Advances in

TISSUE ENGINEERING

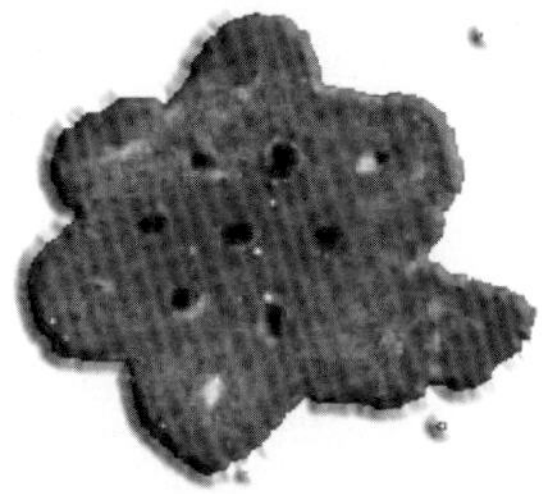

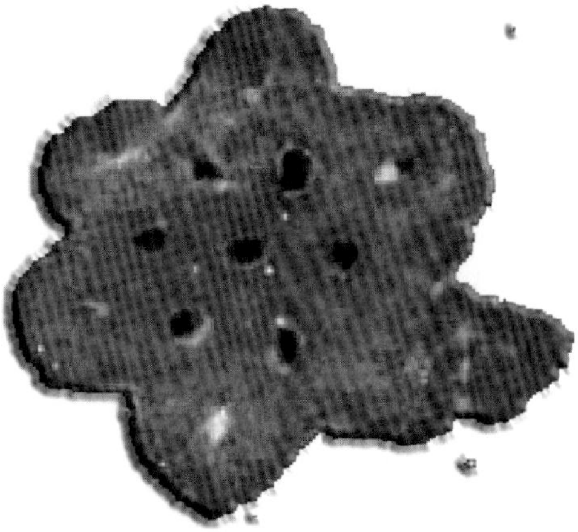

Advances in TISSUE ENGINEERING

Editor

Julia Polak

Imperial College, London, UK

Co-editors

Sakis Mantalaris

Sian E. Harding

Imperial College, London, UK

ICP Imperial College Press

Published by

Imperial College Press
57 Shelton Street
Covent Garden
London WC2H 9HE

Distributed by

World Scientific Publishing Co. Pte. Ltd.
5 Toh Tuck Link, Singapore 596224
USA office: 27 Warren Street, Suite 401-402, Hackensack, NJ 07601
UK office: 57 Shelton Street, Covent Garden, London WC2H 9HE

British Library Cataloguing-in-Publication Data
A catalogue record for this book is available from the British Library.

ADVANCES IN TISSUE ENGINEERING

ISBN-13 978-1-84816-182-5
ISBN-10 1-84816-182-4

Typeset by Stallion Press
Email: enquiries@stallionpress.com

Printed in Singapore by Mainland Press Pte Ltd

Contents

Contributors xix

Foreword xxxv

Introduction xxxvii

Part I: Tissue Engineering: Past, Present, and Future **1**

Chapter 1: An Introduction 3

Robert M. Nerem

1. Introduction 3
2. The Early Years 4
3. The 1990s 6
4. 2000 to Present 6
5. What About the Future? 8
6. Concluding Discussion 10

Part II: Cells for Tissue Engineering **13**

Chapter 2: A Brief Introduction to Different Cell Types 15

Lee Buttery and Kevin M. Shakesheff

1. Introduction 16
2. Cells and Tissue Engineering 16
3. Mature or Primary Cells 18
4. Stem Cells 19
5. Sources of Stem Cells 21
6. Immortalised Cell Lines 29
7. Reprogramming 29
8. Differentiation of Cells 30
9. Regulatory Issues 32

Chapter 3: Human Embryonic Stem Cells: International Policy and Regulation 43

Megan Allyse and Stephen Minger

1. Introduction 44
2. Controversy 45
3. International Guidelines 47
4. National Policy Systems 49
5. Conclusion 59

Chapter 4: Human Embryonic Stem Cells: Derivation and Culture 63

Emma L. Stephenson, Peter R. Braude and Chris Mason

1. Introduction 64
2. The Emergence of Human Embryonic Stem Cell Research 64
3. Human Embryonic Stem Cells 67
4. Culture of hESC lines 71
5. Reporting of Derivation 76
6. Concluding Remarks 77

Chapter 5: Stem Cells Differentiation 83

Pascale V. Guillot and Wei Cui

1. Introduction 83
2. Differentiation of Embryonic Stem Cells 84
3. Somatic Stem Cells 86
4. Conclusion 89

Chapter 6: Marrow Stem Cells 95

Donald G. Phinney

1. Introduction 96
2. Hematopoietic Stem Cells: Discovery, Phenotype, and Function 97
3. Mesenchymal Stem Cells: Discovery, Phenotype, and Function 100
4. Endothelial Progenitor Cells: Discovery, 103
Phenotype, and Function 103
5. A Common Origin for Bone Marrow Stem Cells 105
6. Functional Interdependency of Bone Marrow Stem Cells 108
7. Summary 109

Chapter 7: Cord Blood Stem Cells — Potentials and Realities 123

Colin P. McGuckin and Nicolas Forraz

1. Introduction to the Concept of Umbilical Cord Blood Stem Cells 124
2. Cord Blood Current Clinical Uses 126
3. Cord Blood Processing and Cryopreservation 129
4. Cord Blood Banking 132
5. Cord Blood Research and Where the Future Lies 134

Chapter 8: Fat Stem Cells 143

Jeffrey M. Gimble, Bruce A. Bunnell and Farshid Guilak

1. Introduction 144
2. Types of Adipose Tissue 144
3. Isolation Procedures 145
4. Immunophenotype and Cytokine Profile of ASCs 146
5. Immunogenicity of ASCs 147
6. Differentiation Potential 149
7. Mechanisms of Potential Utility: Genetic Engineering and Gene Delivery 158
8. Conclusions and Future Directions 158

Chapter 9: Control of Adult Stem Cell Function in Bioengineered Artificial Niches 175

Matthias P. Lutolf and Helen M. Blau

1. Introduction 176
2. Adult Stem Cells Reside in Niches 177
3. Common Structures and Components of Stem Cell Niches 177
4. Key Functions of Stem Cell Niches 180
5. Niches Control the Fate of Individual Stem Cells 181
6. The HSC Niche 183
7. Prospects for Using Engineered Artificial Niches as Novel Model Systems to Probe and Manipulate Adult Stem Cell Fate 185

Chapter 10: Stem Cell Immunology 199

Anthony P. Hollander and David C. Wraith

1. Why is Stem Cell Immunology Important for Tissue Engineering? 200
2. Evolutionary Context of the Mammalian Immune System 201
3. Materno-Foetal Tolerance as a Model for Understanding Stem Cell Immune Privilege 204
4. Are Embryonic Stem Cells Immune Privileged? 206
5. Are Mesenchymal Stem Cells (MSCs) Immune Privileged? 209
6. Finding a Way Forward for the Use of Allogeneic Stem Cells 210

Chapter 11: Development of a Design of Experiment Methodology: Applications to the Design and Analysis of Experiments 215

Mayasari Lim and Athanasios Mantalaris

1. Analysis of Factors 216
2. Design Strategy 219
3. DOE Designs 220
4. A DOE Example: Investigating the Influence of Cytokines on Erythropoiesis 223
5. Conclusions 226

Chapter 12: Banking Stem Cell Lines for Future Therapies 229

Glyn N. Stacey and Charles J. Hunt

1. Introduction 230
2. The Rationale for Centralised Banks of Human Cell Lines for Clinical Use 232
3. Fundamental Issues for *in Vitro* Cell Culture 233
4. Cell Culture Processes 234
5. Quality Assurance and Quality Control 242
6. International Perspectives 246
7. Future Developments and Expectations 247

Part III: Materials **253**

Chapter 13: Synthetic Biomaterials as Cell-Responsive Artificial Extracellular Matrices 255

Matthias P. Lutolf and Jeffrey A. Hubbell

1. Introduction 256
2. ECMs Instruct Cell Fates and Respond to Cell-Secreted Signals 257
3. Design Principles for Cell-Responsive Artificial ECMs 259
4. Implementation: Classes and Applications of Cell-Responsive Artificial ECMs 263
5. Future Challenges 271

Chapter 14: Bioactive Composite Materials for Bone Tissue Engineering Scaffolds 279

Sophie Verrier and Aldo R. Boccaccini

1. Introduction 280
2. Scaffolds Requirements 281
3. Composite Materials Approach for Tissue Engineering Scaffolds 283
4. *In Vitro* and *In Vivo* Evaluation 288
5. Discussion 302
6. Conclusions and Future Work 303

Chapter 15: Aggregation of Cells Using Biomaterials and Bioreactors 313

Zahia Bayoussef and Kevin M. Shakesheff

1. Introduction 314
2. Cell Adhesion and Natural Cell Aggregation 314
3. Methods of Cell Aggregation 315
4. Synthetic Cell Aggregation 316
5. Cell Aggregation on Scaffolds 323
6. Bioreactors and Cell Aggregation 325
7. Summary and Conclusion 327

Chapter 16: Nanotechnology for Tissue Engineering 333

Jean S. Stephens-Altus and Jennifer L. West

1. Introduction 334
2. Nanostructured Scaffolds 334
3. Nanoparticles for Cellular Imaging 340
4. Conclusions 343

Chapter 17: Microscale Technologies for Tissue Engineering 349

Ali Khademhosseini, Yanan Du, Bimal Rajalingam, Joseph P. Vacanti and Robert S. Langer

1. Introduction 350
2. Microscale Technologies for Controlling Stem Cell Fate 351
3. Microscale Technologies for Engineering Complex Tissues Containing Different Cell Types and Vasculature 357
4. Conclusion 363

Part IV: Non-Invasive Methods to Monitor Tissue Re-Modelling 371

Chapter 18: Biosensors 373

Tony Cass

1. Introduction to Sensor Technology 374
2. The Importance of Mass Transport in Sensor Performance 376
3. Electrochemical Biosensors 378
4. Optical Biosensors 384
5. Mass Sensors 390
6. Cell Sensing Strategies 391
7. Tissue Sensing Strategies 393
8. Conclusions and Outlook 394

Chapter 19: Tissue-Engineering Monitoring Using Microdialysis 401

Zhaohui Li, Olga Boubriak, Jill Urban and Zhanfeng Cui

1. Introduction 402
2. Methodology of Microdialysis 403

3. Microdialysis for Tissue Engineering Monitoring — Case Studies 409
4. Summary 415

Chapter 20: Characterisation of Tissue Engineering Constructs by Raman Spectroscopy and X-ray Micro-Computed Tomography (μCT) 421

Ioan Notingher and Julian R. Jones

1. Introduction 422
2. Principles and Instrumentation 426
3. Applications of Raman Micro-Spectroscopy to Cells 430
4. Application of μCT to the Quantification of Scaffolds 435
5. Conclusions 438

Chapter 21: Role of Stem Cell Imaging in Regenerative Medicine 443

Gabriella Passacquale and Kishore Bhakoo

1. Introduction 444
2. Ideal Imaging Technology for Non-Invasive Stem Cell Tracking 445
3. Non-Invasive Tracking of Stem Cells Using MRI 448
4. Role of Imaging in Stem Cell-Based Therapy for the Central Nervous System 453
5. Multimodality 456
6. Conclusions 456

Part V: Biotechnology Sector **467**

Chapter 22: Lessons Learnt 469

Nancy L. Parenteau, Susan J. Sullivan, Kelvin G. M. Brockbank and Janet Hardin Young

1. Background: The Marriage of Biology and Engineering 470
2. The Evolving Emphasis from Engineering to Biology 471

	3. Vascular Tissue Engineering	472
	4. Cartilage	475
	5. Extracorporeal Devices	477
	6. Skin	481
	7. Conclusion	483
Chapter 23:	The Promise of Stem Cells: A Venture Capital Perspective	491
	Cathy Prescott	
	1. Introduction	492
	2. Venture Capital — Balancing Risk	492
	3. The Value Proposition	494
	4. "Watch and Wait"	494
	5. The European Regulatory Environment	497
	6. An Evolving Patent Landscape	498
	7. Future Prospects?	499
Part VI:	**Tissue Engineering Products**	**501**
Chapter 24:	Cell Expansion, Cell Encapsulation, 3D Cultures	503
	Julia M. Polak and Athanasios Mantalaris	
	1. Introduction	504
	2. Controlled Differentiation	505
	3. Generation of Clinically Relevant Number of Cells in 3D Cultures as an Integrated and Scalable Process	507
	4. Encapsulation	508
	5. Bioprocessing for Regenerative Medicine	508
	6. Discussion	511
Chapter 25:	Bioreactor Engineering: Regenerating the Dynamic Cell Microenvironment	517
	Tal Dvir and Smadar Cohen	
	1. Tissue Engineering — The Introduction of 3D Cell Cultures	518
	2. Mass Transport Challenges in 3D Cell Cultures	518
	3. First Generation of Tissue Engineering Bioreactors	520

4. Perfusion Bioreactors — Theory and Practice 522
5. Examples of Perfusion Bioreactors in TE 524
6. Bioreactors Providing Physical Signals 526
7. Microfabricated Bioreactors 530
8. Concluding Remarks and Future Aspects 531

Chapter 26: UK Regulatory Issues: The View from the Researcher 537

Caroline Munro and Neil Harris

1. Introduction 538
2. The Product Life Cycle of a Cell-Based Therapeutic 539
3. Stage 1: Procurement — The Obtaining of Cells or Tissue Components from Donors Under cGCP 542
4. Stage 2: Analysis — Initial Isolation, Screening, Characterisation and Manipulation of Cells/Other Components and Storage 543
5. Stage 3: Confirmed Proof of Product and Process — Initial to Final Screening for Potential Use 544
6. Stage 4: Product Manufacturing — Production of Clinical Grade Material Under cGMP 546
7. Stage 5: Pre-Clinical Trials — Assessment of Safety and Performance for Regulatory Submission 546
8. Stage 6: Clinical Trials — Clinical Assessments of Product Safety and Performance 549
9. Stage 7: Launch — Commercially Available Product 551
10. Stage 8: Post-Market — Ongoing Processes Following Commercialisation of Product 552
11. Regulations, Guidelines and Codes of Practice 555

Part VII: Tissue Repair **559**

Chapter 27: Stem Cell Therapy: Past, Present, and Future 561

Frédéric Baron and Rainer Storb

1. Hematopoietic Stem Cells and Hematopoietic Stem Cell Transplantation 562
2. Plasticity of Adult Hematopoietic Stem Cells: Lessons Learnt from a Canine Model of Duchenne Muscular Dystrophy 571
3. Mesenchymal Stromal Cells 573
4. Embryonic Stem Cells 575
5. Future of Stem Cell Therapy 575

Chapter 28: Tissue Engineered Skin Comes of Age? 593

Sheila MacNeil

1. To What Extent can Tissue Engineered Skin Deliver Normal Skin Structure and Function for Clinical Use? 594
2. Which Patients can Benefit from Tissue Engineered Products? 595
3. History of the Development of Tissue Engineered Skin for Burns Injuries 595
4. Development of Tissue Engineered Skin for Chronic Wounds 603
5. Development of Tissue Engineered Skin for Reconstructive Surgery 605
6. The Design Process for Tissue Engineered Products 605
7. Keratinocyte Stem Cells — Where are We? 606
8. Clinical and Future Development Issues 607
9. Laboratory Uses of Tissue Engineered Skin 609
10. Conclusion 612

Chapter 29: Liver Repair 619

Nataša Levičar, Madhava Pai and Nagy A. Habib

1. Introduction 620
2. Haematopoietic Stem Cells and Liver Regeneration 620
3. Clinical Studies 622
4. Conclusions 625

Chapter 30: Tissue Engineering for Tooth Regeneration 633

Ivan. A. Diakonov and Paul Sharpe

1. Tooth Development and Morphology 634
2. Stem Cells in the Tooth 636
3. Two Strategies for Tissue Engineering Tooth Germs *De Novo* 641
4. Challenges 646
5. Conclusions 646

Chapter 31: Urogenital Repair 655

Anthony Atala

1. Introduction 656
2. Tissue Engineering Strategies for Urogenital Repair 656
3. Tissue Engineering of Specific Urologic and Genital Structures 660
4. Kidney 665
5. Genital Tissues 667
6. Other Applications of Genitourinary Tissue Engineering 669
7. Conclusion 670

Part VIII: Cardiac Repair **677**

Chapter 32: Basic Science 679

Sian E. Harding

1. Introduction 680
2. Conclusion 688

Chapter 33: Cardiac Repair Clinical Trials 695

Amanda Green and Eric Alton

1. Introduction 696
2. A New Treatment Option 696
3. Acute Myocardial Infarction 697
4. Chronic Ischaemic Heart Disease 715

5. Which is the Optimal Method of Delivery? 719
6. The Future for Clinical Trials 721

Chapter 34: Myocardial Recovery Following LVAD Support 733

Robert S. George and Emma J. Birks

1. Introduction 734
2. Overview of LVADs 735
3. LVAD and Myocardial Recovery — Clinical Implications 741
4. Remodelling versus Reverse Remodelling 745
5. Conclusion 750

Part IX: Osteoarticular Repair **761**

Chapter 35: Animal Models 763

Elizabeth A. Horner, Jennifer Kirkham and Xuebin B. Yang

1. Introduction 764
2. The Choice of Animal 765
3. *In Vivo* Models for Bone and Cartilage Tissue Engineering 766
4. Conclusion 774

Chapter 36: *In Vitro* 3D Human Tissue Models for Osteochondral Diseases 781

Sourabh Ghosh and David L. Kaplan

1. Introduction 782
2. Factors Governing the Simulation of Tissue Microenvironments 784
3. Osteochondral Tissue 792
4. Conclusions 805

Chapter 37: Application of Tissue Engineering for Craniofacial Reconstruction 821

Deepak M. Gupta, Matthew D. Kwan, Bethany J. Slater and Michael T. Longaker

1. Introduction 822
2. Craniosynostosis: A Case for Mechanisms Underlying Bone Formation 823
3. Distraction Osteogenesis: Endogenous Tissue Engineering 827
4. Cellular-Based Tissue Engineering: Regenerative Medicine 830
5. Conclusion 834

Chapter 38: Clinical Trials 843

Anne K. Haudenschild and Marc H. Hedrick

1. Introduction to Osteoarticular Repair 844
2. Impact 845
3. Current Medical Treatment of OA 845
4. Current Trials 848
5. Promising Future Technologies 850
6. Discussion 851

Part X: Lung Repair **859**

Chapter 39: Tissue Engineering for the Respiratory Epithelium: Is There a Future for Stem Cell Therapy in the Lung? 861

Valérie Besnard and Jeffrey A. Whitsett

1. Introduction: The Challenges Facing Cell-Based Therapy for Treatment of Lung Disease 862
2. Lung Morphogenesis 864
3. Sources of Stem Cells 870
4. Do Bone Marrow-Derived Cells Contribute to Repair? 872
5. Use of Intrinsic Pulmonary Progenitor Cells 875
6. The Hope of Stem Cell Therapy for Treatment of Lung Disease 877
7. Conclusion 879

Chapter 40: The Artificial Lung 887

Andreas Nikolas Maurer and Georg Matheis

1. Introduction 888
2. Mechanical Ventilation 889

3. Development of Membrane Ventilators (Artificial Lungs, Medical Devices) 889
4. Near Future Tasks to Enhance Membrane Ventilators 890
5. From Medical Devices to Biohybrid Lungs 892
6. Organoid Structures 894
7. Conclusion 895

Index 903

Contributors

Megan Allyse, MA
Wellcome Trust PhD Candidate
Institute for Science and Society
University of Nottingham
West Wing, Law and Social Science Building
University Park, Nottingham, NG7 2RD, UK

Eric Alton, MA, MD, FRCP, FMedSci
Professor of Gene Therapy and Respiratory Medicine
Honorary Consultant Physician, The Royal Brompton Hospital
Department of Gene Therapy
The National Heart and Lung Institute
Imperial College London
Emmanuel Kaye Building
Manresa Road, London, SW3 6LR, UK

Anthony Atala, MD
W. Boyce Professor and Chair, Department of Urology
Director, Wake Forest Institute for Regenerative Medicine
Wake Forest University Health Sciences
A1 Bldg., 4th Floor
Medical Center Boulevard
Winston-Salem, NC 27157, USA

Frédéric Baron, MD
Department of Hematology
University of Liege, Belgium
National Fund for Scientific Research Belgium
Hematology, CHU Sart-Tilman, B4000 Liege, Belgium

Zahia Bayoussef, MPharm
Wolfson Centre for Stem Cells, Tissue Engineering,
and Modelling (STEM)
Centre for Biomolecular Sciences, School of Pharmacy
University of Nottingham
University Park, Nottingham, NG7 2RD, UK

Valérie Besnard, PhD
Division of Pulmonary Biology
Cincinnati Children's Hospital Medical Center
3333 Burnet Avenue, Cincinnati, OH 45229-3039, USA

Kishore Bhakoo, PhD
Head, Stem Cell Imaging
MRC Clinical Sciences Centre
Imperial College London
Hammersmith Campus
Du Cane Road, London, W12 0NN, UK

Emma J. Birks, MRCP, PhD
Royal Brompton and Harefield NHS Trust
Harefield, Middlesex, UB9 6JH, UK

Helen M. Blau, PhD
Professor and Director
Baxter Laboratory in Genetic Pharmacology
Stanford University School of Medicine
269 Campus Drive, Stanford, CA 94305-5175, USA

Aldo R. Boccaccini, Dr-Ing. Habil.
Reader in Materials Science
Department of Materials
Imperial College London
Prince Consort Road
London, SW7 2BP, UK

Olga Boubriak, BSc, PhD
Department of Physiology, Anatomy and Genetics
University of Oxford
South Parks Road, Oxford, OX1 3QX, UK

Peter R. Braude, MA, PhD, FRCOG, FMedSci
Head, Department of Women's Health
Division of Reproduction and Endocrinology
King's College London

Assisted Conception Unit, St. Thomas' Hospital
Guy's and St. Thomas' NHS Foundation Trust
10th Floor, North Wing
Westminster Bridge Road, London, SE1 7EH, UK

Kelvin G. M. Brockbank, PhD
Principal Research Scientist, Georgia Tech/Emory Center
for the Engineering of Living Tissues (Atlanta, Georgia, USA)
President and Chief Science Officer, Cell and Tissue Systems, Inc.
2231 Technical Parkway, North Charleston, SC 29406, USA

Bruce A. Bunnell, PhD
Professor, Department of Pharmacology
Center for Gene Therapy
Chairman, Division of Gene Therapy
Tulane National Primate Research Center
Tulane University Health Sciences Center
SL83, Tulane University School of Medicine
1430 Tulane Avenue, New Orleans, LA 70112-2699, USA

Lee Buttery, MSc, PhD
Wolfson Centre for Stem Cells, Tissue Engineering
and Modelling (STEM)
Centre for Biomolecular Sciences, School of Pharmacy
University of Nottingham
University Park, Nottingham, NG7 2RD, UK

Tony Cass, DPhil (Oxon), FRSC, Cchem, Csci, MIET
Deputy Director and Research Director
Institute of Biomedical Engineering
Faculty of Engineering
Imperial College London
South Kensington Campus
Level 4, Bessemer Building
London, SW7 2AZ, UK

Smadar Cohen, PhD
Claire and Harold Oshry Professor Chair in Biotechnology
Department of Biotechnology Engineering
Ben-Gurion University of the Negev
Marcus Campus, Building 39, Room 222
Beer-Sheva, 84105, Israel

Wei Cui, MD, PhD
Institute of Reproductive and Developmental Biology
Faculty of Medicine, Imperial College London
Hammersmith Campus
Du Cane Road, London, W12 0NN, UK

Zhanfeng Cui, BSc, MSc, MA, PhD, FIChemE
Department of Engineering Science
University of Oxford
Parks Road, Oxford, OX1 3PJ, UK

Ivan A. Diakonov
Department of Craniofacial Development
King's College London Dental Institute at Guy's
King's College and St. Thomas' Hospitals
Floor 27 Guy's Tower, St. Thomas St.
London, SE1 9RT, UK

Yanan Du, PhD
Postdoctoral Associate
Harvard-Massachusetts Institute of Technology
Division of Health Sciences and Technology
Harvard Medical School/Brigham and Women's Hospital
Room 265, 65 Landsdowne Street
Cambridge, MA 02139, USA

Tal Dvir, MSc
Department of Biotechnology Engineering
Ben-Gurion University of the Negev
Marcus Campus, Building 39, Room 223
Beer-Sheva, 84105, Israel

Nicolas Forraz, PhD, BSc(Hons)
Senior Research Associate
Manager Research and Development, "Novus Sanguis" charity, funding Responsible Regenerative Medicine
Newcastle Centre for Cord Blood
Institute of Human Genetics
North East England Stem Cell Institute
International Centre for Life, Central Parkway
Newcastle University
Newcastle-upon-Tyne, NE1 3BZ, UK

Robert S. George, BSc, BM, MRCS
Royal Brompton and Harefield NHS Trust
Harefield, Middlesex, UB9 6JH, UK

Sourabh Ghosh, PhD
Department of Biomedical Engineering
Tufts University
Medford, Massachusetts 02155, USA

Jeffrey M. Gimble, MD, PhD
Stem Cell Biology Laboratory
Pennington Biomedical Research Center
Louisiana State University System
6400 Perkins Rd, Baton Rouge, LA 70808, USA

Amanda Green, Grad Dip Phys, MSc
Department of Gene Therapy
The National Heart and Lung Institute
Imperial College London and The Royal Brompton Hospital
The Emmanuel Kaye Building
Manresa Road, London, SW3 6LR, UK

Farshid Guilak, PhD
Director, Orthopaedic Research Laboratories
Department of Surgery
Duke University Medical Center
375 MSRB, Box 3093
Durham, NC 27710, USA

Deepak M. Gupta, MD
Postdoctoral Research Fellow
Hagey Laboratory for Pediatric Regenerative Medicine
Stanford University School of Medicine
257 Campus Drive, Stanford, CA 94305-5148, USA

Pascale V. Guillot, BSc, MSc, PhD
Group Leader in Stem Cell Therapy
Division of Surgery, Oncology, Reproductive Biology
and Anaesthetics (SORA)
Institute of Reproductive and Developmental Biology
Faculty of Medicine, Imperial College London
Hammersmith Campus
Du Cane Road, London, W12 0NN, UK

Nagy A. Habib, FRCS, MD
Department of Biosurgery and Surgical Technology (B Block)
Division of Surgery, Oncology, Reproductive Biology
and Anaesthetics (SORA)
Imperial College London
Hammersmith Hospital Campus
Du Cane Rd, London, W12 0NN, UK

Sian E. Harding, PhD, FAHA
Professor of Cardiac Pharmacology
National Heart and Lung Institute (NHLI), Faculty of Medicine
Imperial College London
Guy Scadding Building
Dovehouse St., London, SW3 6LY, UK

Neil Harris, PhD
LGC Limited
Queens Road, Teddington
Middlesex, TW11 0LY, UK

Anne K. Haudenschild, PhD
Cytori Therapeutics Inc.
3020 Callan Road
San Diego, CA 92121, USA

Marc H. Hedrick, MD
Cytori Therapeutics Inc.
3020 Callan Road
San Diego, CA 92121, USA

Anthony P. Hollander, PhD
ARC Professor of Rheumatology and Tissue Engineering
Department of Cellular and Molecular Medicine
School of Medical Sciences, University of Bristol
University Walk, Clifton, Bristol, BS8 1TD, UK

Elizabeth A. Horner, PhD
Skeletal Tissue Engineering Laboratory
Department of Oral Biology
Leeds Dental Institute, University of Leeds
Leeds, LS2 9LU, UK

Jeffrey A. Hubbell, PhD
Laboratory for Regenerative Medicine and Pharmacobiology
Institute of Bioengineering
Swiss Federal Institute of Technology (EPFL)
Bldg. AAB039, Station 15
CH-1015 Lausanne, Switzerland

Charles J. Hunt, PhD
Operations Manager, UK Stem Cell Bank
National Institute for Biological Standards and Control
Blanche Lane, South Mimms, Herts, EN6 3QG, UK

Julian R. Jones, PhD, DIC, Meng (Oxon.)
Royal Academy of Engineering/EPSRC
Research Fellow and Lecturer Designate
Department of Materials
Imperial College London
South Kensington Campus
London, SW7 2AZ, UK

David L. Kaplan, PhD
Professor and Chair, Department of Biomedical Engineering
Director, Bioengineering and Biotechnology Center
Professor, Department of Chemical and
Biological Engineering, Tufts University
4 Colby Street, Medford, Massachusetts 02155, USA

Ali Khademhosseini, PhD
Assistant Professor
Harvard-Massachusetts Institute of Technology
Division of Health Sciences and Technology
Harvard Medical School/Brigham and Women's Hospital
Room 265, 65 Landsdowne Street
Cambridge, MA 02139, USA

Jennifer Kirkham, PhD
Professor of Biomineralization
Head, Department of Oral Biology
Leeds Dental Institute, University of Leeds
Leeds, LS2 9LU, UK

Matthew D. Kwan, MD
Postdoctoral Research Fellow
Hagey Laboratory for Pediatric Regenerative Medicine
Stanford University School of Medicine
257 Campus Drive, Stanford, CA 94305-5148, USA

Robert S. Langer, PhD
Institute Professor, Department of Chemical Engineering
Massachusetts Institute of Technology
Room E25-342, 77 Massachusetts Avenue
Cambridge, MA 02139-4307, USA

Nataša Levičar, PhD
Department of Surgical Oncology and Technology (B Block)
Imperial College London
Hammersmith Hospital Campus
Du Cane Road, London, W12 0NN, UK

Zhaohui Li, BSc, Dphil
Department of Engineering Science
University of Oxford
Parks Road, Oxford, OX1 3PJ, UK

Mayasari Lim
Biological Systems Engineering Laboratory
Department of Chemical Engineering and
Chemical Technology
Imperial College London
South Kensington Campus
Room 132, Ground Floor, ACEX Building
London, SW7 2AZ, UK

Michael T. Longaker, MD, MBA, FACS
Deane P. and Louise Mitchell Professor
Director, Hagey Laboratory for Pediatric
Regenerative Medicine
Director, Program in Regenerative Medicine
Deputy Director, Institute of Stem Cell Biology and
Regenerative Medicine
Lucile Packard Children's Hospital
Stanford University School of Medicine
257 Campus Drive, Stanford, CA 94305-5148, USA

Matthias P. Lutolf, PhD
Laboratory of Stem Cell Bioengineering (LSCB)
Institute of Bioengineering
Swiss Federal Institute of Technology (EPFL)
Bldg. AI 3138, Station 15
CH-1015 Lausanne, Switzerland

Sheila MacNeil, BSc, PhD
Professor of Tissue Engineering
Department of Engineering Materials
Kroto Research Institute, University of Sheffield
North Campus, Broad Lane, Sheffield, S3 7HQ, UK

Athanasios Mantalaris, PhD, MSc, BSc (Hons), Reader in Biological Systems Engineering
Biological Systems Engineering Laboratory
Department of Chemical Engineering and Chemical Technology
Imperial College London
South Kensington Campus
Room 515, ACE Extension
London, SW7 2AZ, UK

Chris Mason, MBBS, PhD, FRCS
Stem Cell and Regenerative Medicine Bioprocessing Unit
Advanced Centre for Biochemical Engineering
University College London
Roberts Building, Torrington Place
London, WC1E 7JE, UK

Georg Matheis, PhD
Novalung GmbH
Lotzenaecker 3
72379 Germany

Andreas Nikolas Maurer, PhD
Novalung GmbH
Lotzenaecker 3
72379 Germany

Colin P. McGuckin, DPhil, BSc(Hons) PGCTLHE, CBiol, MIBiol, MRSC, EurProBiol
Professor of Regenerative Medicine
President, "Novus Sanguis" Charity, funding Responsible Regenerative Medicine
Newcastle Centre for Cord Blood
Institute of Human Genetics
North East England Stem Cell Institute
International Centre for Life, Central Parkway
Newcastle University
Newcastle-upon-Tyne, NE1 3BZ, UK

Stephen Minger, PhD
Senior Lecturer and Director of King's Stem Cell Biology Laboratory
Wolfson Centre for Age-Related Diseases
King's College London
Strand, London, WC2R 2LS, UK

Caroline Munro, PhD
2, Hillgarth Court
Elvington, York, YO41 4BD, UK

Robert M. Nerem, PhD
Professor and Director
Institute for Bioengineering and Bioscience
Georgia Institute of Technology
315 Ferst Drive, Atlanta, GA 30332-0363, USA

Ioan Notingher, PhD
Nanoscience Group
School of Physics and Astronomy
The University of Nottingham
University Park, Nottingham, NG7 2RD, UK

Madhava Pai, MS, MSc, DIC, MRCS
Department of Surgical Oncology and Technology (B Block)
Imperial College London
Hammersmith Hospital Campus
Du Cane Road, London, W12 0NN, UK

Nancy L. Parenteau, PhD
Co-founder and Managing Member
Parenteau BioConsultants, LLC
PO Box 448
Fair Haven, VT 05743-0448, USA

Gabriella Passacquale, MD
Stem Cell Imaging
MRC Clinical Sciences Centre
Imperial College London
Hammersmith Campus
Du Cane Road, London, W12 0NN, UK

Donald G. Phinney, PhD
Co-Editor, *Stem Cells* (AlphaMed Press)
Associate Director of Research, Center for Gene Therapy
Professor of Microbiology and Immunology
Tulane University Health Sciences Center
1430 Tulane Avenue
Room 672, J. Bennett Johnston Bldg. (SL-99)
New Orleans, LA 70112, USA

Julia M. Polak, DBE, MD, DSc, FRCP, FRCPath, FMedSci, ILT
Department of Chemical Engineering
Tissue Engineering and Regenerative Medicine
Imperial College London
South Kensington Campus
Room 144, Roderic Hill Building
London, SW7 2AZ, UK

Cathy Prescott, DPhil (Oxon)
Biolatris Ltd.
72 Kimberley Road
Cambridge, CB4 1HJ, UK

Bimal Rajalingam, MD
Postdoctoral Associate
Harvard-Massachusetts Institute of Technology
Division of Health Sciences and Technology
Harvard Medical School/Brigham and Women's Hospital
Room 265, 65 Landsdowne Street
Cambridge, MA 02139, USA

Kevin M. Shakesheff, MPharmS, PhD
Wolfson Centre for Stem Cells, Tissue Engineering,
and Modelling (STEM)
Centre for Biomolecular Sciences, School of Pharmacy
University of Nottingham
University Park, Nottingham, NG7 2RD, UK

Paul Sharpe
Department of Craniofacial Development
King's College London Dental Institute at Guy's
King's College and St. Thomas' Hospitals
Floor 27 Guy's Tower, St. Thomas St.
London, SE1 9RT, UK

Bethany J. Slater, MD
Postdoctoral Research Fellow
Hagey Laboratory for Pediatric Regenerative Medicine
Stanford University School of Medicine
257 Campus Drive, Stanford, CA 94305-5148, USA

Glyn N. Stacey, PhD
Head, Division of Cell Biology and Imaging
Director, UK Stem Cell Bank
National Institute for Biological Standards and Control
Blanche Lane, South Mimms, Herts, EN6 3QG, UK

Jean S. Stephens-Altus, PhD
Department of Bioengineering
Rice University
6100 Main St., MS-142
Houston, TX 77005, USA

Emma L. Stephenson, MA
Advanced Centre for Biochemical Engineering
University College London
Department of Women's Health
Division of Reproduction and Endocrinology
King's College London
Stem Cell Research Lab
Assisted Conception Unit, Guy's Hospital
Guy's and St. Thomas' NHS Foundation Trust
4th Floor, Southwark Wing
Great Maze Pond, London, SE1 9RT, UK

Rainer Storb, MD
Head and Member, Transplantation Biology Program
Fred Hutchinson Cancer Research Center
Professor of Medicine, University of Washington
PO Box 19024
1100 Fairview Avenue North, D1-100
Seattle, WA 98109-1024, USA

Susan J. Sullivan, PhD
Senior Director, Alliance Management
ImmunoGen, Incorporated
830 Winter Street
Waltham, MA 02451-1477, USA

Jill Urban, BSc, PhD
Department of Physiology, Anatomy and Genetics
University of Oxford
South Parks Road, Oxford, OX1 3QX, UK

Joseph P. Vacanti, MD
John Homans Professor of Surgery, Harvard Medical School
Chief of Pediatric Surgery
Director, Laboratory for Tissue Engineering and Organ Fabrication
Massachusetts General Hospital
Warren Building 1157, 55 Fruit Street
Boston, MA 02114, USA

Sophie Verrier, PhD
Research Leader
Biomaterials and Tissue Engineering
AO Foundation
Clavadelerstrasse 8
7270 Davos, Switzerland

Jennifer L. West, PhD
Department of Bioengineering
Rice University
6100 Main St., MS-142
Houston, TX 77005, USA

Jeffrey A. Whitsett, MD
Chief, Section of Neonatology, Perinatal and Pulmonary Biology
Cincinnati Children's Hospital Medical Center
3333 Burnet Avenue, Cincinnati, OH 45229-3039, USA

David C. Wraith, PhD
Professor of Experimental Pathology
Department of Cellular and Molecular Medicine
School of Medical Sciences, University of Bristol
University Walk, Clifton, Bristol, BS8 1TD, UK

Xuebin B. Yang, MD, MSc, PhD
Head of Tissue Engineering Research
Skeletal Tissue Engineering Laboratory
Department of Oral Biology
Leeds Dental Institute, University of Leeds
Leeds, LS2 9LU, UK

Janet Hardin Young, PhD
Co-founder
Parenteau BioConsultants, LLC
PO Box 448
Fair Haven, VT 05743-0448, USA

Foreword

It is a pleasure and an honour to write this foreword to a book edited by one of the most respected and foremost authorities on tissue engineering, Professor Dame Julia Polak, with contributions by internationally known leaders in individual fields of tissue engineering and stem cell research. Both disciplines are crucial to the delivery of new treatment of diseases. The UK National Stem Cell Network (UKNSCN) is working hard to unite and facilitate the initiatives in Stem Cells for Regenerative Medicine in the UK, and it is encouraging to see so many of the key figures contributing to this book. It is an authoritative document of the current status of research and developments, and as such a great contribution to the current knowledge base. I have no doubt it will be widely read by all working closely in the field, and by others wishing to know the current status of the research. I think the editor and authors are to be congratulated for bringing it together. I enjoyed reading it, even as a non-tissue or stem cell scientist.

Lord Naren Patel of Dunkeld
Chair, UK National Stem Cell Network (UKNSCN)

Introduction

Tissue engineering is a relatively young science, especially in terms of the soft tissues of the body. The present intense and growing interest can be seen not only in a scientific but also a sociological context, as the focus in a population with increased longevity now turns to the paradigm of "healthy old age". Evident successes in the field of skin, bone and dental replacements also fuel enthusiasm as treatments are seen to become clinical reality. The burgeoning science of stem cells is following a similar trajectory, and the two strands have such clear complementarity that they are frequently brought together both at scientific meetings and in the regrouping of university departments. In the present volume we have reflected this with a detailed consideration of cell types for tissue engineering, which *de facto* concentrate on the many varieties of stem cells that are being discovered and characterised.

A striking feature of tissue engineering and stem cell research is that the problems being addressed have an unexpected degree of commonality between different organ systems. Stem cell researchers in different fields are asking the same questions about directed differentiation of stem cells, genetic modification, the extent to which exogenous cells will remain after implantation and the role of the niche in controlling cell fate. They are also revising views on the plasticity of organs including brain and heart, as a result of the advances in stem cell biology. Engineers are tailoring the design of materials to the particular organ in terms of physical properties, but issues of biocompatibility and biodegradability have shared goals. Here we have presented together the latest information from the pulmonary, cardiac, skin, osteoarticular, liver, urogenital and dental areas, to draw out the differences and similarities of the approaches.

This volume also attempts to define the path from basic science to practical application. The issue of scale-up of stem cell numbers is fundamental to all the organs systems, with the role of bioreactors central to the success of translation. Importantly, the regulatory environment is considered as an integral part of the selection of stem cell types, and the variations in the international landscape are drawn out. A contribution from the UK Stem Cell Bank shows how the control and standardisation of lines will be crucial to underpin both research and translational efforts. The full discussion of the regulatory hurdles includes not only the stem cells but the fabrication of final approved products. We have brought in the opinion of venture capitalists to provide a perspective on the economics of the process by which these products will be carried through to the clinic.

Looking to the future, we highlight exciting new areas of stem cell biology, and the cross-fertilisation between the investigation of signalling in authentic stem cells with the understanding of intrinsic tissue regeneration in health and disease. We showcase new developments in the customisation and functionalisation of biomaterials for both bioreactor design and *in vivo* implantation. The vital contribution of novel sophisticated imaging technologies to assess the success of tissue engineering strategies *in vitro* and *in vivo* is emphasised. Above all, this volume shows the tremendous inventiveness and synergy that comes when biologists and physical scientists join together in a focused effort to address human disease.

PART I

TISSUE ENGINEERING: PAST, PRESENT, AND FUTURE

Chapter 1

An Introduction

Robert M. Nerem

Abstract

In the last few decades tissue engineering has emerged as a technology, and this has now evolved into what is called regenerative medicine, including not only the replacement of tissues and organs, but also repair and regeneration. This field is an outgrowth of the biological revolution. Although research in this field goes back well into the 20th century, it was in the 1990s that research accelerated and that an industry began to emerge. In this same period the Tissue Engineering Society was formed, and this has evolved into what is now the Tissue Engineering and Regenerative Medicine International Society. Although the industry still may be characterized as being in a fledgling state, there are some positive signs appearing. Furthermore, the field has become a global activity, one that has the potential to alter the practice of medicine as we know it. Thus, building on the past, the future of tissue engineering and regenerative medicine remains extremely promising.

Keywords: Tissue Replacement and Repair; Regeneration; Stem Cell Technology; Industry.

Outline

1. Introduction
2. The Early Years
3. The 1990s
4. 2000 to Present
5. What About the Future?
6. Concluding Discussion

Acknowledgements

References

1. Introduction

Over the last few decades tissue engineering has emerged as a technology, and this now has evolved into what is called regenerative medicine. For many tissue engineering has had a broad meaning, one that includes not only the replacement of tissues and organs, but also repair and regeneration. For others, however, it is the term regenerative medicine that has the broader meaning. Whatever the use of these terms, it is a more biologic approach to treating tissues and organs that represents the future, and what one in general means is the triad of replacement, repair, and/or regeneration of tissues and organs. It is this combination of approaches that represents a new era in our efforts to treat problems associated with failing tissues and organs, an era where our goal is to harness the intrinsic power of biology and where our thinking is driven by the desire to in some way imitate nature.[1]

It should be noted that for this author replacement involves the growing/fabricating of tissues and organs outside the body and these then being implanted. Repair is not what has been afforded by the electromechanical devices/implants of the 20th century where there was no biological function. Rather, repair is a biological one at the cell and molecular level, including the repair of DNA. Finally, regeneration means to literally grow *in vivo* a new tissue or organ, where the challenge is to cause this regeneration without scarring.

The medical device/implant industry started to emerge in the middle of the last century, and this industry and the technology it has fostered has made enormous contributions to healthcare. All of us know individuals who have a medical device/implant and who thus have benefited from the products of this industry.[2] These have reduced pain and alleviated symptoms, but they have not been cures. Thus, as will be discussed later, the biological revolution is spawning a revolution in this industry, one where the products will be different and the way products are engineered will be different. This is due, at least in part, to the advent of tissue engineering and regenerative medicine. This more biologic approach brings the potential of revolutionizing treatments and therapies for patients, and it is this emerging science and technology that will lead to the next generation of medical implants. It also will lead to strategies that bypass the need for replacement through the fostering of repair and regeneration. All of this will in turn offer hope to millions of patients who today have conditions where existing treatments and therapies are inadequate or in many cases do not exist at all.

2. The Early Years

The field of tissue engineering and regenerative medicine is a product of the biological revolution. This revolution in biology dates back to the early part of the 20th century and the advent of cell culture, i.e. the ability to grow cells in the laboratory, outside of the body. The next major step was the identification of DNA as a double helix, this in the early 1950s, and by the 1970s recombinant DNA technology had been established. More recently we have the Human Genome Project, and this will be followed by the proteome and the physiome, areas now of significant activity. It was the advent of cell culture, however, that in a very real way initiated the biological revolution.

Another important development in these early years was the field of transplantation, i.e. the ability to transplant an organ from one individual to another individual. The earliest report of an attempt to carry out such a transplantation was reported in a *New York Times* article published November 14, 1911. This attempt apparently was not successful, and it was not until four decades later that such an operation was successfully performed, this at Massachusetts General Hospital in Boston. The transplantation of organs such as the heart, kidney, liver, and pancreas are now all routinely carried out and have helped thousands of patients. Even so, there are more than 90,000 individuals in the United States alone waiting for an organ for transplantation. The patient need far exceeds the availability of organs from donors, and one cannot foresee this transplantation crisis going away without there being an alternative to donated organs. Tissue engineering and regenerate medicine has the potential to be that alternative.

The earliest mention of the concept of a more biological approach dates back to a book published in 1938.[3] It was not until the second half of the 20th century, however, that one began to see more and more research in this emerging field. This is when many of the pioneers in research began to make their mark. Leaders that emerged in the 1970s and 1980s include Anthony Atala, Francois Auger, Stephen Badylak, Eugene Bell, Robert Langer, Gail Naughton, Julia Polak, Charles Vacanti, his brother Joseph Vacanti, and Ioannis Yannas. In fact, one of the early publications was that of Professor Yannas, who with his collaborators described in 1982 the partial regeneration of skin following the grafting of an animal with a cell seeded scaffold.[4]

The term tissue engineering was in fact not "coined" until Autumn 1987, this at a meeting at the National Science Foundation. This led to the first meeting called "tissue engineering" which was held in early 1988 at Lake Tahoe, California,[5] and it was that same year that the first research grants were awarded through a Federal program labeled tissue engineering. This was a National Science Foundation program, and I and my collaborator were fortunate enough to get one of these first six grants. Our goal in 1988 was to create a tissue engineered blood vessel; however, the application was not for clinical use, but rather to develop a better *in vitro* model for vascular biology studies.

The 1980s also was a time when startups were being formed. This included two pioneering companies, Organogenesis in 1986 and Marrow Tech in 1987, with the latter becoming later Advanced Tissue Sciences. More will be said about these companies in a later section.

3. The 1990s

Although as indicated previously, research in tissue engineering goes back well into the 20th century, it was in the 1990s that research in the field accelerated. This thus was a decade of considerable excitement. Not only was research in this field rapidly expanding in the academic arena, but it was a time when there was considerable activity on the commercial front, and there were bold visions of the future.[6,7]

This also was the decade when tissue engineering as a community began to organize. In 1995 the Tissue Engineering Society was formed, and it was Charles Vacanti who was the first president of the society. The journal *Tissue Engineering* was launched in 1996, and in that same year the first meeting of the Tissue Engineering Society was held in Orlando, Florida. In addition, this was the decade when major centers of activity began to emerge including our own center in Atlanta, the Georgia Tech/Emory Center for the Engineering of Living Tissues, established in 1998 as a National Science Foundation Engineering Research Center.

With all this there was considerable hype in the media. An example of this is the respected business journal *Barron's* touting the future of tissue engineering in an article entitled "Spare Body Parts" and that headlined a US$100 billion industry.[8] The television media also contributed to the excitement and the hype. An example of this is Dr. Michael Guillen, an ABC science correspondent, stating in a September 29, 1999 broadcast that "when historians look back at the 20th century [...] the greatest achievement will not be space travel or computers [...] but will be in the fields of tissue engineering and genetic medicine." Furthermore, the fault for all of this does not entirely lie with the media as scientists have also contributed to the hype by overstating the potential benefit for patients and/or by talking about unrealistic timelines for a product or a treatment to reach the patient bedside. The result, however, was that pioneering companies were caught in a circle where the hype helped with the need to continue to raise funds but also lead to increased expectations by investors and by the public.

It also was in the 1990s when the term "regenerative medicine" came into use. For some this term is synonymous with stem cell technology; however, regenerative medicine must be more than just stem cells if one is to move from basic biology to clinical applications. The two areas of tissue engineering and regenerative medicine in fact are very much complementary, and there are many who use the two terms interchangeably.

4. 2000 to Present

The industry that grew up in the 1990s did bring some products to the market. These were largely skin substitutes, with the one exception being Carticel, an autologous cell procedure for treating cartilage defects. These were developed by pioneering individuals such as Professor Yannas whose dermal regeneration template received regulatory approval in 1996. Then there were other skin substitutes, TransCyte, Appligraf, and Dermagraft, developed by Advanced Tissue Sciences and Organogenesis.[9,10] These pioneering companies, however, had their own set of problems. These were addressed as part of a meeting at Georgia Tech held under the auspices of the Medical Technology Leadership Forum (MTLF) in 2007.[11] These problems ranged from an insufficient science base, e.g. not understanding the mechanism of action, to overestimating the patient need, i.e. market size, to underestimating the disruptiveness of this new technology. It was hampered by delays in regulatory approval and in reimbursement approval, and there were a variety of less than optimal business/management decisions. The "bottom line" was that the time from the benchtop to a marketable product was far too long, with a result that several of these pioneering companies, including both Advanced Tissue Sciences and Organogenesis, ran into severe financial problems.

This industry continued into this century in a fledgling state. This is an industry that in some ways is still in the process of being born; however, it is an industry that continues to progress. Whereas *Barron's* talked about a US$100 billion industry, Michael Lysaght[12] estimates that in 2006 total sales was US$240 million, this as compared to less than US$100 million in 2000. Furthermore, in an expanded study using a broader definition that attempts to include all aspects of tissue engineering, regenerative medicine, and stem cell therapeutics, Lysaght and co-workers estimate annual sales at US$1.5 billion for 2007 with another US$860 million in development stage spending, a total of 170 companies, and more than 6000 employees.[13] Of those using cellular approaches, 62% are employing allogeneic cells, and of the stem cell companies, 61% are based on adult stem cells, 27% cord blood-derived cells, and only 12% embryonic stem cells. Whereas this industry in the 1990s appeared to be dominated by US companies, it clearly is now becoming a global industry, with a high percentage of new firms being located outside of the US. Also many of the new firms are in the stem cell area.

There thus are some positive signs appearing. Organogenesis, one of the pioneering skin substitute companies with their product Apligraf, is now turning a profit, and Advanced Tissue Sciences, having gone bankrupt and then acquired by Smith & Nephew, now has its skin substitute products manufactured and marketed by Advanced Biohealing. One of the real successes in bringing tissue engineering to the patient bedside has been the use of the acellular small intestine submucosa (SIS) for a variety of purposes, including musculoskeletal repairs.[14] It thus would be wrong to describe the 1990s as a decade of failures as there are products on the market and these products are helping patients.

Furthermore, the development of these products in the 1990s represented learning experiences, and they have provided insight into what some of the critical issues were and continue to be. Some even say that the industry has come of age, that it has moved into a new era.[15] This may be an overstatement; however, the good news is that the "big boys" of the medical device industry are increasingly investing in tissue engineering and regenerative medicine. They realize that the convergence of biologics with medical devices eventually will have an enormous impact on the industry.[11] To ultimately commercialize a technology, however, requires addressing a variety of other issues that go beyond the science. These will be addressed in various chapters of this publication; however, clearly, scientific success does not ensure commercial success as the issues are very different.

Finally, it should also be noted that the organization of the community has continued to evolve. The Tissue Engineering Society became the Tissue Engineering Society International (TESI) and held meetings in Freiburg, Germany and in Kobe, Japan in 2001 and 2002 respectively, and then in 2003 TESI returned to Orlando, Florida. At that time discussions began to take place to make the society a truly international organization, and in 2005 out of TESI the Tissue Engineering and Regenerative Medicine International Society (TERMIS) was established. As part of this three regional chapters were formed, one in North America, one in Europe, and one in Asia. The plan included holding a World Congress once every three years, with 2006 being in Pittsburgh, 2009 in Korea, and 2012 somewhere in Europe. In years not corresponding to a TERMIS World Congress, the chapters are each authorized to hold a regional meeting.

5. What About the Future?

Even though tissue engineering has been over promised and under delivered in the past, and the same can be said for the more recent development of regenerative medicine, the potential is still there. Advances that are envisioned include the following:[16]

- *in vitro* models for the study of basic biology and for use in drug discovery;
- blood cells derived from stem cells and expanded *in vitro*, thus reducing the need for blood donors;
- an insulin-secreting, glucose responsive bioartificial pancreas;
- heart valves that when implanted into an infant grow as the child grows; and
- repair/regeneration of the central nervous system.

What is necessary in order to create this brighter future? To start with, as with many new technologies, it is only when the second generation of products comes along that there develops a strong scientific foundation upon which the product development/treatment strategy can be established. This is true of tissue engineering and regenerative medicine, and this foundation is very much needed. The critical issues that need to be addressed have been identified recently through two different studies. One of these is a report published by the Multi-Agency Tissue Engineering Science (MATES) Interagency Working Group, a consortium of six Federal agencies.[17] The other is published in late 2007.[18] These represent a road map, and if we address what is identified in these two separate studies, we will be laying the foundation. To do this, however, will require the expanded and accelerated efforts of the tissue engineering and regenerative medicine community. As part of the expansion of this effort, more attention will need to be given to the biology. We need to understand basic biological mechanisms and this includes developmental biology. This in turn requires that more biologists join the multi-disciplinary efforts now in progress.

It also must be recognized that the science and the basic technology represents only the beginning. As one moves from the benchtop of the research laboratory ultimately to the patient bedside, there are a variety of other issues that will need to be addressed.[19] One of these is the manufacturing process as it must be recognized that it is one thing to make one of a kind of a product, a substitute or delivery vehicle in a research laboratory, it is quite different to make 1000 per week with the reproducible quality that would be required to obtain regulatory approval. There also are problems that arise in the testing of the initial cells, in the growth and handling of large volumes of cells, and in maintaining sterility and phenotype. In this the optimization of bioreactor design is critical to meeting the needs of cell expansion and tissue production. Furthermore, the manufacturing process of a living cell product needs to be critically controlled, and the development of new manufacturing techniques will be needed. There will need to be greater attention given to the quality control of what has been engineered. For long-term success a better understanding of the mechanism of action will be needed for use in setting product specifications and for quality control. Once manufactured, how is off-the-shelf availability of a cell-based product to be provided? Is it to be stored fresh in which case its shelf life will be limited or is it to be cryopreserved so as to have extended shelf life? What is the best approach if a particular product or cell-based strategy is to be implemented in the variety of hospitals that make up our healthcare system?

The ultimate success of a tissue-engineered product is also dependent on the market for the product. This is influenced by its uniqueness, accessibility, and cost, as well as the market competition and physician acceptance. There needs to be appropriate business models. In developing these, we need to learn from the problems encountered by the early pioneering companies. Their experiences should be viewed as case studies from which we all can learn as we move towards the next generation of products, and the MTLF meeting held at Georgia Tech in June 2007 represented a start in this.[11]

6. Concluding Discussion

In summary, the advances in the science and the technology have been and continue to be exciting; however, to date tissue engineering and regenerative medicine have been over-promised and under-delivered. Still, if the tremendous potential that exists could be realized, this would dramatically alter the practice of medicine in the future.[20] If we are able to address the critical issues, then there will be a whole new generation of totally biologic products and strategies.

It should be noted that there are significant, multi-disciplinary national initiatives that have been or are being established in countries in Europe and also in Asia. In some cases these have resulted or are resulting in the formation of centers and in others these are taking the form of a broader type of initiative. Example countries in Europe include Germany, Ireland, the Netherlands, and the United Kingdom and in Asia examples are China, Japan, and Korea. The US itself needs a national initiative. Stem cell technology[21] has become an important part of tissue engineering and regenerative medicine, and the US should not let the controversy surrounding human embryonic stem cells prevent a major expansion and acceleration of the US effort in tissue engineering and regenerative medicine. Now that there is the road map provided by the recent studies, how is this to be implemented in the US and elsewhere?

The fact that this field has become a global activity bodes well for the future, in particular with the current constraints on human embryonic stem cell research in the US where as a result other countries are taking the lead. It is this global effort that will write the future history of tissue engineering and regenerative medicine. As we look into this future, in some cases the advances in a treatment or therapy will be made through the development of a replacement tissue or organ. In other cases, for example where the biological complexity prevents us from growing replacement parts in the laboratory, the solution in the future may be the fostering of repair or regeneration. Perhaps we will even be able to diagnose disease at what we now consider to be a preclinical stage and to induce a biological repair and/or regeneration at an early state. All of this will be part of the future of this still emerging field of tissue engineering and regenerative medicine, a future that has the potential of altering the practice of medicine as we know it today.

Acknowledgements

The author is a member of the Georgia Tech/Emory Center for Engineering of Living Tissues, funded as a National Science Foundation Engineering Research Center under Award EEC-9731643.

References

1. Nerem, R.M. The challenge of imitating nature. In: Lanza, R., Langer, R. and Vacanti, J. (eds.), *Principles of Tissue Engineering*, 3rd Ed. (Elsevier/Academic Press, New York, NY, 2007), pp. 7–14.
2. Citron, P. and Nerem, R.M. Bioengineering: 25 years of progress — but still only a beginning. *Technol. Soc.* **26**, 415–431 (2004).
3. Carrel, A. and Lindbergh, C. *The Culture of Organs* (Paul B. Hoeber, Inc., Harper Brothers, New York, NY, 1938).
4. Yannas, I.V., Burke, J.F., Orgill, D.P. and Skrabut, E.M. Wound tissue can utilize a polymeric template to synthesize a functional extension of skin. *Science* **215**, 174–176 (1982).
5. Skalak, R. and Fox, C. (eds.). *NSF Workshop, UCLA Symposia on Molecular and Cellular Biology, 1988* (Alan R. Liss, Inc., 1988).
6. Langer, R. and Vacanti, J.P. Tissue engineering. *Science* **260**, 920–926 (1993).
7. Nerem, R.M. and Sambanis, A. Tissue engineering: from biology to biological substitutes. *Tissue Eng.* **1**, 3–13 (1995).
8. Palmer, J. Spare body parts. *Barron's* (15 May 2000).
9. Parenteau, N. Skin: the first tissue-engineered products — the organogenesis story. *Sci. Am.* **280**(4), 83–84 (1999).
10. Naughton, G. Skin: the first tissue-engineered products — the advanced tissue sciences story. *Sci. Am.* **280**(4), 84–85 (1999).
11. *The Impact of the Convergence of Biologic with the Medical Device Industry*, MTLF Report, 4–5 June 2007. Forum held at Georgia Institute of Technology, Medical Technology Leadership Forum (2007).
12. Lysaght, M.J., Deweerd, E.S. and Jaklenec, A. The tissue engineering industry. In: Lanza, R., Langer, R. and Vacanti, J. (eds.), *Principles of Tissue Engineering*, 3rd Ed. (Elsevier Academic Press, 2007), pp. 1265–1270.
13. Lysaght, M.J., Jaklenec, A. and Deweerd, E. Great expectations: private sector activity in tissue engineering, regenerative medicine, and stem cell therapeutics. *Tissue Eng. Part A* **14**(2), 305–315 (2008).

14. Badylak, S., Arnoczsky, S., Plouhar, P., Haut, R., Mendenhall, V., Clarke, R. and Horvath, C. Naturally occurring extracellular matrix as a scaffold for musculoskeletal repair. *Clin. Orthop. Relat. Res.* **3675**, 333–343 (1989).
15. Mason, C. Regenerative medicine: the industry comes of age. *Med. Dev. Technol.* **18**(2), 25–28 (2007).
16. Nerem, R.M. Cell-based therapies: from basic biology to replacement, repair, and regeneration. *Biomaterials* **28**, 5074–5077 (2007).
17. *Advancing Tissue Science and Engineering: A Multi-Agency Strategic Plan,* The Multi-Agency Tissue Engineering Science (MATES) Interagency Workshop Group, Washington, D.C. (2007).
18. Johnson, P.C., Mikos, A.G., Fisher, J.P. and Jansen, J.A. Strategic directions in tissue engineering. *Tissue Eng.* **13**(12), 2827–2837 (2007).
19. Ahsan, T. and Nerem, R.M. Bioengineered tissues: the science, the technology, and the industry. *Orthod. Craniofac. Res.* **8**, 134–140 (2005).
20. Nerem, R.M. Tissue engineering: the hope, the hype, and the future. *Tissue Eng.* **12**(5), 1143–1150 (2006).
21. Committee on the Biological and Biomedical Applications of Stem Cell Research, Commission on Life Sciences, National Research Council, Board on Neuroscience and Behavioral Health, Institute of Medicine. *Stem Cells and the Future of Regenerative Medicine* (National Academies Press, Washington, D.C., 2001).

PART II

CELLS FOR TISSUE ENGINEERING

Chapter 2

A Brief Introduction to Different Cell Types

Lee Buttery and Kevin M. Shakesheff

Abstract

For any tissue engineering or regenerative medicine strategy its success is dependent largely on controlling the biology of the cells at the site of repair or regeneration, since it is the cells that constitute and co-ordinate the basic structure and function of tissues. The ideal situation would be to augment intrinsic self-repair mechanisms by stimulating the mobilisation, recruitment and activity of cells within the body. At present, however, this approach is quite limited, but as our knowledge of cell biology, cell environments, cell signalling and cell trafficking increases such activation of self-repair mechanism might become possible. The alternative is to supply cells exogenously and this raises a number of questions and challenges, such as what are the most appropriate sources and types of cell, how to control the growth and differentiation of the cells and how to deliver the cells to site of repair. This chapter provides a brief summary of some of the various cell types, including differentiated somatic adult cells, somatic stem cells, foetal cells and embryonic stem cells that are being and which might be used in promoting and understanding tissue repair and regeneration.

Keywords: Cell Sources; Stem Cell; Embryonic Stem Cell; Adult Somatic Stem Cell; Differentiation.

Outline

1. Introduction
2. Cells and Tissue Engineering
3. Mature or Primary Cells
4. Stem Cells
5. Sources of Stem Cells
    5.1. Adult (somatic) stem cells (ASC)

5.2. The adult stem cell niche
5.3. Bone marrow stem cells
5.4. Haematopoietic stem cells (HSCs)
5.5. Bone marrow stromal stem cells (BMSCs)/mesenchymal stem cells (MSCs)
5.6. Multipotent adult stem cells (MAPCs)
5.7. ASCs from other tissues
5.8. Cord blood stem cells and foetal stem cells
5.9. Embryonic stem cells
5.10. Epiblast stem cells
6. Immortalised Cell Lines
7. Reprogramming
8. Differentiation of Cells
9. Regulatory Issues
9.1. Cells
9.2. Animal studies
References

1. Introduction

The processes and methodology behind developing a tissue engineering or regenerative medicine product/therapy involves three core starting components — cells, scaffolds and signals, which can be applied singly and in various combinations. Cells must be able to form the desired tissue and this is ultimately dependent on the type and source cell. The scaffolds play an important role in creating an environment that is compatible with supporting the growth of the cells leading to development of the mature tissue and also interacting and integrating with the surrounding tissues after implantation. Finally, signals are delivered by molecules such as growth factors or extracellular matrix components (ECM), which need to be present at the site to stimulate the correct response from the cells.[1–4]

This chapter focuses on the uses and sources of cells for tissue engineering and will also provide a brief discussion of approaches to control their growth and differentiation and also regulatory and ethical issues associated with use of cells. More extensive discussions of different types of cells and their application in tissue engineering and regenerative medicine are presented in other chapters of this book.

2. Cells and Tissue Engineering

Many tissue engineering and regenerative medicine strategies are dependent on cells supplied from an exogenous source and for those cells to contribute to achieving an effective, long-lasting and stable repair of damaged or diseased tissues there are a number of important criteria that should be considered and ideally met.[1–11] These include:

(i) Obtaining sufficient numbers of cells to be able to achieve the repair — even small amounts of tissue may require tens of millions of cells.
(ii) Deciding how best to promote proliferation and expansion of cell numbers and whether that should be done *in vitro* and/or after implantation.
(iii) Accessibility — how easy is it to take samples of the relevant tissue from which to isolate cells?
(iv) Differentiating the cells to the correct cell type and ensuring the cells perform the necessary functions, such as secreting molecules like extracelluar matrix proteins, hormones, cytokines, etc.
(v) Ensuring that the cells adopt the appropriate two-/three-dimensional organisation and tissue architecture and that they are structurally and mechanically compliant with the normal demands of native tissue.
(vi) Integrating with the native cells and tissues (including vascularisation and innervation, if required) and overcoming or minimising the risk of immune rejection.
(vii) Delivering cells — are the cells going to be delivered directly or be combined with a scaffold and how does that influence other factors and considerations listed?

To some extent the ability to satisfy these criteria is dependent on the qualities and sources of the cells. There are a number of different sources of cells that can be and have been used for tissue repair and regeneration and include:

(i) Mature, differentiated cells isolated from the patients' own tissues,[9,12] e.g. skin.
(ii) "Adult" stem cells isolated from specific tissue sites/compartments within the patient,[8,13–16] e.g. haematopoietic stem cells (HSCs) from the bone marrow.
(iii) Cells isolated from (elective) aborted foetuses, e.g. embryonic germ (EG) cells derived from the developing gonad region of first trimester foetuses or tissue-specific cell types from later foetuses.[17]
(iv) Cells isolated from the very early stages of the developing embryo, e.g. embryonic stem (ES) cells derived from the pre-implantation blastocyst.[18,19]

Mature, differentiated cells and "adult" stem cells are integral components of our bodies and are also often referred to as somatic cells and somatic stem cells. Conversely, ES cells are derived from small clusters of cells that exist only transiently during the very early stages of development and in some respects they can be regarded as being "man-made".

3. Mature or Primary Cells

Whilst mature cells isolated from tissue biopsies can be potentially used for re-implantation into the same donor or a related or immunologically matched recipient, thus overcoming or limiting problems of immunocompatibility, they are probably not the best source of cells for tissue repair. These are generally differentiated and differentiating cells that inherently have a low potential to proliferate, which makes generating sufficient cells to promote tissue repair potentially difficult. Moreover, these cells are usually committed to a particular cell type that is restricted to the tissue type from where they were harvested. This therefore raises issues of accessibility of tissue sites from which cells can be harvested. For example, while samples or biopsies of skin tissues can be quite readily harvested with minimal risk to the patient, harvesting cells and tissues like heart and brain is more challenging and poses a far greater risk to the patient.

Despite these potential limitations there are some clinical examples where primary/mature cells isolated from a patient have been used to promote repair of their own tissues. A good example is the Carticel® autologous chondrocyte product for articular cartilage repair. This involves first collecting a small biopsy of normal cartilage tissue from the patient, which is then shipped to the company where chondrocytes are isolated and cultured for several weeks to increase cell numbers, before being shipped back to the surgeon for re-implantation into sites of articular cartilage injury and damage. This system appears to work well, especially since articular damage while debilitating is not normally life threatening, so surgery can be planned well in advance and can be performed by minimally invasive "keyhole" surgical techniques.[20–24] There are several other products used in clinical orthopaedics based on variations of this approach.

One goal of tissue engineering is to supply ready to use functional tissue and there are examples of where mature cells have been used to produce "fully formed" tissues in the laboratory that are supplied as a ready for use product. Probably the best example of this approach is tissue engineered skin, e.g. Apligraf®, which is a living skin substitute comprising skin cells (keratinocytes and fibroblasts) isolated from neonatal foreskin incorporated with a natural scaffold (collagen type 1) and is used clinically as a living bandage to treat burns and non-healing ulcers.[12] The product is stable for a number of weeks and thus can be used "off the shelf", which greatly benefits the patient who requires immediate treatment. Surprisingly this product does not invoke an obvious immune response and may relate to how the product is processed to remove dendritic cells, macrophages and other immune response cells and the fact that the fibroblasts and keratinocytes used in generating the product do not appear to express HLA molecules.[25]

4. Stem Cells

A stem cell can be described as an immature, primal, or undifferentiated cell that is usually capable of dividing to produce at least one daughter cell that is identical to the mother cell, a process called self-renewal, which may be perpetuated over a number of cell divisions or possibly even indefinitely.[26–28] Figure 1 illustrates some of the general concepts of self-renewal. All three mechanisms illustrated in Fig. 1 may be relevant and may, to some extent, occur in the same stem cell population, but overall asymmetric division contributing to both self-renewal and differentiation is probably the predominant mechanism. Evidence for symmetric

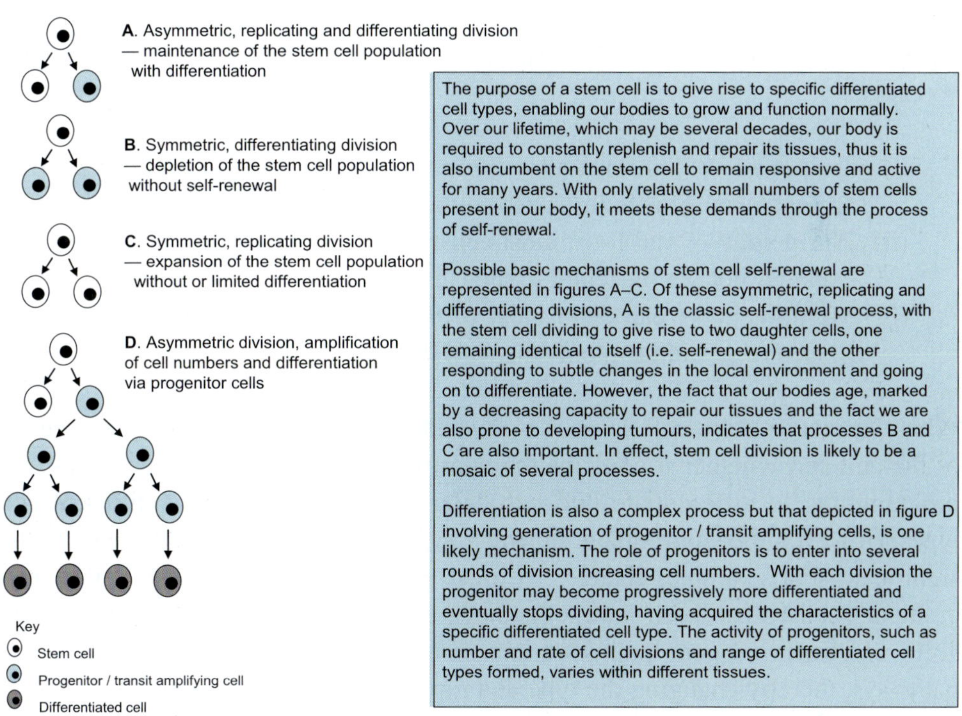

Fig. 1. A general overview of the concept of stem cell self-renewal and differentiation.

differentiating division where a stem cell population is reduced and eventually exhausted is supplied by the fact that our bodies age and our capacity to repair tissues, where stem cells are known to be present like bone and skin also diminishes with age. Evidence for symmetric, replicating division, where a stem population expands and potentially overgrows the tissue is somewhat more controversial but there is evidence to suggest that cancers may be the result of stem cell self-renewal.[29,30]

Stem cells can be isolated from various tissues and grown in the culture dish and in this environment self-renewal of stem cells can be often readily stimulated and maintained for several weeks, months or even years resulting in considerable amplification of stem cell numbers, which in terms of cell therapy and tissue engineering can be extremely useful. In the body however, stem cells may divide relatively infrequently, remaining dormant or quiescent for prolonged periods until they receive the appropriate set of signals to start and stop dividing. This tight control of stem cell self-renewal *in vivo* is necessary to ensure that these cells do not divide indeterminately and as discussed above potentially overgrow the tissue and in effect become a cancer. For these reasons stems cells are also usually quite rare with only small numbers of cells being found at defined locations within our tissues, for example within the bone marrow.[7,8,13,14,26]

Another important concept is stem cell potency, plasticity or range of cell types to which it can give rise. There are three basic measures of stem cell potency:

(i) Totipotent — can form all cell types that contribute to the formation of an organism. This is restricted to the fertilised egg or zygote.
(ii) Pluripotent — can form most cell types of an organism including germ cells but not the placental tissues (e.g. ES cells and EG cells).
(iii) Multipotent — can form most cells issues within a particular tissue or tissues (e.g. "adult" stem cells — HSCs).

What determines stem cell potency is dependent to a large extent on the genetics of the cell and whether it contains the appropriate active or activated genes and programming to differentiate into particular cell type or range of different cell types. However, the environment in which the stem cell is located or placed is also significant. For example, changes in local growth factor/cytokine/hormone etc. gradients, cell-cell and cell-matrix contact are important in the switching "on" and "off" of genes and gene pathways and possibly even re-programming of those pathways, thereby changing the type and range of cells that are generated.[26,27,31,32] As will be discussed in the succeeding sections of this chapter, this classification of stem cell potency is not rigid. Indeed, it is evident that the distinction between pluripotent and multipotent is becoming increasingly more blurred, with some cells seemingly having greater plasticity than previously realised.

The process of differentiation where a cell acquires a particular set of characteristics enabling it to perform a specific function (e.g. insulin production by pancreatic beta cells, dopamine production by neural cells, bone formation by osteoblasts, etc.) usually involves formation of an intermediate progenitor cell, sometimes also called a transit amplifying cell. To some extent progenitor cells are similar to stem cells, particularly "adult" stem cells, being able to divide a number of times, and may even have a very limited capacity for self-renewal, but usually with each successive division the cells become progressively more differentiated. The main function of progenitor cells is to increase cell numbers (often exponentially) and in response to morphogenic stimuli contribute to the formation of tissue and organs with the appropriate size, shape and mass. Progenitor cells may also exhibit a degree of plasticity, ranging from being unipotent and restricted to differentiating to one specific cell type through to being bi- or tripotent and giving rise to several differentiated cell types. For example in skeletal tissue, cartilage cells (chondrocytes), fat cells (adipocytes) and bone forming cells (osteoblasts) are believed to arise from a common tri-potential progenitor and this progenitor is itself derived from a stem cell located within the bone marrow microenvironment.[14] Apoptosis or programmed cell death is also an integral part of the process of cell proliferation and differentiation, although it is not discussed here.

5. Sources of Stem Cells

5.1. Adult (somatic) stem cells (ASC)

ASCs are found in many tissues throughout the body. These cells divide and differentiate to replenish the supply of differentiated cells that die as part of the natural "life cycle" of the tissue or to repair damaged tissue. Tissues like the blood, skin, liver, gut and bone are replenished and repaired almost constantly and while the ability to maintain and repair our tissues diminishes as we get older, the fact that we can live for several decades demonstrates clearly the capacity of ASCs for self-renewal.

In the body, ASCs usually differentiate to a particular cell type or range of cell types and these are usually associated with the tissue in which they are located. In this regard ASCs are considered to be multipotent cells. However, evidence is accumulating to suggest that ASCs might in fact have pluripotency and this is supported by several lines of experimental investigation:

(i) Stem cells from the bone marrow have been differentiated into neural cells[33] and likewise stem cells from the brain have been differentiated into blood cells.[34,35]

(ii) Studies on bone marrow transplantation in rodents and humans have demonstrated that small numbers of marrow-derived stem cells can migrate to various tissues and organs around the body and at least colonize, but possibly also differentiate into or stimulate cells in those tissues.[36,37]

(iii) Bone marrow stem cells have been used in clinical trials to help promote repair of heart tissue damaged after a heart attack.[38,39]

(iv) The expression of genes and proteins associated with pluripotency (e.g. those expressed by ES cells) have been detected in some ASCs.[40,41]

5.2. The adult stem cell niche

The capacity of ASC to self-renew and to differentiate into mature cell types is a tightly regulated process which is controlled by a number of tissue-specific environmental factors. In order to maintain tissue homeostasis, these cells often reside in what is referred to as a "stem cell niche".[8,26,27,42] This "niche" is essentially a microenvironment that limits the exposure of stem cells to differentiation, apoptotic and other signalling events that would otherwise deplete stem cell reserves. In addition, the "niche" tightly controls stem cell division to avoid overpopulation of a tissue with these cell types, which would give rise to cancer. When required, stem cells can be activated to produce transit amplifying or progenitor cells that are at that point committed to producing the mature cell types of that tissue. Hence, the dynamic interplay between stem cells and their niche is essential in the maintenance of healthy tissues and so it is important to understand these relationships in order to utilise these cell types effectively for therapeutic applications.

5.3. Bone marrow stem cells

The bone marrow is one of the most abundant sites for ASCs and these are also probably the most studied and best understood of all ASC types. Part of the reason why marrow stem cells have been studied so extensively, compared to other ASCs is accessibility — samples of bone marrow can be collected relatively easily by introducing a needle directly into the bone marrow, usually at the iliac crest and aspirating marrow tissue. The marrow tissue can be then cultured for further investigation. Stem cells from the bone marrow often enter the circulation and can be also found in peripheral blood samples.[14,16,43–47]

There are in fact two distinct types of ASC within the bone marrow — the haematopoietic stem cell (HSC) that classically gives rise to the entire blood cell lineage[16,44] and the bone marrow stromal stem cell (BMSC), also called the mesenchymal stem cell (MSC), that classically gives rise to various connective tissues notably bone, cartilage and adipose tissue.[14,45–48] Other cell types are also found in bone marrow aspirates, notably red blood cells, endothelial cells, fibroblasts, adipocytes and osteoblasts. There is also evidence for distinct subpopulations of stem cells with the marrow environment, notably multipotent adult progenitor cells (MAPCs), which appear to share similar properties to ES cells.[40,41] With such a heterogeneous mix of cells it is necessary to apply specific methods to isolate and characterise the stem cells, such as density centrifugation, cell sorting techniques like FACS or MACS and differential adhesion.[14,16,40–48]

In terms of characterisation of a stem cell population such as from the bone marrow, the classic experiment is the colony forming unit (CFU) assay,[14,16,40–53] where the cell fraction is diluted to specific densities or number of cells, including down to a single cell (determined from a total mononuclear cell count). The cells are cultured for several days and then observations made for growth of distinctive colonies of differentiating cells which will be the progeny of any stem cells present the culture. The colonies can be characterised by morphology and also by more detailed molecular and biochemical assays.[14,53] The colonies can be also counted and collectively with data from the other assays this can be used to assess the relative abundance or frequency of a particular stem cell type within the bone marrow. It is believed that the stem cell population within the marrow is somewhere between one cell in 10,000 down to 100,000 cells.[14,16,40–53]

There are some more rudimentary differences between the HSCs and BMSCs, which are that HSCs tend not to adhere directly to cell culture plates (require co-culture or a semi-solid matrix, see below), whereas BMSCs will, in general adhere to cell culture plates. Thus, a simple adhesion assay performed over an hour or so can be used to initially differentiate (termed a differential adhesion assay) between the HSCs[54,55] and BMSCs fractions.[14,16,40,53]

5.4. Haematopoietic stem cells (HSCs)

Of all the different types of stem cell, HSCs are the best studied and are also an example of a very successful stem cell therapy, forming the basis of bone marrow transplantation to treat various types of leukaemia. Within the body, HSC are found mainly in the bone marrow (may also circulate in the blood and occupy other tissues like the spleen) and they function to generate the entire blood cell lineage (e.g. red blood cells, leucocytes and lymphocytes).[16,44] Although the marrow is a diffuse tissue, the proliferation and differentiation of HSCs is tightly controlled within microenvironments or niches.[8,16,26,27,44] The exact nature of the HSCs niche is not known, but physical contact with stromal cells (including BMSCs) and osteoblasts is important, as is the presence of various growth factors, in particular stem cell factor (SCF) and also a number of specific cytokines (colony stimulating factors).[16,44]

There are a number of "markers" expressed on the surface of HSC such as CD34 (and many others that are used in varying combinations) which can be used sort fractions of HSC.[16,44,56,57] Once collected, HSC populations can be cultured and stimulated with various growth factors to study mechanisms of differentiation, e.g. CFU assays, or expanded to increase numbers of HSCs for clinical application. HSCs do not generally adhere to cell culture plates and so are either co-cultured on a feeder layer of growth-arrested stromal cells or fibroblasts or on a semi-solid support matrix such as methylcellulose, which is a gel-like material. In both cases biochemicals such as SCF and CSFs need to be added to the culture medium to help stimulate and maintain HSCs proliferation. Since only small numbers of HSCs are usually present in the body it is also possible to treat a patient with CSFs, which stimulate and mobilise HSCs within the body and potentially can increase the "yield" of stem cells from bone marrow biopsies.[16,44,58,59]

5.5. Bone marrow stromal stem cells (BMSCs)/mesenchymal stem cells (MSCs)

This stem cell type "shares" the marrow environment with HSCs (and several other cell types) and as indicated above there is a certain amount of interaction between the different cell populations. Again the exact nature of the microenvironment or niche in which BMSCs reside is not known, but the marrow and the bone tissue that confines it are dynamic environments and the key function of BMSCs is to help maintain these environments, differentiating into cells such as osteoblasts, adipocytes and stromal cells/fibroblasts.[14,36,40,46,48,53] For example, throughout our life, bone is constantly being resorbed by osteoclasts (derived from HSCs) and replaced by osteoblasts in response to normal physiological demand, e.g. mechanical loading associated with walking, running, etc. and metabolic turnover, with bone tissue serving as a reservoir of essential mineral ions and cytokines. Our ability to repair fractured or broken bones also serves as powerful illustration of the function of BMSCs.

Selection of BMSCs from marrow aspirates is probably less efficient than that for HSCs because fewer specific "markers" exist for BMSCs.[14,48,53] Initial isolation can be achieved by differential adhesion assays, which will remove the HSC fraction but other cells present in the marrow like adipocytes, and stromal cells/fibroblasts will also adhere to cell culture plates. There are some "markers" that are useful for selection of BMSCs including Stro-1, an as yet uncharacterised cell surface antigen, CD44, a glycoprotein that binds hyaluronan, a component of the extracellular matrix and CD105, also called endoglin, and a receptor for the cytokine transforming growth factor beta.[14,48,53] Both CD44 and CD105 can be also used for selecting HSC, but some distinction can be achieved by subsequent adhesion of the selected cells.[53,60–63]

The BMSC can, in general, be maintained and expanded in culture quite easily, without the need to add specific growth factors, such as required for HSC. The differentiation potentials of BSMCs can be assessed by CFU assays and also by adding specific growth factors to stimulate differentiation into a particular cell type. Such experiments reveal that BMSCs can be readily differentiated *in vitro* to osteoblasts, chondrocytes, adipocytes, stromal cells, tendon cells and muscle cells and possibly several other cell types.[14,48,53] Based on the potential to differentiate into these various connective tissue cell types, BMSCs have been used in a number of tissue engineering studies, for example to repair damaged bones and joints in both animal models and also humans.

There are also a few clinical examples where BMSCs have been used to help treat genetic diseases of the bone, such as osteogenesis imperfecta,[64–66] where the osteoblasts have an impaired capacity to synthesise bone matrix protein (collagen type 1), resulting in brittle bones. Implantation of donor BMSCs can help restore differentiation of normal osteoblasts, leading to synthesis of normal bone tissue.

5.6. Multipotent adult stem cells (MAPCs)

MAPCs were isolated as a subset of BMSCs/MSCs and so are extremely rare cells. They express cell surface "markers" that are distinct from BMSCs and also express "markers" associated with ES cells and pluripotency.[40,41] Relatively few experiments have been performed to date to explore and confirm the biology of these cells.

5.7. ASCs from other tissues

Many tissues have now been shown to contain a population of stem cells (or cells with stem cell-like characteristics) that serve to replace or repair cells that are lost due to normal wear and tear or injury. Notable examples of such tissue include skin,[67–69] gut[70,71] and liver,[72,73] which as part of their normal function have high cell attrition rates. However, many other tissues, which appear to a have a much more limited capacity for normal repair, such as cartilage,[74] heart,[75] teeth[76] and brain[18,77,78] have also been shown to contain stem cells or putative stem cells. However, it is not known why these tissues are unable to effect self repair as efficiently as tissues like the skin.

Clearly, the isolation and therapeutic application of ASC from some tissues is going to be challenging, however, their study using animal models or human surgical and cadaveric samples is still important for understanding their biology and potentially designing methods, such as scaffold delivery systems to stimulate these cells in the body.

5.8. Cord blood stem cells and foetal stem cells

Blood collected from the umbilical cords of newborn babies has been shown to contain populations of stem cells with characteristics similar to both HSCs and BMSCs and could potentially be useful in allogenic transplants.[79,80]

Stem cells have also been isolated from the cord blood during early pregnancy, using a needle guided by ultrasound. Again these foetal cord blood stem cells have characteristics similar to both HSCs and BMSCs, but unlike cord blood collected from term pregnancies, these foetal stem cells may be less developed immunologically and potentially would cause fewer problems with immune rejection.[81,82] More controversially foetal stem cells can be potentially isolated from elective abortions.

5.9. Embryonic stem cells

Embryonic stem cells are derived from the inner cell mass of the pre-implantation blastocyst and have been obtained from a number of different species including mice, non-human primates and humans.[83–85] Manipulation of the human embryo is restricted to within the first 14 days of its creation.[86] This time point marks a specific stage in embryonic development with the onset of organogenesis and in particular the formation of the primitive streak, which is the rudimentary central nervous system. ES cells are usually isolated before day 5, when the entire embryo is comprised of few hundred cells.

Experiments *in vitro* and *in vivo* show that while ES cells retain the capacity to form any and all foetal and adult cell types including the germ cells, they seem unable to form tissues like the placenta and for this reason ES cells are described as being pluripotent and are not totipotent.[87–90] If implanted into immunodeficient mice a single, undifferentiated ES cell has the capacity to form a teratoma (also called a germ cell tumour), which is a type of tumour comprised of ectodermal, mesodermal and endodermal cells (classic examples can have perfectly formed teeth, hair or kidney tubules, etc. inside the tumour). The ability to induce formation of a teratoma is often used as a measure of stem cell pluripotency.[87–90] Further evidence of pluripotency is provided by injecting ES cells into fresh blastocyts and implanting the blastocyst into a surrogate mother and allowing the embryo to develop naturally — the resulting chimera demonstrates the ability of the injected ES cells to contribute to normal development.[87–90] If the ES cell has been genetically manipulated either to over-express or lack a specific gene or expresses a fluorescent "marker" like GFP then specific developmental processes can be investigated and observed in the resulting chimera (or in the cultured cells).

Most ES cell lines to date (including mouse and human) have been derived by co-culture on growth arrested mouse embryonic fibroblasts.[91] Subsequently, it has been shown that ES cell lines can be also maintained on human fibroblasts derived from tissues like neonatal foreskin. More innovative approaches have utilised the pluripotent characteristic of ES cells to differentiate them into fibroblasts, which have then been used to support culture of ES cells.[92] And most recently human ES cell lines have been derived using human foetal fibroblasts and culture media that are completely free from contact with animal cells or products[93,94] — this is an important step in developing the clinical potential of ES cells and minimising transfer of potentially harmful agents (e.g. animal pathogens). It is also possible to grow ES cells independent of co-culture with another cell type on culture surfaces coated with ECM preparations like gelatin or Matrigel[95] and also purified ECM proteins like laminin[96] have been shown to support ES cell growth. How co-culture and/or coating with ECM proteins helps maintain ES cell pluriopotency and growth is not known, but it may fulfil a niche or microenvironment role.

There are a number of fundamental differences in the biology of mouse and human ES cells and these may be relevant to transferring results from mouse experiments to human.[87–90] One of the most apparent differences is responsiveness to specific cytokines which are required, in addition to co-culture or culture on treated surfaces, to help maintain pluripotency of all ES cells. Mouse ES cells require leukaemia inhibitory factor (LIF), whereas human ES cells appear to be unresponsive to LIF and instead require basic fibroblast growth factor (bFGF). Such differences highlight key difference in the regulation and maintenance of pluripotency.[87–90]

Undifferentiated ES cells express specific transcription factors such as octamer binding protein-4 (Oct-4) and Nanog and also cell surface "markers" including stage specific embryonic antigens (SSEA1-4).[87–90] These characteristic markers are common to both mouse and human ES cells although they vary slightly in their expression of SSEA (SSEA1 on mouse ES cells and SSEA3/4 on human ES cells). ES cells will normally express the enzyme telomerase, which maintains specific cap regions (telomeres) on the chromosomes and function to prevent mutations or damage when the chromosomes are segregated during normal division. Telomerase is therefore associated with a capacity for a cell to divide. Most somatic cells do not express telomerase and with cell division the telomeres progressively shorten and the cell stops dividing when the telomeres research a certain minimum length.

Related to chromosome integrity is the cell karyotype and it is important with any cell type, but particularly stem cells to perform regular karyotype analyses to confirm a normal complement of chromosomes (euploidy for all cells expect the gametes). Any change in chromosome number (aneuploidy) can have deleterious effects on cell behaviour and function. Such changes have been reported in some human ES cell lines.[87–90]

Differentiation of ES cells is induced almost immediately when they are removed from contact with feeder cells and the cytokines LIF or bFGF. The ES cells are removed from the feeders or culture plates either by physical scraping methods and/or enzymatic treatment with collagenase or trypsin. The dispersed ES cells (now differentiating cells) are usually maintained in suspension culture (seeding density can be varied) where they proliferate and differentiate to form distinctive cellular aggregates or embryoid bodies, which contain differentiating cells of ectodermal, endodermal and mesodermal lineage.[87–90] It is important to note that not all ES cell lines form EBs and there are some human ES cells that appear incapable of forming these structures, at least in their classic free-floating aggregate form.[87–90] This again highlights fundamental differences in the basic biology of ES cells, and not all ES cells are alike. Other methods for forming embryoid bodies include culturing a single or small number of ES cells in a small droplet of culture medium suspended from the lid of the culture plate (hanging drop) or similarly in small tube (pellet culture).[87–90] Both the hanging drop and pellet methods produce single embryoid bodies and while being much more labour-intensive (as compared to suspension culture) these methods offer better control over embryoid body growth and size and this can potentially improve subsequent control of differentiation. Another method that combines elements of the hanging drop and suspension methods is to encapsulate single or small numbers of ES cells within alginate beads, which are then cultured in a bioreactor.[97]

To some extent embryoid bodies mimic the process of gastrulation seen in early development. Embryoid bodies can be left intact or dispersed by mechanical and enzymatic methods to single cells and used in a variety of culture experiments, such as CFU assays or stimulation with specific growth factors to study and control differentiation to particular cell types.

5.10. Epiblast stem cells

The epiblast is a tissue that forms at a later stage than ES cells, after the developing embryo implants into the uterus. Cells derived from the epiblast of the mouse and rat and have been shown to behave as pluripotent stem cells.[98,99] Perhaps more interesting is the fact that rodent epiblast stem cells have characteristics that are more similar to human ES cells. As discussed above mouse and human ES cells differ in the factors regulating pluripotency and this has resulted in much debate over similarities and differences between ES cells from different species and the implications for understanding mechanisms of development and their therapeutic applications. However, the rodent epiblast stem cells appear to be regulated by factors and mechanisms that are similar to human ES cells, in particular activin and nodal. These cells therefore provide a valuable cell source for further investigation of the mechanisms of early development and for exploring and understanding the fundamental properties and differences of ES cell from different species.

6. Immortalised Cell Lines

To overcome the limitations of obtaining sufficient numbers of cells with which to develop a tissue engineered product it is possible to engineer immortalised cell lines.[100,101] This is often useful for primary cells that typically have low proliferating potential (e.g. neurons) but is also applicable to ASC where there might be issues of harvesting sufficient numbers of cells (e.g. mesenchymal stem cells). There are a number of approaches for immortalising cells, such as transfecting them with genes encoding growth factors or enzymes involved in regulating cell division such as telomerase. The immortalised cells remain competent to respond to mitogenic signals and will also differentiate, however, there is potential for the growth cells to be become completely disrupted, resulting in overt proliferation and potentially resulting in a tumour or tumour-like growth. Therefore it may be also necessary to further engineer the cells, particularly if they are to be used clinically, so that they can selectively killed or inactivated.[100,101]

7. Reprogramming

In the light of the success of cloning Dolly the sheep[102] and the various other animals that have followed, much interest has been generated in understanding the mechanisms of nuclear cloning/reprogramming and potentially harnessing them for tissue repair strategies. The process of somatic cell nuclear transfer involves the removal of the genetic material of an oocyte and the introduction of the nucleus from a somatic cell into the enucleated oocyte. Under appropriate conditions the oocyte-somatic nucleus "hybrid" can be stimulated to start dividing and has the potential to generate an intact embryo and produce a viable organism that is genetically identical to the donor of the somatic nucleus. The mechanisms that induce such a remarkable transformation are not fully understood, but it seems likely that factors present in the enucleated oocyte can initiate reprogramming of the somatic nucleus to generate a totipotent cell, with the capacity to form a viable embryo.[103–105]

In terms of humans, while ethical, moral and other societal laws and issues prevent cloning of a human infant (e.g. reproductive cloning), the cloning of an embryo (for up to 14 days) is currently allowed in some countries, like the UK.

It is hoped that this can be used to generate cloned human ES cells, which could then be used to generate cells and tissues matched to a specific patient (e.g. therapeutic cloning).[106] However, success has so far been very limited due, in part, to the technical challenges of cloning and also an incomplete understanding of the mechanisms — efficiency of cloning is often very low (requiring large numbers of oocytes).

To overcome some of the technical challenges and in particular the demand for large numbers of donor oocytes, studies are being performed on analysing and using oocyte extracts to try and identify the factors and mechanisms that stimulate reprogramming of a somatic nucleus.[103,104] It is hoped that such factors could be used to reprogramme a somatic cell to become pluripotent without the need to create/destroy an embryo.[103–105] Interspecies nuclear transfer experiments are also being investigated, with the prospect that more abundant animal oocytes such as from cows and rabbits could be used to reprogramme/clone human ES cells.[105,107]

Experiments on reprogramming of somatic cells to a pluirotent, ES cell-like state are also being performed by transduction with transcription factors associated with pluripotency. Such studies have successfully reprogrammed rodent and human fibroblasts to an induced pluripotent state, with the cells exhibiting ES cell-like morphology and expressing ES cell markers.[108–111] Such experiments are important in understanding the mechanisms of pluripotency and in the longer term could replace the use of ES cells and overcome ethical issues associated with ES cells.

8. Differentiation of Cells

Achieving effective and efficient differentiation of cells is a key goal in any tissue engineering strategy and remains a significant challenge (see Fig. 2). There are a number of approaches which can be applied either singly or more commonly in combination.

Stimulation with specific growth factors is a useful approach for directing differentiation of cells to a particular cell type but it is rarely, if ever, 100% effective, with some cells differentiating to other (unwanted) cell types. If using serum-containing medium, it is also useful to batch test a range of samples as many cells are fastidious in their serum requirements and responsiveness. This random differentiation occurs with all cell types/sources and in terms of studying mechanisms of differentiation and/or the potential clinical applications, where a pure cell population is required or desirable, the unwanted/contaminating cell types need to be eliminated. To this end, methods like FACS or MACS are extremely useful for sorting a particular cell type (positive selection or negative depletion) although it does depend on the cell type of interest expressing a suitable cell surface "marker". Similar to this approach is the differential adhesion assay, which is based on the ability of cells to interact with specific ligands or motifs within extracellular

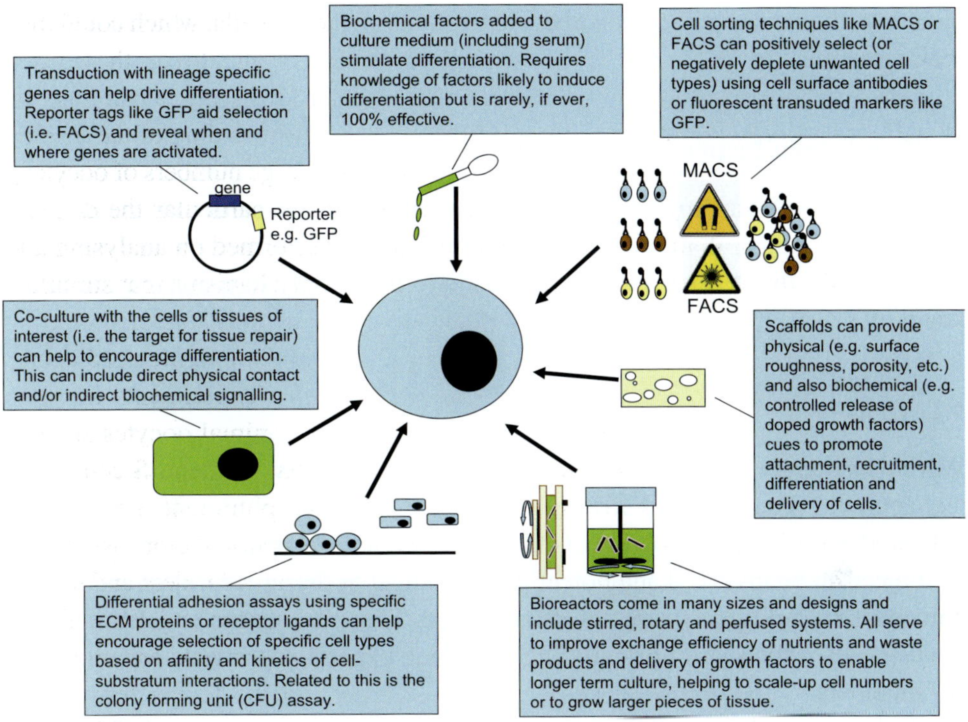

Fig. 2. A general overview of approaches and methods for controlling stem cell growth and differentiation.

matrix proteins, like collagen, fibronectin, laminin, etc. This is usually performed on tissue culture plates and concentration of protein coatings and duration of exposure of cells to the surface can all influence selection of specific cell types.

Many cells types, but especially stem cells (both ES cells and ASCs) are amenable to gene manipulation techniques, which can be used to generate cultures that are near 100% pure for the desired cell type.[108–113] Engineered gene constructs carrying fluorescent tags like green fluorescent protein are also useful for cell sorting and also imaging gene induction and localisation of cells. These experiments obviously require construction or access to an expression vector with the appropriate gene and are also influenced by transfection efficiency, which can be highly variable.

In addition to genetic and biochemical signals, cells and tissues are also influenced by biophysical signals. This includes cell density and factors like compliance of the substratum, which are known to influence differentiation.

Finally, other factors that influence differentiation include scaffolds which can vary in their intrinsic chemistry (e.g. natural or synthetic), whether they are engineered to be biodegradable and also whether they include growth factors or other bioactive molecules either adsorbed onto their surfaces or encapsulated within their structure.

9. Regulatory Issues

Research and clinical application of tissue engineering and stem cells is subjected to a number of ethical, moral and societal rules and regulations. It is not the intention to debate those rules and regulations here but simply to present some of the key issues.

9.1. Cells

Work on virtually any human cell, especially if it involves a patient, requires ethical approval. This involves writing a detailed description of how the cells will be collected and for what experiments they will be used. Justification for the number of subjects to be involved and potential biomedical or therapeutic applications and outcomes, including any risks, is also required. This information is debated by the ethics panel, which will normally include some experts in the field together with lay members.

If the cells are going to be transplanted into a patient, then as part of the ethical review process it is important to demonstrate aseptic, clean processing of the cells to avoid any risk of transferring infections. These are described as Good Laboratory Practice (GLP) and Good Manufacturing Practice (GMP) and usually require laboratories dedicated solely for clinical application of cells — this is very expensive to implement and maintain.[94,98,114] Most therapies require several small-scale studies (clinical trials) where patients are voluntarily recruited before any approval is given for the product to enter the clinic fully.

Use of ES cells and associated applications like cloning is particularly emotive since their derivation involves destruction of an embryo. The debate on the use of these cells and techniques involves a similar ethics review process, but is conducted at a governmental level. Permission to use human ES cells varies in countries around the world.[115] The UK has some of the most liberal laws and permits work on human ES cells, including derivation of new lines. The UK is also one of the first countries to establish a human stem cell bank (including ASCs and ES cells).[116] Samples of human stem cells that have been isolated or derived by researchers in the UK must be sent to this central facility, which not only banks the stem cells, it also performs quality assurance tests including GMP and develops standardised protocols to expand these cell lines and distribute them to other researchers. To receive stem cells from the UK Stem Cell Bank a researcher must make a request that is processed in a manner similar to ethical review.

9.2. Animal studies

Prior to clinical application, many tissue engineering and stem cell therapies are tested and developed using animals. Any experiment performed on a live animal requires a Home Office Project Licence issued under the Animals (Scientific Procedures) Act 1986. Again, this is an ethical review-type process where the animal species, experimental protocol, expected outcomes, potential risks and level of suffering and numbers of animals to be used have be fully explained and justified.

References

1. Stock, U.A. and Vacanti, J.P. Tissue engineering: current state and prospects. *Annu. Rev. Med.* **52**, 443–451 (2001).
2. Atala, A. Engineering tissues, organs and cells. *J. Tissue Eng. Regen. Med.* **1**, 83–96 (2007).
3. Shieh, S.J. and Vacanti, J.P. State-of-the-art tissue engineering: from tissue engineering to organ building. *Surgery* **137**, 1–7 (2005).
4. Naughton, G.K. From lab bench to market: critical issues in tissue engineering. *Ann. N. Y. Acad. Sci.* **961**, 372–385 (2002).
5. Gurtner, G.C., Callaghan, M.J. and Longaker, M.T. Progress and potential for regenerative medicine. *Annu. Rev. Med.* **58**, 299–312 (2007).
6. Parenteau, N.L., Rosenberg, L. and Hardin-Young, J. The engineering of tissues using progenitor cells. *Curr. Top. Dev. Biol.* **64**, 101–139 (2004).
7. Polak, J.M. and Bishop, A.E. Stem cells and tissue engineering: past, present, and future. *Ann. N. Y. Acad. Sci.* **1068**, 352–366 (2006).
8. Mimeault, M. and Batra, S.K. Concise review: recent advances on the significance of stem cells in tissue regeneration and cancer therapies. *Stem Cells* **24**, 2319–2345 (2006).
9. Fodor, W.L. Tissue engineering and cell based therapies, from the bench to the clinic: the potential to replace, repair and regenerate. *Reprod. Biol. Endocrinol.* **1**, 102 (2003).
10. Jimenez, P.A. and Jimenez, S.E. Tissue and cellular approaches to wound repair. *Am. J. Surg.* **187,** 56S–64S (2004).
11. Yamada, K.M. and Cukierman, E. Modelling tissue morphogenesis and cancer in 3D. *Cell* **130**, 601–610 (2007).
12. Eaglstein, W.H. and Falanga, V. Tissue engineering and the development of Apligraf, a human skin equivalent. *Clin. Ther.* **19**, 894–905 (1997).
13. Robey, P.G. Stem cells near the century mark. *J. Clin. Invest.* **105**, 1489–1491 (2000).

14. Biancho, P. and Robey, P.G. Marrow stromal stem cells. *J. Clin. Invest.* **105**, 1663–1668 (2000).
15. Ting, A.E., Mays, R.W., Frey, M.R., Hof, W.V., Medicetty, S. and Deans, R. Therapeutic pathways of adult stem cell repair. *Crit. Rev. Oncol. Hematol.* **65**, 81–93 (2008).
16. Morrison, S.J., Uchida, N. and Weissman, I.L. The biology of hematopoietic stem cells. *Annu. Rev. Cell. Dev. Biol.* **11**, 35–71 (1995).
17. Shamblott, M.J., Axelman, J., Littlefield, J.W., Blumenthal, P.D., Huggins, G.R., Cui, Y., Cheng, L. and Gearhart, J.D. Human embryonic germ cell derivatives express a broad range of developmentally distinct markers and proliferate extensively *in vitro*. *Proc. Natl. Acad. Sci. USA* **98**, 113–118 (2001).
18. Martin, G.R. Isolation of a pluripotent cell line from early mouse embryos cultured in medium conditioned by teratocarcinoma stem cells. *Proc. Natl. Acad. Sci. USA* **78**, 7634–7638 (1981).
19. Thomson, J.A., Itskovitz-Eldor, J., Shapiro, S.S., Waknitz, M.A., Swiergiel, J.J., Marshall, V.S. and Jones, J.M. Embryonic stem cell lines derived from human blastocysts. *Science* **282**, 1145–1147 (1998).
20. Brittberg, M., Lindahl, A., Nilsson, A., Ohlsson, C., Isaksson, O. and Peterson, L. Treatment of deep cartilage defects in the knee with autologous chondrocyte transplantation. *N. Engl. J. Med.* **331**, 889–895 (1994).
21. Boulas, H.J., Herren, A. and Buchler, U. Osteochondral metatarsophalangeal autografts for traumatic articular metacarpophalangeal defects: a preliminary report. *J. Hand Surg. [Am]* **18**, 1086–1092 (1993).
22. Meyers, M.H. Resurfacing of the femoral head with fresh osteochondral allografts. Long-term results. *Clin. Orthop. Relat. Res.* **197**, 111–114 (1985).
23. Peterson, L., Brittberg, M., Kiviranta, I., Akerlund, E.L. and Lindahl, A. Autologous chondrocyte transplantation. Biomechanics and long-term durability. *Am. J. Sports Med.* **30**, 2–12 (2002).
24. Redman, S.N., Oldfield, S.F. and Archer, C.W. Current strategies for articular cartilage repair. *Eur. Cell Mater.* **9**, 23–32 (2005).
25. Briscoe, D.M., Dharnidharka, V.R., Isaacs, C., Downing, G., Prosky, S., Shaw, P., Parenteau, N.L. and Hardin-Young, J. The allogeneic response to cultured human skin equivalent in the hu-PBL-SCID mouse model of skin rejection. *Transplantation* **67**, 1590–1599 (1999).
26. http://stemcells.nih.gov/info/basics/defaultpage.asp.
27. Watt, F.M. and Hogan, B.L.M. Out of Eden: stem cells and their niches. *Science* **287**, 1427–1430 (2000).
28. McKay, R. Stem cells — hype and hope. *Nature* **406**, 361–364 (2000).
29. Alison, M.R., Murphy, G. and Leedham, S. Stem cells and cancer: a deadly mix. *Cell Tissue Res.* **331**, 109–124 (2008).

30. Shibata, D. Stem cells as common ancestors in a colorectal cancer ancestral tree. *Curr. Opin. Gastroenterol.* **24**, 59–63 (2008).
31. Barrilleaux, B., Phinney, D.G., Prockop, D.J. and O'Connor, K.C. Review: *ex vivo* engineering of living tissues with adult stem cells. *Tissue Eng.* **12**, 3007–3019 (2006).
32. Adams, G.B., Martin, R.P., Alley, I.R., Chabner, K.T., Cohen, K.S., Calvi, L.M., Kronenberg, H.M. and Scadden, D.T. Therapeutic targeting of a stem cell niche. *Nat. Biotechnol.* **25**, 238–243 (2007).
33. Bjornson, C.R., Rietze, R.L., Reynolds, B.A., Magli, M.C. and Vescovi, A.C. Turning brain into blood: a hematopoietic fate adopted by adult neural stem cells *in vivo*. *Science* **283,** 534–537 (1999).
34. Mezey, E., Chandross, K.J., Harta, G., Maki, R.A. and McKercher, S.R. Turning blood into brain: cells bearing neuronal antigens generated *in vivo* from bone marrow. *Science* **290**, 1779–1782 (2000).
35. Long, Y. and Yang, K.Y. Bone marrow derived cells for brain repair: recent findings and current controversies. *Curr. Mol. Med.* **3**(8), 719–725 (2003).
36. Krause, D.S., Theise, N.D., Collector, M.I. *et al.* Multi-organ, multi-lineage engraftment by a single bone marrow-derived stem cell. *Cell* **105**, 369–377 (2001).
37. Grove, J.E., Bruscia, E. and Krause, D.S. Plasticity of bone marrow-derived stem cells. *Stem Cells* **22**, 487–500 (2004).
38. Orlic, D., Kajstura, J., Chimenti, S., Jakoniuk, I., Anderson, S.M. and Li, B. Bone marrow cells regenerate infacrted myocardium. *Nature* **410**, 701–704 (2001).
39. Yoon, Y.S., Wecker, A., Heyd, L., Park, J.S., Tkebuchava, T., Kusano, K., Hanley, A., Scadova, H., Qin, G., Cha, D.H., Johnson, K.L., Aikawa, R., Asahara, T. and Losordo, D.W. Clonally expanded novel multipotent stem cells from human bone marrow regenerate myocardium after myocardial infarction. *J. Clin. Invest.* **115**, 326–338 (2005).
40. Jiang, Y., Jahagirdar, B.N., Reinhardt, R.L., Schwartz, R.E., Keene, C.D., Ortiz-Gonzalez, X.R., Reyes, M., Lenvik, T., Lund, T., Blackstad, M., Du, J., Aldrich, S., Lisberg, A., Low, W.C., Largaespada, D.A. and Verfaillie, C.M. Pluripotency of mesenchymal stem cells derived from adult marrow. *Nature* **418**, 41–49 (2002).
41. Verfaillie, C.M., Schwartz, R., Reyes, M. and Jiang, Y. Unexpected potential of adult stem cells. *Ann. N. Y. Acad. Sci.* **996**, 231–234 (2003).
42. Mitsiadis, T.A., Barrandon, O., Rochat, A., Barrandon, Y. and De Bari, C. Stem cell niches in mammals. *Exp. Cell. Res.* **313**, 3377–3385 (2007).
43. He, Q., Wan, C. and Li, G. Concise review: multipotent mesenchymal stromal cells in blood. *Stem Cells* **25**, 69–77 (2007).

44. Verfaillie, C.M. Hematopoietic stem cells for transplantation. *Nat. Immunol.* **3**, 314–317 (2002).
45. Chamberlain, G., Fox, J., Ashton, B. and Middleton, J. Concise review: mesenchymal stem cells: their phenotype, differentiation capacity, immunological features, and potential for homing. *Stem Cells* **25**, 2739–2749 (2007).
46. Caplan, A.I. Adult mesenchymal stem cells for tissue engineering versus regenerative medicine. *J. Cell Physiol.* **213**, 341–347 (2007).
47. Delorme, B., Chateauvieux, S. and Charbord, P. The concept of mesenchymal stem cells. *Regen. Med.* **1**(4), 497–509 (2006).
48. Oreffo, R.O., Cooper, C., Mason, C. and Clements, M. Mesenchymal stem cells: lineage, plasticity, and skeletal therapeutic potential. *Stem Cell Rev.* **1**, 169–178 (2005).
49. Friedenstein, A.J., Petrakova, K.V., Kurolesova, A.I. and Folova, G.P. Heterotropic transplants of bone marrow. Analysis of precursor cells for osteogenic and hematopoietic tissues. *Transplantation* **6**, 230–247 (1968).
50. Leary, A.G. and Ogawa, M. Blast cell colony assay for umbilical cord blood and adult bone marrow progenitors. *Blood* **69**, 953–956 (1987).
51. Emerson, S.G. *Ex vivo* expansion of hematopoietic precursors, progenitors, and stem cells: the next generation of cellular therapeutics. *Blood* **87**, 3082–3088 (1996).
52. Akashi, K., Traver, D., Miyamoto, T. and Weissman, I.L. A clonogenic common myeloid progenitor that gives rise to all myeloid lineages. *Nature* **404**, 193–197 (2000).
53. Pittenger, M.F., Mackay, A.M., Beck, S.C., Jaiswal, R.K., Douglas, R., Mosca, J.D., Moorman, M.A., Simonetti, D.W., Craig, S. and Marshak, D.R. Multilineage potential of adult human mesenchymal stem cells. *Science* **284**, 143–147 (1999).
54. van der Sluijs, J.P., Baert, M.R. and Ploemacher, R.E. Differential adherence of murine hematopoietic stem cell subsets to fibronectin. *Exp. Hematol.* **22**, 1236–1243 (1994).
55. Kerk, D.K., Henry, E.A., Eaves, A.C. and Eaves, C.J. Two classes of primitive pluripotent hemopoietic progenitor cells: separation by adherence. *J. Cell Physiol.* **125**, 127–134 (1985).
56. Sutherland, D.R., Stewart, A.K. and Keating, A. CD34 antigen: molecular features and potential clinical applications. *Stem Cells* **11**(Suppl. 3), 50–57 (1993).
57. Krause, D.S., Ito, T., Fackler, M.J., Smith, O.M., Collector, M.I., Sharkis, S.J. and May, W.S. Characterization of murine CD34, a marker for hematopoietic progenitor and stem cells. *Blood* **84**, 691–701 (1994).

58. Wright, D.E., Wagers, A.J., Gulati, A.P., Johnson, F.L. and Weissman, I.L. Physiological migration of hematopoietic stem and progenitor cells. *Science* **294**, 1933–1936 (2001).
59. Abkowitz, J.L., Robinson, A.E., Kale, S., Long, M.W. and Chen, J. Mobilization of hematopoietic stem cells during homeostasis and after cytokine exposure. *Blood* **102**(4), 1249–1253 (2003).
60. Aslan, H., Zilberman, Y., Kandel, L., Liebergall, M., Oskouian, R.J., Gazit, D. and Gazit, Z. Osteogenic differentiation of noncultured immunoisolated bone marrow-derived CD105+ cells. *Stem Cells* **24**, 1728–1737 (2006).
61. Wagner, W., Wein, F., Seckinger, A., Frankhauser, M., Wirkner, U., Krause, U., Blake, J., Schwager, C., Eckstein, V., Ansorge, W. and Ho, A.D. Comparative characteristics of mesenchymal stem cells from human bone marrow, adipose tissue, and umbilical cord blood. *Exp. Hematol.* **33**, 1402–1416 (2005).
62. Lee, R.H., Kim, B., Choi, I., Kim, H., Choi, H.S., Suh, K., Bae, Y.C. and Jung, J.S. Characterization and expression analysis of mesenchymal stem cells from human bone marrow and adipose tissue. *Cell Physiol. Biochem.* **14**, 311–324 (2004).
63. Bühring, H.J., Battula, V.L., Treml, S., Schewe, B., Kanz, L. and Vogel, W. Novel markers for the prospective isolation of human MSC. *Ann. N. Y. Acad. Sci.* **1106**, 262–271 (2007).
64. Horwitz, E.M., Prockop, D.J., Fitzpatrick, L.A., Koo, W.W., Gordon, P.L., Neel, M., Sussman, M., Orchard, P., Marx, J.C., Pyeritz, R.E. and Brenner, M.K. Transplantability and therapeutic effects of bone marrow-derived mesenchymal cells in children with osteogenesis imperfecta. *Nat. Med.* **5**, 309–313 (1999).
65. Chamberlain, J.R., Schwarze, U., Wang, P.R., Hirata, R.K., Hankenson, K.D., Pace, J.M., Underwood, R.A., Song, K.M., Sussman, M., Byers, P.H. and Russell, D.W. Gene targeting in stem cells from individuals with osteogenesis imperfecta. *Science* **303**, 1198–1201 (2004).
66. Guillot, P.V., Abass, O., Bassett, J.H., Shefelbine, S.J., Bou-Gharios, G., Chan, J., Kurata, H., Williams, G.R., Polak, J. and Fisk, N.M. Intrauterine transplantation of human fetal mesenchymal stem cells from first trimester blood repairs bone and reduces fractures in osteogenesis imperfecta mice. *Blood* **111**, 1717–1725 (2008).
67. Jones, P.H. and Watt, F.M. Separation of human epidermal stem cells from transit amplifying cells on the basis of differences in integrin function and expression. *Cell* **73**, 713–724 (1993).
68. Tumbar, T., Guasch, G., Greco, V., Blanpain, C., Lowry, W.E., Rendl, M. and Fuchs, E. Defining the epithelial stem cell niche in skin. *Science* **303**, 359–363 (2004).

69. Fuchs, E., Tumbar, T. and Guasch, G. Socializing with the neighbors: stem cells and their niche. *Cell* **116**, 769–778 (2004).
70. Brittan, M. and Wright, N.A. Gastrointestinal stem cells. *J. Pathol.* **197**, 492–509 (2002).
71. Reya, T. and Clevers, H. Wnt signalling in stem cells and cancer. *Nature* **434**, 843–850 (2005).
72. Thorgeirsson, S.S. Hepatic stem cells in liver regeneration. *FASEB J.* **10**, 1249–1256 (1996).
73. Fausto, N. Liver regeneration and repair: hepatocytes, progenitor cells, and stem cells. *Hepatology* **39**, 1477–1487 (2004).
74. Dowthwaite, G.P., Bishop, J.C., Redman, S.N., Khan, I.M., Rooney, P., Evans, D.J., Haughton, L., Bayram, Z., Boyer, S., Thomson, B., Wolfe, M.S. and Archer, C.W. The surface of articular cartilage contains a progenitor cell population. **117**, 889–897 (2004).
75. Bearzi, C., Rota, M., Hosoda, T., Tillmanns, J., Nascimbene, A., De Angelis, A., Yasuzawa-Amano, S., Trofimova, I., Siggins, R.W., Lecapitaine, N., Cascapera, S., Beltrami, A.P., D'Alessandro, D.A., Zias, E., Quaini, F., Urbanek, K., Michler, R.E., Bolli, R., Kajstura, J., Leri, A. and Anversa, P. Human cardiac stem cells. *Proc. Natl. Acad. Sci. USA* **104**, 14068–14073 (2007).
76. Yen, A.H. and Sharpe, P.T. Stem cells and tooth tissue engineering. *Cell Tissue Res.* **331**, 359–372.
77. Suhonen, J.O., Peterson, D.A., Ray, J. and Gage, F.H. Differentiation of adult hippocampus-derived progenitors into olfactory neurons *in vivo*. *Nature* **383**, 624–627 (1996).
78. Merkle, F.T., Mirzadeh, Z. and Alvarez-Buylla, A. Mosaic organization of neural stem cells in the adult brain. *Science* **317**(5836), 381–384 (2007).
79. Weiss, M.L. and Troyer, D.L. Stem cells in the umbilical cord. *Stem Cell Rev.* **2**, 155–162 (2006).
80. Harris, D.T., Badowski, M., Ahmad, N. and Gaballa, M.A. The potential of cord blood stem cells for use in regenerative medicine. *Expert Opin. Biol. Ther.* **7**, 1311–1322 (2007).
81. Guillot, P.V., Gotherstrom, C., Chan, J., Kurata, H. and Fisk, N.M. Human first-trimester fetal MSC express pluripotency markers and grow faster and have longer telomeres than adult MSC. *Stem Cells* **25**, 646–654 (2007).
82. Guillot, P.V., O'Donoghue, K., Kurata, H. and Fisk, N.M. Fetal stem cells: betwixt and between. *Semin. Reprod. Med.* **24**, 340–347 (2006).
83. Martin, G.R. Isolation of a pluripotent cell line from early mouse embryos cultured in medium conditioned by teratocarcinoma stem cells. *Proc. Natl. Acad. Sci. USA* **78**, 7634–7638 (1981).

84. Thomson, J.A., Kalishman, J., Golos, T.G., Durning, M., Harris, C.P., Becker, R.A. and Hearn, J.P. Isolation of a primate embryonic stem cell line. *Proc. Natl. Acad. Sci. USA* **92**, 7844–7848 (1995).
85. Thomson, J.A., Itskovitz-Eldor, J., Shapiro S.S., Waknitz, M.A., Swiergiel, J.J., Marshall, V.S. and Jones, J.M. Embryonic stem cell lines derived from human blastocysts. *Science* **282**, 1145–1147 (1998).
86. Fischbach, G.D. and Fischbach, R.L. Stem cells: science, policy, and ethics. *J. Clin. Invest.* **114**, 1364–1370 (2004).
87. Smith, A.G. Culture and differentiation of embryonic stem cells. *J. Tissue Cult. Methods* **13**, 89–94 (1991).
88. Keller, G.M. *In vitro* differentiation of embryonic stem cells. *Curr. Opin. Cell Biol.* **7**, 862–869 (1995).
89. Keller, G. Embryonic stem cell differentiation: emergence of a new era in biology and medicine. *Genes Dev.* **19**, 1129–1155 (2005).
90. Trounson, A. The production and directed differentiation of human embryonic stem cells. *Endocr. Rev.* **27**, 208–219 (2006).
91. Skottman, H. and Hovatta, O. Culture conditions for human embryonic stem cells. *Reproduction* **132**, 691–698 (2006).
92. Stojkovic, P., Lako, M., Stewart, R., Przyborski, S., Armstrong, L., Evans, J., Murdoch, A., Strachan, T. and Stojkovic, M. An autogeneic feeder cell system that efficiently supports growth of undifferentiated human embryonic stem cells. *Stem Cells* **23**, 306–314 (2005).
93. Rodríguez, C.I., Galán, A., Valbuena, D. and Simón, C. Derivation of clinical-grade human embryonic stem cells. *Reprod. Biomed. Online* **12**, 112–118 (2006).
94. Hovatta, O. Derivation of human embryonic stem cell lines, towards clinical quality. *Reprod. Fertil. Dev.* **18**, 823–828 (2006).
95. Xu, C., Inokuma, M.S., Denham, J., Golds, K., Kundu, P., Gold, J.D. and Carpenter, M.K. Feeder-free growth of undifferentiated human embryonic stem cells. *Nat. Biotechnol.* **19**, 971–974 (2001).
96. Draper, J.S., Moore, H.D., Ruban, L.N., Gokhale, P.J. and Andrews, P.W. Culture and characterization of human embryonic stem cells. *Stem Cells Dev.* **13**, 325–336 (2004).
97. Gerecht-Nir, S., Cohen, S., Ziskind, A. and Itskovitz-Eldor, J. Three-dimensional porous alginate scaffolds provide a conducive environment for generation of well-vascularized embryoid bodies from human embryonic stem cells. *Biotechnol. Bioeng.* **88**, 313–320 (2004).
98. Brons, I.G., Smithers, L.E., Trotter, M.W., Rugg-Gunn, P., Sun, B., Chuva de Sousa Lopes, S.M., Howlett, S.K., Clarkson, A., Ahrlund-Richter, L.,

Pedersen, R.A. and Vallier, L. Derivation of pluripotent epiblast stem cells from mammalian embryos. *Nature* **448**, 191–195 (2004).

99. Tesar, P.J., Chenoweth, J.G., Brook, F.A., Davies, T.J., Evans, E.P., Mack, D.L., Gardner, R.L. and McKay, R.D. New cell lines from mouse epiblast share defining features with human embryonic stem cells. *Nature* **448**, 196–199 (2007).
100. Hodges, H., Pollock, K., Stroemer, P., Patel, S., Stevanato, L., Reuter, I. and Sinden, J. Making stem cell lines suitable for transplantation. *Cell Transplant.* **16**, 101–115 (2007).
101. Obinata, M. The immortalized cell lines with differentiation potentials: their establishment and possible application. *Cancer Sci.* **98**, 275–283 (2007).
102. Wilmut, I., Schnieke, A.E., McWhir, J., Kind, A.J. and Campbell, K.H. Viable offspring derived from fetal and adult mammalian cells. *Nature* **385**, 810–813 (1997).
103. Agarwal, S. Cellular reprogramming. *Methods Enzymol.* **420**, 265–283 (2006).
104. Alberio, R., Campbell, K.H. and Johnson, A.D. Reprogramming somatic cells into stem cells. *Reproduction* **132**, 709–720 (2006).
105. Hochedlinger, K. and Jaenisch, R. Nuclear reprogramming and pluripotency. *Nature* **441**, 1061–1067 (2006).
106. Lanza, R.P., Cibelli, J.B. and West, M.D. Prospects for the use of nuclear transfer in human transplantation. *Nat. Biotechnol.* **17**, 1171–1174 (1999).
107. Tecirlioglu, R.T., Guo, J. and Trounson, A.O. Interspecies somatic cell nuclear transfer and preliminary data for horse-cow/mouse iSCNT. *Stem Cell Rev.* **2**, 277–287 (2006).
108. Wernig, M., Meissner, A., Foreman, R., Brambrink, T., Ku, M., Hochedlinger, K., Bernstein, B.E. and Jaenisch, R. *In vitro* reprogramming of fibroblasts into a pluripotent ES-cell-like state. *Nature* **448**, 318–324 (2007).
109. Takahashi, K. and Yamanaka, S. Induction of pluripotent stem cells from mouse embryonic and adult fibroblast cultures by defined factors. *Cell* **126**(4), 663–676 (2006).
110. Takahashi, K., Tanabe, K., Ohnuki, M., Narita, M., Ichisaka, T., Tomoda, K. and Yamanaka, S. Induction of pluripotent stem cells from adult human fibroblasts by defined factors. *Cell* **131**, 861–872 (2007).
111. Yu, J., Vodyanik, M.A., Smuga-Otto, K., Antosiewicz-Bourget, J., Frane, J.L., Tian, S., Nie, J., Jonsdottir, G.A., Ruotti, V., Stewart, R., Slukvin, II. and Thomson, J.A. Induced pluripotent stem cell lines derived from human somatic cells. *Science* **318**, 1917–1920 (2007).
112. Conrad, C., Gupta, R., Mohan, H., Niess, H., Bruns, C.J., Kopp, R., von Luettichau, I., Guba, M., Heeschen, C., Jauch, K.W., Huss, R. and

Nelson, P.J. Genetically engineered stem cells for therapeutic gene delivery. *Curr. Gene Ther.* **7**, 249–260 (2007).
113. Menendez, P., Wang, L. and Bhatia, M. Genetic manipulation of human embryonic stem cells: a system to study early human development and potential therapeutic applications. *Curr. Gene Ther.* **5**(4), 375–385 (2005).
114. Bosse, R., Singhofer-Wowra, M., Rosenthal, F. and Schulz, G. Good manufacturing practice production of human stem cells for somatic cell and gene therapy. *Stem Cells* **15**(Suppl. 1), 275–280 (1997).
115. http://www.dh.gov.uk/en/Policyandguidance/Healthandsocialcaretopics/Stemcell/index.htm.
116. http://www.ukstemcellbank.org.uk.

Chapter 3

Human Embryonic Stem Cells: International Policy and Regulation

Megan Allyse and Stephen Minger

Abstract

Although human embryonic stem cells may have enormous potential for the treatment of degenerative diseases, their origins and derivation have raised unprecedented controversy in many societies. Many organisations, on both international and national levels, have responded with regulatory systems that seek to manage and supervise such work. These systems vary from voluntary professional guidelines and international treaties to government agencies and binding national legislation. This chapter briefly summarises the main controversies surrounding human embryonic stem cell research and the existing international mechanisms that address its conduct. We then offer a more detailed overview of the relevant regulatory structures of five nations — the US, the UK, China, India and South Korea — where research with human embryonic stem cells is commonly practiced. A review of the responsible state and professional bodies in each country is included, along with a brief discussion of any relevant legislation. Regulations relevant to the conduct of international collaboration with national research groups are included where possible.

Keywords: Ethics Committee; International Collaboration; Somatic Cell Nuclear Transfer; Cytoplasmic Hybrid; Chimera; Informed Consent.

Outline

1. Introduction
2. Controversy
3. International Guidelines
4. National Policy Systems
   4.1. India
   4.2. South Korea

4.3. China
4.4. United Kingdom
4.5. United States
5. Conclusion
References

1. Introduction

Although the existence and potential properties of stem cells have been recognised by the scientific community since the 1960s the isolation of human embryonic stem cells (hESC) in 1998 nevertheless exploded onto the cultural and political scene with relatively little warning. The resulting whirlwind of debate has had many consequences, both for the profile of medical research and therapy in the public sphere and the increasingly complex relationship between the political and scientific spheres. Very few areas of medical research, with the exception of those conceived of as having military applications, receive as much political and public attention as hESC have garnered in the last decade.

In particular, hESC seem to have contributed to an increased emphasis on public mediation of scientific research by means of State regulation. In many countries, hESC research has given rise to a host of new regulations, including several relatively novel forms of state supervision. Indeed, between decisions on whether or not to permit stem cell research at all, whether to permit embryonic stem cell research specifically, whether to fund such research and how it can be regulated many governments have found themselves devoting disproportionate amounts of time to the contemplation of this relatively narrow but highly potent field. The resulting patchwork of guidelines and statutory controls has had implications for the practice of the science itself as an increasingly global scientific community attempts to navigate an ever more complex regulatory environment. In this chapter, we offer a brief survey of the current state of affairs in the form of the international and national policy and regulation surrounding hESC research in areas of the world where it is practiced most intensely.

In an attempt to offer some insight on why it has garnered such unprecedented attention, we begin with a brief introduction to the nature of the controversy which surrounds hESC research. After a brief exploration of developments on an international level we turn to the national systems presently in place in five of the current world leaders in hESC research. That the countries profiled here — India, South Korea, China, the United States (US) and the United Kingdom (UK) — represent some of the most permissive regulatory environments with respect to hESC research is neither coincidental nor intended to indicate a *de facto* endorsement of the practice of hESC research *per se*. Rather, these countries represent something of a self-selecting sample. In countries that have placed an outright ban on the practice of hESC research the issue of how such research interacts with the political system is somewhat moot, although in some nations, such as Italy, there exists ongoing debate which represents a constant tension between the perceived economic and health-related benefits of hESC research and the views of a relatively traditional political system with strong ties to the Catholic Church, arguably the strongest opponent of hESC research. On the other hand, there exist many nations who, for a variety of reasons, have made no conclusive decisions on the practice of hESC research, leaving us little data on which to base an analysis.

Nevertheless, there exists sufficient variation between even these five nations to offer a relatively comprehensive survey of the way in which hESC research can be managed and supervised on a national level. Between those countries, such as the US, which have passed no national legislation on the matter, although the government nevertheless manages to express strong opinions through the allocation of funding, and South Korea, whose explicit legislation on the practice of stem cell research incorporates penal provisions for violation of procedural and ethical requirements, there exists a continuum which encompasses India's system of non-binding guidelines and the UK's highly specialised bureaucratic approach.

2. Controversy

A major contributor to the amount of attention hESC have received is undoubtedly their strong, and often strongly opposed, association with two extremes of therapeutic medicine: the extension and cessation of human life. Historically, the most intractable, and often unresolved, political debates — abortion, contraception, human experimentation — have centred around technologies designed to bring about one of these two fundamental outcomes. As such, the combined social and political weight accumulated during such debates lurks in the background of discussions over hESC research.

On the one hand, hESC offer the potential, still largely theoretical, to ameliorate, if not totally reverse, the effects of some of the most damaging degenerative conditions in human medicine. As the basis of the new field of regenerative medicine, they have been hailed as the "cure" for everything from spinal cord injury-based paraplegia to Parkinson's disease. In many societies that increasingly value health and well-being above almost any other social service,[1] this promise has led to a level of public interest which allows support for the development of research into hESC-based therapies to, as it were, punch above its political weight. As a former member of Germany's Genetics Commission explained:

> It is simply not possible to be against the healing or prevention of diseases. In general, one might rightfully criticise that it is precisely the impending medical

> applications of modern biology which are going to strengthen the tendency, in a haphazard manner, towards a technicalisation of human life. In a concrete case the right of those affected to protection of their physical soundness also fundamentally includes access to precisely those technologies.[2]

In many areas, this translates to an argument that governments are failing in a fundamental duty to promote the health of their citizens by not permitting or actively encouraging hESC research.[1]

The second major attraction of the field, which stems inevitably from the first, is the extent to which it is seen as offering the potential for significant commercial and economic returns on the investment of whoever manages to bring it to the open market.[3] The potential for a "stem cell gold rush",[4] a term used to imply that large numbers of potentially lucrative biotechnology companies are likely to move to wherever the legislation is sufficiently permissive for them to operate profitably, has been cited as a strong argument in favour of permitting the practice of hESC research. This argument also carries an internationally competitive aspect. In the words of the UK's Department of Health, "[if] we fail to capitalise on our current position of strength in stem cell research to develop stem cell therapies and technology, there is a danger that [we] will have to pay significantly greater amounts than [we] otherwise would have done in order to import stem cell expertise and products from overseas".[5] The anticipated economic potential of hESC has generated such intense interest and excitement that some observers have accused it of crossing the line from "stem cell hope" to "stem cell hype".[6] This has not stopped many from embracing the development of hESC technology wholesale. Several nations and American states have staked considerable investment in the growth of the stem cell field.

It must be pointed out, however, that at present this commercial success remains strictly theoretical. One of the more secular arguments against the practice of hESC is the fact that it represents a largely unproven technology. In addition to the many, and not insignificant, practical barriers to its use in commercial medicine there remains the fact that only one, very early stage, clinical trial of embryonic stem cells has been announced to date.[7] Opponents of the research use this lack of results to argue that society would more effectively fulfil its duty to alleviate suffering by financing other, less controversial, research with a stronger track record of success — such as adult stem cell (ASC)-based therapies.[8] Another relatively secular line of thought uses a slippery-slope argument to assert that if we allow experiments on embryos, and in particular permit the creation of embryos specifically for hESC research, we are contributing to a societal attitude which views human embryos, and by extension foetuses, children and adults, in an instrumental light as raw materials for an industrial process. No matter how altruistic the motives of the commercial process, once we submit to the commodification of human materials we are opening ourselves to a future in which human beings and human body parts may be bought and sold.[9]

These arguments, which generally derive from a belief in the moral significance of the human embryo, are notably more common in countries which have a history of contentious debates over abortion, particularly the US, the UK and Germany, all of which exhibit strong and well-funded anti-hESC lobbies. All three share cultural roots in the Christian Church, which maintains that the embryo effectively becomes a person at the moment fertilisation occurs. As such, the derivation of embryonic stem cell lines, which destroys the embryo as a reproductive entity, is tantamount to the practice of abortion, to which the Church maintains a fierce opposition.[10] The secular corollary to this argument uses a human rights approach to argue that any form of embryonic stem cell research constitutes a violation of research ethics and human rights since the embryo is incapable of giving consent to experimentation.[11] These arguments form the basis of the ban on the derivation of hESC in several countries including Costa Rica,[12] El Salvador,[13] Italy[14] and Poland.[15]

A final area of controversy surrounds the practice of somatic cell nuclear transfer (SCNT), sometimes referred to by the more incendiary term "human cloning", and cytoplasmic hybrid ("cybrid") research. The possibility of using SCNT to create the cloned offspring of adult human beings has engendered an extraordinary level of rejection, with respect to its unity and strength, on the part of a wide variety of interest groups across nations, religions and cultures. Similarly, the prospect of mixing genetic material from animals and humans also appears to arouse revulsion in some sectors, particularly when it is imagined that the process might be used for reproductive purposes. However, these technologies also offer the possibility of isolating two specific forms of hESC, disease-specific and patient-specific lines, which may be unobtainable through the use of surplus *in vitro* fertilisation (IVF) embryos. And while only lines derived from purely human material could ever be used in clinical practice, cybrid lines offer potentially valuable research tools.

Perhaps the best way to understand the conflict over cloning is to skim through the process that brought about the *United Nations Declaration on Human Cloning*, which represents the closest thing to an international law pertaining to hESC research currently in effect.

3. International Guidelines

Following the birth of Dolly the sheep in 1997, concerns arose over the application of human cloning technology to human beings. In response, the UN General Assembly formed the *Ad Hoc* Committee on an International Convention against the Reproductive Cloning of Human Beings in 2001. In point of fact, it was not until 2003 that Costa Rica first introduced a "draft international convention for the prohibition of all forms of human cloning". The proposed convention stated that human cloning "for any [...] purpose whatsoever, is morally repugnant, unethical and contrary to respect for the person and constitutes a grave violation of fundamental human rights which cannot under any circumstances be justified or accepted".[16] This language immediately drew strong criticism from supporters of stem cell research for its lumping of the use of SCNT technologies for therapeutic purposes into the same moral category as reproductive cloning. In response, another draft convention was presented which also called for an international ban on reproductive human cloning but, significantly, required that member states put in place legislation "to *control* other forms of human cloning [...] by means of national legislation",[17] leaving the legal field open to the practice of therapeutic cloning provided it was regulated by the state. Despite the fact that this language would have allowed those member states that objected to cloning technology to ban both therapeutic and reproductive cloning it faced heavy opposition, particularly from the US. So intense was the debate that the Committee was forced to table the issue for a year.[18]

Finally, in 2005, a watered-down draft submitted by Italy was accepted and became the *United Nations Declaration on Human Cloning*. It is worth noting, however, that this version is non-binding. Member states are "called upon to prohibit any attempts to create human life through cloning processes"[19] through the use of national legislation but no binding requirement is stated. A majority of nations have transposed the Declaration through the implementation of national laws banning cloning technologies but, significantly, several such laws, including those in effect in the UK, China, Belgium, South Korea and several states of the US, specify a ban on reproductive cloning, making the practice of therapeutic cloning a separate issue.

Other UN instruments, including the *Declaration on the Human Genome and Human Rights* (1997), the *Universal Declaration on Bioethics and Human Rights* (2005) and the International Bioethics Committee's (IBC's) *Human Embryo Research and International Solidarity and Cooperation*, may be applicable to national governments but do not require the passing of specific legislation. On a non-governmental level, the World Medial Association's (WMA's) *Declaration of Helsinki* (2000) and the *International Ethical Guidelines for Biomedical Research Involving Human Subjects* (2002) ("CIOMS Guidelines"), prepared by the Council for International Organizations of Medical Sciences (CIOMS) in collaboration with the World Health Organization (WHO), also offer specific recommendation on the practice of medical research — particularly in the areas of informed consent and the preservation of human rights — but do not address hESC research directly.[a] Although such instruments are not technically binding on either governments or individuals, many institutions and governments have incorporated them into appropriate research conduct requirements through the use of "compliance with accepted ethical standards" language that may directly influence research practice. Initial investigations into the conduct of South Korea's Professor Hwang Woo Suk, for instance, were spurred not by suspicions that he had falsified research in publication but that he had violated "international ethical standards"[20] in obtaining human oocytes for his research on cloned human embryos. The investigating Institutional Review Board (IRB) on Human Subjects Research and Ethics at Hanyang University Hospital cited the *Declaration of Helsinki* in both their initial finding that, given the evidence available at the time, Professor Hwang did not appear to have violated existing standards of research ethics and might therefore have kept his post and their later determination that he should be stripped of both his funding and his professional license.[21] Notwithstanding subsequent revelations of misappropriated government research funds and the publication of false research results, the incident stands as an indication that demonstrated compliance with international ethical standards serves an increasingly central role in the legitimacy of scientific research today.

4. National Policy Systems

International influence aside, however, the primary instrument for regulating scientific research remains national policy. In the following sections, we explore five different national regulatory systems with respect to hESC and how they affect the conduct of research.

4.1. India

The responsible body for stem cell research in India is the Indian Council of Medical Research (ICMR), which proudly announces itself as one of the oldest research bodies in the world. Initially designated the Indian Research Fund Association, it was founded in 1911 to provide national support for medical

[a] Other potentially relevant documents include the *Nuremburg Code* (1947), the *Belmont Report* (1978), the *European Forum for Good Human Practice* (1997), the *Ethics of Research Related to Healthcare in Developing Countries* (2002), the *European Convention on Human Rights and Biomedicine* (last updated 2000), the *European Directive 2004/23/EC* (2004), the *International Declaration on Human Genetic Data* (2003) and the *National Research Council's Guidelines for Human Embryonic Stem Cell Research* (2005).

research in India. After independence from Britain was declared, the Council was renamed and placed under the Ministry of Health and Family Welfare (MHFW), where it serves as the apex body for the "formulation, coordination and promotion of biomedical research"[22] in India. As such, ICMR has been responsible for the development and promulgation of the *Draft Guidelines for Stem Cell Research/Regulation* ("2004 Guidelines"). Although the guidelines were completed in 2004 they have not, at the time of writing, been ratified by the Indian Parliament. We examine them here as an example of an approach to national science policy, although the chance exists that they may never become binding legislation. The ICMR is also responsible for the *2006 Ethical Guidelines for Biomedical Research on Human Subjects* ("2006 Guidelines"), which include a more detailed description of permitted areas of research. At present, these Guidelines are also non-binding.

Together, these two documents outline a multi-tiered regulatory system with several zones of ethical permissibility. The chief regulatory body envisaged under the proposed legislation is the National Apex Committee for Stem Cell Research and Therapy (NAC-SCRT), formed from both professional and lay members and attached to the ICMR. The NAC-SCRT is intended to both create and enforce policy towards stem cell research and therapy and to serve as a central repository for information about the state of stem cell research in India. Centres that wish to conduct research or clinical trials involving human stem cells must be registered with the NAC-SCRT (although clinical trials may also require the approval of the Drug Controller General) and all newly-derived hESC lines must be registered in its database.

Below the NAC-SCRT sit individual Institutional Committees for Stem Cell Research and Therapy (IC-SCRT), which act in parallel with existing Institutional Ethics Committees (IEC) to approve and monitor research proposals and conduct. Research classified as "Permitted", which encompasses research on any established stem cell line, derivation of new stem lines from supernumerary IVF embryos (donated with appropriate informed consent and not cultured beyond 14 days after fertilisation) and clinical trials using proven clinical grade stem cells, may be carried out with only the approval of the IC-SRTC and IEC. Research classified as "Restricted", including the creation of embryos for the purpose of hESC derivation (incorporating the creation of embryos through SCNT), international clinical trials, the introduction of human stem cells into animals at a germ line level, chimera research and any study in which the anonymity of donors cannot be guaranteed, must first gain approval from the relevant IC-SCRT and then from the NAC-SCRT, which has discretion to approve restricted research on a case-by-case basis. Finally, human germ line modification or reproductive cloning, the transfer of any cloned or genetically modified embryo into a human womb, the breeding of animal/human hybrids and the directed non-antilogous donation of stem cells would be "Prohibited" under this system.

Of particular note to international researchers is the existence of the Division of International Health, tasked with coordinating research between Indian and foreign institutions. Such collaborations must accord with MHFW requirements and are limited to those countries with which India has signed a Memorandum of Understanding. Under Memorandum No. L20025/90-90-F, the transfer of biological materials, such as embryos or established stem cell lines, abroad may also require the approval of the ICMR as does any clinical trial which is sponsored or funded by an international pharmaceutical company.[22] NAC-SCRT also carries the authority to reject international collaboration in clinical trials if they are not satisfied that the materials provided from the collaborating country are properly certified. In the event of a conflict between the ethical procedures and requirements in place in the collaborating country and those in India, Indian researchers are required to ensure that Indian ethical guidelines and law prevail.

Derived stem cells and stem cell-based therapies may be patented in India, although foetal cells may not. The Guidelines state that should a stem cell-based therapy result in significant commercial benefit "a proportion of benefits shall be ploughed into the community which has [...] contributed to the IPR".[23]

On the whole, the proposed system in India is moderately permissive, although allowing the NAC-SCNT discretion in approving restricted research proposals means that the permissibility of the research environment may be subject to change depending on the members of the Committee. Although international collaboration must now be monitored — there is some concern about India's poorer populations being exploited by international clinical drugs trials[24] — it remains encouraged. However, the voluntary nature of the system is a serious weakness. Reports have emerged from India in recent years of researchers conducting inadequately robust therapeutic trials with hESC and concern remains that there is no monitoring or enforcement ability within the system as it stands.

4.2. South Korea

South Korea has worked hard to remain at the forefront of the stem cell research frontier, not least through large commitments of government funds to academic labs working in the hESC field. Although many claimed that the repercussions of the revelations surrounding the conduct of Dr. Hwang would have a permanent effect on the credibility of hESC science as a whole, and of South Korea in particular, the long term effects do not appear to have been as severe as predicted. Somewhat ironically, had Dr. Hwang's misconduct occurred a mere year later than it did he would have faced imprisonment rather than merely losing his professional standing. South Korea's *Bioethics and Biosafety Act*, which took effect on January 1, 2005, is one of the more comprehensive pieces of national legislation on the subject of hESC and one of the few to incorporate penal provisions.

At the centre of the Act stands the National Bioethics Committee (NBC), responsible directly to the President, which serves as the central regulatory and adjudicatory body on matters of embryo research, SCNT research, DNA testing and gene therapy as well as peripheral matters which are deemed to fall under the aegis of bioethics, biosafety or "issues of social or moral significance concerning the research, development and utilisation of life sciences and biotechnologies".[25] The Committee's membership features several government ministers, granting it a more directly political nature than comparable bodies in other countries. The remainder of the body is divided between representatives of the natural and moral sciences and industry. In addition to the NBC, any institution which engages in embryo research, DNA banking, gene therapy or areas of biotechnology deemed to have moral significance is required to maintain an IRB tasked with reviewing the ethical implications of proposed research, including matters of patient or subject safety and the maintenance of proper informed consent procedures. Such institutions must also be registered with the Ministry of Health and Welfare (MHW), which monitors their quality and must approve the research protocol for all work with embryos. The MHW is also empowered to inspect, fine or close any institution which it finds unfit to continue research.

Supernumerary embryos from the IVF process may be frozen for up to five years after which they are available for hESC research in fields approved by the President. The provision of supernumerary embryos is subject to lengthy informed consent procedures and requires that neither the donor nor the IVF institution receive financial compensation for the donation. Like many countries, South Korea has outlawed the reproductive cloning of human beings, although they remain at the forefront of the field of reproductive animal cloning. Chimera research involving human and animal genetic material is also prohibited, as is the creation of embryos for the purposes of research. However, human SCNT research for therapeutic research purposes is permitted.

Penal provisions are provided of up to ten years in prison for offences relating to reproductive cloning, five years for offences relating to the creation of human/animal hybrids and three years for creating embryos for purposes other than pregnancy, demanding or receiving financial compensation for embryos or other genetic material, violating informed consent procedures or conducting SCNT research for purposes not approved by the Ministry. One-year prison sentences or fines may be levied for procedural violations such as storing genetic material in unlicensed facilities or conducting research without a license.

Negligence fines also apply to those who neglect to report relevant details of their research to the appropriate bodies or who inappropriately disclose confidential information.

In 2005, South Korea announced that it was forming the World Stem Cell Hub, an international collaborative body that would assist in coordinating and disseminating resources and information relating to stem cell research. Unfortunately, Dr. Hwang was the hub's first president; after he was forced to resign in 2005 the hub dissolved. The *Bioethics and Biosafety Act* makes no specific provisions regarding the conduct of international collaboration so it is unclear to what extent its provisions apply to international researchers. Nor does it address the issue of patentability, although South Korea does recognise patents for inventions involving hESC.[26] In general, however, the binding nature of South Korea's legislation, especially the provision of penal provisions, mark it as potentially the most tightly regulated system in hESC research.

4.3. China

China has received increasing international attention in recent years as its economic growth and competitiveness, despite being founded on what remains a largely communist base, continue to grow at astonishing rates. The Chinese government has affirmed a commitment to becoming a world leader in the field of biomedicine, with a particular focus on clinical treatments using stem cells; a significant number of experimental treatments using autologous stem cells are already in use in various medical centres in China. Much like India, however, China has struggled to keep the pace of regulation commensurate with scientific and structural development. Although China was a co-author of the *Universal Declaration on Human Rights and Biomedicine* (1998), the IBC's *Human Embryo Research and International Solidarity and Cooperation* and at least one draft of the *Declaration on Human Cloning*, national law on these subjects has been slow to develop. The Chinese government is aware of the concerns raised by this legislative silence, both in terms of potential misconduct of researchers working in China and the damage a lack of clearly enforced ethical standards may do to China's international scientific reputation.[27] Recently, at least partially in response to unfavourable reports of scientific[28] and research conduct,[29] considerable focus has been given to the development of tougher systems of national regulation with regards to areas of research such as hESC.

For instance, the *Regulations on Human Assisted Reproductive Technologies* (2001) ("ART Regulations") now regulate the supply side of hESC research in the form of embryo formation and donation. And in January 2007, after nine years of deliberation, the new *Regulation on Ethical Review of Biomedical*

Research Involving Human Subjects (2007) ("Human Subjects Regulation") finally went into effect. There remains a gap, however, between the IVF clinic and the therapeutic clinic that some critics argue requires more explicit legislation. At present, hESC is only explicitly addressed by the "voluntary" *Ethical Guiding Principles for Research on Human Embryonic Stem Cells* (2003) ("Guiding Principles"), a joint publication of the Ministry of Health (MoH) and the Ministry of Science and Technology (MoST) who share responsibility for funding and managing stem cell research in China. The Biomedical Ethics Committee of the MoH acts as the chief coordinating body for ethical review bodies on a national level but its interaction with regional and local ethical boards is on a "professional guidance" basis only. Under the new Human Subjects Regulation, ethical review boards are required at ministry, provincial and institutional levels and tasked with considering proposed research against such ethical standards as the *Declaration of Helsinki* and the CIOMS Guidelines. While this level of scrutiny is potentially formidable, anecdotal evidence suggests that, as of yet, training in formal ethical reasoning is not widespread.[30]

The MoH is also responsible for the management of IVF clinics and the enforcement of the ART Regulations, which inevitably impacts on some areas of hESC research. The ART Regulations prohibit the buying and selling of human gametes and embryos, reproductive cloning and germ line genetic manipulation. They also mandate the existence of informed consent procedures and the requirement of voluntary donation. Once an embryo is donated, however, it leaves the purview of the MOH and enters the jurisdiction of MoST and the Guiding Principles. The principles reiterate the ban on reproductive cloning and the creation of human animal chimeras but permit therapeutic cloning. Research may be conducted on surplus IVF embryos or gametes and donated germ cells provided it is restricted to within 14 days of fertilisation. Institutions who wish to conduct embryo research are advised to maintain standing ethics review committees specifically to consider work with hESC.

Although the Guiding Principles are non-binding this does not translate, as it might in more free market societies, to being optional. The principles are explicitly intended to apply to all stem cell research that takes place within the territorial boundaries of China, regardless of whether the research is privately or publicly funded, and this applies to the conduct of international research as well. Although Chinese researchers have historically been reticent about publishing in Western journals, international collaboration is becoming more common. It is worth noting, however, that receiving administrative or government approval of such projects may take a significant amount of time; authorisation for new clinical trials from the State Food and Drug Administration can take up to a year. According to the Human Subjects Regulation, research will not be permitted where evidence of properly obtained informed consent from the donor or research subject cannot be provided.

Finally, under Chinese patent law scientific inventions, strategies for the diagnosis and treatment of disease and any creation that is contrary to "social morality" are unpatentable.[31] There is no evidence as of yet which indicates whether any of these exceptions may be construed to apply to stem cell therapies.

4.4. United Kingdom

The UK is proud to have the most comprehensive regulatory system in the world with respect to IVF and hESC. The relative sophistication of the UK system is most likely attributable to the length of the deliberation that led to its enactment. Initial calls for regulation of embryological science began in 1979, with the birth of Louise Brown, but it was not until 1990 that the *Human Fertilisation and Embryology Act* ("HFE Act") was passed. The main purpose of the act was to create the Human Fertilisation and Embryology Authority (HFEA), tasked with regulating the practice of assisted reproductive technologies and embryological research. Although the possibility of cytoplasmic hybrid research and cloning were dismissed as "science fiction possibilities"[32] at the time of the HFEA's formation, the HFE Act is phrased in such a way that authority over the practice of stem cell research automatically devolved to the HFEA when it became a reality almost a decade later.

The HFE Act was supplemented twice in 2001, with the *Human Reproductive Cloning Act* and the *Human Fertilisation and Embryology (Research Purposes) Regulations*, but in general it has remained unchanged through 17 years of scientific development. The reason for this remarkable longevity is that the Act is not, in fact, prescriptive legislation as it is generally conceived. Merely, it defines the HFEA as a statutory, non-governmental body composed of a small core of employees and a rotating committee of appointed specialists. The body is conceived of primarily as a licensing authority; no one in the UK, whether privately or publicly funded, may engage in the creation, treatment, research or storage of embryos without an explicit license from the HFEA. From this relatively simple formula springs the entirety of the UK policy towards hESC. Because ART centres are licensed on a premises basis but research laboratories are licensed on a project basis the HFEA is in a position to make an analysis of the separate legality of every line of embryo-based research undertaken in the UK on a case-by-case basis. In this way, policy evolves at precisely the same speed as scientific research in the form of HFEA decisions on whether or not to grant specific licenses.

There are interesting implications to the system. As a statutory body, the HFEA's primary mandate is to determine the legality of proposed strands of research. They are not, interestingly, overly tasked with decisions of ethics, although ethical reasoning does feature in many discussions over new research projects. The *Code of Conduct*, which the HFEA issues as a guideline to all facilities under its jurisdiction, focuses primarily on issues of safety, accountability and reporting. Then too, the HFEA deals only with embryos. Once a stem cell line is established, it becomes human tissue and leaves the HFEA's jurisdiction. At this point it enters the purview of the Human Tissue Authority (HTA), created in 2004 to harmonise the UK with *EU Directive 2004/23/EC*. The UK also has a system of regional and institutional Research Ethics Committees (REC), which must also issue approval of research protocols before they can go forward.

For such a closely regulated system, the UK research environment remains one of the most permissive in the world. Reproductive cloning is forbidden but SCNT is allowed. It is one of the only countries in the world to explicitly allow both the creation of embryos for research purposes, if the HFEA is convinced that the research cannot be done with existing embryos, and the donation of human oocytes to SCNT research. Supernumerary embryos from IVF, an increasingly common procedure in the UK, remain the most common source of hESC but because pre-implantation genetic diagnosis (PGD) is also permitted, although only for early-onset, fatal genetic conditions, rejected PGD embryos may also be available for research. The HFEA has recently concluded that the creation of cybrids is both under their remit (some doubt initially arose over whether they qualified as human embryos) and permitted by law, provided they are used to investigate the causes and treatment of degenerative diseases.

As a member of the EU, the UK is bound by EU regulations on the movement of human tissues and genetic materials across national borders; those wishing to import or export embryos or gametes must gain approval from the HFEA, who will wish to ascertain that the materials have been sourced in accordance with UK regulations on informed consent and anonymity. International collaboration is otherwise unrestricted. The UK is also subject to the decisions of the European Patent Office (EPO) which has twice declined to recognise patents on hESC under Article 53a of the *European Patent Code* which prohibits patents on inventions whose exploitation would run counter to "order public or morality".[33]

One unique feature of the UK system is the existence of the government-funded UK Stem Cell Bank (UKSCB.) As a condition of every HFEA license to derive hESC, a living colony from any resulting line must be deposited in the bank. The UKSCB also accepts submissions from international sources provided the appropriate paperwork is provided to demonstrate that the derivation of the line accords with UK standards of informed consent, safety and quality.

The bank makes research samples of any line it holds available to qualified research laboratories at no cost provided they can supply evidence of appropriate protocols. In conjunction with the UK's flexible and stable regulatory system, the Stem Cell Bank is likely to ensure that the UK remains a hub for international hESC research despite its comparatively modest financial investment in the field.

4.5. United States

As the largest funder of medical research in the world, and the birthplace of hESC isolation, the approach of the US to stem cell research is potentially of enormous influence on international policy. However, the US has struggled for decades with violent conflicts over opposing views on abortion that coloured views on embryo research even before the derivation of hESC. The federalist nature of the US government system tends to discourage federal policy on issues which are seen as containing moral dimensions; even the right to an abortion is enshrined not in federal legislation but in case law in the form of the landmark *Roe vs. Wade* case. Instead, the majority of the federal government's influence comes through the allocation or withholding of funds. Embryo research was not eligible for federal funding for many years under the relatively conservative administrations of Presidents Reagan and Bush Sr. Under President Clinton, the National Bioethics Advisory Committee (NBAC) issued a recommendation that "research involving the derivation and use of ES cells from embryos remaining after infertility treatments should be eligible for federal funding, given an appropriate framework for public oversight and review"[34] but in 2000 the administration changed and the recommendation was never implemented. In 2001, President George W. Bush announced that, due to moral concerns about the destruction of embryos, federal funding would be allocated only to research on hESC lines that had been created before January 2001. At the time, the administration announced that more than 60 usable lines were in existence,[35] although subsequent investigation indicated that this number was closer to 22.

Research in the US tends to operate on a "funders ethics carry" basis. Since it does not fund hESC research the National Institutes of Health (NIH) has no direct supervisory role in its conduct. It does, however, issue guidelines to researchers and institutional review boards, including the *Guidance for Investigators and Institutional Review Boards Regarding Research Involving Human Embryonic Stem Cells, Germ Cells and Stem Cell-Derived Test Articles,*[36] which attempt to clarify the point at which unregulated research ends and research on human subjects, which is subject to regulation by the Department of Health and Human Services (DHHS) and the Food and Drug Administration (FDA), begins.

Notwithstanding the use of federal funds or live human subjects, however, there are no federal laws pertaining to stem cell research.

The regulation of hESC research thus devolves to individual states, if they choose to implement it. The response on the part of the 50 states has been varied. Louisiana, Michigan, Minnesota, North Dakota, Pennsylvania, South Dakota and Indiana have passed state legislation specifically banning the conduct of hESC research. Arizona, Kansas, Virginia, Missouri and Georgia have followed the lead of the federal government in saying that they will not allow state funds to support such research, but they have not forbidden the practice *per se*. Florida and Maine have existing laws forbidding research on the foetus, but because "foetus" is undefined in the legislation it remains unclear whether this would apply to embryo research as well. Iowa, Missouri and New Jersey have specifically allowed the practice of stem cell research but do not offer state funding. California, Massachusetts, Connecticut, Illinois, Maryland, Washington, New Jersey, New York, Wisconsin and New Mexico provide state funding of various amounts, most notably California's US$3 billion in state bonds specifically earmarked for stem cell research. Texas and Oregon are currently considering pending legislation that would create similar funds. In the remaining states, as no specific legislation has been passed, federal policies stand.[b]

It is important to note that restrictions on state and federal funding do not constitute legal regulation of the conduct of hESC research. Privately funded companies in many states that have not explicitly banned the practice of hESC are free to conduct any kind of research, potentially including reproductive cloning. By contrast, California has amended its Health and Safety Code[37] to require ethical review by both a Stem Cell Research Oversight Committee and an IRB. Most other states subscribe to the "funders ethics" principle.

In theory, this regulatory sparseness makes the US ideal for international collaboration. There are no existing restrictions on the conduct of international research collaboration provided the import and export of human tissue and stem cells does not violate the policies of the Centers for Disease Control and Prevention (CDC) or the United States Department of Agriculture (USDA). Those wishing to develop cells for clinical use will also be subject to the regulations of the FDA. In addition, patent law in the US tends to be highly favourable to inventors; the patenting of stem cell techniques and products is permitted. However, the US represents a volatile regulatory environment in which permission or restrictions regarding stem cell research may be granted and withdrawn with comparatively little notice. For institutions located in states other than

[b] For updated information regarding the current state of legislation related to hESC in the US, see www.hinxtongroup.org.

those, like California, that have amended their constitution to explicitly permit stem cell research, long-term collaborations, at least on projects which involve the derivation of new hESC lines, retain an element of risk. Although the overall trend in the US does appear to be towards a permissive regulatory environment, it seems unlikely that a stable and uniform research policy will emerge any time soon.

5. Conclusion

It is hard, after reviewing chapters like this one, not to concede the huge instability that marks the interface between modern science and politics. By its nature, human embryonic stem cell science represents a polarising influence in an already uncertain field; the disparity between those who see human embryonic stem cells as the answer to previously insurmountable medical problems and those who see them as a violation of fundamental spiritual principles is so wide that it is tempting to dismiss the development of any kind of unified regulatory response as fantasy. The US is a telling example. Although it is widely acknowledged that the lack of coherent economic and legal governance has a negative effect on scientific progress, the ideological divisions are sufficiently strong to create one of the few historical situations where the advantages promised by medical science are insufficient to overrule more conservative social resistance.

Then too, there is the problem of pace. Effective regulatory environments are often complex and require both time and political capital to build and maintain. Where insufficient quantities of such capital have accumulated, as in India, regulation is wont to remain either theoretical or voluntary. And the pace of scientific advances does not stop to allow the political and social realms to catch up; although the non-democratic nature of the Chinese government means that it should theoretically be capable of enacting state regulation more quickly, the pace of scientific development has far outstripped any efforts at ethical control.

Interestingly, it in the two centralised democracies, not always noted for their legislative efficiency, that the regulatory regime is most stable and legally supported. It must be conceded, however, that the UK was merely successful at subsuming the issue of embryonic stem cells under existing legislation, while South Korea had considerable incentive, as a result of the Hwang scandal, to demonstrate a strong stance on ethical regulation. Nevertheless, both stand as proof that it is possible to implement comprehensive, enforceable regulation that supports, not restricts, the progress of scientific research and its promises for the future.

References

1. Callahan, D. *What Price Better Health? Hazards of the Research Imperative* (University of California Press, California, 2003).
2. van den Daele, W. As quoted by Beck-Gernsheim, E. Health and responsibility: from social change to technological change and vice versa. In: Adam, B., Beck, U. and Van Loom, J. (eds.), *Risk Society and Beyond: Critical Issues for Social Theory* (Sage Publication, London, 1989).
3. Marshall, E. The business of stem cells. *Science* **287**(5457), 1419–1421 (2000).
4. Philipkoski, K. The stem cell gold rush. *Wired* (4 November 2004), http://www.wired.com.
5. UK Department of Health and UK Department of Trade and Industry. *UK Stem Cell Initiative: Report and Recommendations* (November 2005), http://www.advisorybodies.doh.gov.uk/uksci.
6. Amos, J. Winston warns of stem cell "hype". *BBC News* (5 September 2005), http://www.bbc.co.uk.
7. http://www.geron.com.
8. Sherley, J. Human embryonic stem cell research: no way around a scientific bottleneck. *J. Biomed. Biotechnol.* **2**, 71–72 (2004).
9. Waldby, C. and Squier, S. Ontogeny, ontology and phylogeny: embryonic life and stem cell technologies. *Configurations* **11**, 27–46 (2003).
10. De Dios Vial Correa, J. and Sgreccia, E. *Declaration on the Production and the Scientific and Therapeutic Use of Human Embryonic Stem Cells* (Pontifical Academy for Life, Vatican, 2000), http://www.vatican.va/roman_curia/pontifical_academies/acdlife/documents/rc_pa_acdlife_doc_20000824_cellule-staminali_en.html.
11. Doerflinger, R. The ethics of funding embryonic stem cell research: a Catholic viewpoint. *Kennedy Inst. Ethics J.* **9**(2), 137–150 (1999).
12. Costa Rica. *Decreto de la Regulaci? n de la Reproducci? n Asistida.* Legislación Nacional, No. 24029-S, El Presidente de la Republica y el Ministro de Salud.
13. El Salvador. *La Constituci? n de la Republica*. Decreto No. 38 (1983), http://www.asamblea.gob.sv/constitucion/index1983.htm.
14. Fineschi, V., Neri, M. and Turillazzi, E. The new Italian law on assisted reproduction technology (law 40/2004). *J. Med. Ethics* **31**, 536–539 (2005).
15. Poland Law of 7 January 1993 on family planning, protection of human fetuses and the conditions under which pregnancy termination is permissible. Translated in *Int. Dig. Health Legis.* **44**(2), 253–255 (1993), http://annualreview.law.harvard.edu/population/abortion/POLAND.abo.htm.

16. Permanent representative of Costa Rica. *Draft International Convention on the Prohibition of All Forms of Human Cloning*. Presented to the *Ad Hoc* Committee (2003).
17. United Nations General Assembly. *International Convention Against the Reproductive Cloning of Human Beings*. Document A/C.6/58/L.8 (2003).
18. United Nations. *Legal Committee Recommends UN Declaration on Human Cloning To General Assembly*. Press Release GA/L/3271 (2005).
19. United Nations. *United Nations Declaration on Human Cloning*. Document A/C.6/59/L.27/Add.1 (2005).
20. Sun, J.-Y. Stem cell research must march on. *The Korean Times* (28 November 2005).
21. Xinhua. South Korean stem cell pioneer admits irregularities in research. *People's Daily Online,* China (25 November 2005), http://english.people daily.com.cn/200511/25/eng20051125_223824.html.
22. http://www.icmr.nic.in.
23. Draft Guidelines for Stem Cell Research/Regulation by the Indian Council of Medical Research (2006), http://www.icmr.nic.in.
24. Kahn, J. A nation of Guinea pigs. *Wired* (Issue 14.03, March 2006).
25. South Korea. *Bioethics and Biosafety Act, 2005* (unofficial translation), http://www.ruhr-uni-bochum.de/kbe/Bioethics&BiosafetyAct-South Korea-v1.0.pdf.
26. Schubert, S. Europe halts decisions on stem-cell patents. *Nature Online* (8 June 2005), http://www.nature.com/nature/journal/v435/n7043/full/ 435720b.html.
27. Salter, B., Cooper, M. and Dickens, A. China and the global stem cell economy: an emerging political strategy? *Regenerative Med.* **1**(5), 671–683 (2006).
28. Cyranoski, D. Named and shamed. *Nature Online* **441**, 392–393 (2006).
29. Cyranoski, D. Chinese clinical trials: consenting adults? Not necessarily… *Nature Online* **435**(138) (2005).
30. Hennig, W. Bioethics in China: although national guidelines are in place, their implementation remains difficult. *EMBO Rep.* **17**(9), 850–854 (2006).
31. Article 5 and Article 25. Chinese Patent Law as reprinted by the European Patent Office, http://www.epo.org/patents/patent-information/east-asian/ helpdesk/china/faq.html#new1.
32. Clarke, K. *Lord Hansard* (2 April 1990), Column 920.
33. Press release of the European Patent Office. *The EPO Follows the EU's Directive on Biotechnology Patents* (27 October 2005), http://www.epo. org/about-us/press/releases/archive/2005/27102005.html.
34. National Bioethics Advisory Committee. *The Ethical Use of Human Stem Cells in Research*. Draft Report (9 July 1999) Rockville, MD, NBAC.

35. Speech of President Bush on Stem Cell Research Funding (11 August 2001), http://www.whitehouse.gov/news/releases/2001/08/20010811-1.html.
36. http://www.stemcells.nih.gov.
37. See California Codes — Health and Safety Code125300–125355, http://www.leginfo.ca.gov/calaw.html.

Chapter 4

Human Embryonic Stem Cells: Derivation and Culture

Emma L. Stephenson, Peter R. Braude and Chris Mason

Abstract

There has been continued substantial interest from both scientists and the public in the therapeutic and scientific potential of stem cells since the first isolation of human embryonic stem cells (hESC) in 1998.[1] Pluripotent hESCs derived from the inner cell mass of preimplantation embryos following fertilisation *in vitro* (IVF) have been well studied, and proposed not only as potentially useful in treating degenerative diseases, but invaluable clinically relevant alternatives to animal models for studying early development, and for identifying novel pharmaceuticals with high throughput drug screens *in vitro*.[2] In addition, due to ethical controversy surrounding the use of embryos in stem cell research, there has been a paradigm shift in some research groups who have reported alternative methods of obtaining embryonic stem-like cells without the use of embryos. Most recently there has been some enthusiasm for exploring the use of induced pluripotent stem cells (iPS) which may be able to be derived from somatic cells by manipulation of transcription factors.[3] The derivation, culture and characterisation of hESC are currently a labour intensive and time consuming process. Emerging tissue engineering technology such as robotic control of culture will overcome such hurdles and facilitate the scale-up needed for clinical therapies.

Keywords: IVF; Human Embryo; Inner Cell Mass; Derivation; Human Embryonic Stem Cell; Standards.

Outline

1. Introduction
2. The Emergence of Human Embryonic Stem Cell Research
   2.1. Regulation of human embryo research
3. Human Embryonic Stem Cells

3.1. Definition
3.2. Embryonic stem cell derivation methods
3.2.1. *Embryonic stem cells without the destruction of embryos*
3.3. Preimplantation genetic diagnosis and stem cell derivation
4. Culture of hESC lines
4.1. Feeder cells
4.2. Media composition
4.2.1. *Oxygen tension*
5. Reporting of Derivation
6. Concluding Remarks
References

1. Introduction

In 1978, after almost two decades of both animal and human research, the August edition of *The Lancet*[4] published a letter that heralded great change in the field of infertility — the birth of Louise Brown, the first baby born following *in vitro* fertilisation (IVF). Professor Robert Edwards and Mr. Patrick Steptoe had together marked the moment when science and medicine entered the arena of human conception, offering hope to the multitude of couples around the world previously told they had no chance of conceiving their own genetic child. In the following 25 years, 68,000 IVF babies were born in the UK, with an estimated one million worldwide. The numbers continue to rise rapidly as diagnosis, treatment and follow-up care are constantly researched and improved. It is this great success story that laid the foundations for the field of human embryonic stem cell research.

2. The Emergence of Human Embryonic Stem Cell Research

The first report concerning embryonic stem cells was in 1967, when Cole, Edwards and Paul[5] described long-lived and stable cells obtained from either whole or dissected rabbit blastocysts. A number of additional landmark papers soon followed from Cambridge, with Gardner describing the generation of the world's first chimeric mouse with its distinctive coat-colour pattern in 1968[6] by the injection of inner cell mass (ICM) cells into the blastocoelic cavity, and the much cited paper by Evans and Kaufman in 1981[7] describing the establishment in culture of pluripotential cells from mouse embryos. Hormonal manipulation of the mice and the use of serum-supplemented medium resulted in the growth of karyotypically normal, pluripotent cells from whole mouse blastocysts on mouse fibroblast feeder layers. Fishel, Edwards and Evans reported in 1984[8] that whole human embryos cultured *in vitro* produced outgrowths that survived for several days and secreted human chorionic gonadotrophin, but that were not suitable for stem cell colonies. Pluripotent cell lines were derived from several non-rodent species in the following years, but the developmental potential of these cells were poorly characterised. In 1994 Bongso and colleagues described the isolation and culture of ICM cells from whole human blastocysts[9] producing cells with typical stem cell morphology, that were alkaline phosphatase positive, karyotypically normal and which could be maintained for two passages. This was a significant step forward, as major differences exist between early human and mouse development, such as the timing of embryonic genome expression and the structure and function of foetal membranes. Thomson and co-workers reported the isolation of an ES cell line from a rhesus monkey blastocyst in 1995.[10] This primate line showed the morphological and differential features of ES cells on complete characterisation. These results were followed by the first report in 1998[1] of embryonic stem cell lines derived from human blastocysts. Thomson and colleagues used 36 human embryos to derive five hESC lines, using immunosurgery to isolate the ICMs. The lines were fully characterised, including expression of cell-surface markers that characterise undifferentiated cells, an ability to differentiate into all three germ layers, and teratoma formation on transplantation to SCID mice.

There was therefore a 13-year gap (1981–1994) between the derivation of the first mESC line and the first report of human ICM cells cultured for two passages, a further year before the demonstration of a primate embryonic cell line (1995), and an additional three years before the report of the first fully-characterised hESC line (1998). The technical challenges associated with translation of stem cell derivation methods between species, and the increasing use by IVF units of cryopreservation of surplus embryos for patient use resulting in scarce supply of human blastocysts for research, both contributed to this time interval. In addition it took several years for the development of culture media able to reliably and appropriately support the growth of human embryos to the blastocyst stage. The development of specialist sequential culture media tailored to the changing requirements of embryo metabolism was a significant step in the success of infertility treatment and hence stem cell isolation. The most common developmental stage of embryo used for derivation attempts is the blastocyst — between five and eight days after fertilisation. It is at this stage where the pluripotent population of cells can be distinguished morphologically and isolated for stem cell culture. However, attempts have also been successful with morula stage embryos,[11] single blastomeres from cleavage stage embryos,[12] and from apparently arrested embryos.[13]

Since Thomson's 1998 paper, the hESC field has expanded rapidly, with different research groups working on derivation methods and efficiency, on striving to remove animal products from culture to enable cells to be used in medical therapy, on differentiation and genetic modification studies, and much more.[14–16] As stem cell research touches so acutely the fundamental processes of reproduction and development, it attracts interest from social scientists and ethicists, and also draws responses from concerned public, and religious and pro-life groups.[17]

2.1. Regulation of human embryo research

The birth of Louise Brown in 1978 prompted intense debate as to whether it was morally acceptable to allow research to be undertaken on human embryos, which resulted in the establishment of a UK framework for regulating such research (see also Chapter 3 of this book). The Committee of Enquiry into Human Fertilisation and Embryology, the Warnock Committee (1982–1984)[18] took a gradualistic view on the status of the human embryo, deciding that the degree of protection afforded increased with development of the embryo. They recommended that research on human embryos should be permitted up to 14 days after fertilisation (or the primitive streak stage, whichever is the sooner) but only under closely regulated licences by a statutory regulator. The Human Fertilisation and Embryology (HFEA) was set up following the passage of the 1990 Human Fertilisation and Embryology Act, and charged with regulation and inspection of all clinics offering *in vitro* fertilisation and the storage of any gametes or embryos. It also was to regulate the use of human embryos in research allowed for five specified purposes, deemed necessary or desirable. In January 2001, driven largely by the potential of hESC in regenerative therapy, the HFE (Research Purposes) Regulations were extended to allow research on embryos to be licensed for three additional purposes. The HFEA Code of Practice (Sixth Edition)[19] states that the HFEA may only grant licences for research projects if it appears to the HFEA that the activity to be authorised by the licence is necessary or desirable for one or more of the following purposes:

(i) to promote advances in the treatment of infertility;
(ii) to increase knowledge about the causes of congenital disease;
(iii) to increase knowledge about the causes of miscarriages;
(iv) to develop more effective techniques of contraception;
(v) to develop methods for detecting the presence of gene or chromosome abnormalities in embryos before implantation;
(vi) to increase knowledge about the development of embryos;
(vii) to increase knowledge about serious disease; and
(viii) to enable such knowledge to be applied in the development of treatments to combat serious disease.

Thus, with these amendments to the HFE Act, the King's College London and the Roslin Institute in Scotland were granted the first licenses for hESC research in early 2002. Subsequently, the King's College London deposited one the first cell lines into the UK Stem Cell Bank coinciding with the official opening of the facility in May 2004.

Additional legislation, the Human Reproductive Cloning Act was also introduced in 2001 following a judicial review requested by the ProLife Alliance, which prohibits the uterine transfer of a human embryo created otherwise than by fertilisation. Hence although under an appropriate license centres can research using somatic cell nuclear transfer, embryos resulting from this procedure cannot be transferred to a woman, thus allaying fears that the HFE Act wording permitted human reproductive cloning.

The EU directive on tissues and cells became mandatory in April 2007[20] and requires a stringent standard of quality for IVF laboratories and laboratories deriving stem cells intended for therapeutic use, more rigorous than that required by the HFEA, covering quality and safety for the donation, procurement, testing, processing, preservation, storage and distribution of human tissues and cells. This swift rate of scientific change and discovery requires legislation keep pace so that appropriate guidance and regulation is always in place. The Human Fertilisation and Embryology Act is now over 15 years old and requires modification[21] especially in the requirement to incorporate the European Tissues and Cells Directive. A draft new bill including establishment of a combined HFEA and Human Tissue authority as a Regularity Authority for Tissues and Embryos (RATE)[22,23] will be considered by parliament in 2008.

In an attempt to ensure that the UK remains one of the global leaders in stem cell research, the government established the UK Stem Cell Initiative in 2005, with a remit to develop a ten-year vision and costed strategy for UK stem cell research, for implementation between 2006–2015. The Pattison report listed 11 recommendations designed to consolidate the UK position of strength and develop this over the decade.

3. Human Embryonic Stem Cells

3.1. Definition

Thomson and colleagues in their landmark paper of 1998[1] listed their proposed essential characteristics to define primate ES cells. In the broadest sense human ES cells are defined as: cells derived from the pre- or peri-implantation human embryo, capable of prolonged undifferentiated proliferation and demonstrating stable developmental potential to form derivatives of all three embryonic germ layers even after prolonged culture.

3.2. Embryonic stem cell derivation methods

Following the first identification of embryonic stem cells from intact blastocysts in 1967[5] success in establishing hESC lines was achieved by using immunosurgery to isolate the inner cell mass (ICM). This method, developed by Solter and Knowles in 1975[24] uses a complement reaction after removal of the zona pellucida to selectively lyse the trophectoderm cells, leaving the ICM intact. Although this method has proven successful in many laboratories, there is a real need to develop an alternative method for isolating the ICM, that does not require the use of animal serum and complement, which could preclude their clinical use. Amit and Itskovitz-Eldor in 2002[25] reported the derivation of two hESC lines using immunosurgery from five cryopreserved blastocysts, and a further line by gentle removal of the trophectoderm with 27G needles, hence without contamination with animal products. In 2004, Heins and colleagues[26] successfully established four lines by simply plating whole blastocysts on irradiated mouse embryonic fibroblasts (MEF), which had either spontaneously hatched or had pronase-removal of the zona pellucida. Such success was replicated by Suss-Toby and co-workers[27] in 2004, Findikli *et al.*[28] in 2005, and is now a commonly used technique. However, there is no consensus on whether the efficiency is comparable to immunosurgery, since some groups have not been able to derive any lines using this method. Stojkovic and colleagues[29] reported in 2004 a novel three-step culture system for human blastocysts, which were used on day eight after culture in buffalo rat liver (BRL)-conditioned media for two days. This resulted in significantly more ICM cells than the traditional six-day culture, and immunosurgery was then used to produce one hESC. Eight morula-derived hESC lines were reported in 2004 by Strelchenko *et al.*[11] by injecting zona-free morulae under a MEF feeder layer using a micromanipulator. Whilst success using early stage mouse embryos had previously been reported, this was the first incidence of success with human embryos. Kim and colleagues[30] tailored their derivation attempts to the quality of the blastocysts, only performing immunosurgery on good quality blastocysts with large, distinct ICMs, partial culture on blastocysts with small ICMs, and whole blastocyst culture for those with poorly defined ICMs, achieving hESC lines with all methods, although with differing efficiency. Mechanical isolation has proven at least as successful as immunosurgery in personal experience (unpublished results). This group used 29 G needles to gently separate the ICM whilst stabilising the blastocyst in a minute volume of

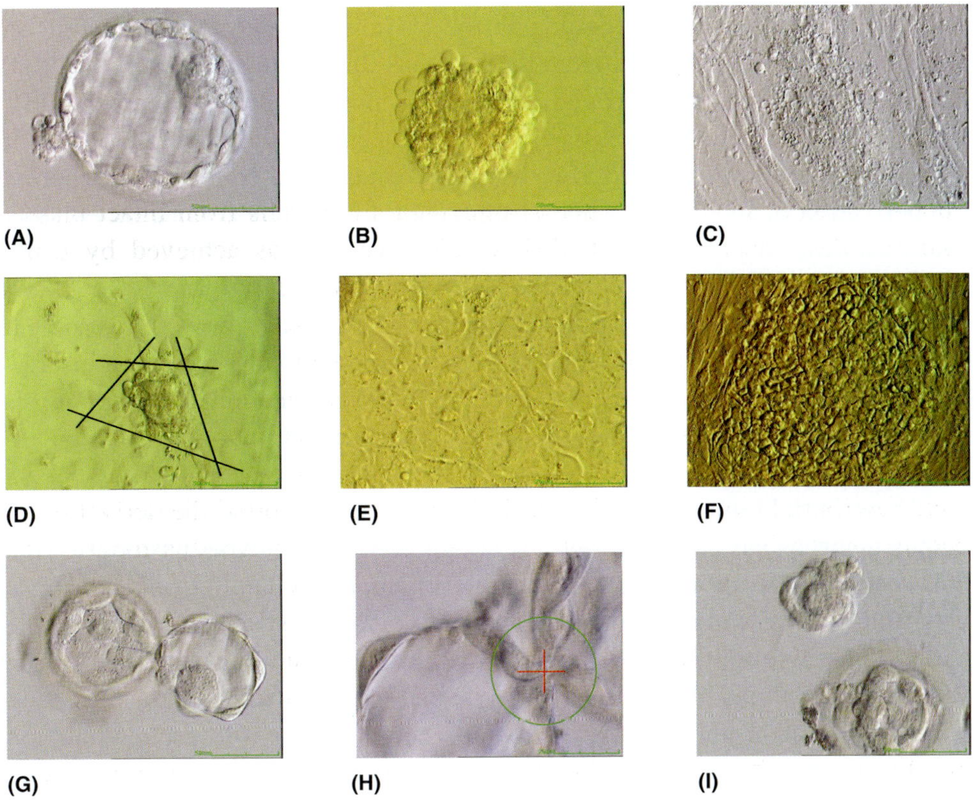

Fig. 1. **(A)** Day 6 blastocyst before immunosurgery. **(B)** Blastocyst during complement reaction of immunosurgery. The blebbing structures are lysing trophectoderm cells. **(C)** Initial cell outgrowth one day after plating the isolated inner cell mass. **(D)** Day 7 inner cell mass of expanded blastocyst. The superimposed lines show the cuts made with needles to isolate the inner cell mass. **(E)** Initial appearance of stem cell-like cells 11 days after plating the inner cell mass. **(F)** Passage 1 of stem cell line. **(G)** Day 5 blastocyst before laser isolation. **(H)** Laser overlay to show target of pulse on trophectoderm cells. **(I)** Isolated inner cell mass with a small number of surrounding trophectoderm cells, the majority of the trophectoderm remains in zona pellucida.

media — an acceptable method for isolating the ICM for obtaining clinical grade hESC lines (see Fig. 1).

3.2.1. *Embryonic stem cells without the destruction of embryos*

One of the main objections to hESC derivation is that human embryos have to be destroyed in order to culture the ICM. Despite some clinics only using embryos that are truly surplus to therapeutic need, this public feeling combined with the restrictive federal funding in the US has driven attempts to find an alternative methodology. Although the NIH has funded approximately US$129 million of research on approved hESC lines derived before October 2001, US research teams have been trying to develop means that would qualify for NIH funding on lines derived after that date. Chung and colleagues[31] reported the derivation of mouse ESC lines from single blastomeres, using an existing green fluorescent protein (GFP) positive stem cell line as a feeder layer, and showed that the biopsied embryos could produce term births. Meissner *et al.*[32] used a different approach by generating mouse ESC lines from Cdx2-deficient blastocysts. Cdx2 encodes the earliest known trophectoderm-specific transcription factor activated in the eight-cell embryo and hence embryos lacking this gene are morphologically abnormal, lack functional trophoblast and fail to implant.

These initial reports were with the use of mouse embryos, but success with human embryos has been reported. Although Fong and co-workers reported the unsuccessful derivation of hESC lines from pairs of human blastomeres,[33] this was followed by a manuscript from Klimanskaya *et al.*[34] who announced the derivation of two stable hESC lines with 17 further hESC-like outgrowths from 91 individual human blastomeres. A single cell was removed from an eight-cell embryo just as if the embryo was undergoing biopsy for preimplantation genetic diagnosis, and then allowed to develop in culture for two days. The resulting blastomere-derived vesicle was co-cultured with a GFP — hESC line to attempt derivation. The parental embryo was kept in culture until the blastocyst stage and then cryopreserved as if for future use. Revazova and colleagues recently reported the derivation of six pluripotent human embryonic stem cell (hESC) lines from blastocysts of parthenogenetic origin.[35]

Another avenue being explored is that of reprogramming somatic cells. Cowan and colleagues[36] in 2005 and Strelchenko *et al.*[37] in 2006 both reported variations on the theme. Cowan showed nuclear reprogramming of somatic cells after fusion with hESCs, resulting in tetraploid cells with the morphology and behaviour of hES cells. Strelchenko described reprogramming of human somatic cells by embryonic stem cell cytoplast producing a mixed population of cells where, in a proportion, the nucleus of the hES cell was replaced by the somatic cell nucleus thus giving rise to a cell with the genotype of the somatic donor, but the "stemness" of the recipient hES cell. In a groundbreaking publication Takahashi and Yamanaka described the induction of pluripotent stem cells from mouse embryonic or adult fibroblasts by introducing four factors, Oct3/4, Sox2, c-Myc, and Klf4, under ES cell culture conditions.[38] These cells, designated induced pluripotent stem (iPS) cells, exhibited the morphology and growth properties of ES cells and expressed ES cell marker genes. They rapidly followed this[3] with the demonstration of the generation of iPS cells from adult human dermal fibroblasts with the same four factors. Human iPS cells were similar to human embryonic stem cells in morphology, proliferation, surface antigens, gene expression, epigenetic status of pluripotent cell-specific genes, and telomerase activity. Furthermore, these cells could differentiate into cell types of the three germ layers *in vitro* and in teratomas. Following this, Yu and colleagues[39] showed that a different combination of four factors, Oct4, Sox2, Nanog, and LIN28, were sufficient to reprogramme human somatic cells to pluripotent stem cells that exhibited the essential characteristics of embryonic stem (ES) cells.

3.3. Preimplantation genetic diagnosis and stem cell derivation

hESCs with mutations significant to human disease provide a powerful *in vitro* tool for modelling the disease, identifying molecular mechanisms that may be blocked or prevent expression, for studying pathogenesis in particular differentiated cell types as well as providing an ideal system for investigating *in vitro* efficacy and toxicity of drugs. Preimplantation genetic diagnosis (PGD) provides access to embryos that carry genetic disorders and being often from fertile couples the embryos are more likely to be good quality. Pickering *et al.*[40] reported the generation and characterisation of a hESC line carrying the most common cystic fibrosis mutation Δ508. The line showed the same morphology and protein expression as unaffected lines, and provides the opportunity to test therapies relevant to cystic fibrosis, as well as investigating the progression of the disease with the differentiation of the cells to pulmonary phenotype. At around the same time, Materiel and colleagues[41] reported the derivation of three hESC lines carrying genetic disorders, and Verlinsky *et al.* reported[42] the derivation of 18 hESC lines with genetic disorders, although most were cryopreserved at early passage and not fully characterised. These hESC lines with genetic disorders can provide an unlimited source of cells for the study of the disease, and the "disease-in-a-dish" technology is likely to be fruitful in the study of clinically significant genetic disease.

4. Culture of hESC Lines

The first reports of successful derivation of hESCs used very similar growth conditions to those used for deriving mESC lines, and contained animal derived products for almost all stages of culture. Ideally these products will be eliminated from culture for use of the cells in clinical application. Although the MHRA and FDA have approved the use of some animal products in therapy in a limited number of cases, comprehensive quality assurance, safety testing and validation is required. Furthermore, Martin and colleagues[43] reported that hESCs cultures in animal-derived serum replacement medium express an immunogenic non-human sialic acid. If these cells were transplanted, antibodies in the recipient serum could cause immunoglobulin binding and deposition of complement, which would lead to the destruction of the stem cells *in vivo*. Animal products are used at many stages of *in vitro* culture — cumulus cells surrounding the oocyte are removed using a porcine hyaluronidase before microinjection of sperm; many proprietary IVF media contain pooled human serum albumin as the protein source; immunosurgery to isolate the ICM uses animal antibody and complement; and the isolated ICM cells are placed on mouse feeder layers, culture and preparation of which uses bovine serum and a porcine enzyme. The media used to culture the ICM and then stem cells contain calf serum, and occasionally the medium is conditioned by buffalo rat liver cells. Each stage of culture must be considered and assessed for suitability for clinical application of hESCs.

4.1. Feeder cells

Unlike mESCs, hESCs lose pluripotency and differentiate rapidly when grown on plastic tissue-culture dishes, even in the presence of leukaemia inhibitory factor (LIF). The culture of ICM or hES cells appears to require support from a feeder layer either as a contact layer or a source of conditioned medium. There have been reports on the use of BD Matrigel™ Matrix (BD Biosciences, USA) as a synthetic non-cellular matrix, however this also contains components of animal origin. Reports on the successful modification of this culture system soon appeared after derived lines became available for research. Xu and co-workers first reported feeder-free growth of undifferentiated hESCs[44] where they were able to maintain hESCs for at least 130 population doublings on Matrigel or laminin, but required media conditioned by MEFs. Amit *et al.*, continued with the feeder-free option, by culturing hESCs on human or bovine fibronectin with a serum replacement medium supplemented with various combination of growth factors.[45] Transforming growth factor-beta (TGFβ) and basic fibroblast growth factor (bFGF) with or without LIF proved most effective, but they noted that all three parameters proved somewhat inferior to the traditional MEF culture. Another variation for feeder-free growth was described by Klimanskaya *et al.*, in 2005,[46] who prepared extracellular matrix from MEFs and media with serum replacement and plasmanate®.

Taking a different route, Richards and colleagues[47] described the prolonged undifferentiated growth of human ICM and hESCs on human fibroblast feeders, using a human serum source. However it was observed that the differentiation rates of the hESCs increased beyond ten passages. The same group subsequently published a comparative evaluation of various human feeders for hESC culture in 2003.[48] In 2003, Hovatta and co-workers described a culture system using post-natal commercially available human foreskin fibroblasts as feeder cells,[49] and reported derivation of two new lines in these conditions, but with bovine serum as the protein source. They followed this with a report in 2005 of derivation of hESCs on placental fibroblast feeders[50] using serum replacement medium. In 2003, Amit *et al.* also described the use of human foreskin feeders for hESCs in Knock-Out Dulbecco's Modified Eagles Medium (KO-DMEM, Invitrogen) with serum replacement[51] stating that this resulted in a completely animal free culture system. Unfortunately, it seems that the serum replacement contains animal derivatives, and the establishment of the foreskin cell line was achieved using foetal calf serum, resulting in direct and indirect exposure of the hESCs to products of animal origin.

Another approach to avoid use of MEFs has been to derive fibroblasts from existing hESC lines. This approach offers a number of advantages; both xenogeneic and allogeneic feeders carry the risk of pathogen transmission; not all human feeder systems or cell-free matrices support the culture of undifferentiated hESCs equally well, like hESCs, stem cell-derived feeder layers would be immortal, allowing scale-up and mass-production. Xu and co-workers reported that immortalised fibroblast-like cells derived from hESCs supported undifferentiated cell growth.[52] They derived a fibroblast-like cell type and transfected them with a retrovirus expressing hTERT to extend their replicative capacity, resulting in immortalised cells. Conditioned medium from this source was able to support hESCs on Matrigel.

Ludwig *et al.* described the derivation of two hESC lines in defined conditions in early 2006.[53] The culture system was feeder-free and contained protein components derived solely from recombinant sources or purified from human material. The only undefined component was human serum albumin (HSA), but the authors argued that standard IVF media for culture of embryos contains HSA so they were not introducing any further risk into the culture system. Unfortunately the concentration of growth factors and proteins used in the media were such to make the system prohibitively expensive for the majority of research laboratories. However, the group must be commended for publishing the exact formulations of their culture system and their protocols. It is only through the sharing of information in this way that the field will move forward efficiently.

4.2. Media composition

To try to characterise the milieu provided by feeder cells, Lim and Bodnar[54] performed proteome analysis of MEF-CM resulting in the identification of

136 unique proteins species. This served to highlight the complexity of the environment provided by the feeder cells and provided some guidance as to the types of proteins which should be investigated for their effects on growth and differentiation of hESCs in a defined culture environment. Koivisto and colleagues[55] investigated the effect of bFGF, finding that it supported non-differentiated growth of hESCs with the best result at 8 ng/ml. Beattie *et al.*[56] reported that pluripotency of hESCs could be maintained by Activin A in the absence of feeder layers or conditioned medium although keratinocyte growth factor and nicotinamide were needed to return the proliferation rate to normal. Xiao and co-workers[57] developed these findings further by showing that Activin A induces the expression of bFGF, Oct4, Nanog, Nodal, Wnt3 and FGF8 and suppresses bone morphogenic protein (BMP) pathways in hESCs. Brain-derived neurotrophic factor (BDNF), neurotrophin (NT)3 and (NT)4 also mediate hESC survival, and their addition to cell cultures resulted in a 36-fold improvement in their clonal survival.[58] A comprehensive study of various xeno-free media[59] unfortunately concluded that none of those tested were as effective at culturing hESC as the conventional KOSR system. Several commercially available proprietary media which claim various levels of clinical suitability and efficacy are becoming available, e.g. hEScGRO™ (Chemicon), StemPro® (Invitrogen) and mTeSR™ (Stem Cell Technologies).

4.2.1. *Oxygen tension*

The dissolved oxygen tension (DOT) in the oviduct and uterine cavity lies between 3% and 5%, at the implantation site 2%–4%, and in foetal blood during pregnancy 4%–5%.[60] Hence as hESCs derive from the ICM of the blastocyst, the cells from which they originate develop *in vivo* at physiological hypoxia. It is still the norm to culture human embryos for therapeutic use *in vitro* in media with a gas phase of 5% CO_2 in air (21% O_2) despite ample evidence of a potentially detrimental effect of oxidative stress on development.[61,62] Although these culture conditions are clearly sufficient for embryo viability, and will generate successful and healthy pregnancies, the longer-term effects of oxygen concentration are uncertain. Of particular concern is the long-term effect of enhanced oxygen tension on the function and enduring potential of mitochondria.

Prompted by the evidence from IVF units that embryos cultured at 4%–6% oxygen are of improved viability[63] and that decreasing the dissolved oxygen tension during all phases of embryo production increases pregnancy success rates, several stem cell laboratories have also begun to investigate the effect of DOT on stem cell growth and fate. Ezashi and co-workers[64] demonstrated that hESCs grew as well under 3% and 5% oxygen as at 21% in routine culture, with growth rates only slightly reduced at 1% oxygen. Forsyth and colleagues reported that culture in 2% oxygen enhanced hESC clonal recovery sixfold in three different hESC lines and significantly reduced the acquisition of spontaneous chromosomal abnormalities.[65] These data suggest, perhaps unsurprisingly, that physiological

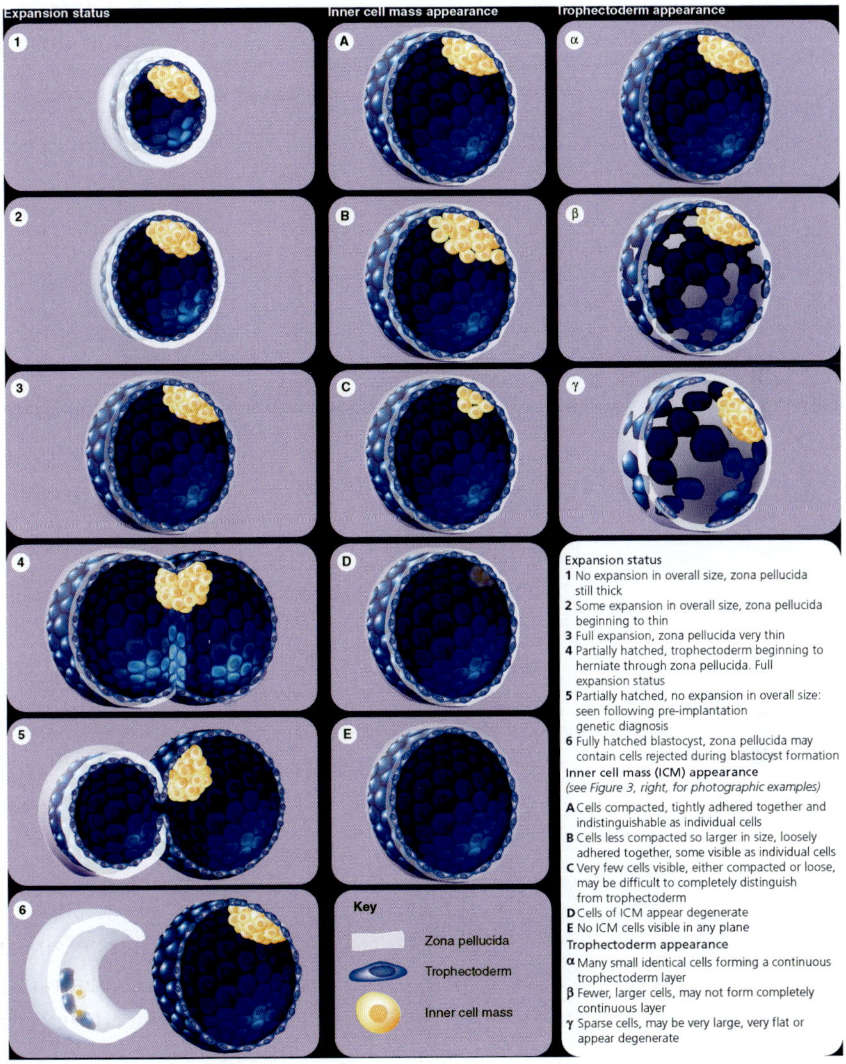

Fig. 2. Classification system for human embryonic stem cell (hESC) derivation. Diagrammatic representation of the grading system for blastocysts used in hESC derivation. Column 1 shows expansion status only; the ICM and trophectoderm are standard to highlight the variable. Similarly, Column 2 shows ICM appearance only, and Column 3 trophectoderm morphology only. Copyright of Future Medicine. Published in *Regen. Med.* **1**(6), 739–750 (2006).[66]

oxygen is more appropriate for hESC culture, and likely to reduce or prevent the accumulation of karyotypic abnormalities over prolonged time in culture.

5. Reporting of Derivation

The interpretation of results and implementation of new methods in the stem cell field is fraught with difficulties, not least of which is the highly variable reporting

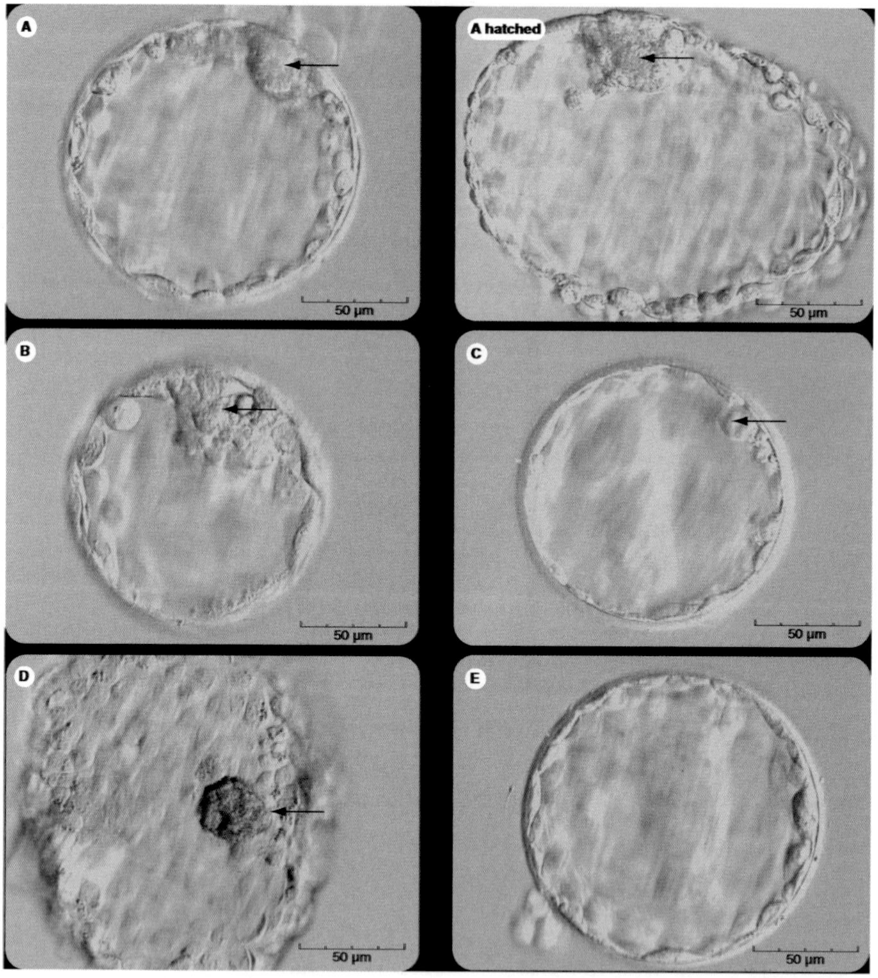

Fig. 3. Photographic examples of inner cell mass (ICM) grades A–E. Photographic examples of the ICM appearance represented in diagrammatic form in Fig. 2. These pictures demonstrate ICM grade A–E only, and do not consider expansion or trophectoderm appearance. It is important to scan through every plane of the blastocyst to confer an accurate grade. ICM indicated by arrow. Copyright of Future Medicine. Published in *Regen. Med.* **1**(6), 739–750 (2006).[66]

of the number and quality of embryos used to derive each new hESC line.[66] It is almost impossible to ascertain from the literature a true efficiency for deriving a hESC line. Until there is a clear minimum information (MI) convention amongst the derivation teams, understanding and comparing novel derivation methodologies and their potential impact on the resulting stem cell line will continue to be impossible or at best extremely difficult.

Open and transparent reporting of results will move the human embryonic stem cell field forward more rapidly towards clinical translation and application. There is little debate over the value of the development of standards for other biological areas of research, and the hESC field must be included.[67,68] The requirement for standards in derivation and stem cell science originates in part because the field is new and therefore needs structure. Also, it involves expertise from diverse scientific backgrounds, from embryology to tissue engineering. The challenge faced in embryology and stem cell derivation is the attempt to overlay discrete defined standards on a continuous spectrum of developmental variance. However, the use of MI datasets in other biological fields have shown that standards can greatly help avoid misunderstandings, duplication of work and aid clear, precise communication.[69,70]

With this in mind, Stephenson *et al.* published a draft proposal for a universal minimum information convention for the reporting on the derivation of human embryonic stem cell lines.[66] This was followed by the results of the community consultation as an international community consensus.[71] The articles are open-access publications in order to make the convention freely available to the international community and encourage universal participation (see Figs. 2 and 3).

6. Concluding Remarks

Despite ethical debate and political interference, and concerns over systems and reporting, results from hESC so far are very encouraging. From one line in 1998, there is now anecdotal evidence of 300–350 hESC lines internationally. Similarly substantial progress has been made towards the clinical application of these cells, with moves toward Good Manufacturing Practice (GMP) compatible culture systems and early understanding of the processes and control of differentiation to the desired cell type. Although caution must still be maintained when a projecting time scale to therapy, the concerted effort and cooperative multi-disciplinary approach of the stem cell community will in no doubt reach the goal of the clinic by following a safe, efficacious and regulated route.

References

1. Thomson, J.A. *et al.* Embryonic stem cell lines derived from human blastocysts. *Science* **282**(5391), 1145–1147 (1998).
2. Sinha, G. Cell biology. Human embryonic stem cells may be toxicology's new best friends. *Science* **308**(5728), 1538 (2005).
3. Takahashi, K. *et al.* Induction of pluripotent stem cells from adult human fibroblasts by defined factors. *Cell* **131**(5), 861–872 (2007).
4. Steptoe, P.C. and Edwards, R.G. Birth after the reimplantation of a human embryo. *Lancet* **2**(8085), 366 (1978).
5. Edwards, R.G., IVF and the history of stem cells. *Nature* **413**(6854), 349–351 (2001).
6. Gardner, R.L. Mouse chimeras obtained by the injection of cells into the blastocyst. *Nature* **220**(167), 596–597 (1968).
7. Evans, M.J. and Kaufman, M.H. Establishment in culture of pluripotential cells from mouse embryos. *Nature* **292**(5819), 154–156 (1981).
8. Fishel, S.B., Edwards, R.G. and Evans, C.J. Human chorionic gonadotropin secreted by preimplantation embryos cultured *in vitro*. *Science* **223**(4638), 816–818 (1984).
9. Bongso, A. *et al.* Isolation and culture of inner cell mass cells from human blastocysts. *Hum. Reprod.* **9**(11), 2110–2117 (1994).
10. Thomson, J.A. *et al.* Isolation of a primate embryonic stem cell line. *Proc. Natl. Acad. Sci. USA* **92**(17), 7844–7848 (1995).
11. Strelchenko, N. *et al.* Morula-derived human embryonic stem cells. *Reprod. Biomed. Online* **9**(6), 623–629 (2004).
12. Klimanskaya, I. *et al.* Derivation of human embryonic stem cells from single blastomeres. *Nat. Protoc.* **2**(8), 1963–1972 (2007).
13. Zhang, X. *et al.* Derivation of human embryonic stem cells from developing and arrested embryos. *Stem Cells* **24**, 2669–2676 (2006).
14. Hewitt, Z.A., Amps, K.J. and Moore, H.D. Derivation of GMP raw materials for use in regenerative medicine: hESC-based therapies, progress toward clinical application. *Clin. Pharmacol. Ther.* **82**(4), 448–452 (2007).
15. Goswami, J. and Rao, M. Embryonic stem cell therapy. *IDrugs* **10**(10), 713–719 (2007).
16. Puceat, M. and Ballis, A. Embryonic stem cells: from bench to bedside. *Clin. Pharmacol. Ther.* **82**(3), 337–339 (2007).
17. http://www.corethics.org/.
18. Committee of Inquiry into Human Fertilisation and Embryology. Report. HMSO, London, 1984 (Cmnd 9314; Warnock report).

19. HFEA. *The Code of Practice*, 6th Ed. (2003).
20. 2004/23/EC, D., Of the European Communities and of the Councils of 31 March 2004 on setting standards of quality and safety for the donation, procurement, testing, processing, preservation, storage and distribution of human tissues and cells. Official Journal of the European Union 2004. http://europa.eu.int/eur-lex/pri/en/oj/dat/2004/1_102/1_10220040407en00480058.pdf (accessed 6 September 2006).
21. http://www.hfea.gov.uk/docs/2005-05-18_review_of_the_act_annex_A.pdf.
22. http://www.hfea.gov.uk/docs/White_Paper_Dec_06_web_version.pdf.
23. Human Tissues and Embryos (Draft) Bill. Department of Health HMSO (Cm 7087) 2007.
24. Solter, D. and Knowles, B.B. Immunosurgery of mouse blastocyst. *Proc. Natl. Acad. Sci. USA* **72**(12), 5099–5102 (1975).
25. Amit, M. and Itskovitz-Eldor, J. Derivation and spontaneous differentiation of human embryonic stem cells. *J. Anat.* **200**(Pt. 3), 225–232 (2002).
26. Heins, N. *et al.* Derivation, characterization, and differentiation of human embryonic stem cells. *Stem Cells* **22**(3), 367–376 (2004).
27. Suss-Toby, E. *et al.* Derivation of a diploid human embryonic stem cell line from a mononuclear zygote. *Hum. Reprod.* **19**(3), 670–675 (2004).
28. Findikli, N. *et al.* Establishment and characterization of new human embryonic stem cell lines. *Reprod. Biomed. Online* **10**(5), 617–627 (2005).
29. Stojkovic, M. *et al.* Derivation of human embryonic stem cells from day-8 blastocysts recovered after three-step *in vitro* culture. *Stem Cells* **22**(5), 790–797 (2004).
30. Kim, H.S. *et al.* Methods for derivation of human embryonic stem cells. *Stem Cells* **23**(9), 1228–1233 (2005).
31. Chung, Y. *et al.* Embryonic and extraembryonic stem cell lines derived from single mouse blastomeres. *Nature* **439**(7073), 216–219 (2006).
32. Meissner, A. and Jaenisch, R. Generation of nuclear transfer-derived pluripotent ES cells from cloned Cdx2-deficient blastocysts. *Nature* **439**(7073), 212–215 (2006).
33. Fong, C.-Y. and Ariff Bongso, M.R. Unsuccessful derivation of human embryonic stem cell lines from pairs of human blastomeres. *Reprod. Biomed.* **13**(2), 295–300 (2006).
34. Klimanskaya, I. *et al.* Human embryonic stem cell lines derived from single blastomeres. *Nature* **444**, 481–485 (2006).
35. Revazova, E.S. *et al.* Patient-specific stem cell lines derived from human parthenogenetic blastocysts. *Cloning Stem Cells* **9**, 432–449 (2007).

36. Cowan, C.A. *et al.* Nuclear reprogramming of somatic cells after fusion with human embryonic stem cells. *Science* **309**(5739), 1369–1373 (2005).
37. Strelchenko, N. *et al.* Reprogramming of human somatic cells by embryonic stem cell cytoplast. *Reprod. Biomed. Online* **12**(1), 107–111 (2006).
38. Takahashi, K. and Yamanaka, S. Induction of pluripotent stem cells from mouse embryonic and adult fibroblast cultures by defined factors. *Cell* **126**(4), 663–676 (2006).
39. Yu, J. *et al.* Induced pluripotent stem cell lines derived from human somatic cells. *Science* **318**(5858), 1917–1920 (2007).
40. Pickering, S.J. *et al.* Generation of a human embryonic stem cell line encoding the cystic fibrosis mutation deltaF508, using preimplantation genetic diagnosis. *Reprod. Biomed. Online* **10**(3), 390–397 (2005).
41. Mateizel, I. *et al.* Derivation of human embryonic stem cell lines from embryos obtained after IVF and after PGD for monogenic disorders. *Hum. Reprod.* **21**(2), 503–511 (2006).
42. Verlinsky, Y. *et al.* Human embryonic stem cell lines with genetic disorders. *Reprod. Biomed. Online* **10**(1), 105–110 (2005).
43. Martin, M.J. *et al.* Human embryonic stem cells express an immunogenic nonhuman sialic acid. *Nat. Med.* **11**(2), 228–232 (2005).
44. Xu, C. *et al.* Feeder-free growth of undifferentiated human embryonic stem cells. *Nat. Biotechnol.* **19**(10), 971–974 (2001).
45. Amit, M. *et al.* Feeder layer- and serum-free culture of human embryonic stem cells. *Biol. Reprod.* **70**(3), 837–845 (2004).
46. Klimanskaya, I. *et al.* Human embryonic stem cells derived without feeder cells. *Lancet* **365**(9471), 1636–1641 (2005).
47. Richards, M. *et al.* Human feeders support prolonged undifferentiated growth of human inner cell masses and embryonic stem cells. *Nat. Biotechnol.* **20**(9), 933–936 (2002).
48. Richards, M. *et al.* Comparative evaluation of various human feeders for prolonged undifferentiated growth of human embryonic stem cells. *Stem Cells* **21**(5), 546–556 (2003).
49. Hovatta, O. *et al.* A culture system using human foreskin fibroblasts as feeder cells allows production of human embryonic stem cells. *Hum. Reprod.* **18**(7), 1404–1409 (2003).
50. Genbacev, O. *et al.* Serum-free derivation of human embryonic stem cell lines on human placental fibroblast feeders. *Fertil. Steril.* **83**(5), 1517–1529 (2005).
51. Amit, M. *et al.* Human feeder layers for human embryonic stem cells. *Biol. Reprod.* **68**(6), 2150–2156 (2003).

52. Xu, C. *et al.* Immortalized fibroblast-like cells derived from human embryonic stem cells support undifferentiated cell growth. *Stem Cells* **22**(6), 972–980 (2004).
53. Ludwig, T.E. *et al.* Derivation of human embryonic stem cells in defined conditions. *Nat. Biotechnol.* **24**(2), 185–187 (2006).
54. Lim, J.W. and Bodnar, A. Proteome analysis of conditioned medium from mouse embryonic fibroblast feeder layers which support the growth of human embryonic stem cells. *Proteomics* **2**(9), 1187–1203 (2002).
55. Koivisto, H. *et al.* Cultures of human embryonic stem cells: serum replacement medium or serum-containing media and the effect of basic fibroblast growth factor. *Reprod. Biomed. Online* **9**(3), 330–337 (2004).
56. Beattie, G.M. *et al.* Activin A maintains pluripotency of human embryonic stem cells in the absence of feeder layers. *Stem Cells* **23**(4), 489–495 (2005).
57. Xiao, L., Yuan, X. and Sharkis, S.J. Activin A maintains self-renewal and regulates fibroblast growth factor, Wnt, and bone morphogenic protein pathways in human embryonic stem cells. *Stem Cells* **24**(6), 1476–1486 (2006).
58. Pyle, A.D., Lock, L.F. and Donovan, P.J. Neurotrophins mediate human embryonic stem cell survival. *Nat. Biotechnol.* **24**(3), 344–350 (2006).
59. Rajala, K. *et al.* Testing of nine different xeno-free culture media for human embryonic stem cell cultures. *Hum. Reprod.* **22**(5), 1231–1238 (2007).
60. Fischer, B. and Bavister, B.D. Oxygen tension in the oviduct and uterus of rhesus monkeys, hamsters and rabbits. *J. Reprod. Fertil.* **99**(2), 673–679 (1993).
61. Guerin, P., El Mouatassim, S. and Menezo, Y. Oxidative stress and protection against reactive oxygen species in the pre-implantation embryo and its surroundings. *Hum. Reprod.* **7**(2), 175–189 (2001).
62. Harvey, A.J., Kind, K.L. and Thompson, J.G. REDOX regulation of early embryo development. *Reproduction* **123**(4), 479–486 (2002).
63. Catt, J.W. and Henman, M. Toxic effects of oxygen on human embryo development. *Hum. Reprod.* **15**(Suppl. 2), 199–206 (2000).
64. Ezashi, T., Das, P. and Roberts, R.M. Low O_2 tensions and the prevention of differentiation of hES cells. *Proc. Natl. Acad. Sci. USA* **102**(13), 4783–4788 (2005).
65. Forsyth, N.R. *et al.* Physiologic oxygen enhances human embryonic stem cell clonal recovery and reduces chromosomal abnormalities. *Cloning Stem Cells* **8**(1), 16–23 (2006).
66. Stephenson, E.L., Braude, P.R. and Mason, C. Proposal for a universal minimum information convention for the reporting on the derivation

of human embryonic stem cell lines. *Regen. Med.* **1**(6), 739–750 (2006).
67. Brivanlou, A.H., Gage, F.G., Jaenisch, R., Jessell, T., Melton, D. and Rossant, J. Setting standards for human embryonic stem cells. *Science* **300**, 913–916 (2003).
68. Veiga, A. *et al.* Selection of embryos for stem cell derivation: can we optimize the process? In: Pellicer, A. and Simon, C. (eds.), *Stem Cells in Human Reproduction: Basic Science and Therapeutic Potential* (Informa Healthcare, Abingdon, UK, 2007).
69. Le Novere, N. *et al.* Minimum information requested in the annotation of biochemical models (MIRIAM). *Nat. Biotechnol.* **23**(12), 1509–1515 (2005).
70. Hopwood, N. A history of normal plates, tables and stages in vertebrate embryology. *Int. J. Dev. Biol.* **51**(1), 1–26 (2007).
71. Stephenson, E.L. *et al.* International community consensus standard for reporting derivation of human embryonic stem cell lines. *Regen. Med.* **2**(4), 349–362 (2007).

Chapter 5

Stem Cells Differentiation

Pascale V. Guillot and Wei Cui

Abstract

Stem cells (both embryonic and somatic) have the capacity to self-renew and differentiate into specific lineage under permissive conditions. Although progress has been made on differentiation of pluripotent ES to a defined cell type, it is far from clear which are the signalling pathways controlling differentiation of a defined cell type and even less known about how to differentiate ES cells to be a functional cell type. Along with ES cells, foetal mesenchymal stem cells (MSC) are the only stem cells to be pluritpotent, expressing Oct-4 and Nanog, while adult MSC remain multipotent.

Keywords: Embryonic Stem (ES) Cells; Mesenchymal Stem Cells (MSC); Somatic Stem Cells; Adult Mesenchymal Stem Cells; Foetal Mesenchymal Stem Cells; Embryonic Stem Cells; Differentiation.

Outline

1. Introduction
2. Differentiation of Embryonic Stem Cells
3. Somatic Stem Cells
   3.1. Adult mesenchymal stem cells
   3.2. Foetal mesenchymal stem cells
4. Conclusion
   References

1. Introduction

Stem cells can be classified into two main categories; embryonic stem (ES) cells, which are derived from the inner cell mass of pre-implantation embryos, and somatic stem cells found in adult, neonatal and foetal tissues or organs, such as mesenchymal stem cells (MSC) and haematopoietic stem cells (HSC) from bone marrow. All stem cells regardless of their source share two common properties: self-renewal cells can divide and produce at least one daughter cell equivalent to the mother cell for a long period of time, and differentiation cells are able to give rise to a specialised cell type.[1] These properties of stem cells make them a good cell source for medical and biomedical applications, including regenerative medicine and treatment of certain diseases.

Most of the applications of stem cells require differentiation of stem cells to particular cell lineages. The signals that trigger stem cell differentiation are complex and the potential of stem cells to differentiate into specific cell types, a phenomenon known as plasticity, is largely dependent on the type of stem cells. For example, embryonic stem cells are pluripotent, being able to generate the cell types from all three germ layers in the body, while somatic stem cells are considered multipotent, as they usually generate the cell types of their tissue of origin. However, over the last years, evidence suggests that somatic stem cells from foetal tissues may be more plastic than originally stated and may have the capacity to differentiate into cells from non-mesenchymal lineages.

2. Differentiation of Embryonic Stem Cells

Embryonic stem (ES) cells can self-renew continuously and differentiate into all cell types in the body, including germ cells. For example, injection of mouse ES cells into a blastocyst is able to produce chimera offspring in which all the tissues carry the cells differentiated from the ES cells[2] and injection of human ES cells into severe combined immunodeficiency (SCID) mice produce teratoma containing cell types from all three germ layers.[3] The unique features of ES cells enable them to provide an unlimited cell source for the production of most, if not all, required cell types. However, they also create a challenge in how to differentiate the pluripotent cells into a particular cell lineage or cell type *in vitro*.

The most commonly used method for differentiating ES cells to a particular cell type is to grow ES cells initially as aggregates in suspension culture, which leads to formation of multi-differentiated structures called embryoid bodies (EBs). This technique, to a certain degree, resembles early embryo development *in vivo* but without proper axial organisation or elaboration of a body plan. This method has some advantages: it is easy to perform and likely to obtain the required cell type because cellular differentiation with this method proceeds on a schedule similar to that in the embryo. Several cell types have been successfully differentiated by this method such as neural cells,[4] cardiomyocytes,[5] pancreatic cells[6] and blood cells.[7] However, the main limitation of this method is that the differentiated cells are often heterogeneous as individual cells may receive different signals depending on their locations in an EB. In addition, it is difficult to dissect out the signalling pathways required for differentiation of particular cell types as often serum is applied in the culture medium. Therefore, several strategies have been applied to selectively enrich required cell population, including using serum-free culture medium with specific growth factors, identifying specific cell surface markers for further purification of required cell types, and generation of reporter and/or drug resistant cell lines for isolating a subset of cells.

Another method to differentiate ES cells into defined cell types is by co-culture with other cells. The best known example of this type is differentiation of ES cells to dopaminergic neurons by culturing ES cells on PA6 stromal cells derived from skull bone marrow.[8] However, it is unclear what factors produced by PA6 cells drive ES cells to the neural cell type.

In recent years, progress has been made in several laboratories to differentiate ES cells to defined cell types in chemically defined media without EB formation. In 2003, scientists at the University of Edinburgh first differentiated mouse ES cells in a chemically defined medium, N2B27, to neural progenitors without EB formation.[9] Based on this system, Gerrard *et al.* efficiently differentiate human ES cells to neural progenitors by blocking BMP signalling pathway.[10] Differentiation of ES cells in defined culture system can increase the efficiency to generate particular cell types as ES cells in the culture receive similar paracrine and autocrine signals. Thus, if a culture condition is optimal for a defined cell type, it will lead to more efficient differentiation. In addition, differentiation in a defined culture condition will provide a system to dissect out effects of certain factors or signalling pathways on differentiation. Subsequent studies have identified that FGF-Erk signalling pathway is critical for mouse ES cell differentiation[11] and activin/nodal are important for differentiation of ES cells to definitive endoderm further leading to the production of hepatic and pancreatic cell types.[12] However, although adherent cultures in a defined medium can enhance ES cell differentiation to defined cell types, it may not be sufficient for differentiated cell types to obtain mature functional properties as physiological functions of a particular cell type may require cell-cell interactions. Culture-differentiated cells with defined cell types in a three-dimensional system may facilitate their functionality.

Since the establishment of human ES cells in 1998, tremendous progresses have been made on differentiation of ES to defined cell types. However, it is far from clear which are the signalling pathways controlling differentiation of a defined cell type and even less is known about how to differentiate ES cells into a functional cell type.

3. Somatic Stem Cells

3.1. Adult mesenchymal stem cells

Adult MSC are easy to isolate from bone marrow, although MSCs make up only a minor fraction of bone marrow and other tissues. The exact proportion is difficult to establish since different methods used to collect the bone marrow affect the harvest, but it is estimated that MSC comprise 0.001% to 0.01% of the total adult bone marrow, and are hence about ten-fold less abundant than haematopoietic stem cells (HSC).[13]

HSC can reconstitute the haemopoietic system and are characterised by the expression of CD34 and CD45 antigens, and the absence of markers such as CD38 and HLA-DRE.[14,15]

All MSCs can be maintained in an undifferentiated state *in vitro*, and can differentiate down the main three mesodermal lineages bone, fat and cartilage, making, under permissive conditions, good candidates for protein and gene delivery in musculoskeletal diseases.[16] When cultured in the presence of defined permissive medium, *ex vivo* expanded adult bone marrow MSC can also give rise to muscle and other connective tissues and represent an invaluable tool for cell and gene therapy.[17] However, one pitfall of adult bone marrow MSCs is the decrease in proliferation rate, capacity to differentiate into multiple mesenchymal lineages after *in vitro* successive passaging, reflecting the difficulty on maintaining stem cell plasticity *in vitro*.[17] This is particularly challenging, considering that stem cells have to be expanded *in vitro* to a number suitable for therapeutic use. Other important parameters that have to be taken into consideration for cell therapy include the kinetics of differentiation, differentiation potency and functionality. Cell culture conditions affect differentiation potency, for example cell confluence is known to decrease osteogenic efficiency.[17] Studies have indicated that adult MSCs may either be multipotent, or consist in a mixed population of committed preogenitor cells, each with different differentiation potential.[18] For example, Pittinger *et al.* reported that individual colonies derived from MSC precursors are heterogeneous in terms of differentiation capacity, with only one-third of the original population being able to differentiate down the tri-osteogenic, adipogenic and chondrogenic lineages.[19] This observation is consistent with Muglia *et al.*, who showed that only 30% of *in vitro* expanded clones have the tri-lineage differentiation potential.[20]

Nevertheless, a number of publications document that adult bone marrow MSCs can differentiate down non-mesenchymal lineages both *in vivo* and *in vitro*. However, as recently discussed by Phinney *et al.*, the description of MSC fate *in vitro* when cultured under specific conditions really reflect true differentiation or artifact associated with the methods used for cell cuture.[21] For example, several groups have documented their capacity to undergo differentiation down the endodermal-derived hepatic[22,23] and endothelial[24] lineages and the ectodermal-derived neuronal lineage[25,26] in culture. Adult MSCs can also undergo multipotent differentiation *in vivo*, although results are limited by their poor ability to engraft following transplantation due to their loss of expression of adhesion molecules after expansion in culture,[18] adult bone marrow MSCs lead to osteoblast differentiation after transplantation into children with osteogenesis imperfecta.[27,28] Prockop *et al.* were the first to demonstrate that adult bone marrow MSCs can migrate to the brain when transplanted into newborn mice and undergo astrocyte and neuronal differentiation *in vivo.*[29] Other examples include differentiation down the epithelial lineage when transplanted in mice with lung injury.[30]

Other sources of adult MSC with multilineage potential are under investigation and can be found within the connective tissue of most organs, i.e. adipose tissue,[31] synovial membrane and periosteum.[32]

3.2. Foetal mesenchymal stem cells

The prevalence of MSCs declines with age: in the marrow of a newborn, one MSC is found among 10,000 nucleated marrow cells, compared to one MSC per 250,000 nucleated cells in adult bone marrow and one per 2×10^6 in that of an 80-year-old.[13] In contrast, the foetus is relatively rich in MSCs, first trimester foetal blood containing one MSC among every 3000 nucleated cells (which then decline rapidly with advancing gestation) and second trimester foetal bone marrow one MSC among every 400 cells.[33] The capacity of foetal and adult MSCs to differentiate also seems to differ. D'Ippolito *et al.* examined the osteogenic differentiation of postnatal MSCs of different ages and found that younger individuals exhibited increased osteogenic potential than older individuals.[34]

Umbilical cord blood is a source of stem cells that can be isolated by a non-invasive method. The number of HSC circulating in foetal blood increases from the first to peak in the second trimester, probably due to cells migrating from the foetal liver to establish haemopoiesis in the foetal bone marrow.[35] Some HSCs remain in the umbilical cord at delivery where they can be collected for allogeneic or occasionally autologous cell transplantation. MSCs found in the umbilical cord blood have some advantages over adult bone marrow MSCs. They are more immature, expressing pluripotency markers TRA-1-60, TRA-1-81, SSEA-4, SSEA-3 and Oct-4,[36] and elicit a lower level of graft rejection, the basis if this is thought to be linked to their longer telomeres status.[37] In addition to bone, fat and cartilage, umbilical cord blood MSC can differentiate into skeletal myogenic,[38] neural,[39] glial,[37] and hepatic cells[22] (see also Chapter 7).

Human MSCs found in various foetal tissues, amniotic fluid and placenta are also more primitive[40] than human adult bone marrow MSCs and have a higher differentiation potential. Taylor *et al.* showed a ten-fold competitive engraftment advantage of foetal liver relative to adult bone marrow cells in foetal recipients.[15] In mid-gestation foetal sheep, allogeneic foetal MSCs transplantation led to low-level multi-organ engraftment,[41] while foetal liver cell transplants in *mdx* (muscular dystrophy) mice resulted in multi-compartment engraftment and myogenic differentiation.[42]

The differentiation capacity of foetal MSCs depends on their tissue of origin, i.e. blood, liver, bone marrow, lung, pancreas, kidney, neural system, skeletal muscle,[43] spleen,[44] amniotic fluid or placenta. For example, we recently demonstrated that human first trimester foetal bone and bone marrow have higher *in vitro* and *in vivo* osteogenic potency than foetal liver and adult bone marrow MSC.[45] Hu *et al.* isolated a population of MSC from foetal pancreas at four to six months' gestation, and termed these cells "pancreas-derived mesenchymal stem cells" (PMSCs).[46] Foetal PMSCs show a multipotent capacity, 80% of the cells differentiating into chondrogenic, osteogenic or adipogenic lineages under appropriate conditions. In terms of clinical applications, nestin-positive precursors derived from human foetal pancreas have shown to differentiate into pancreatic endocrine cells, which reversed hyperglycaemia in diabetic mice.[47] Alternatively, Yao *et al.* isolated foetal pancreatic ductal stem cells to show differentiation into insulin-producing cells *in vitro*, a property which can also be induced by glucose.[48] The kidney represents another source of foetal epithelial and stromal cell lineages.[49] Culturing foetal kidney in an identical way to bone marrow MSCs yields an adherent population of cells with similar phenotype. When transplanted in a foetal lamb model, human foetal kidney MSCs showed long-term persistence and site-specific differentiation type.[50] Stem cells from the central nervous system are more limited in their differentiation potential than ES cells, but still give rise to the three major neural cell types: neurons, astrocytes and oligodendrocytes. After transplantation in rat brain-injury models, foetal cortical grafts survive and function. Experience already exists with human foetal neural tissue transplants in many patients with Parkinson's disease who have received foetal mesencephalic precursors grafted into the striatum. Although these grafts resulted in symptomatic improvement, they were associated with unacceptable side effects. These problems, and the limited supply of foetal tissue (up to six foetal brains needed to treat one patient), have led to a search for alternatives. Foetal neural stem cells were recently transplanted in a foetal sheep model to show long term engraftment, and gave rise to multi-lineage haemopoiesis.[51]

Amniotic and placental MSC differentiate along adipogenic and osteogenic mesenchymal lineages.

4. Conclusion

The molecular mechanisms that govern the maintenance of the undifferentiated state and stem cell differentiation are not yet fully understood. These processes involved various factors and depend in part on the stem cell niche and *ex vivo* expansion conditions. The full and functional differentiation of stem cells to specific lineages is still in its infancy and is increasingly challenged by the number of various stem cell sources and their environmental conditions.

References

1. Smith, A. A glossary for stem-cell biology. *Nature* **441**, 1060 (2006).
2. Smith, A.G. Embryo-derived stem cells: of mice and men. *Annu. Rev. Cell Dev. Biol.* **17**, 435–462 (2001).
3. Thomson, J.A., Itskovitz-Eldor, J., Shapiro, S.S. *et al.* Embryonic stem cell lines derived from human blastocysts. *Science* **282**(5391), 1145–1147 (1998).
4. Zhang, S.C., Wernig, M., Duncan, I.D., Brustle, O. and Thomson, J.A. *In vitro* differentiation of transplantable neural precursors from human embryonic stem cells. *Nat. Biotechnol.* **19**(12), 1129–1133 (2001).
5. Maltsev, V.A., Wobus, A.M., Rohwedel, J., Bader, M. and Hescheler, J. Cardiomyocytes differentiated *in vitro* from embryonic stem cells developmentally express cardiac-specific genes and ionic currents. *Circ. Res.* **75**(2), 233–244 (1994).
6. Schroeder, I.S., Rolletschek, A., Blyszczuk, P., Kania, G. and Wobus, A.M. Differentiation of mouse embryonic stem cells to insulin-producing cells. *Nat. Protoc.* **1**(2), 495–507 (2006).
7. Wiles, M.V. and Keller, G. Multiple hematopoietic lineages develop from embryonic stem (ES) cells in culture. *Development* **111**(2), 259–267 (1991).
8. Kawasaki, H., Mizuseki, K., Nishikawa, S. *et al.* Induction of midbrain dopaminergic neurons from ES cells by stromal cell-derived inducing activity. *Neuron* **28**(1), 31–40 (2000).
9. Ying, Q.L., Stavridis, M., Griffiths, D., Li, M. and Smith, A. Conversion of embryonic stem cells into neuroectodermal precursors in adherent monoculture. *Nat. Biotechnol.* **21**(2), 183–186 (2003).
10. Gerrard, L., Rodgers, L. and Cui, W. Differentiation of human embryonic stem cells to neural lineages in adherent culture by blocking bone morphogenetic protein signaling. *Stem Cells* **23**(9), 1234–1241 (2005).
11. Kunath, T., Saba-El-Leil, M.K., Almousailleakh, M., Wray, J., Meloche, S. and Smith, A. FGF stimulation of the Erk1/2 signalling cascade triggers transition

of pluripotent embryonic stem cells from self-renewal to lineage commitment. *Development* **134**(16), 2895–2902 (2007).
12. D'Amour, K.A., Agulnick, A.D., Eliazer, S., Kelly, O.G., Kroon, E. and Baetge, E.E. Efficient differentiation of human embryonic stem cells to definitive endoderm. *Nat. Biotechnol.* **23**(12), 1534–1541 (2005).
13. Caplan, A.I. The mesengenic process. *Clin. Plast. Surg.* **3**, 429–435 (1994).
14. Huss, R. Isolation of primary and immortalized CD34-hematopoietic and mesenchymal stem cells from various sources. *Stem Cells* **18**(1), 1–9 (2000).
15. Taylor, P.A., McElmurry, R.T., Lees, C.J., Harrison, D.E. and Blazar, B.R. Allogenic fetal liver cells have a distinct competitive engraftment advantage over adult bone marrow cells when infused into fetal as compared with adult severe combined immunodeficient recipients. *Blood* **99**(5), 1870–1872 (2002).
16. Friedenstein, A.J., Deriglasova, U.F. and Kulagina, N.N. Precursors for fibroblasts in different populations of hematopoietic cells as detected by the *in vitro* colony assay method. *Exp. Hematol.* **2**, 83–92 (1974).
17. Caplan, A.I. Mesenchymal stem cells. *J. Orthop. Res.* **5**, 641–650 (1991).
18. Kemp, K.C., Hows, J. and Donaldson, C. Bone marrow-derived mesenchymal stem cells. *Leuk. Lymphoma* **46**(11), 1531–1544 (2005).
19. Pittenger, M.F., Mackay, A.M., Beck, S.C. *et al.* Multilineage potential of adult human mesenchymal stem cells. *Science* **284**(5411), 143–147 (1999).
20. Muraglia, A., Cancedda, R. and Quarto, R. Clonal mesenchymal progenitors from human bone marrow differentiate *in vitro* according to a hierarchical model. *J. Cell. Sci.* **113**, 1161–1166 (2000).
21. Phinney, D.G. and Prockop, D.J. Concise review: mesenchymal stem/multipotent stromal cells: the state of transdifferentiation and modes of tissue repair — current views. *Stem Cells* **25**(11), 2896–2902 (2007).
22. Lee, K.D., Kuo, T.K., Whang-Peng, J. *et al. In vitro* hepatic differentiation of human mesenchymal stem cells. *Hepatology* **40**(6), 1275–1284 (2004).
23. Snykers, S., Vanhaecke, T., De Becker, A. *et al.* Chromatin remodeling agent trichostatin A: a key-factor in the hepatic differentiation of human mesenchymal stem cells derived of adult bone marrow. *BMC Dev. Biol.* **7**, 24 (2007).
24. Oswald, J., Boxberger, S., Jorgensen, B. *et al.* Mesenchymal stem cells can be differentiated into endothelial cells *in vitro*. *Stem Cells* **22**, 377–384 (2004).
25. Ortiz-Gonzalez, X.R., Keene, C.D., Verfaillie, C.M. and Low, W.C. Neural induction of adult bone marrow and umbilical cord stem cells. *Curr. Neurovasc. Res.* **1**(3), 207–213 (2004).
26. Kim, S., Honmou, O., Kato, K. *et al.* Neural differentiation potential of peripheral blood and bone marrow-derived precursor cells. *Brain Res.* **1123**(1), 27–33 (2006).

27. Horwitz, E.M., Prockop, D.J., Fitzpatrick, L.A. *et al.* Transplantability and therapeutic effects of bone marrow-derived mesenchymal cells in children with osteogenesis imperfecta. *Nat. Med.* **5**(3), 309–313 (1999).
28. Horwitz, E.M., Gordon, P.L., Koo, W.K. *et al.* Isolated allogeneic bone marrow-derived mesenchymal cells engraft and stimulate growth in children with osteogenesis imperfecta: implications for cell therapy of bone. *Proc. Natl. Acad. Sci. USA* **99**(13), 8932–8937 (2002).
29. Kopen, G.C., Prockop, D.J. and Phinney, D.G. Marrow stromal cells migrate throughout forebrain and cerebellum, and they differentiate into astrocytes after injection into neonatal mouse brains. *Proc. Natl. Acad. Sci. USA* **96**(19), 10711–10716 (1999).
30. Ortiz, L.A., Gambelli, F., McBride, C. *et al.* Mesenchymal stem cell engraftment in lung is enhanced in response to bleomycin exposure and ameliorates its fibrotic effects. *Proc. Natl. Acad. Sci. USA* **100**(14), 8407–8411 (2003).
31. Astori, G., Vignati, F., Bardelli, S. *et al.* "*In vitro*" and multicolor phenotypic characterization of cell subpopulations identified in fresh human adipose tissue stromal vascular fraction and in the derived mesenchymal stem cells. *J. Transl. Med.* **5**(1), 55 (2007).
32. De Bari, C., Dell'Accio, F., Karystinou, A., Guillot, P.V., Fisk, N.M., Jones, E.A., McGonagle, D., Khan, I.M., Archer, C.W., Misiadis, T.A., Donaldson, N., Luyten, F.P. and Pitzalis, C. A biomarker-based mathematical model to predict bone forming potency of human synovial and periosteal mesenchymal stem cells. *Arthritis Rheum.* **58**(1), 240–250 (2007).
33. in 't Anker, P.S., Noort, W.A., Scherjon, S.A. *et al.* Mesenchymal stem cells in human second-trimester bone marrow, liver, lung, and spleen exhibit a similar immunophenotype but a heterogeneous multilineage differentiation potential. *Haematologica* **88**(8), 845–852 (2003).
34. D'Ippolito, G., Schiller, P.C., Ricordi, C., Roos, B.A. and Howard, G.A. Age-related osteogenic potential of mesenchymal stromal stem cells from human vertebral bone marrow. *J. Bone Miner. Res.* **14**(7), 1115–1122 (1999).
35. Clapp, D.W., Freie, B., Lee, W.H. and Zhang, Y.Y. Molecular evidence that *in situ*-transduced fetal liver hematopoietic stem/progenitor cells give rise to medullary hematopoiesis in adult rats. *Blood* **86**(6), 2113–2122 (1995).
36. McGuckin, C.P., Forraz, N., Baradez, M.O. *et al.* Production of stem cells with embryonic characteristics from human umbilical cord blood. *Cell Prolif.* **38**(4), 245–255 (2005).
37. El-Badri, N.S., Hakki, A., Saporta, S. *et al.* Cord blood mesenchymal stem cells: potential use in neurological disorders. *Stem Cells Dev.* **15**(4), 497–506 (2006).

38. Gang, E.J., Jeong, J.A., Hong, S.H. *et al.* Skeletal myogenic differentiation of mesenchymal stem cells isolated from human umbilical cord blood. *Stem Cells* **22**(4), 617–624 (2004).
39. Jeong, J.A., Gang, E.J., Hong, S.H. *et al.* Rapid neural differentiation of human cord blood-derived mesenchymal stem cells. *Neuroreport* **15**(11), 1731–1734 (2004).
40. Guillot, P.V., Gotherstrom, C., Chan, J., Kurata, H. and Fisk, N.M. Human first trimester fetal mesenchymal stem cells (MSC) express pluripotency markers, grow faster and have longer telomeres compared to adult MSC. *Stem Cells* **25**(3), 646–654 (2006).
41. Schoeberlein, A., Holzgreve, W., Dudler, L., Hahn, S. and Surbek, D.V. Tissue-specific engraftment after *in utero* transplantation of allogeneic mesenchymal stem cells into sheep fetuses. *Am. J. Obstet. Gynecol.* **192**(4), 1044–1052 (2005).
42. Mackenzie, T.C., Shaaban, A.F., Radu, A. and Flake, A.W. Engraftment of bone marrow and fetal liver cells after *in utero* transplantation in MDX mice. *J. Pediatr. Surg.* **37**(7), 1058–1064 (2002).
43. Young, H.E., Steele, T.A., Bray, R.A. *et al.* Human reserve pluripotent mesenchymal stem cells are present in the connective tissues of skeletal muscle and dermis derived from fetal, adult, and geriatric donors. *Anat. Rec.* **264**(1), 51–62 (2001).
44. in 't Anker, P.S., Noort, W.A., Kruisselbrink, A.B. *et al.* Non-expanded primary lung and bone marrow-derived mesenchymal cells promote the engraftment of umbilical cord blood-derived CD34(+) cells in NOD/SCID mice. *Exp. Hematol.* **31**(10), 881–889 (2003).
45. Guillot, P.V., De Bari, C., Dell'Accio, F., Kurata, H., Polak, J. and Fisk, N.M. Comparative osteogenic transcription profiling of various fetal and adult mesenchymal stem cell sources. Differentiation. *Stem Cells*. E pub.
46. Hu, Y., Liao, L., Wang, Q. *et al.* Isolation and identification of mesenchymal stem cells from human fetal pancreas. *J. Lab. Clin. Med.* **141**(5), 342–349 (2003).
47. Huang, H. and Tang, X. Phenotypic determination and characterization of nestin-positive precursors derived from human fetal pancreas. *Lab. Invest.* **83**(4), 539–547 (2003).
48. Yao, Z.X., Qin, M.L., Liu, J.J., Chen, X.S. and Zhou, D.S. *In vitro* cultivation of human fetal pancreatic ductal stem cells and their differentiation into insulin-producing cells. *World J. Gastroenterol.* **10**(10), 1452–1456 (2004).
49. Al-Awqati, Q. and Oliver, J.A. Stem cells in the kidney. *Kidney Int.* **61**(2), 387–395 (2002).

50. Almeida-Porada, G., Porada, C. and Esmail, D. Differentiation potential of human metanephric stem cells: from mesenchyme to blood. *Blood* **96**, 494a (2000).
51. Almeida-Porada, G., Crapnell, K., Porada, C. *et al. In vivo* haematopoietic potential of human neural stem cells. *Br. J. Haematol.* **130**, 276–283 (2005).

Chapter 6

Marrow Stem Cells

Donald G. Phinney

Abstract

The skeletal system of adult mammals fulfills many roles including providing structural support for the body musculature and functioning as a storehouse for calcium. Additionally, it also serves as a repository for the bone marrow, which represents a rich source of stem cells. The hematopoietic stem cell (HSC), which generates all lineages of the immune system and blood, was the first stem cell shown to reside within adult bone marrow and remains the best characterized with respect to phenotype and function. Soon after the discovery of the HSC a separate cell population in marrow was described with the capacity to form ectopic bone tissue *in vivo*. This stem cell population, referred to as mesenchymal stem cells or multipotent marrow stromal cells (MSCs) has since been shown to differentiate into various connective tissue cell types including adipocytes, chondrocytes, myoblasts, and osteoblasts. Most recently, adult bone marrow has also been shown to harbor endothelial progenitor cells (EPCs) that participate in postnatal vasculogenesis. Therefore, bone marrow appears unique with respect to the fact that it harbors three physically and functionally distinct adult stem cell populations. Herein I review the discovery of these stem cells and emphasize their characterization and function. Moreover, I describe recent findings that indicate these stem cells are derived from a common precursor during embryonic development. This fact may account for their phenotypic similarities and functional interdependency.

Keywords: Hematopoietic Stem Cells; Mesenchymal Stem Cells; Marrow Stromal Cells; Endothelial Progenitor Cells; Bone Marrow.

Outline

1. Introduction
2. Hematopoietic Stem Cells: Discovery, Phenotype, and Function
3. Mesenchymal Stem Cells: Discovery, Phenotype, and Function

4. Endothelial Progenitor Cells: Discovery, Phenotype, and Function
5. A Common Origin for Bone Marrow Stem Cells
6. Functional Interdependency of Bone Marrow Stem Cells
7. Summary
References

1. Introduction

Over the past decade several groundbreaking discoveries have challenged long-standing tenets of stem cell biology. These include the discovery of stem and progenitor cells resident in most adult mammalian tissues, the unexpected ability of adult stem cells to differentiate across germinal boundaries, and the identification of a small subset of genes whose ectopic expression can reprogram adult somatic cells into embryonic stem-like cells. While these advances are remarkable in their own right, they were achieved by building upon fundamental contributions made to the field of stem cell biology over many decades. Most prominent among these was the discovery, isolation, and characterization of the hematopoietic stem cell (HSC) resident in adult bone marrow. Seminal discoveries made by pursuing the characterization of the HSC have forged many fundamental concepts about adult stem cell biology that remain highly relevant in the field today.

Initially, it was thought that stem cells resided only in adult tissues that undergo a high rate of turnover and as such functioned to maintain tissue homeostasis. The mammalian hematopoietic system, wherein a billion red blood cells are replaced every day, serves as one example. However, in tissues that traditionally are not thought to exhibit high turnover verifying the existence of a resident tissue-specific stem cell has been more difficult. This bias made the general scientific community slow to accept the concept of a mesenchymal stem cell (MSC) as the precursor of connective tissue cell types. In fact, it was first demonstrated over 40 years ago that whole bone marrow contained a cell population capable of generating heterotopic bone tissue that supported hematopoiesis and exhibited a measurable capacity for self-renewal *in vivo*, thereby fulfilling the criteria of a stem cell. Nevertheless, several decades passed before the concept of the MSC for connective tissues gained widespread acceptance. Subsequently, MSCs have attracted interest as therapeutic vectors to treat numerous diseases despite the fact that many aspects of their basic biology remain indeterminate. Over the past several decades, stem and progenitor cells resident in other adult tissues including the brain, heart, lungs, and liver have also been described.

Another recently identified adult stem cell population is endothelial progenitor cells (EPCs) that participate in postnatal vasculogenesis. Although the cellular processes that govern vasculogenesis during embryonic development are well described, a long-standing paradigm was that new vessel formation in adults occurred via angiogenesis, a process wherein existing endothelial cells (ECs) at the proximal ends of vessels proliferate in response to local cues to generate new vessels. However, recent experimental evidence has demonstrated that EPCs differentiate into ECs *in situ* and contribute to postnatal vasculogenesis under physiological conditions and following pathological tissue damage. EPCs also reside in adult bone marrow and share many characteristics with HSCs.

Therefore, bone marrow is unique in that it represents a reservoir for three physically and functionally distinct stem cell populations; HSCs, EPCs, and MSCs. The HSC is the best described adult stem cell with respect to its ontogeny, phenotype, and function. In contrast, many aspects of MSC biology remain indeterminate and the field of EPC biology continues to evolve at a rapid pace. Herein, I describe the isolation and characterization of these stem cell populations from marrow. I also discuss recent evidence that indicates the three populations derive from a common ancestor during development, which may explain their functional interdependency.

2. Hematopoietic Stem Cells: Discovery, Phenotype, and Function

The existence of the HSC was inferred from studies showing that transplantation of normal bone marrow cells *in vivo* could rescue animals from the lethal effects of radiation (see also Chapter 27 of this book).[1,2] Soon thereafter, Till and McCulloch[3,4] demonstrated that a proportion of transplanted marrow cells colonized the spleen and formed macroscopic colonies of proliferating cells that differentiated into erythrocytes, granulocytes, and megakaryocytes. The subsequent development of experimental methods to generate radiation-induced chromosomal aberrations in transplanted donor cells confirmed that these macroscopic spleen colonies were clonal in nature and therefore derived from single cells, referred to as colony forming-unit spleen (CFU-S). This work provided evidence for a common myeloid progenitor cell within bone marrow.[5–7] Thereafter, several groups provided cytological evidence in radiation-induced chimeras that a single donor marrow cell could yield myeloid and lymphoid cell lineages *in vivo*.[8–10] In one such study the distribution pattern of genetic markers in the hematopoietic system of primary transplant recipients was shown to be preserved after serial passage of cells into secondary recipients.[10] This result validated the existence of a self-renewing HSC in marrow. Moreover, the identification of restricted progenitors that yielded only myeloid or lymphoid cell lineages *in vivo* spurred the concept of a hierarchical model for HSC differentiation.[10]

The capacity to reconstitute the hematopoietic system of irradiated animals provided an invaluable tool to explore the nature of the repopulating HSC. Accordingly, many groups began to fractionate bone marrow based on physical or biological parameters to enrich for the CFU-S, or fractions that were radioprotective *in vivo*.[11–14] For example, Visser *et al.* showed that fractionation of marrow cells by equilibrium density centrifugation and affinity for wheat germ lectin provided greater than a 100-fold enrichment in the number of CFU-S in marrow and a similar enrichment for the cells that provided radioprotection *in vivo*, indicating that these two properties co-purified in marrow.[14] Subsequently, the advent of fluorescence activated cell sorting provided a means to fractionate populations based on the expression of cell surface antigens such as Thy-1.[15,16] These advances, together with the use of *in vitro* colony forming assays to quantify the number of committed hematopoietic progenitors in marrow,[17–21] facilitated dissection of the hematopoietic process and its cellular components.

The ability to study B- and T-cell lymphopoiesis *in vitro* revealed that stem/progenitor cells typically lacked expression of cell surface antigens used to delineate mature hematopoietic cell types.[22] Accordingly, an immunodepletion scheme was devised to remove committed lymphoid and myeloid lineages from bone marrow, which yielded a rare population of cells that expressed low levels of Thy-1 and were highly enriched in clonal progenitors for spleen, thymic, and pre-B-cell colonies.[23] Moreover, these cells were also capable of regenerating the hemato-lymphoid system of lethally irradiated mice and therefore were enriched for HSCs. Further refinement of the fractionation procedure yielded a population defined phenotypically as $Thy^{lo}Lin^{-}Sca\text{-}1^{+}$ that was shown to differentiate with approximately unit efficiency into myelo-monocytic, pre-B-, and T-cells.[24] Moreover, transplantation of 30 $Thy^{lo}Lin^{-}Sca1^{+}$ cells was shown to be sufficient to rescue 50% of lethally irradiated mice while transfer of up to 900 $Thy^{lo}Lin^{-}Sca1^{-}$ cells were not radioprotective *in vivo*. Collectively, these studies suggested that all HSCs were defined phenotypically as $Thy^{lo}Lin^{-}Sca1^{+}$.

Subsequent *in vivo* repopulation assays using highly purified fractions of $Thy^{lo}Lin^{-}Sca1^{+}$ cells revealed that only a subset of these cells were capable of long-term self-renewal *in vivo*.[25] Moreover, use of the vital dye rhodamine-123 (Rh-123) resolved the $Thy^{lo}Lin^{-}Sca1^{+}$ population into two fractions including a $Rh\text{-}123^{lo}$ subset that contained a 20-fold higher level of precursors for CFU-S and was enriched in long-term repopulating cells as measured in serial transplants *in vivo*.[26] Consistent with these findings, other groups reported that the $Thy^{lo}Lin^{-}Sca1^{+}$ population was also heterogeneous with respect to its cell cycle status[27] and expression of specific hematopoietic lineage markers.[28] In the latter case, Morrison and Weissman[29] showed that fractionation based on expression of CD4 and Mac-1 resolved three populations with distinct repopulating activities *in vivo* and that only the Lin$^-$Mac-1$^-$CD4$^-$ population was highly enriched for long-term reconstituting HSCs. Other studies showed that the receptor tyrosine kinases c-kit[30,31] was also expressed on long-term repopulating HSCs and that Flk-2[32] expression discriminated long-term from short-term repopulating HSCs. Herein, the Thy-1^{lo}Flk-2$^-$ subset was shown to contain the long-term repopulating HSCs while the Thy-1^{lo}Flk-2$^+$ subset contained predominantly short-term repopulating stem cells. A third fraction characterized as Thy-1$^-$Flk-2$^+$ was shown to be enriched for multipotent progenitors that exhibited only transient engraftment *in vivo*. Therefore, loss of Thy-1 expression and gain of Flk-2 expression marked the loss of self-renewal during HSC maturation. Collectively, these studies defined the long-term repopulating HSC in mice as c-kit$^+$Thy1^{lo}Lin$^-$Sca-1$^+$ (KTLS) and demonstrated that lack of Flk-2 expression could replace Thy-1 during the isolation procedure (Fig. 1).

Related studies demonstrated that all human colony-forming progenitors in marrow that were detectable by *in vitro* assays expressed the CD34 antigen and

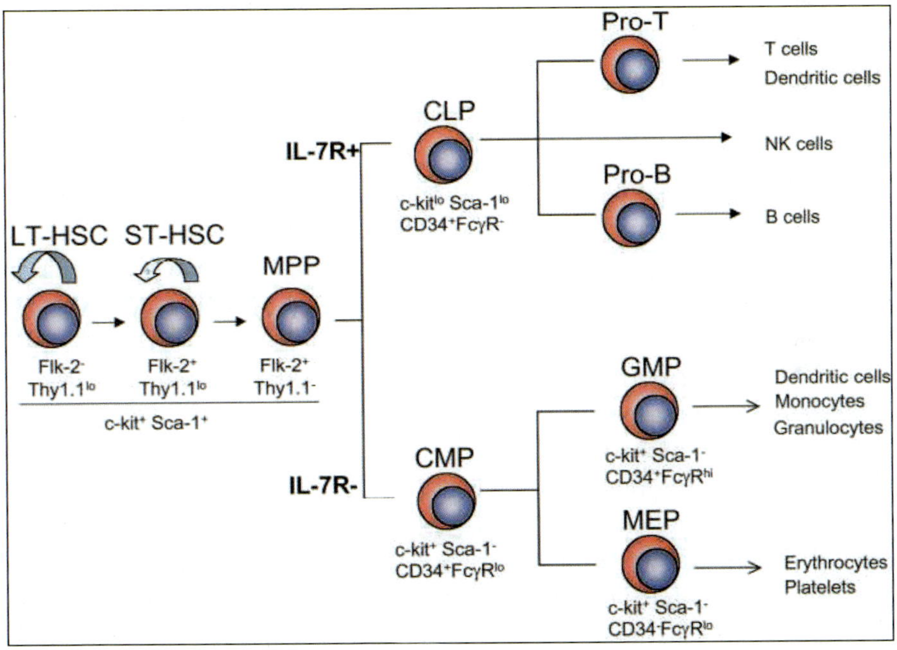

Fig. 1. A current map of the hematopoietic lineage. Long-term (LT) HSC self-renew for life, while their downstream ST-HSC self-renew for six to eight weeks. The other progenitors have been prospectively isolated to phenotypic, functional, and for the more mature cells, gene expression profile homogeneity. CMP, common myeloid progenitor; CLP, common lymphoid progenitor; GMP, granulocyte/monocyte progenitor; MEP, megakaryocyte/erythrocyte progenitor. Reprinted with permission from *Immunological Reviews*.[22]

that populations enriched for this antigen exhibited durable engraftment when transplanted to lethally irradiated baboons[33] or human breast cancer patients following marrow ablative therapy.[34] Baum *et al.*[35] further showed that Thy1^{+}CD34^{+} human fetal bone marrow cells were competent to establish long-term cultures *in vitro* that generated both myeloid and lymphoid cell lineages. Moreover, Thy1^{+}CD34^{+} but not Thy1^{-}CD34^{+} cells were shown to sustain myelopoiesis and B-cell lymphopoiesis for several months in an *in vivo* stem cell assay employing SCID mice. Additional markers used to further enrich the human CD34 marrow population for long-term engrafting cells included CD38, CD90, and CD133.[36–38] Importantly, although CD34 was also shown to be expressed by murine HSCs,[39,40] several groups have recently isolated HSCs from mouse and man that lack expression of this protein but still exhibit long-term engraftment *in vivo.*[41,42] Variability in expression of CD34 has been reported to distinguish activated early progenitors mobilized from marrow from that of quiescent stem cells.[43] However, the exact function of CD34 in HSCs remains unclear.

3. Mesenchymal Stem Cells: Discovery, Phenotype, and Function

The discovery of the MSC is generally attributed to Alexander Friedenstein and co-workers, who in the 1960s were the first to demonstrate that whole bone marrow transplanted under the kidney capsule of mice generated a heterotopic ossicle that was invaded by host-derived hematopoietic cells and functioned as a hematopoietic organ.[44,45] This osseous tissue was shown to persist within animals for no less than 14 months following a single heterotopic transplant.[46] Friedenstein further showed through a series of semi-syngeneic and allogeneic transplants that the marrow-derived osteogenic progenitors responsible for heterotopic bone formation *in vivo* were functionally distinct from hematopoietic cells.[47] These studies also revealed that the osteogenic capacity of marrow was maintained *in vivo* through four serial transplants before being exhausted. Therefore, the *in vivo* repopulating ability of these cells was commensurate with that demonstrated for the HSC. Based on these results, which demonstrated protracted self-maintenance and multi-lineage differentiation *in vivo*, Friedenstein postulated the existence of an osteogenic stem cell in bone marrow. Friedenstein and co-workers also demonstrated that this "osteogenic" activity segregated with the fibroblastoid (stromal) cell fraction that was enriched from bone marrow by plastic adherence.[48] Moreover, Friedenstein characterized these stromal cells as a rare population in marrow with a high proliferative capacity *in vitro*. The cells were also shown to be able to grow at clonal density but formed colonies that were heterogeneous with respect to their size and morphology. In accordance with nomenclature developed in the hematopoietic field, Friedenstein labeled these cells as colony forming-unit fibroblasts (CFU-F).

Friedenstein's elegant studies prompted two divergent lines of investigation into the biology of marrow stromal cells. One avenue of research stemmed from the realization that stromal cells may be exploited to develop culture systems that recapitulate aspects of hematopoiesis *in vitro*. In the 1970s, Dexter and co-workers[18] showed that long-term bone marrow cultures (LT-BMCs), which were prepared by charging an established monolayer of stromal cells with fresh bone marrow, supported production of CFU-S, GM-CFC, granulocytes, and monocytes. Subsequently, human LT-BMCs were established that could sustain hematopoiesis for up to 20 weeks.[49] Use of LT-BMCs to identify soluble factors and cell mediated interactions necessary for sustained hematopoiesis provided a wealth of information regarding the phenotype and function of marrow stromal cells. Development of these culture systems also played an invaluable role in the identification and isolation of the HSC.

Alternatively, various groups extended Friedenstein's landmark studies by showing that marrow stromal cells transplanted *in vivo* in a closed system formed a mixture of osseous, cartilaginous, and fibrous tissue[50–52] or could be induced *in vitro* to differentiate into adipocytes, osteoblasts, and chondrocytes under the appropriate culture conditions.[53–58] One such study revealed an inverse relationship between osteogenic and adipogenic differentiation of stromal cells in culture, which suggested the existence of a common precursor for both cell lineages.[58] Additionally, Grigoriadis *et al.* reported that a cell line derived from fetal rat calvarial tissue was capable of differentiating into four distinct connective tissue lineages.[59] These and other studies led to the formulation of the mesengenic process, which postulated the existence of a multipotent stem cell for connective tissues in bone marrow that differentiated according to a hierarchical model similar to that described for the HSC.[60] Ensuing work demonstrating that single cell-derived populations of human[61] and rodent[62] stromal cells were capable of multilineage differentiation *in vitro* substantiated, in part, the mesengenic process. Accordingly, the term "mesenchymal stem cell" was adopted to describe this population.

Initial characterizations revealed that MSCs isolated by plastic adherence lacked expression of most markers common to hematopoietic cell types including CD2, CD3, CD4, CD8, Mac-1/CD11b, CD14, CD15, CD19, CD20, B220, CD45, Thy-1, and myeloperoxidase.[63] Populations were also shown to express the LDL receptor and alkaline phosphatase,[64] smooth muscle actin,[65] type IV collagen and laminin,[66] factor VIII,[67] and MUC18[68] and as such were categorized as marrow myoid, vascular smooth muscle, or endothelial-like cells. MSCs enriched from bone marrow by plastic adherence were also shown to be functionally heterogeneous with respect to their differentiation potential.[69–72] Subsequently, human and murine MSCs were shown to express a compliment of integrin receptors and adhesion molecules including CD29, CD44, CD49a-f, CD51, CD73, CD105, CD106, and CD166 that aided in refining the isolation process.[61,73–76] Additionally, the antibody Stro-1, which was originally produced by immunizing mice against human bone marrow cells,[77] was shown to immuno-select a cell population capable of sustaining hematopoiesis and differentiating into adipocytes, osteoblasts, and chondrocytes *in vitro* as well as forming ectopic bone tissue when implanted into athymic mice *in vivo*.[78]

More recent studies have shown cells phenotypically and functionally equivalent to MSCs can be enriched from bone marrow by selection for the nerve growth factor receptor (CD271),[79] SSEA-1[80] or SSEA-4.[81] Moreover, CD271-expressing populations have been shown to co-express CD140b (platelet-derived growth factor receptor β), CD340 (HER-2/erbB2), and CD349 (frizzled-9).[82] In one such study, SSEA-4$^+$ but not SSEA-4$^-$ cells sorted from human bone marrow aspirates were shown to form a homogeneous cell monolayer and differentiate into adipocytes, osteoblasts, and chondrocytes *in vitro* as well as form well organized and vascularized ectopic bone tissue when implanted into immunodeficient mice.

Other groups have reported that bone marrow-derived MSCs also express CD146 and 3G5,[83] consistent with their characterization as specialized vascular pericytes (see below).

Despite these advances, there still remains no consensus opinion in the field as to how best to define MSCs. In most laboratories, MSCs are defined by expression of a subset of cell surface receptors, such as CD29, CD44, CD73, CD105, and CD106, lack of expression of hematopoietic markers, and the capacity to differentiate into adipocytes, osteoblasts, and chondrocytes *in vitro*, although the chondrogenic potential of cells is less regularly evaluated. However, it is well established that *in vivo* assays provide a more stringent assessment of the overall differentiation potential of MSCs[70] even though such assays are rarely used. Analysis of the murine MSC transcriptome has also revealed that cells express a diverse repertoire of proteins that regulate angiogenesis, immunity and defense, neural activities, and hematopoiesis.[84] Moreover, many of these proteins are expressed by distinct subpopulations of cells. Therefore, the composition of this population is much more complex at a biochemical level then previously envisioned and may account for the broad therapeutic efficacy of MSCs in experimental models of disease. However, in the absence of specific evidence for self-renewal *in vivo* or asymmetric division *in vitro*, it is impossible to discriminate between the *bona fide* MSC and multipotent progenitors that lack the capacity for self-renewal. A better understanding of the molecular signaling networks that regulate fate decisions in cells will undoubtedly aid in refining the definition of the MSC.

4. Endothelial Progenitor Cells: Discovery, Phenotype, and Function

The discovery of adult EPCs is credited to Asahara *et al.*,[85] who showed that $CD34^+$ or $Flk\text{-}1^+$ (VEFR-2) populations isolated from human peripheral blood grew as spindle-shaped cells and formed blood island-like clusters or tube-like structures *in vitro* and integrated into capillary vessel walls within neovascular zones of the hind limbs in rabbit and mouse models of unilateral hind limb ischemia. In addition to CD34 and Flk-1, these EPCs were also shown to express CD31, Tie2, and E-selectin, expression of which increased as cells were cultured *in vitro*. Cultured cells were also shown to express factor VIII, endothelial constitutive nitric oxide synthase (eNOS), and demonstrated the ability to uptake acetylated LDL. Hence, these studies identified progenitor cells in adults capable of contributing to new vessel formation and therefore challenged the long-standing tenet that vasculogenesis, the formation of new blood vessels in the absence of pre-existing ones via *de novo* production of ECs, occurs exclusively during embryogenesis.[86]

In related studies, Asahara *et al.*[87] performed bone marrow transplantation in immune-deficient mice using donor marrow cells that express the *β*-galactosidase gene regulated from an endothelial cell-specific promoter, e.g. Flk-1 or Tie-2. Analyses of the transplant recipients revealed β-galactosidase mRNA and protein expression at sites of normal and pathological neovascularization including in reproductive tissue of hormonally induced animals, granulation tissues of cutaneous wounds, ischemic limbs and heart, and tumors produced by implantation of colon cancer cells. These findings, together with the fact that differentiated ECs were shown to rarely incorporate into foci of neovascularization *in vivo*,[88] suggested that circulating EPCs recruited from the bone marrow constitute the majority of cells that contribute to postnatal vasculogenesis. These findings were also consistent with early observations showing that artificial vascular grafts implanted into human patients with cardiovascular disease[89,90] or experimental animals[91] were remodeled into an endothelium by circulating cells, which also appeared to originate from the bone marrow.[92]

Subsequent studies have identified several interrelated classes of EPCs in adults. These include the $CD34^+Flk\text{-}1^+$ cells described above, which also have been shown to co-express the HSC maker CD133.[93,94] In mice, these EPCs co-purify with the $c\text{-}kit^+Lin^-Sca1^+$ fraction of bone marrow. Accordingly, these populations are believed to be derived from a common precursor of hematopoietic and ECs, referred to as the hemangioblast[95,96] (see below). In contrast, several groups have reported that mature ECs can be derived from $CD14^+$ monocytes cultured in the presence of angiogenic growth factors such as VEGF.[97–99]

The monocyte-derived cells were characterized as ECs based on their expression of von Willebrand factor, VE-cadherin, acetylated LDL receptor, VEGFR-1, VEGFR-2, eNOS, and their ability to form tubular-like structures *in vitro*. In a related study, mononuclear cells isolated from human peripheral blood and differentiated into ECs *in vitro* were shown to express very low levels of CD34 and CD117 (c-kit) and high levels of the monocyte markers CD11b, CD11c, CD14 and the pan-leukocyte marker CD45.[100] Additionally, the cells failed to proliferate to a significant extent *in vitro* but did secrete large quantities of VEGF and HGF, which was postulated to account for their pro-angiogenic affects *in vivo*. Elsheikh *et al.*[101] further demonstrated that only a specific subset of $CD14^+$ monocytes that expressed VEGFR-2 had the ability to differentiate into ECs and contribute to re-endothelialization of injured femoral arteries when transplanted *in vivo*. In contrast, $CD14^+VEGFR\text{-}2^-$ cells did not exhibit any of these characteristics. Therefore, expression of VEGFR-2 on peripheral blood monocytes appeared essential for their endothelial-like function. Based on these data, circulating ECs arise from two distinct sources; monocytes and authentic EPCs resident in bone marrow.

A recent study by Yoder *et al.*[102] further validates this hypothesis by demonstrating that EPCs and monocyte-derived ECs can be discriminated based on their growth characteristics and functional properties *in vitro* and *in vivo*. In this study expansion of peripheral blood mononuclear cells was performed using two methods. In the first method cells were expanded on fibronectin-coated plates in the presence of endothelial cell growth factors. After 48 hours the non-adherent cells were collected, replated under the same conditions, and cultured an additional five to seven days after which colonies, referred to as colony forming unit-endothelial cells (CFU-ECs), became visible. In the second method, cells were expanded on collagen-coated plates using the same medium formulation. However, the non-adherent cells were discarded and the adherent cells fed with fresh medium daily. Under these conditions colonies of adherent cells, termed endothelial colony-forming cells (ECFCs), became visible after 14 to 21 days. Importantly, CFU-ECs could not be generated from ECFCs and *vice versa*, suggesting that the two types of colonies, which were morphologically distinct, were derived from separate cell populations. Further analysis confirmed that both populations expressed a variety of endothelial lineage specific markers but only the CFU-ECs expressed CD45 and the monocyte/macrophage markers CD14 and CD115. In addition, the CFU-ECs were shown to express non-specific esterase activity and phagocytose *E. coli*, a property characteristic of macrophages. Additionally, the CFU-ECs failed to generate secondary colonies when plated at limiting dilution but yielded granulocytic/monocytic colonies (CFU-GMs) at a low frequency. Finally, isolation of CFU-ECs and ECFCs from a patient with polycythemia vera proved that the cells were not clonally related *in vivo*. Furthermore, only ECFCs were shown to generate chimeric blood vessels when transplanted into immunodeficient mice. Collectively, these results demonstrated that monocyte-derived endothelial cells and EPCs are distinct entities and that only EPCs participate in postnatal vasculogenesis.

5. A Common Origin for Bone Marrow Stem Cells

It is well established that hematopoietic and endothelial lineages follow a similar temporal and spatial developmental course during embryogenesis. For example, during mouse development hematopoietic and endothelial progenitors arise simultaneously at embryonic day 7.5 (E7.5) within blood islands of the yolk sac and the para-aortic splanchnopleura.[103] The para-aortic splanchnopleura gives rise to the aorta-gonad-mesonephros (AGM) region, which by day E10.5 contains definitive HSCs.[104] Concurrently, ECs within the splanchnopleura colonize the floor of the aorta and differentiate *in situ* to produce the vasculature of the body wall, kidney, visceral organs, and limbs. Heterotopic transplantation of these splanchnopleural cells into quail/chicken chimeras has shown that they also give rise to hematopoietic precursors.[105] These findings are consistent with other reports demonstrating that hematopoietic cells arise from the ventral wall of the dorsal aorta and other arterial regions.[106–108] Collectively, these and other studies support the long-standing notion that hematopoietic and endothelial lineages arise form a common precursor, termed the hemangioblast, during development.[109]

Genetic and molecular studies have also pointed to the existence of the hemangioblast. As noted above, HSCs and EPCs express a number of common phenotypic markers including VEGFR-2 (Flk-1)[85] as well as several transcription factors such as *SCL/Tal1*[110] and *Runx1*.[107] Deletion of VEGFR-2 has been shown to arrest both hematopoietic and endothelial differentiation in developing mouse embryos[111,112] and *Runx1*-deficient embryos develop primitive hematopoiesis and early vascular networks but die between E11–E12.5 due to a lack of definitive hematopoiesis and defects in vascular development.[113] Finally, embryonic stem (ES) cells have been identified *in vitro* that transiently generate blast forming cells capable of differentiating into both hematopoietic and endothelial cell lineages, and as such possess hemangioblast activity *in vitro*.[114,115]

In contrast, the developmental origin of the MSC remains enigmatic. Intuitively, one may speculate MSCs that produce bone and cartilage derive from the lateral plate mesoderm (somatopleura), which forms the appendicular skeleton. However, several lines of evidence suggest that the closest *in vivo* approximation of the MSC is a specialized marrow pericyte, referred to in the early literature as an adventitial reticular cell, which lines the abluminal surface of the endothelium of venous sinusoids. First, a number of studies have shown that marrow pericytes are capable of differentiating into adipocytes, chondrocytes, and osteoblasts *in vitro*.[116–118] Second, the antibody Stro1, which has been used to enrich MSCs from bone marrow, specifically stains ECs as well as perivascular cells of blood vessel walls in a variety of tissues.[119] Other studies have shown that MSCs share characteristics with vascular smooth muscle cells,[65] marrow myoid cells,[120] vascular pericytes,[121] and endothelial-like cells[66] and also express the pericyte markers CD146 and 3G5.[83] Most recently, Brachvogel *et al.*[122] showed by lineage tracing that annexin A5-positive cells that arise as angioblasts in the primary vascular plexus generate perivascular (mural) cells in most blood vessels. These perivascular cells were shown to express stem cell markers, generate a calcified matrix, and differentiate into adipocytes. These studies are consistent with more recent findings that indicate MSCs or MSC-like cells isolated from most adult tissue reside within a perivascular niche.[123]

Intriguingly, although mural cells are known to have distinct origins during development depending upon their location and function *in vivo*,[124] a recent paper by Yamashita *et al.*[125] has provided new evidence that mural cells may arise from a common vascular stem cell that also generates ECs (Fig. 2). Herein, Flk1$^+$ ES cells, which were previously shown to possess hemangioblast activity,[115] were shown to generate mural cells or endothelial cells following exposure to PDGF-BB or VEGF, respectively. Moreover, clonal analysis showed both cell lineages were derived from a single Flk1$^+$ ES cell *in vitro* and from Flk1$^+$VE-cadherin$^-$ cells derived from day E8.5 embryos. Finally, intracardiac injection of Flk1$^+$ ES cells into developing chick embryos resulted in their incorporation as both endothelial and mural cells into the vasculature of the head, yolk sac, heart, and intersomitic regions. Therefore, if MSCs exist as perivascular cells *in vivo*, they may derive from the angioblasts. This notion is supported by another study showing that mouse embryonic dorsal aorta explants grafted into chick embryos generate blood vessels but also muscle and other connective tissues.[126] Interestingly, these cells were shown to express Flk-1 and c-kit when cultured *in vitro*. Due to their broad developmental potential they were categorized as meso-angioblasts. Collectively, these data suggest that HSCs, EPCs, and MSCs may derive from a common progenitor during development (Fig. 3).

The aforementioned findings are consistent with the fact that MSCs exhibit many functional characteristics of mural cells. For example, during the early phases of vessel formation, mural cells migrate along capillary sprouts that express PDGF-BB, which functions as a recruitment factor, thereby covering the outer surface of the developing endothelium, and supporting capillary structure.[127,128]

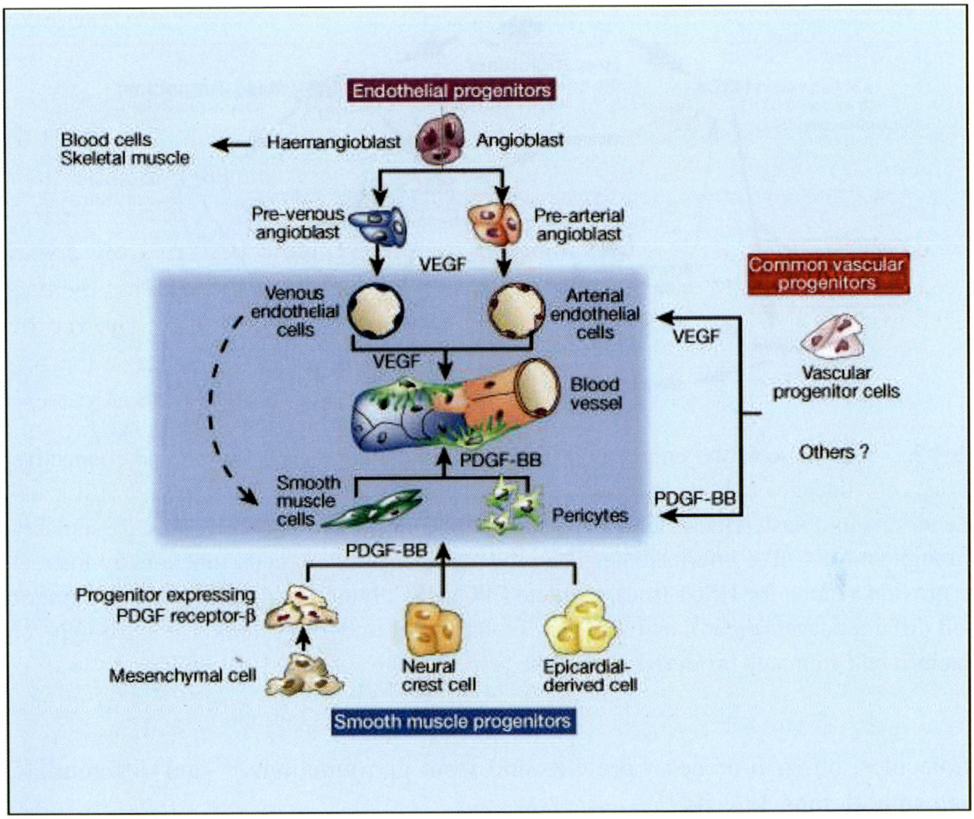

Fig. 2. Endothelial cells arise from precursors called angioblasts or hemangioblasts in the embryo (top). Smooth muscle cells and pericytes are derived from a variety of progenitors (bottom). These include mesenchymal cells in the embryo and adult, embryonic neural crest cells, and progenitors in the embryonic epicardium (the outer layer of the heart). Smooth muscle cells also form from endothelial cells. Recent studies indicate that pericytes and smooth muscle cells may also originate from a common vascular precursor (right). VEGF promotes the development of endothelial cells from this precursor and PDGF-BB stimulates their development into smooth muscle cells and pericytes. Reprinted with permission from *Nature*.[149]

Additionally, pericytes physically interact with ECs via binding to adhesion molecules, integrins or formation of gap junctions, which enable the cells to regulate endothelial cell proliferation and migration, thereby controlling vascular growth.[129] Pericytes are also thought to play a role in regulating blood flow and produce extracellular matrix proteins that protect newly formed vessels from rupture or regression. Similarly, MSCs express PDGF receptors,[130] secrete various extracellular matrix and basement membrane proteins,[131] express adhesion

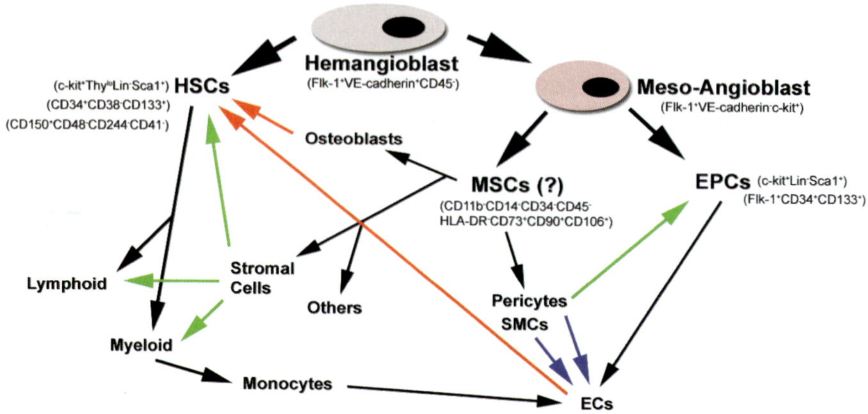

Fig. 3. Schematic of the ontogeny of hematopoietic, endothelial, mural, and connective tissue cells lineages in bone marrow. The hemangioblast gives rise to definitive HSCs and the meso-angioblast, which yields EPCs and other connective tissue cell types, presumably through an MSC-like intermediate. The progeny of these stem cells functionally interact to provide a niche for HSCs (red), regulate EPC, HSC, lymphoid, and myeloid progenitor cell differentiation (green), and maintain blood vessel integrity (blue). The phenotype of the different stem cells is derived from the text.

molecules, integrin receptor proteins and from gap junctions,[132] and differentiate into smooth muscle cells.[133]

6. Functional Interdependency of Bone Marrow Stem Cells

In addition to their common origin during development, various studies have demonstrated that HSCs, EPCs, and MSCs and/or their progeny are functionally interdependent. For example, MSCs and their progeny are known to secrete an array of matrix and adhesion molecules as well as numerous cytokines that regulates various aspects of hematopoiesis, including maintenance of the HSC.[134] In addition, recent studies have shown that osteoblasts play an important role in regulating HSC self-renewal *in vivo*. Specifically, Zhang *et al.*[135] showed that inactivation of the BMPR1A receptor in mice led to an increase in the number of long-term repopulating HSCs in bone marrow. Loss of this receptor also led to increased formation of trabecular-like bone, which was populated at its surface by spindle-shaped bone lining cells that expressed N-cadherin. Subsequent studies revealed that N-cadherin, in a complex with β-catenin, was asymmetrically localized to the cell surface of HSCs adjacent to the bone lining cells. Simultaneously, Calvi *et al.*[136] showed that over-expression of parathyroid hormone receptors within osteoblasts led to increased formation of trabeculae and trabecular osteoblastic cells in the long bones of mice and a commensurate increase in the number of long-term repopulating HSCs. The osteoblasts also exhibited increased production of Jag1, a ligand for the Notch receptor, which in turn activated Notch signaling in HSCs resulting in expansion of the stem cell compartment *in vivo*. Collectively, these data indicate that specific subsets of osteoblasts constitute a niche *in vivo* that regulates the size of the HSC pool. These findings are consistent with earlier studies showing that osteoblasts play an important role in maintaining HSCs in marrow.[137,138]

Studies have also indicated that ECs may regulate aspects of hematopoiesis. For example, several groups have shown that ECs maintain long-term repopulating HSCs cultured *in vitro*.[139,140] Moreover, ablation of ECs *in vivo* by administration of an anti-VE-cadherin antibody leads to hematopoietic failure.[141] These studies suggest that a vascular niche for HSCs may also exist and play a role in maintaining HSCs in extramedullary sites *in vivo* such as the spleen and liver.[142] Recently, Keil *et al.*[143] demonstrated that HSC activity was contained within the $CD150^+CD48^-CD244^-CD41^-$ negative fraction of bone marrow. Moreover, since $CD150^+CD48^-$ cells were uniformly negative for CD244 expression, these data indicated that CD150, CD48, and CD41 can substitute for Thy-1, Sca1, c-kit, and lineage-specific antibodies to identify the definitive HSC *in vivo*. Immunostaining of spleen tissue from mice administered cyclophospamide and G-CSF revealed that $CD150^+CD48^-CD41^-$ cells were mainly associated with the sinusoidal endothelium. In bone marrow, some of these cells were found associated with the endosteum, but most were observed in contact with the endothelium lining of the sinusoids. These data reveal a separate vascular HSC niche *in vivo*. Moreover, they provide experimental evidence to support a previously proposed model wherein the osteoblast niche maintains quiescent HSCs and the vascular niche supports mobilized HSCs that proliferate and differentiate during their journey out of bone marrow.[144]

Finally, MSCs and their progeny have recently been shown to secrete various factors that regulate endothelial cell growth, mobility, and angiogenesis.[84] MSCs have also been shown to regulate endothelial cell function *in vitro* and *in vivo*,[145–148] consistent with their pericyte-like characteristics.

7. Summary

The bone marrow of adult mammals is unique in that it harbors three physically and functionally distinct stem cell populations; HSCs, EPCs, and MSCs. Developmental and molecular studies provide strong evidence that these stem cells derive from a common precursor, the hemangioblast, during development. In adult bone marrow, complex functional interrelationships exist between these stem cells and their progeny, which serve to regulate the size of the stem cell pool, maintain hematopoiesis and bone homeostasis. These interrelationships exemplify the complex architecture and function of bone and marrow as an organ system.

References

1. Ford, C.E., Hamerton, J.L., Barnes, D.W. and Loutit, J.F. Cytological identification of radiation-chimeras. *Nature* **177**, 452–454 (1956).
2. Micklem, H.S., Ford, C.E., Evans, E.P. and Gray, J. Interrelationships of myeloid and lymphoid cells: studies with chromosome-marked cells transplanted into lethally irradiated mice. *Proc. R. Soc. Lond. Ser. B.* **165**, 78–102 (1966).
3. Till, J.E. and McCulloch, E.A. A direct measurement of the radiation sensitivity of normal mouse bone marrow cells. *Radiat. Res.* **14**, 213–222 (1961).
4. McCulloch, E.A. and Till, J.E. The sensitivity of cells from normal mouse bone marrow to gamma radiation *in vitro* and *in vivo*. *Radiat. Res.* **16**, 822–832 (1962).
5. Becker, A.J., McCulloch, E.A. and Till, J.E. Cytological demonstration of the clonal nature of spleen colonies derived from transplanted mouse marrow cells. *Nature* **197**, 452–454 (1963).
6. Siminovitch, L., McCulloch, E.A. and Till, J.E. The distribution of colony-forming cells among spleen colonies. *J. Cell. Physiol.* **62**, 327–336 (1963).
7. Wu, A.M., Till, J.E., Siminovitch, L. and McCulloch, E.A. A cytological study of the capacity for differentiation of normal hematopoietic colony-forming cells. *J. Cell. Physiol.* **69**, 177–184 (1967).
8. Nowell, P.C., Hirsch, B.E., Fox, D.H. and Wilson, D.B. Evidence for the existence of multi-potential lympho-hematopoietic stem cells in the adult rat. *J. Cell. Physiol.* **75**, 151–158 (1970).
9. Wu, A.M., Till, J.E., Siminovitch, L. and McCulloch, E.A. Cytological evidence for a relationship between normal hematopoietic colony-forming cells and cells of the lymphoid system. *J. Exp. Med.* **127**, 455–464 (1968).
10. Abramson, S., Miller, R.G. and Phillips, R.A. The identification in adult bone marrow of pluripotent and restricted stem cells of the myeloid and lymphoid systems. *J. Exp. Med.* **145**, 1567–1579 (1977).
11. Metcalf, D., Moore, M.A.S. and Shortman, K. Adherence column and buoyant density separation of bone marrow stem cells and more differentiated cells. *J. Cell. Physiol.* **78**, 441–450 (1971).
12. Murphy, M.J., Bertles, J.F. and Gordon, A.S. Identifying characteristics of the hematopoietic precursor cell. *J. Cell Sci.* **9**, 23–47 (1971).

13. Nicola, N.A., Metcalf, D., von Melcher, H. and Burgess, A.W. Isolation of murine fetal hematopoietic progenitor cells and selective fractionation of various erythroid precursors. *J. Blood* **58**, 376–386 (1982).
14. Visser, J.W., Bauman, J.G., Mulder, A.H., Eliason, J.F. and de Leeuw, A.M. Isolation of murine pluripotent hemopoietic stem cells. *J. Exp. Med.* **159**, 1576–1590 (1984).
15. Castagnola, C., Visser, J., Boersam, W. and Bekkum, D.W. Purification of rat pluripotent hematopoietic stem cells. *Stem Cells* **1**, 250–260 (1982).
16. Goldschneider, I., Gordon, L.K. and Morris, R.J. Demonstration of Thy-1 antigen on pluripotent hemopoietic stem cells in the rat. *J. Exp. Med.* **148**, 1351–1366 (1978).
17. Dexter, T.M., Allen, T.D., Lajtha, L.G., Schofield, R. and Lord, B.I. Stimulation of differentiation and proliferation of haemopoietic cells *in vitro*. *J. Cell. Physiol.* **82**, 461–474 (1973).
18. Dexter, T.M., Allen, T.D. and Lajtha, L.G. Conditions controlling the proliferation of haemopoietic stem cells *in vitro*. *J. Cell. Physiol.* **91**, 335–344 (1977).
19. Humphries, R.K., Eaves, A.C. and Eaves, C.J. Characterization of a primitive erythropoietic progenitor found in mouse marrow before and after several weeks in culture. *Blood* **53**, 746–763 (1979).
20. Humphries, R.K., Eaves, A.C. and Eaves, C.J. Self-renewal of hematopoietic stem cells during mixed colony formation *in vitro*. *Proc. Natl. Acad. Sci. USA* **78**, 3629–3633 (1981).
21. Whitlock, C.A. and Witte, O.N. Long-term culture of B lymphocytes and their precursors from murine bone marrow. *Proc. Natl. Acad. Sci. USA* **79**, 3608–3612 (1982).
22. Weissman, I.L. The road ended up at stem cells. *Immunol. Rev.* **185**, 159–174 (2002).
23. Muller-Sieburg, C., Whittlock, C.A. and Weissman, I.L. Isolation of two early B lymphocyte progenitors from mouse marrow: a committed pre-B cell and a clonogenic Thy-1^{lo} hematopoietic stem cell. *Cell* **44**, 653–662 (1986).
24. Spangrude, G.J., Heifeld, S. and Weissman, I.L. Purification and characterization of mouse hematopoietic stem cells. *Science* **241**, 58–62 (1988).
25. Smith, L.G., Weissman, I.L. and Heimfeld, S. Clonal analysis of hematopoietic stem-cell differentiation *in vivo*. *Proc. Natl. Acad. Sci. USA* **88**, 2788–2792 (1991).
26. Spangrude, G.J. and Johnson, G.R. Resting and activated subsets of mouse multipotent hematopoietic stem cells. *Proc. Natl. Acad. Sci. USA* **87**, 7433–7437 (1990).

27. Fleming, W.H., Alpern, E.J., Uchida, N., Ikuta, K., Spangrude, G.J. and Weissman, I.L. Functional heterogenity is associated with the cell cycle status of murine hematopoietic stem cells. *J. Cell Biol.* **122**, 897–902 (1993).
28. Wineman, J.P., Gilmore, G.L., Gritzmacher, C., Torbett, B.E. and Muller-Sieburg, C.E. CD4 is expressed on murine pluripotent hematopoietic stem cells. *Blood* **80**, 1717–1724 (1992).
29. Morrison, S.J. and Weissman, I.L. The long-term repopulating subset of hematopoietic stem cells is deterministic and isolatable by phenotype. *Immunity* **1**, 661–673 (1994).
30. Okada, S., Nakauchi, H., Nagayoshi, K., Nishikawa, S., Nishikawa, S.-I., Miura, Y. and Suda, T. Enrichment and characterization of murine hematopoietic stem cells that express c-kit molecule. *Blood* **78**, 1706–1712 (1991).
31. Ikuta, K. and Weissman, I.L. Evidence that hematopoietic stem cells express mouse c-kit but do not depend on Steel factor for their generation. *Proc. Natl. Acad. Sci. USA* **89**, 1502–1506 (1992).
32. Christensen, J.L. and Weissman, I.L. Flk-2 is a marker in hematopoietic stem cell differentiation: a simple method to isolate long-term stem cells. *Proc. Natl. Acad. Sci. USA* **98**, 14541–14546 (2001).
33. Berenson, R.J., Andrews, R.G., Bensinger, W.I., Kalamasz, D., Knitter, G., Buchner, C.D. and Bernstein, D. Antigen CD34+ marrow cells engraft lethally irradiated baboons. *J. Clin. Invest.* **81**, 951–955 (1988).
34. Berenson, R.J., Bensinger, W.I., Hill, R.S., Andrews, R.G., Garcia-Lopez, J., Kalamasz, D.F., Still, B.J., Spitzer, G., Buckner, C.D., Bemstein, I.D. and Thomas, E.D. Engraftment after infusion of CD34+ marrow cells in patients with breast cancer or neuroblastoma. *Blood* **77**, 1717–1722 (1991).
35. Baume, C.M., Weissman, I.L., Tsukamoto, A.S., Buckle, A. and Peault, B. Isolation of a candidate human hematopoietic stem-cell population. *Proc. Natl. Acad. Sci. USA* **89**, 2804–2808 (1992).
36. Murray, L., Chen, B., Galy, A., Chen, S., Tushinski, R., Uchida, N., Negrin, R., Tricot, G., Jagannath, S., Vesole, D., Barlogie, B., Hoffman, R. and Tsukamoto, A. Enrichment of human hematopoietic stem cell activity in the CD34+Thy-1+Lin$^-$ subpopulation from mobilized peripheral blood. *Blood* **85**, 368–378 (1995).
37. Terstappen, L.W., Huang, S., Safford, M., Lansdorp, P.M. and Loken, M.R. Sequential generations of hematopoietic colonies derived from single non-lineage-committed CD34+CD38- progenitor cells. *Blood* **77**, 1218–1227 (1991).
38. Yin, A.H., Miraglia, S., Zanjani, E.D., Almeida-Porada, G., Ogawa, M., Leary, A.G., Olweus, J., Kearney, J. and Buck, D.W. AC133, a novel marker

for human hematopoietic stem and progenitor cells. *Blood* **90**, 5002–5012 (1997).
39. Krause, D., Ito, T., Fackler, M., Collector, M., Sharkis, S. and May, W. Characterization of murine CD34, a marker for hematopoietic progenitor and stem cells. *Blood* **84**, 691–701 (1994).
40. Morel, F., Szilvassy, S.J., Travis, M., Chen, B. and Galy, A. Primitive hematopoietic cells in murine bone marrow express the CD34 antigen. *Blood* **88**, 3774–3784 (1996).
41. Osawa, M., Hanada, K., Hamada, H. and Nakauchi, H. Long-term lympho-hematopoietic reconstitution by a single CD34-low/negative hematopoietic stem cell. *Science* **273**, 242–245 (1996).
42. Goodell, M.A., Rosenzweig, M., Kim, H., Marks, D.F., DeMaria, M., Paradis, G., Grupp, S.A., Sieff, C.A., Mulligan, R.C. and Johnson, R.P. Dye efflux studies suggest that hematopoietic stem cells expressing low or undetectable levels of CD34 antigen exist in multiple species. *Nat. Med.* **3**, 1337–1345 (1997).
43. Furness, S.G. and McNagny, K. Beyond mere markers: functions for CD34 family of sialomucins in hematopoiesis. *Immunol. Res.* **34**, 13–32 (2006).
44. Petrakova, K.V., Tolmacheva, A.A. and Friedenstein, A.J. Bone formation occurring in bone marrow transplantation in diffusion chambers. *Bull. Exp. Biol. Med.* **56**, 87–91 (1963).
45. Friedenstein, A.J., Piatetzky-Shapiro, I.I. and Petrakova, K.V. Osteogenesis in transplants of bone marrow cells. *J. Embryol. Exp. Morph.* **16**, 381–390 (1966).
46. Friedenstein, A.J., Petrakova, K.V., Kurolesova, A.I. and Frolova, G.P. Heterotopic transplants of bone marrow. *Transplantation* **6**, 230–247 (1968).
47. Friedenstein, A.J. and Kuralesova, A.I. Osteogenic precursor cells of bone marrow in radiation chimeras. *Transplantation* **12**, 99–108 (1971).
48. Friedenstein, A.J., Chailakhjan, R.K. and Lalykina, K.S. The development of fibroblast colonies in monolayer cultures of guinea-pig bone marrow and spleen cells. *Cell Tissue Kinet.* **3**, 393–403 (1970).
49. Gartner, S. and Kaplan, H. Long-term culture of human bone marrow cells. *Proc. Natl. Acad. Sci. USA* **77**, 4756–4759 (1980).
50. Ashton, B.A., Allen, T.D., Howlett, C.R., Eaglesom, C.C., Hattori, A. and Owen, M. Formation of bone and cartilage by marrow stromal cells in diffusion chambers *in vivo*. *Clin. Orthop. Relat. Res.* **151**, 294–307 (1980).
51. Bab, I., Ashton, B.A., Gazit, D., Marx, G., Williamson, M.C. and Owen, M.E. Kinetics and differentiation of marrow stromal cells in diffusion chambers *in vivo*. *J. Cell Sci.* **84**, 139–151 (1986).

52. Grundle, R., Joyner, C.J. and Triffitt, J.T. Human bone tissue formation in diffusion chamber culture *in vivo* by bone-derived cells and marrow stromal fibroblastic cells. *Bone* **16**, 597–601 (1995).
53. Owen, M.E., Cave, J. and Joyner, C.J. Clonal analysis *in vitro* of osteogenic differentiation of marrow CFU-F. *J. Cell Sci.* **87**, 731–738 (1987).
54. Leboy, P.S., Beresford, J.N., Devlin, C. and Owen, M.E. Dexamethasone induction of osteoblast mRNAs in rat marrow stromal cell cultures. *J. Cell. Physiol.* **146**, 370–378 (1991).
55. Gimble, J.M., Dorheim, M.A., Cheng, Q., Median, K., Wang, C.S., Jones, R., Koren, E., Pietrangeli, C. and Kincade, P.W. Adipogenesis in a murine bone marrow stromal cell line capable of supporting B lineage lymphocyte growth and proliferation: biochemical and molecular characterization. *Eur. J. Immunol.* **20**, 379–387 (1990).
56. Berry, L., Grant, M.E., McClure, J. and Rooney, P. Bone-marrow-derived chondrogenesis *in vitro*. *J. Cell Sci.* **101**, 333–342 (1992).
57. Johnstone, B., Hering, T.M., Caplan, A.I., Goldberg, V.M. and Yoo, J.U. *In vitro* chondrogenesis of bone marrow-derived mesenchymal progenitor cells. *Exp. Cell Res.* **238**, 265–272 (1998).
58. Beresford, J.N., Bennet, J.H., Devlin, C., Leboy, P.S. and Owen, M.E. Evidence for an inverse relationship between the differentiation of adipocytic and osteogenic cells in rat marrow stromal cell cultures. *J. Cell Sci.* **102**, 341–351 (1992).
59. Grigoriadis, A.E., Heersche, J.N. and Aubin, J.E. Differentiation of muscle, fat, cartilage, and bone from progenitor cells present in a bone-derived clonal cell population: effect of dexamethasone. *J. Cell Biol.* **106**, 2139–2151 (1988).
60. Caplan, A.I. The mesengenic process. *Clin. Plast. Surg.* **21**, 429–435 (1994).
61. Pittenger, M.F., MacKay, A.M., Beck, S.C., Jaiswal, R.K., Douglas, R., Mosca, J.D., Moorman, M.A., Simonetti, D.W., Craig, S. and Marshak, D.R. Multilineage potential of adult human mesenchymal stem cells. *Science* **284**, 143–147 (1999).
62. Dennis, J.E., Merriam, A., Awadallah, A., Yoo, J.U., Johnstone, B. and Caplan, A.I. A quadripotential mesenchymal progenitor cell isolated from the marrow of an adult mouse. *J. Bone Miner. Res.* **14**, 700–709 (1999).
63. Deryugina, E.I. and Muller-Sieburg, C.E. Stromal cells in long-term cultures: keys to the elucidation of hematopoietic development. *Crit. Rev. Immun.* **13**, 115–150 (1993).
64. Liesveld, J.L., Abboud, C.N., Duerst, R.E., Ryan, D.H., Brennan, J.K. and Lichtman, M.A. Characterization of human marrow stromal cells: role in progenitor cell binding and granulopoiesis. *Blood* **73**, 1794–1800 (1989).

65. Charbord, P., Lerat, H., Newton, I., Tamayo, E., Gown, A.M., Singer, J.W. and Herve, P. The cytoskeleton of stromal cells from human bone marrow cultures resembles that of cultured smooth muscle cells. *Exp. Hematol.* **18**, 276–282 (1990).
66. Perkins, S. and Fleishman, R.A. Stromal cell progeny of murine bone marrow fibroblast colony-forming units are clonal endothelial-like cells that express collagen IV and laminin. *Blood* **75**, 620–725 (1990).
67. Hasthorpe, S., Bogdanovski, M., Rogerson, J. and Radley, J.M. Characterization of endothelial cells in murine long-term marrow culture. Implication for hemopoietic regulation. *Exp. Hematol.* **20**, 476–481 (1992).
68. Gottlieb, D.J., Simmons, P.J. and Bradstock, K.F. MUC18, a member of the immunoglobulin superfamily, is expressed on bone marrow fibroblasts and a subset of hematological malignancies. *Leukemia* **12**, 414–421 (1998).
69. Friedenstein, A.J., Latzinik, N.W., Grosheva, A.G. and Gorskaya, U.F. Marrow microenvironment transfer by heterotopic transplantation of freshly isolated and cultured cells in porous sponges. *Exp. Hematol.* **10**, 217–227 (1982).
70. Kuznetsov, S.A., Krebsbach, P.H., Satomura, K., Kerr, J., Riminucci, M., Benayahu, D. and Robey, P.G. Single-colony derived strains of human marrow stromal fibroblasts form bone after transplantation *in vivo*. *J. Bone Miner. Res.* **12**, 1335–1347 (1997).
71. Phinney, D.G., Kopen, G., Righter, W., Webster, S., Tremain, N. and Prockop, D.J. Donor variation in the growth properties of human marrow stromal cells. *J. Cell. Biochem.* **75**, 424–436 (1999).
72. Mauraglia, A., Cancedda, R. and Quarto, R. Clonal mesenchymal progenitors from human bone marrow differentiate *in vitro* according to a hierarchical model. *J. Cell Sci.* **113**, 1161–1166 (2000).
73. Bruder, S.P., Ricalton, N.S., Boynton, R.E., Connolly, T.J., Jaiswal, N., Zaia, J. and Barry, F.P. Mesenchymal stem cell surface antigen SB-10 corresponds to activated leukocyte cell adhesion molecule and is involved in osteogenic differentiation. *J. Bone Miner. Res.* **13**, 655–663 (1998).
74. Barry, F.P., Boynton, R.E., Haynesworth, S., Murphy, J.M. and Zaia, J. The monoclonal antibody SH-2, raised against human mesenchymal stem cells, recognizes an epitope on Endoglin (CD105). *Biochem. Biophys. Res. Commun.* **265**, 134–139 (1999).
75. Barry, F., Boynton, R., Murphy, M., Haynesworth, S. and Zaia, J. The SH-3 and SH-4 antibodies recognize distinct epitopes on CD73 from human mesenchymal stem cells. *Biochem. Biophys. Res. Commun.* **289**, 519–524 (2001).

76. Pittenger, M.F., Mbalaviele, G., Black, M., Mosca, J.D. and Marshak, D.R. Mesenchymal stem cells. In: Koller, M.R., Palsson, B.O. and Master, J.R.W. (eds.), *Human Cell Culture* (Kluwer Academic Publishers, Great Britain, 2001), pp. 189–207.
77. Simmons, P.J. and Torok-Storb, B. Identification of stromal cell precursors in human bone marrow by a novel monoclonal antibody STRO-1. *Blood* **78**, 55–62 (1991).
78. Dennis, J.E., Carbillet, J.P., Caplan, A.I. and Charbord, P. The STRO-1+ marrow cell population is multipotential. *Cells Tissues Organs* **170**, 73–82 (2002).
79. Quirici, N., Soligo, D., Bossolasco, P., Servida, F., Lumini, C. and Deliliers, G.L. Isolation of bone marrow mesenchymal stem cells by anti-nerve growth factor receptor antibodies. *Exp. Hematol.* **30**, 783–791 (2002).
80. Anjos-Afonso, F. and Bonnet, D. Nonhematopoietic/endothelial SSEA-1+ cells define the most primitive progenitors in the adult murine bone marrow mesenchymal compartment. *Blood* **109**, 1298–1306 (2007).
81. Gang, E.J., Bosnakovski, D., Figueiredo, C.A., Visser, J.W. and Perlingeiro, R.C. SSEA-4 identified mesenchymal stem cells from bone marrow. *Blood* **109**, 1743–1751 (2007).
82. Buhring, H.-J., Battula, V.L., Trem, S., Schewe, B., Kanz, L. and Vogel, W. Novel markers for the prospective isolation of human MSC. *Ann. N. Y. Acad. Sci.* **1106**, 262–271 (2007).
83. Shi, S. and Gronthos, S. Perivascular niche of postnatal mesenchymal stem cells in human bone marrow and dental pulp. *J. Bone Miner. Res.* **18**, 696–704 (2003).
84. Phinney, D.G., Hill, K., Michelson, C., DuTreil, M., Hughes, C., Humphries, S., Wilkinson, R., Baddoo, M. and Bayly, E. Biological activities encoded by the murine mesenchymal stem cell transcriptome provide a basis for their developmental plasticity and broad clinical efficacy. *Stem Cells* **24**, 186–198 (2006).
85. Asahara, T., Murohara, T., Sullivan, A., Silver, M., van der Zee, R., Li, T., Witzenbichler, B., Schatteman, G. and Isner, J.M. Isolation of putative progenitor endothelial cells for angiogenesis. *Science* **275**, 964–967 (1997).
86. Risau, W. and Flamme, I. Vasculogenesis. *Annu. Rev. Cell Dev. Biol.* **11**, 73–91 (1995).
87. Asahara, T., Masuda, H., Takahashi, T., Kalka, C., Pastore, C., Silver, M., Kearne, M., Magner, M., and Isner, J.M. Bone marrow origin of endothelial progenitor cells responsible for postnatal vasculogenesis in physiological and pathological neovascularization. *Circ. Res.* **85**, 221–228 (1999).

88. Kalka, C., Masuda, H., Takahashi, T., Li, T. and Asahara, T. Administration of culture-expanded endothelial progenitor cells (EPC) augments therapeutic neovascularization. *Circulation* **97**(Suppl. I), I-455 (1998).
89. Mackenzie, D.C. and Loewenthal, J. Endothelial growth in nylon vascular grafts. *Br. J. Surg.* **48**, 212–217 (1960).
90. Poole, C.J., Sabiston Jr., D.C., Florey, H.W. and Allison, P.R. Growth of endothelium in arterial prosthetic grafts and following endarterectomy. *Surg. Forum* **13**, 225–227 (1962).
91. Stump, M.M., Jordan Jr., G.L., Debakey, M.E. and Halpert, B. Endothelium grown from circulating blood on isolated intravascular Dacron hub. *Am. J. Pathol.* **43**, 361–367 (1963).
92. Shi, Q., Rafii, S., Wu, M.H., Wijelath, E.S., Yu, C., Ishida, A., Fujita, Y., Kothari, S., Mohle, R., Sauvage, L.R., Moore, M.A., Storb, R.F. and Hammond, W.P. Evidence for circulating bone marrow-derived endothelial cells. *Blood* **92**, 362–367 (1998).
93. Peichev, M., Naiyer, A.J., Pereira, D., Zhu, Z., Lane, W.J., Williams, M., Oz, M.C., Hicklin, D.J., Witte, L., Moore, M.A.S. and Rafii, S. Expression of VEGFR-2 and AC133 by circulating human $CD34^+$ cells identifies a population of functional endothelial precursors. *Blood* **95**, 952–958 (2000).
94. Gehling, U.M., Ergun, S., Schumacher, U., Wagener, C., Pantel, K., Otte, M., Schuch, G., Schafhausen, P., Mende, T., Kilic, N., Kluge, K., Schafer, B., Hossfeld, D.K. and Fiedler, W. *In vitro* differentiation of endothelial cells from AC133-positive progenitor cells. *Blood* **95**, 3106–3112 (2000).
95. Jackson, K.A., Majka, S.M., Wang, H., Pocius, J., Hartley, C.J., Majesky, N.W., Entman, M.L., Hirschi, K.K. and Goodell, M.A. Regeneration of ischemic cardiac muscle and vascular endothelium by adult stem cells. *J. Clin. Invest.* **107**, 1395–1402 (2001).
96. Grant, M.B., May, W.S.,Caballero, S., Brown, G.A., Guthrie, S.M., Mames, R.N., Byrne, B.J., Vaught, T., Spoerri, P.E., Peck, A.B. and Scott, E.W. Adult hematopoietic stem cells provide functional hemangioblast activity during retinal neovascularization. *Nat. Med.* **8**, 607–612 (2002).
97. Fernandez Pujol, B., Lucibello, F.C., Gehling, U.M., Lindemann, K., Weidner, N., Zuzarte, M.L., Adamkiewicz, J., Elsasser, H.P., Muller, R. and Havemann, K. Endothelial-like cells derived from human CD14 positive monocytes. *Differentiation* **65**, 287–300 (2000).
98. Schmeisser, A., Garlichs, C.D., Zhang, H. Eskafi, S., Graffy, C., Ludwig, J., Strasser, R.H. and Daniel, W.G. Monocytes coexpress endothelial and macrophagocytic lineage markers and form cord-like structures in Matrigel under angiogenic conditions. *Cardiovasc. Res.* **49**, 671–680 (2001).

99. Schmeisser, A., Graffy, C., Daniel, W.G. and Strasser, R.H. Phenotypic overlap between monocytes and vascular endothelial cells. *Adv. Exp. Med. Biol.* **522**, 59–74 (2003).
100. Rehman, J., Li, J., Orschell, C.M. and March, K.L. Peripheral blood "endothelial progenitor cells" are derived from monocyte/macrophages and secrete angiogenic growth factors. *Circulation* **107**, 1164–1169 (2003).
101. Elsheikh, E., Uzunel, M., He, Z., Holgersson, J., Nowak, G. and Sumitran-Holgersson, S. Only a specific subset of human peripheral-blood monocytes has endothelial-like functional capacity. *Blood* **106**, 2347–2355 (2005).
102. Yoder, M.C., Mead, L.E., Prater, D., Krier, T.R., Mroueh, K.N., Li. F., Krasich, R., Temm, C.J., Prchal, J.T. and Ingram, D.A. Redefining endothelial progenitor cells via clonal analysis and hematopoietic stem/progenitor cell principals. *Blood* **109**, 1801–1809 (2007).
103. Keller, G., Lacause, G. and Robertson, S. Development of the hematopoietic system in the mouse. *Exp. Hematol.* **27**, 777–787 (1999).
104. Muller, A.M., Medvinsky, A., Strouboulis, J., Grosveld, F. and Dzierzak, E. Development of hematopoietic stem cell activity in the mouse embryo. *Immunity* **1**, 219–301 (1994).
105. Pardanaud, L., Lutton, D., Prigent, M., Bourcheix, L.M., Catala, M. and Dieterlen-Lievre, F. Tow distinct endothelial lineages in ontogeny, one of them related to hemopoiesis. *Development* **122**, 1363–1371 (1996).
106. Jaffredo, T., Gautier, R., Eichmann, A. and Dieterlen-Lievre, F. Intra-aortic hemopoietic cells are derived from endothelial cells during ontogeny. *Development* **125**, 4575–4583 (1998).
107. North, T., Gu, T.L., Stacy, T., Wang, Q., Howard, L., Binder, M., Marin-Padilla, M. and Speck, N.A. Cbfa2 is required for the formation of intra-aortic hematopoietic clusters. *Development* **126**, 2563–2575 (1999).
108. Tavian, M., Coulombel, L., Luton, D., Clemente, H.S., Dieterlen-Lievre, F. and Peault, B. Aorta-associated CD34+ hematopoietic cells in the early human embryo. *Blood* **87**, 67–72 (1996).
109. Sabin, F.R. Studies on the origin of blood vessels and of red corpuscles as seen in the living blastoderm of the chick during the second day of incubation. *Contrib. Embryol.* **9**, 213–262 (1920).
110. Sanchez, M., Gottgens, B., Sinclair, A.M., Stanley, M., Begley, C.G., Huner, S. and Green, A.R. An SCL 3′ enhancer targets developing endothelium together with embryonic and adult haematopoietic progenitors. *Development* **126**, 3891–3904 (1999).
111. Shalaby, F., Rossant, J., Yamaguchi, T.P., Gertsenstein, M., Wu, X.F., Breitman, M.L. and Schuh, A.C. Failure of blood-island formation and vasculogenesis in Flk-1-deficient mice. *Nature* **376**, 62–66 (1995).

112. Carmeliet, P., Ferreira, V., Breier, G., Pollefeyt, S., Kieckens, L., Gertsenstein, M., Fahrig, M., Vandenhoeck, A., Harpal, K., Eberhardt, C., Declercq, C., Pawling, J., Moons, L., Collen, D., Risau, W. and Nagy, A. Abnormal blood vessel development and lethality in embryos lacking a single VEGF allele. *Nature* **380**, 435–439 (1996).
113. Okuda, T., van Deursen, J., Hiebert, S.W., Grosveld, G. and Downing, J.R. AML1, thc targct of multiplc chromosomal translocations in human leukemia, is essential for normal fetal liver hematopoiesis. *Cell* **84**, 321–330 (1996).
114. Choi, K., Kennedy, M., Kazarov, A., Papadimitriou, J.C. and Keller, G.A. Common precursor for hematopoietic and endothelial cells. *Development* **125**, 725–732 (1998).
115. Nishikawa, S.I., Nishikawa, S., Hirashima, M., Matsuyoshi, N. and Kodama, H. Progressive lineage analysis by cell sorting and culture identifies FLK1+VE-cadherin+ cells at a diverging point of endothelial and hematopoietic lineages. *Development* **125**, 1747–1757 (1998).
116. Doherty, M.J., Ashton, B.A., Walsh, S., Beresford, J.N., Grant, M.E. and Canfield, A.E. Vascular pericytes express osteogenic potential *in vitro* and *in vivo*. *J. Bone Miner. Res.* **13**, 828–838 (1988).
117. Bianco, P. and Cossu, G. Uno, nessuno e centomila: searching for the identity of mesodermal progenitors. *Exp. Cell Res.* **251**, 257–263 (1999).
118. Farrington-Rock, C., Crofts, N.J., Doherty, M.J., Ashton, B.A., Griffin-Jones, C. and Canfield, A.E. Chondrogenic and adipogenic potential of microvascular pericytes. *Circulation* **110**, 2226–2232 (2004).
119. Bianco, P., Riminucci, M., Gronthos, S. and Robey, P.G. Bone marrow stromal stem cells: nature, biology, and potential applications. *Stem Cells* **19**, 180–192 (2001).
120. Bonanno, E., Ercoli, L., Missori, P., Rocchi, G. and Spagnoli, L.G. Homogeneous stromal cell population from normal human adult bone marrow expressing α-smooth muscle actin filaments. *Lab. Invest.* **71**, 308–310 (1994).
121. Weiss, L. The hematopoietic microenvironment of bone marrow: an ultra structural study of the stroma in rats. *Anat. Rec.* **186**, 161–184 (1976).
122. Brachvogel, B., Moch, H., Pausch, F., Schlotzer-Schrehardt, U., Hofmann, C., Hallmann, R., von der Mark, K., Winkler, T. and Poschl, E. Perivascular cells expressing annexin A5 define a novel mesenchymal stem cell-like population with the capacity to differentiate into multiple mesenchymal lineages. *Development* **132**, 2657–2668 (2005).

123. da Silva Meirelles, L., Chagastelles, P.C. and Nardi, N.B. Mesenchymal stem cells reside in virtually all post-natal organs and tissues. *J. Cell Sci.* **119**, 2204–2213 (2006).
124. Gittenberger-de Groot, A.C., DeRuiter, M.C., Bergwerff, M. and Poelmann, R.E. Smooth muscle cell origin and its relation to heterogeneity in development and disease. *Arterioscler. Thromb. Vasc. Biol.* **19**, 1589–1594 (1999).
125. Yamashita, J., Itoh, H., Hirashima, M., Ogawa, M., Nishikawa, S., Yurugi, T., Naito, M., Nakao, K. and Nishikawa, S. Flk1-positive cells derived from embryonic stem cells serve as vascular progenitors. *Nature* **408**, 92–96 (2000).
126. Minasi, M.G., Riminucci, M., De Angelis, L., Borello, U., Berarducci, B., Innocenzi, A., Caprioli, A., Sirabella, D., Baiocchi, M., De Maria, R., Boratto, R., Jaffredo, T., Broccoli, V., Bianco, P. and Cossu, G. The mesoangioblast: a multipotent, self-renewing cell that originates from the dorsal aorta and differentiates into most mesodermal tissues. *Development* **129**, 2773–2783 (2002).
127. Lindahl, P., Johansson, B.R., Leveen, P. and Betsholtz, C. Pericyte loss and micro-aneurysm formation in PDGF-B deficient mice. *Science* **277**, 242–245 (1997).
128. Hellstrom, M., Kalen, M., Lindahl, P., Abramsson, A. and Betsholtz, C. Role of PDGF-B and PDGFR-beta in recruitment of vascular smooth muscle cells and pericytes during embryonic blood vessel formation in the mouse. *Development* **126**, 3047–3055 (1999).
129. Hirschi, K.K. and D'Amore, P.A. Pericytes in the microvasculature. *Cardiovasc. Res.* **32**, 687–698 (1996).
130. Satomura, K., Derubeis, A.R., Fedarko, N.S., Ibaraki-O'Connor, K., Kuznetsov, S.A., Rowe, D.W., Young, M.F. and Gehron Robey, P. Receptor tyrosine kinase expression in human bone marrow stromal cells. *J. Cell. Physiol.* **177**, 426–438 (1998).
131. Clark, B.R. and Keating, A. Biology of bone marrow stroma. *Ann. N. Y. Acad. Sci.* **770**, 70–78 (1995).
132. Rosendaal, M., Gregan, A. and Green, C.R. Direct cell–cell communication in the blood-forming system. *Tissue Cell* **23**, 457–470 (1991).
133. Galmiche, M.C., Koteliansky, V.E., Briere, J., Herve, P. and Charbord, P. Stromal cells from human long-term marrow cultures are mesenchymal cells that differentiate following a vascular smooth muscle differentiation pathway. *Blood* **82**, 66–76 (1993).
134. Phinney, D.G. Gene expression profiles of mesenchymal stem cells. In: Nolta, J.A. (ed.), *Genetic Engineering of Mesenchymal Stem Cells* (Springer, 2006).

135. Zhang, J., Niu, C., Ye, L., Huang, H., He, X., Tong, W.G., Ross, J., Haug, J., Johnson, T., Feng, J.Q., Harris, S., Wiedemann, L.M., Mishina, Y. and Li, L. Identification of the haematopoietic stem cell niche and control of the niche size. *Nature* **425**, 836–841 (2003).
136. Calvi, L.M., Adams, G.B., Weibrecht, K.W., Weber, J.M., Olson, D.P., Knight, M.C., Martin, R.P., Schipani, E., Divieti, P., Bringhurst, F.R., Milner, L.A., Kronenberg, H.M. and Scadden, D.T. Osteoblastic cells regulate the haematopoietic stem cell niche. *Nature* **425**, 841–846 (2003).
137. Visnjic, D., Kalajzic, Z., Rowe, D.W., Katavic, V., Lorenzo, J. and Aguila, H.L. Hematopoiesis is severely altered in mice with an induced osteoblast deficiency. *Blood* **103**, 3258–3264 (2004).
138. El-Badri, N.S., Wang, B.Y., Cherry and Good, R.A. Osteoblasts promote engraftment of allogeneic hematopoietic stem cells. *Exp. Hematol.* **26**, 110–116 (1998).
139. Ohneda, O., Fennie, C., Zheng, Z., Donahue, C., La, H., Villacorta, R., Cairns, B. and Lasky, L.A. Hematopoietic stem cell maintenance and differentiation are supported by embryonic aorta-gonad-mesonephros region-derived endothelium. *Blood* **92**, 908–919 (1998).
140. Li, W., Johnson, S.A., Shelley, W.C. and Yoder, M.C. Hematopoietic stem cell repopulating ability can be maintained *in vitro* by some primary endothelial cells. *Exp. Hematol.* **32**, 1226–1237 (2004).
141. Avecilla, S.T., Hattori, K., Heissig, B., Tejada, R., Liao, F., Shido, K., Jin, D.K., Dias, S., Zhang, F., Hartman, T.E., Hackett, N.R., Crystal, R.G., Witte, L., Hicklin, D.J., Bohlen, P., Eaton, D., Lyden, D., de Sauvage, F. and Rafii, S. Chemokine-mediated interaction of hematopoietic progenitors with the bone marrow vascular niche is required for thrombopoiesis. *Nat. Med.* **10**, 64–71 (2004).
142. Taniguchi, H., Toyoshima, T., Fukao, K. and Nakauchi, H. Prescence of hematopoietic stem cells in the adult liver. *Nat. Med.* **2**, 198–203 (1996).
143. Kiel, M.J., Yilmaz, O.H., Iwashita, T., Yilmaz, O.H., Terhorst, C. and Morrison, S.J. SLAM family receptors distinguish hematopoietic stem and progenitor cells and reveal endothelial niches for stem cells. *Cell* **121**, 1109–1121 (2005).
144. Kopp, H.G., Avecilla, S.T., Hooper, A.T. and Rafii, S. The bone marrow vascular niche: home of HSC differentiation and mobilization. *Physiology* **20**, 349–356 (2005).
145. Kinnaird, T., Stabile, E., Burnett, M.S., Lee, C.W., Barr, S., Fuchs, S. and Epstein, S.E. Marrow-derived stromal cells express genes encoding a broad spectrum of arteriogenic cytokines and promote *in vitro* and *in vivo* arteriogenesis through paracrine mechanisms. *Circ. Res.* **94**, 678–685 (2004).

146. Hashi, C.K., Zhu, Y., Yang, G.Y., Young, W.L., Hsiao, B.S., Wang, K., Chu, B. and Li, S. Antithrombogenic property of bone marrow mesenchymal stem cells in nanofibrous vascular grafts. *Proc. Natl. Acad. Sci. USA* **104**, 11915–11920 (2007).
147. Ladage, D., Brixius, K., Steingen, C., Mehlhorn, U., Schwinger, R.H., Bloch, W. and Schmidt, A. Mesenchymal stem cells induce endothelial activation via paracine mechanisms. *Endothelium* **14**, 53–63 (2007).
148. Zacharek, A., Chen, J., Cui, X., Li, A., Li, Y., Roberts, C., Feng, Y., Gao, Q. and Chopp, M. Angiopoietin1/Tie2 and VEGF/Flk1 induced by MSC treatment amplifies angiogenesis and vascular stabilization after stroke. *J. Cereb. Blood Flow Metab.* **27**, 1684–1691 (2007).
149. Carmeliet, P. One cell, two fates. *Nature* **408**, 43–45 (2000).

Chapter 7

Cord Blood Stem Cells — Potentials and Realities

Colin P. McGuckin and Nicolas Forraz

Abstract

With approximately 130 million babies born worldwide every year umbilical cord blood represents perhaps the largest potential source of stem cells for regenerative medicine. Between 1972 and 2008, it is estimated that over 10,000 patients would have been treated by cord blood cells for over 80 different clinical conditions. Cord blood stem cells are used clinically mostly to support patients suffering from haematological and immunological diseases but they also provide emerging therapeutic solutions for limited cases of type 1 diabetes or infant cerebral injuries. Cord blood samples are collected after birth and bioprocessed before cryopreservation in either public biobanks for unrelated allogenic use or private family biobanks for related allogenic and autologous uses. Regenerative medicine research demonstrated the existence of multipotent stem cells with embryonic characteristics in cord blood, which can produce over 20 tissue types including liver, neural or insulin-secreting cells. Cord blood stem cells not only offer therapeutic benefits at present but also show real potential for the advancement of regenerative medicine.

Keywords: Stem Cells; Cord Blood; Regenerative Medicine; Cryopreservation and Banking; Cell Therapy.

Outline

1. Introduction to the Concept of Umbilical Cord Blood Stem Cells
2. Cord Blood Current Clinical Uses
3. Cord Blood Processing and Cryopreservation
4. Cord Blood Banking
5. Cord Blood Research and Where the Future Lies

References

1. Introduction to the Concept of Umbilical Cord Blood Stem Cells

Treatment of leukaemia using chemotherapy and radiotherapy led to one of the earliest understandings that cancer therapy is a balance between keeping the good cells alive and killing the bad cells. Unfortunately this was a balance that sometimes was lost, resulting in the death of the bone marrow. Since bone marrow produces not only blood but also the immune system, fatalities were common. The only possibility for rescuing the patient is, therefore, a bone marrow transplant. Transplant immunology in the first half of the 20th century was a little understood science, but it became clear that close human leucocyte antigen (HLA) code matches were required. Unfortunately, registries for such codes did not much exist and only links from the blood transfusion services created any hope of finding the elusive match.

Whilst today this is still a problem, the haematologists who were tasked with such searches were often reduced to asking the local population to be tested for compatibility. It rarely found a match.

In 1972 doctors Ende and Ende, USA, required such transplant for a patient with acute lymphoblastic leukaemia. Using umbilical cord blood of several donors they were able to transfuse the cells into a 16-year-old Afro-American male successfully although long-term follow-up data on the patient is not available.[1]

In fact, as early as from the end of World War II, placental/cord blood was considered as a potential source of cells for blood transfusion.[2] In the 1970s and 1980s umbilical cord blood was further characterised as containing a source of haematopoietic progenitor cells that could be used in bone marrow transplantation.[3–5] In 1988, Gluckman and colleagues confirmed this potential by transplanting a six-year-old boy from North Carolina, USA, suffering from Fanconi anaemia with HLA-matched umbilical cord blood from his baby sister in Paris.[6]

So what is cord blood? Originally, thanks to the haematology community the fluid inside the physical umbilical cord was only known as "blood". What started to change the landscape was the growing, badly documented and anecdotal information that patients who had undergone bone marrow replacement with cord blood also had improvements in other organ-based systems, including heart and kidney. Was this wishful thinking? Could it be real? No one really believed the patients, instead choosing to think that they were just happy to have recovered from a serious illness. Other sceptics also said that the appearance of bone marrow related cells in systemic organs such as heart and liver, were simply either fused cells, or cells that had become lost.[7–9]

Indeed anecdotal information without proper evidence is probably responsible for setting back the promotion of cord blood as a potential therapy, especially as was sometimes the case, it included speculation that hair was re-growing on bald patients!

Now we understand that the fluidic umbilical cord blood is fundamentally different to that found in adult human peripheral blood. In the adult human, the majority of circulating cells are the red blood cells with the white cell components of platelets, monocytes, lymphocytes, neutrophils and rare numbers of other immune cells being in the minority.

So, what is different about umbilical cord blood? Beyond the observation that nucleated red blood cells are predominant in cord blood, early studies in our research group noted that CD56 natural killer cell progenitors were more in abundance. Building on this we have been able to characterise the different levels of T-regulatory cells and other immune system cells available in the cord blood.[10] What we are now categorically able to say is that umbilical cord blood is a richer source of both stem cells and early immune system cells than both adult human peripheral blood and bone marrow. The purpose of this abundance is much debated. For us, we postulate that the circulating cells in cord blood exist for two main reasons during the gestational period. (i) They are there to help repair mistakes in the developing foetus; this is not hard to understand, since many immune system cells in the adult human exist to help remove dying and damaged cells, including — to varying degrees of success — microbially infected and cancer transformed cells. While the child is protected to a degree by the mother, the need for an immune system increases as the child grows in size. Further, mistakes in gestational cell production can only be removed by the immune system. (ii) Circulating stem cells also help to replenish organs undergoing rapid development and also represents the redistribution of such cells around the bone marrow system, which in the child, is present in most hollow bones (c/f adult bone marrow is restricted to the ends of long bones, ribs and cranial regions, of the major skeletal system). To this degree, we therefore believe that the bone marrow of the child, and to a lesser extent the adult, is composed of extremely responsive stem cell groups, which, although 99% of the time make blood/immune cells, have the capability to do other things, including helping repair and replenish damaged organs. Treatment of myocardial infarction with bone marrow cells has effectively demonstrated that the bone marrow has capabilities beyond standard blood production.[11–12] Differentiation of cells of all three primordial germ layers, further demonstrates that cells in the umbilical cord have capacities far beyond blood.[12–24]

Therefore, as this chapter will show, there is now every reason to believe that the anecdotal testimony of cord blood-transplanted patients was not in fact deluded, and they may well have benefited from their transplant in ways that would never have previously been believed. Could it be therefore, that the bone marrow is home to specialised cells which can aid in organ regeneration? Why then is it that systemic disease is still so common — could it be that there are simply not enough in adult bone marrow — and is this a therapeutically viable potential worth pursuing?

2. Cord Blood Current Clinical Uses

After Ende and Ende, and later Gluckman and colleagues led the use of umbilical cord blood transplants, the application which became more common was of course the standard replacement of the bone marrow. In certain types of childhood blood diseases, notably acute leukaemia, the success rate with a unit of cord blood was eventually found to be higher than with standard bone marrow or the cytokine-mobilised peripheral blood leukapheresis processes.[25,26]

The main limitation to the use of cord blood transplants was that the yield of cells in a single cord blood was generally too small to use for an adult bone marrow transplant. This limitation aside, the Japanese haematological community was the quickest to take on board the potential of cord transplants, albeit with a slower uptake in other health care systems.[27] It is expected in 2008 that 10,000 successful cord blood transplants across the world will have been achieved.

In our own centre we were first to realise that it might be possible to use more than one cord blood unit to help an adult human patient. Steve Proctor from Newcastle pioneered the use of no less than seven units of cord blood for a relapsed chemotherapy resistant adult male with acute lymphoblastic leukaemia (ALL). An ablative stem cell intervention transplant was carried out using melphalan, total body irradiation, and rituximab preconditioning followed with a single matched umbilical cord blood and co-supported by six mismatched cords (one unit per 10 kg recipient weight). Despite the matched cord being sub-optimal in size, there was found to be no adverse reactions to the multi-cord process and the matched cord blood quickly engrafted, with no molecular evidence of the mismatched cords engrafting. Unfortunately, as can be the case with any leukaemia patient, the patient's remission only lasted eight months before "minimal-residual disease" occurred and the leukaemia resurfaced (any leukaemia cells which managed to hide from the chemotherapy ablation can re-multiply, despite the new bone marrow).

Other centres, notably the University of Minnesota Medical Center, have pioneered the routine transplant of what has become know as the "Double Cord" protocol, and Professor Wagner of that centre has reputedly developed a "Triple Cord" also.[28,29]

The prospect, therefore that cord blood can be useful for adults, as well as children, has been one of the biggest advances for the use of this stem cell source. However, it is worth noting that the issue of cord bloods not being "big" enough comes from the simple haematological fact that for a bone marrow replacement a huge cell dose is required. Data from the work on treatment of patients with myocardial infarct and the trial using bone marrow for liver disease[30] indicate that smaller doses are required. Therefore, it is possible to predict a lower number of cells will be needed when using cord blood stem cells in the future. For cord blood banking (noted later) this is a primary issue.

Table 1 lists the current diseases treatable or at least supportable with cord blood-derived cells. Rather than simply calling these "stem cell therapies" it is important to highlight one trial in particular as an example of how the cord blood is likely to be more than stem cells and have potential immunotherapy, see also Refs. 31 to 33.

The trial — coded NCT00305344 — from the University of Florida at Gainesville uses autologous umbilical cord blood mononucleated cell infusions into newly diagnosed type 1 diabetes patients. The trial highlights the importance of pioneers not only in science but also in medicine. Faced with a child who was diagnosed with type 1 diabetes, the parents noted that their cord blood had been stored in a cord blood bank. Upon thought, the doctors in Gainesville decided that an autologous transfusion would not require preconditioning with chemotherapy, nor with immunosuppressive therapy. Atkinson, Haller and Schatz proceeded and found that cord blood infusion had protected the pancreas, which was in a state of autoimmune attack.[34] It is likely that cord blood is a great deal more than blood cells and stem cells — that it also contains unique immune cells, potentially capable of immune modulation. This begs the question — why do you need immune modulation in a developing baby? Since the baby has only half the HLA types similar to the mother, the placental tissues attached to the wall of the mother's womb are going to be in constant conflict with the immune system of the mother. The only way for the placenta not to be rejected by the mother is immunomodulation. A number of different systems for this have been suggested, but a lot more investigation is required to really understand it. A possible explanation could be in involvement of HLA-G expressed on cytotrophoblasts and modulating immune tolerance of the placenta and the foetus by the mother. Interestingly the interaction of HLA-G with regulatory T-cells is increasingly studied during cell and tissue transplantation.[35] Indeed research on the placental tissues has identified a number of interesting mesenchymal-like stem cells. Such cells have been postulated to have immunomodulatory capabilities and this may be part of the process (see also Chapter 10). Nevertheless, following this success, Atkinson and colleagues managed to get authorisation to start a clinical trial for 23 patients, including laboratory investigations which are showing that the immune system, not least the regulatory T-cell system is modulated in these patients upon transfusion of the cord blood cells.

Less than 20 years ago, the use of cord blood as a clinical intervention was accepted in only one therapy. Today almost 80 diseases and disorders could use

Table 1. Disorders treatable/supportable by cord blood cells.

Oncologic Disorders	**Immune Deficiencies**
Acute lymphoblastic leukaemia	Ataxia telangectasia
Acute myeloid leukaemia	Cartilage-hair hypoplasia
Autoimmune lymphoproliferative syndromes	Chronic granulomatous disease
Burkitt lymphoma	DiGeorge syndrome
Chronic myeloid leukaemia	Hypogammaglobulinaemia
Cytopenia related to monosomy	IKK gamma deficiency
Familial histocytosis	Immune dysregulation polyendocrinopathy
Haemaphagocytic lymphohistiocytosis	Mucolipidosis type II
Hodgkin's disease	Myelokathesis
Juvenile myelomonocytic leukaemia	Severe combined immunodeficiency
Langerhans cell histocytosis	Wiscott-Aldrich syndrome
Myelodysplastic syndromes	x-linked agammaglobulinaemia
Non-Hodgkin's lymphoma	x-linked immunodeficiency
	x-linked lymphoproliferative syndrome
Haematological Disorders	**Metabolic Disorders**
Amegarakarocytic thrombocytopenia	Adrenoleukodystrophy
Autoimmune neutropenia	Alpha mannosidosis
Congenital dyserythropoietic anaemia	Diabetes mellitus, Type 1
Congenital sideroblastic anaemia	Gaucher's disease (infantile)
Cyclic neutropenia	Globoid cell leukodystrophy
Diamond Blackfran anaemia	Gunther disease
Evan's syndrome	Hermansky-Pudlak syndrome
Fanconi anaemia	Hurler syndrome
Glanzmann's disease	Hurler-Scheie syndrome
Hypoproliferative anaemia	Krabbe's disease
Juvenile dermatomyositis	Maroteau-lamy syndrome
Juvenile xanthogranulomas	Metachromatic leukodystrophy
Kostmann's syndrome	Mucolipidosis Types II, III
Pancytopenia	Neimann Pick syndrome, Types A and B
Red cell aplasia	Sandoff syndrome
Refractory anaemia	Sanfilippo syndrome
Severe aplastic anaemia	Tay Sachs disease
Shwachman syndrome	
Severe neonatal thrombocytopenia	
Sickle cell disorders	
Severe neonatal thrombocytopenia	
Sickle cell disorders	
Systemic mastocytosis	
Thalassaemia	
Thrombocytopenias with absent radius	

cord blood as part of patient therapy. Table 1 shows that haematological cancers are no longer the only successful therapies using cord blood. This exponential rise in less than 20 years has taken the stem cell community somewhat by surprise and whilst other stem cell sources continue to gain more media attention, the number of cord blood therapies increases each year.

3. Cord Blood Processing and Cryopreservation

Given that clinical intervention for haematological disorders requires most of the white cells available, and that we do not yet fully understand what else is in the cord blood that is useful, the need to process cord blood has increased in importance. Cord blood banking has become a financially viable industry (with an estimated 500,000 units in the US and 135,000 units stored in private family cord blood banks, see Fig. 1). However, not all of the cord blood companies that have

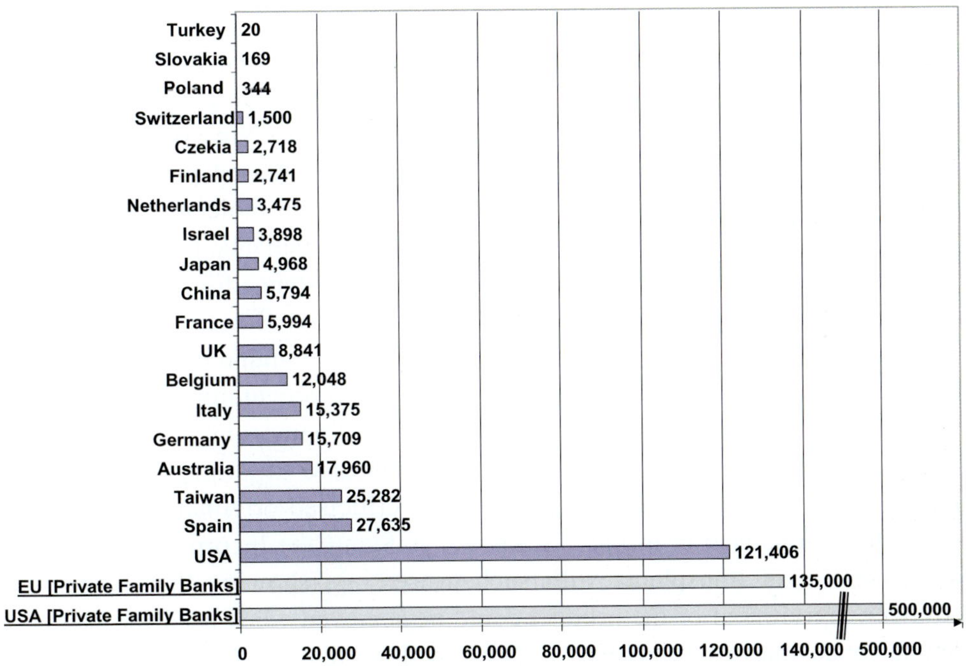

Fig. 1. Cord blood units stored in different countries. Around 276,000 cord blood samples are stored in public cord blood bank as of 23 October 2007 as highlighted by the blue columns. An estimated 135,000 cord blood units in Europe and 500,000 cord blood units in the US are stored in private family cord blood banks. *Source*: Bone Marrow Donors Worldwide and personal research.

set up to store — for a fee — a child's cord blood are to be recommended. Some have given a bad name to the industry and indeed to cord blood in general.

Originally cord blood was stored in the same way that a bag of donated peripheral blood was stored — at room temperature and for a short period. Unfortunately, cord blood is fundamentally different from peripheral blood; primarily the cell content and the potential for activating platelets and clotting can be a problem. Therefore, processing the cord blood is the recommendation of most experts. Some major private cord blood banks do not, however, process, preferring instead to use a simpler and more economical method by adding cryopreservant dimethylsulphoxide (DMSO) and freeze. The authors are not currently aware of any public bank in the world that still carries out direct freezing and the vast majority of successful cord blood transplants already carried out have been with processed cord blood. In our hands, cord blood which has not been processed tends to have major clotting problems and low cell viability if stored whole and then thawed.[36]

(i) Methods

Right now, different organisations process in different ways. The most popular methods for processing focus on red cell depletion (including the use of Ficoll solution, Hetastarch, Lymphoprep solution, PrepaCyte solution (BioE Corporation, MN, USA) and centrifugal elutriation). Based primarily on the weight of the red blood cell due to the content of iron molecules, the system assumes that the red cells are not required for later transplants. Indeed, given the very effective blood transfusion banks already in existence for red cell transfusion, there is little need to store these cells currently. However, many of the weight-defined processing systems also lose certain immune cells which could be useful, not least the neutrophillic populations and progenitors. An alternative system is known as plasma depletion, which as the name suggests, reduces the volume of the cord blood to a size more manageable for storage, helping cell loss to a minimum. This system was heavily criticised originally, but a public cord blood organisation in the US, known as StemCyte Incorporated, championed the use of transplants with such processing, showing that earlier and stronger engraftment of the product could be achieved when used for haematological bone marrow recovery.[37,38] Now, with well over 500 such successful transplants, critics of this system of processing are having to re-evaluate their opposition, since it clearly works.

Bioprocessing in itself, means nothing without proper evaluation of what is in the final product. One cannot assume that all cord bloods are the same.[39,40]

(ii) Contamination

The variability in quality, volume, cellularity, stem cell content and infection rate, is one of the main barriers to using cord blood at all. Although the birth of a Caesarean child is more controlled, it is not without infection rates of the outer materials of the umbilical cord and these can be transferred to the cord blood itself if not properly processed.[41]

It is quite easy to see how a cord blood can become contaminated from a normal vaginal birth. Any cord blood, therefore, which is intended for storage or transplant, requires analysis for cellularity, infection risk and preferably some level of analysis of the content of the stem cell compartment. The latter analysis is not easy to choose given that the blood stem cell antigen CD34 is only part of the story. Our own research shows that many cells which are CD34-negative are still useful. The CD34 analysis is, however, easy to do and widely accepted. Some private bioprocessing centres do not analyse routinely.[10,42–45]

(iii) Cryopreservation

Current practice is for cryopreservation and storage of cord blood for 20 years. The usefulness of such a long storage period will need to be confirmed in humans even if animal data are available. Currently the oldest cord blood to be thawed and used successfully is 11 years old in a human and 17 years old in an animal model.[46] On the bright side, this year it will be 19 years stored, and next year 20 years. Each passing year the debate that cord blood will not last in the liquid nitrogen freezer becomes less and less relevant. How cord blood is frozen, at what rate and with which cryopreservant is, however, hugely important. Simply taking the cord blood and freezing at −80° Celcius, results in massive cell death. Similarly, storing at −196° Celcius in liquid nitrogen cannot also be achieved without the addition of DMSO. The controlled rate freezing process is central to the quality of the stored cells as it can be set to provide the optimum freezing curve for the cord blood cells and it also produces a record of this process for control and quality monitoring.[47]

Our own published work has shown that the success of thawing out the cord blood will also be affected by the processing methods used. Indeed the use of lymphoprep-based blood processing, whilst interesting for research, yields a low quality product after thawing that it is not recommended.[10,36]

A great deal more cord blood bioprocessing and cryopreservation is required. One such study, funded by the European Commission — called CRYSTAL is underway and is investigating novel methodologies for cryopreservation of stem cells for human applications.

(iv) Storage vessels

Some partially automated storage systems use bag storage and some use vials. Both are generally polypropylene or similar strength plastics designed to withstand extreme temperature changes. In our experience double layering of the vial or bag, is the best way to ensure lack of breakages inside a liquid nitrogen tank. However, the main barrier to infection of the cord blood sample is for it to be totally processed inside a GMP (good manufacturing practice)-regulated clean room, with appropriate certification and staff training.[48] We do not recommend bioprocessing in any other environment, even in an automated system that is manufacturer-designated as a "closed system". Such systems (e.g. the Biosafe SEPAX system) do not guarantee to remove operator's error.[48]

4. Cord Blood Banking

In an ideal world a cord blood bank should be set up in every metropolitan city, with the samples all tissue-typed for HLA specification and each bank linked by computer network. The reality is different and is unlikely to happen for a long time.[49] Cord blood banking is expensive to set up and the health care system still thinks this new way of treating diseases needs to be more firmly established. However, Table 1 shows that an explosion has taken place in the applications of this novel resource: 20 years ago — one type of transplant; ten years ago, a handful; today about 80 clinical interventions.

Western governments have taken some of this on board, and even in the UK, one government-funded bank does exist — the NHS Cord Blood Bank. Whilst this is a step in the right direction, it does not cover even a small fraction of the necessary tissue types of the UK. One interim solution could be for the private cord blood banks to work more closely with the public systems to help co-fund cord blood banking, research and transplants, e.g. Future Health Ltd. (UK), CryoSave (Switzerland) and the Virgin Healthbank Ltd. (UK) are three companies that work with the public sector, albeit to varying degrees. In the European ideal of "Freedom of Choice", both ethically and legally, parents should be given a choice to store cord blood both privately or publicly, or, indeed, not at all, for the resource belongs to the child, under guardianship of the parents. As with many human resources, not all are exploited.

There is a wider issue however, of bad companies operating under the guise of "helping" people. In the relatively short history of cord blood banking, once set up, private cord blood banks eventually reach their operating costs within two to six years depending on how they process (GMP-processed profit attainment being in the later years), and profitability being undeniably lucrative. Several cord blood companies have proven so successful they have even been floated on stock markets, leading to significant investment in the private side of this industry. The public side, however, has been seriously under-funded — on a global scale. It is unfortunate that for a small industry, notoriety has come from some bad blood banks, who advertise on websites that the cord blood can be used to cure all manner of diseases (when it cannot yet) and that it will be properly processed (when it is not always). The only way to stop this is to have governments legislate with regard to this and other potential human tissues useable for cellular therapies. The European Parliament has organised in 2006, under the leadership of Member of Parliament Mikolasik, an Advanced Cell Therapies panel to address this issue and the variability around the European Union, since some countries are better than others at addressing this urgent issue[50] (see also Chapter 3). The UK is one of the most advanced in the world at moving towards better practice, but even so, some companies operate outside the recommendations, even pretending to operate in the heavily legislated UK, when in fact they are in another country.

Public opinion on cord blood and banking varies widely from country to country. Figure 1 shows the number of cord blood units stored in public banks across the world with approximately 276,000 units stored.

It is further estimated that over 135,000 cord blood units in Europe and 500,000 cord blood units in the US are stored in private family cord blood banks. Korea is perhaps the most pro-cord blood banking and transplant country in the world with over 15% of live births storing umbilical cord blood. Awareness of cord blood also changes in unusual ways. In Spain, in 2006, the grandchild of King Juan-Carlos had her cord blood stored privately, reigniting the debate on cord blood in that country and increasing storage in Spanish-speaking countries.

In the US it is now law in over 12 states, including New York, that pregnant women are made aware of cord blood banking and their options. Further a recent voted guideline of the American Medical Association said doctors should encourage women willing to donate their infants' cord blood to donate to public banks.[51,52]

In the UK, in 2006, the Royal College of Obstetricians and Gynaecologists produced a report relatively unsupportive against private cord blood banking, but pro-public cord blood banking for bone marrow therapy. Unfortunately, public cord blood banking is not currently financially viable in a free health care system. Given that hospital trusts put themselves at significant legal liability if they refuse to help with cord blood collections and given that the American Medical Association (2007) has come out in favour of support for this, perhaps in the future the Royal Colleges would rewrite the report. Many obstetrics units cannot be fully supportive of the college's recommendation since their primary role is to care for the safety and health of the child and mother. In free health care systems such as the UK, midwifery units are severely understaffed and taking part in cord blood collections is an extra piece of work to consider. What then must be done? In Newcastle, we provide extra staff for this purpose to ensure that the obstetrics unit are not expected to do this extra work. For countrywide cord blood collections this would be an onerous and expensive process. The solution therefore, maybe for obstetric units to work more closely with cord blood banks, so that in the future collections are more straightforward.

5. Cord Blood Research and Where the Future Lies

In 2005, the cord blood centre staged on the stem cell agenda the report of cells possessing embryonic stem cell markers from cord blood[16] (Fig. 2). Our intention was not to rival embryonic stem cells, but rather to point out that the so-called "embryonic antigens" were in fact not so. We believe that these intra- and extra-cellular antigens are in fact markers of pluripotency, indicating potential for development. The question, as with all stem cell types, is, can that potential be controlled? The cells we characterised were also highly expandable and when put into culture grew in a similar way as ES cells. Now we have achieved differentiation of cord blood stem cells into tissues of all three germ layers. Figure 3 shows examples of such differentiation. Being the first in the world to produce hepatic cells and insulin-secreting C-peptide-producing cells from cord blood, we feel that this is a reasonable indication that cord blood has potential beyond simple bone marrow therapy. Clearly, no aspect of stem cell research is easy and much work

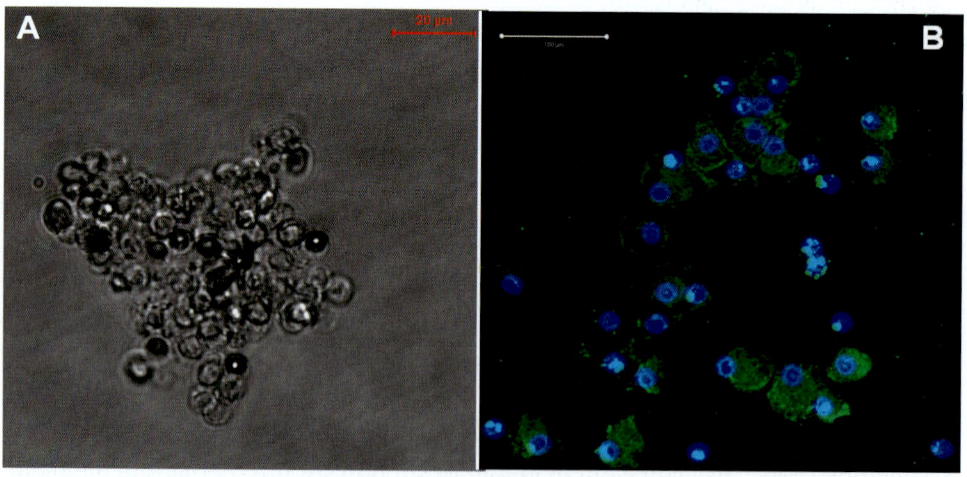

Fig. 2. Cord blood embryonic (CBE)-like stem cells **(A)** showing their typical "cluster"-like morphology upon isolation and expansion. **(B)** is a fluorescent confocal micrograph showing immunoreactivity for the pluripotent transcription factor Oct-4 (in green) expressed either in the cytoplasm or in the nucleus (blue DAPI stain).

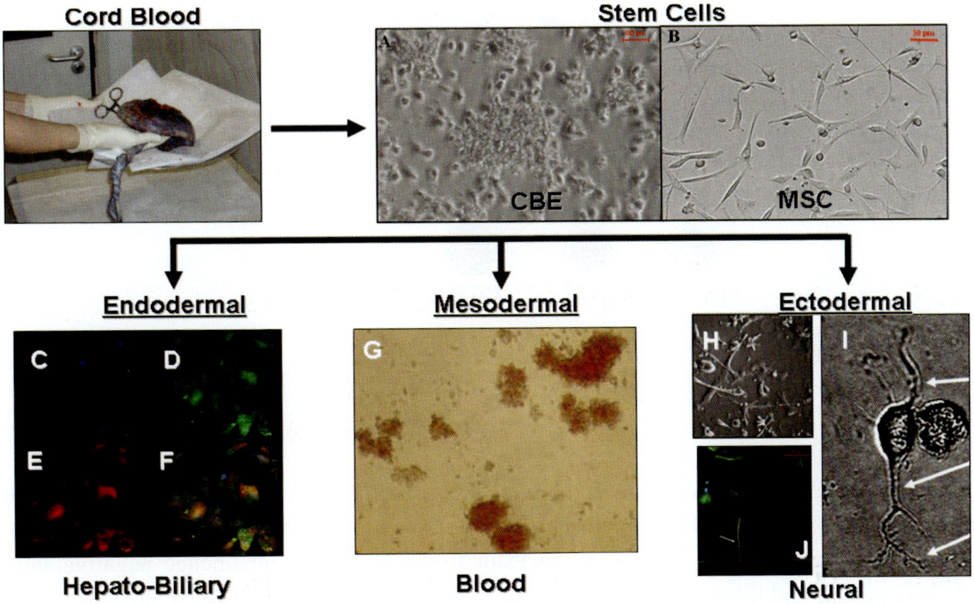

Fig. 3. Pluripotency of cord blood stem cells. Cord blood samples can be processed to derive different stem cell subpopulations **(A)** CBE and/or **(B)** MSC. Cord blood stem cells are pluripotent and can give rise to tissues of all three germs layers: endodermal — hepatobiliary cells **(C)** DAPI nuclear stain of hepatobiliary cells positive for cytokeratin-7 **(D)** and cytokeratin-19 **(E)**, overlay **(F)**; mesodermal — blood colonies; and ectodermal — neural cells **(H)** including bipolar neurons **(I)** with dendrites, arrows expressing GFAP in red and NF-200 in green **(J)**.

remains to be done, but given that certain embryonic researchers are moving away from their technologies to adult stem cell-defined technologies, cord blood is likely to be one of the main sources in regenerative medicine. Our research group has spent a number of years characterising stem and progenitor cell groups in umbilical cord blood samples.[10,17,20] Our experience tells us that different cell populations can be isolated in cord blood as presented in Fig. 4.

Cord blood stem cells are multipotent.[15,18] Our work was also replicated by a leading group in Poland, led by Domanska, Lukomska and colleagues.[53,54] Their work was instrumental in showing that neural progenitors derived from cord blood could produce true neurospheric lines and could be integrated into the brain of animal models with evidence of integration. They were also the first to demonstrate the formation of action potentials in their cells.

Liver is a major organ for drug metabolism and detoxification. The human liver is well known for being capable of major self-regeneration notably by the rapid division of mature cells in response to tissue loss of injury. Resident hepatic

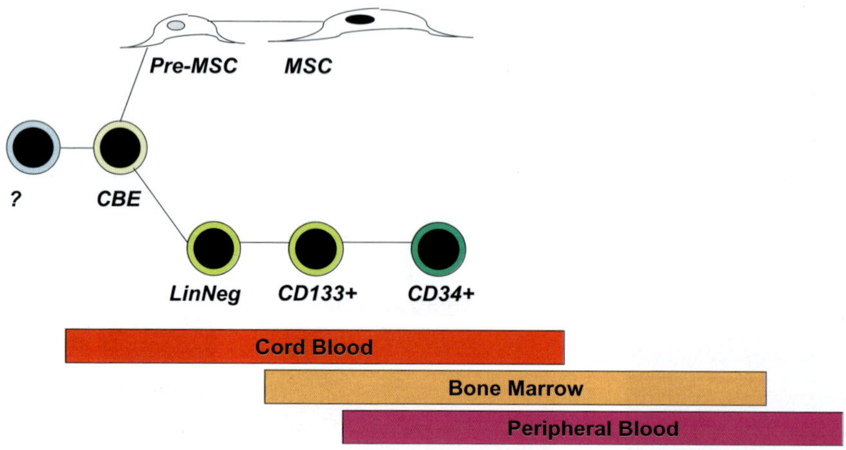

Fig. 4. Stem cell population in cord blood. This simplified model compares the different type of stem cells available from cord blood. We have identified the cord blood embryonic (CBE)-like stem cells as a common ancestor to the non-adherent lineage negative stem cells (LinNeg) and the adherent pre-mesenchymal stem cell (MSC). Bone marrow is a host to MSC and LinNeg stem cells which give rise to CD133+ and CD34+ progenitors, whereas adult peripheral blood has minimal quantities of CD133+ progenitors and minimal amount of CD34+ cells.

stem/progenitor cells (also called oval cells) have been implicated in a small way in liver repair but their role is poorly understood.[55] Liver diseases are significant and can be caused by a range of disorders including excess consumption of drugs or alcohol, medications, and infections. It is estimated that over 500,000 people suffer from hepatitis worldwide. When considering that over 70% of patients awaiting a liver transplant never find a suitable donor, stem cell-based therapies or support appear extremely attractive for these patients (see also Chapter 29).

Our group is working on a large liver tissue engineering programme from cord blood stem cells. Our strategy includes the combination of cord blood stem cells to soluble factors (cytokines and proteins) as well as biomaterials (e.g. collagen IV and other bio-scaffolds) to gradually induce liver tissue differentiation. In collaboration with NASA and its spin-off corporation Synthecon (Houston, Texas, USA), we have been provided with a custom designed unique Rotating Cell Culture System (RCCS) bioreactor developed by NASA, specifically designed for our stem cell requirements. Our work with NASA-engineered RCCS bioreactors has demonstrated that they allow continuous rotation, expansion and endogenous extracellular matrix production in a shear stress-free environment favouring the development of functional 3D tissues. A key finding in this was our ability to expand, maintain and differentiate cord blood embryonic (CBE)-like stem cells into hepatic progenitor in three-dimensions with and without supporting bio-scaffolds. For instance we have been able to differentiate cord blood stem cells into hepatic-biliary tissue constructs expressing tissue specific markers (including cytokeratin-7, -18, -19, alpha-fetoprotein) but also capable of functional activity (albumin production, cytochrome P450's indocyanine green uptake, Periodic Acid Schiff) (Fig. 3).[14,16]

Our research suggests that cord blood stem cells can produce over 20 tissues *in vitro*. In addition to neural and liver tissue engineering we focus our current efforts on cardiovascular, pancreatic and orthopaedic applications.[12–24]

An increasing number of research groups around the world are also developing methodologies to differentiate cord blood stem cells for regenerative medicine. Kang and colleagues in Korea, and Young and colleagues in USA are exploring the potential of cord blood stem cells for spinal cord repair and regeneration.[56] Lazzari and colleagues in Italy are investigating cord blood stem cells as a source of cells for acute renal failure.[57] Stamm and colleagues in Germany are investigating cord blood cells for cardiovascular diseases.[58]

With over 130 million live births every year around the world, umbilical cord blood offers a vast ethnically and genetically diverse source of stem cells. We believe the way forward is to increase the overall inventory of cord blood units stored around the country and worldwide. This will require co-operation of both public and private family banks within a scientifically and ethically sound legal framework to guarantee best practice in collection and bioprocessing.

References

1. Ende, M. and Ende, N. Hematopoietic transplantation by means of fetal (cord) blood. A new method. *Va Med. Mon.* **99**, 276–280 (1972).
2. Halbrecht, J. Fresh and stored placental blood. *Lancet* **2**, 1263–1271 (1939).
3. Knudtzon, S. *In vitro* growth of granulocytic colonies from circulating cells in human cord blood. *Blood* **43**, 357–361 (1974).
4. Fauser, A.A. and Messner, H.A. Granuloerythropoietic colonies in human bone marrow, peripheral blood, and cord blood. *Blood* **52**, 1243–1248 (1978).
5. Broxmeyer, H.E., Douglas, G.W. and Hangoc, G. Human umbilical cord blood as a potential source of transplantable hematopoietic stem/progenitor cells. *Proc. Natl. Acad. Sci. USA* **86**, 3828–3832 (1989).
6. Gluckman, E., Broxmeyer, H.A., Auerbach, A.D. *et al.* Hematopoietic reconstitution in a patient with Fanconi's anemia by means of umbilical-cord blood from an HLA-identical sibling. *N. Engl. J. Med.* **321**, 1174–1178 (1989).

7. Körbling, M., Katz, R.L., Khanna, A., Ruifrok, A.C., Rondon, G., Albitar, M., Champlin, R.E. and Estrov, Z. Hepatocytes and epithelial cells of donor origin in recipients of peripheral-blood stem cells. *N. Engl. J. Med.* **346,** 738–746 (2002).
8. Jang, Y.Y. and Sharkis, S.J. Stem cell plasticity: a rare cell, not a rare event. *Stem Cell Rev.* **1**, 45–51 (2005).
9. Wagers, A.J., Sherwood, R.I., Christensen, J.L. and Weissman, I.L. Little evidence for developmental plasticity of adult hematopoietic stem cells. *Science* **297**, 2256–2259 (2002).
10. McGuckin, C.P., Basford, C., Hanger, K., Habibollah, S. and Forraz, N. Cord blood revelations — the importance of being a first born girl, big, on time and to a young mother! *Early Hum. Dev.* **83**, 733–741 (2007).
11. Strauer, B.E., Brehm, M., Zeus, T., Köstering, M., Hernandez, A., Sorg, R.V., Kögler, G. and Wernet, P. Repair of infarcted myocardium by autologous intracoronary mononuclear bone marrow cell transplantation in humans. *Circulation* **106**, 1913–1918 (2002).
12. Stamm, C., Westphal, B., Kleine, H.D., Petzsch, M., Kittner, C., Klinge, H., Schümichen, C., Nienaber, C.A., Freund, M. and Steinhoff, G. Autologous bone-marrow stem-cell transplantation for myocardial regeneration. *Lancet* **361**, 45–46 (2003).
13. Denner, L., Bodenburg, Y., Zhao, J.G., Howe, M., Cappo, J., Tilton, R.G., Copland, J.A., Forraz, N., McGuckin, C. and Urban, R. Directed engineering of umbilical cord blood stem cells to produce C-peptide and insulin. *Cell Prolif.* **40**, 367–380 (2007).
14. Russomano, T., Cardoso, R., Falcao, F., Dalmarco, G., V Dos Santos, C., Dos Santos, L., G de Azevedo, D., Dos Santos, M., Martinelli, L., Motta, J., Forraz, N. and McGuckin, C. Development and validation of a 3D clinostat for the study of cells during microgravity simulation. *Conf. Proc. IEEE Eng. Med. Biol. Soc.* **1,** 564–566 (2005).
15. McGuckin, C., Forraz, N., Baradez, M.O., Basford, C., Dickinson, A.M., Navran, S. and Hartgerink, J.D. Embryonic-like stem cells from umbilical cord blood and potential for neural modeling. *Acta Neurobiol. Exp.* (*Wars*), **66**, 321–329 (2006).
16. McGuckin, C.P., Forraz, N., Baradez, M.O., Navran, S., Zhao, J., Urban, R., Tilton, R. and Denner, L. Production of stem cells with embryonic characteristics from human umbilical cord blood. *Cell Prolif.* **38**, 245–255 (2005).
17. McGuckin, C.P., Forraz, N., Pettengell, R. and Thompson, A. Thrombopoietin, flt3-ligand and c-kit-ligand modulate HOX gene expression in expanding cord blood CD133 cells. *Cell Prolif.* **37**, 295–306 (2004).

18. McGuckin, C.P., Forraz, N., Allouard, Q. and Pettengell, R. Umbilical cord blood stem cells can expand hematopoietic and neuroglial progenitors *in vitro*. *Exp. Cell Res.* **295**, 350–359 (2004).
19. Forraz, N., Pettengell, R. and McGuckin, C.P. Characterization of a lineage-negative stem-progenitor cell population optimized for *ex vivo* expansion and enriched for LTC-IC. *Stem Cells* **22**, 100–108 (2004).
20. McGuckin, C.P., Pearce, D., Forraz, N., Tooze, J.A., Watt, S.M. and Pettengell, R. Multiparametric analysis of immature cell populations in umbilical cord blood and bone marrow. *Eur. J. Haematol.* **71**, 341–350 (2003).
21. McGuckin, C.P., Forraz, N., Baradez, M.O., Lojo-Rial, C., Wertheim, D., Whiting, K., Watt, S.M. and Pettengell, R. Co-localization analysis of sialomucins CD34 and CD164. *Stem Cells* **21**, 162–170 (2003).
22. McGuckin, C.P., Forraz, N. and Liu, W.M. Diaminofluorene stain detects erythroid differentiation in immature haemopoietic cells treated with EPO, IL-3, SCF, TGFbeta1, MIP-1alpha and IFNgamma. *Eur. J. Haematol.* **70**, 106–114 (2003).
23. Forraz, N., Pettengell, R. and McGuckin, C.P. Haemopoietic and neuroglial progenitors are promoted during cord blood *ex vivo* expansion. *Br. J. Haematol.* **119**, 888 (2002).
24. Forraz, N., Pettengell, R., Deglesne, P.A. and McGuckin, C.P. AC133+ umbilical cord blood progenitors demonstrate rapid self-renewal and low apoptosis. *Br. J. Haematol.* **119**, 516–524 (2002).
25. Fasouliotis, S.J. and Schenker, J.G. Human umbilical cord blood banking and transplantation: a state of the art. *Eur. J. Obstet. Gynecol. Reprod. Biol.* **90**, 13–25 (2005).
26. Grewal, S.S., Barker, J.N., Davies, S.M. and Wagner, J.E. Unrelated donor hematopoietic cell transplantation: marrow or umbilical cord blood? *Blood* **101**, 4233–4244 (2003).
27. Ooi, J., Iseki, T., Takahashi, S., Tomonari, A., Takasugi, K., Shimohakamada, Y., Yamada, T., Ishii, K., Ohno, N., Nagamura, F., Uchimaru, K., Tojo, A. and Asano, S. Unrelated cord blood transplantation for adult patients with *de novo* acute myeloid leukemia. *Blood* **103**, 489–491 (2004).
28. Brunstein, C.G. and Wagner, J.E. Cord blood transplantation for adults. *Vox Sang.* **91**, 195–205 (2006).
29. Majhail, N.S., Brunstein, C.G. and Wagner, J.E. Double umbilical cord blood transplantation. *Curr. Opin. Immunol.* **18**, 571–575 (2006).
30. Gordon, M.Y., Levicar, N., Pai, M., Bachellier, P., Dimarakis, I., Al-Allaf, F., M'Hamdi, H., Thalji, T., Welsh, J.P., Marley, S.B., Davies, J., Dazzi, F., Marelli-Berg, F., Tait, P., Playford, R., Jiao, L., Jensen, S., Nicholls, J.P.,

Ayav, A., Nohandani, M., Farzaneh, F., Gaken, J., Dodge, R., Alison, M., Apperley, J.F., Lechler, R. and Habib, N.A. Characterization and clinical application of human CD34+ stem/progenitor cell populations mobilized into the blood by granulocyte colony-stimulating factor. *Stem Cells* **24**, 1822–1830 (2006).

31. Escolar, M.L., Poe, M.D., Provenzale, J.M., Richards, K.C., Allison, J., Wood, S., Wenger, D.A., Pietryga, D., Wall, D., Champagne, M., Morse, R., Krivit, W. and Kurtzberg, J. Transplantation of umbilical-cord blood in babies with infantile Krabbe's disease. *N. Engl. J. Med.* **352**, 2069–2081 (2005).
32. Slatter, M.A. and Gennery, A.R. Umbilical cord stem cell transplantation for primary immunodeficiencies. *Expert Opin. Biol. Ther.* **6**, 555–565 (2006).
33. Talvensaari, K., Clave, E., Douay, C., Rabian, C., Garderet, L., Busson, M., Garnier, F., Douek, D., Gluckman, E., Charron, D. and Toubert, A. A broad T-cell repertoire diversity and an efficient thymic function indicate a favorable long-term immune reconstitution after cord blood stem cell transplantation. *Blood* **99**, 1458–1464 (2002).
34. Haller, M., Cooper, S., Putnam, A., Freed, B., Brusko, T., Chase, P., Atkinson, M. and Schatz, D. Autologous cord blood transfusion associated with prolonged honeymoon in a child with type 1 diabetes (T1D). *Diabetes* **54**(Suppl. 1), A485 (2005).
35. Rouas-Freiss, N., Naji, A., Durrbach, A. and Carosella, E.D. Tolerogenic functions of human leukocyte antigen G: from pregnancy to organ and cell transplantation. *Transplantation* **84**, S21–S25 (2007).
36. Jones, J., Stevens, C.E., Rubinstein, P., Robertazzi, R.R., Kerr, A. and Cabbad, M.F. Obstetric predictors of placental/umbilical cord blood volume for transplantation. *Am. J. Obstet. Gynecol.* **188**, 503–509 (2003).
37. Chow, R., Nademanee, A., Rosenthal, J., Karanes, C., Jaing, T.H., Graham, M.L., Tsukahara, E., Wang, B., Gjertson, D., Tan, P., Forman, S. and Petz, L.D. Analysis of hematopoietic cell transplants using plasma-depleted cord blood products that are not red blood cell reduced. *Biol. Blood Marrow Transplant.* **13**, 1346–1357 (2007).
38. Rosenthal, J., Jaing, T., Chan, L., Eames, G., Graham, M., Tan, A., Lin, H., Nademanee, A., Karanes, C. and Wang, B. Unrelated cord blood transplantation (UCBT) for 59 pediatric patients with benign indications using plasma depleted cord blood (PD CB) — an audited retrospective analysis. *Biol. Blood Marrow Transplant.* **13**, 7–17 (2007).
39. Dimitriou, H., Perdikogianni, C., Stiakaki, E., Vorgia, P., Hatzidaki, E. and Kalmanti, M. The impact of mode of delivery and gestational age on cord blood hematopoietic stem/progenitor cells. *Ann. Hematol.* **85**, 381–385 (2006).

40. Tommiska, V., Heinonen, K., Ikonen, S., Kero, P., Pokela, M.L., Renlund, M., Virtanen, M. and Fellman, V. A national short-term follow-up study of extremely low birth weight infants born in Finland in 1996–1997. *Pediatrics* **107**, E2 (2001).
41. Donaldson, C., Armitage, W.J., Laundy, V., Barron, C., Buchanan, R., Webster, J., Bradley, B. and Hows, J. Impact of obstetric factors on cord blood donation for transplantation. *Br. J. Haematol.* **106**, 128–132 (1999).
42. Ballen, K.K., Wilson, M., Wuu, J., Ceredona, A.M., Hsieh, C., Stewart, F.M., Popovsky, M.A. and Quesenberry, P.J. Bigger is better: maternal and neonatal predictors of hematopoietic potential of umbilical cord blood units. *Bone Marrow Transplant.* **27**, 7–14 (2001).
43. Mohyeddin Bonab, M.A., Alimoghaddam, K.A., Goliaei, Z.A. and Ghavamzadeh, A.R. Which factors can affect cord blood variables? *Transfusion* **44**, 690–693 (2004).
44. Opie, T.M., Shields, L.E. and Andrews, R.G. Cell-surface antigen expression in early and term gestation fetal hematopoietic progenitor cells. *Stem Cells* **16**, 343–348 (1998).
45. Nakagawa, R., Watanabe, T., Kawano, Y., Kanai, S., Suzuya, H., Kaneko, M., Watanabe, H., Okamoto, Y., Kuroda, Y. and Nakayama, T. Analysis of maternal and neonatal factors that influence the nucleated and CD34+ cell yield for cord blood banking. *Transfusion* **44**, 262–267 (2004).
46. Broxmeyer, H.E., Srour, E., Orschell, C., Ingram, D.A., Cooper, S., Plett, P.A., Mead, L.E. and Yoder, M.C. Cord blood stem and progenitor cells. *Meth. Enzymol.* **419**, 439–473 (2006).
47. Hunt, C.J., Armitage, S.E. and Pegg, D.E. Cryopreservation of umbilical cord blood: tolerance of CD34(+) cells to multimolar dimethyl sulphoxide and the effect of cooling rate on recovery after freezing and thawing. *Cryobiology* **46**, 76–87 (2003).
48. Chivu Mihaela, S.N., Camelia, S., Bleotu, C., Alexiu, I. and Hudita, D. Optimizing donor selection in order to establish a cord blood banking facility: maternal and obstetric factors impact. *Central Eur. J. Med.* **2**, 180–189 (2007).
49. Katz-Benichou, G. Umbilical cord blood banking: economic and therapeutic challenges. *Int. J. Healthcare Technol. Manag.* **8**, 464–477 (2007).
50. Proposal for Regulation of the European Parliament and of the Council on Advanced Therapy Medicinal Products {SEC(2005) 1444}.
51. AMA press release on *Guidelines for Cord Blood Banking* (Associated Press, 16 November 2007).
52. Umbilical Cord Blood Bill into law by Governor Eliot Spitzer, State of New York Chapter 427, 2007.

53. Jurga, M., Buzańska, L., Małecki, M., Habich, A. and Domańska-Janik, K. Function of ID1 protein in human cord blood-derived neural stem-like cells. *J. Neurosci. Res.* **84**, 993–1002 (2006).
54. Kozłowska, H., Jabłonka, J., Janowski, M., Jurga, M., Kossut, M. and Domańska-Janik, K. Transplantation of a novel human cord blood-derived neural-like stem cell line in a rat model of cortical infarct. *Stem Cells Dev.* **16**, 481–488 (2007).
55. Cantz, T., Manns, M.P. and Ott, M. Stem cells in liver regeneration and therapy. *Cell Tissue Res.* **331**, 271–282 (2008).
56. Lim, J.H., Byeon, Y.E., Ryu, H.H., Jeong, Y.H., Lee, Y.W., Kim, W.H., Kang, K.S. and Kweon, O.K. Transplantation of canine umbilical cord blood-derived mesenchymal stem cells in experimentally induced spinal cord injured dogs. *J. Vet. Sci.* **8**, 275–282 (2007).
57. Lo Cicero, V., Montemurro, T., Morigi, M., Corna, D., Lecchi, L., Giordano, R., Rebulla, P., Remuzzi, G. and Lazzari, L. Cord blood mesenchymal stem cells for acute renal failure repair. *Blood* **108**(11), 87A (2006).
58. Ma, N., Ladilov, Y., Kaminski, A., Piechaczek, C., Choi, Y.H., Li, W., Steinhoff, G. and Stamm, C. Umbilical cord blood cell transplantation for myocardial regeneration. *Transplant Proc.* **38**, 771–773 (2006).

Chapter 8

Fat Stem Cells

Jeffrey M. Gimble, Bruce A. Bunnell and Farshid Guilak

Abstract

Adipose tissue is an abundant, accessible, and replenishable source of adult stem cells. Adipose-derived stem cells (ASCs) can be isolated from human lipoaspirates by collagenase digestion, differential centrifugation, and plastic adherence. The ASCs display a consistent immunophenotype that is similar, but not identical, to that of bone marrow-derived mesenchymal stem cells (BMSCs). Like BMSCs, ASCs inhibit mixed lymphocyte reactions *in vitro*, suggesting that it will be possible to perform allogeneic transplants in clinical settings. The expanded ASCs are multipotent and can differentiate into adipocytes, chondrocytes, endothelial cells, myocytes, neuronal-like cells, and osteoblasts, among other lineages. Furthermore, genetic engineering methods can be applied to ASCs, allowing their potential use as gene delivery vehicles *in vivo*. This chapter reviews the expanding literature base relating to ASC applications for regenerative medicine.

Keywords: Adipose-Derived Stem Cells (ASCs); Cytokine Profile; Differentiation; Immunophenotype; Immunogenicity.

Outline

1. Introduction
2. Types of Adipose Tissue
3. Isolation Procedures
4. Immunophenotype and Cytokine Profile of ASCs
5. Immunogenicity of ASCs
6. Differentiation Potential
   6.1. Adipocyte
   6.2. Cardiac myocytes
   6.3. Chondrocyte
   6.4. Endodermal and ectodermal lineages

6.5. Endothelial and smooth muscle cells
6.6. Hematopoietic support
6.7. Neuronal
6.8. Osteoblast
6.9. Skeletal myocyte
7. Mechanisms of Potential Utility: Genetic Engineering and Gene Delivery
8. Conclusions and Future Directions
References

1. Introduction

Adipose tissue is an abundant and accessible source of adult or somatic stem cells termed "adipose-derived stem cells" (ASCs). There has been increased interest in ASCs for tissue engineering applications and the subject has been reviewed by multiple authors.[1–5] This chapter will present a concise synopsis of the recent literature relating to ASCs and will end with a discussion of the questions and challenges to be addressed in the future.

2. Types of Adipose Tissue

Adipose tissue is present in multiple depots throughout the body (Table 1). The relative abundance of each can change as a function of age, gender, nutritional, and pathological status of the individual. Bone marrow adipose tissue is not present in the newborn state but, with advancing age, presents itself first in the distal skeleton before appearing throughout the marrow cavity.[6,7] In contrast, brown adipose tissue (BAT) is most abundant in the newborn human, where it serves as a thermogenic organ protecting the major vessels and vital organs. With advancing age, BAT disappears at these sites, where it is replaced by white adipose tissue (WAT). In conditions where energy intake exceeds utilization, WAT accumulates and results in obesity when its levels are disproportionate to the lean body mass. While this was once thought to be an age-dependent pathology, the epidemic rise in childhood obesity worldwide indicates that it can occur throughout one's lifetime. Mammary adipose tissue is relatively gender specific, appearing in post-pubertal women and often increasing following parturition and lactation. Mechanical adipose depots, found in the soles of the feet and palms of the hand as well as other sites of frequent physical impact, serve a protective function and tend to be spared during periods of starvation. Finally, under pathologic conditions, adipose tissue appears in ectopic depots. Patients with metabolic syndrome store excess lipids within their liver, leading to a condition known as hepatosteatosis;

Table 1. Adipose tissue depots.

Type	Function
Bone marrow	Occupy space no longer required for hematopoiesis; provide cytokines and growth factors required for hematopoiesis and osteogenesis.
Brown adipose tissue (BAT)	Thermogenesis; protection of vital organs in the newborn state.
Ectopic adipose tissue	Fat accumulation occurring abnormally within the liver, skeletal, or cardiac muscles due to improper energy utilization and storage under conditions such as the metabolic syndrome.
Mammary	Energy and nutrient source during lactation.
Mechanical	Cushion for protection of joints (knee), extremities (palms, soles), and vital organs (eye) from mechanical trauma.
White adipose tissue (WAT)	Insulation; energy storage and reservoir; endocrine organ for the production of adipokines with systemic influences.

similar changes can occur in skeletal muscle. Each of these depots can serve as a source of ASCs or equivalent stromal cells. For example, bone marrow has proven to be an abundant source of mesenchymal stem cells (MSCs) or stromal cells with multipotent capabilities as reviewed elsewhere in this volume. However, the depots can differ; direct comparison of murine BAT and WAT determined that the plasticity of BAT derived ASCs was reduced relative WAT.[8]

3. Isolation Procedures

The procedures for the isolation of ASCs were first pioneered by Rodbell and colleagues using rat epididymal fat.[9–12] These methods were later adapted and refined for human tissues by multiple groups.[13–15] The method relies on a combination of physical disruption of the tissue, collagenase digestion, differential centrifugation, and final selection of the ASCs based on their plastic adherence (Table 2 and Fig. 1). This final step enriches the ASCs from the heterogeneous stromal vascular fraction (SVF) cells by ~30-fold.[16] The cell yield per unit volume of tissue is reproducible between laboratories and indicates that a litre of lipoaspirate could yield up to 250 million ASCs within a week of culture[16–18] (Table 3). It should be noted that the surgical approach to liposuction can impact the cell recovery; yields using ultrasound-assisted liposuction are lower than those of tumescent liposuction.[18]

Table 2. Adherence-based isolation of ASCs.[175]

1.	Obtain specimen as intact tissue or lipoaspirate.
2.	Wash minced tissue in PBS warmed to 37°C.
3.	Incubate tissue in PBS containing 0.1% collagenase (type I) and 1% bovine serum albumin for up to 1 hr at 37°C with gentle rocking/agitation.
4.	Centrifuge the tissue digest for 5 min at 300 × g at room temperature (× 2).
5.	Aspirate the supernatant containing the floating mature adipocytes and the collagenase digestion solution.
6.	Retain the pellet containing the stromal vascular fraction (SVF) cells.
7.	Plate the SVF cells at a density corresponding to 0.16 ml of tissue per cm^2 in DMEM/F12 medium containing 10% fetal bovine serum and 1% antibiotic/antimycotic.
8.	After an overnight incubation to promote adherence, wash the flask with PBS warmed to 37°C and add fresh DMEM/F12 medium containing 10% fetal bovine serum and 1% antibiotic/antimycotic.
9.	Feed cells every 2nd to 3rd day until they achieve >80%.
10.	Trypsin digest and passage at 5×10^3 cells per cm^2.

The ASCs have been further isolated by limit dilution[19] and ring-cloning methodologies.[20] These analyses have determined that the progeny of a single ASC are capable of multiple differentiation lineages, consistent with the definition of a stem cell. Recently, a number of groups have begun to exploit antibody-based selection methods to isolate ASCs, using both flow cytometry and magnetic immunobead approaches effectively.[21,22] While antibody-based methods have the potential to select ASCs more precisely, it remains to be determined if they can be adapted for the high throughput and high volume necessary for most tissue engineering applications.

4. Immunophenotype and Cytokine Profile of ASCs

Multiple laboratories have used flow cytometry and immunohistochemical methods to characterize the surface protein expression profile of human ASCs (Table 4 and Fig. 2).[16,17,19,23,24] Likewise, investigators have explored the cytokine expression profile of undifferentiated, differentiated, and endotoxin stimulated ASCs (Table 5).[25–29] Overall, the expression profiles of ASC cytokines and surface proteins are similar to those reported for human bone marrow-derived MSCs. Recent studies have begun to utilize proteomic and transcriptomic tools to more fully characterize the ASC phenotype[24,29,30] (Bunnell, manuscript in preparation). It is likely that these unbiased global expression analyses will identify additional proteins relevant to the isolation and manipulation of ASCs for regenerative medicine.

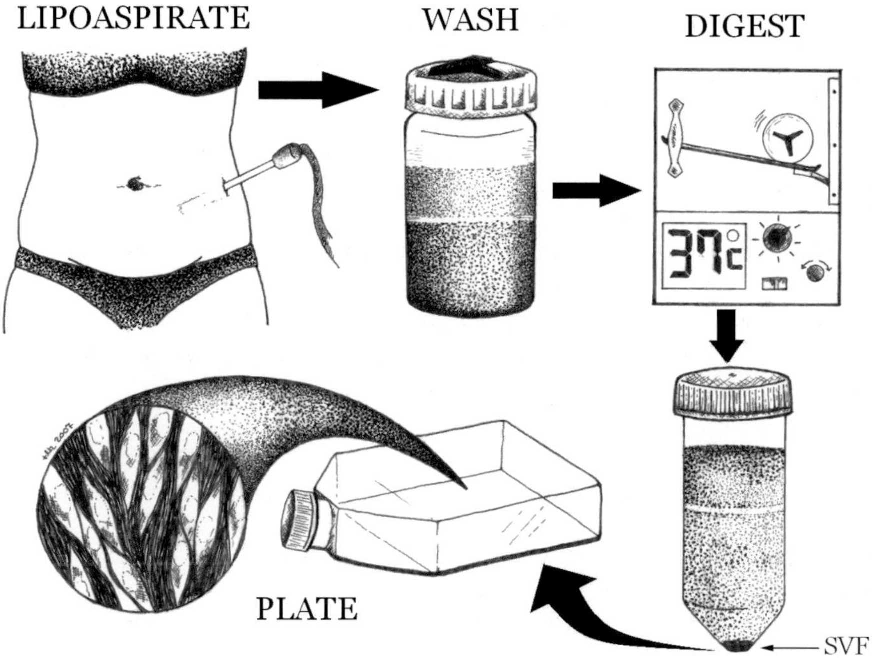

Fig. 1. Isolation and expansion of ASCs. Human lipoaspirates are processed by first washing in phosphate buffered saline, followed by collagenase digestion and differential centrifugation to precipitate the stromal vascular fraction cells. Next, these are plated in the presence of fetal bovine serum (10%), allowed to adhere for 24 hours, washed with phosphate buffered saline, and then culture expanded for analysis and subsequent differentiation along lineage pathways.

Table 3. Adipose stem cell yield.[16]

Culture period	6.0 ± 2.4 days
Cells/ml tissue	247 K ± 136 K
Cells/cm^2	38.4 K ± 21.2 K
Mean age	41 ± 10 years
Mean BMI	26.1 ± 4.8 kg/m^2
Subjects	$n = 44$

5. Immunogenicity of ASCs

A number of laboratories have examined the immunogenicity of ASCs and SVF cells.[31–35] In mixed lymphocyte reactions, the SVF cells elicit a robust proliferative response in allogeneic peripheral blood monocytes; however, with progressive passage, the ASCs fail to do so.[32,33] Moreover, when added to an active mixed

Table 4. Human ASC immunophenotype.[16,17,19,23,24]

Positive Markers	
CD9	Tetraspan
CD10	Common acute lymphocytic leukemia antigen
CD13	Aminopeptidase
CD29	β_1-integrin
CD34	Sialomucin
CD44	Hyaluronate receptor
CD49d	α_4-integrin
CD54	Intracellular adhesion molecule-1
CD55	Decay accelerating factor
CD59	Protectin
CD71	Transferrin
CD73	5′-ectonucleotidase
CD90	Thy1
CD105	Endoglin
CD146	Muc-18
CD166	Activated leukocyte cell adhesion molecule
HLA-ABC	Histocompatability locus antigen — ABC
α-SMA	α-smooth muscle actin
Negative Markers	
CD11b	α_b-integrin
CD14	LPS receptor
CD16	Fc receptor
CD18	β_2-integrin
CD45	Common leukocyte antigen
CD50	Intracellular adhesion molecule-3
CD56	Neural cell adhesion molecule
CD62	E-selectin
CD104	β_4-integrin
HLA-DR	Histocompatability locus antigen — DR

lymphocyte reaction between allogeneic responder and stimulator peripheral blood monocytes, the ASCs are immunosuppressive.[31–33] These properties are retained following osteogenic differentiation of the ASCs, in a manner similar to that observed with bone marrow-derived mesenchymal stem cells.[35,36] The ASCs' lack of immunogenicity has been linked to the absence of the major histocompability class II antigens (HLA-DR) on their surface.[31–33,35] Furthermore, their immunosuppressive properties have been tied to their production of prostaglandin E2 rather than hepatocyte growth factor or transforming growth factor-β.[31] These

A single study has examined the ability of human ASCs to display biochemical characteristics of pancreatic cells.[101] Under serum-free conditions in the presence of activin, extendin, HGF, nicotinamide, and pentagastrin, ASCs expressed mRNAs for the pancreatic hormones glucagon, insulin, and somatostatin as well as the transcription factors isl-1, Ipf-1, Ngn-3, and Pax-6.[101] It remains to be determined if the differentiated ASCs are glucose-responsive with respect to their secretion of insulin and additional analyses, *in vitro* and *in vivo*, will be required to document their utility for tissue engineering applications.

A comparable study has examined the ability of human ASCs to undergo epithelial cell differentiation.[102] In the presence of all-trans retinoic acid, the ASCs underwent morphological changes characteristic of epithelial cells and expressed the lineage specific marker, cytokeratin-18.[102] Human ASCs have been used clinically to promote epithelial tissue repair. In a Spanish Phase I clinical trial, autologous human ASCs were used to treat chronic fistulas in Crohn's disease patients.[103,104] In 75% of the cases, the fistulas closed and were covered with new epithelial tissue.[104] These preliminary studies warrant further investigation.

6.5. Endothelial and smooth muscle cells

Adipose tissue was identified as a source of endothelial progenitors even before it was appreciated as a source of multipotent stem cells. In the late 1980s and early 1990s, Williams and colleagues demonstrated that adipose-derived endothelial cells could be seeded onto vascular grafts to improve arterial function.[105,106] More recently, three independent groups published nearly simultaneous papers documenting the ability of ASCs to express angiogenic cytokines and to improve vascular recovery following peripheral arterial ischemic injuries in rodent models.[26,27,107] This has been associated with conversion of the ASC from a CD31-negative to CD31-positive surface profile, likely due to the influence of VEGF.[107–109] Similarly, the ASCs are known to express markers consistent with a smooth muscle phenotype, such as calponin and α-smooth muscle actin.[110,111] Together, these results indicate that ASCs have potential value for regenerative medicine approaches to vascular and urogenital trauma.

6.6. Hematopoietic support

The immunophenotype of human ASCs includes many of the cell adhesion molecules found on bone marrow MSCs that have proved to be responsible for their hematopoietic support function.[16,19,23,24] These include CD9, CD29, and CD44; antibodies to each of these proteins interferes with hematopoietic support in murine hematopoietic co-culture models.[112–115] In addition, the ASCs express and secrete cytokines associated with hematopoietic support, including M-CSF, GM-CSF, TNFα, IL-6, IL-7, IL-8, IL-11, and stem cell factor.[116,117] *In vitro*, ASCs support the differentiation of $CD34^+$ hematopoietic progenitor cells. Co-cultures established with a human ASC feeder layer support the expansion of early B-cell, myeloid, natural killer cell, and T-cell populations.[116,117] Indeed, the unfractionated SVF cells from adipose tissue have been used to reconstitute the hematopoietic population of lethally irradiated mice, suggesting that adipose tissue contains hematopoietic stem cells.[118] Ongoing studies have demonstrated some benefit to providing bone marrow-derived MSCs in combination with hematopoietic stem cells to aid engraftment.[119–121] These results suggest that human ASCs transplantation might also accelerate recovery in patients requiring hematopoietic stem cell reconstitution following high-dose chemotherapy.[117]

6.7. Neuronal

Human, non-human primate, and murine ASCs display biochemical and morphological characteristics of neuronal- and oligodendrocyte-like cells when cultured in the presence of antioxidants and the absence of serum.[19,122–126] Similar induction has been obtained in response to indomethacin, insulin, and methylisobutylxanthine.[127,128] The ASCs express the neuronal-associated markers nestin, NeuN, and intermediate filament M.[19,122–124,127] Likewise, they also express the oligodendrocyte marker, glial fibrillary acidic protein (GFAP).[122,124] Further studies with murine ASCs have detected the neuronal-associated glutamate receptor subunits NR1 and NR2, MAP2, S-100, and β-III tubulin; however, the dopaminergic markers, dopa decarboxylase and tyrosine hydroxylase, were not present.[124] The differentiated murine ASCs exhibited reduced viability in the presence of N-methyl D-aspartate (NMDA), consistent with a neuronal phenotype.[124] To date, no studies have demonstrated that ASCs exhibit the electrophysiological profile of a mature neuronal cell.[129] This area of research merits further investigation since studies of bone marrow MSCs suggest that the expression of neuronal biochemical features alone is not sufficient evidence to document true differentiation.[130,131] Some studies suggest that MSCs express nestin and other markers in response to cytoskeletal disruption and extreme stress.[130,131]

In vivo, ASCs display a beneficial effect in central nervous system injury models. In a rat spinal cord injury model, intravenous injection of ASCs improved recovery of motor function.[132] The ASCs migrated to the injury site and displayed biochemical markers of both neuronal and oligodendroglial cells.[132] Intraventricular delivery of human ASCs exerted a similar positive effect on the recovery of rats from a middle cerebral artery occlusion (stroke) model.[133] Multiple tracking techniques have demonstrated that the ASCs migrate to the ischemic injury site within the brain.[134,135] The mechanisms underlying the ASC actions may involve the release of growth factors. *In vitro* co-culture models provide complementary findings. Human ASCs promoted the growth and differentiation of isolated murine neural stem cells in a cell-contact dependent manner[136] while murine 3T3-L1 adipocytes supported neurite outgrowths through the release of angiopoietin 1.[137] Thus, paracrine factors released by ASCs or stimulated by their presence may account for the beneficial effects of ASC transplants on central nervous system injury models.

6.8. Osteoblast

Human ASCs display osteogenic characteristics *in vitro* following exposure to ascorbate, dexamethasone, 1,25-dihydroxy vitamin D_3, and β-glycerophosphate.[19,38,138–140] This is associated with the generation of a calcified extracellular matrix, as demonstrated by the presence of calcium phosphate by alizarin red or von Kossa staining and the expression of osteoblast-associated mRNAs.[19,38,138–140] Further studies have been undertaken to optimize the osteogenic cocktail. Using murine ASCs, these have found that the substitution of retinoic acid for dexamethasone exerts maximal osteogenic effects.[141] Additionally, the presence of growth factors such as BMP-2[84] and growth differentiation factor-5 (GDF-5)[142] can further promote osteogenesis by goat and rat ASCs while bFGF reversibly inhibits osteogenesis in murine ASCs.[43] Likewise, the histone deacetylase inhibitors, valproic acid and trichostatin, both exerted osteogenic effects on both human ASCs and MSCs *in vitro.*[143] This latter finding suggests that epigenetic mechanisms may participate in the control of ASC differentiation. Finally, genetic engineering approaches, such as the introduction of genes encoding BMP-2,[144] BMP-7,[145] Runx2,[146] or telomerase,[147,148] can induce or promote osteogenesis in human or non-human primate ASCs.

In bone marrow of the aging human, osteogenic MSCs routinely undergo adipogenic differentiation in response to physiological and environmental cues.[6,149] In part, this may reflect mechanical as well as biochemical inputs. *In vitro*, lineage commitment between the adipocyte and osteoblast pathways can be manipulated by altering the density of adhesion sites on a plastic surface.[150] Low density conditions favor adipogenesis while high density conditions favor osteogenesis.[150] The Rho GTPase pathway and cytoskeletal tension underlie these observations, since the lineages can be determined by dominant negative and constitutively active forms of the RhoA protein.[150] Consistent with these observations, mechanical and environmental mechanisms can further modulate ASC osteogenesis. *In vitro,* shear stresses generated by pulsating fluid flow induced osteogenesis differentiation in ASCs.[151] In contrast, exposure of murine ASCs to low oxygen tensions (2%) inhibited osteogenic differentiation.[86,152] The seeding of human ASCs into three-dimensional matrices as opposed to two-dimensional conventional cultures enhanced their osteogenic capacity.[153,154] Human ASCs seeded onto osteoinductive hydroxyapatite or tricalcium phosphate scaffolds formed new bone six weeks after *in vivo* implantation into immunodeficient mice or rats;[155–157] in at least one study, osteogenic pre-induction of the cells prior to implantation enhanced new bone formation.[157] The adipose tissue depot may determine the osteogenic features of the ASCs; in rabbits, ASCs from visceral, as opposed to subcutaneous, adipose tissue exhibited greater osteogenic potential *in vitro.*[158] Comparison of human ASCs to bone marrow MSCs have led some to conclude that their osteogenic capabilities are similar[89,159] while others argue that ASCs are inferior to MSCs.[91,94] Nevertheless, others have pursued studies on the use of ASCs to repair calvarial[160–163] or palatal[163] bone defects in mice and rabbits. While the murine ASCs form new bone within a two- to four-week period,[160] osteoclasts remodel and resorb the bone by 12 weeks following implantation.[161] This suggests that further optimization of ASC osteogenic induction may be required. In humans, a single case has been reported where autologous ASCs, in combination with a bone graft from the patient's iliac crest, were used to repair a critical-sized calvarial defect in a seven-year-old.[164] Obviously, further preclinical and clinical studies will be necessary to validate safety and efficacy before these methods can be considered routine clinical practice.

6.9. Skeletal myocyte

Human ASCs can display biochemical features consistent with skeletal myocytes when maintained under classical myogenic culture conditions. In the presence of horse serum, under low fetal bovine serum conditions, and/or in the presence of the murine myocyte cell line, C2C12, ASCs form multinucleated myotubes.[19,38,152,165] Furthermore, the ASCs express mRNA and/or proteins for the myogenic transcription factors, myoD and myogenin, as well as myosin light chain kinase.[19,38,165] While the mechanisms regulating lineage determination remain to be determined, several candidate pathways exist. The introduction of the myoD transcription factor is known to transdifferentiate pre-adipoctyes into myoblasts,[166] just as PPARγ and C/EBPα can convert myoblasts into adipocytes.[167] A similar myogenic outcome can be achieved by the fusion of skeletal myoblasts with pre-adipocytes.[168] The Rho GTPase family may regulate selection of the adipogenic and myogenic pathway.[169] Cells isolated from mice

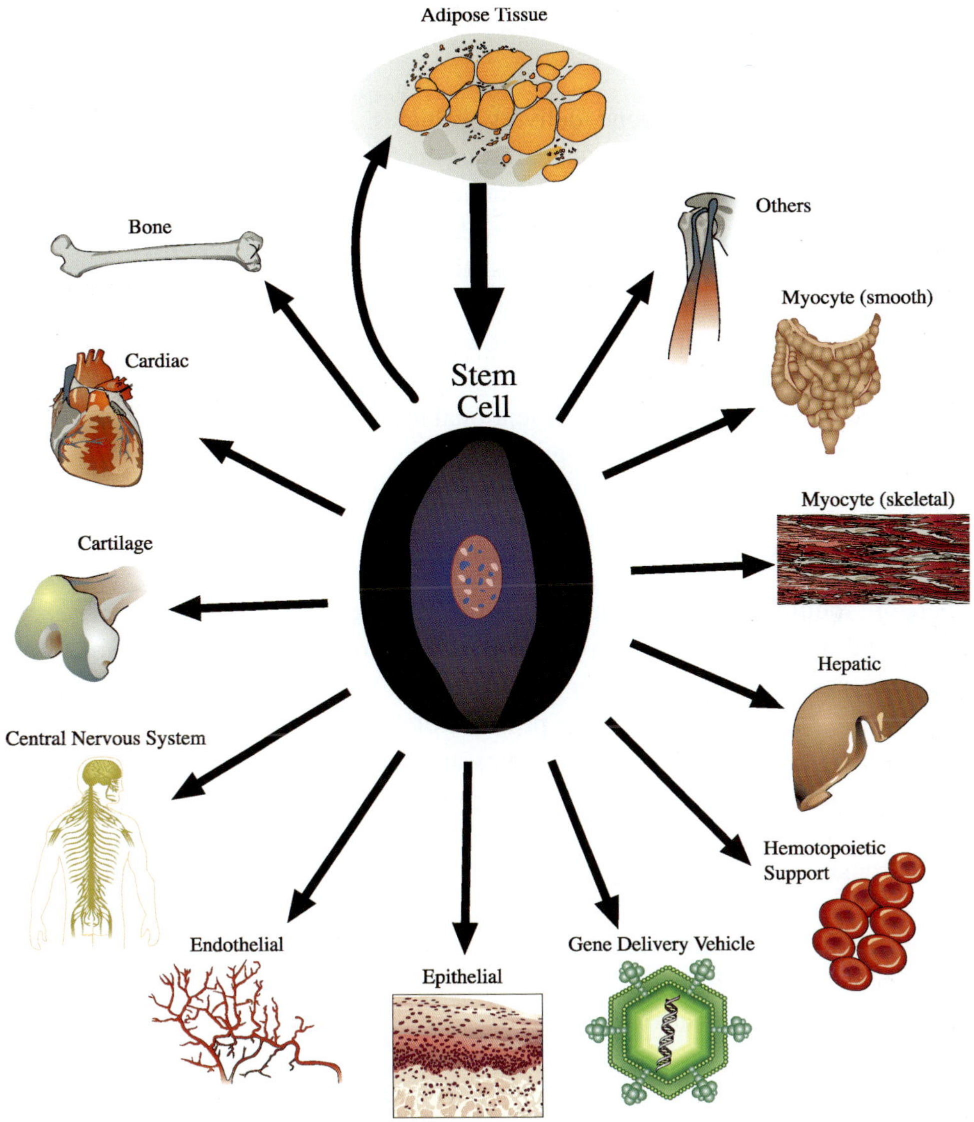

Fig. 3. Differentiation potential of ASCs and their future clinical applications.

lacking the regulatory protein of the Rho GTPase fail to undergo adipogenesis and display enhanced myogenesis.[169] This defect interferes with the cellular response to insulin-like growth factor (IGF-1) and exposure to this agent induces the myogenic response.[169] Thus, as with osteoblast differentiation, the Rho signal transduction pathway may prove to be a target for manipulation of ASC myogenic commitment.

7. Mechanisms of Potential Utility: Genetic Engineering and Gene Delivery

The ASCs can serve as gene delivery vehicles. Adenoviral, herpes simplex, lentiviral, and retroviral vectors can all transduce ASCs.[144,147,148,170,171] To date, the most efficient transduction has been obtained using lentiviral vectors.[171] Since the ASCs are manipulated *in vitro*, it is possible to use high titers of virus while minimizing the risk to patients. Nevertheless, alternative gene delivery systems, using nucleofection and plasmid DNA constructs, merit evaluation and development. Such approaches using both bone marrow-derived MSCs[172] and human ASCs[173] are gaining momentum. Already, luciferase reporter genes have been successfully introduced into human ASCs by conventional transfection methods and tracked *in vivo* using non-invasive monitoring.[174]

8. Conclusions and Future Directions

There are multiple potential sources of adult stem cells and all merit further evaluation and development; however, the advantage of adipose tissue relates to its relatively easy access, abundance, and presence throughout the lifespan of the individual. Moreover, there is the promise of "off-the-shelf" use of allogeneic ASCs for transplantation. Further studies are needed to document the safety of ASCs in animal models before moving to clinical studies. Specifically, studies examining the tumor potential of ASCs at various passages of expansion need to be rigorously conducted in appropriate transplant models. These should focus on both ASCs alone and in combination with various scaffolds and growth factors. Likewise, various modes of delivery should be assessed. It is possible that ASCs will behave differently when administered intravenously, intraperitoneally, or in the context of a solid matrix.

References

1. Gimble, J. and Guilak, F. Adipose-derived adult stem cells: isolation, characterization, and differentiation potential. *Cytotherapy* **5**(5), 362–369 (2003).
2. Fraser, J.K., Schreiber, R., Strem, B., Zhu, M., Alfonso, Z., Wulur, I. and Hedrick, M.H. Plasticity of human adipose stem cells toward endothelial cells and cardiomyocytes. *Nat. Clin. Pract. Cardiovasc. Med.* **3**(Suppl. 1), S33–S37 (2006).
3. Gimble, J.M., Katz, A.J. and Bunnell, B.A. Adipose-derived stem cells for regenerative medicine. *Circ. Res.* **100**(9), 1249–1260 (2007).

4. Moseley, T.A., Zhu, M. and Hedrick, M.H. Adipose-derived stem and progenitor cells as fillers in plastic and reconstructive surgery. *Plast. Reconstr. Surg.* **118**(Suppl. 3), 121S–128S (2006).
5. Schaffler, A. and Buchler, C. Concise review: adipose tissue-derived stromal cells — basic and clinical implications for novel cell-based therapies. *Stem Cells* **25**(4), 818–827 (2007).
6. Gimble, J.M., Robinson, C.E., Wu, X. and Kelly, K.A. The function of adipocytes in the bone marrow stroma: an update. *Bone* **19**(5), 421–428 (1996).
7. Gimble, J.M., Zvonic, S., Floyd, Z.E., Kassem, M. and Nuttall, M.E. Playing with bone and fat. *J. Cell Biochem.* **98**(2), 251–266 (2006).
8. Prunet-Marcassus, B., Cousin, B., Caton, D., Andre, M., Penicaud, L. and Casteilla, L. From heterogeneity to plasticity in adipose tissues: site-specific differences. *Exp. Cell Res.* **312**(6), 727–736 (2006).
9. Rodbell, M. Metabolism of isolated fat cells. I. Effects of hormones on glucose metabolism and lipolysis. *J. Biol. Chem.* **239**, 375–380 (1964).
10. Rodbell, M. Metabolism of isolated fat cells. II. The similar effects of phospholipase C (*Clostridium perfringens* alpha toxin) and of insulin on glucose and amino acid metabolism. *J. Biol. Chem.* **241**(1), 130–139 (1966).
11. Rodbell, M. The metabolism of isolated fat cells. IV. Regulation of release of protein by lipolytic hormones and insulin. *J. Biol. Chem.* **241**(17), 3909–3917 (1966).
12. Rodbell, M. and Jones, A.B. Metabolism of isolated fat cells. The similar inhibitory action of phospholipase C (*Clostridium perfringens* alpha toxin) and of insulin on lipolysis stimulated by lipolytic hormones and theophylline. *J. Biol. Chem.* **241**(1), 140–142 (1966).
13. Engfeldt, P., Arner, P. and Ostman, J. Influence of adipocyte isolation by collagenase on phosphodiesterase activity and lipolysis in man. *J. Lipid Res.* **21**(4), 443–448 (1980).
14. Deslex, S., Negrel, R., Vannier, C., Etienne, J. and Ailhaud, G. Differentiation of human adipocyte precursors in a chemically defined serum-free medium. *Int. J. Obes.* **11**(1), 19–27 (1987).
15. Hauner, H., Entenmann, G., Wabitsch, M., Gaillard, D., Ailhaud, G., Negrel, R. and Pfeiffer, E.F. Promoting effect of glucocorticoids on the differentiation of human adipocyte precursor cells cultured in a chemically defined medium. *J. Clin. Invest.* **84**(5), 1663–1670 (1989).
16. Mitchell, J.B. McIntosh, K., Zvonic, S., Garrett, S., Floyd, Z.E., Kloster, A., Halvorsen, Y.D., Storms, R.W., Goh, B., Kilroy, G.S., Wu, X. and Gimble, J.M. The immunophenotype of human adipose derived cells: temporal

changes in stromal- and stem cell-associated markers. *Stem Cells* **24**, 376–385 (2006).

17. Aust, L., Devlin, B., Foster, S.J., Halvorsen, Y.D., Hicok, K., du Laney, T., Sen, A., Willingmyre, G.D. and Gimble, J.M. Yield of human adipose-derived adult stem cells from liposuction aspirates. *Cytotherapy* **6**(1), 7–14 (2004).
18. Oedayrajsingh-Varma, M.J., van Ham, S.M., Knippenberg, M., Helder, M.N., Klein-Nulend, J., Schouten, T.E., Ritt, M.J. and van Milligen, F.J. Adipose tissue-derived mesenchymal stem cell yield and growth characteristics are affected by the tissue-harvesting procedure. *Cytotherapy* **8**(2), 166–177 (2006).
19. Zuk, P.A., Zhu, M., Ashjian, P., De Ugarte, D.A., Huang, J.I., Mizuno, H., Alfonso, Z.C., Fraser, J.K., Benhaim, P. and Hedrick, M.H. Human adipose tissue is a source of multipotent stem cells. *Mol. Biol. Cell.* **13**(12), 4279–4295 (2002).
20. Guilak, F., Lott, K.E., Awad, H.A., Cao, Q., Hicok, K.C., Fermor, B. and Gimble, J.M. Clonal analysis of the differentiation potential of human adipose-derived adult stem cells. *J. Cell. Physiol.* **206**, 229–237 (2006).
21. Sengenes, C., Lolmede, K., Zakaroff-Girard, A., Busse, R. and Bouloumie, A. Pre-adipocytes in the human subcutaneous adipose tissue display distinct features from the adult mesenchymal and hematopoietic stem cells. *J. Cell. Physiol.* **205**(1), 114–122 (2005).
22. Zannettino, A.C.W., PS, Khor, F., Itescu, S., Gimble, J.M. and Gronthos, S. Multipotential human adipose-derived stromal stem cells exhibit a perivascular phenotype *in vitro* and *in vivo*. *J. Cell. Physiol.* **214**(2), 413–421 (2008).
23. Gronthos, S., Franklin, D.M., Leddy, H.A., Robey, P.G., Storms, R.W. and Gimble, J.M. Surface protein characterization of human adipose tissue-derived stromal cells. *J. Cell. Physiol.* **189**(1), 54–63 (2001).
24. Katz, A.J., Tholpady, A., Tholpady, S.S., Shang, H. and Ogle, R.C. Cell surface and transcriptional characterization of human adipose-derived adherent stromal (hADAS) cells. *Stem Cells* **23**(3), 412–423 (2005).
25. Halvorsen, Y.D., Bond, A., Sen, A., Franklin, D.M., Lea-Currie, Y.R., Sujkowski, D., Ellis, P.N., Wilkison, W.O. and Gimble, J.M. Thiazolidinediones and glucocorticoids synergistically induce differentiation of human adipose tissue stromal cells: biochemical, cellular, and molecular analysis. *Metabolism* **50**(4), 407–413 (2001).
26. Miranville, A., Heeschen, C., Sengenes, C., Curat, C.A., Busse, R. and Bouloumie, A. Improvement of postnatal neovascularization by human adipose tissue-derived stem cells. *Circulation* **110**(3), 349–355 (2004).

27. Rehman, J., Traktuev, D., Li, J., Merfeld-Clauss, S., Temm-Grove, C.J., Bovenkerk, J.E., Pell, C.L., Johnstone, B.H., Considine, R.V. and March, K.L. Secretion of angiogenic and antiapoptotic factors by human adipose stromal cells. *Circulation* **109**(10), 1292–1298 (2004).
28. Kilroy, G.E., Foster, S., Wu, X., Ruiz, J., Sherwood, S., Heifetz, A., Ludlow, J.W., Stricker, D.M., Potiny, S., Green, P., Halvorsen, Y.D.C., Cheatham, B., Storms, R.W. and Gimble, J.M. Cytokine profile of human adipose-derived stem cells: expression of angiogenic, hematopoietic, and pro-inflammatory factors. *J. Cell. Physiol.* **212**(3), 702–709 (2007).
29. Zvonic, S., Lefevre, M., Kilroy, G., Floyd, Z.E., Delany, J.P., Kheterpal, I., Gravois, A., Dow, R., White, A., Wu, X. and Gimble, J.M. Secretome of primary cultures of human adipose-derived stem cells (ASCs): modulation of serpins by adipogenesis. *Mol. Cell Proteomics* **6**, 18–28 (2007).
30. Delany, J., Floyd, Z.E., Zvonic, S., Smith, A., Gravois, A., Reiners, E., Wu, X., Kilroy, G., Lefevre, M. and Gimble, J.M. Proteomic analysis of primary cultures of human adipose derived stem cells: modulation by adipogenesis. *Mol. Cell Proteomics* **4**, 731–740 (2005).
31. Cui, L., Yin, S., Liu,W., Li, N., Zhang, W. and Cao, Y. Expanded adipose-derived stem cells suppress mixed lymphocyte reaction by secretion of prostaglandin E2. *Tissue Eng.* **13**(6), 1185–1195 (2007).
32. Puissant, B., Barreau, C., Bourin, P., Clavel, C., Corre, J., Bousquet, C., Taureau, C., Cousin, B., Abbal, M., Laharrague, P., Penicaud, L., Casteilla, L. and Blancher, A. Immunomodulatory effect of human adipose tissue-derived adult stem cells: comparison with bone marrow mesenchymal stem cells. *Br. J. Haematol.* **129**(1), 118–129 (2005).
33. McIntosh, K., Zvonic, S., Garrett, S., Mitchell, J.B., Floyd, Z.E., Hammill, L., Kloster, A., Halvorsen, Y.D., Ting, J.P., Storms, R.W., Goh, B., Kilroy, G., Wu, X. and Gimble, J.M. The immunogenicity of human adipose derived cells: temporal changes *in vitro*. *Stem Cells* **24**, 1245–1253 (2006).
34. Wolbank, S., Peterbauer, A., Fahrner, M., Hennerbichler, S., van Griensven, M., Stadler, G., Redl, H. and Gabriel, C. Dose-dependent immunomodulatory effect of human stem cells from amniotic membrane: a comparison with human mesenchymal stem cells from adipose tissue. *Tissue Eng.* **13**(6), 1173–1183 (2007).
35. Niemeyer, P., Kornacker, M., Mehlhorn, A., Seckinger, A., Vohrer, J., Schmal, H., Kasten, P., Eckstein, V., Sudkamp, N.P. and Krause, U. Comparison of immunological properties of bone marrow stromal cells and adipose tissue-derived stem cells before and after osteogenic differentiation *in vitro*. *Tissue Eng.* **13**(1), 111–121 (2007).

36. Bartholomew, A., Sturgeon, C., Siatskas, M., Ferrer, K., McIntosh, K., Patil, S., Hardy, W., Devine, S., Ucker, D., Deans, R., Moseley, A. and Hoffman, R. Mesenchymal stem cells suppress lymphocyte proliferation *in vitro* and prolong skin graft survival *in vivo*. *Exp. Hematol.* **30**(1), 42–48 (2002).
37. Sen, A., Lea-Currie, Y.R., Sujkowska, D., Franklin, D.M., Wilkison, W.O., Halvorsen, Y.D. and Gimble, J.M. Adipogenic potential of human adipose derived stromal cells from multiple donors is heterogeneous. *J. Cell Biochem.* **81**(2), 312–319 (2001).
38. Zuk, P.A., Zhu, M., Mizuno, H., Huang, J., Futrell, J.W., Katz, A.J., Benhaim, P., Lorenz, H.P. and Hedrick, M.H. Multilineage cells from human adipose tissue: implications for cell-based therapies. *Tissue Eng.* **7**(2), 211–228 (2001).
39. Neubauer, M., Hacker, M., Bauer-Kreisel, P., Weiser, B., Fischbach, C., Schulz, M.B., Goepferich, A. and Blunk, T. Adipose tissue engineering based on mesenchymal stem cells and basic fibroblast growth factor *in vitro*. *Tissue Eng.* **11**(11–12), 1840–1851 (2005).
40. Vashi, A.V., Abberton, K.M., Thomas, G.P., Morrison, W.A., O'Connor, A.J., Cooper-White, J.J. and Thompson, E.W. Adipose tissue engineering based on the controlled release of fibroblast growth factor-2 in a collagen matrix. *Tissue Eng.* **12**(11), 3035–3043 (2006).
41. Butterwith, S.C., Peddie, C.D. and Goddard, C. Regulation of adipocyte precursor DNA synthesis by acidic and basic fibroblast growth factors: interaction with heparin and other growth factors. *J. Endocrinol.* **137**(3), 369–374 (1993).
42. Yamashiro, H., Inamoto, T., Yagi, M., Ueno, M., Kato, H., Takeuchi, M., Miyatake, S., Tabata, Y. and Yamaoka, Y. Efficient proliferation and adipose differentiation of human adipose tissue-derived vascular stromal cells transfected with basic fibroblast growth factor gene. *Tissue Eng.* **9**(5), 881–892 (2003).
43. Quarto, N. and Longaker, M.T. FGF-2 inhibits osteogenesis in mouse adipose tissue-derived stromal cells and sustains their proliferative and osteogenic potential state. *Tissue Eng.* **12**(6), 1405–1418 (2006).
44. Koellensperger, E., von Heimburg, D., Markowicz, M. and Pallua, N. Human serum from platelet-poor plasma for the culture of primary human preadipocytes. *Stem Cells* **24**(5), 1218–1225 (2006).
45. Zaragosi, L.E., Ailhaud, G. and Dani, C. Autocrine fibroblast growth factor 2 signaling is critical for self-renewal of human multipotent adipose-derived stem cells. *Stem Cells* **24**(11), 2412–2419 (2006).
46. Halberstadt, C., Austin, C., Rowley, J., Culberson, C., Loebsack, A., Wyatt, S., Coleman, S., Blacksten, L., Burg, K., Mooney, D. and Holder Jr., W. A hydrogel

material for plastic and reconstructive applications injected into the subcutaneous space of a sheep. *Tissue Eng.* **8**(2), 309–319 (2002).

47. Kral, J.G. and Crandall, D.L. Development of a human adipocyte synthetic polymer scaffold. *Plast. Reconstr. Surg.* **104**(6), 1732–1738 (1999).
48. Patrick, Jr., C.W. Adipose tissue engineering: the future of breast and soft tissue reconstruction following tumor resection. *Semin. Surg. Oncol.* **19**(3), 302–311 (2000).
49. Patrick Jr., C.W., Zheng, B., Johnston, C. and Reece, G.P. Long-term implantation of preadipocyte-seeded PLGA scaffolds. *Tissue Eng.* **8**(2), 283–293 (2002).
50. Patrick, Jr., C.W., Chauvin, P.B., Hobley, J. and Reece, G.P. Preadipocyte seeded PLGA scaffolds for adipose tissue engineering. *Tissue Eng.* **5**(2), 139–151 (1999).
51. Schoeller, T., Lille, S., Wechselberger, G., Otto, A., Mowlavi, A. and Piza-Katzer, H. Histomorphologic and volumetric analysis of implanted autologous preadipocyte cultures suspended in fibrin glue: a potential new source for tissue augmentation. *Aesthet. Plast. Surg.* **25**(1), 57–63 (2001).
52. von Heimburg, D., Zachariah, S., Low, A. and Pallua, N. Influence of different biodegradable carriers on the *in vivo* behavior of human adipose precursor cells. *Plast. Reconstr. Surg.* **108**(2), 411–420 (2001); discussion 421–422.
53. von Heimburg, D., Zachariah, S., Heschel, I., Kuhling, H., Schoof, H., Hafemann, B. and Pallua, N. Human preadipocytes seeded on freeze-dried collagen scaffolds investigated *in vitro* and *in vivo*. *Biomaterials* **22**(5), 429–438 (2001).
54. Mauney, J.R., Volloch, V. and Kaplan, D.L. Matrix-mediated retention of adipogenic differentiation potential by human adult bone marrow-derived mesenchymal stem cells during *ex vivo* expansion. *Biomaterials* **26**(31), 6167–6175 (2005).
55. Mauney Jr., N.T., Gillen, K., Kirker-Head, C., Gimble, J.M. and Kaplan, D.L. Engineering adipose-like tissue *in vitro* and *in vivo* utilizing human bone marrow and adipose-derived mesenchymal stem cells with silk fibroin 3D scaffolds. *Biomaterials* **28**(35), 5280–5290 (2007).
56. Eppley, B.L., Snyders Jr., R.V., Winkelmann, T. and Delfino, J.J. Autologous facial fat transplantation: improved graft maintenance by microbead bioactivation. *J. Oral Maxillofac. Surg.* **50**(5), 477–482 (1992); discussion 482–483.
57. Kawaguchi, N., Toriyama, K., Nicodemou-Lena, E., Inou, K., Torii, S. and Kitagawa, Y. *De novo* adipogenesis in mice at the site of injection of basement membrane and basic fibroblast growth factor. *Proc. Natl. Acad. Sci. USA* **95**(3), 1062–1066 (1998).

58. Tabata, Y., Miyao, M., Inamoto, T., Ishii, T., Hirano, Y., Yamaoki, Y. and Ikada, Y. *De novo* formation of adipose tissue by controlled release of basic fibroblast growth factor. *Tissue Eng.* **6**(3), 279–289 (2000).
59. Toriyama, K., Kawaguchi, N., Kitoh, J., Tajima, R., Inou, K., Kitagawa, Y. and Torii, S. Endogenous adipocyte precursor cells for regenerative soft-tissue engineering. *Tissue Eng.* **8**(1), 157–165 (2002).
60. Kimura, Y., Ozeki, M., Inamoto, T. and Tabata, Y. Time course of *de novo* adipogenesis in matrigel by gelatin microspheres incorporating basic fibroblast growth factor. *Tissue Eng.* **8**(4), 603–613 (2002).
61. Kimura, Y., Ozeki, M., Inamoto, T. and Tabata, Y. Adipose tissue engineering based on human preadipocytes combined with gelatin microspheres containing basic fibroblast growth factor. *Biomaterials* **24**(14), 2513–2521 (2003).
62. Inoue, S., Hori, Y., Hirano, Y., Inamoto, T. and Tabata, Y. Effect of culture substrate and fibroblast growth factor addition on the proliferation and differentiation of human adipo-stromal cells. *J. Biomater. Sci. Polym. Ed.* **16**(1), 57–77 (2005).
63. Hiraoka, Y., Yamashiro, H., Yasuda, K., Kimura, Y., Inamoto, T. and Tabata, Y. *In situ* regeneration of adipose tissue in rat fat pad by combining a collagen scaffold with gelatin microspheres containing basic fibroblast growth factor. *Tissue Eng.* **12**(6), 1475–1487 (2006).
64. Coleman, S.R. Facial recontouring with lipostructure. *Clin. Plast. Surg.* **24**(2), 347–367 (1997).
65. Matsumoto, D., Sato, K., Gonda, K., Takaki, Y., Shigeura, T., Sato, T., Aiba-Kojima, E., Iizuka, F., Inoue, I., Suga, H. and Yoshimura, K. Cell-assisted lipotransfer (CAL): supportive use of human adipose-derived cells for soft tissue augmentation with lipoinjection. *Tissue Eng.* **12**(12), 3375–3382 (2006).
66. Rangappa, S., Fen, C., Lee, E.H., Bongso, A. and Sim, E.K. Transformation of adult mesenchymal stem cells isolated from the fatty tissue into cardiomyocytes. *Ann. Thorac. Surg.* **75**(3), 775–779 (2003).
67. Gaustad, K.G., Boquest, A.C., Anderson, B.E., Gerdes, A.M. and Collas, P. Differentiation of human adipose tissue stem cells using extracts of rat cardiomyocytes. *Biochem. Biophys. Res. Commun.* **314**(2), 420–427 (2004).
68. Planat-Benard, V., Menard, C., Andre, M., Puceat, M., Perez, A., Garcia-Verdugo, J.M., Penicaud, L. and Casteilla, L. Spontaneous cardiomyocyte differentiation from adipose tissue stroma cells. *Circ. Res.* **94**(2), 223–229 (2004).
69. Yamada, Y., Wang, X.D., Yokoyama, S., Fukuda, N. and Takakura, N. Cardiac progenitor cells in brown adipose tissue repaired damaged myocardium. *Biochem. Biophys. Res. Commun.* **342**(2), 662–670 (2006).

70. Yamada, Y., Yokoyama, S., Wang, X.D., Fukuda, N. and Takakura, N. Cardiac stem cells in brown adipose tissue express CD133 and induce bone marrow non-hematopoietic cells to differentiate into cardiomyocytes. *Stem Cells* **25**(5), 1326–1333 (2007).
71. Katz, A.J., Zang, Z., Shang, H., Chamberlain, A.T., Berr, S.S., Roy, R.J., Khurgel, M., Epstein, F.H. and French, B.A. Serial MRI assessment of human adipose-derived stem cells (HASCS) in a murine model of reperfused myocardial infarction. *Adipocytes* **2**(1), 1–10 (2006).
72. Strem, B.M., Zhu, M., Alfonso, Z., Daniels, E.J., Schreiber, R., Beygui, R., MacLellan, W.R., Hedrick, M.H. and Fraser, J.K. Expression of cardiomyocytic markers on adipose tissue-derived cells in a murine model of acute myocardial injury. *Cytotherapy* **7**(3), 282–291 (2005).
73. Miyahara, Y., Nagaya, N., Kataoka, M., Yanagawa, B., Tanaka, K., Hao, H., Ishino, K., Ishida, H., Shimizu, T., Kangawa, K., Sano, S., Okano, T., Kitamura, S. and Mori, H. Monolayered mesenchymal stem cells repair scarred myocardium after myocardial infarction. *Nat. Med.* **12**(4), 459–465 (2006).
74. Song, Y.H., Gehmert, S., Sadat, S., Pinkernell, K., Bai, X., Matthias, N. and Alt, E. VEGF is critical for spontaneous differentiation of stem cells into cardiomyocytes. *Biochem. Biophys. Res. Commun.* **354**(4), 999–1003 (2007).
75. Awad, H.A., Halvorsen, Y.D., Gimble, J.M. and Guilak, F. Effects of transforming growth factor beta1 and dexamethasone on the growth and chondrogenic differentiation of adipose-derived stromal cells. *Tissue Eng.* **9**(6), 1301–1312 (2003).
76. Awad, H.A., Wickham, M.Q., Leddy, H.A., Gimble, J.M. and Guilak, F. Chondrogenic differentiation of adipose-derived adult stem cells in agarose, alginate, and gelatin scaffolds. *Biomaterials* **25**(16), 3211–3222 (2004).
77. Erickson, G.R., Gimble, J.M., Franklin, D.M., Rice, H.E., Awad, H. and Guilak, F. Chondrogenic potential of adipose tissue-derived stromal cells *in vitro* and *in vivo*. *Biochem. Biophys. Res. Commun.* **290**(2), 763–769 (2002).
78. Wickham, M.Q., Erickson, G.R., Gimble, J.M., Vail, T.P. and Guilak, F. Multipotent stromal cells derived from the infrapatellar fat pad of the knee. *Clin. Orthop.* **412**, 196–212 (2003).
79. Huang, J.I., Zuk, P.A., Jones, N.F., Zhu, M., Lorenz, H.P., Hedrick, M.H. and Benhaim, P. Chondrogenic potential of multipotential cells from human adipose tissue. *Plast. Reconstr. Surg.* **113**(2), 585–594 (2004).
80. Guilak, F., Lott, K.E., Awad, H.A., Cao, Q., Hicok, K.C., Fermor, B. and Gimble, J.M. Clonal analysis of the differentiation potential of human adipose-derived adult stem cells. *J. Cell Physiol.* **206**(1), 229–237 (2006).

81. Betre, H., Ong, S.R., Guilak, F., Chilkoti, A., Fermor, B. and Setton, L.A. Chondrocytic differentiation of human adipose-derived adult stem cells in elastin-like polypeptide. *Biomaterials* **27**(1), 91–99 (2006).
82. Estes, B.T., Wu, A.W., Storms, R.W. and Guilak, F. Extended passaging, but not aldehyde dehydrogenase activity, increases the chondrogenic potential of human adipose-derived adult stem cells. *J. Cell Physiol.* **209**(3), 987–995 (2006).
83. Estes, B.T., Wu, A.W. and Guilak, F. Potent induction of chondrocytic differentiation of human adipose-derived adult stem cells by bone morphogenetic protein 6. *Arthritis Rheum.* **54**(4), 1222–1232 (2006).
84. Knippenberg, M., Helder, M.N., Zandieh Doulabi, B., Wuisman, P.I. and Klein-Nulend, J. Osteogenesis versus chondrogenesis by BMP-2 and BMP-7 in adipose stem cells. *Biochem. Biophys. Res. Commun.* **342**(3), 902–908 (2006).
85. Chiou, M., Xu, Y. and Longaker, M.T. Mitogenic and chondrogenic effects of fibroblast growth factor-2 in adipose-derived mesenchymal cells. *Biochem. Biophys. Res. Commun.* **343**(2), 644–652 (2006).
86. Malladi, P., Xu, Y., Chiou, M., Giaccia, A.J. and Longaker, M.T. Effect of reduced oxygen tension on chondrogenesis and osteogenesis in adipose-derived mesenchymal cells. *Am. J. Physiol. Cell. Physiol.* **290**(4), C1139–C1146 (2006).
87. Wang, D.W., Fermor, B., Gimble, J.M., Awad, H.A. and Guilak, F. Influence of oxygen on the proliferation and metabolism of adipose derived adult stem cells. *J. Cell. Physiol.* **204**, 184–191 (2005).
88. Winter, A., Breit, S., Parsch, D., Benz, K., Steck, E., Hauner, H., Weber, R.M., Ewerbeck, V. and Richter, W. Cartilage-like gene expression in differentiated human stem cell spheroids: a comparison of bone marrow-derived and adipose tissue-derived stromal cells. *Arthritis Rheum.* **48**(2), 418–429 (2003).
89. Lee, R.H., Kim, B., Choi, I., Kim, H., Choi, H.S., Suh, K., Bae, Y.C. and Jung, J.S. Characterization and expression analysis of mesenchymal stem cells from human bone marrow and adipose tissue. *Cell Physiol. Biochem.* **14**(4–6), 311–324 (2004).
90. Huang, J.I., Kazmi, N., Durbhakula, M.M., Hering, T.M., Yoo, J.U. and Johnstone, B. Chondrogenic potential of progenitor cells derived from human bone marrow and adipose tissue: a patient-matched comparison. *J. Orthop. Res.* **23**(6), 1383–1389 (2005).
91. Im, G.I., Shin, Y.W. and Lee, K.B. Do adipose tissue-derived mesenchymal stem cells have the same osteogenic and chondrogenic potential as bone marrow-derived cells? *Osteoarthr. Cartil.* **13**(10), 845–853 (2005).

92. Liu, T.M., Martina, M., Hutmacher, D.W., Hoi, J., Hui, P., Lee, E.H. and Lim, B. Identification of common pathways mediating differentiation of bone marrow and adipose tissues derived human mesenchymal stem cells (MSCs) into three mesenchymal lineages. *Stem Cells* **25**(3), 750–760 (2007).
93. Mehlhorn, A.T., Niemeyer, P., Kaiser, S., Finkenzeller, G., Stark, G.B., Sudkamp, N.P. and Schmal, H. Differential expression pattern of extracellular matrix molecules during chondrogenesis of mesenchymal stem cells from bone marrow and adipose tissue. *Tissue Eng.* **12**(10), 2853–2862 (2006).
94. Sakaguchi, Y., Sekiya, I., Yagishita, K. and Muneta, T. Comparison of human stem cells derived from various mesenchymal tissues: superiority of synovium as a cell source. *Arthritis Rheum.* **52**(8), 2521–2529 (2005).
95. Yoshimura, H., Muneta, T., Nimura, A., Yokoyama, A., Koga, H. and Sekiya, I. Comparison of rat mesenchymal stem cells derived from bone marrow, synovium, periosteum, adipose tissue, and muscle. *Cell Tissue Res.* **327**(3), 449–462 (2007).
96. Mochizuki, T., Muneta, T., Sakaguchi, Y., Nimura, A., Yokoyama, A., Koga, H. and Sekiya, I. Higher chondrogenic potential of fibrous synovium- and adipose synovium-derived cells compared with subcutaneous fat-derived cells: distinguishing properties of mesenchymal stem cells in humans. *Arthritis Rheum.* **54**(3), 843–853 (2006).
97. Lee, K.D., Kuo, T.K., Whang-Peng, J., Chung, Y.F., Lin, C.T., Chou, S.H., Chen, J.R., Chen, Y.P. and Lee, O.K. *In vitro* hepatic differentiation of human mesenchymal stem cells. *Hepatology* **40**(6), 1275–1284 (2004).
98. Seo, M.J., Suh, S.Y., Bae, Y.C. and Jung, J.S. Differentiation of human adipose stromal cells into hepatic lineage *in vitro* and *in vivo*. *Biochem. Biophys. Res. Commun.* **328**(1), 258–264 (2005).
99. Talens-Visconti, R., Bonora, A., Jover, R., Mirabet, V., Carbonell, F., Castell, J.V. and Gomez-Lechon, M.J. Human mesenchymal stem cells from adipose tissue: differentiation into hepatic lineage. *Toxicol. In Vitro* **21**(2), 324–329 (2007).
100. Talens-Visconti, R., Bonora, A., Jover, R., Mirabet, V., Carbonell, F., Castell, J.V. and Gomez-Lechon, M.J. Hepatogenic differentiation of human mesenchymal stem cells from adipose tissue in comparison with bone marrow mesenchymal stem cells. *World J. Gastroenterol.* **12**(36), 5834–5845 (2006).
101. Timper, K., Seboek, D., Eberhardt, M., Linscheid, P., Christ-Crain, M., Keller, U., Muller, B. and Zulewski, H. Human adipose tissue-derived mesenchymal stem cells differentiate into insulin, somatostatin, and glucagon expressing cells. *Biochem. Biophys. Res. Commun.* **341**(4), 1135–1140 (2006).
102. Brzoska, M., Geiger, H., Gauer, S. and Baer, P. Epithelial differentiation of human adipose tissue-derived adult stem cells. *Biochem. Biophys. Res. Commun.* **330**(1), 142–150 (2005).

103. Garcia-Olmo, D., Garcia-Arranz, M., Garcia, L.G., Cuellar, E.S., Blanco, I.F., Prianes, L.A., Montes, J.A., Pinto, F.L., Marcos, D.H. and Garcia-Sancho, L. Autologous stem cell transplantation for treatment of rectovaginal fistula in perianal Crohn's disease: a new cell-based therapy. *Int. J. Colorectal Dis.* **18**(5), 451–454 (2003).
104. Garcia-Olmo, D., Garcia-Arranz, M., Herreros, D., Pascual, I., Peiro, C. and Rodriguez-Montes, J.A. A phase I clinical trial of the treatment of Crohn's fistula by adipose mesenchymal stem cell transplantation. *Dis. Colon Rectum* **48**(7), 1416–1423 (2005).
105. Williams, S.K., Jarrell, B.E., Rose, D.G., Pontell, J., Kapelan, B.A., Park, P.K. and Carter, T.L. Human microvessel endothelial cell isolation and vascular graft sodding in the operating room. *Ann. Vasc. Surg.* **3**(2), 146–152 (1989).
106. Williams, S.K., Wang, T.F., Castrillo, R. and Jarrell, B.E. Liposuction-derived human fat used for vascular graft sodding contains endothelial cells and not mesothelial cells as the major cell type. *J. Vasc. Surg.* **19**(5), 916–923 (1994).
107. Planat-Benard, V., Silvestre, J.S., Cousin, B., Andre, M., Nibbelink, M., Tamarat, R., Clergue, M., Manneville, C., Saillan-Barreau, C., Duriez, M., Tedgui, A., Levy, B., Penicaud, L. and Casteilla, L. Plasticity of human adipose lineage cells toward endothelial cells: physiological and therapeutic perspectives. *Circulation* **109**(5), 656–663 (2004).
108. Cao, Y., Sun, Z., Liao, L., Meng, Y., Han, Q. and Zhao, R.C. Human adipose tissue-derived stem cells differentiate into endothelial cells *in vitro* and improve postnatal neovascularization *in vivo*. *Biochem. Biophys. Res. Commun.* **332**(2), 370–379 (2005).
109. Martinez-Estrada, O.M., Munoz-Santos, Y., Julve, J., Reina, M. and Vilaro, S. Human adipose tissue as a source of Flk-1+ cells: new method of differentiation and expansion. *Cardiovasc. Res.* **65**(2), 328–333 (2005).
110. Lee, W.C., Rubin, J.P. and Marra, K.G. Regulation of alpha-smooth muscle actin protein expression in adipose-derived stem cells. *Cells Tissues Organs* **183**(2), 80–86 (2006).
111. Jeon, E.S., Moon, H.J., Lee, M.J., Song, H.Y., Kim, Y.M., Bae, Y.C., Jung, J.S. and Kim, J.H. Sphingosylphosphorylcholine induces differentiation of human mesenchymal stem cells into smooth-muscle-like cells through a TGF-{beta}-dependent mechanism. *J. Cell Sci.* **119**(Pt. 23), 4994–5005 (2006).
112. Oritani, K., Wu, X., Medina, K., Hudson, J., Miyake, K., Gimble, J.M., Burstein, S.A. and Kincade, P.W. Antibody ligation of CD9 modifies production of myeloid cells in long-term cultures. *Blood* **87**(6), 2252–2261 (1996).

113. Jacobsen, K., Miyake, K., Kincade, P.W. and Osmond, D.G. Highly restricted expression of a stromal cell determinant in mouse bone marrow *in vivo*. *J. Exp. Med.* **176**(4), 927–935 (1992).
114. Wu, X., Miyake, K., Medina, K.L., Kincade, P.W. and Gimble, J.M. Recognition of murine integrin beta 1 by a rat anti-stromal cell monoclonal antibody. *Hybridoma* **13**(5), 409–416 (1994).
115. Miyake, K., Underhill, C.B., Lesley, J. and Kincade, P.W. Hyaluronate can function as a cell adhesion molecule and CD44 participates in hyaluronate recognition. *J. Exp. Med.* **172**(1), 69–75 (1990).
116. Kilroy, G.E., Foster, S.J., Wu, X., Ruiz, J., Sherwood, S., Heifetz, A., Ludlow, J.W., Stricker, D.M., Potiny, S., Green, P., Halvorsen, Y.-D.C., Cheatham, B., Storms, R.W. and Gimble, J.M. Cytokine profile of human adipose-derived stem cells: expression of angiogenic, hematopoietic, and pro-inflammatory factors. *J. Cell. Physiol.* **212**, 702–709 (2007).
117. Corre, J., Barreau, C., Cousin, B., Chavoin, J.P., Caton, D., Fournial, G., Penicaud, L., Casteilla, L. and Laharrague, P. Human subcutaneous adipose cells support complete differentiation but not self-renewal of hematopoietic progenitors. *J. Cell. Physiol.* **208**(2), 282–288 (2006).
118. Cousin, B., Andre, M., Arnaud, E., Penicaud, L. and Casteilla, L. Reconstitution of lethally irradiated mice by cells isolated from adipose tissue. *Biochem. Biophys. Res. Commun.* **301**(4), 1016–1022 (2003).
119. Koc, O.N., Gerson, S.L., Cooper, B.W., Dyhouse, S.M. and Haynesworth, S.E. Rapid hematopoietic recovery after coinfusion of autologous-blood stem cells and culture-expanded marrow mesenchymal stem cells in advanced breast cancer patients receiving high dose chemotherapy. *J. Clin. Oncol.* **18**(2), 307 (2000).
120. Koç, O.N. and Lazarus, H.M. Mesenchymal stem cells: heading into the clinic. *Bone Marrow Transplant* **27**(3), 235–239 (2001).
121. Lazarus, H.M., Koc, O.N., Devine, S.M., Curtin, P., Maziarz, R.T., Holland, H.K., Shpall, E.J., McCarthy, P., Atkinson, K., Cooper, B.W., Gerson, S.L., Laughlin, M.J., Loberiza Jr., F.R., Moseley, A.B. and Bacigalupo, A. Cotransplantation of HLA-identical sibling culture-expanded mesenchymal stem cells and hematopoietic stem cells in hematologic malignancy patients. *Biol. Blood Marrow Transplant* **11**(5), 389–398 (2005).
122. Safford, K.M., Hicok, K.C., Safford, S.D., Halvorsen, Y.D., Wilkison, W.O., Gimble, J.M. and Rice, H.E. Neurogenic differentiation of murine and human adipose-derived stromal cells. *Biochem. Biophys. Res. Commun.* **294**(2), 371–379 (2002).

123. Safford, K.M. and Rice, H.E. Stem cell therapy for neurologic disorders: therapeutic potential of adipose-derived stem cells. *Curr. Drug Targets* **6**(1), 57–62 (2005).
124. Safford, K.M., Safford, S.D., Gimble, J.M., Shetty, A.K. and Rice, H.E. Characterization of neuronal/glial differentiation of murine adipose-derived adult stromal cells. *Exp. Neurol.* **187**(2), 319–328 (2004).
125. Kang, S.K., Putnam, L.A., Ylostalo, J., Popescu, I.R., Dufour, J., Belousov, A. and Bunnell, B.A. Neurogenesis of rhesus adipose stromal cells. *J. Cell Sci.* **117**(Pt. 18), 4289–4299 (2004).
126. Krampera, M., Marconi, S., Pasini, A., Galie, M., Rigotti, G., Mosna, F., Tinelli, M., Lovato, L., Anghileri, E., Andreini, A., Pizzolo, G., Sbarbati, A. and Bonetti, B. Induction of neural-like differentiation in human mesenchymal stem cells derived from bone marrow, fat, spleen and thymus. *Bone* **40**, 382–390 (2006).
127. Ashjian, P.H., Elbarbary, A.S., Edmonds, B., DeUgarte, D., Zhu, M., Zuk, P.A., Lorenz, H.P., Benhaim, P. and Hedrick, M.H. *In vitro* differentiation of human processed lipoaspirate cells into early neural progenitors. *Plast. Reconstr. Surg.* **111**(6), 1922–1931 (2003).
128. Fujimura, J., Ogawa, R., Mizuno, H., Fukunaga, Y. and Suzuki, H. Neural differentiation of adipose-derived stem cells isolated from GFP transgenic mice. *Biochem. Biophys. Res. Commun.* **333**(1), 116–121 (2005).
129. Kokai, L.E., Rubin, J.P. and Marra, K.G. The potential of adipose-derived adult stem cells as a source of neuronal progenitor cells. *Plast. Reconstr. Surg.* **116**(5), 1453–1460 (2005).
130. Neuhuber, B., Gallo, G., Howard, L., Kostura, L., Mackay, A. and Fischer, I. Re-evaluation of *in vitro* differentiation protocols for bone marrow stromal cells: disruption of actin cytoskeleton induces rapid morphological changes and mimics neuronal phenotype. *J. Neurosci. Res.* **77**(2), 192–204 (2004).
131. Lu, P., Blesch, A. and Tuszynski, M.H. Induction of bone marrow stromal cells to neurons: differentiation, transdifferentiation, or artifact? *J. Neurosci. Res.* **77**(2), 174–191 (2004).
132. Kang, S.K., Shin, M.J., Jung, J.S., Kim, Y.G. and Kim, C.H. Autologous adipose tissue-derived stromal cells for treatment of spinal cord injury. *Stem Cells Dev.* **15**(4), 583–594 (2006).
133. Kang, S.K., Lee, D.H., Bae, Y.C., Kim, H.K., Baik, S.Y. and Jung, J.S. Improvement of neurological deficits by intracerebral transplantation of human adipose tissue-derived stromal cells after cerebral ischemia in rats. *Exp. Neurol.* **183**(2), 355–366 (2003).

134. Kubis, N., Tomita, Y., Tran-Dinh, A., Planat-Benard, V., Andre, M., Karaszewski, B., Waeckel, L., Penicaud, L., Silvestre, J.S., Casteilla, L., Seylaz, J. and Pinard, E.Vascular fate of adipose tissue-derived adult stromal cells in the ischemic murine brain: a combined imaging-histological study. *Neuroimage* **34**(1), 1–11 (2007).
135. Rice, H.E., Sheng, H., Evenson, D.A., Freememan, A.J., Safford, K.M., Provenzale, J.M., Warner, D.S. and Johnson, G.A. SPIO-labeling and transplantation of adipose-derived stem cells in MCAO-injured mice. *Am. J. Roetgenol.* **188**(4), 1101–1108 (2007).
136. Kang, S.K., Jun, E.S., Bae, Y.C. and Jung, J.S. Interactions between human adipose stromal cells and mouse neural stem cells *in vitro*. *Brain Res. Dev. Brain Res.* **145**(1), 141–149 (2003).
137. Kosacka, J., Nowicki, M., Kacza, J., Borlak, J., Engele, J. and Spanel-Borowski, K. Adipocyte-derived angiopoietin-1 supports neurite outgrowth and synaptogenesis of sensory neurons. *J. Neurosci. Res.* **83**(7), 1160–1169 (2006).
138. Halvorsen, Y.C., Wilkison, W.O. and Gimble, J.M. Adipose-derived stromal cells — their utility and potential in bone formation. *Int. J. Obes. Relat. Metab. Disord.* **24**(Suppl. 4), S41–S44 (2000).
139. Halvorsen, Y.D., Franklin, D., Bond, A.L., Hitt, D.C., Auchter, C., Boskey, A.L., Paschalis, E.P., Wilkison, W.O. and Gimble, J.M. Extracellular matrix mineralization and osteoblast gene expression by human adipose tissue-derived stromal cells. *Tissue Eng.* **7**(6), 729–741 (2001).
140. Huang, J.I., Beanes, S.R., Zhu, M., Lorenz, H.P., Hedrick, M.H. and Benhaim, P. Rat extramedullary adipose tissue as a source of osteochondrogenic progenitor cells. *Plast. Reconstr. Surg.* **109**(3), 1033–1041 (2002); discussion 1042–1043.
141. Malladi, P., Xu, Y., Yang, G.P. and Longaker, M.T. Functions of vitamin D, retinoic acid, and dexamethasone in mouse adipose-derived mesenchymal cells. *Tissue Eng.* **12**(7), 2031–2040 (2006).
142. Zeng, Q., Li, X., Choi, L., Beck, G., Balian, G. and Shen, F.H. Recombinant growth/differentiation factor-5 stimulates osteogenic differentiation of fat-derived stromal cells *in vitro*. *Connect. Tissue Res.* **47**(5), 264–270 (2006).
143. Cho, H.H., Park, H.T., Kim, Y.J., Bae, Y.C., Suh K.T. and Jung, J.S. Induction of osteogenic differentiation of human mesenchymal stem cells by histone deacetylase inhibitors. *J. Cell Biochem.* **96**(3), 533–542 (2005).
144. Dragoo, J.L., Choi, J.Y., Lieberman, J.R., Huang, J., Zuk, P.A., Zhang, J., Hedrick, M.H. and Benhaim, P. Bone induction by BMP-2 transduced stem cells derived from human fat. *J. Orthop. Res.* **21**(4), 622–629 (2003).

145. Yang, M., Ma, Q.J., Dang, G.T., Ma, K., Chen, P. and Zhou, C.Y. *In vitro* and *in vivo* induction of bone formation based on *ex vivo* gene therapy using rat adipose-derived adult stem cells expressing BMP-7. *Cytotherapy* **7**(3), 273–281 (2005).
146. Zhang, X., Yang, M., Lin, L., Chen, P., Ma, K.T., Zhou, C.Y. and Ao, Y.F. Runx2 overexpression enhances osteoblastic differentiation and mineralization in adipose — derived stem cells *in vitro* and *in vivo*. *Calcif. Tissue Int.* **79**(3), 169–178 (2006).
147. Kang, S.K., Putnam, L., Dufour, J., Ylostalo, J., Jung, J.S. and Bunnell, B.A. Expression of telomerase extends the lifespan and enhances osteogenic differentiation of adipose tissue-derived stromal cells. *Stem Cells* **22**(7), 1356–1372 (2004).
148. Jun, E.S., Lee, T.H., Cho, H.H., Suh, S.Y. and Jung, J.S. Expression of telomerase extends longevity and enhances differentiation in human adipose tissue-derived stromal cells. *Cell Physiol. Biochem.* **14**(4–6), 261–268 (2004).
149. Nuttall, M.E. and Gimble, J.M. Controlling the balance between osteoblastogenesis and adipogenesis and the consequent therapeutic implications. *Curr. Opin. Pharmacol.* **4**(3), 290–294 (2004).
150. McBeath, R., Pirone, D.M., Nelson, C.M., Bhadriraju, K. and Chen, C.S. Cell shape, cytoskeletal tension, and RhoA regulate stem cell lineage commitment. *Dev. Cell* **6**(4), 483–495 (2004).
151. Knippenberg, M., Helder, M.N., Doulabi, B.Z., Semeins, C.M., Wuisman, P.I. and Klein-Nulend, J. Adipose tissue-derived mesenchymal stem cells acquire bone cell-like responsiveness to fluid shear stress on osteogenic stimulation. *Tissue Eng.* **11**(11–12), 1780–1788 (2005).
152. Lee, J.H. and Kemp, D.M. Human adipose-derived stem cells display myogenic potential and perturbed function in hypoxic conditions. *Biochem. Biophys. Res. Commun.* **341**(3), 882–888 (2006).
153. Gabbay, J.S., Heller, J.B., Mitchell, S.A., Zuk, P.A., Spoon, D.B., Wasson, K.L., Jarrahy, R., Benhaim, P. and Bradley, J.P. Osteogenic potentiation of human adipose-derived stem cells in a 3-dimensional matrix. *Ann. Plast. Surg.* **57**(1), 89–93 (2006).
154. Leong, D.T., Khor, W.M., Chew, F.T., Lim, T.C. and Hutmacher, D.W. Characterization of osteogenically induced adipose tissue-derived precursor cells in 2-dimensional and 3-dimensional environments. *Cells Tissues Organs* **182**(1), 1–11 (2006).
155. Justesen, J., Pedersen, S.B., Stenderup, K. and Kassem, M. Subcutaneous adipocytes can differentiate into bone-forming cells *in vitro* and *in vivo*. *Tissue Eng.* **10**(3–4), 381–391 (2004).

156. Hicok, K.C., Du Laney, T.V., Zhou, Y.S., Halvorsen, Y.D., Hitt, D.C., Cooper, L.F. and Gimble, J.M. Human adipose-derived adult stem cells produce osteoid *in vivo*. *Tissue Eng.* **10**(3–4), 371–380 (2004).
157. Leong, D.T., Abraham, M.C., Rath, S.N., Lim, T.-C., Chew, F.T. and Hutmacher, D.W. Investigating the effects of preinduction on human adipose-derived precursor cells in an athymic rat model. *Differentiation* **74**, 519–529 (2006).
158. Peptan, I.A., Hong, L. and Mao, J.J. Comparison of osteogenic potentials of visceral and subcutaneous adipose-derived cells of rabbits. *Plast. Reconstr. Surg.* **117**(5), 1462–1470 (2006).
159. Hattori, H., Sato, M., Masuoka, K., Ishihara, M., Kikuchi, T., Matsui, T., Takase, B., Ishizuka, T., Kikuchi, M., Fujikawa, K. and Ishihara, M. Osteogenic potential of human adipose tissue-derived stromal cells as an alternative stem cell source. *Cells Tissues Organs* **178**(1), 2–12 (2004).
160. Cowan, C.M., Shi, Y.Y., Aalami, O.O., Chou, Y.F., Mari, C., Thomas, R., Quarto, N., Contag, C.H., Wu, B. and Longaker, M.T. Adipose-derived adult stromal cells heal critical-size mouse calvarial defects. *Nat. Biotechnol.* **22**(5), 560–567 (2004).
161. Cowan, C.M., Aalami, O.O., Shi, Y.Y., Chou, Y.F., Mari, C., Thomas, R., Quarto, N., Nacamuli, R.P., Contag, C.H., Wu, B. and Longaker, M.T. Bone morphogenetic protein 2 and retinoic acid accelerate *in vivo* bone formation, osteoclast recruitment, and bone turnover. *Tissue Eng.* **11**(3–4), 645–658 (2005).
162. Dudas, J.R., Marra, K.G., Cooper, G.M., Penascino, V.M., Mooney, M.P., Jiang, S., Rubin, J.P. and Losee, J.E. The osteogenic potential of adipose-derived stem cells for the repair of rabbit calvarial defects. *Ann. Plast. Surg.* **56**(5), 543–548 (2006).
163. Conejero, J.A., Lee, J.A., Parrett, B.M., Terry, M., Wear-Maggitti, K., Grant, R.T. and Breitbart A.S. Repair of palatal bone defects using osteogenically differentiated fat-derived stem cells. *Plast. Reconstr. Surg.* **117**(3), 857–863 (2006).
164. Lendeckel, S., Jodicke, A., Christophis, P., Heidinger, K., Wolff, J., Fraser, J.K., Hedrick, M.H., Berthold, L. and Howaldt, H.P. Autologous stem cells (adipose) and fibrin glue used to treat widespread traumatic calvarial defects: case report. *J. Craniomaxillofac. Surg.* **32**(6), 370–3 (2004).
165. Mizuno, H., Zuk, P.A., Zhu, M., Lorenz, H.P., Benhaim, P. and Hedrick, M.H. Myogenic differentiation by human processed lipoaspirate cells. *Plast. Reconstr. Surg.* **109**(1), 199–209 (2002); discussion 210–211.
166. Weintraub, H., Tapscott, S.J., Davis, R.L., Thayer, M.J., Adam, M.A., Lassar, A.B. and Miller, A.D. Activation of muscle-specific genes in pigment, nerve, fat, liver, and fibroblast cell lines by forced expression of MyoD. *Proc. Natl. Acad. Sci. USA* **86**(14), 5434–5438 (1989).

167. Hu, E., Tontonoz, P. and Spiegelman, B.M. Transdifferentiation of myoblasts by the adipogenic transcription factors PPAR gamma and C/EBP alpha. *Proc. Natl. Acad. Sci. USA* **92**(21), 9856–9860 (1995).
168. Sasao, N., Hirayama, E. and Kim, J. Characterization of heterokaryons between skeletal myoblasts and preadipocytes: myogenic potential of 3T3-L1 preadipocytes. *Eur. J. Cell. Biol.* **82**(2), 97–103 (2003).
169. Sordella, R., Jiang, W., Chen, G.C., Curto, M. and Settleman, J. Modulation of Rho GTPase signaling regulates a switch between adipogenesis and myogenesis. *Cell* **113**(2), 147–158 (2003).
170. Katz, A.J., Llull, R., Hedrick, M.H. and Futrell, J.W. Emerging approaches to the tissue engineering of fat. *Clin. Plast. Surg.* **26**(4), 587–603, viii (1999).
171. Morizono, K., De Ugarte, D.A., Zhu, M., Zuk, P., Elbarbary, A., Ashjian, P., Benhaim, P., Chen, I.S. and Hedrick, M.H. Multilineage cells from adipose tissue as gene delivery vehicles. *Hum. Gene Ther.* **14**(1), 59–66 (2003).
172. Aslan, H., Zilberman, Y., Arbeli, V., Sheyn, D., Matan, Y., Liebergall, M., Li, J.Z., Helm, G.A., Gazit, D. and Gazit, Z. Nucleofection-based *ex vivo* non-viral gene delivery to human stem cells as a platform for tissue regeneration. *Tissue Eng.* **12**(4), 877–889 (2006).
173. Zaragosi, L.E., Billon, N., Ailhaud, G. and Dani, C. Nucleofection is a valuable transfection method for transient and stable transgene expression in adipose tissue-derived stem cells. *Stem Cells* **25**, 790–797 (2007).
174. Wolbank, S., Peterbauer, A., Wassermann, E., Hennerbichler, S., Voglauer, R., van Griensven, M., Duba, H.C., Gabriel, C. and Redl, H. Labelling of human adipose-derived stem cells for non-invasive *in vivo* cell tracking. *Cell Tissue Bank.* **8**(3), 163–177 (2007).
175. Dubois, S.G., Halvorsen, Y.-D.C., Ravussin, E. and Gimble, J.M. Primary stromal cell culture from adipose tissue: from liposuction to needle biopsy. *Adipocytes* **1**(3), 139–144 (2005).

Chapter 9

Control of Adult Stem Cell Function in Bioengineered Artificial Niches

Matthias P. Lutolf and Helen M. Blau

Stem cells are characterized by their dual ability to self-renew and differentiate, yielding essentially unlimited numbers of progeny that can replenish tissues with either high turnover such as blood and skin or contribute to the regeneration of organs with less frequent remodeling such as muscle. In contrast to their embryonic counterparts, adult stem cells can only preserve their unique functions if they are in intimate contact with an instructive microenvironment, termed niche. Stem cells integrate a complex array of niche signals that regulate their fate, keeping them in a relatively quicscent state during homeostasis, or controlling their numbers via symmetric or asymmetric divisions in response to the regenerative demands of a tissue. This chapter provides an overview of the current state of knowledge of structural and functional hallmarks of mammalian stem cell niches and offers a perspective on how bioengineering principles could be used to deconstruct the niche and providing novel insights into the role of its specific components in the regulation of stem cell fate. Such "artificial niches" constitute powerful tools for elucidating stem cell regulatory mechanisms that should fuel the development of novel therapeutic strategies for tissue regeneration.

Keywords: Adult Stem Cells; Niches; Self-Renewal; Differentiation; Artificial Niches.

Outline

1. Introduction
2. Adult Stem Cells Reside in Niches
3. Common Structures and Components of Stem Cell Niches
4. Key Functions of Stem Cell Niches
5. Niches Control the Fate of Individual Stem Cells

6. The HSC Niche
7. Prospects for Using Engineered Artificial Niches as Novel Model Systems to Probe and Manipulate Adult Stem Cell Fate
References

1. Introduction

While embryonic stem cells form tissues, adult stem cells are essential for tissue maintenance and repair throughout life. In adulthood, tissue homeostasis and regeneration are critically dependent on both the self-renewal and the differentiation capacity of stem cells. Due to these unique properties, the potential of adult stem cells for use in the treatment of various genetic diseases or injuries due to trauma is enormous. However, to tap this clinical reservoir and fully exploit its potential, we must increase our knowledge of the regulatory mechanisms that govern stem cell behavior. To date only a few adult stem cell types are approved for clinical use. Bone marrow transplants containing hematopoietic stem cells (HSCs) have saved the lives of numerous leukemia and lymphoma patients and skin transplants have significantly alleviated disfigurement and increased function as burn victims grow. Moreover, recent findings suggest that cells with stem cell-like properties may give rise to and maintain some cancers, including acute leukemia, brain and breast. Thus, an increased understanding of stem cell regulatory mechanisms may not only augment regenerative medicine but also spawn new strategies for the treatment of cancer.

To overcome the hurdles inherent to enlisting adult stem cells therapeutically or to targeting cancer stem cells for destruction, fundamental questions regarding the precise cell-intrinsic and -extrinsic regulation of quiescence, self-renewal and differentiation must be addressed. Which genes determine the unique properties of a stem cell? Do adult stem cells of diverse tissues share genes in common? How can the regulation of these genes be manipulated to advantage? What genetic determinants distinguish stem cells from their more specialized progenitors and can this specialization be reversed? Stem cells are exposed to a multitude of diverse converging biochemical and biophysical cues present in their vicinity. To what extent is stem cell fate predetermined or subject to such extrinsic influences? How can extrinsic factors direct stem cell divisions, altering their timing and numbers, leading to either maintenance or expansion of the stem cell pool? Can perturbation of certain soluble or immobilized factors typical of the niche lead either to the promotion or to the arrest of cancer?

To address these questions, methodologies for controlling adult stem cell behavior outside of tissues that allow simplification of the numerous variables typical of their niches would be beneficial. Bioengineering technologies could greatly facilitate this type of deconstruction enabling analyses of stem cell behavior in a manner previously not possible. In this chapter, after highlighting the fundamentals of adult stem cell function in niches, we present some ways in which bioengineering can be combined with stem cell biology to generate new *in vitro* model systems that will be useful in devising novel strategies for tissue regeneration.

2. Adult Stem Cells Reside in Niches

Adult stem cells reside in anatomically well-defined locales, termed niches, comprised of complex mixtures of extracellular cues delivered by support cells in close proximity[1–6] (Fig. 1A). Niches, in concert with cell-intrinsic regulatory networks, control multiple functions of adult stem cells, as discussed in detail below. The niche concept was postulated by Schofield more than two decades ago,[7] but concrete evidence in support of this hypothesis is relatively recent. Germline stem cell niches in *Drosophila melanogaster* were the first to be rigorously studied. These discrete niches were optimal for study as they could be visualized microscopically and manipulated genetically, serving as paradigms for other niches. Seminal studies led by the groups of Spradling, Fuller and others revealed that stromal cells can provide instructive microenvironments that support stem cell self-renewal in the testis and ovary via direct cell-cell interactions (e.g. Refs. 8–11).

As described in several excellent recent reviews[3,5,6,12] mammalian niches have recently been identified and characterized in multiple self-renewing tissues including the skin (in the bulge region of the hair follicle), intestine (in the epithelium), brain (in the subventricular zone), bone marrow (on the endosteal surface and near blood vessels) and muscle (beneath the muscle fiber basal lamina). Although most mammalian niches remain somewhat ill-defined because they are poorly accessible and therefore difficult to manipulate experimentally, niche factors are being identified at a rapid pace and as described below, evidence is accumulating that these factors play a critical role in regulating stem cell fate.

3. Common Structures and Components of Stem Cell Niches

Recent studies have demonstrated that features characteristic of stem cell niches are conserved from *Drosophila* to mammals, including several aspects of niche structure and biochemical composition. Figure 1A illustrates the prototypical niche architecture with its key components. Stem cells are in intimate physical contact with stromal support cells providing short-range signals via soluble factors and transmembrane cell-cell adhesion proteins, and they are surrounded by an extracellular matrix which provides structure and organization

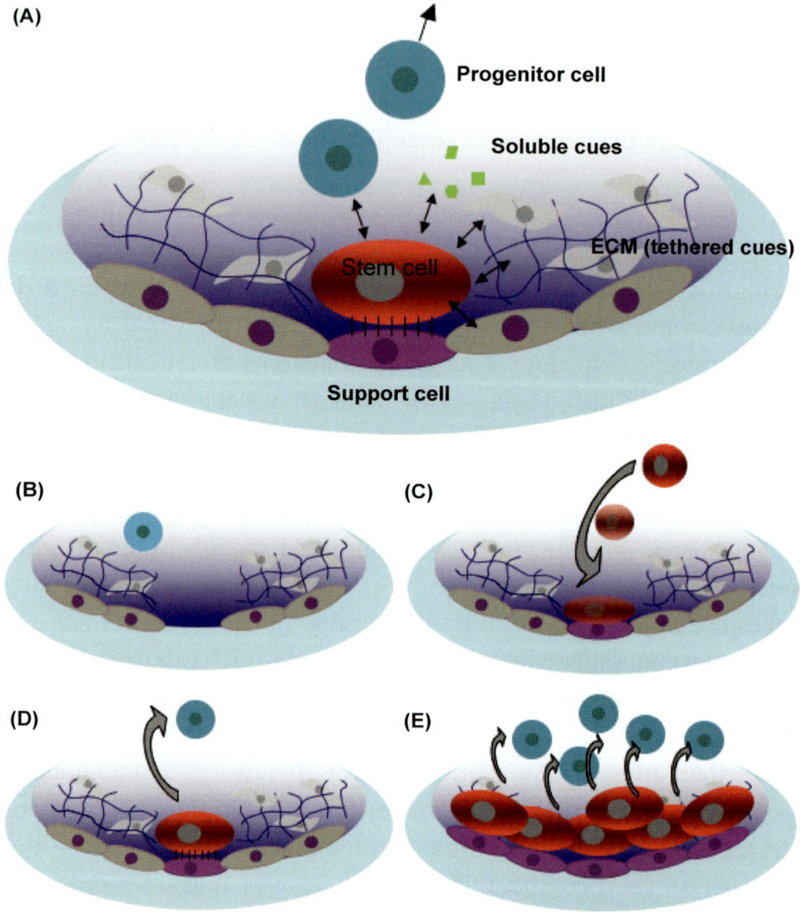

Fig. 1. Structure and function of the prototypical stem cell niche. Adult stem cells are located within instructive microenvironments comprised of complex mixtures of extracellular cues delivered by support cells in close proximity. The main components of a niche are support cells and their secreted transmembrane cell-cell adhesion proteins, soluble factors and the surrounding extracellular matrix **(A)**. Absence of a niche depletes the stem cell pool, highlighting the importance of the niche for stem cell function **(B)**. Stem cells can "home" back to their niche such as is the case when HSCs repopulate their bone marrow niches after transplantation **(C)**. Under homeostatic conditions, the stem cell numbers remain constant **(D)**, while a misregulated balance between self-renewal and differentiation can lead to cancer **(E)**.

as well as biochemical and mechanical signals. Intriguingly, blood vessels are often found near niches (not shown in Fig. 1A), presumably to transport long-range signals and serve as a conduit for recruitment of circulating cells into the niche.

Niche components can be categorized into three main types of molecular signals that are presented to stem cells in a precise spatial organization:

(1) *Membrane-anchored signals arising from support cells.* Support cells anchor the stem cells to their niches. Adhesion proteins on the cell surface restrict the movement of the stem cells and keep them within the niche. These adhesive interactions also help to position and orient the stem cells within the niche in close proximity to self-renewal signals emanating from the support cells. Cell-cell interactions of this kind are governed for example by the family of cadherins forming adherens junctions. The extracellular domain of these transmembrane receptors on support cells can interact either with other cadherins of the same kind (i.e. homophilic binding) or with different cadherins (i.e. heterophilic binding) on stem cells.[13] The importance of such interactions is highlighted in *Drosophila* where the removal of cadherin function in germline stem cells niches leads to stem cell loss.[14] How cell-cell interactions via transmembrane proteins such as cadherins, are involved not only in adhesion, but also in directing stem cell self-renewal, is at great interest and remains to be determined. As described later in this chapter, the development of bioengineering approaches may aid in elucidating their function.

    Support cells within the niche also provide transmembrane cues unrelated to adhesion. A case in point is Notch signaling, which plays a role in controlling cell fate throughout embryonic development[15] as well as in adult stem cell self-renewal and differentiation. Notch receptors are highly conserved transmembrane proteins that are expressed by a number of stem cells or support cells. Binding to their corresponding jagged or delta ligands induces proteolytic cleavage and release of the intracellular domain of the receptor, which translocates to the cell nucleus to alter expression of target genes.

(2) *Physically tethered ligands from the surrounding extracellular matrix (ECM).* Stem cells in the niche are in contact with a protein- and sugar-rich network that comprises the ECM. The ECM can be either in the form of a two-dimensional sheet-like basal lamina, as is the case for epithelial stem cells or muscle satellite cells, or a highly hydrated three-dimensional fibrillar polymer network that fully encompasses the cells, as in the case of HSCs. Many adult stem cells express receptors for domains of these ECM proteins such as fibronectin, or glycosaminoglycans such as hyaluronic acid. Stem cells interact with these ECM components via adhesion protein receptors such as the integrins. These integral membrane proteins define cell shape, motility, and regulate the cell cycle.

Moreover, signal transduction induced via integrins appears to play a role in stem cell maintenance and differentiation. For example, beta-1 integrin signaling is involved in maintenance of epidermal stem cells[16] or neural stem/progenitors[17] in the niche.

(3) *"Soluble" effectors that include protein morphogens, growth factors, chemokines and cytokines* These molecules play an important function in directing stem cell fate. Of particular relevance are morphogen proteins such as Wnts, hedgehog proteins, fibroblast growth factors (FGFs) or bone morphogenetic proteins (BMPs). These signals are conserved and can be found in many niches across different species from *Drosophila* to mammals. Many soluble proteins are bound to the ECM via electrostatic interactions involving heparan sulfate proteoglycans such as heparin, localizing their response to the niche and protecting them from rapid proteolytic inactivation.[18] In addition to secreted proteins, it should be noted that small molecules or high local concentrations of ions, for example Ca^{2+} in HSC niches, can also provide important regulatory cues in stem cell niches.[19]

4. Key Functions of Stem Cell Niches

The complex ensemble of localized signals described above constitutes niches and is postulated to act in covert on stem cells in close proximity to physically tether the relatively quiescent stem cells, protect them from rapid differentiation, and regulate their self-renewal. The importance of the niche as a stem cell regulatory network is exemplified by the fact that loss of contact with the niche or a disruption of the physical structure of the niche results in loss of stem cell function (Fig. 1B). This is most evident when adult stem cells are removed from their microenvironment and cultured *in vitro* where they tend to quickly differentiate. Further evidence that, molecular niche components are required to maintain stem cell function is provided by findings that empty niches are repopulated when stem cells home back to their protective milieu (Fig. 1C). A point in case is the recruitment of circulating HSCs upon transplantation back to their bone marrow niches where they proliferate to re-establish homeostatic conditions (Fig. 1D). Intriguingly, upon complete loss of stem cells, the niche has been demonstrated in *Drosophila*[20] and mice[21] to instruct progenitor cells with a reduced self-renewal capacity to revert to a stem cell state. This observation strongly further implicates niches as locales of the key regulators that establish stem cell function. Moreover, these findings suggest that self-renewal is not exclusively an intrinsic function of stem cells, but that multipotency can be induced in progenitors by extrinsic cues.

The interplay between stem cells and their niches creates a dynamic and reciprocal system that balances stem cell numbers in response to the physiological demands of a tissue. The size of the stem cell pool has been shown to correlate with the size of the niche.[5] Specific niche signals keep the number of stem cells in the niche constant during homeostasis and proliferation and allows the stem cell pool to expand under physiological stress and in pathological situations such as cancer (Fig. 1E). It is conceivable that multiple types of niches exist for each stem cell type, and that niches can be formed *de novo* or be destroyed in response to physiological needs. This characteristic responsiveness makes the niche a prime target for regenerative medicine and tissue engineering. For example, rather than targeting the stem cells, support cells in the niche could be manipulated or transplanted to indirectly alter endogenous stem cell numbers in order to improve tissue function and overcome adverse effects of disease or aging.[22,23] Moreover, if tumor-propagating cancer stem cells are dependent on signals from a niche,[24,25] therapeutic ablation of components of this cancer stem cell niche could provide a promising path forward to fight cancer.

5. Niches Control the Fate of Individual Stem Cells

In principle, an individual adult stem cell in its niche can undergo four different fates (excluding death) as schematically depicted in Fig. 2A: it can remain quiescent, undergo self-renewal divisions that result in two daughter stem cells (termed *symmetric* divisions), self-renewal divisions that result in one daughter stem cell and one differentiated cell (termed *asymmetric* divisions; asymmetric with regards to the identity of the two daughter cells), or divisions in the absence of self-renewal resulting in two differentiated progeny. All of these fates have to be actively regulated and coordinated by the niche in order to assure an appropriate size of the stem cell pool during homeostasis and regeneration over the course of a lifetime of an organism. Studies on stem cell niches in invertebrates have helped to shed light on the mechanisms by which the delicate balance between self-renewal and differentiation is regulated, as described in detail in several excellent reviews.[26–29]

At the crux of these cell fate choices is regulation of cell division. Under homeostatic conditions, stem cell numbers in the niche can be kept constant via asymmetric divisions (Fig. 2B). The decision of a stem cell to undergo an asymmetric division may be controlled by cell-intrinsic and/or cell-extrinsic mechanisms. On the one hand, intracellular partitioning of protein components, for example the cell fate determinant protein Numb or cell polarity factors such as PAR proteins[27] may drive asymmetry. On the other hand, the asymmetric positioning of daughter cells relative to external niche components may induce

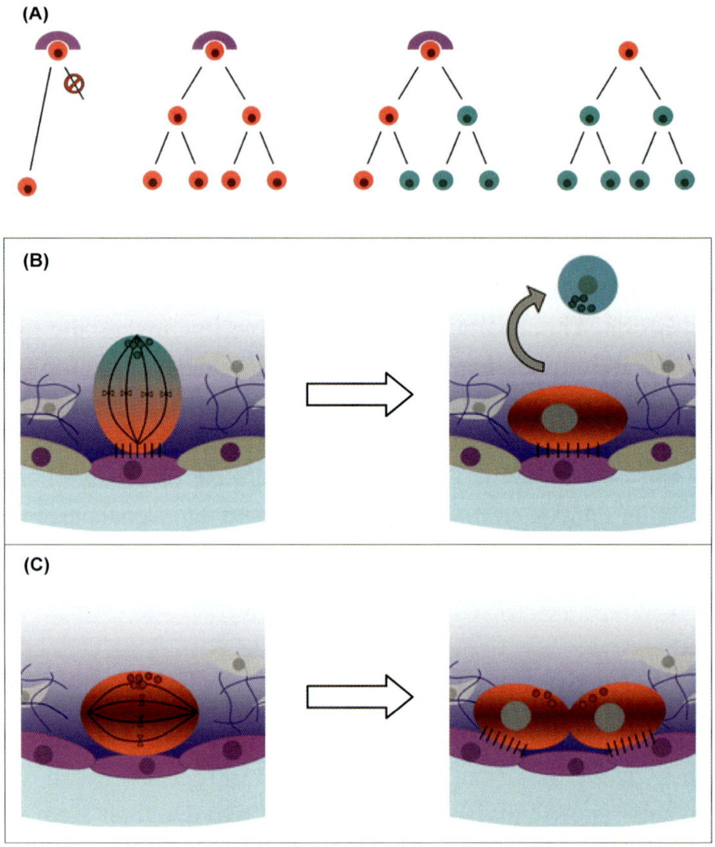

Fig. 2. Stem cell fates in the niche **(A)**. Stem cells are red, differentiated cells are blue. Adult stem cells are generally considered to be quiescent in their niches (fate 1) unless they are activated to proliferate in response to the demands of a tissue (e.g. stress or trauma). Quiescence is actively regulated by niche proteins such as cell-cell adhesion proteins. During development and/or stress the stem cell pool needs to expand via symmetric self-renewal divisions (fate 2), while during homeostatic conditions self-renewal divisions are of asymmetric nature, generating one daughter cell retaining stem cell identity and another one being already partially differentiated (fate 3). A fourth possible fate is that of (symmetric) differentiation division, whereby both daughter cells lose stem cell function. This fate can be observed in many cases during *in vitro* culture of adult stem cells. Possible mechanisms of stem cell divisions in the niche (**B** and **C**). Asymmetric divisions can be controlled inside the cell via localization of cell fate determinants segregated to the cytoplasm of just one daughter cell **(B)**. A hallmark of this type of division is the regulated orientation of the mitotic spindle that retains only one daughter in the niche. Both daugher cells are exposed to a different microenvironment: that of the original niche (the daughter that maintains stem cell identity) and that outside of the niche (the daughter that loses stem cell function). In a symmetric self-renewal division both daughter cells are exposed to the same niche environment and therefore keep their stem cell identity **(B)**.

asymmetric fates. In this case, both daughter cells may initially be equivalent, but their different microenvironment may impose two disparate fates. A hallmark of this type of extrinsically controlled asymmetric division is the retention of the mitotic spindle orientation perpendicular to the niche, as evidenced in asymmetric divisions of *Drosophila* germline stem cells.[28] One daughter remains in the niche, comprised of support cells termed "cap" cells in the *Drosophila* ovary and "hub" cells in the *Drosophila* testis, and retains stem cell identity, while the other daughter moves away from the niche and starts to differentiate. Accordingly, adhesion junctions at the interface between the niche and germline stem cells may be involved in controlling orientation of the mitotic spindle.[30] In mice, asymmetric segregation of Numb or regulated spindle orientation has also been detected in neural progenitors in the developing cortex,[31] satellite muscle stem cells[32,33] and epidermal progenitors.[34] These findings suggest that asymmetric division may serve as an evolutionarily conserved mechanism to control, or limit, stem cell numbers.[29]

In development or during regeneration, adult stem cells in the niche must increase in numbers. This occurs via symmetric self-renewal divisions giving rise to two daughter stem cells (Fig. 2C). Consistent with the above mechanism of asymmetric division (Fig. 2B), a symmetric division could for example occur by cell-extrinsic mechanisms, if both daughter cells encounter the same niche environment after division and remain in the niche as stem cells. In this case divisions would be predicted to have a mitotic spindle orientation parallel to the niche, a prediction borne out for the epithelium of the skin.[34] Of course, this process must also be tightly orchestrated, as overproduction of stem cells could lead to tumors (Fig. 1E). Of note, the same stem cell populations may switch between symmetric during embryonic development or regeneration in the adult, and asymmetric cell divisions during late fetal development and homeostasis as has for example been observed in the developing forebrain.[35,36]

6. The HSC Niche

In the following paragraphs, we illustrate the concepts described above using one specific niche example, namely that of the hematopoietic stem cell (HSC) niche (Fig. 3). The elucidation of the regulatory function of the HSC niche is particularly relevant, as the limited number of stem cells for transplantation to treat blood cancers poses a significant problem.

Throughout the entire life span of organisms, HSCs have the capacity to undergo self-renewal divisions to generate daughter stem cells and progeny that differentiate into all lineages of lymphoid and myeloid blood cells.[37] A wealth of recent studies using elegant knockout or transgenic *in vivo* strategies have

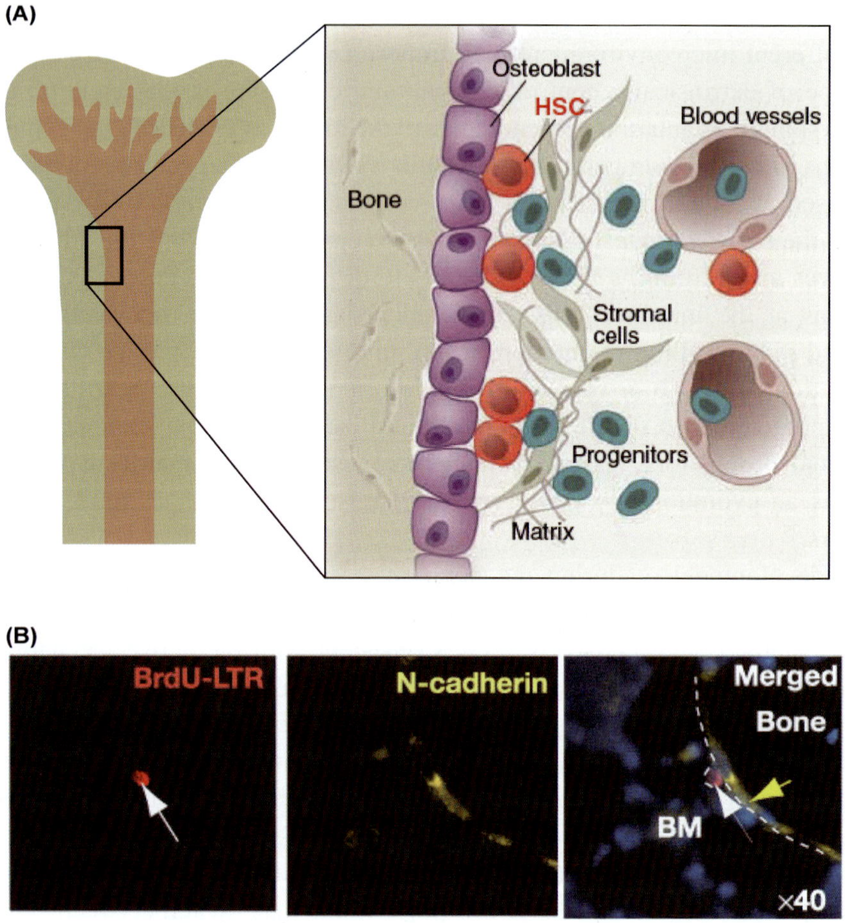

Fig. 3. The endosteal HSC niche. HSCs reside along the endosteal surface of trabecular bone in close proximity to osteoblasts that provide critical regulatory cues (**A**, adapted with permission from Refs. 40 and 103). Identification of HSCs in their endosteal niche via label-retaining strategies and immunohistochemistry (**B**, adapted with permission from Ref. 40). The nucleotide analogue bromodeoxyuridine (BrdU) is retained in HSCs because they divide very infrequently (left panel). Osteoblasts express the transmembrane cell-cell adhesion protein N-cadherin (middle and right panel).

highlighted that HSC function is under the control of niches, consistent with Schofield's hypothesis.[7] Two types of niches located in the bone marrow have been proposed thus far[38]: an endosteal niche[39–43] and a vascular niche,[44–47] wherein HSCs are thought to adhere to osteoblast support cells in the endosteum (Fig. 3), or reticular cells in the vasculature, respectively. For example, when an increase in numbers of a subset of osteoblast in the bone marrow was achieved by expressing a constitutively active form of parathyroid hormone (PTH),[39] or by knocking-down bone morphogenetic protein receptor-1A (BMPR-1A),[40] an accompanying increase in the HSC pool was observed, defining these osteoblasts as key components of the HSC niche. HSCs may transit from one niche to the other depending on their state of activity,[48–51] or the two may be so closely juxtaposed that they function as a single entity. Regardless, the niche localizes stem cells via cell-cell interactions presumably involving N-cadherin[40,52,53] (note: that the role of N-cadherin as a niche factor is still controversial[54]), extracellular matrix proteins such as osteopontin,[42,55] fibronectin[56] and members of cell adhesion molecules (CAMs) family;[57] or heparan sulfate proteoglycans[58,59] such as the glycosaminoglycan hyaluronic acid.[60] Many adhesion molecules also play a key role in the homing of HSCs to these niches.[61–63]

Most HSCs in the niche are in a quiescent G0 state[50,64] which is maintained by putative niche cues such as osteoblast-secreted angiopoietin-1 binding to the receptor tyrosine kinase Tie2,[41] endothelial cell-expressed CXCL12[46] and some additional proteins involved in the above cell-cell or cell-ECM proteins. Several soluble proteins have emerged as candidates for regulating HSC self-renewal, including the stem cell factor or c-kit ligand,[65] Wnt-3a,[66,67] Sonic hedgehog (Shh),[68,69] stromal cell-derived factor 1 (sdf-1)/CXCL12[70] or FGF-1.[71] Notably, hypoxic conditions[72] and a high calcium concentration[43] in the bone marrow are also important parameters for keeping HSCs in a multipotent state.

Although convincing evidence for asymmetric division of HSCs is lacking, a series of *in vitro* studies employing micromanipulation suggest that initial divisions of mouse or human HSCs can generate daughter cells with different developmental potential.[73–79] However, to date there is no evidence that HSCs asymmetrically segregate chromosomes or retain BrdU[54] like other mammalian adult stem cells.[80,81]

7. Prospects for Using Engineered Artificial Niches as Novel Model Systems to Probe and Manipulate Adult Stem Cell Fate

The unique self-renewal and differentiation potential of stem cells drives the current intense interest in harnessing these cells for applications in regenerative medicine. However, progress in exploiting this potential has been impeded by an unyielding fundamental biological problem: their function cannot be well controlled *in vitro*. More robust protocols are required to direct the *in vitro* differentiation of embryonic stem cells and most adult stem cell types are highly refractory to undergoing self-renewal divisions and thus cannot be maintained or expanded in culture. This is crucial for HSCs, for example, as it would overcome the shortage of donor cells available for transplantation from adults and facilitate the future use of autologous cord blood HSCs. A promising route to overcoming these hurdles is the amalgamation of stem cell biology with biomolecular materials engineering and microfabrication technologies, in order to generate novel cell culture platforms that mimic crucial biochemical or structural aspects of the niche and allow an assessment of stem cell function in high-throughput.[82] Several lines of research that have emerged recently appear particularly promising and are briefly discussed in the following paragraphs.

(1) *Engineering artificial niches via biomolecular materials.* Synthetic approaches that mimic specific physicochemical and biochemical characteristics of the niche using well-defined and tunable biomolecular materials[83,84] may provide a means of controlling adult stem cell fate outside the niche. Polymer hydrogels based on synthetic building blocks may be particularly well suited for this purpose, as they are inert, and can be readily functionalized with desired protein niche cues, as described in detail in Chapter 13 of this book. In contrast to the ubiquitous tissue culture plastic dish, soft hydrogel substrates mimic some of the physicochemical aspects of natural tissues imbibing typically between 90% and 99% water. Notably, it was recently demonstrated that adult stem cells change their fate in response to the elasticity of a substrates,[85] aberrant rigid plastic surfaces likely favoring the commitment of stem cells to undesired lineages. Thus, the choice of a substrate that can be molecularly engineered and that recapitulates the physiochemical characteristics of niches may be critical in controlling stem cell behavior outside of their natural microenvironment.

Current culture systems rely primarily on the use of soluble growth factors and cytokines as medium supplements. Many of these factors are not components of the natural niche and it is therefore not surprising that they fail to promote stem cell maintenance or self-renewing divisions.

As a result, over the last four decades, despite numerous attempts to establish *in vitro* conditions to achieve HSC expansion without loss of differentiation potential, none currently meets that goal satisfactorily. The majority of these efforts have either involved the use of cocktails containing "classical" hematopoietic cytokines such as interleukins (e.g. IL-3 or -6), or the co-culture of HSCs in contact with a layer of stromal "feeder" cells. Cytokines have in many cases induced unwanted differentiation in conjunction with proliferation. On the other hand, the co-culture approach has yielded some promising results, for example using the stromal cell clone from murine fetal liver, AFT024,[86] but is limited by its ill-defined composition. An alternative approach involves genetic engineering to overexpress or silence critical regulatory genes in HSCs. HSC expansion has been achieved by manipulation of genes including bcl-2,[87] the polycomb group gene Bmi-1,[88] the transcription factor HOXB4,[89] the proto-oncogene c-myc,[52] or genes involved in cell cycle control.[90–92] Although in many cases successful and informative, the genetic engineering of stem cells for clinical purposes faces challenges shared by gene therapy. Thus, if similarly effective, manipulation of stem cells from the "outside" using protein cues would seem advantageous. Notably, when HOXB4 was taken up by cells from the medium as a TAT-fusion protein and therefore only transiently present, substantial stem cell expansion was observed.[93] Niche proteins may achieve similar effects acting via cell surface receptors to alter intracellular signaling thereby directing cell fate towards quiescence or self-renewal, just as in the native niche.

(2) *Microfabrication technologies to explore stem cell biology at the single cell level and in high-throughput.* In many cases, stem cell regulation is best explored *in vitro* by interrogating a large number of individual, spatially confined stem cells. The reason is that, to date, adult stem cells can only be isolated with limited purity, even when the most advanced phenotypic marker combinations and flow cytometry tools are utilized. The resulting cell heterogeneity hampers any conventional population-based ("bulk") *in vitro* analyses. As a result, analyses of rare stem cells may be skewed by the behavior of overgrowing progenitors, for example in cases where stem cells grow much slower than progenitors. Unicellular systems employing standard multiwell well plates allow stem cell progeny to be analyzed and followed over time at the single cell level as clones, but are cumbersome and highly inefficient. Micronscale technologies have been successfully employed to generate microwell arrays for to culture stem cells (e.g. Refs. 94–97) as described in detail in Chapter 17 of this book. We expect that these tools will be efficient to dissect the behavior of many rare stem cell populations. A further integration of hydrogel technologies and protein patterning into microfabrication platforms will be crucial to optimally mimic biochemical and physicochemical aspects of the natural niche of a stem cell.

(3) *Protein microarrays to dissect the complexity of the niche.* A major challenge in stem cell research is to unveil the complexity of signaling that governs adult stem cell behavior. Conventional experimental paradigms, i.e. the testing of one signal at a time, fail to address this complexity. Several groups have recently successfully addressed this challenge.[98–100] Mixtures of protein signaling cues such as ECM components, morphogens,

and other signaling proteins were microarrayed on flat substrates using robotic spotting technology. Stem cells can be exposed to such multi-component artificial microenvironments, and their response quantified at the single cell level via multi-parameter analysis involving for example immunocytochemistry-based read-outs. These high-throughput analyses of signaling networks have allowed elucidation of the effect of combinations of stem cell regulatory proteins on self-renewal and differentiation. In one pioneering study by Soen *et al.*,[99] primary human neural precursor cells were cultured on printed "microenvironmental" arrays to explore the extent and direction of differentiation into neurons and glias. Co-stimulation via Wnt and Notch maintained stem cells in an undifferentiated-like state, in contrast to BMP-4 that led to expression of differentiation markers. The application of such spotting technologies to hydrogels could allow the combinatorial protein effects to be studied in stem cells grown under physico-chemical conditions more closely mimicking tissues, as described above.

(4) *Novel technologies to mimic spatial and temporal complexity of the niche.* As described above, niche signals are presented in a complex but spatially well-defined "polar" manner, wherein each side of a stem cell may be exposed to a different microenvironment. This polarization of signals may lead to intracellular protein segregation with important consequences for the outcome of cell divisions, as we have discussed earlier. Clever protein spotting approaches with control of protein deposition at the subcellular scale could be explored to mimic the intricate physiologic spatial organization of niche cues.

Moreover, signal gradients play an important role in regulating stem cell function, for example by controlling "homing" towards a cytokine gradient via chemotaxis. It is likely that the niche comprises gradients of soluble and ECM-tethered biomolecules, whereby fate decisions of self-renewal versus differentiation are controlled by the distance from the cell to the niche, just as occurs during morphogenesis in development. Microfluidics technologies offer elegant means to recapitulate spatial complexity such as in biomolecule gradients. Numerous examples of soluble and surface adsorbed protein gradients formed via microfluidics-based systems have been described in the last few years, most of which have employed rigid glass or plastic surfaces. Burdick *et al.* have demonstrated that adhesion ligand gradients can be formed on hydrogels as well.[101] Recently, Choi and colleagues have presented an elegant microfluidics-based approach whereby cells within alginate gels could be exposed to desirable *soluble* gradients in 3D.[102] Applied to adult stem cell culture, such intricate control over the biochemical microenvironment in 3D, constitutes and important step towards recapitulation of the niche to prevent stem cells from rapid differentiation or to selectively differentiate them towards "desired" lineages. Future approaches should address the *immobilization* of biomolecule gradients in 3D hydrogel microenvironments.

In summary, two decades ago it was widely believed that stem cells are regulated intrinsically, impervious to extrinsic signals. Powerful approaches employing genetic engineering in developmental organisms and transgenic mouse models have shed light on the importance of the niche and extrinsic factors in stem cell regulation. In the foreseeable future, the amalgamation of biomaterials science, microfabrication technology and stem cell biology will provide tools that have the potential to revolutionize our understanding of how stem cell fate is controlled. By simplifying the niche and analyzing the dynamic responses of stem cells to well-defined artificial microenvironments, the role of both specific niche components and niche architecture in regulating fundamental behaviors such as mechanisms of cell division, self-renewal and differention can be elucidated. This may well lead to the generation of adequate numbers of stem cells and the ability to control of their differentiation as needed in order to maximize their utility not only as cell-based therapeutics for tissue regeneration and replacement, but also in the treatment of some cancers that arise due to defects in stem cell regulatory mechanisms.

References

1. Watt, F.M. and Hogan, B.L. Out of Eden: stem cells and their niches. *Science* **287**(5457), 1427–1430 (2000).
2. Spradling, A., Drummond-Barbosa, D. and Kai, T. Stem cells find their niche. *Nature* **414**(6859), 98–104 (2001).
3. Fuchs, E., Tumbar, T. and Guasch, G. Socializing with the neighbors: stem cells and their niche. *Cell* **116**(6), 769–778 (2004).
4. Yamashita, Y.M., Fuller, M.T. and Jones, D.L. Signaling in stem cell niches: lessons from the Drosophila germline. *J. Cell Sci.* **118**(4), 665–672 (2005).
5. Scadden, D.T. The stem-cell niche as an entity of action. *Nature* **441**(7097), 1075–1079 (2006).
6. Moore, K.A. and Lemischka, I.R. Stem cells and their niches. *Science* **311**(5769), 1880–1885 (2006).
7. Schofield, R. Relationship between spleen colony-forming cell and hematopoietic stem-cell — hypothesis. *Blood Cells* **4**(1–2), 7–25 (1978).

8. Xie, T. and Spradling, A.C. A niche maintaining germ line stem cells in the Drosophila ovary. *Science* **290**(5490), 328–330 (2000).
9. Tran, J., Brenner, T.J. and DiNardo, S. Somatic control over the germline stem cell lineage during Drosophila spermatogenesis. *Nature* **407**(6805), 754–757 (2000).
10. Kiger, A.A., White-Cooper, H. and Fuller, M.T. Somatic support cells restrict germline stem cell self-renewal and promote differentiation. *Nature* **407**(6805), 750–754 (2000).
11. Kiger, A.A., Jones, D.L., Schulz, C., Rogers, M.B. and Fuller, M.T. Stem cell self-renewal specified by JAK-STAT activation in response to a support cell cue. *Science* **294**(5551), 2542–2545 (2001).
12. Li, L. and Xie, T. Stem cell niche: structure and function. *Annu. Rev. Cell Dev. Biol.* **21**, 605–631 (2005).
13. Halbleib, J.M. and Nelson, W.J. Cadherins in development: cell adhesion, sorting, and tissue morphogenesis. *Genes Dev.* **20**(23), 3199–3214 (2006).
14. Song, X., Zhu, C.H., Doan, C. and Xie, T. Germline stem cells anchored by adherens junctions in the Drosophila ovary niches. *Science* **296**(5574), 1855–1857 (2002).
15. Artavanis-Tsakonas, S., Rand, M.D. and Lake, R.J. Notch signaling: cell fate control and signal integration in development. *Science* **284**(5415), 770–776 (1999).
16. Zhu, A.J., Haase, I. and Watt, F.M. Signaling via beta 1 integrins and mitogen-activated protein kinase determines human epidermal stem cell fate *in vitro*. *Proc. Natl. Acad. Sci. USA* **96**(12), 6728–6733 (1999).
17. Campos, L.S., Leone, D.P., Relvas, J.B., Brakebusch, C., Fassler, R., Suter, U. *et al.* Beta 1 integrins activate a MAPK signalling pathway in neural stem cells that contributes to their maintenance. *Development* **131**(14), 3433–3444 (2004).
18. Ramirez, F. and Rifkin, D.B. Cell signaling events: a view from the matrix. *Matrix Biol.* **22**(2), 101–107 (2003).
19. Adams, G.B., Chabner, K.T., Alley, I.R., Olson, D.P., Szczepiorkowski, Z.M., Poznansky, M.C. *et al.* Stem cell engraftment at the endosteal niche is specified by the calcium-sensing receptor. *Nature* **439**(7076), 599–603 (2006).
20. Kai, T. and Spradling, A. Differentiating germ cells can revert into functional stem cells in *Drosophila melanogaster* ovaries. *Nature* **428**(6982), 564–569 (2004).
21. Nakagawa, T., Nabeshima, Y.I. and Yoshida, S. Functional identification of the actual and potential stem cell compartments in mouse spermatogenesis. *Dev. Cell.* **12**(2), 195–206 (2007).

22. Conboy, I.M., Conboy, M.J., Wagers, A.J., Girma, E.R., Weissman, I.L. and Rando, T.A. Rejuvenation of aged progenitor cells by exposure to a young systemic environment. *Nature* **433**(7027), 760–764 (2005).
23. Adams, G.B., Martin, R.P., Alley, I.R., Chabner, K.T., Cohen, K.S., Calvi, L.M. *et al.* Therapeutic targeting of a stem cell niche. *Nat. Biotechnol.* **25**(2), 238–243 (2007).
24. Kaplan, R.N., Riba, R.D., Zacharoulis, S., Bramley, A.H., Vincent, L., Costa, C. *et al.* VEGFR1-positive haematopoietic bone marrow progenitors initiate the pre-metastatic niche. *Nature* **438**(7069), 820–827 (2005).
25. Sneddon, J.B., Zhen, H.H., Montgomery, K., van de Rijn, M.V., Tward, A.D., West, R. *et al.* Bone morphogenetic protein antagonist gremlin 1 is widely expressed by cancer-associated stromal cells and can promote tumor cell proliferation. *Proc. Natl. Acad. Sci. USA* **103**(40), 14842–14847 (2006).
26. Doe, C.Q. and Bowerman, B. Asymmetric cell division: fly neuroblast meets worm zygote. *Curr. Opin. Cell Biol.* **13**(1), 68–75 (2001).
27. Betschinger, J. and Knoblich, J.A. Dare to be different: asymmetric cell division in Drosophila, C-elegans and vertebrates. *Curr. Biol.* **14**(16), R674–R685 (2004).
28. Yamashita, Y.M. and Fuller, M.T. Asymmetric stem cell division and function of the niche in the Drosophila male germ line. *Int. J. Hematol.* **82**(5), 377–380 (2005).
29. Morrison, S.J. and Kimble, J. Asymmetric and symmetric stem-cell divisions in development and cancer. *Nature* **441**(7097), 1068–1074 (2006).
30. Yamashita, Y.M., Mahowald, A.P., Perlin, J.R. and Fuller, M.T. Asymmetric inheritance of mother versus daughter centrosome in stem cell division. *Science* **315**(5811), 518–521 (2007).
31. Cayouette, M. and Raff, M. Asymmetric segregation of Numb: a mechanism for neural specification from Drosophila to mammals. *Nat. Neurosci.* **5**(12), 1265–1269 (2002).
32. Conboy, I.M. and Rando, T.A. The regulation of notch signaling controls satellite cell activation and cell fate determination in postnatal myogenesis. *Dev. Cell.* **3**(3), 397–409 (2002).
33. Kuang, S.H., Kuroda, K., Le Grand, F. and Rudnicki, M.A. Asymmetric self-renewal and commitment of satellite stem cells in muscle. *Cell* **129**(5), 999–1010 (2007).
34. Lechler, T. and Fuchs, E. Asymmetric cell divisions promote stratification and differentiation of mammalian skin. *Nature* **437**(7056), 275–280 (2005).
35. Chenn, A. and McConnell, S.K. Cleavage orientation and the asymmetric inheritance of notch1 immunoreactivity in mammalian neurogenesis. *Cell* **82**(4), 631–641 (1995).

36. Noctor, S.C., Martinez-Cerdeno, V., Ivic, L. and Kriegstein, A.R. Cortical neurons arise in symmetric and asymmetric division zones and migrate through specific phases. *Nat. Neurosci.* **7**(2), 136–144 (2004).
37. Kondo, M., Wagers, A.J., Manz, M.G., Prohaska, S.S., Scherer, D.C., Beilhack, G.F. *et al.* Biology of hematopoietic stem cells and progenitors: implications for clinical application. *Annu. Rev. Immunol.* **21**, 759–806 (2003).
38. Wilson, A. and Trumpp, A. Bone-marrow haematopoietic-stem-cell niches. *Nat. Rev. Immunol.* **6**(2), 93–106 (2006).
39. Calvi, L.M., Adams, G.B., Weibrecht, K.W., Weber, J.M., Olson, D.P., Knight, M.C. *et al.* Osteoblastic cells regulate the haematopoietic stem cell niche. *Nature* **425**(6960), 841–846 (2003).
40. Zhang, J., Niu, C., Ye, L., Huang, H., He, X., Tong, W.G. *et al.* Identification of the haematopoietic stem cell niche and control of the niche size. *Nature* **425**(6960), 836–841 (2003).
41. Arai, F., Hirao, A., Ohmura, M., Sato, H., Matsuoka, S., Takubo, K. *et al.* Tie2/angiopoietin-1 signaling regulates hematopoietic stem cell quiescence in the bone marrow niche. *Cell* **118**(2), 149–161 (2004).
42. Nilsson, S.K., Johnston, H.M., Whitty, G.A., Williams, B., Webb, R.J., Denhardt, D.T. *et al.* Osteopontin, a key component of the hematopoietic stem cell niche and regulator of primitive hematopoietic progenitor cells. *Blood* **106**(4), 1232–1239 (2005).
43. Adams, G.B. and Scadden, D.T. The hematopoietic stem cell in its place. *Nat. Immunol.* **7**(4), 333–337 (2006).
44. Avecilla, S.T., Hattori, K., Heissig, B., Tejada, R., Liao, F., Shido, K. *et al.* Chemokine-mediated interaction of hematopoietic progenitors with the bone marrow vascular niche is required for thrombopoiesis. *Nat. Med.* **10**(1), 64–71 (2004).
45. Kiel, M.J., Yilmaz, O.H., Iwashita, T., Yilmaz, O.H., Terhorst, C. and Morrison, S.J. SLAM family receptors distinguish hematopoietic stem and progenitor cells and reveal endothelial niches for stem cells. *Cell* **121**(7), 1109–1121 (2005).
46. Sugiyama, T., Kohara, H., Noda, M. and Nagasawa, T. Maintenance of the hematopoietic stem cell pool by CXCL12-CXCR4 chemokine signaling in bone marrow stromal cell niches. *Immunity* **25**(6), 977–988 (2006).
47. Arai, F. and Suda, T. Maintenance of quiescent hematopoietic stem cells in the osteoblastic niche. *Ann. N. Y. Acad. Sci.* **1106**, 41–53 (2007).
48. Kiel, M.J. and Morrison, S.J. Maintaining hematopoietic stem cells in the vascular niche. *Immunity* **25**(6), 862–864 (2006).

49. Bryder, D., Rossi, D.J. and Weissman, I.L. Hematopoietic stem cells — the aradigmatic tissue-specific stem cell. *Am. J. Pathol.* **169**(2), 338–346 (2006).
50. Wilson, A., Oser, G.M., Jaworski, M., Blanco-Bose, W.E., Laurenti, E., Adolphe, C. *et al.* Dormant and self-renewing hematopoietic stem cells and their niches. *Ann. N. Y. Acad. Sci.* **1106**, 64–75 (2007).
51. Kaplan, R.N., Psaila, B. and Lyden, D. Niche-to-niche migration of bone-marrow-derived cells. *Trends Mol. Med.* **13**(2), 72–81 (2007).
52. Wilson, A., Murphy, M.J., Oskarsson, T., Kaloulis, K., Bettess, M.D., Oser, G.M. *et al.* c-Myc controls the balance between hematopoietic stem cell self-renewal and differentiation. *Genes Dev.* **18**(22), 2747–2763 (2004).
53. Jang, Y.-Y. and Sharkis, S.J. A low level of reactive oxygen species selects for primitive hematopoietic stem cells that may reside in the low-oxygenic niche. *Blood* **110**(8), 3056–3063 (2007).
54. Kiel, M.J., Radice, G.L. and Morrison, S.J. Lack of evidence that hematopoietic stem cells depend on N-cadherin-mediated adhesion to osteoblasts for their maintenance. *Cell Stem Cell* **1**(2), 204–217 (2007).
55. Stier, S., Ko, Y., Forkert, F., Lutz, C., Neuhaus, T., Grunewald, E. *et al.* Osteopontin is a hematopoietic stem cell niche component that negatively regulates stem cell pool size. *J. Exp. Med.* **201**(11), 1781–1791 (2005).
56. Simmons, P.J., Levesque, J.P. and Zannettino, A.C. Adhesion molecules in haemopoiesis. *Baillieres Clin Haematol.* **10**(3), 485–505 (1997).
57. Taichman, R.S. Blood and bone: two tissues whose fates are intertwined to create the hematopoietic stem-cell niche. *Blood* **105**(7), 2631–2639 (2005).
58. Gupta, P., McCarthy, J.B. and Verfaillie, C.M. Stromal fibroblast heparan sulfate is required for cytokine-mediated *ex vivo* maintenance of human long-term culture-initiating cells. *Blood* **87**(8), 3229–3236 (1996).
59. Gupta, P., Oegema, T.R. Jr., Brazil, J.J., Dudek, A.Z., Slungaard, A. and Verfaillie, C.M. Structurally specific heparan sulfates support primitive human hematopoiesis by formation of a multimolecular stem cell niche. *Blood* **92**(12), 4641–4651 (1998).
60. Nilsson, S.K., Haylock, D.N., Johnston, H.M., Occhiodoro, T., Brown, T.J. and Simmons, P.J. Hyaluronan is synthesized by primitive hemopoietic cells, participates in their lodgment at the endosteum following transplantation, and is involved in the regulation of their proliferation and differentiation *in vitro*. *Blood* **101**(3), 856–862 (2003).
61. Scott, L.M., Priestley, G.V. and Papayannopoulou, T. Deletion of alpha4 integrins from adult hematopoietic cells reveals roles in homeostasis, regeneration, and homing. *Mol. Cell Biol.* **23**(24), 9349–9360 (2003).

62. Whetton, A.D. and Graham, G.J. Homing and mobilization in the stem cell niche. *Trends Cell Biol.* **9**(6), 233–238 (1999).
63. Prosper, F. and Verfaillie, C.M. Regulation of hematopoiesis through adhesion receptors. *J. Leukoc. Biol.* **69**(3), 307–316 (2001).
64. Cheshier, S.H., Morrison, S.J., Liao, X. and Weissman, I.L. *In vivo* proliferation and cell cycle kinetics of long-term self-renewing hematopoietic stem cells. *Proc. Natl. Acad. Sci. USA* **96**(6), 3120–3125 (1999).
65. Bowie, M.B., Kent, D.G., Copley, M.R. and Eaves, C.J. Steel factor responsiveness regulates the high self-renewal phenotype of fetal hematopoietic stem cells. *Blood* **109**(11), 5043–5048 (2007).
66. Reya, T., Duncan, A.W., Ailles, L., Domen, J., Scherer, D.C., Willert, K. *et al.* A role for Wnt signalling in self-renewal of haematopoietic stem cells. *Nature* **423**(6938), 409–414 (2004).
67. Willert, K., Brown, J.D., Danenberg, E., Duncan, A.W., Weissman, I.L., Reya, T. *et al.* Wnt proteins are lipid-modified and can act as stem cell growth factors. *Nature* **423**(6938), 448–452 (2003).
68. Bhardwaj, G., Murdoch, B., Wu, D., Baker, D.P., Williams, K.P., Chadwick, K. *et al.* Sonic hedgehog induces the proliferation of primitive human hematopoietic cells via BMP regulation. *Nat Immunol.* **2**(2), 172–180 (2001).
69. Trowbridge, J.J., Scott, M.P. and Bhatia, M. Hedgehog modulates cell cycle regulators in stem cells to control hematopoietic regeneration. *Proc. Natl. Acad. Sci. USA* **103**(38), 14134–14139 (2006).
70. Bowie, M.B., McKnight, K.D., Kent, D.G., McCaffrey, L., Hoodless, P.A. and Eaves, C.J. Hematopoietic stem cells proliferate until after birth and show a reversible phase-specific engraftment defect. *J. Clin. Invest.* **116**(10), 2808–2816 (2006).
71. de Haan, G., Weersing, E., Dontje, B., van Os, R., Bystrykh, L.V., Vellenga, E. *et al. In vitro* generation of long-term repopulating hematopoietic stem cells by fibroblast growth factor-1. *Dev. Cell.* **4**(2), 241–251 (2003).
72. Ito, K., Hirao, A., Arai, F., Matsuoka, S., Takubo, K., Hamaguchi, I. *et al.* Regulation of oxidative stress by ATM is required for self-renewal of haematopoietic stem cells. *Nature* **431**(7011), 997–1002 (2004).
73. Suda, J., Suda, T. and Ogawa, M. Analysis of differentiation of mouse hematopoietic stem-cells in culture by sequential replating of paired progenitors. *Blood* **64**(2), 393–399 (1984).
74. Suda, T., Suda, J. and Ogawa, M. Disparate differentiation in mouse hematopoietic colonies derived from paired progenitors. *Proc. Natl. Acad. Sci. USA Biol. Sci.* **81**(8), 2520–2524 (1984).

75. Leary, A.G., Strauss, L.C., Civin, C.I. and Ogawa, M. Disparate differentiation in hematopoietic colonies derived from human paired progenitors. *Blood* **66**(2), 327–332 (1985).
76. Ema, H., Takano, H., Sudo, K. and Nakauchi, H. *In vitro* self-renewal division of hematopoietic stem cells. *J. Exp. Med.* **192**(9), 1281–1288 (2000).
77. Takano, H., Ema, H., Sudo, K. and Nakauchi, H. Asymmetric division and lineage commitment at the level of hematopoietic stem cells: inference from differentiation in daughter cell and granddaughter cell pairs. *J. Exp. Med.* **199**(3), 295–302.
78. Giebel, B., Zhang, T., Beckmann, J., Spanholtz, J., Wernet, P., Ho, A.D. *et al.* Primitive human hematopoietic cells give rise to differentially specified daughter cells upon their initial cell division. *Blood* **107**(5), 2146–2152 (2006).
79. Ho, A.D. and Wagner, W. The beauty of asymmetry: asymmetric divisions and self-renewal in the haematopoietic system. *Curr. Opin. Hematol.* **14**(4), 330–336 (2007).
80. Smith, G.H. Label-retaining epithelial cells in mouse mammary gland divide asymmetrically and retain their template DNA strands. *Development* **132**(4), 681–687 (2005).
81. Shinin, V., Gayraud-Morel, B., Gomes, D. and Tajbakhsh, S. Asymmetric division and cosegregation of template DNA strands in adult muscle satellite cells. *Nat. Cell Biol.* **8**(7), 677–687 (2006).
82. Underhill, G.H. and Bhatia, S.N. High-throughput analysis of signals regulating stem cell fate and function. *Curr. Opin. Chem. Biol.* **11**, 1–10 (2007).
83. Langer, R. and Tirrell, D.A. Designing materials for biology and medicine. *Nature* **428**(6982), 487–492 (2004).
84. Lutolf, M.P. and Hubbell, J.A. Synthetic biomaterials as instructive extracellular microenvironments for morphogenesis in tissue engineering. *Nat. Biotechnol.* **23**(1), 47–55 (2005).
85. Engler, A.J., Sen, S., Sweeney, H.L. and Discher, D.E. Matrix elasticity directs stem cell lineage specification. *Cell* **126**(4), 677–689 (2006).
86. Moore, K.A., Ema, H. and Lemischka, I.R. *In vitro* maintenance of highly purified, transplantable hematopoietic stem cells. *Blood* **89**(12), 4337–4347 (1997).
87. Domen, J., Cheshier, S.H, and Weissman, I.L. The role of apoptosis in the regulation of hematopoietic stem cells: overexpression of Bcl-2 increases both their number and repopulation potential. *J. Exp. Med.* **191**(2), 253–264 (2000).
88. Iwama, A., Oguro, H., Negishi, M., Kato, Y., Morita, Y., Tsukui, H. *et al.* Enhanced self-renewal of hematopoietic stem cells mediated by the polycomb gene product Bmi-1. *Immunity* **21**(6), 843–851 (2001).

89. Sauvageau, G., Thorsteinsdottir, U., Eaves, C.J., Lawrence, H.J., Largman, C., Lansdorp, P.M. *et al.* Overexpression of HOXB4 in hematopoietic cells causes the selective expansion of more primitive populations *in vitro* and *in vivo*. *Genes Dev.* **9**(14), 1753–1765 (1995).
90. Cheng, T., Rodrigues, N., Shen, H., Yang, Y., Dombkowski, D., Sykes, M. *et al.* Hematopoietic stem cell quiescence maintained by p21cip1/waf1. *Science* **287**(5459), 1804–1808 (2000).
91. Stier, S., Cheng, T., Forkert, R., Lutz, C., Dombkowski, D.M., Zhang, J.L. *et al.* *Ex vivo* targeting of p21Cip1/Waf1 permits relative expansion of human hematopoietic stem cells. *Blood* **102**(4), 1260–1266 (2003).
92. Yuan, Y., Shen, H., Franklin, D.S., Scadden, D.T. and Cheng, T. *In vivo* self-renewing divisions of haematopoietic stem cells are increased in the absence of the early G1-phase inhibitor, p18INK4C. *Nat. Cell. Biol.* **6**(5), 436–442 (2004).
93. Krosl, J., Austin, P., Beslu, N., Kroon, E., Humphries, R.K. and Sauvageau, G. *In vitro* expansion of hematopoietic stem cells by recombinant TAT-HOXB4 protein. *Nat. Med.* **9**(11), 1428–1432 (2003).
94. Chin, V.I., Taupin, P., Sanga, S., Scheel, J., Gage, F.H. and Bhatia, S.N. Microfabricated platform for studying stem cell fates. *Biotechnol. Bioeng.* **88**(3), 399–415 (2004).
95. Falconnet, D., Csucs, G., Grandin, H.M. and Textor, M. Surface engineering approaches to micropattern surfaces for cell-based assays. *Biomaterials* **27**(16), 3044–3063 (2006).
96. Dykstra, B., Ramunas, J., Kent, D., McCaffrey, L., Szumsky, E., Kelly, L. *et al.* High-resolution video monitoring of hematopoietic stem cells cultured in single-cell arrays identifies new features of self-renewal. *Proc. Natl. Acad. Sci. USA* **103**(21), 8185–8190 (2006).
97. Khademhosseini, A., Ferreira, L., Blumling, J., Yeh, J., Karp, J.M., Fukuda, J. *et al.* Co-culture of human embryonic stem cells with murine embryonic fibroblasts on microwell-patterned substrates. *Biomaterials* **27**(36), 5968–5977 (2006).
98. Flaim, C.J., Chien, S. and Bhatia, S.N. An extracellular matrix microarray for probing cellular differentiation. *Nat. Methods* **2**(2), 119–125 (2005).
99. Soen, Y., Mori, A., Palmer, T.D., Brown, P.O. Exploring the regulation of human neural precursor cell differentiation using arrays of signaling microenvironments. *Mol. Syst. Biol.* **2**, 37 (2006).
100. Derda, R., Li, L.Y., Orner, B.P., Lewis, R.L., Thomson, J.A. and Kiessling, L.L. Defined substrates for human embryonic stem cell growth identified from surface arrays. *ACS Chem. Biol.* **2**(5), 347–355 (2007).

101. Burdick, J.A., Khademhosseini, A. and Langer, R. Fabrication of gradient hydrogels using a microfluidics/photopolymerization process. *Langmuir* **20**(13), 5153–5156 (2004).
102. Choi, N.W., Cabodi, M., Held, B., Gleghorn, J.P., Bonassar, L.J. and Stroock, A.D. Microfluidic scaffolds for tissue engineering. *Nat. Mater.* **6**(11), 908–915 (2007).
103. Moore, K.A. and Lemischka, I.R. Stem cells and their niches. *Science* **311**(5769), 1880–1885 (2006).

Chapter 10

Stem Cell Immunology

Anthony P. Hollander and David C. Wraith

Abstract

The field of stem cell biology is fast approaching a degree of maturity that will allow novel therapies to be exploited both clinically and commercially. The first such treatments are likely to rely on the use of a patient's own (autologous) stem cells in order to avoid immune rejection. However if these approaches are to have any long-term viability then the use of donor (allogeneic) cells instead of individualised therapies will be essential in order to allow scale-up of production. There is some evidence that stem cells are hypoimmunogenic as they express only very low levels of major histocompatibility antigens and they are therefore likely to avoid immune rejection. In this chapter we review the evidence for hypoimmunogenicity of stem cells and highlight some differences between adult and embryonic stem cells in their immune stimulatory properties. We also review data demonstrating that the hypoimmunogenic properties of stem cells may sometimes be lost once they have differentiated or after they have been exposed to inflammatory cytokines.

Keywords: Embryonic Stem Cells; Adult Stem Cells; Immune Rejection; Immune Privilege; Immune Suppression.

Outline

1. Why is Stem Cell Immunology Important for Tissue Engineering?
2. Evolutionary Context of the Mammalian Immune System
3. Materno-Foetal Tolerance as a Model for Understanding Stem Cell Immune Privilege
   3.1. Low expression by placental cells of MHC-I
   3.2. Absence of MHC-II molecules
   3.3. Synthesis by the placental cells of indolamine-2,3-dioxygenase (IDO)
   3.4. Expression of Fas ligand (FasL)
   3.5. Secretion of soluble immunosuppressive factors

4. Are Embryonic Stem Cells Immune Privileged?
5. Are Mesenchymal Stem Cells (MSCs) Immune Privileged?
6. Finding a Way Forward for the Use of Allogeneic Stem Cells
References

1. Why is Stem Cell Immunology Important for Tissue Engineering?

One of the most exciting new opportunities in medicine is the use of stem cells, which promise to provide cures for diseases such as Parkinson's, diabetes, osteoarthritis and heart failure. However, they will only be used to treat large numbers of people if they can be turned into commercially successful products. The majority of existing cell therapy techniques are based on the use of autologous (i.e. the patient's own) cells. This is a very expensive and limited approach because of the need to culture cells on a case-by-case basis. Indeed it is clear that for this reason alone, those companies marketing tissue engineering and cell therapies are struggling to make any profit from them. If we are to realise the full potential of stem cells and establish a flourishing cell therapy sector then an alternative allogeneic (donor) cell source must be used to create safe, economically viable, off-the-shelf cell and tissue engineering implants.

Despite the huge therapeutic potential of all types of donor stem cells (adult, foetal, and embryonic) their use in clinical applications will only become a reality once we are able effectively to manage the possibility of immune rejection of implanted cells and tissues. Aggressive immunosuppressive regimes have been used for conventional organ transplantation for many years.[1,2] There are three reasons why this approach may be inappropriate for preventing the rejection of implanted stem cells, their differentiated progeny or engineered tissues derived from them. (1) The range of diseases that will be treated by the new stem cell techniques will include many that are not life-threatening but which have a significant negative impact on the patients' quality of life. These patients are unlikely to tolerate the risk of serious side-effects that are associated with aggressive immunosuppression. (2) There is some evidence that both adult and embryonic stem cells are hypoimmunogenic and so, even after differentiation, they may survive in the recipient with only mild or no immunosuppression. This possibility is discussed in detail later in the chapter. (3) These implants, unlike donated organs, will not be contaminated by donor dendritic cells and so will be unable to stimulate an immune response directly, only indirectly through the host's own antigen presenting cells.[3] This more limited immune stimulation may be suppressed more easily than is the case for organ transplantation where both direct and indirect pathways are activated.

One potential means of providing full cell compatibility would be therapeutic nuclear transfer.[4] However, there are serious concerns with the efficiency and long-term viability of this approach. In addition, as it is an individualised treatment the therapy would be expensive and impractical for treatment of large numbers of patients. Therefore new technologies are required that will provide affordable immunosuppression with fewer side effects than current regimes, making the risks acceptable to most patients irrespective of their medical condition. Alternatively we need to document under which conditions, if any, stem cells might be administered without immunosuppression, relying instead on their innate immune privileged state.

2. Evolutionary Context of the Mammalian Immune System

The immune system has evolved to protect us from infection.[5] In mammals, however, the sophisticated immune system can cause complications and these are collectively known as hypersensitivity conditions.[6]

The immune system is a perfect model of evolutionary advance. Invertebrates have their own immune defence mechanisms and these have been retained throughout evolution.[7] Invertebrate immune mechanisms include physical barriers, the production of molecules designed to sequester invading agents, such as lectins, the secretion of degradative enzymes and the presence of phagocytes. These scavenging cells first evolved to carry foodstuffs around the body of acoelomates. This function became redundant with the evolution of coelomates and the phagocytic cells then evolved a protective role. All of these mechanisms can be observed in vertebrates. Humans retain various protective barriers including skin and mucus whilst lysozyme is a prime example of an enzyme involved in so-called innate immunity. Phagocytic cells are retained in the form of neutrophils and macrophages and these cells provide a first line of defence against bacterial infection as part of the innate immune response.

The difference between vertebrates and invertebrates is the presence of an adaptive or acquired immune response.[8] Acquired immunity refers to the fact that vertebrates respond more quickly and strongly to an antigen on second encounter. Features of acquired immunity include antigen specificity, memory and a gradual strengthening of the immune response to a particular antigen, referred to as affinity maturation. It is thanks to the evolution of acquired immunity that humans can be protected by vaccination. The acquired immune response is possible because of the evolution of lymphocytes bearing antigen-specific receptors. In fact, a primitive form of antigen receptor can be found in jawless fish and this demonstrates an early form of receptor diversity arising from somatic rearrangement.[9] Jawed fish display immunoglobulin gene rearrangement and also have examples of lymphoid tissue organisation. Progression from bony fish through amphibians to birds and mammals shows a clear increase in complexity of the immune system. There are more and more immunoglobulin types, effective separation of lymphocyte function and more complex organisation of lymphoid structures.

A major evolutionary step forward has been the generation of Fc receptors. These molecules are present on the surface of cells in the innate immune system and are able to bind the immunoglobulin/antibody effector molecules of the adaptive immune system.[10] In this way, cells of the innate immune system can indirectly display antigen specificity and hence greatly improve their protective role. Furthermore, evolution has increased the diversity of function of cells with innate immune properties. Neutrophils are best suited to rapid bacterial eradication whereas eosinophils are particularly effective in killing helminth parasites. These cells are suitably armed with Fc receptors appropriate for anti-bacterial (IgG) and anti-helminth (IgE) immune responses respectively.

Fc receptors bind to the constant domains of immunoglobulin molecules. These antibody molecules have highly variable antigen binding regions (FAb) and distinct Fc domains. Different Fc structures define the function of antibodies. For example, IgA molecules are designed to cross mucosal epithelium and hence are functional against mucosal pathogens. IgG antibodies are relatively small, can cross the placenta and hence are important in transferring maternal immunity to the foetus. IgG and IgM antibodies are also effective in triggering the complement cascade and hence are a vital defence mechanism in bacterial infection. IgE antibodies bind to Fc receptors on eosinophils and hence target these cells to the surface of parasites. Mast cells and basophils also carry high affinity Fc receptors for IgE and hence are involved in immediate hypersensitivity, allergic reactions.

The adaptive immune system is distinct from the innate immune system in having highly specific antigen receptors on lymphocytes. B-lymphocytes secrete antibodies and carry immunoglobulin molecules as antigen receptors at the cell surface. T-lymphocytes carry a structurally similar but functionally distinct surface receptor for antigen, the T-cell receptor. There is further complexity in that T-cells are divided into α:β and γ:δ subsets. The α:β cells are further divided into CD4-helper cells and CD8-cytotoxic cells. Effectively the two α:β cell subsets have arisen under evolutionary pressure from infectious agents. CD8 cells are designed to detect viral antigens at the earliest time during virus infection of a cell. CD4-cells have an even more diverse role and have been described as the orchestrator of the immune response. These cells can detect pathogens at different sites in the body and within distinct subcellular compartments and can then dictate the appropriate generation of an immune response against the pathogen. CD4 T-cells are directed to secrete cytokines in response to pathogens at different sites. Hence, Th1 cells, secreting interferon gamma, are effective against intracellular bacteria, e.g. *Mycobacterium tuberculosis*, whereas Th2 cells, secreting IL-4 and IL-5, protect against parasite infections. Th1, Th2 and Th3 cells all control secretion of immunoglobulins by B-cells. Importantly, the isotype of immunoglobulin correlates with the role of the specific antibody in protection from infection. For example, Th3 cells protect against mucosal pathogens and direct B-cells to secrete IgA, the only antibody isotype capable of crossing mucosal epithelium.

B-lymphocytes respond to antigen directly and their receptor interacts with conformational determinants of the antigen. T-cells, on the other hand, respond to small fragments of antigen bound to major histocompatibility complex (MHC) molecules, which in humans are also called human leukocyte antigen (HLA) molecules.[11] The immune system utilises distinct MHC molecules for CD4 and CD8 cells. This division of labour allows the immune system to direct appropriate immune effector mechanisms against pathogens at distinct tissue and subcellular compartments. Class I MHC binds antigenic epitopes within the endoplasmic reticulum and hence alerts the immune system that the cell contains foreign proteins. This alarm signal then triggers CD8 cytotoxic T-cells to attack the virus-infected cell and halt spread of the infection. Class II MHC, on the other hand, binds antigen in intracellular vesicles and hence generates alarm signals from antigens such as extracellular pathogens and microbes that dwell within the vesicle itself. This signal then activates CD4 helper T-cells and they orchestrate the immune response as discussed above. A further class of molecules, class Ib, inhibits cell killing by natural killer (NK) cells, e.g. in humans, cells lacking MHC-I but expressing HLA-G avoid killing by NK cells.

The immune system has clearly evolved to protect against infection. The sophisticated immune system of human beings, however, can still become overly responsive to non-infectious antigens. So-called hypersensitivity diseases include allergic responses directed to otherwise innocuous foreign antigens and autoimmune diseases against the individual's own tissues. Furthermore, the unnatural act of tissue transplantation can lead to a hypersensitivity reaction leading ultimately to graft rejection in the case of a solid organ transplant or alternatively graft versus host disease (GVHD) in the case of a bone marrow transplant.[2] Tissue transplantation confronts the immune system with an extraordinary antigenic challenge and requires long-term use of extremely potent immunosuppressive drugs to overcome rejection. It is important to appreciate, however, that graft rejection may be one of the most primitive immune mechanisms. So, for example, when sponges from two different colonies are held together for a number of days they begin to kill each other whereas sponges from the same colony continue to grow happily. Furthermore, earthworms and starfish will reject tissue grafts from other species. Such rejection reactions are mediated by phagocytic cells and tell us that even lowly invertebrates carry receptors capable of distinguishing self from non-self antigens.

Some of the standard terminology used to describe immune responses is defined in Table 1 and used throughout this chapter.

Table 1. Definitions and explanations of immunological terminology.

Word/Phrase	Definition
Autologous	Self.
Allogeneic	Non-self (different at a major or minor histocompatibility antigen).
Xenogeneic	Different species.
Helper T-cell	Supports humoral as well as cell-mediated immunity.
Tc cell	Cytotoxic T-cell.
NK cell	Natural killer cell.
MHC	Major histocompatibility antigens.
HLA	Human leukocyte antigens.
MHC-I/HLA-I	Molecules that activate cell killing by Tc cells and NK cells.
MHC-II/HLA-II	Molecules that activate Th cells.
HLA-G	Molecule that inhibits NK cells from killing target cells.
Tolerance	Non-reactivity of the host's immune system to a set of antigens while maintaining full immunocompetence and reactivity and to other antigens.
IDO	Indolamine-2,3-dioxygenase — depletes tryptophan, thereby inhibiting T-cell proliferation.
FasL	Fas ligand — binds to fas on activated T-cells, inducing T-cell killing.
B-cells	Antibody-producing cells.
APCs	Antigen-presenting cells — present peptide fragments of antigen to T-cells. Antigen can be presented to either CD8 or CD4 T-cells depending on whether the antigen is presented by either class I or II MHC.
DCs	Dendritic cells — professional antigen presenting cells.
Co-stimulatory molecules	Signals, together with MHC, to drive lymphocyte activation and proliferation.

3. Materno-Foetal Tolerance as a Model for Understanding Stem Cell Immune Privilege

The growing embryo and foetus in the uterus of placental mammals presents the maternal immune system with foreign MHC antigens of paternal origin and yet this allogeneic tissue is not usually rejected by the mother's immune system. This state of natural tolerance is at least in part a result of a variety of immunosuppressive properties of the placental cells at the interface between mother and foetus. A range of mechanisms have been proposed as an explanation of materno-foetal tolerance.[12,13] These warrant detailed consideration as there is growing evidence that some of the pathways involved are recapitulated by both embryonic stem cells (ESCs) and mesenchymal stem cells (MSCs).

3.1. Low expression by placental cells of MHC-I

MHC-I molecules evolved to bind viral particles present in the cytosolic compartment of infected cells. Newly synthesised MHC-I α-units are retained in the endoplasmic reticulum where they bind to the co-stimulatory molecule, β_2-microglobulin (β_2m). Once a viral peptide has bound to the MHC-I, the MHC-peptide-β_2m complex translocates to the cell surface where it activates cytotoxic T-cells through their T-cell receptor, resulting in killing of the virus-infected cell. Since viruses require the host's intracellular machinery to produce their own proteins, this mechanism has evolved to deal with any cell that may be infected by virus rather than to mop up extracellular viral particles. Therefore almost all cells can express MHC-I.

MHC-I molecules are highly polymorphic, reflecting the capacity of the immune system to mount an adaptive response to infection. However this same polymorphism means that in the foetus, MHC-I molecules are not identical to the mother's and so could trigger an immune response resulting in foetal rejection. Reduced expression of MHC-I by the foetal and placental cells can be seen as a mechanism for the avoidance of activation of Tc cells.

3.2. Absence of MHC-II molecules

Unlike MHC-I, MHC-II molecules are only expressed by a restricted number of cell types, namely those antigen-presenting cells (APCs) that are essential for activation of other cells involved in the adaptive immune response. These include macrophages, immature dendritic cells (DCs) and B-cells. The role of MHC-II is to bind peptide that has been generated in the intracellular vesicles of APCs after phagocytic ingestion of bacterium or protein. Presentation of the MHC-II-peptide complex at the cell surface results in activation of helper T-cells through their T-cell receptor as well as activation of specific B-cells when they serve as APC. This results in the release of a range of soluble factors (antibodies and cytokines) that contribute to clearance of bacterium or foreign protein from the body.

Like MHC-I, the MHC-II molecules are highly polymorphic and after transplantation they can be recognised as non-self by the host's immune system. Direct stimulation of the host's own immune system results in a rejection reaction. Thus if an organ is transplanted into a recipient, the greater its contamination with APC's the greater the risk of rejection. The absence of MHC-II expression by foetal and placental cells ensures that the foetus is not rejected as part of an allogeneic response.

3.3. Synthesis by the placental cells of indolamine-2,3-dioxygenase (IDO)

IDO is a key regulatory molecule that catabolises conversion of tryptophan to kynurenine. In the placenta, IDO expression creates a zone of tryptophan depletion and this inhibits T-cell proliferation, which is tryptophan-dependent (thus in mice, T-cell-mediated abortion can be induced using 1-methyl tryptophan to inhibit endogenous IDO).

3.4. Expression of Fas ligand (FasL)

FasL (CD178) is a member of the TNF family. It is expressed on the membranes of Tc cells. Any cell bearing the Fas receptor (CD95) is a target for induction of apoptosis by FasL. It is a natural mechanism for killing and removal of activated Fas-bearing lymphocytes, thereby limiting an immune response. Expression of FasL by placental cells can lead to killing of the mother's activated Tc/lymphocytes in the vicinity of the foetus.

3.5. Secretion of soluble immunosuppressive factors

IL-10 synthesis by placental cells is an important regulatory mechanism as this cytokine has immunosuppressive functions. It may be responsible for downregulation of MHC-I (see above). It may also upregulate HLA-G, part of the MHC-Ib complex. HLA-G is itself an immunosuppressive molecule as it inhibits cell killing by natural killer (NK) cells. Other soluble factors that might contribute to localised immunosuppression include prostaglandin E_2 (PGE_2) and TGFβ.

4. Are Embryonic Stem Cells Immune Privileged?

The concept that ESCs, like the blastocyst from which they were derived, may exhibit immuno-privileged properties was first postulated by Fändrich *et al.*,[12] who noted the lack of expression of MHC molecules by ESCs. A number of subsequent *in vitro* studies have noted the low expression of both MHC-I and MHC-II as well as co-stimulatory molecules by undifferentiated ESCs of both human and rodent origin.[3,14–17]

Where these studies diverge is in their prediction of how ESCs will change their immune status after differentiation or after exposure to inflammatory mediators such as interferon-γ (IFNγ). Drukker *et al.* undertook a detailed study[3,14] in which they followed changes in MHC expression after ESC differentiation in embryoid bodies (EBs) and then further differentiation to teratomas. There was increased MHC-I expression by EB cells and even higher expression by teratomas, though not as high as HeLa cells. However, IFNγ increased MHC-I expression by ESCs to levels that were similar to that of HeLA cells. In contrast MHC-II expression was minimal at all stages of differentiation and even after exposure to IFNγ. These authors conclude that in the absence of MHC-II expression by ESCs and with no contaminating DCs, there is unlikely to be any immune rejection of the embryonic cells through the direct pathway. However if MHC-I is upregulated *in vivo* then there may be immune rejection through activation of CD8-cytotoxic cells assuming activation of CD4 cells by the indirect pathway of antigen presentation. The indirect pathway involves uptake of donor cell antigens by host APCs. Indirect presentation to T-cells and cognate recognition by B-cells is predicted to result in allo-reactive antibody generation, antibody binding to the ESCs/tissue grafts and rejection through activation of complement. In a later report[15] the same authors quantified MHC-I expression more carefully. They found that the inner cell mass of the blastocyst had negligible expression of MHC-I but levels were increased two- to four-fold after EB formation and eight- to ten-fold after teratoma formation. They also noted the general absence of ligands for NK cells and the resistance of ESCs to killing by NK cells.

Grinnemo *et al.*[18] have taken a more pessimistic view. They do not see ESCs as being any different to the majority of cells, namely lacking MHC-II and expressing some MHC-I that can be upregulated by IFNγ. Whilst this interpretation is essentially the same as Drukker *et al.*, these authors do not believe that the low basal expression of MHC-I is of any real consequence. Therefore the authors conclude that in order to avoid immune rejection of ESCs, a strategy of inducing tolerance induction will be required.

In their review of the literature, Fairchild *et al.*[19] are equally pessimistic, stating that activation of the indirect pathway (alloantigen presentation by host DCs) is inevitable. Indeed these authors suggest that even if rejection through MHC molecules can be avoided, there will still be a risk of rejection through minor histocompatibility antigens.

There have been a small number of *in vivo* studies of immune responses to ESCs and these have produced contradictory results. Burt *et al.*[16] differentiated mouse ESCs to haematopoetic precursor cells *in vitro* and then implanted them into allogeneic mice. There was good engraftment of the implanted cells with no evidence of GVHD and bi-directional tolerance (i.e. host for donor and donor for host). Similarly, Menard *et al.*[20] implanted cardiac-committed mouse ESCs into sheep myocardium after initiation of infarction. In this xenogeneic model the nine controls (medium only) showed reduced ventricular function with time. The nine treated animals mostly improved with time even in the absence of immunosuppression by cyclosporine. There was no evidence of an inflammatory infiltrate in any animals. The authors postulated that the lack of rejection could be due to reduced MHC expression or tolerance induction because of the aortic route of administration. Hildebrand *et al.*[21] implanted mouse ESCs and the same cells differentiated to ectodemal/neuroectodermal phenotype, in the cochlea of guinea without any immunosuppression. These cells survived for at least nine weeks with no signs of immune response. The authors postulated that there could have been be some protection of the graft at this site by the blood-labyrinth-barrier. Li *et al.*[17] injected human ESCs into muscles of immune-competent mice and showed no evidence of rejection, after 48 hours.

In contrast, Drukker *et al.*[22] found clear evidence of immune rejection of human ESCs implanted into immune competent mice after one month, whilst, Bonnevie *et al.*[23] found evidence of xenogeneic immune rejection after implantation of mouse ESCs into baboon myocardium. There was evidence for upregulation of MHC-I on the implanted ESCs in this model. In another study, Kofidis *et al.*[24] found that mouse ESCs injected into allogeneic mouse infarcted myocardium led to good engraftment of the cells but clear evidence for immune rejection, with accumulation of immune cells and gradual reduction in graft size. In this study, immune rejection correlated with a progressive increase in MHC-I expression and evidence of differentiation of the mESCs along the myocardial lineage (expression of connexin-43 and α-sarcomeric actin).

Taken together the *in vitro* and *in vivo* studies outlined above seem to indicate that whilst undifferentiated ESCs are likely to be only weakly immunogenic, after differentiation and/or exposure to inflammatory cytokines this hypoimmunogenic state will be gradually lost through upregulation of MHC-I, leading to immune rejection. However, administration of cells by certain routes or into certain immune privileged sites of the body may sometimes be enough to prevent immune rejection.

Strategies for prevention of immune rejection in the clinical setting will need to include some degree of MHC matching combined with induction of tolerance. Taylor *et al.*[25] have attempted to estimate how many human ESC lines would be required to provide cells for the UK population. They have concluded that 150 cell lines would provide a complete match in just 8% of recipients, but a beneficial match in 31% of recipients and an acceptable match in 58%–80% of recipients. Their analysis suggested that there would be minimal benefit from increasing the cell pool further. This paper did not, however, account for the fact that not all ESC lines can be effectively differentiated to all lineages and it is therefore likely that a much larger number will be required to account for differentiation potential as well as avoidance of immune rejection. Transplantation to immune privileged sites such as brain, testis, or cornea remains a possibility. Non-matched foetal dopaminergic neurones have been implanted successfully in Parkinson's disease patients whilst non-matched corneal transplants are also often used, demonstrating the feasibility of utilising site-specific immune privilege. Generation of isogenic human ESC lines by somatic nuclear transfer may be technically possible, but no method for human cells has been developed and even if it were possible the costs would prohibit its widespread use.

5. Are Mesenchymal Stem Cells (MSCs) Immune Privileged?

The earliest evidence that MSCs may avoid immune rejection came from a xenogeneic *in vivo* study in which engraftment of human MSCs occurred after intrauterine transplantation into sheep, even when the transplant was performed after the foetuses had developed and become immune competent.[26] This led to a large number of *in vitro* studies investigating the potential mechanism of hypoimmunogenicity.[13,27–32] These collectively demonstrated that MSCs express a moderate level of MHC-I (greater than ESCs but less than many other cell types), but no MHC-II or co-stimulatory molecules. In mixed lymphocyte reactions (MLRs) in which MSCs are used to drive T-cell proliferation there is little or no T-cell response. Importantly, in third party MLRs where DCs are used to drive T-cell proliferation, MSCs suppress the proliferative response. This demonstrates that they have both immune privilege and tolerance-inducing properties. Intriguingly, exposure of MSCs to IFNγ increases expression of MHC-II but the cells remain both hypoimmunogenic (MLRs) and immunosuppressive (third party MLRs).[27] It has been proposed that the mechanism for this is a paradoxical effect involving upregulation of IDO production by MSCs via IFNγ. This would suggest that in an inflammatory environment, increased MSC production of IDO leads to depletion of tryptophan and therefore suppression of T-cell proliferation (despite the increased MHC-II expression).

Studies of the effects of differentiation (chondrogenesis, adipogenesis and osteogenesis) on MSC immune privilege and immunosuppression are less clear cut. One study has shown no effect of differentiation on the capacity of MSC to reduce immune responses by any of the suggested pathways.[27] Another study of rabbit MSC osteogenesis showed no loss of immune privilege/immunosuppression after *in vitro* differentiation although there was some gradual loss of these properties *in vivo*.[32] A third study has suggested that rat MSCs retain their immunosuppressive properties after adipogenic and osteogenic differentiation but not after chondrogenesis.[33] Data from a range of pre-clinical *in vivo* studies are equally difficult to interpret.[34,35] In baboons, MSCs prolong the time to rejection of histoincompatible skin grafts. In other animals MSCs improve the outcome of tissue (renal/neural/lung) injury, apparently through local anti-inflammatory effects. One study in mice has failed to demonstrate immunosuppression by allogeneic MSCs.[36]

The early recognition that MSCs may sometimes induce tolerance through suppression of T-cell proliferation has led to the experimental use of these cells in some clinical trials. More than 100 patients have received MSCs. These have mostly been allogeneic but HLA-matched and generally given together with haematopoetic stem cell transplantation for the treatment of malignancy or innate metabolic diseases. No side effects have been noted to date and there is some evidence for therapeutic immunosuppression, which has led to new trials of MSCs in treating GVHD.[31,35,37]

Overall, some authors are very optimistic about the potential for use of allogeneic MSCs without the need for immunosuppression.[13] Others are far more cautious. Uccelli *et al.*[38] point out that the mechanism of immunosuppression is still not well delineated and that it is likely that key molecules must be identified through functional molecular profiling of the MSC transcriptome. This molecular profiling approach may best be undertaken using IFNγ-activated MSCs since it is only after exposure to this cytokine that they express their most effective immunosuppressive activity. There remains an urgent need to understand the effect of differentiation on immunosuppression, both *in vitro* and *in vivo*. It seems probable that differentiation will ultimately lead to loss of immunosuppression by these cells and if that is the case then we will need to learn how to recapitulate the effect by activating the embryonic/foetal pathways that seem to be so central to the function of MSCs.

6. Finding a Way Forward for the Use of Allogeneic Stem Cells

There is growing evidence that both ESCs and MSCs share with embryonic and foetal cells the characteristic of very low expression of MHC-I and MHC-II molecules and low immunogenicity. MSCs appear to have the additional capacity to suppress third party immune responses. One new study by Koch *et al.*[39] has demonstrated that murine ESCs may also have the capacity to suppress third party responses, however this one report needs to be substantiated by other studies. Further exploration of the mechanisms underlying these distinct patterns of immunological behaviour are vital if we are to harness the pathways in order to induce tolerance in other types of transplanted cell.

References

1. Lechler, R.I., Sykes, M., Thomson, A.W. and Turka, L.A. Organ transplantation — how much of the promise has been realized? *Nat. Med.* **11**(6), 605–613 (2005).

2. Lechler, R. and George, A.J.T. Transplantation and rejection. In: Male, D., Brostoff, J., Roth, D.B. and Roitt, I. (eds.), *Immunology* (Mosby/Elsevier, Philadelphia, 2006), pp. 383–400.
3. Drukker, M. and Benvenisty, N. The immunogenicity of human embryonic stem-derived cells. *Trends Biotechnol.* **22**(3), 136–141 (2004).
4. Mayhall, E.A., Paffett-Lugassy, N. and Zon, L.I. The clinical potential of stem cells. *Curr. Opin. Cell. Biol.* **16**(6), 713–720 (2004).
5. Male, D. Introduction to the immune system. In: Male, D., Brostoff, J., Roth, D.B. and Roitt, I. (eds.), *Immunology* (Mosby/Elsevier, Philadelphia, 2006), pp. 3–18.
6. Platts-Mills, T.A.E., Male, D., Hay, F., Westwood, O.M.R. and Britton, W. Hypersensitivity in immunology. In: Male, D., Brostoff, J., Roth, D.B. and Roitt, I. (eds.), *Immunology* (Mosby/Elsevier, Philadelphia, 2006), pp. 423–492.
7. Du Pasquier, L. Meeting the demand for innate and adaptive immunities during evolution. *Scand J. Immunol.* **62**(Suppl. 1), 39–48 (2005).
8. Pancer, Z. and Cooper, M.D. The evolution of adaptive immunity. *Annu. Rev. Immunol.* **24**, 497–518 (2006).
9. Pancer, Z., Amemiya, C.T., Ehrhardt, G.R., Ceitlin, J., Gartland, G.L. and Cooper, M.D. Somatic diversification of variable lymphocyte receptors in the agnathan sea lamprey. *Nature* **430**(6996), 174–180 (2004).
10. Jefferis, R. Antibodies in immunology. In: Male, D., Brostoff, J., Roth, D.B. and Roitt, I. (eds.), *Immunology* (Mosby/Elsevier; Philadelphia, 2006), pp. 59–86.
11. Roth, D.B. T cell receptors and MHC molecules. In: Male, D., Brostoff, J., Roth, D.B. and Roitt, I. (eds.), *Immunology* (Mosby/Elsevier, Philadelphia, 2006), pp. 105–126.
12. Fandrich, F., Dresske, B., Bader, M. and Schulze, M. Embryonic stem cells share immune-privileged features relevant for tolerance induction. *J. Mol. Med.* **80**(6), 343–350 (2002).
13. Barry, F.P., Murphy, J.M., English, K. and Mahon, B.P. Immunogenicity of adult mesenchymal stem cells: lessons from the fetal allograft. *Stem Cells Dev.* **14**(3), 252–265 (2005).
14. Drukker, M., Katz, G., Urbach, A., Schuldiner, M., Markel, G., Itskovitz-Eldor, J. *et al.* Characterization of the expression of MHC proteins in human embryonic stem cells. *Proc. Natl. Acad. Sci. USA* **99**(15), 9864–9869 (2002).
15. Drukker, M. Immunogenicity of human embryonic stem cells: can we achieve tolerance? *Springer Semin. Immunopathol.* **26**(1–2), 201–213 (2004).
16. Burt, R.K., Verda, L., Kim, D.A., Oyama, Y., Luo, K. and Link, C. embryonic stem cells as an alternate marrow donor source: engraftment without graft-versus-host disease. *J. Exp. Med.* **199**(7), 895–904 (2004).

17. Li, L., Baroja, M.L., Majumdar, A., Chadwick, K., Rouleau, A., Gallacher, L. *et al.* Human embryonic stem cells possess immune-privileged properties. *Stem Cells* **22**(4), 448–456 (2004).
18. Grinnemo, K.H., Sylven, C., Hovatta, O., Dellgren, G. and Corbascio, M. Immunogenicity of human embryonic stem cells. *Cell Tissue Res.* **331**, 67–78 (2008).
19. Fairchild, P.J., Robertson, N.J., Minger, S.L. and Waldmann, H. Embryonic stem cells: protecting pluripotency from alloreactivity. *Curr. Opin. Immunol.* **19**, 596–602 (2007).
20. Menard, C., Hagege, A.A., Agbulut, O., Barro, M., Morichetti, M.C., Brasselet. C. *et al.* Transplantation of cardiac-committed mouse embryonic stem cells to infarcted sheep myocardium: a preclinical study. *Lancet* **366**(9490), 1005–1012 (2005).
21. Hildebrand, M.S., Dahl, H.H., Hardman, J., Coleman, B., Shepherd, R.K. and de Silva, M.G. Survival of partially differentiated mouse embryonic stem cells in the scala media of the guinea pig cochlea. *J. Assoc. Res. Otolaryngol.* **6**(4), 341–354 (2005).
22. Drukker, M., Katchman, H., Katz, G., Even-Tov Friedman, S., Shezen, E., Hornstein, E. *et al.* Human embryonic stem cells and their differentiated derivatives are less susceptible to immune rejection than adult cells. *Stem Cells* **24**(2), 221–229 (2006).
23. Bonnevie, L., Bel, A., Sabbah, L., Al Attar, N., Pradeau, P., Weill, B. *et al.* Is xenotransplantation of embryonic stem cells a realistic option? *Transplantation* **83**(3), 333–335 (2007).
24. Kofidis, T., deBruin, J.L., Tanaka, M., Zwierzchoniewska, M., Weissman, I., Fedoseyeva, E. *et al.* They are not stealthy in the heart: embryonic stem cells trigger cell infiltration, humoral and T-lymphocyte-based host immune response. *Eur. J. Cardiothorac. Surg.* **28**(3), 461–466 (2005).
25. Taylor, C.J., Bolton, E.M., Pocock, S., Sharples, L.D., Pedersen, R.A. and Bradley, J.A. Banking on human embryonic stem cells: estimating the number of donor cell lines needed for HLA matching. *Lancet* **366**, 2019–2025 (2005).
26. Liechty, K.W., MacKenzie, T.C., Shaaban, A.F., Radu, A., Moseley, A.M., Deans, R. *et al.* Human mesenchymal stem cells engraft and demonstrate site-specific differentiation after *in utero* transplantation in sheep. *Nat. Med.* **6**(11), 1282–1286 (2000).
27. Le Blanc, K., Tammik, C., Rosendahl, K., Zetterberg, E. and Ringden, O. HLA expression and immunologic properties of differentiated and undifferentiated mesenchymal stem cells. *Exp. Hematol.* **31**(10), 890–896 (2003).
28. Le Blanc, K. Immunomodulatory effects of fetal and adult mesenchymal stem cells. *Cytotherapy* **5**(6), 485–489 (2003).

29. Ilancheran, S., Michalska, A., Peh, G., Wallace, E.M., Pera, M. and Manuelpillai, U. Stem cells derived from human fetal membranes display multilineage differentiation potential. *Biol. Reprod.* **77**(3), 577–588 (2007).
30. Aggarwal, S. and Pittenger, MF. Human mesenchymal stem cells modulate allogeneic immune cell responses. *Blood* **105**(4), 1815–1822 (2005).
31. Dazzi, F., Ramasamy, R., Glennie, S., Jones, S.P. and Roberts, I. The role of mesenchymal stem cells in haemopoiesis. *Blood Rev.* **20**(3), 161–171 (2006).
32. Liu, H., Kemeny, D.M., Heng, B.C., Ouyang, H.W., Melendez, A.J. and Cao, T. The immunogenicity and immunomodulatory function of osteogenic cells differentiated from mesenchymal stem cells. *J. Immunol.* **176**(5), 2864–2871 (2006).
33. Chen, X., McClurg, A., Zhou, G.Q., McCaigue, M., Armstrong, M.A. and Li, G. Chondrogenic differentiation alters the immunosuppressive property of bone marrow-derived mesenchymal stem cells, and the effect is partially due to the upregulated expression of B7 molecules. *Stem Cells* **25**(2), 364–370 (2007).
34. Le Blanc, K. and Pittenger, M. Mesenchymal stem cells: progress toward promise. *Cytotherapy* **7**(1), 36–45 (2005).
35. Le Blanc, K. Mesenchymal stromal cells: tissue repair and immune modulation. *Cytotherapy* **8**(6), 559–561 (2006).
36. Fibbe, W.E., Nauta, A.J. and Roelofs, H. Modulation of immune responses by mesenchymal stem cells. *Ann. N. Y. Acad. Sci.* **1106**, 272–278 (2007).
37. van Laar, J.M. and Tyndall, A. Adult stem cells in the treatment of autoimmune diseases. *Rheumatology (Oxford)* **45**(10), 1187–1193 (2006).
38. Uccelli, A., Moretta, L. and Pistoia, V. Immunoregulatory function of mesenchymal stem cells. *Eur. J. Immunol.* **36**, 2566–2573 (2006).
39. Koch, C.A., Geraldes, P. and Platt, J.L. Immunosuppression by embryonic stem cells. *Stem Cells* **26**, 89–98 (2008).

Chapter 11

Development of a Design of Experiment Methodology: Applications to the Design and Analysis of Experiments

Mayasari Lim and Athanasios Mantalaris

Abstract

Stem cell cultures are complex and intricate processes. There are many contributing factors that affect and influence the outcome of the culture, including (1) factors affecting the physicochemical environment, (2) nutrients and metabolites, (3) growth factors, and (4) the various cell types that exist within the cell culture system, as shown in Fig. 1A.[1–8] Physicochemical factors, such as pH, temperature, dissolved oxygen and carbon dioxide levels, determine the cellular environment and affect cellular behaviour and functionality. Slight changes in the levels of these physicochemical parameters will result in large changes in the cellular output. For instance, the optimal pH for the process of megakaryopoiesis in haematopoietic cell cultures has been reported to be at pH 7.60 while for the process of granulopoiesis it has been reported to be at pH 7.21.[1,2] Such reports highlight the transient nature of cell bioprocesses and the existence of gradients occurring in the *in vivo* microenvironment. Specifically, the inherent heterogeneous and transient nature of cell bioprocesses, which includes cells at different maturation and differentiation stages (stem cells, progenitors, precursors, and mature cells), necessitate that culture conditions are dynamic during the culture time. Consequently, control of cell culture bioprocesses is difficult to achieve and culture conditions are generally not optimal.

Keywords: Design of Experiments; Bioprocess.

Outline

1. Analysis of Factors
2. Design Strategy
3. DOE Designs
4. A DOE Example: Investigating the Influence of Cytokines on Erythropoiesis
5. Conclusions

   References

1. Analysis of Factors

When investigating a process, two main considerations need to be recognised, namely the input(s) and output(s) of a process (Fig. 1B). The input(s) of a process is often referred as the process variable(s), including both controllable and uncontrollable variables.[9] Controllable variables are those which can be manipulated and adjusted accordingly. Uncontrollable variables are often related to noise or nuisances that are beyond control yet still affect the process. The output of a process is a response variable; it represents the result of the experiment. Response variables are dependent variables as they are affected by process variables and will change according to changes in the process variables. In stem cell culture bioprocesses, for example, the four key categories of process variables that need to be considered are (1) factors determining the physicochemical environment, (2) nutrients and metabolites, (3) growth factors, and (4) different cell types existing in the culture. Currently, when investigating the effects of controllable process variables, traditional methods such as dose-response experimentation are being utilised. In dose-response experiments, each process factor is being investigated one factor at a time while the other factor(s) is kept at a constant value. This is often called the one-factor analysis. Unfortunately, this type of experimentation method is highly inefficient, particularly when many process factors are involved. When factors are investigated singly, the study in unable to reveal process interactions between different factors and may, therefore, not yield accurate process information. In many cases, a false optimal is obtained as a result of this type of experimentation methodology. A much more effective and efficient way of experimentation is through the design of experiments (DOE). DOE maximises the use of process data while requiring the minimal amount of experimentation and time to produce relationships that accurately and quantitatively describe the process.

As an example, let us investigate a process using two factors, *A* and *B*. When dose-response/one-factor analysis is used, two experiments have to be performed; the first experiment is to manipulate factor *A* while keeping factor *B* at a constant

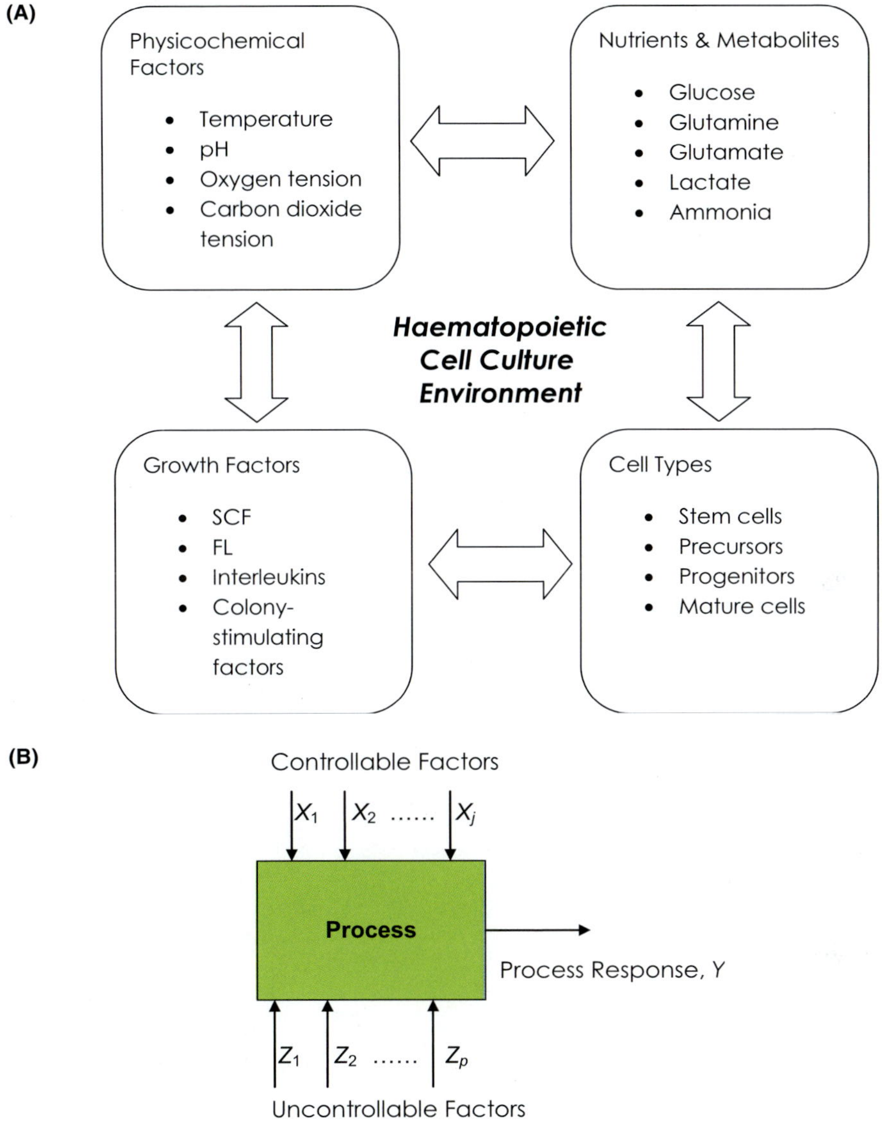

Fig. 1. **(A)** Factors affecting cell culture bioprocesses. **(B)** Process variables influencing a process.

value (often the mid-value of the range is chosen). The result is a line profile showing the influence of A on the process (Fig. 2A). From this plot, the optimal value of A is determined for the maximum output (in this case, $A = 10$). Subsequently, the second experiment is performed by keeping factor A constant (at the optimal value determined from the previous experiment; $A = 10$) while manipulating factor B. The result is another line profile showing the influence of

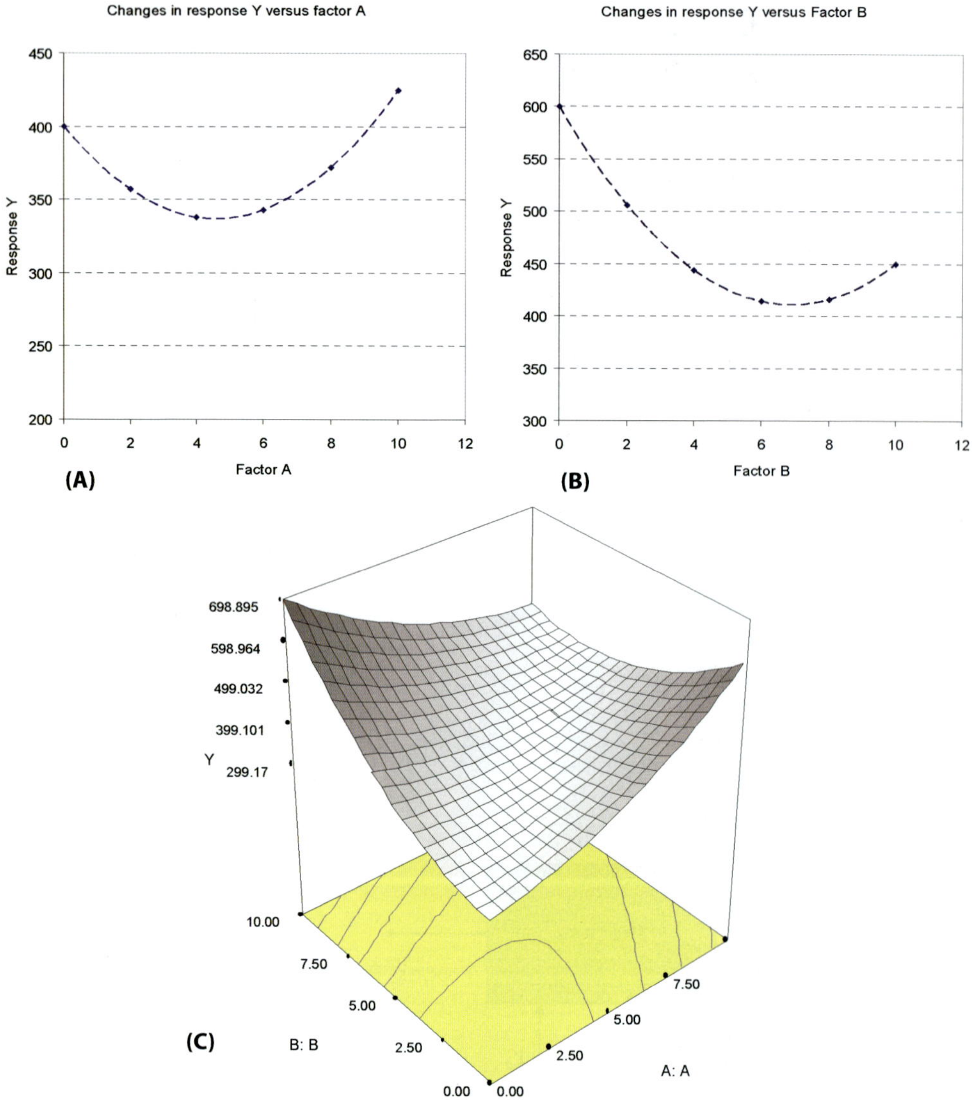

Fig. 2. Example of one-factor (dose-response) and DOE analysis. **(A)** Response *Y* when varying factor *A* only. **(B)** Response *Y* when varying Factor *B* only. **(C)** Response *Y* when varying both *A* and *B* using DOE.

B on the process (Fig. 2B). From this plot, the optimal value of *B* for maximum output is determined to $B = 0$. Therefore, the final outcome from the one-factor analysis is found to be $Y = 600$ with $A = 10$ and $B = 0$. In contrast, when using a DOE approach, where both factors *A* and *B* are manipulated simultaneously in one experiment to generate a result showing the effects of *A* and *B* on the process

(Fig. 2C), a completely different result — the correct one — is obtained. From the DOE plot, the optimal values of A and B for maximum production are actually at $A = 0$ and $B = 10$, giving an output $Y = 700$. This example clearly illustrates that using a dose-response method to study a process is not always ideal as it can result in false conclusions about the process. Furthermore, one-factor analysis is a longer procedure, requiring more experiments, thus wasting time and labour. From this example, it is clear that DOE is a much more efficient and effective way of experimentation, which, in addition, is more capable of dealing with a large number of factors compared to the dose-response method. Through the use of DOE methodology, complete three-dimensional (3D) response surface plots or contour maps can be generated that reveal process characteristics and factor relationships accurately. From these plots, one can easily determine the optimal operating conditions for improved bioprocesses. It is, therefore, a very powerful and highly efficient method for studying complex systems, such as stem cell cultures.

2. Design Strategy

When dealing with a complex system, a systematic approach should be taken to obtain useful information for defining the bioprocess of interest (Fig. 3). Typically this would involve the following three steps: screening, characterisation, and optimisation using the appropriate DOE design(s) in each stage. In the first stage, *process screening*, a large number of factors (more than five) is typically involved. The objective, therefore, is to identify factors that have a significant influence on the process against factors that have little or no influence at all. Process screening allows the quick identification of critical factors that influence the process amongst the many possible process variables in minimal time and effort (number of experiments). Some of the commonly used designs for such investigations are

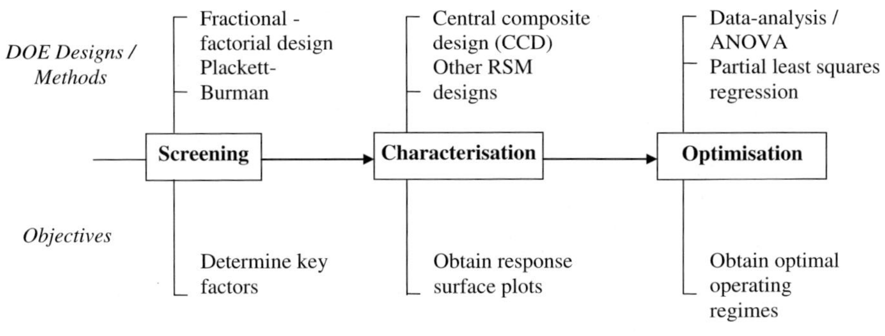

Fig. 3. Systematic approach to DOE methodology for the design and analysis of experiments.

the fractional factorial design and the Plackett-Burman design. These designs allow rapid screening of a large number of factors using a very small number of experiments, yet also providing statistical significances. In the Plackett-Burman design, for example, a 12-run experiment can screen up to 11 factors. In the second stage, *process characterisation*, the goal is to obtain a more detailed and quantitative description about the process by way of a quadratic model and 3D surface response plots to define process characteristics. This is achieved by designing experiments for the factors identified from *process screening* so that full process interactions of contributing factors are revealed. Common designs include the central composite design (CCD) and Box-Benhken design, which reveal process relationships that can be described using a quadratic or cubic model. In the third stage, *process optimisation*, proper analysis of the results and the generation of 3D plots will reveal optimal operating regimes for the important process parameters, yielding the best conditions of an optimised process. Other methods of *process optimisation* include the path of steepest ascent in which the rate of change due to a factor is investigated to determine the optimal process conditions. The use of this systematic framework to study an unknown process alleviates many of the unnecessary experiments, saving significant time, effort, and money, while ensuring that proper characterisation of the process will be achieved. The outcome is a reproducible and statistically valid elucidation of bioprocesses that do not require *a priori* assumptions.

To date, the use of DOE in stem cell bioprocess characterisation has been somewhat limited; much of the research still utilises traditional dose-response methods. However, successful application of various DOE designs for haematopoietic stem cell culture studies has been demonstrated by various investigators. These include the use of fractional factorial designs to perform screening experiments followed by optimisation via the path of steepest ascent,[10,11] and the use of central-composite designs (CCD) to perform process characterisation.[12–14] However, full process characterisation, which considers the totality of process factors, including physicochemical parameters, nutrients, metabolites and growth factor concentrations, has not yet been established. The successful application of DOE methodologies to resolve the complexity in stem cell bioprocessing will yield invaluable information and result in reproducible, well-characterised cultures.

3. DOE Designs

Two-level factorial designs are the basic building blocks for most DOE designs and are, therefore, the most commonly applied designs. Factorial designs are very effective in studying both main and interactive effects of a process rendering them extremely useful for response surface studies or modelling.[15] Two-level factorial designs study each factor in a process at two levels, a high and a low. A basic two-level full factorial design involves all possible combinations of high and low values of the various process factors under investigation. This yields a total number of 2^n experiments, where n is the number of factors in the experiment.[9,16] Specifically, in a two-factor two-level factorial design, the design points relate to four corners of a square, representing full coverage of the experimental space, and yields four (2^2) experimental runs. Similarly, a three-factor factorial design yields eight (2^3) experimental runs and covers all corners of a cube (Fig. 4A). Basic two-level factorial designs are mainly used to study linear relationships between process factors and their responses, yet are still capable of revealing both main and interactive effects independently. Modifications can be made to the full factorial design to generate other DOE designs suited for different applications.

One of the most common and widely applied designs for characterisation studies is the central composite design. CCD is used to build second-order response surface models and derives from two-level factorial designs, but also includes a centre-point and axial points in the basic 2^n design. The centre-point is used to estimate curvature in a second-order model and its replicates are used for estimating pure experimental error, while axial points are used mainly to estimate quadratic terms in a second-order model.[15] The position of the axial point depends on α, the axial distance, which varies according to the extent of the region of operation. The value of α can range from 1 to $\sqrt{n}$; where n is the number of factors in the experiment. The choice of the α value affects design rotatability, which relates to robustness and stability of the design. A rotatable design is one that provides equal variance in the estimation of predicted values at any two locations in the design space.[15] A CCD with an α value of $\sqrt{n}$ is referred to as a spherical design and is considered rotatable. A CCD with an α value of 1 is a cubodial or face-centred design (Fig. 4B). Though they are not rotatable, face-centred designs are often applied in many experimental studies where design ranges are restricted and are still very effective in providing full process characterisation for a defined process regime. Other available designs include the Box-Behnken design, spherical response surface method (RSM) designs, and Equiradial designs.[15]

When a large number of factors is involved in a process, full factorial designs are not practical since the number of experimental runs become significantly large. In an eight-factor experiment, for example, the required number of experimental runs for a full factorial design would be 256 (2^8). Running this design would be costly and extremely time-consuming, thus making the full factorial design an impractical option for studying this process problem. In general, when a large number of factors (more than five) are involved, screening is performed and two-level fractional factorial designs can be used to identify the influencing

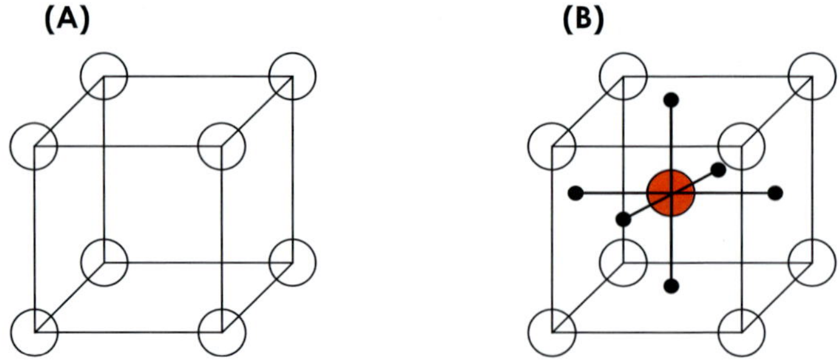

(C)

Number of Factors

	2	3	4	5	6	7	8	9
4	Full	1/2 Fract.						
8		Full	1/2 Fract.	1/4 Fract.	1/8 Fract.	1/16 Fract.		
16			Full	1/2 Fract.	1/4 Fract.	1/8 Fract.	1/16 Fract.	1/32 Fract.
32				Full	1/2 Fract.	1/4 Fract.	1/8 Fract.	1/16 Fract.
64					Full	1/2 Fract.	1/4 Fract.	1/8 Fract.
128						Full	1/2 Fract.	1/4 Fract.

Fig. 4. **(A)** Three-factor two-level factorial design. **(B)** Face-centred central composite design. **(C)** Experiment design matrix for full and fractional factorial designs (red = Res III; yellow = Res IV; green = Res V). Adopted from Design Expert software, StatEase.

parameters. Fractional factorial designs derive from full factorial designs, where only a fraction of the full design is used to perform process screening. The assumption made in these designs is that higher-order interactions (involving three or more factors) are often negligible, therefore only the main factors and low-order interactions (two-factor interactions) are estimated in this design. The number of experimental runs is reduced as both the main and interactive effects are aliased with each other. The effectiveness of a fractional factorial design depends on the resolution of the design; these include Resolution III, IV, and V.[15,16] Resolution III designs are ones that have no main effects aliased with another main effect; they are aliased with two-factor interactions. They provide the smallest number of experimental runs possible but only allow a rough estimation of the main effects since they are biased by two-factor interactions. Resolution IV designs have no main effects aliased with another main effect or a two-factor interaction; they therefore allow a good estimation of each main factor. Two-factor interactions are aliased with each other and are therefore not estimable. Finally, Resolution V designs are ones that have no main effects or two-factor interactions aliased with each other, so they can estimate both main effects and two-factor interactions but they tend to have too many experimental runs. In conclusion, a design with a higher resolution represents a more thorough analysis since fewer factors are aliased with each other, but this also means that it requires a greater number of experiments. For most screening purposes, a Resolution IV design is sufficient to perform accurate initial screening. Figure 4C shows a matrix of the various fractional factorial and full factorial designs and their corresponding number of experimental runs required.

4. A DOE Example: Investigating the Influence of Cytokines on Erythropoiesis

To demonstrate the application of the DOE strategy developed herein in stem cell bioprocessing, cord blood CD34+ cells were cultured and differentiated towards the erythroid lineage. Specifically, the effects of different growth factors for the directed erythropoietic differentiation were evaluated. The process of erythropoiesis is influenced by a number of early- and late-acting growth factors. EPO, for example, has been known to be the lineage-specific cytokine responsible for differentiation towards the erythroid lineage. Based on literature, a number of cytokines were selected to be investigated using the DOE methodology, including both early- and late-acting ones. Specifically, SCF, FL, IL-3, IL-6, granulocyte-macrophage colony stimulating factor (GM-CSF), TPO, and EPO were chosen. The goal of the DOE methodology was to identify and optimise the cytokines that would enable the greatest cell expansion (total cell number) coupled with the maximum yield of red blood cell formation (% red blood cells).

Cryopreserved human umbilical cord blood units used in the DOE experiments were supplied by the London Cord Blood Bank in accordance with regulatory and ethical policies. First, mononuclear cells (MNCs) were separated and isolated from the cord blood using Ficoll-Hypaque (GE-Healthcare) via density centrifugation according to the manufacturer's instructions. MNCs were then incubated with CD34 direct microbeads (Miltenyi Biotec) for 30 minutes at 4°C before being subjected to a high-gradient magnetic field in midi- and mini-MACS columns for positive microbead selection (Miltenyi Biotec). Cells were subjected to two-column selections and a yield of >94% purity in CD34+ cells was obtained in all cell samples. Freshly isolated CD34+ cells were then cultured in culture medium containing 10% (v/v) foetal bovine serum (FBS; Gibco) in Iscove's Modified Dulbecco's Medium (IMDM; Gibco) plus the growth factor cocktail at concentrations specified by the DOE. Cells were cultured in 24-well plate systems

(Corning) at 5×10^4 cells/mL and fed every three days starting from day 4 after initial seeding of the cord blood CD34+ cells. Culture conditions were maintained at 37°C, 5% carbon dioxide and 20% oxygen in a fully humidified incubator. Recombinant human SCF, IL-3, IL-6, FL, TPO, GM-CSF (all from Biosource), and EPO (R&D Systems) were used in the cell culture DOE studies.

The first step in elucidating the effects of the seven chosen growth factors was process screening by using a resolution IV fractional factorial design. This design yields 16 independent experiments, as a result of $2^{(7-3)}$ experimental runs. The output, total cell number and % red blood cells, was evaluated at day 10 of the culture. Specifically, cell enumeration of viable cells determined the total cell expansion by staining the cells with eryothrosin-B stain solution (ATCC) and counting using a haemocytometer. Flow cytometry was employed to evaluate the expression of CD45, CD71, and glycophorin-A (GPA), thus determining the percentage (%) of red cells. Specifically, during each passage, 0.5–1.0 $\times 10^6$ cells were collected and stained with CD45-FITC, CD71-PE and GPA-PC5 (all from BD Biosciences) for 30 minutes at 4°C. Following incubation, the cells were washed twice with PBS + 0.1% BSA and 0.1% sodium azide before being fixed with 1% para-formaldehyde. Analysis using the Epics-Altra (Beckman Coulter) cytometer was performed within three days of staining; 30,000 events were collected for each cell sample and the data is analysed using WinList. The results from the screening experiment were then used in the next step, process characterisation, where a face-centred CCD was employed to further characterise process effects and interactions. Three centre replicates were included in this design for process repeatability and estimation of pure error. Data obtained from the characterisation study were analysed; the ANOVA table was evaluated for model significance and lack of fit, and 3D surface response plots for each factor variable were generated to determine optimal cytokine operating conditions. Cell cultures were repeated at least three times using the optimal cytokine concentrations obtained from the characterisation study and the culture period was extended to 16 days. The design software used to generate the DOE experiments and analyse the data was Design Expert (StatEase) and MODDE7 (Umetrics).

From the screening experiment, two coefficient plots were obtained for total cell expansion and the percentage of red blood cells (%RBC), as shown in Fig. 5A. Both plots conclude that only SCF and EPO play a significant role in affecting cell expansion and differentiation towards the erythroid lineage in cord blood CD34+ cell cultures. Both coefficients of SCF and EPO have a p-value of < 0.05, which indicates that their effects are significant. These two cytokines were subsequently used in the characterisation study by performing a two-factor CCD experiment. A total of 11 experiments were generated from this design resulting in two contour maps for each parameter of interest (Fig. 5B). For cell expansion,

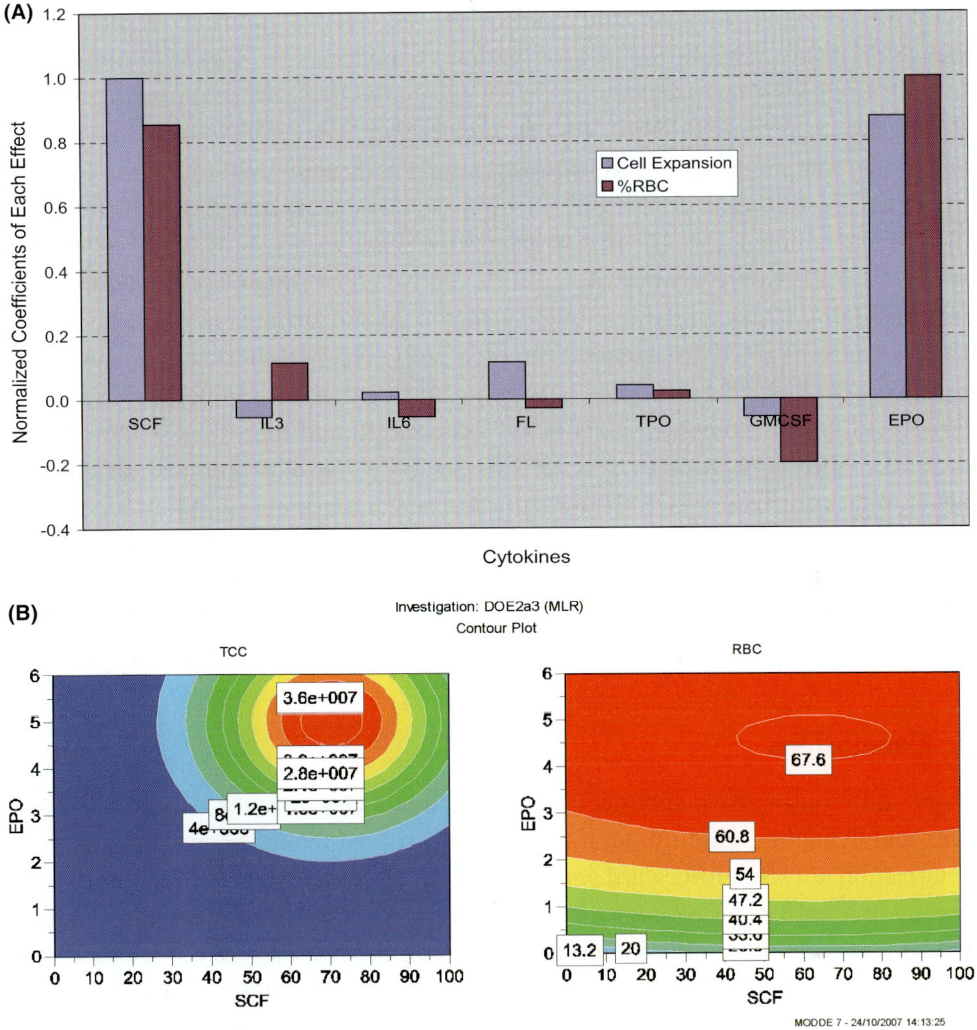

Fig. 5. **(A)** Process screening results. Normalised coefficient plot for total cell expansion and %RBC produced after ten days in culture. **(B)** Process characterisation results. Contour maps for total cell expansion and %RBC produced after ten days in culture.

the results demonstrate that both SCF and EPO play an important role, revealing a combination of main and interactive effects from each factor (Fig. 5B). However, for differentiation into the erythroid lineage, the results demonstrate that only EPO, but not SCF, significantly affects this process (Fig. 5B) since the contour lines are nearly parallel to the x-axis for SCF showing there is little/no change as SCF concentration increases or decreases. These results are consistent with the literature since EPO has been identified as an erythroid lineage-dominant cytokine for the initiation and control of red cell maturation. Furthermore, the DOE results enable the determination of optimal cytokine concentrations for cell expansion and differentiation. Specifically, these were determined to be 75 ng/ml of SCF and 4.5 U/ml of EPO, respectively. The DOE results were further verified in a separate optimisation study, where cells were cultured at the optimal cytokine concentration conditions for up to 16 days. Cell growth and differentiation were evaluated and repeated at least three times, confirming high cell expansion (>20,000) by the end of the culture and accelerated red cell maturation by way of high GPA expression (>90%) at day 7 of the culture. In conclusion, the DOE results enabled the generation of a fully characterised and quantitative process identifying relationships between significant growth factors affecting the process of erythropoiesis in cord blood cultures. In addition, they demonstrated the successful application of a systematic strategy that is simple and efficient in generating information-rich data. Using the screening method, quick identification of the important growth factors was accomplished. The characterisation analysis resulted in a quantitative model describing the effects of each contributing factor. Finally, the optimal cytokine concentrations were determined from the 3D surface response plots.

5. Conclusions

Stem cell bioprocesses, such as in haematopoietic cell cultures, are highly complex and dynamic. A multitude of factors are involved rendering control and characterisation not straightforward and incomplete. However, in order to achieve high quality and high yield cellular products and bioprocesses that meet clinical standards, a good understanding and tight control of the cell culture is required. It is, therefore, necessary to utilise an efficient and effective method for analysing such complex bioprocesses in order for us to achieve maximum process information with minimal time and effort. DOE methodologies provide such tools that facilitate the efficient and effective analysis of complex bioprocesses. In this chapter, we have demonstrated the feasibility of DOE systematic framework and design strategy that enables identification of important process parameters, elucidation of interaction effects, and optimisation of the process conditions.

References

1. Yang, H., Miller, W.M. and Papoutsakis, E.T. Higher pH promotes megakaryocytic maturation and apoptosis. *Stem Cells* **20**, 320–328 (2002).
2. Hevehan, D.L., Papoutsakis, T.E. and Miller, W.M. Physiologically significant effects of pH and oxygen on granulopoiesis. *Exp. Hematol.* **28**, 267–275 (2000).

3. McAdams, T.A., Miller, W.M. and Papoutsakis, T.E. pH is a potent modulator of erythroid differentiation. *Br. J. Haematol.* **103**, 317–325 (1998).
4. Hevehan, D.L., Wenning, L.A., Miller, W.M. and Papoutsakis, T.E. Dynamic model of *ex vivo* granulocytic kinetics to examine the effects of oxygen tension, pH and IL-3. *Exp. Hematol.* **28**, 1016–1028 (2000).
5. Proulx, C., Dupuis, N., St-Amour, I., Boyer, L. and Lemieux, R. Increased megakaryopoiesis in cultures in CD34-enriched cord blood cells maintained at 39°C. *Biotechnol. Bioeng.* **88**(6), 675–680 (2004).
6. Alcorn, M.J. and Holyoake, T.L. *Ex vivo* expansion of haematopoietic progenitor cells. *Blood Rev.* **10**(3), 167–176 (1996).
7. Banu, N., Deng, B., Lyman, S.D. and Avraham, H. Modulation of haematopoietic progenitor development by Flt3-ligand. *Cytokine* **11**(9), 679–688 (1998).
8. Lazzari, L., Lucchi, S., Rebulla, P., Porretti, L., Puglisi, G., Lecchi, L. and Sirchia, G. Long-term expansion and maintenance of cord blood haematopoietic stem cells using thrombopoietin, Flt3-ligand, interleukin (IL)-6 and IL-11 in a serum-free and stroma-free culture system. *Br. J. Haematol.* **112**(2) (2001), 397–404.
9. Natrella, M. In: Croarkin, C. and Tobias, P. (eds.), *NIST/SEMATECH e-Handbook of Statistical Methods* (National Institute of Standards and Technology, USA).
10. Yao, C.-L., Chu, I.-M., Hsieh, T.-B. and Hwang, S.-M. A systematic strategy to optimize *ex vivo* expansion medium for human hematopoietic stem cells derived from umbilical cord blood mononuclear cells. *Exp. Hematol.* **32**, 720–727 (2004).
11. Yao, C.-L., Liu, C.-H., Chu, I.-M., Hsieh, T.-B. and Hwang, S.-M. Factorial designs combined with the steepest ascent method to optimize serum-free media for *ex vivo* expansion of human hematopoietic stem cells. *Enzy. Microbial Technol.* **33**(4), 343–352 (2003).
12. Cortin, V., Garnier, A., Pineault, N., Lemieux, R., Boyer, L. and Proulx, C. Efficient *in vitro* megakaryocyte maturation using cytokine cocktails optimized by statistical experimental design. *Exp. Hematol.* **33**, 1182–1191 (2005).
13. Zandstra, P.W., Conneally, E., Petzer, A.L., Piret, J.M. and Eaves, C.J. Cytokine manipulation of primitive human hematopoietic cell self-renewal. *Proc. Natl. Acad. Sci. USA* **94**, 4698–4703 (1997).
14. Zandstra, P.W., Petzer, A.L., Eaves, C.J. and Piret, J.M. Cellular determinants affecting the rate of cytokine depletion in cultures of human hematopoietic cells. *Biotechnol. Bioeng.* **54**(1), 58–66 (1997).
15. Myers, R.H. and Montgomery, D.C. *Response Surface methodology: Process and Product Optimization Using Designed Experiments* (John Wiley and Sons, Inc., 2002).
16. Montgomery, D.C. *Design and Analysis of Experiments* (John Wiley and Sons, Inc., Arizona, 2001).

Chapter 12

Banking Stem Cell Lines for Future Therapies

Glyn N. Stacey and Charles J. Hunt

Abstract

The provision of central resources of human stem cell lines will be an important element in enabling progress in stem cell research and the development of safe and effective cell therapies. These resource centres, commonly referred to as "stem cell banks", could promote advances in the field of stem cell research, by providing access to well-characterised and quality controlled seed stocks of human stem cell lines that have been checked for appropriate ethical provenance. Such banks can also deliver benefits for the field through establishing international collaboration and standardisation between the developing national stem cell banking centres. This chapter will review the key issues that centres banking and distributing stem cells must address to support the research community in the development of exciting cell therapies for the future.

Keywords: Quality Assurance; Safety; Standardisation; Cell Lines; Stem Cells; Quality Control.

Outline

1. Introduction
2. The Rationale for Centralised Banks of Human Cell Lines for Clinical Use
3. Fundamental Issues for *in Vitro* Cell Culture
4. Cell Culture Processes
   4.1. The cell banking process
   4.2. Scaleability of stem cell culture
   4.3. Cryopreservation and low-temperature storage
5. Quality Assurance and Quality Control
   5.1. Risk from donor selection/harvest

5.2. Cell Line Master File (CLMF): principles and potential content
5.3. Ethical issues and the UK model
6. International Perspectives
6.1. The "International Stem Cell Banking Initiative"
7. Future Developments and Expectations
References

1. Introduction

The viability and general quality of cell cultures are highly susceptible to the adverse effects of laboratory accidents and microbial contamination. They are also prone to genetic change and persistent microbial infections that may not be evident from microscopic examination but which can have critical effects on their properties and characteristics. In the 1940s, the discovery that fowl spermatozoa could be treated with cryoprotective agents, frozen to ultra-low temperatures and subsequently recovered in a viable state,[1] led to the adoption of this technique for a range of cell cultures. This critical development has led to the establishment of cell and tissue banks worldwide.

Although definitions may vary slightly depending on the particular application, a cell bank in its simplest form is a collection of containers of cryopreserved cells that are aliquots of homogenous cell suspension derived from a single preparation/culture of cells. Such stocks of cells deliver a number of key benefits. Cells are generally held in the gaseous phase of liquid nitrogen (typically around $-160^{\circ}C$ to $-180^{\circ}C$) at which temperature the cells are in "suspended animation", in which state no biological processes can occur, thus assuring stability in the long term. Cells have been recovered into culture from the cryopreserved state over decades (summarised in Ref. 2) and theoretically should remain viable indefinitely. In addition, a cryopreserved bank of cells can be held in cold storage until samples have been taken and quality control, characterisation and safety testing completed, thereby ensuring an appropriate level of testing as well as addressing regulatory issues before cells are used, providing greater confidence and increased levels of safety for banked cells.

Distribution of frozen cells from a central bank significantly promotes standardisation of experimental research, testing and production activities utilising cell cultures. National and international cell banks have been established for a variety of purposes and there are international bodies that coordinate these activities in different fields (Table 1).

With the successful isolation of human embryonic stem (hES) cell lines in 1998,[3] and the demonstration of their ability to develop into each of the three germ layers required to provide all the cells of the human body, there was an

Table 1. National and international biological resource centre activity.

Activity type	Description	Examples
Cell line banking organisations	General cell line collections	ATCC (www.atcc.org/), DSMZ (www.dsmz.de/), ECACC (www.ecacc.org.uk/), ICLC (http://www.iclc.it/), JCRB/HSRRB (http://cellbank.nibio.go.jp/)
	Cell banks servicing genetic research	Coriell Institute for Medical Research (http://www.coriell.org/) ECACC (www.ecacc.org.uk/)
	Cell banks focused on serving the cancer community	DSMZ (leukaemia and lymphoma cell lines) (www.dsmz.de/) Health Science Research Resources Bank (previously JCRB) (http://cellbank.nibio.go.jp/)
	Biological resource centres focused on specialist applications of cells in production and testing of biologicals	Centers for Disease Control (www.cdc.gov/) NIBSC, UK (www.nibsc.ac.uk/)
	Cell banks focused on stem cell research	UK Stem Cell Bank (www.ukstemcellbank.org.uk/) WiCell (www.wicell.org/)
Biological resource centre coordination	International networks supporting the development of biological resource centres	European Culture Collection Oraganisation (www.eccosite.org/) International Society for Biological and Environmental Repositories (ISBER) (www.isber.org/) International Stem Cell Forum (www.stemcellforum.org/) World Federation for Culture Collections (http://wdcm.nig.ac.jp/wfcc/) World Health Organization (www.who.int/)

immediate recognition of the tremendously exciting clinical potential of these cells for the replacement of damaged or defective tissues. However, for some this also raised significant ethical issues relating to the sanctity of early human life and whether the use of human embryos could be justified from ethical and religious perspectives.

In some countries this may not be considered a serious issue; in a few all work on human embryonic stem cells has been banned. In others, national policy has set a date limit on the derivation and use of hES cell lines and research on hES cells isolated after this cut-off date is prohibited. In the UK, following a period of public consultation and deliberation by a House of Lords select committee, a pragmatic position was taken to extend UK legislation to permit the use of supernumerary embryos for the isolation of, and research with, hES cells. However, the UK Government also acknowledged the ethical concerns and instigated a framework for ethical oversight of all hES cells research. An important part of this framework was the establishment of a national bank for human stem cell lines that could, through its Steering Committee, ensure ethical consent to donation and full traceability as well as providing appropriate quality control for its cell lines. Thus the UK National Stem Cell Bank project was initiated through government funding administered by the Medical Research Council and the Biotechnology and Biological Sciences Research Council.[4]

Since the announcement of the UK Stem Cell Bank as the first public service collection for human stem cell lines a number of countries have taken up the idea of creating national stem cell banks for hES cells. In the remainder of this chapter we describe the key issues and activities involved in establishing a central resource for supply of hES cell lines and in particular the challenges in providing cell banks for the development of therapies that are fit for use as seed stocks in clinical trials.

2. The Rationale for Centralised Banks of Human Cell Lines for Clinical Use

An obvious immediate solution to the provision of hES cells to the research community is for the originating research laboratories which isolated the lines to carry out the distribution and this option was quickly adopted by the NIH who funded individual research groups to distribute their cell lines. Clearly the originating laboratories are best placed to provide instruction on the culture and characterisation of the cells for research purposes. However, there are a number of potential disadvantages with this strategy. Inevitably the quality control and levels of support that can be provided from each centre will vary and most research laboratories are not geared to routine cell banking nor dealing with a customer-focused activity.

After an initial period of active distribution to service key collaborations, researchers will inevitably wish to return to the more challenging task of basic stem cell research. Additionally, handling of intellectual property will also be variable depending on the governing body/ownership of each line.

A centralised bank such as the UK Stem Cell Bank, funded solely to bank and distribute stem cell lines will not be diverted by competing research interests and can focus entirely on assuring appropriate governance, quality control and customer focused delivery and service. In addition, consistent and open procedures can be put in place, that ensure that any lines made available by the bank are available under reasonable conditions for research purposes. Centralised banking also offers the great advantage that all cells are subjected to the same system of quality control and safety testing and can provide a central resource with expertise in appropriate quality standards, for researchers wishing to establish stocks of cells for clinical use. The ongoing ability of such a centralised resource able to facilitate access to cells on an international basis will have a long term and significant influence in avoiding the need for individual laboratories to derive their own lines thus helping to avoid unnecessary use of fresh donated embryos and other tissues. Centralised banks also permit more efficient international transfer of cells, by providing a central point for information on regulation, thus helping to avoid cases of unwitting transgression of ethical and legal requirements in the country to which cells are transferred.

3. Fundamental Issues for *in Vitro* Cell Culture

In the management of cell culture systems there are three well-recognised technical aspects that are vitally important to assuring reliable delivery of cells for any purpose. Firstly it is important that the cells should be free of microbial contamination, as apart from any safety issues the characteristics of cells can be significantly affected by persistent and non-cytopathic microbial infection. Secondly, authenticity (i.e. "are the cells what they are thought to be") is vital for the validity of research based on cell line data. There is a long history of cell line cross-contamination that has been the cause of much wasted research effort.[5] Identity testing techniques for human cell lines, such as Short Tandem Repeat analysis, are readily accessible from service providers and consensus profiles of common cell lines are available in literature.[6,7] Thus there is no excuse for a modern laboratory not to check and assure the identity of human cell lines. Finally, stability of genotype and phenotype following multiple passages *in vitro* can be critical to the value of a culture.

A range of techniques have been used to evaluate stability, depending on the application to which the cells will be put. A diverse range of monoclonal antibodies have been developed for surface and intracellular antigens on different cell types and comprehensive gene expression studies are now possible using microarray and microfluidics techniques to analyse cell mRNA. However, the presence of mRNA does not always prescribe the presence of the associated antigen and it is wise to complement RNA expression studies with investigation of antigen for key markers. Karyology generally will not give useful information to differentiate diploid cell lines (except where derived from different sexes or where there is a characteristic chromosome abnormality), but can be used as a general measure of genetic stability and to identify the appearance of altered or transformed clones of altered karyotype. Multiple single nucleotide polymorphism and comparative genome hybridisation (CGH) microarray technologies are also now able to provide valuable additional data on genetic stability.

In addition to these technical procedures to evaluate cell cultures it is also vital that workers using the cells are aware of the conditions and events than can have significant impact on the biology of cells as underlined in Good Cell Culture Practice (GCCP).[8] Common hazards affecting cell performance in day to day cell culture include change in the composition of the culture medium or the source of serum, suboptimal passaging and effects due to fluctuation in temperature or gas environment in multiuser cell culture incubators. Good levels of education relating to cell biology and cell culture processes are therefore vital to ensure good reliability of cell cultures. Stem cell lines are challenging to culture and are known to be susceptible to genetic and phenotypic alteration on extended passage.[9,10] The principles of GCCP are therefore especially important in the delivery of human stem cell lines for research and the development of therapy.

4. Cell Culture Processes

4.1. The cell banking process

The approach to cell banking taken by an individual stem cell bank will depend on the remit for the work supported by that bank, the resources available and any applicable regulatory demands.[11] Clearly there will be significant differences between a group supplying stocks of cells for local researchers, a group supplying cells for national or international research programmes and a bank providing cells for clinical use.[11]

The early culture of human stem cell lines has depended on the use of bovine serum, supplementary growth factors and mouse embryonic feeder cells and these are still used in many laboratories. This is a largely undefined culture system in which key raw materials, feeder cells and serum, may all show high batch to batch variability. Many laboratories have switched to other media and growth supplements, such as "Knock-out" DMEM™ and "serum replacement" medium (SR™ , Invitrogen), which claim to be better defined. However, many laboratories are still dependent on the use of mouse embryonic feeder cells to culture hES cells, and it appears that alternative growth surface substrates, such as Matrigel ™ and complex mixtures of natural extracellular matrix proteins, do not appear to be adequate and are too costly for routine use. The routine culture and expansion of hES cells for experimental use is still a rapidly developing area and any potential short term and possibly permanent effects of the different culture systems and media in use today have yet to be realised. This dynamic and as yet uncertain environment presents a significant challenge to the establishment of banking procedures and the assurance of reliable stocks of stem cell lines that will be available for future researchers.

In order to address this situation, the UK Stem Cell Bank has established a principle of operation for its stem cell lines. In it, the "pre-master", "master cell bank" and the first "distribution cell bank" are all prepared according to the procedures recommended by the depositor of each line at the time the cell line is deposited. Subsequently, as new media become widely accepted for use with stem cell lines; the Bank will adapt cells from the master cell bank and provide banks of cells established under new culture conditions which are relevant to current research. Depositors are encouraged to lodge with the Bank stocks of cells from the earliest passage known by the depositor to be consistent with reasonable cell recovery and proliferation. In this way material representing the earliest cultures will be retained in case they are needed for future scientific reference or to enable return to early pristine material to avoid the cumulative effects of multiple adaptations to new culture conditions (see Fig. 1).

Homogeneity of each cell bank is also a key challenge for assuring reliability of cell stocks. This is difficult to establish for hES cells passaged by individual colony dissection, as with current methodologies a single pool of cells cannot be homogenised before cryopreservation without significant loss of viability. However, new developments in the maintenance of hES cells, including the ability to passage cells as single cell suspensions will no doubt provide solutions to this issue in due course (e.g. Ref. 12). Even when such methods become widely validated it will still be necessary to carry out tests on cell banks to test for homogeneity throughout each bank of cells, sampling as a minimum vials from early, mid and late aliquots and checking their viability and characteristics.

Reliability of stem cell lines provided from a cell bank is dependent on adopting good cell culture practice (see Ref. 8) and the reliability and reproducibility of the processes, reagents and facilities used. Documenting and recording all procedures and reagents provides traceability which enables troubleshooting, should cells fail quality control, or if there are customer complaints regarding released

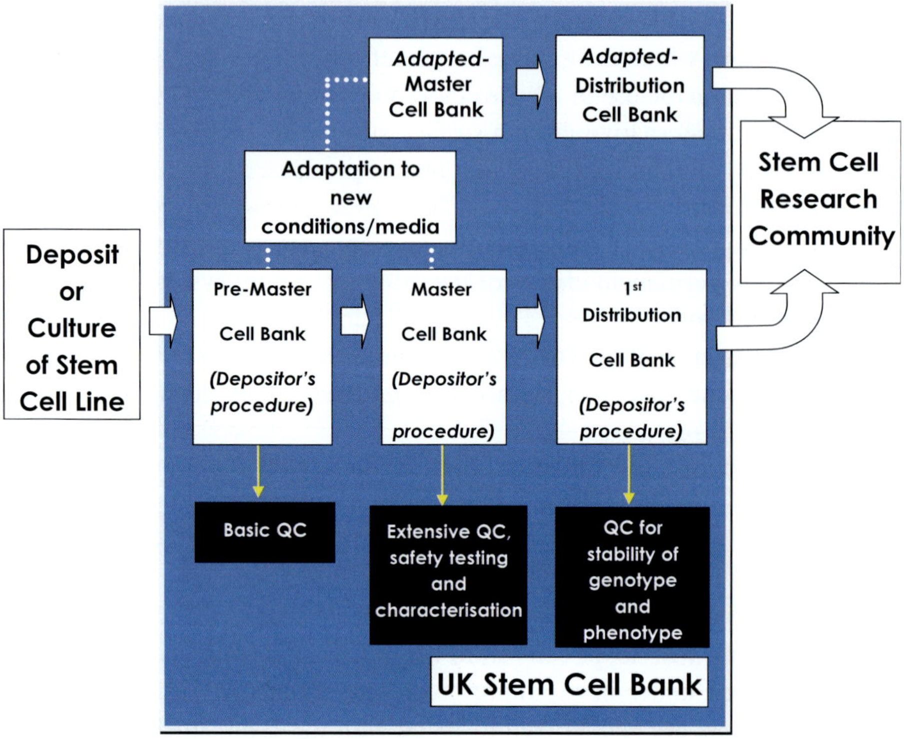

Fig. 1. UK Stem Cell Bank Scheme for assuring long term security of early cells and adapting cell line to current best practice (Quality Control (QC) is performed on "adapted" cell banks as for those established using the depositors' procedures).

cells and also provides independent auditing to give the bank and its customers confidence that all procedures claimed to be used for each cell banking process are effective. Such traceability and independent auditing is a formal requirement where the cells are ultimately intended for clinical use and will become particularly critical for the credibility of the bank should there be adverse events in future clinical trials.

Contamination of cell culture reagents is a crucial issue for cell lines used for both research and therapy. Any form of viral contamination, whether pathogenic for humans or not, will be undesirable in a clinical product from the perspectives of safety and efficacy. Virus infection in cells used for research purposes, if not evident from a clear cytopathic effect, can easily lead to the production of unreliable or misleading data due to the inevitable influence of virus infection on cells. Thus, screening of cell line for evidence of viral infection and preventing introduction of viral contamination in cell culture reagents of biological origin are important activities for stem cell banks. However virus screening is costly and over-zealous pursuit of virus contamination could potentially divert resources from the main aim of these centres which is to enable researchers to gain access to stem cell lines. Accordingly, it is helpful to try to evaluate the likelihood of viral contamination in cell culture reagents of animal origin and reduce this risk to acceptable limits wherever practicable.

The infectious risk due to microbial contamination of cell culture reagents can be evaluated in the light of a number of factors including:

- species and tissue of origin,
- geographical origin,
- the ability of the processing, purification and formulation methods to reduce or eliminate viable organisms,
- the effectiveness and reliability of any specific sterilisation or disinfection steps, and
- microbiological testing performed on the product.

Viral testing of animal-derived products will be most appropriate when no sterilisation process can be used. Care should be taken to assess viral testing and to ensure that appropriate methods were employed to give defined and acceptable levels of sensitivity and specificity. In the case of material destined for clinical application, tests should only be performed by a qualified laboratory with appropriate experience.

Unless stringent efforts have been made to avoid it, any cell line established in a research laboratory, unless very carefully controlled, may well have been exposed to bovine and possibly murine viruses. The longer the cell lines history of *in vitro* passage the more likely it is to have had multiple exposures to microorganisms.

The presence of viral contamination in serum has become a serious issue for manufacturers of biomedical products since the 1970s when an outbreak of foot-and-mouth disease in the UK focused attention on the need for "clean" sources of serum. More recently the outbreak of bovine spongiform encephalopathy (BSE) in cattle and subsequently new variant Creutzfeldt-Jakob disease (CJD) in humans again focused attention on the sourcing and traceability of serum. Whilst the usual sources of serum used in cell culture are either human or bovine, other species may also be used particularly in the derivation of hES cells including rodent serum for complement. Serum manufacturers now offer irradiated serum which offers a further reduction in the risk of viral contamination from this source.

The subculture of some hES cell lines has been achieved using trypsin which has been identified as a source of infection of cells in culture due to porcine virus.[13] Until very recently, there have been no preparations of these key culture reagents of animal origin that had received any kind of viral evaluation or inactivation. However, irradiated trypsin is now readily available for cell culture for clinical application and in addition the availability of recombinant trypsin now offers a further solution to avoid viral contamination.

Numerous products used in cell culture are derived from natural bacterial strains (e.g. collagenase, pronase) and recombinant microorganisms (e.g. various cytokines, growth factors, collagen). Where the cultured cells are intended for use in humans it is obvious that the product processing and purification methods must remove any residual organisms or spores, but it may also be necessary to assess the risk from other processing intermediates, including the bacteriological growth media, as they may contain raw materials of animal origin.

The risk to laboratory workers from potential prion or viral infection is generally considered to be low. The use of serum from known sources and source animals screened for disease and the adoption of good aseptic technique and containment of cell culture work in a class II microbiological cabinet will provide protection for laboratory workers against most contaminants.

4.2. Scaleability of stem cell culture

In procedures where donor cells are not deliberately replicated, careful optimisation and automation of the process can enable efficient utilisation and delivery of cell/tissue preparations, sometimes referred to as "scale-out". Where cells are replicated *in vitro* the process will obviously have the potential for "scale-up" in bioreactor systems to deliver large homogenous cell lots to provide treatment for multiple patients. Traditionally, the development of large-scale animal cell cultures for biotechnology applications have depended on technology developed for industrial microbial fermentation. However, in recent years a broader range of technologies has become available that has been developed specifically for animal cell culture (for a review see Ref. 14). Whilst the growth of cells in single large lots, maintained for extended periods of time in bioreactors has many advantages over the preparation of cells in many small discrete lots, large scale systems are at high risk of loss of cultures due to adverse culture conditions (e.g. depleted nutrients, increased levels of toxic metabolites) or contamination. Thus, such systems require highly specialised control mechanisms and equipment to sustain the cultures reliably (for a review see Ref. 15). A variety of approaches to scale up specifically for embryonic stem cells have been published in the literature[16,17] (see Table 2 and also elsewhere in this book). Of course the expansion of stem cells within such systems will require special attention to the phenotypic and genotypic characteristics of the stem cell lines in order to sustain the efficacy of the cells for therapy and to give confidence in the safety of the final product.

Table 2. Examples of automated systems for cell culture scale-up and analysis.*

Automated cell line scale-up systems

Automation Partnership

- CompacT SelecT — Expansion and maintenance of multiple cell lines; sub-culturing; expanding cell numbers through the seeding of a number of flasks and performing transient transfections.
- SelecT — Automated cell culture lines are maintained and expanded in T-175 flasks. The system is designed to prevent cross-contamination even when many cell lines (up to 182) are cultured in parallel. The system can be set to harvest, count and seed without operator intervention — at any time, day or night, to help optimise screening productivity.
- Cello™ — Cello is designed to culture multiple cell lines in parallel, from seeding through expansion and sub-cloning. The system is aimed to provide the throughput and capacity needed to support several projects running in parallel, with minimal operator involvement. Cello cultures both adherent and non-adherent cell lines with the capability to utilise 384-, 96-, 24- and six-well plates.
- Piccolo™ — Fully automated cell culture system for the rapid optimisation of recombinant protein production in microbial or insect cells.
- Cellmatc — Full automation of processes needed to culture cells in roller bottles and T-flasks using a robotic arm. It simulates normal cell culture manipulation thus avoiding the need for process change.

Hamilton

- Cellhost — Fully automated system for colony dissection, culture and expansion of ES cells. ES cells propagated with cell host are reported to retain their typical morphology after plating and media change.

MediHealth and NovaThera™ Ltd

- A compact, manual feed, disposable, batch culture bioreactor that allows 3D culture of cells including stem cells in three dimensions. NovaThera's™ encapsulation technology can be used with NovaPod™ and is claimed to be suitable for the production of pluripotent stem cells and their progeny without the need for continuous operator intervention.

Tecan

- Cellerity — Fully automated cell-line maintenance, expansion, harvesting and plating. A modular system that incorporates an integrated CO_2 incubator, cell-counter for determining cell number and viability, refrigerated media storage with in-line warming immediately prior to dispense.

(*Continued*)

Table 2. (*Continued*).

Automated monitoring and analysis systems for cell culture
20/20 Technology
• Bionomic system — Environmental control stage for live cell, real time imaging.
Amnis
• ImageStream® — Combines the visual capability of microscopy and the statistical rigour of flow cytometry in a single platform, enabling quantitation of cell structure features not accessible by traditional flow cytometry.
Beckman Coulter
• Cellomics KineticScan HCS Reader — Temporal and high-resolution spatial analysis of physiological processes, dynamic distribution and activity of cellular constituents, and morphological features in individual live cells.
CompuCyte
• iCyte® Automated Imaging Cytometer — Automated high-content cellular analysers which is also designed for higher throughput. Capability to precisely measure cellular DNA content and simultaneously combine these measurements with other molecular markers and cell morphology.
Luminex Corporation
• Luminex 100 IS System — Designed to enable simultaneous assay of up to 100 analytes in a single well of a microtitre plate. The system can deliver many assay formats including nucleic acid assays, receptor-ligand assays, immunoassays and enzymatic assays.
Nikon
• BioStation IM — Compact cell incubation and monitoring system that permits live cell imaging. Long-term time-lapse experiments (including studies of cell growth, morphology, and protein expression) are enabled through its design to provide consistent environmental control of temperature, humidity and gas concentration and observation by phase and fluorescence imaging.
Nova Biomedical
• BioProfile 400 — Metabolite analyser for mammalian cell culture.

* The authors would like to thank Ms. M. Gillett (UCL) for collating the information in this table.

4.3. Cryopreservation and low-temperature storage

This is a fundamental requirement for long-term banking of cell cultures, but is a much neglected aspect of cell biology that requires careful development and standardisation both in the establishment of new techniques and in their application to new cell/tissue types. Two approaches to the cryopreservation of hES cells have been adopted based on empirical studies which have applied methods used successfully on other, apparently closely related, cells: the mouse ES (mES) cell and the embryo. This has led to the development of protocols that use either conventional freezing or vitrification.

Cryopreservation by slow cooling has been based on methods used successfully for the mES cell, in turn adapted from methods applied to both primary and continuous cell lines. However, the adoption of this technique for hES cells has met with notable failures.[18] The disparity between successful recovery of mES cells and the failures seen with hES cells following slow cooling has been attributed by some to the "cooperative nature" of hES cells and the requirement to passage and cryopreserve many of the cell lines currently available as clusters of cells rather than single cell suspensions. The potential for ice formation within the clusters and consequential disintegration of the cell clusters on thawing are cited as contributory reasons for the poor recovery. However, this does not take into account a number of more recent studies which indicate that high survival is obtainable with conventional cooling particularly with the application of seeding at high subzero temperatures (for discussion see Ref. 19).

The majority of cell lines currently held by stem cell banks have been preserved by the "vitrification" technique despite these recent studies and the practical difficulties associated with this methodology. Vitrification is the solidification of a liquid without crystallisation and the growth of ice. Thus, during cooling, ice nucleation in the system is inhibited through a number of mechanisms with the solution increasing in viscosity until a glass is formed. In this condition the system displays the properties of a solid but retains the molecular structure of a liquid.

The adoption of vitrification as the predominant method of cryopreservation is largely due to three early comparative studies[20–22] which highlighted poor recovery of cells frozen by slow cooling and indicated that vitrified cells yielded a comparatively high degree of viable, undifferentiated cells capable of proliferation. However, in one of these studies, recovery of undifferentiated colonies was low (<30%) and there has been much anecdotal evidence to suggest that the yield of undifferentiated, proliferative cells is substantially lower than in the remaining reported studies.

The vitrification method, while capable of producing relatively good recovery, is both technically challenging and potentially susceptible to microbiological contamination. The use of very narrow bore, *open* pulled straws in order to facilitate the ultra-rapid cooling rates necessary to vitrify the low solute concentration cryoprotectant solution used in this method, and their subsequent storage often under liquid nitrogen, exposes the material to an increased risk of contamination. The transmission of infections through various routes including liquid nitrogen storage has been the subject of debate both in the context of IVF and cell/tissue banking with a number of demonstrated instances of contamination and infection.[23,24] The use of open pulled straws, even where the material is stored in the vapour phase above liquid nitrogen to reduce cross-contamination, is unlikely to be regarded favourably by regulatory agencies for material destined for therapeutic applications.

Logistically, it is difficult to imagine this method being applied outside of the laboratory as the process is difficult, if not impossible, to scale-up. Whilst the UK Stem Cell Bank has been able to apply this method to banks of 100–200 straws, the process is labour-intensive, operator dependent and variable. In addition, the meta-stable nature of the vitrified "glass" makes the material extremely sensitive to changes in storage temperature and both storage and transportation requires the maintenance of ultra-low temperatures to prevent ice nucleation and ice crystal growth during storage and distribution: events which can lead to a severe reduction in viability.

Addressing the fundamental cryobiological issues in the preservation, storage and distribution of stem cell lines will be important in the delivery of good quality stem cell lines to researchers and to clinical trials.

5. Quality Assurance and Quality Control

5.1. Risk from donor selection/harvest

For any therapy based on the use of human cells the most obvious microbiological risk is that of viral contamination of the donors' cells. The most highly pathogenic organisms present in significant numbers in donors would be expected to be readily detected through donor screening for the most prevalent serious human pathogens in the donor population. Accordingly, donor screening for cells and tissues for transplantation focus on a relatively small but significant group of viruses for "safety" testing purposes (e.g. Ref. 25). Of course such testing regimes are based on evaluation of risk in the donor population, but where stem cell lines are concerned the original donor may be from various countries with stem cell research programmes and accordingly geographical variation in prevalence of most common viral and other pathogens is a significant issue.

Another key factor is the need to evaluate the impact of differences between procedures established for processing of tissue and cells for transplantation versus clinical applications involving *in vitro* cell culture. Under the well-developed controls on cells and tissues for transplantation,[26,27] there is an emphasis on minimising the potential for replication of any microbial contamination during processing. However, *in vitro* cultures of cells are highly susceptible to the expansion of endogenous or introduced viral infection. Of course the overall risk will depend on the tissue and species tropism of the contaminating virus but even for non-human pathogens adaptation to growth in human cells is known and clearly the presence of any virus or any other potential pathogen must be considered undesirable and a potential risk. Viral testing regimes for stem cell lines have been reviewed[28] and it is clear that for stem cell banks charged with supplying cells for use in clinical trials, very careful review of the history of cell lines proposed for clinical use must be carried out. This should include an evaluation of the medical hazards associated with the donor history and the potential exposure of the cells to contaminants during the process of cell line derivation. Mapping the history of the cell line thus becomes a central issue in establishing its fitness for clinical application. An important and valuable tool is that of "process mapping" which has been used widely in risk assessment to identify all the steps in the preparation and processing of cells and other inputs during the derivation process. An example of such a process map for the derivation of a hES cell line is given in Fig. 2, and illustrates some of the steps at which contamination may be introduced.

5.2. Cell Line Master File (CLMF): principles and potential content

Having constructed a process map in collaboration with the scientists that established the stem cell line it is then possible to evaluate the nature and level of risks of contamination. However it is too easy to assume that the only real risks to a potential therapeutic product based on stem cells lines are virological, but in fact the modern world is acutely attuned to the need for appropriate ethical governance of the use of donated cells and these ethical issues are therefore an additional aspect that must be clearly documented for any cell line intended for therapeutic application. Thus, it is a critical step in the progression of a stem cell line towards utility in cell therapy, that a body of information is developed that demonstrates ethical sourcing of the original cells used to derive the line and exclusion of key risks in the cell line derivation and culture history. Since the original sampling of donated tissue cannot be expected to meet the requirements of pharmaceutical GMP; the process of gathering information for is initially

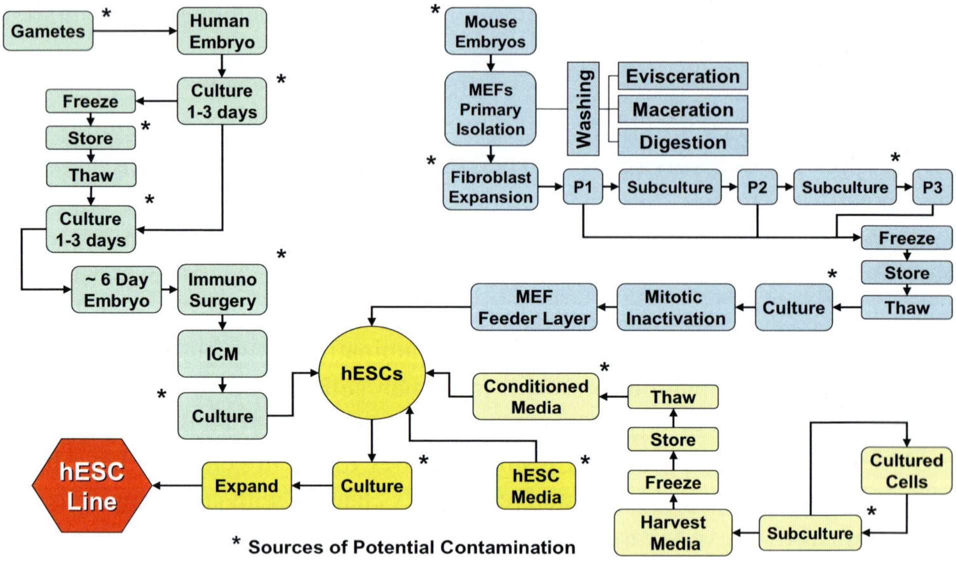

Fig. 2. Flowchart showing the process of cell line derivation and stages where there is the potential for the introduction of microbiological contamination.* Courtesy of the National Institute for Biological Standards and Control (NIBSC).

focused on demonstrating traceability to a microbiologically and ethically acceptable source, but is progressively developed to show increasing control and definition of the environment in which the cells are grown, prepared and stored. It is important that this process is inclusive and that any evidence, information or data is recorded and traceability to the cell line established. A key reference point in this process is the establishment of the first frozen stocks of the cell line that might be considered suitable for clinical use. Thus, a collation of all data available on the cells up to that point will be highly valuable in enabling the cells to be qualified for clinical use and for their acceptability under new regulatory environments that apply due to new geographical situations or temporal change. At the UK Stem Cell Bank this collation of data is referred to as a "Cell Line Master File" and the some of the key elements that may be required for this dossier are given in Table 3.

5.3. Ethical issues and the UK model

High standards of ethical governance are necessary to protect the use of altruistically donated tissues and cells, in order to maintain public support for stem

Table 3. Some central elements that may be required for a cell line master file used in support of proposed clinical application of a human stem cell line.*

Item	Type of documentation
Donor tissue procurement	• Traceability to fully informed consent. • Copies of informed consent protocol. • Blank consent form. • Details of centre/contact that obtained and hold original consent documents. • Details and record of harvesting and primary storage of donated cells.
Derivation of cell line	• Record of transmission to use of cells/tissue for R&D or clinical use. • Record of thawing of cells from primary storage containers for use in attempted cell lines derivation. • Records of all materials and reagents potentially in contact with the cells (type, source, composition, traceability). • Record of cell line isolation environment (SOPs, records of air quality and environmental control, staff involved and their training status, identification and history and validation of laboratories and equipment used, record of other work performed historically and concurrently in the derivation lab environment. • Records of material procedures, environment used to establish the first frozen stock of the cell line.
Subsequent centralised cell banking procedures	• GMP compliant records of expansion, preservation, quality control, "safety testing" and characterisation of master and working cell banks. • Records of bank facility processes and validation for staff training/competency, technical procedures, external and internal contractor service agreements and testing/release data.
Manufacturers cell banks and production	• cGMP compliance for preparation, validation and testing of production cell banks according to the national/regional guidelines/requirements.

* In the development of this information the authors would like to thank Ms. Penny Carter (NIBSC) and representatives of the UK Medicines and Healthcare Products Regulatory Agency (www.mhra.gsi.gov.uk/) for helpful discussions.

cell therapy and consequently the supply of tissue donated for research purposes.

Good ethical and scientific governance has been established as a key factor in the use of all human tissue; irrespective of the anatomical site of origin. International standards are sure to follow to facilitate cross-border delivery of new therapies. Efforts to achieve this are already underway[29,30] and in due course the international organisations engaged in the establishment of international standards, such as the World Health Organization[31] and the International Conference on Harmonisation (ICH),[32] will be able to assist in harmonisation of the requirements for clinical grade cell products.

The future of stem cell research is bound to be challenged by very rare but significant events such as unexpected adverse reactions in clinical trials, occasional failure to carry out the necessary safety and efficacy testing, the consequence of "quack doctor" operations and the attention of the media to all of these. A strong yet permissive regulatory environment is important to sustain scientific and clinical progress.

6. International Perspectives

Much activity has also been focused on the delivery of registries of hES cells which are ethically approved. In the US the long established NIH registry[33] lists all hES cell lines that were established before June 2001 and are therefore acceptable for federally funded research. In the UK the Medical Research Council also hosts a registry of hES cell lines that are approved for use in the UK. In this case approval is primarily focused on demonstration of informed consent from donors and the value of the cell lines for research into serious human disease, and there is otherwise no regulatory constraint on the use of such cells in the UK. The International Stem Cell Initiative[34] described below has also established a registry based on technically standardised studies of hES cell lines. In 2007 a registry project was launched by the European Commission which aims to establish a comprehensive database of human stem cell lines that are fit for use in European laboratories. This project, called the European Human Embryonic Stem Cell Registry,[35] has oversight from an international Scientific Advisory Board and Steering Committee and addresses a broad range of scientific, technical, regulatory, ethical and public interest aspects of the use of human stem cell lines.

A particularly effective and truly international initiative has emerged from a collaboration of national stem cell funding bodies which have adopted the title the International Stem Cell Forum.[36] Some 21 countries are currently registered as members of this organisation and their primary focus is to fund initiatives that can only be achieved through broad international collaboration. This group has funded a number of projects including:

- an Ethics Working Party that has reviewed and published on the international spectrum of regulation on fully informed consent,
- an Intellectual Property working group to evaluate the international landscape of commercial issues for stem cell research and therapy,
- the International Stem Cell Initiative which to date has performed a standardised evaluation of the features of hES cell lines across the world[37] with a second phase now focused on evaluation of defined growth media, and
- the International Stem Cell Banking Initiative described below.

6.1. The "International Stem Cell Banking Initiative"

Quality control procedures adopted in different research groups are likely to vary considerably even though minimum criteria for scientific characterisation of human stem cell lines have been published (for example Ref. 38). International guidance on cell banking for cell lines has been published for application to cells for research and testing[8] and for cells intended for use in the manufacture of medicinal products.[39,40] National guidance has been published for the use of stem cell lines in the UK and for the operation of the UK Stem Cell Bank.[41] More recently, the International Cell Banking Initiative, funded by the International Stem Cell Forum has coordinated input from 17 countries on guidance on best practice for the banking and supply of human stem cell lines for research purposes which is available on the ISCF website.[37]

7. Future Developments and Expectations

The field of stem cell research has experienced an extremely high degree of interest from the public and governments, due to the great promise stem cells hold for future medicine. Despite the exciting prospects, it is vital that the expectations of the public are handled with care to enable the public to have a realistic view on what can be achieved and over what time period these achievements can be realistically expected. Establishment of reliable methods for the very basic technologies of growing, characterising and preserving stem cells has yet to be properly realised. Complex biological questions have yet to be answered relating to reliable differentiation of stem cells to deliver products that may have real clinical potential.

Most biological medicines, such as vaccines and monoclonal antibodies, take decades to become established as accepted therapies. On the road to products for stem cell therapy it is clear there will be lines of research that meet dead ends and unfortunate problems that arise in early clinical applications that may stall progress. In this situation there is a clear need for public confidence in the scientific and clinical communities who are responsible for delivering new therapies. This will be assisted by appropriate and open ethical governance and regulation to match the permissive approach to stem cell research in countries like the UK. The professional, transparent and reliable operation of stem cell banks can play important part in supporting both the quality of research and public confidence and thus help to promote timely delivery of safe and reliable cell therapies.

References

1. Polge, C., Smith, A.U. and Parkes, A.S. Revival of spermatozoa after vitrification and dehydration at low temperatures. *Nature* **164**, 666 (1949).
2. Stacey, G.N. and Day, J.G. Long-term *ex situ* conservation of biological resources and the role of biological resource centres. In: Day, J.G. and Stacey, G.N. (eds.), *Cryopreservation and Freeze-Drying Protocols* (Humana Press, 2007), pp. 1–14.
3. Thomson, J., Itskovitz-Eldor, J., Shapiro, S. *et al.*, Embryonic stem cell lines derived from human blastocysts. *Science* **282**, 1145–1147 (1998).
4. The UK Stem Cell Bank, http://www.ukstemcellbank.org/.
5. National Institutes of Health. Notice Regarding Authentication of Cultured Cell Lines Notice No NOT-OD-08-017 (2007), http://grants.nih.gov/grants/guide/notice-files/NOT-OD-08-017.html/.
6. MacLeod, R.A.F., Dirks, W.G., Matsuo, Y., Kaufman, M., Milch, H. and Drexler, H.G. Widespread intra-species cross-contamination of human tumour cell lines arising at source. *Int. J. Cancer* **83**, 555–563 (1999).
7. Masters, J.R., Thomson, J.A., Daly-Burns, B., Reid, Y.A., Dirks, W.G., Packer, P., Toji, L.H., Ohno, T., Tanabe, H., Arlett, C.F., Kelland, L.R., Harrison, M., Virmani, A., Ward, T.H., Ayres, K.L. and Debenham, P.G. Short tandem repeat profiling provides an international reference standard for human cell lines. *Proc. Natl. Acad. Sci. USA* **98**, 8012–8017 (2001).
8. Coecke, S., Balls, M., Bowe, G., Davis, J., Gstraunthaler, G., Hartung, T., Hay, R., Merten, O.-W., Price, A., Shechtman, L., Stacey, G.N. and Stokes, W. Guidance on good cell culture practice. A report of the second ECVAM task force on good cell culture practice. *Altern. Lab. Anim.* **33**, 1–27 (2005).
9. Draper, J.S., Smith, K., Gokhale, P., Moore, H.D., Maltby, E., Johnson, J., Meisner, L., Zwaka, T.P., Thomson, J.A. and Andrews, P.W. Recurrent gain of chromosomes 17q and 12 in cultured human embryonic stem cells. *Nat. Biotechnol.* **22**, 53–54 (2004).

10. Baker, D.E., Harrison, N.J., Maltby, E., Smith, K., Moore, H.D., Shaw, P.J., Heath, P.R., Holden. H. and Andrews, P.W. Adaptation to culture of human embryonic stem cells and oncogenesis *in vivo*. *Nat. Biotechnol.* **25**, 207–215 (2007).
11. Stacey, G.N. Sourcing established human embryonic stem cell lines. In: Sullivan, S., Cowan, C. and Eggan, K. (eds.), *Human Embryonic Stem Cells: A Practical Handbook* (John Wiley & Sons Ltd, 2007), pp. 11–24.
12. Ellerström, C., Strehl, R., Noaksson, K., Hyllner, J. and Semb, H. Facilitated expansion of human embryonic stem cells by single-cell enzymatic dissociation. *Stem Cells* **25**,1690–1696 (2007).
13. Hallauer, C., Kronauer, G. and Siegl, G. Parvovirus contaminants of permanent human cell lines I virus isolation from 1960-1970. *Arch. Gesamte Virusforsch.* **35**, 80–90 (1971).
14. Davis, J. Systems for cell culture scale-up. In: Stacey, G. and Davis, J. (eds.), *Medicines from Animal Cell Culture* (Wiley, 2007), Chapter 10, pp. 145–171.
15. Stoll, T.S. and Grabarek, P. Monitoring control and automation in upstream processing. In: Stacey, G. and Davis, J. (eds.), *Medicines from Animal Cell Culture* (Wiley, 2007), Chapter 13, pp. 203–244.
16. Fernandes, A.M., Fernandes, T.G., Diogo, M.M., da Silva, C.L., Henrique, D. and Cabral, J.M. Mouse embryonic stem cell expansion in a microcarrier-based stirred culture system. *J. Biotechnol.* **132**, 227–236 (2007).
17. Fong, W.J., Tan, H.L., Choo, A. and Oh, S.K. Perfusion cultures of human embryonic stem cells. *Bioproc. Biosyst. Eng.* **27**, 381–387 (2005).
18. Heng, B.C., Kuleshova, L.L., Bested, S.M., Liu, H. and Cao, T. The cryopreservation of human embryonic stem cells. *Biotechnol. Appl. Biochem.* **41**, 97–104 (2005).
19. Hunt, C.J. The banking and cryopreservation of human embryonic stem cells. *Transfus. Med. Hemother.* **34**, 293–304 (2007).
20. Reubinoff, B.E., Pera, M.F., Vajta, G. and Trounson, A.O. Effective cryopreservation of human embryonic stem cells by the open pulled straw vitrification method. *Human Reprod.* **16**, 2187–2194 (2001).
21. Zhou, C.Q., Mai, Q.Y., Li, T. and Zhaung, G.J. Cryopreservation of human embryonic stem cells by vitrification. *Chin. Med. J. Engl.* **117**, 1050–1055 (2004).
22. Richards, M., Fong, C.Y., Tan, S., Chan, W.K. and Bongso, A. An efficient and safe xeno-free cryopreservation method for the storage of human embryonic stem cells. *Stem Cells* **22**, 779–789 (2004).
23. Tedder, R.S., Zuckerman, M.A., Goldstone, A.H., Hawkins, A.E., Fieldin, A., Briggs, E.M., Irwin, D., Blair, S., Gorman, A.M., Patterson, K.G., Linch,

D.C., Hepenstall, J. and Brink, N.S. Hepatitis B transmission from contaminated cryopreservation tank. *Lancet* **346**, 137–140 (1995).
24. Steyaert, S.R., Leroux-Rouls, G.G. and Dhont, M. Infections in IVF: review and guidelines. *Hum. Reprod. Update* **6**, 432–441 (2000).
25. EU Directive 2006/17/EC. Implementing Directive 2004/23/EC of the European Parliament and the Council as regards certain technical requirements for the donation, procurement, and testing of human tissues and cells, http://europa.eu.int/eur-lex/lex/LexUriServ/site/en/oj/2006/l_038/l_03820060209en00400052.pdf.
26. EU Directive 2004/23/EC. On setting standards of quality and safety for the donation, procurement, testing, processing, preservation, storage and distribution of human tissues and cells, http://europa.eu.int/eur-lex/pri/en/oj/dat/2004/l_102/l_10220040407en00480058.pdf.
27. EU Directive 2006/86/EC. Implementing Directive 2004/23/EC of the European Parliament and the Council as regards traceability requirements, notification of serious adverse reactions and events and certain technical requirements for the coding, processing, preservation, storage and distribution of human tissues and cells. http://eur-lex.europa.eu/LexUriServ/site/en/oj/2006/l_294/l_29420061025en00320050.pdf.
28. Cobo, F., Stacey, G.N., Hunt, C., Cabrera, C., Nieto, A., Montes, R., Cortes, J.L., Catalina, P., Barnie, A. and Concha, A. Microbiological control in stem cell banks: approaches to standardisation. *Appl. Microbiol. Biotechnol.* **68**, 456–466 (2005).
29. International Society for Stem Cell Research. Draft guidelines for the conduct of human stem cell research. Draft for comment 6-30-06 (2006), http://www.isscr.org/StaticContent/StaticPages/ISSCRTaskForceGuidelinesDRAFT6-30-06.pdf.
30. International Stem Cell Banking Initiative website, http://www.stemcellforum.org/forum_initiatives/international_stem_cell_banking_initiative.cfm.
31. WHO website, http://who.int/.
32. International Conference on Harmonisation website, http://www.ich.org/.
33. US National Institutes for Health, http://stemcells.nih.gov/research/registry/.
34. Andrews, P.A., Benvenisty, N., McKay, R., Pera, M.F., Rossant, J., Semb, H. and Stacey, G.N. Commentary: the international stem cell initiative — towards benchmarks for human embryonic stem cell research. *Nat. Biotechnol.* **23**, 795–797 (2005).
35. European hESC Registry website, http://www.hescreg.eu/.
36. ISCF website, http://www.stemcellforum.org/.
37. International Stem Cell Initiative, Adewumi, O., Aflatoonian, B., Ahrlund-Richter, L., Amit, M., Andrews, P.W., Beighton, G., Bello, P.A., Benvenisty, N.,

Berry, L.S., Bevan, S., Blum, B., Brooking, J., Chen, K.G., Choo, A.B., Churchill, G.A., Corbel, M., Damjanov, I., Draper, J.S., Dvorak, P., Emanuelsson, K., Fleck, R.A., Ford, A., Gertow, K., Gertsenstein, M., Gokhale, P.J., Hamilton, R.S., Hampl, A., Healy, L.E., Hovatta, O., Hyllner, J., Imreh, M.P., Itskovitz-Eldor, J., Jackson, J., Johnson, J.L., Jones, M., Kee, K., King, B.L., Knowles, B.B., Lako, M., Lebrin, F., Mallon, B.S., Manning, D., Mayshar, Y., McKay, R.D., Michalska, A.E., Mikkola, M., Mileikovsky, M., Minger, S.L., Moore, H.D., Mummery, C.L., Nagy, A., Nakatsuji, N., O'Brien, C.M., Oh, S.K., Olsson, C., Otonkoski, T., Park, K.Y., Passier, R., Patel, H., Patel, M., Pedersen, R., Pera, M.F., Piekarczyk, M.S., Pera, R.A., Reubinoff, B.E., Robins, A.J., Rossant, J., Rugg-Gunn, P., Schulz, T.C., Semb, H., Sherrer, E.S., Siemen, H., Stacey, G.N., Stojkovic, M., Suemori, H., Szatkiewicz, J., Turetsky, T., Tuuri, T., van den Brink, S., Vintersten, K., Vuoristo, S., Ward, D., Weaver, T.A., Young, L.A. and Zhang, W. Characterization of human embryonic stem cell lines by the International Stem Cell Initiative. *Nat. Biotechnol.* **25**, 803–816 (2007).

38. Loring, J.F. and Rao, M.S. Establishing standards for the characterization of human embryonic stem cell lines. *Stem Cells* **24**, 145–150 (2006).
39. ICH. Human Medicines Evaluation Unit: ICH Topic Q 5 D — Quality of biotechnological products. Derivation and characterisation of cell substrates used for production of biotechnological/biological products. ICH Technical Co-ordination, European Agency for the Evaluation of Medicinal Products, London (1997), http://www.eudra.org/emea.html.
40. WHO. WHO expert committee on biological standardization (47th Report): requirements for the use of animal cells as *in vitro* substrates for the production of biologicals. WHO Technical Report Series 878, World Health Organization, Geneva (1998).
41. UK Stem Cell Bank, http://www.ukstemcellbank.org.uk/.

PART III

MATERIALS

Chapter 13

Synthetic Biomaterials as Cell-Responsive Artificial Extracellular Matrices

Matthias P. Lutolf and Jeffrey A. Hubbell

Abstract

Cells in our tissues are exposed to complex arrays of biochemical and biophysical cues from their protein- and sugar-rich extracellular matrix (ECM). In concert with cell-intrinsic regulatory cascades, these temporally and spatially coordinated signals instruct cells to acquire specific fates, controlling, for example, cell division, differentiation, migration or apoptosis. Conversely, cells are constantly secreting signals that can trigger structural and biochemical microenvironmental changes, as is most evident during proteolytic remodeling of the ECM. The resulting reciprocal and dynamic cell-matrix interaction is crucial for tissue development, maintenance and regeneration and, if gone awry, it can be involved in disease progression such as tumor metastasis. Recent efforts in the development of synthetic biomaterials for tissue engineering aimed to mimic the cell-instructive *and* cell-responsive function of ECMs. This chapter focuses on the molecular design, function and application of such smart biomaterials as cell-responsive artificial ECMs that can for example actively participate in cascades of morphogenesis during tissue regeneration (see also other chapters in this book).

Keywords: Cell-Instructive; Cell-Responsive; Tissue Engineering; Artificial ECM; Hydrogel; Hybrid; Protein-Polymer; Peptide-Polymer; Nanofibrillar; Self-Assembly.

Outline

1. Introduction
2. ECMs Instruct Cell Fates and Respond to Cell-Secreted Signals

3. Design Principles for Cell-Responsive Artificial ECMs
   3.1. Selecting biophysically relevant materials structures
   3.2. Selecting mild chemistries
   3.3. Designing responsiveness to cell-secreted biomolecules
   3.4 Designing building block modularity
4. Implementation: Classes and Applications of Cell-Responsive Artificial ECMs
   4.1. Hybrid protein-polymer and saccharide-polymer systems
   4.2. Hybrid peptide-polymer systems
   4.3. Nanofibrillar-based gels
5. Future Challenges
   References

1. Introduction

Modern tissue engineering strategies rely on the performance of synthetic biomaterials.[1–3] They are conceived to play a role as tunable three-dimensional (3D) scaffolds to attract or support cells, to control their fate and ultimately guide cellular assembly (i.e. morphogenesis) into functional tissues after trauma or disease.

The state-of-the-art in biomaterials design has continuously evolved over the past few decades. After an early empirical phase of biomaterials selection based on availability, design attempts were primarily focused on either achieving structural/mechanical performance, or on rendering biomaterials inert and thus unrecognizable as "foreign bodies" by the immune system. Later, an increased appreciation of the importance of molecular extracellular matrix (ECM) constituents in controlling cell functions has spurred a paradigm shift towards molecular design. A first generation of such molecular biomaterials, termed "bioactive", was designed to incorporate bioactivity through biological recognition, for example by presenting ECM-derived biochemical signals such as adhesion-promoting oligopeptides or growth factor proteins to control specific cell fates.[4] Since then, the field has seen an ever more profound integration of fundamental cell and molecular biology principles into biomaterials design, resulting in synthetic biomimetic materials that start to rival their naturally derived counterparts in biochemical functionality.[5] As tissue engineering advances into the clinic, these smart materials may become key players in guiding tissue-specific morphogenetic processes in regeneration. In the following two paragraphs, we briefly recapitulate some important concepts of cell-matrix interactions, focusing on their role in regulating cell fates during natural processes of tissue dynamics.

2. ECMs Instruct Cell Fates and Respond to Cell-Secreted Signals

The 3D extracellular microenvironment is in essence a complex and highly hydrated polymer gel comprised of several types of macromolecular components secreted by cells (Fig. 1; middle panel, left): (i) physically immobilized protein and sugar components that are crosslinked into a fibrillar network that provides space for diffusion of nutrients and metabolites to and from the cells (we use the term "ECM" for this component), (ii) "soluble" effectors that include growth factors, chemokines and cytokines with important signaling functions, and (iii) membrane-anchored molecules presented from neighboring cells enabling cell-cell communication that is crucial in tissue morphogenesis.

Two major types of ECM macromolecules are the fibrous proteins, such as collagens, elastin, fibronectin, and laminins, and the hydrophilic proteoglycans that contain large glycosaminoglycan side chains such as hyaluronic acid. Although most ECMs share these components, the organization, form, biochemical and mechanical properties vary widely in different tissues. Beyond their most elemental function as solid scaffolds to organize cells into 3D tissues or to provide physical boundaries between neighboring tissues, the ECM, via specific receptor-ligand interactions, regulates multiple cellular fates such as cell adhesion, migration, proliferation and division, all of which arc driving morphogenesis.[6,7] Moreover, the ECM can modulate tissue dynamics through its ability to bind, store and sequester soluble growth factor proteins. ECM binding occurs primarily via electrostatic interactions involving heparan sulfate proteoglycans such as heparin. Binding can protect morphogens from inactivation via enzymatic degradation and in some cases, has been reported to increase biological activity through conformational changes that optimize receptor-ligand interaction.

Conversely, extracellular microenvironments respond to signals secreted by cells, establishing a dynamic and bidirectional cell-matrix crosstalk. This responsiveness may be most evident in processes that involve proteolytic degradation of the ECM (Fig. 1; middle panel, right). Cell-secreted proteolytic enzymes can selectively cleave peptide bonds of domains of macromolecular ECM components in close proximity. As dense ECMs often act as a biophysical barrier for cells within it, this matrix responsiveness is a cornerstone of 3D migration and matrix remodeling that occurs during tissue formation, regeneration and many pathological processes. Notably, 3D migration can also occur via non-proteolytic strategies and the selection of a particular strategy depends on the cell type and microenvironmental context.[8] In essence, proteolytic migration is enabled through integrin-binding to the ECM coupled with a highly localized degradation of pericellular ECM proteins by cell-secreted and cell-activated proteolytic enzymes, involving matrix metalloproteinases (MMPs) as the main family,[9] as well as other enzyme families such as serine proteases.[10]

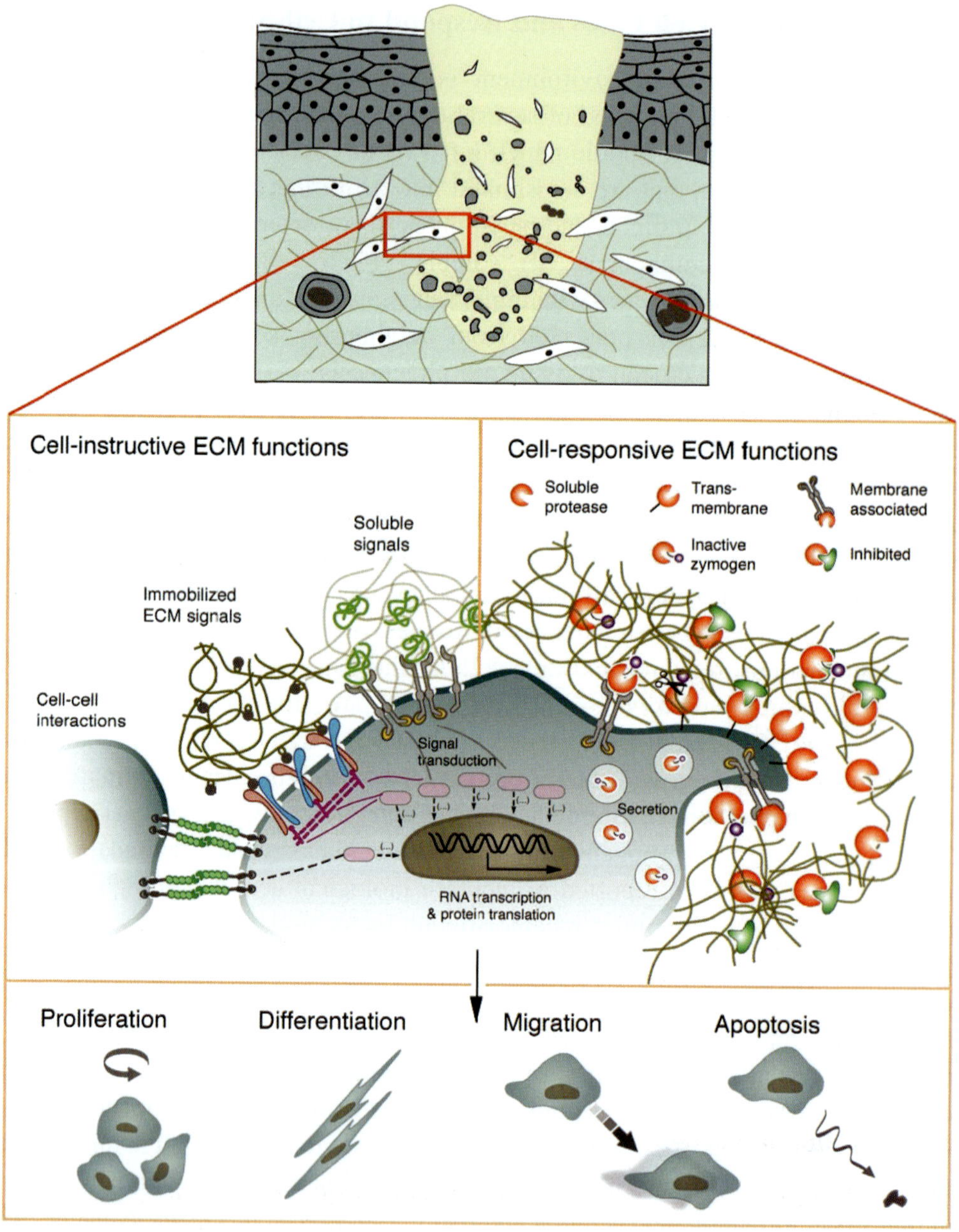

Fig. 1. The environment that surrounds cells regulates their fate and the dynamic state of multicellular tissues (adapted and reproduced with permission from Lutolf and Hubbell;[5] copyright 2005 Nature Publishing Group). Cells are receiving signals from this protein- and proteoglycan-rich gel network, termed ECM, comprised of soluble and physically tethered signals. At the same time, cells are secreting signals that trigger for example enzymatic remodeling of the ECM, generating reciprocal cell-matrix interactions. Localized proteolytic degradation of the ECM is mainly regulated by MMPs.

3. Design Principles for Cell-Responsive Artificial ECMs

One of the emerging avenues in biomaterials for tissue engineering has been the development of synthetic, multifunctional biomaterials that recapitulate both the cell-instructive *and* cell-responsive function of natural ECMs. These materials can be termed artificial ECMs. Although this concept has been pioneered nearly two decades ago by Moghaddam and Matsuda,[11,12] it has only relatively recently captured the imagination of researchers in the biomaterials field, as rapid advances in synthetic chemistry and molecular and cell biology only now permit to overcome the hurdles in its implementation. We believe that functional, synthetic mimics of natural ECMs can only be brought to life via a true fusion of materials engineering with cell/matrix biology. We have identified four main engineering challenges that need to be successfully addressed: (i) materials structures that mimic the physicochemical properties of ECMs, (ii) chemistries to form materials in the presence of proteins, cells or tissues, (iii) strategies to render materials responsive to cell-secreted signals, and (iv) design modularity to create materials with sufficient biological signals to target multiple cell functions that ultimately facilitate tissue morphogenesis.

3.1. Selecting biophysically relevant materials structures

Natural ECMs are mechanically resilient hydrogels of blends of interwoven glycosaminoglycans and protein fibrils and fibers. This molecular architecture results in distinct macroscopic physicochemical properties. As cell behavior not only strongly depends on biochemical ECM signals, but also responds to biophysical cues from the surrounding, designing biophysically appropriate scaffold structures has received considerable attention. This challenge has been successfully addressed in the form of synthetic hydrogel matrices that imbibe typically between 90% and 99% water. There are countless elegant strategies for the synthesis of hydrogels from synthetic building blocks[13] that result in ECM-like viscoelastic, diffusive transport, and interstitial flow characteristics.

3.2. Selecting mild chemistries

The ideal cell-responsive artificial ECM enables the entrappment of cells and biomolecules in the newly formed matrix via a mild and specific crosslinking

These enzymes are synthesized as secreted or transmembrane proenzymes (zymogens) and most of them are processed to an active form by the removal of an amino-terminal propeptide. Apart from association with the membrane (e.g. MT-MMPs), the localization of matrix degradation is controlled via a tightly regulated balance between active MMPs and their natural inhibitors of the family of tissue inhibitors of metalloproteinases (TIMPs), or a complexation of soluble proteinases to cell surface receptors such as integrins.

scheme. This would be advantageous in many clinical applications including delivery systems for protein therapeutics or 3D tissue engineering matrices, in which materials need to be formed *in situ* at the site of a tissue defect. A wealth of mild crosslinking schemes to form hydrogel matrices have been presented in the past few years, both for chemically and physically crosslinked gel systems.[14,15]

3.3. Designing responsiveness to cell-secreted biomolecules

One of the challenges in engineering functional mimetics of natural ECMs is the design of strategies to render materials susceptible to cell-secreted signals. This could possibly establish the aforementioned bidirectional cell-matrix crosstalk and thus permit cell-autonomous processes to occur within the synthetic matrix. We distinguish here between four general modes of cell-responsiveness (Fig. 2) that result in materials belonging to the broader class of bioresponsive, smart hydrogels.[16] First, material building blocks can be synthesized with responsiveness to cell-secreted crosslinking enzymes (Fig. 2A). Second, hydrogels can be functionalized with biomolecules such as enzymes or receptors that selectively bind corresponding substrates or ligands. Upon binding of a cell-secreted enzymatic substrate or ligand, a macroscopic volume change (swelling or collapse) would occur that could for example trigger the release of another soluble effector (Fig. 2B). Several interesting examples of such feedback systems based on responsive hydrogels have been published, for example to control the release of insulin via elegant enzymatic feedback systems. Third, biomolecules can be linked to chains of a hydrogel network via proteolytically labile linkers (Fig. 2C). In the presence of cell-secreted enzymes that diffuse into the gels, the biomolecule would be enzymatically released, generating a biologically controlled drug delivery system with high selectivity. Finally, the polymer backbone in a hybrid hydrogel network could comprise a proteinaceous or oligopeptidic substrate for a protease (Fig. 2D). Upon binding of the enzyme to its substrate and the subsequent cleavage of individual peptide bonds, the otherwise stable gel would start to dissociate to its soluble component. Pioneered by Kopecek and co-workers in the early 1980s,[13] the concept of enzymatically degradable hydrogels is a cornerstone of cell-responsive artificial ECMs, as described in more detail below.

3.4. Designing building block modularity

As the dynamic state of a tissue is regulated via a highly complex temporal and spatial coordination of many different cell-matrix and cell-cell interactions, unifunctional design approaches do not suffice in imitating the natural complexity of the ECM. Rather, the cell-instructive *and* cell-responsive function of natural ECM

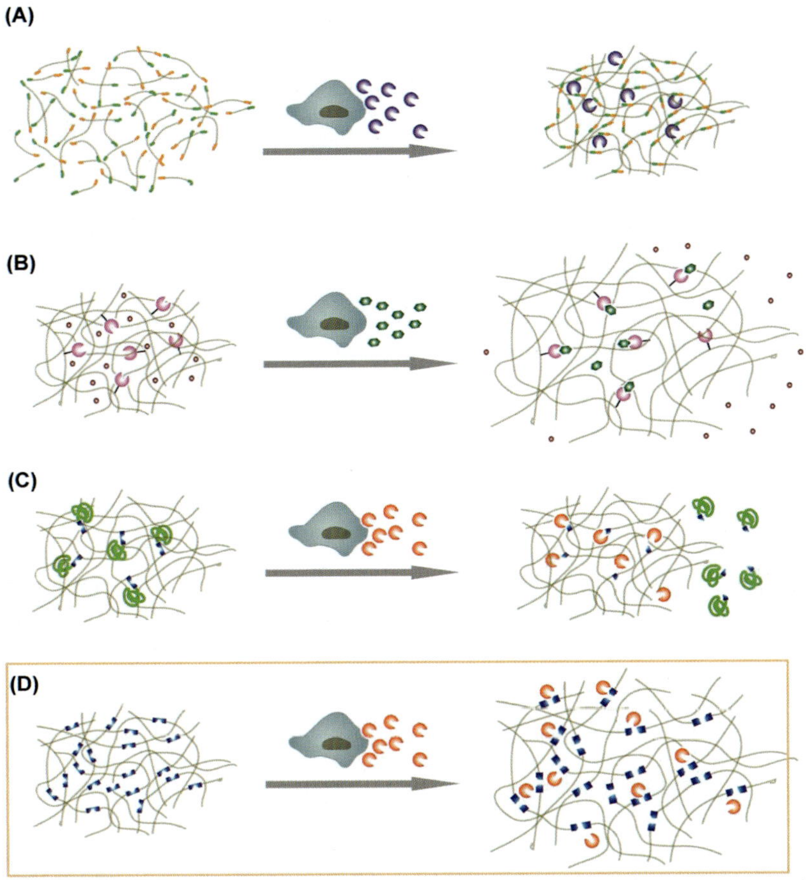

Fig. 2. At least four general concepts of cell-responsiveness, an important feature of artificial ECMs, can be envisioned in hydrogel networks. Material building blocks with substrates for cell-derived crosslinking enzymes can be fabricated allowing crosslinking via biomimetic mechanisms **(A)**. Immobilized biomolecules such as enzymes can be used to generate smart hydrogels in which selective binding induces macroscopic volume changes that can for example trigger the release of drugs **(B)**. Immobilized biomolecules such as growth factors can be linked to gels networks via proteolytically labile bonds to release them by cell-controlled proteolytic mechanisms **(C)**. Gel networks with substrates for proteases in the chain backbone are susceptible to enzymatic breakdown in the presence of the corresponding enzyme **(D)**. The latter concept is central to the fabrication of cell-responsive artificial ECMs discussed in this review.

components needs to be replicated (Fig. 1). Accordingly, we propose modular design strategies for synthetic materials that can account for the multifunctionality of natural ECMs.[5] Essentially, the key ECM proteins could be selected, if desired deconstructed into their peptidic functional subunits such as cell-adhesive,

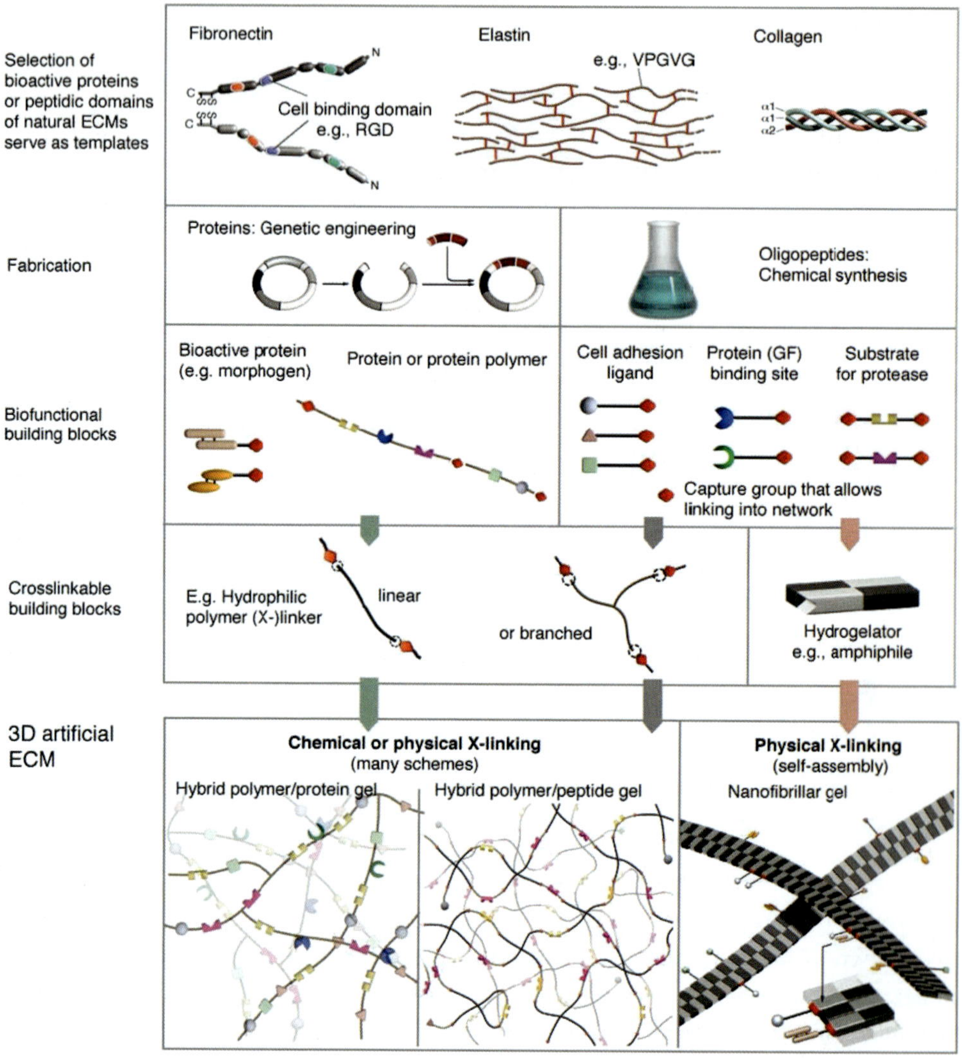

Fig. 3. "Toolbox"-design strategy for the creation of cell-responsive artificial ECMs (adapted and reproduced with permission from Lutolf and Hubbell;[5] copyright 2005 Nature Publishing Group). Crosslinking of biofunctional components from an entire array of building blocks can be used to create tailor-made, cell- or tissue-specific matrices.

protease-responsive, or self-assembling oligopeptide domains, generating an array of building blocks that could be re-assembled into well-defined and tunable scaffolds via mild crosslinking chemistries (Fig. 3). This modular approach would allow matching the matrix functionality exactly to a particular cell- or tissue-type of interest. Of course, the entire selection of mild crosslinking strategies (physical or chemical) is available to crosslink these building blocks into 3D hydrogel networks. Crosslinking should not only be so adaptable as to permit attachment of desired combinations of bioactive building blocks, it should also be sufficiently facile to allow formation of relatively complex materials in the least amount of reaction steps. Ultimately, such designer matrices could be applied in diverse tissue engineering applications, and they could become outstanding discovery tools in cell biology research whereby matrix complexity could be deconstructed and then build up from bottom-up on a biologically inert background.

4. Implementation: Classes and Applications of Cell-Responsive Artificial ECMs

In the following paragraphs, we illustrate the design concepts described above by way of specific implementations. Various research groups have arrived at different approaches by which to realize the concepts of biophysical and biomolecular mimicry, including biological ligand presentation and sensitivity to cell-secreted signals such as proteases, all in the context of materials systems with mild reaction chemistries and substantial building block modularity. We illustrate these, and we also comment on applications that have been explored or foreseen to date.

4.1. Hybrid protein-polymer and saccharide-polymer systems

Biomolecules such as proteins and glycosaminoglycans (GAGs) can confer biofunctionality to matrices and thus serve as important components of the biomaterials tool-kit (Fig. 4). Proteins possess characteristics that are uniquely valuable for implementation into artificial ECMs, most notably the extent to which multiple design functionalities can be designed and the molecular precision with which the materials precursors, i.e. the engineered proteins, can be synthesized using recombinant DNA technology. These proteins, typically designed with some native protein in mind, but not as exact mimetics of the proteins, are often referred to as *protein polymers*, rather than merely *proteins* (a word suggestive of the naturally-occurring protein). Proteins that are isolated from biological sources can also be very valuable, such as fibrinogen, collagen and albumin. In addition to proteins, GAGs can confer interesting functionality, such as enzymatic degradability and affinity for growth factors and in some cases cell surface receptors.

Early pioneering work in engineering protein polymers was from Cappello and co-authors on design of recombinant elastin-like and silk-like copolymers.[17,18] The silk-like domains inter-molecularly self-assemble into a crystalline nanostructure, thus connecting many protein polymer chains into a network linked by the amorphous elastin-like domains. These materials have been widely explored

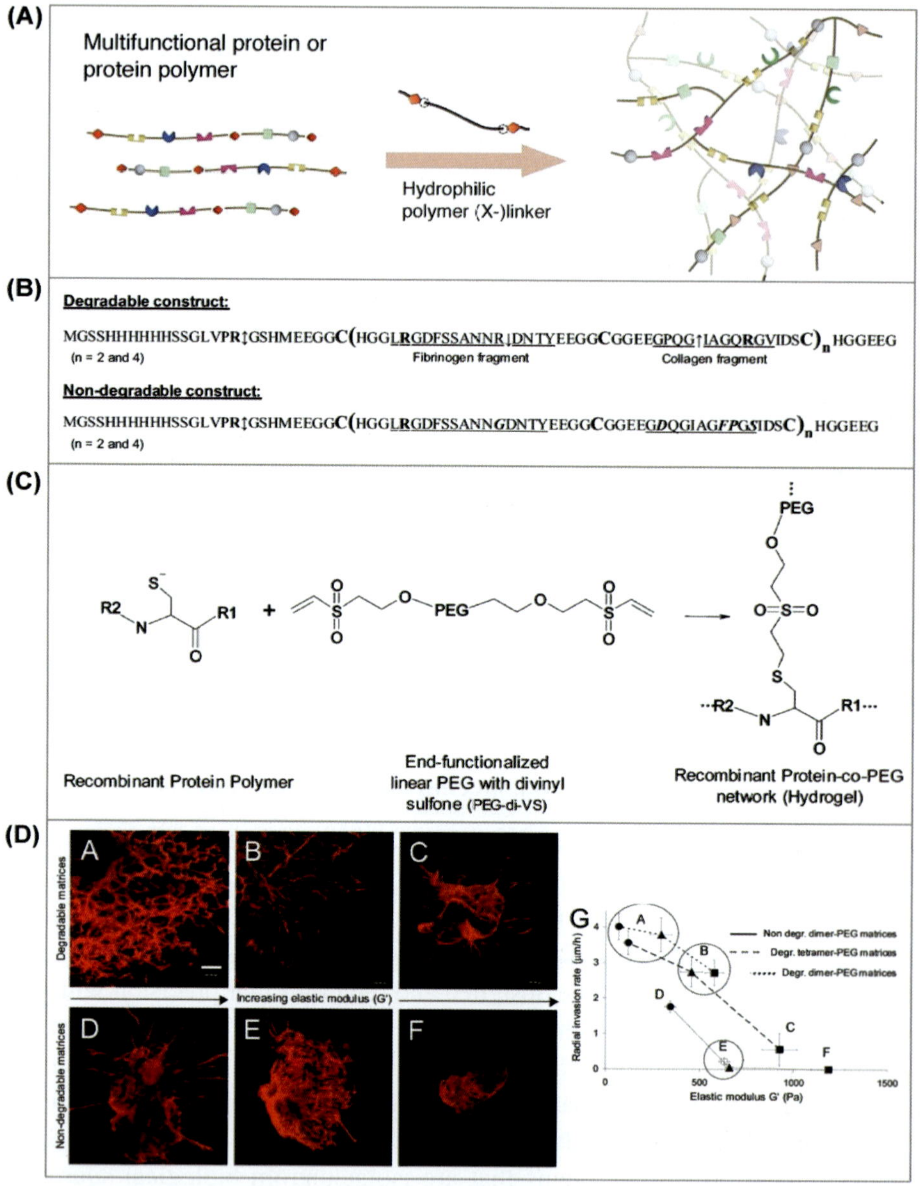

Fig. 4. Biomolecules such as proteins or glycosaminoglycans (GAGs) can be used as building blocks to confer biofunctionality in artificial ECMs **(A)**. An example of the structure of a proteolytically sensitive and cell-adhesive protein polymer produced in *E.coli* by Rizzi *et al.* is shown (**B**, adapted and reproduced with permission from Rizzi *et al.*;[26] copyright 2006 American Chemical Society). Michael-type addition crosslinking chemistry to produce hybrid protein-polymer gels **(C)**. Primary fibroblasts *within* these matrices migrate extensively via proteolytic matrix degradation. Cell invasion rates respond to the elastic moduli (i.e. crosslinking density) of the gels.

as substrates for cell adhesion and, more recently, such as in collaborative work with Gandahari and co-authors, as matrices for enhancing uptake and transfection with plasmid DNA.[19]

Sophisticated molecular engineering has been carried out in protein polymers to confer biological functionality. For example, Tirrell and co-authors have explored incorporation of numerous cell adhesion motifs into elastin-based protein polymers,[20] for example to confer preferential adhesion to endothelial cells in vascular implants.[21] In collaboration with Kornfield, self-assembly approaches have been developed based on coiled-coil interactions, and depending on the details of protein design a wide and controllable range of mechanical properties and erosion rates can be obtained.[22]

Approaches to covalent assembly of protein polymers and protein polymer hybrid materials (i.e. copolymer networks with synthetic polymers) have been developed. For example, recent work from Chilkoti, Setton and co-workers has described design and expression of elastin-based protein polymers comprising substrate domains for tissue transglutaminase; mixing of the soluble protein polymer *in situ*, perhaps also in the presence of cells, leads to rapid formation of elastic gels,[23] which are being explored in repair of articular cartilage. Hubbell and co-authors have described approaches by which to express multifunctional protein polymers, bearing sites for integrin ligation, growth factor binding, and chemical reaction with counter-reactive synthetic polymers for *in situ* gelation, for example exploiting a Michael-type addition reaction between thiols on cysteine residues on the protein polymer and acrylate- or vinyl sulfone-functionalized branched PEG chains.[24,25] These materials have been explored in bone repair,[26] as well as other applications.

Finally, as to protein polymers, one should not overlook the usefulness of naturally occurring proteins in protein-polymer hybrid matrices. The ability to engineer such materials has been illustrated by Seliktar and co-workers, who have developed approaches by which to chemically crosslink partially-denatured (so as to expose cysteine residues) fibrinogen, a blood-derived protein, with reactive acrylate-functionalized branched PEG chains.[27] This approach allows a high level of control of proteolytic remodeling rates *in vivo*, and these materials have been explored in bone repair *in vivo*.[28]

Glycosaminoglycans are also interesting materials in construction of enzymatically-sensitive biomaterial matrices. Prestwich and co-workers have presented chemical schemes by which to functionalize hyaluronic acid and gelatin or other biomolecules with thiol functionalities, to thus allow chemical crosslinking *in situ* with reactive polymers.[29,30] The effectiveness of this novel family of artifical ECMs has been demonstrated in various tissue engineering applications.[31] Hubbell and co-workers have explored functionalization with acrylate groups, to allow photopolymerization *in situ* with visible light initiation.[32] Elisseeff and colleagues have developed schemes by which to functionalize GAGs such as chondroitin sulfate with multiple reactive groups, to allow gelation and also strong bonding to tissue substrates, and have explored this in articular cartilage repair.[33] Of note, other polysaccharides such as alginate can be readily modified to become extremely useful artificial ECM gels for tissue engineering, as demonstrated by Mooney and co-workers over the last couple of years.[34,35] Alginate gels can be fabricated with controlled mechanical properties, degradation characteristics, cell adhesion, and binding or release of bioactive molecules.

As illustrated above, biomolecules such as proteins and GAGs can serve as compelling building blocks of bioactive, biomimetic materials, both naturally occurring biomolecules and those produced by genetic engineering approaches. These materials can be applied as useful tools for basic biological investigation, such as studies of cell adhesion, migration, and self-renewal or lineage selection and differentiation, and can also function as powerful therapeutics in tissue repair alone or in combination of incorporated growth factors.

4.2. Hybrid peptide-polymer systems

Whereas protein polymers are subject to very sophisticated design and very precise production, as described in the preceding section, their production is somewhat laborious and can be costly. This drives interest in biomolecules that could carry similar information as proteins, but be associated with simpler synthesis. Peptides can potentially serve this role, being recognizable by a number of cell-surface receptors and cell-secreted proteases if appropriately designed. As such, hybrid peptide-polymer systems are also a powerful member of the biomaterials toolbox (Fig. 5), although one should recognize that peptides also can be rather expensive, depending on their length and scale of synthesis.

Pioneering work on peptide-mediated crosslinking of biomaterial hydrogel precursors derived from the lab of Griffith. Sperinde and Griffith derivatized branched PEGs with very short peptide substrates for tissue transglutaminase and demonstrated that the enzymatically-mediated amide bond formation between the PEG chain termini could lead to very gentle gelation to form elastic gels.[36,37] Messersmith's laboratory explored longer peptide sequences that served as higher V_{max} substrates and obtained gelation that was considerably faster.[38,39] This work has been taken further by Ehrbar *et al.* in terms of modularity of design, using the coagulation transglutaminase factor XIIIa to crosslink peptide-functionalized PEGs, where the peptide comprises not only a factor XIIIa substrate site for crosslinking but also a protease substrate site for cellular remodeling.[40,41] Using this method, adhesion peptides or engineered growth factors bearing

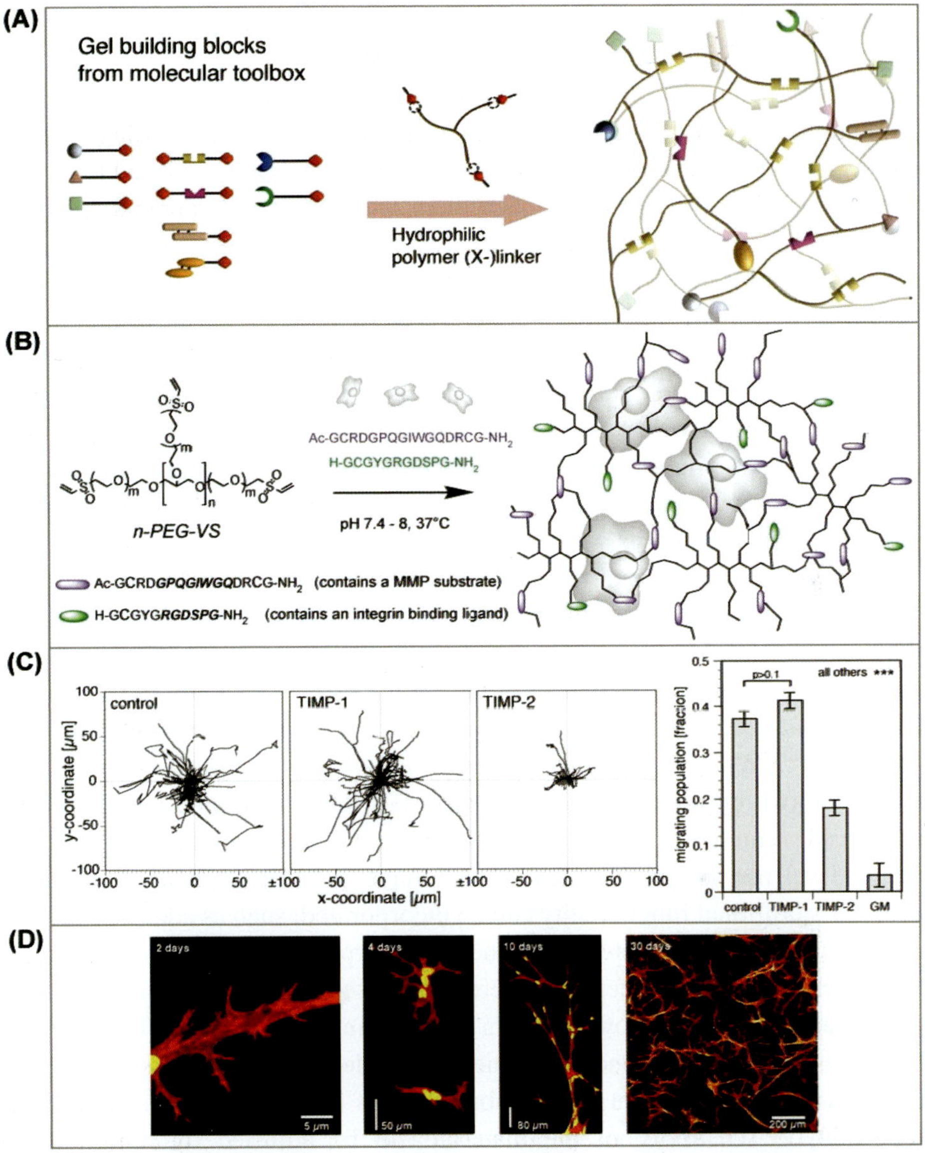

Fig. 5. Peptides can serve as extremely versatile building blocks to confer biofunctionality in artificial ECMs **(A)**. Structure and crosslinking chemistry of a hybrid peptide-polymer gel formed via Michael-type addition from cysteine-bearing cell-adhesive and proteolytically degradable peptides and vinylsulfone-functionalized PEG macromers **(B)**. The 3D migration of primary fibroblast can be blocked via a tissue inhibitor of metalloproteinases (TIMP-2), demonstrating the physiological character of 3D cell behavior with the synthetic matrix (**C**, adapted and reproduced with permission from Raeber *et al.*;[56] copyright 2007 Elsevier). Extensive migration and proliferation in long-term cell culture leads to the formation of multicellular (morphogenetic) networks **(D)**.

a factor XIIIa substrate site can be co-crosslinked into the gels simultaneously with gelation to enable cell-mediated remodeling.

Non-enzymatic chemical reactions have also been employed in peptide-polymer hybrids, including free radical polymerization and chemical addition reactions. An early example was from West and Hubbell, in which acrylate-terminated peptide-PEG-peptide macromers were photopolymerized to induce gelation.[42] The peptide separating the PEG chain from the free-radical polymerizable acrylate end-group was designed to be a substrate for plasmin or MMPs, both families of which are activated on the surfaces of cells as they migrate and remodel their natural environment. Such materials have been explored by West's group in a number of tissue engineering applications, including vascular tissue engineering.[43–45] Healy's group has developed peptide-containing monomers for formation of copolymeric semi-interpenetrating polymer networks that display sensitivity to MMPs involved in cell migration;[46] such materials are being explored in bone repair,[47] as well as other applications.

One important strength of the approach utilized with peptide-polymer hybrids is modularity of design, and this can be illustrated by work from the groups of Hubbell and Lutolf. Michael-type addition reactions have been exploited to induce gelation by mixture of a peptide comprising two cysteine residues and a branched PEG functionalized with a thiol-reactive functionality, such as a vinyl sulfone or an acrylate.[48,49] When the crosslinking peptide contains a protease substrate site between the two cysteine residues, the materials that are formed are sensitive to cell-induced remodeling, allowing migration and invasion, as well as cell-demanded release of growth factors that have been incorporated within the matrix.[50,51] Additional functionality can be incorporated, such as adhesion peptides containing a single free cysteine residue for incorporation, as well as potentially growth factors that have been engineered to possess a free cysteine residue[52] or peptides selected to bind growth factors via heparin-like affinity.[53] The mechanics of such networks can be readily engineered by design of the crosslinking PEG, such as its molecular weight and number of arms in the branched structure, and the proteolytic sensitivity of the material can be adjusted by design of the sequence of the protease substrate that resides within the crosslinking peptide.[54–56] As such, the modules of mechanics, remodeling, adhesion and growth factor incorporation and release can each be designed and selected in nearly independent fashion. Such materials have been explored in applications involving angiogenesis and bone repair, for example.

Peptide-polymer hydrogels have demonstrated high promise already early in their development cycle, both as models for *in vitro* cell biological investigation and as matrices for cell invasion and morphogen delivery in tissue repair in animal models. The materials are powerful in that the principles of their design can be clearly related to a basic understanding of the biochemistry and cell biology of the ECM. It is undoubtful that such materials will continue to play an important role as artificial ECMs for tissue engineering and cell biology.

4.3. Nanofibrillar-based gels

The materials described in the above two sections are biomimetic in their biomolecular recognition features, but not necessarily in their nanomorphology: many components of the ECM are nanofibrillar, such as several of the collagens, or nanofibrillar with a co-continuous amorphous gel phase, such as hyaluronic acid.[57] Biomaterials scientists have begun exploration of means by which to mimic such structures as well, creating gels that are comprised of nanofibrils (Fig. 6).

Stupp and co-workers have long studied amphiphilic block copolymers and their self-assembly in water into ordered structures, including fibrillar gels.[58] Using these approaches, it is possible to form fibrillar structures with fibril diameters in the 10 nm dimension with lengths and branches that lead to macroscopically strong materials. Consistent with the concept of modularity of design, adhesion ligands can be incorporated into these materials, such as the well-studied RGD ligand[59,60] and neurogenic ligands from laminin including the IKVAV sequence, which was demonstrated to enhance neuronal differentiation from stem cells.[61] Affinity for growth factors can be conferred via heparin-binding character, the heparin being an intermediary, binding to the heparan sulfate-binding domains that are frequently found in morphogenetic growth factors.[62] Given that such fibril precursors can be precisely synthesized, and that fibrils can be formed from mixtures of fibril-forming precursors, these methods are very powerful.

Very sophisticated molecular designs can be incorporated into the fibril-forming precursors, as has been demonstrated by Hartgerink and co-workers. Basic understanding of biomolecular affinity can be used to design the self-assembling structures,[63] e.g. exploiting the interactions that are used naturally in self-assembly processes such as triple helix-formation in collagens.[64] With regard to peptide amphiphiles and modular design of functionality, for example, peptide amphiphiles have been designed to incorporate protease substrate sites, to enable the self assembly-inducing hydrophobe to be cleaved from the remainder of the peptide by cell-associated proteases,[65] thus allowing cellular remodeling in these materials as described for the peptide- and protein-polymer hybrids in the two preceding sections.

Sophistication in nanofibril morphology has also been achieved, for example as developed by Woolfson and colleagues[66,67] or Zhang and colleagues, using electrostatic and other interactions to drive peptide self-assembly; this group's extensive structure-function exploration has been reviewed elsewhere.[68] These

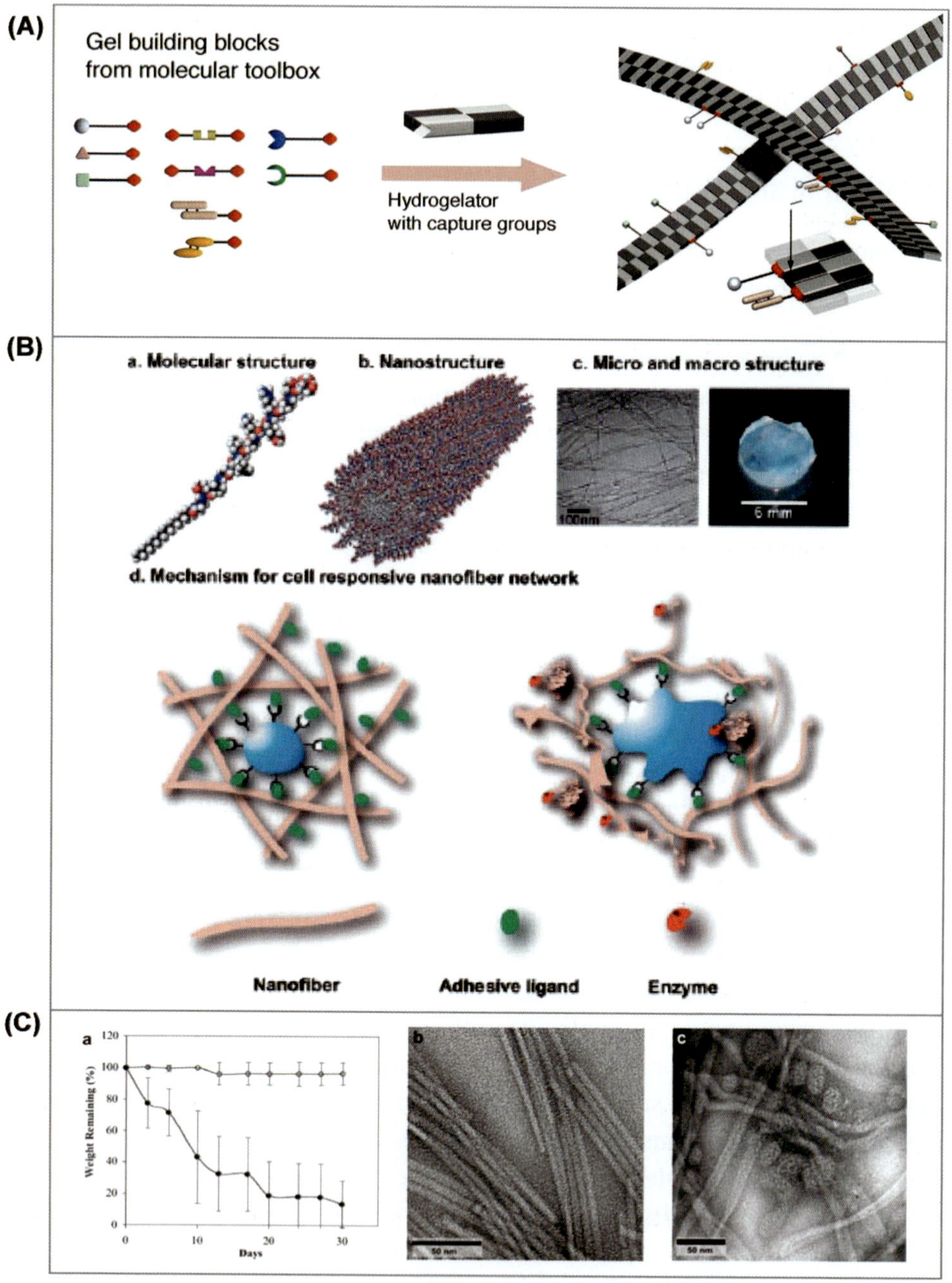

Fig. 6. Small organic hydrogelators, peptide or polymeric amphiphiles can be utilized to generate powerful nanofibrillar artificial ECMs (**A**). The incorporation of protease substrates in self-assembling peptide amphiphiles allows for example to render fibers susceptible to cell-controlled proteolytic gel degradation (**B** and **C**, adapted and reproduced with permission from Jun *et al.*;[65] copyright 2005 Wiley-VCH Verlag GmBH & Co.)

materials have demonstrated very interesting usefulness in basic cell biological investigations,[69] as well as in regenerative medicine applications,[70] including bone,[71] cartilage,[72] and neural[73] tissue engineering. The ability to exploit self-assembly processes, with polymeric amphiphiles, with peptide amphiphiles, and with peptide electrostatic self-assembly offers great potential in being able to engineer both the nanomorphology and the biomolecular recognition characteristics of materials.

5. Future Challenges

Impressive progress has already been made in the design of artificial ECMs. As discussed above, the incorporation of biological recognition principles has enabled researchers to design materials for efficient 3D cell culture and tissue regeneration *in vivo*. However, it should be noted that these systems are highly simplified mimics of natural ECMs that can at best recapitulate a limited set of its key signaling and cell-responsive functions. It is clear that the design complexity and functionality of these smart biomaterials will evolve in the foreseeable future, as bioengineers continue to exploit the nearly inexhaustible toolbox that biology has to offer.

At least five characteristics of natural ECMs remain beyond the realms of possibility in synthetic ECM mimics at the moment. (i) *Near-physiological ECM multifunctionality*. Common artificial ECMs exploit one or two biomolecules classes, while natural ECMs are comprised of many different biochemical cues. Versatile chemical schemes are needed that would allow the site-specific and stable immobilization of *any desired* protein morphogen and morphogen combinations into gels without compromising the protein's bioactivity. This would allow to systematically deconstruct and then reconstruct the natural complexity from bottom-up, addressing difficult biological questions such as signaling crosstalk in cell fate regulation. Alternatively, it may be interesting to deliver molecules (such as small interfering RNA) that are not part of natural ECMs but that can selectively manipulate cell fate by interfering for example with the expression of particular transcription factor genes that regulate cell phenotype. (ii) *Temporal complexity in signal presentation*. The time scales and dynamics of signal presentation from artificial ECMs are rather limited in longevity and complexity. Systems that allow morphogen release from the matrix via cell-controlled mechanisms, or matrices that can capture signals upon secretion by embedded cells in close proximity may be of interest here. (iii) *Spatial complexity in signal presentation*. Matrix-immobilized 3D morphogen gradients play important roles in tissue development and regeneration. For example, several different cell types and patterns can be generated spatially related to the source of a signal, and specific cells migrate to specific locations controlled by such signal gradients. These scenarios could be recapitulated *in vitro* were sophisticated matrices available. While biomolecule gradients have been successfully generated on hydrogel surfaces,[74,75] there are no methodologies to generate them within 3D cell-responsive artificial ECMs. (iv) *Feedback systems controlling cell-matrix interactions*. Feedback systems are common features of natural ECMs. Proteases can for example cleave ECM components and the generated cleavage products can have important signaling function.[76] The exploration of such feedback systems should be an attractive area of research towards higher complexity in smart biomaterials. (v) *Cell-specificity in matrix responsiveness*. Thus far, protease substrates of cell-responsive matrices are short, linear peptides with little specificity for particular proteases, except in matrices composed of natural proteins. The implication of this is that common artificial ECMs are not specific to particular cells or their secreted protease mixtures, respectively. The use of protease substrates with more complex secondary or tertiary structures to achieve selectivity,[77] or else the screening of large numbers of potential substrate sequences via combinatorial peptide libraries to identify specific substrates[78] could provide promising paths forward in this area.

Finally, as pointed out by others,[31,79] depending on the targeted application, the pursuit of ever more complex systems should be balanced against economical and practical constraints. Cell-responsive materials for clinical applications should be cost-effective and robust. In this case, an exact recapitulation of natural ECMs may be unnecessary, as relatively "simple" systems could perform the desired functions. Responsive materials that (i) employ signals to attract progenitor cells and induce desired differentiation pathways and (ii) are *permissive* to give way to subsequent cellular remodeling, could be sufficient to create functional tissues *in situ*. However, if the goal is to engineer tissue *in vitro*,[80] higher matrix complexity appears justified, as the necessary ingredients for morphogenesis are completely absent outside of the body. A fusion of cell-responsive materials technologies with microtechnologies such as microfluidics[81] and bioreactors technologies could one day facilitate the construction of complex tissues with a wealth of applications involving screening tools and powerful models to study fundamental biological questions.

Our understanding of the importance of extracellular microenvironments in regulating cell behavior has led to the design of novel synthetic biomaterials that mimic some of its structural and biochemical key characteristics. Although progress has been made towards recapitulating more complex cellular processes in these materials and much about the molecular and cellular mechanisms of tissue morphogenesis and regeneration is known, the translation of this knowledge into biomaterials design and tissue engineering has only just begun. Once the initial limitations are overcome, the prospects of using such designer microenvironments for tissue engineering and cell biology are compelling.

References

1. Langer, R. and Vacanti, J.P. Tissue engineering. *Science* **260**(5110), 920–926 (1993).
2. Peppas, N.A. and Langer, R. New challenges in biomaterials. *Science* **263**(5154), 1715–1720 (1994).
3. Langer, R. and Tirrell, D.A. Designing materials for biology and medicine. *Nature* **428**(6982), 487–492 (2004).
4. Hubbell, J.A. Bioactive biomaterials. *Curr. Opin. Biotechnol.* **10**(2), 123–129 (1999).
5. Lutolf, M.P. and Hubbell, J.A. Synthetic biomaterials as instructive extracellular microenvironments for morphogenesis in tissue engineering. *Nat. Biotechnol.* **23**(1), 47–55 (2005).
6. Kleinman, H.K., Philp, D. and Hoffman, M.P. Role of the extracellular matrix in morphogenesis. *Curr. Opin. Biotechnol.* **14**(5), 526–532 (2003).
7. Giancotti, F.G. and Ruoslahti, E. Integrin signaling. *Science* **285**(5430), 1028–1032 (1999).
8. Friedl, P. Prespecification and plasticity: shifting mechanisms of cell migration. *Curr. Opin. Cell Biol.* **16**(1), 14–23 (2004).
9. Page-McCaw, A., Ewald, A.J. and Werb, Z. Matrix metalloproteinases and the regulation of tissue remodelling. *Nat. Rev. Mol. Cell Biol.* **8**(3), 221–233 (2007).
10. Lijnen, H.R. Elements of the fibrinolytic system. *Ann. N. Y. Acad. Sci.* **936**, 226–236 (2001).
11. Moghaddam, M.J. and Matsuda, T. Development of a 3-D artificial extracellular matrix. Design concept and artificial vascular media. *ASAIO Trans.* **37**(3), M437–M438 (1991).
12. Moghaddam, M.J. and Matsuda, T. Molecular design of three-dimensional artificial extracellular matrix: photosensitive polymers containing cell adhesive peptides. *J. Polym. Sci. A Polym. Chem.* **31**, 1589–1597 (1993).
13. Kopecek, J. and Yang, J.Y. Revie — hydrogels as smart biomaterials. *Polym. Int.* **56**(9), 1078–1098 (2007).
14. Hennink, W.E. and van Nostrum, C.F. Novel crosslinking methods to design hydrogels. *Adv. Drug Deliv. Rev.* **54**(1), 13–36 (2002).
15. Estroff, L.A. and Hamilton, A.D. Water gelation by small organic molecules. *Chem. Rev.* **104**(3), 1201–1218 (2004).
16. Ulijn. R.V., Bibi, N., Jayawarna, V., Thornton, P.D., Todd, S.J., Mart, R.J. *et al.* Bioresponsive hydrogels. *Mater. Today* **10**(4), 40–48 (2007).
17. Cappello, J., Crissman, J., Dorman, M., Mikolajczak, M., Textor, G., Marquet, M. *et al.* Genetic engineering of structural protein polymers. *Biotechnol. Prog.* **6**, 198–202 (1990).

18. Cappello, J., Crissman, J.W., Crissman, M., Ferrari, F.A., Textor, G., Wallis, O. *et al.* *In-situ* self-assembling protein polymer gel systems for administration, delivery, and release of drugs. *J. Cont. Rel.* **53**, 105–117 (1998).
19. Haider, M., Leung, V., Ferrari, F., Crissman, J., Powell, J., Capello, J. *et al.* Molecular engineering of silk-elastinlike polymers for matrix-mediated gene delivery: biosynthesis and characterization. *Mol. Pharm.* **2**, 139–150 (2005).
20. Maskarinec, S.A. and Tirrell, D.A. Protein engineering approaches to biomaterials design. *Curr. Opin. Biotechnol.* **16**, 422–426 (2005).
21. Heilshorn, S.C., Liu, J.C. and Tirrell, D.A. Cell-binding domain context affects cell behavior on engineered proteins. *Biomacromolecules* **6**, 318–323 (2005).
22. Shen, W., Zhang, K., Kornfield, J.A. and Tirrell, D.A. Tuning the erosion rate of artificial protein hydrogels through control of network topology. *Nat. Mater.* **5**, 153–158 (2006).
23. McHale, M.K., Setton, L.A. and Chilkoti, A. Synthesis and *in vitro* evaluation of enzymatically cross-linked elastin-like polypeptide gels for cartilaginous tissue repair. *Tissue Eng.* **11**, 1768–1779 (2005).
24. Halstenberg, S., Panitch, A., Rizzi, S., Hall, H. and Hubbell, J.A. Biologically engineered protein-graft-poly(ethylene glycol) hydrogels: a cell adhesive and plasm in-degradable biosynthetic material for tissue repair. *Biomacromolecules* **3**(4), 710–723 (2002).
25. Rizzi, S.C. and Hubbell, J.A. Recombinant protein-co-PEG networks as cell-adhesive and proteolytically degradable hydrogel matrixes. Part 1: Development and physicochernical characteristics. *Biomacromolecules* **6**(3), 1226–1238 (2005).
26. Rizzi, S.C., Ehrbar, M., Halstenberg, S., Raeber, G.P., Schmoekel, H.G., Hagenmuller, H. *et al.* Recombinant protein-co-PEG networks as cell-adhesive and proteolytically degradable hydrogel matrixes. Part II: Biofunctional characteristics. *Biomacromolecules* **7**(11), 3019–3029 (2006).
27. Almany, L. and Seliktar, D. Biosynthetic hydrogel scaffolds made from fibrinogen and polyetheylene glycol for 3D cell cultures. *Biomaterials* **26**, 2467–2477 (2005).
28. Peled, E., Boss, J., Bejar, J., Zinman, C. and Seliktar, D. A novel poly(ethylene glycol)-fibrinogen hydrogel for tibial segmental defect repair in a rat model. *J. Biomed. Mater. Res. A* **80**, 874–884 (2007).
29. Shu, X.Z., Ahmad, S., Liu, Y.C. and Prestwich, G.D. Synthesis and evaluation of injectable, *in situ* crosslinkable synthetic extracellular matrices for tissue engineering. *J. Biomed. Mater. Res. Part A* **79**A(4), 902–912 (2006).
30. Vanderhooft, J.L., Mann, B.K. and Prestwich, G.D. Synthesis and characterization of novel thiol-reactive poly(ethylene glycol) cross-linkers for extracellular-matrix-mimetic biomaterials. *Biomacromolecules* **8**(9), 2883–2889 (2007).

31. Prestwich, G.D. Simplifying the extracellular matrix for 3-D cell culture and tissue engineering: a pragmatic approach. *J. Cell. Biochem.* **101**(6), 1370–1383 (2007).
32. Park, Y.D., Tirelli, N. and Hubbell, J.A. Photopolymerized hyaluronic acid-based hydrogels and interpenetrating networks. *Biomaterials* **24**(6), 893–900 (2003).
33. Wang, D.A., Varghese, S., Sharma, B., Strehin, I., Fermanian, S., Gorham, J. *et al.* Multifunctional chondroitin sulphate for cartilage tissue-biomaterial integration. *Nat. Mater.* **6**, 385–392 (2007).
34. Rowley, J.A., Madlambayan, G. and Mooney, D.J. Alginate hydrogels as synthetic extracellular matrix materials. *Biomaterials* **20**(1), 45–53 (1999).
35. Augst, A.D., Kong, H.J. and Mooney, D.J. Alginate hydrogels as biomaterials. *Macromol. Biosci.* **6**(8), 623–633 (2006).
36. Sperinde, J.J. and Griffith, L.G. Synthesis and charcterization of enzymatically-cross-linked poly(ethylene glycol) hydrogels. *Macromolecules* **30**, 5255–5264 (1997).
37. Sperinde, J.J. and Griffith, L.G. Control and prediction of gelation kinetics in enzymatically cross-linked poly(ethylene glycol) hydrogels. *Macromolecules* **33**, 5476–5480 (2000).
38. Hu, B.H. and Messersmith, P.B. Rational design of transglutaminase substrate peptdies for rapid enzymatic formation of hydrogels. *J. Am. Chem. Soc.* **125,** 14298–14299 (2003).
39. Hu, B.H. and Messersmith, P.B. Enzymatically cross-linked hydrogels and their adhesive strength to biosurfaces. *Orthod. Craniofac. Res.* **8**, 145–149 (2005).
40. Ehrbar, M., Rizzi, S.C., Hlushchuk, R., Djonov, V., Zisch, A.H., Hubbell, J.A. *et al.* Enzymatic formation of modular cell-instructive fibrin analogs for tissue engineering. *Biomaterials* **28**(26), 3856–3866 (2007).
41. Ehrbar, M., Rizzi, S.C., Schoenmakers, R.G., San Miguel, B., Hubbell, J.A., Weber, F.E. *et al.* Biomolecular hydrogels formed and degraded via site-specific enzymatic reactions. *Biomacromolecules* **8**(10), 3000–3007 (2007).
42. West, J.L. and Hubbell, J.A. Polymeric biomaterials with degradation sites for proteases involved in cell migration. *Macromolecules* **32**(1), 241–244 (1999).
43. Hahn, M.S., Miller, J.S. and West, J.L. Three-dimensional biochemical and biomechanical patterning of hydrogels for guiding cell behavior. *Adv. Mater.* **18**(20), 2679–2684 (2006).
44. Moon, J.J., Lee, S.H., Hahn, M.S., Nsiah, B.A. and West, J.L. Regulation of endothelial angiogenesis and vasculogenesis in synthetic poly(ethylene glycol) hydrogels modified with biomolecules. *FASEB J.* **21**(6), A748–A749 (2007).

45. Moon, J.J., Lee, S.H. and West, J.L. Synthetic biomimetic hydrogels incorporated with Ephrin-A1 for therapeutic angiogenesis. *Biomacromolecules* **8**(1), 42–49 (2007).
46. Kim, S., Chung, E.H., Gilbert, M. and Healy, K.E. Synthetic MMP-13 degradable ECMs based on poly(N-isopropylacrylamide-co-acrylic acid) semi-interpenetrating polymer networks. I. Degradation and cell migration. *J. Biomed. Mater. Res. Part A* **75A**(1), 73–88 (2005).
47. Chung, E.H., Gilbert, M., Virdi, A.S., Sena, K., Sumner, D.R. and Healy, K.E. Biomimetic artificial ECMs stimulate bone regeneration. *J. Biomed. Mater. Res. Part A* **79A**(4), 815–826 (2006).
48. Lutolf, M.P. and Hubbell, J.A. Synthesis and physicochemical characterization of end-linked poly(ethylene glycol)-co-peptide hydrogels formed by Michael-type addition. *Biomacromolecules* **4**(3), 713–722 (2003).
49. Lutolf, M.P., Tirelli, N., Cerritelli, S., Cavalli, L. and Hubbell, J.A. Systematic modulation of Michael-type reactivity of thiols through the use of charged amino acids. *Bioconjug. Chem.* **12**(6), 1051–1056 (2001).
50. Lutolf, M.P., Lauer-Fields, J.L., Schmoekel, H.G., Metters, A.T., Weber, F.E., Fields, G.B. *et al.* Synthetic matrix metalloproteinase-sensitive hydrogels for the conduction of tissue regeneration: engineering cell-invasion characteristics. *Proc. Natl. Acad. Sci. USA* **100**(9), 5413–5418 (2003).
51. Lutolf, M.P., Raeber, G.P., Zisch, A.H., Tirelli, N. and Hubbell, J.A. Cell-responsive synthetic hydrogels. *Adv. Mater.* **15**(11), 888–892.
52. Zisch, A.H., Lutolf, M.P., Ehrbar, M., Raeber, G.P., Rizzi, S.C., Davies, N. *et al.* Cell-demanded release of VEGF from synthetic, biointeractive cell-ingrowth matrices for vascularized tissue growth. *FASEB J.* **17**(13), 2260–2262 (2003).
53. Maynard, H.D. and Hubbell, J.A. Discovery of a sulfated tetrapeptide that binds to vascular endothelial growth factor. *Acta Biomater.* **1**(4), 451–459 (2005).
54. Lutolf, M.R., Weber, F.E., Schmoekel, H.G., Schense, J.C., Kohler, T., Muller, R. *et al.* Repair of bone defects using synthetic mimetics of collagenous extracellular matrices. *Nat. Biotechnol.* **21**(5), 513–518 (2003).
55. Raeber, G.P., Lutolf, M.P. and Hubbell, J.A. Molecularly engineered PEG hydrogels: a novel model system for proteolytically mediated cell migration. *Biophys. J.* **89**(2), 1374–1388 (2005).
56. Raeber, G.P., Lutolf, M.P. and Hubbell, J.A. Mechanisms of 3-D migration and matrix remodeling of fibroblasts within artificial ECMs. *Acta Biomater.* **3**(5), 615–629 (2007).
57. Pedersen, J.A. and Swartz, M.A. Mechanobiology in the third dimension. *Ann. Biomed. Eng.* **33**(11), 1469–1490 (2005).

58. Palmer, L.C., Velichko, Y.S., de la Cruz, M.O. and Stupp, S.I. Supramolecular self-assembly codes for functional structures. *Philos. Trans. A Math. Phys. Eng. Sci.* **365**(1855), 1417–1433 (2007).
59. Guler, M.O., Hsu, L., Soukasene, S., Harrington, D.A., Hulvat, J.F. and Stupp, S.I. Presentation of RGDS epitopes on self-assembled nanofibers of branched peptide amphiphiles. *Biomacromolecules* **7**(6), 1855–1863 (2006).
60. Storrie, H., Guler, M.O., Abu-Amara, S.N., Volberg, T., Rao, M., Geiger, B. *et al.* Supramolecular crafting of cell adhesion. *Biomaterials* **28**(31), 4608–4618 (2007).
61. Silva, G.A., Czeisler, C., Niece, K.L., Beniash, E., Harrington, D.A., Kessler, J.A. *et al.* Selective differentiation of neural progenitor cells by high-epitope density nanofibers. *Science* **303**(5662), 1352–1355 (2004).
62. Rajangam, K., Behanna, H.A., Hui, M.J., Han, X.Q., Hulvat, J.F., Lomasney, J.W. *et al.* Heparin binding nanostructures to promote growth of blood vessels. *Nano Lett.* **6**(9), 2086–2090 (2006).
63. Dong, H., Paramonov, S.E., Aulisa, L., Bakota, E.L. and Hartgerink, J.D. Self-assembly of multidomain peptides: balancing molecular frustration controls conformation and nanostructure. *J. Am. Chem. Soc.* **129**(41), 2468–12472 (2007).
64. Gauba, V. and Hartgerink, J.D. Self-assembled heterotrimeric collagen triple helices directed through electrostatic interactions. *J. Am. Chem. Soc.* **129**(9), 2683–2690 (2007).
65. Jun, H.W., Yuwono, V., Paramonov, S.E. and Hartgerink, J.D. Enzyme-mediated degradation of peptide-amphiphile nanofiber networks. *Adv. Mater.* **17**(21), 2612–2617 (2005).
66. Ryadnov, M.G. and Woolfson, D.N. Engineering the morphology of a self-assembling protein fibre. *Nat. Mater.* **2**(5), 329–332 (2003).
67. Ryadnov, M.G. and Woolfson, D.N. Introducing branches into a self-assembling peptide fiber. *Angew. Chem. Int. Ed. Engl.* **42**(26), 3021–3023 (2003).
68. Zhao, X. and Zhang, S. Fabrication of molecular materials using peptide construction motifs. *Trends Biotechnol.* **22,** 470–476 (2004).
69. Zhang, S.G., Gelain, F. and Zhao, X.J. Designer self-assembling peptide nanofiber scaffolds for 3D tissue cell cultures. *Semin. Cancer Biol.* **15**(5), 413–420 (2005).
70. Gelain, F., Horii, A. and Zhang, S.G. Designer self-assembling peptide scaffolds for 3-D tissue cell cultures and regenerative medicine. *Macromol. Biosci.* **7**(5), 544–551 (2007).
71. Horii, A. Biological designer self-assembling peptide nanofiber scaffolds significantly enhance osteoblast proliferation, differentiation and 3-D migration. *PLoS ONE* **2**(2), e190 (2007).

72. Kisiday, J., Jin, M., Kurz, B., Hung, H., Semino, C., Zhang, S. *et al.* Self-assembling peptide hydrogel fosters chondrocyte extracellular matrix production and cell division: implications for cartilage tissue repair. *Proc. Natl. Acad. Sci. USA* **99**(15), 9996–10001 (2002).
73. Gelain, F. Designer self-assembling peptide nanofiber scaffolds for adult mouse neural stem cell 3-dimensional cultures. *PLoS ONE* **1**(1), e119 (2006).
74. Burdick, J.A., Khademhosseini, A. and Langer, R. Fabrication of gradient hydrogels using a microfluidics/photopolymerization process. *Langmuir* **20**(13), 5153–5156 (2004).
75. DeLong, S.A., Moon, J.J. and West, J.L. Covalently immobilized gradients of bFGF on hydrogel scaffolds for directed cell migration. *Biomaterials* **26**(16), 3227–3234 (2005).
76. Hamano, Y., Zeisberg, M., Sugimoto, H., Lively, J.C., Maeshima, Y., Yang, C.Q. *et al.* Physiological levels of tumstatin, a fragment of collagen IV alpha 3 chain, are generated by MMP-9 proteolysis and suppress angiogenesis via alpha V beta 3 integrin. *Cancer Cell* **3**(6), 589–601 (2003).
77. Lauer-Fields, J.L., Sritharan, T., Stack, M.S., Nagase, H. and Fields, G.B. Selective hydrolysis of triple-helical substrates by matrix metalloproteinase-2 and -9. *J. Biol. Chem.* **278**(20), 18140–18145 (2003).
78. Richardson, P.L. The determination and use of optimized protease substrates in drug discovery and development. *Curr. Pharm. Des.* **8**(28), 2559–2581 (2002).
79. Alexander, C. and Shakesheff, K.M. Responsive polymers at the biology/materials science interface. *Adv. Mater.* **18**(24), 3321–3328 (2006).
80. Khetani, S.R. and Bhatia, S.N. Engineering tissues for *in vitro* applications. *Curr. Opin. Biotechnol.* **17**(5), 524–531 (2006).
81. Khademhosseini, A., Langer, R., Borenstein, J. and Vacanti, J.P. Microscale technologies for tissue engineering and biology. *Proc. Natl. Acad. Sci. USA* **103**(8), 2480–2487 (2006).

Chapter 14

Bioactive Composite Materials for Bone Tissue Engineering Scaffolds

Sophie Verrier and Aldo R. Boccaccini

Abstract

Synthetic bioactive and bioresorbable composite materials are becoming increasingly important as scaffolds for bone tissue engineering. Next generation biomaterials should combine bioactive and bioresorbable properties to activate *in vivo* mechanisms of tissue regeneration, stimulating the body to heal itself and leading to replacement of the scaffold by the regenerating tissue. In the present chapter composite materials based on smart combinations of biodegradable polymers and bioactive ceramics, including hydroxyapatite and bioactive glasses, are discussed as suitable materials for scaffold fabrication. These composites exhibit tailored physical, biological and mechanical properties as well as predictable degradation behaviour. The appropriate selection of a particular composite for a given application requires a detailed understanding of relevant cells and/or tissue response. Knowledge concerning interactions between cells and their immediate local environment in composite scaffolds has deeply improved in the last years. An overview of these findings is presented highlighting the influence of material processing methods, scaffold microstructure as well as the importance of the nature and amount of the bioactive ceramic particulate included in specific polymer matrices. The chapter also emphasises the response diversity according to the cell type used *in vitro* or the chosen *in vivo* models (species and location), suggesting the utility of standardisation in this field of biomaterials science. Bioactive composites discussed in this chapter, enhanced by microstructural optimisation and surface engineering, are suggested as the materials of choice for development of optimal bone tissue engineering scaffolds.

Keywords: Scaffolds; Bioactive Materials; Bioactive Glass; Composite Materials; Osteoblast; Bone Regeneration; Biodegradable Polymers.

Outline

1. Introduction
2. Scaffolds Requirements
3. Composite Materials Approach for Tissue Engineering Scaffolds
   3.1. Advantages of composites materials
   3.2. Mechanical properties
   3.3. Fabrication technologies
4. *In Vitro* and *In Vivo* Evaluation
   4.1. Calcium phosphate-based composites
      4.1.1. *HA containing composites*
      4.1.2. *Other calcium phosphate containing composites*
   4.2. Bioactive glass containing composites
      4.2.1. *Silicate bioactive glass*
      4.2.2. *BG containing composites: new developments*
5. Discussion
6. Conclusions and Future Work
   References

1. Introduction

A promising and well-researched branch of tissue engineering aims to restore function to diseased or damaged tissue using combinations of functional cells and biodegradable scaffolds made from engineered biomaterials.[1,2] Some of the most promising biomaterials for application in bone tissue engineering are hydroxyapatite (HA), calcium phosphates, bioactive silicate glasses and related composite materials combining bioactive inorganic materials with biodegradable polymers.[3,4] Bioactive inorganic materials are capable of reacting with physiological fluids forming tenacious bonds to bone through the formation of bone-like hydroxyapatite layers leading to effective biological interaction of bone tissue with the material surface.[5] Moreover, in the case of silicate bioactive glasses, e.g. Bioglass®,[5] reactions on the material surface induce the release of critical concentrations of soluble Si, Ca, P and Na ions, which has been shown to lead to favourable intra- and extracellular responses promoting rapid bone formation.[6]

Synthetic biodegradable polymers are widely considered for the development of composite scaffolds in combination with bioactive inorganic particles.[3,4,7] In particular polyesters, such as polylactic acid (PLA), polyglycolic acid (PGA) and their co-polymers poly-(D/L-lactic-*co*-glycolic) acid (PLGA)[8,9] are being investigated. PGA, PLA, polydioxane and their co-polymers are synthetic and biodegradable polymers with extensive US Food and Drug Administration (FDA) approval history.

Also polyhydroxyalkanoates (PHAs), a family of microbial polyesters, find increasingly application for composite scaffolds development, in particular poly-(3-hydroxybutyrate) and poly-(3-hydroxybutyrate-*co*-hydroxyvalerate).[10]

Bone tissue engineering scaffolds are generally highly porous, three-dimensional (3D) structures exhibiting tailored porosity, pore size and interconnectivity. Several scaffold fabrication techniques, including foam replication methods, microsphere sintering, salt or sugar leaching, thermally-induced phase separation (TIPS), electrospinning to form nanofibrous structures, computer-assisted rapid prototyping techniques, textile and foam coating methods as well as biomimetic approaches to optimise the structure, properties and mechanical integrity of scaffolds have been reported in the literature and recent comprehensive reviews of the state-of-the-art in scaffold manufacturing and optimisation are available.[3,4,7,11,12] The incorporation of nanotopographic features when designing the scaffold surface architecture, in order to mimic the nanostructure of natural bone, is also becoming a significant area of research in bone tissue engineering.[13–15]

This chapter is organised in the following manner. In Section 2 we discuss scaffold's requirements for bone tissue engineering. In Section 3 the advantages of the composite materials approach to tissue engineering scaffolds are described, including also a summary of several composite scaffold fabrication technologies, and a discussion of scaffolds' microstructure (e.g. porosity and pore structure) and relevant properties. The focus is on synthetic composite scaffolds based on combinations of HA or bioactive glass particles, and biodegradable polymer matrices, investigating also the enhancement of scaffold performance through surface modification by inducing nanotopography. Section 4 reviews the latest developments on cell/tissue response of bioactive composite scaffolds, highlighting the complex effect of the presence of inorganic particulates within different polymer matrices on cell growth and differentiation. Finally, advantages and limitations of presently developed composite materials for bone tissue scaffolds are discussed (Section 5), and areas where further research is needed are identified (Section 6).

2. Scaffolds Requirements

The most important function of a scaffold is its role as the substrate that allows cells to attach, proliferate, differentiate and organise into normal, healthy bone as the scaffold degrades. Figure 1[3] illustrates the most important factors involved in the optimised design of tissue engineering scaffolds and their interdependencies. Scaffolds for bone tissue engineering are subject to many interrelated biological and structural requirements which must be taken into consideration when selecting the suitable biomaterial for their fabrication. Firstly, scaffolds need to foster cell attachment, differentiation and proliferation which are highly dependent on

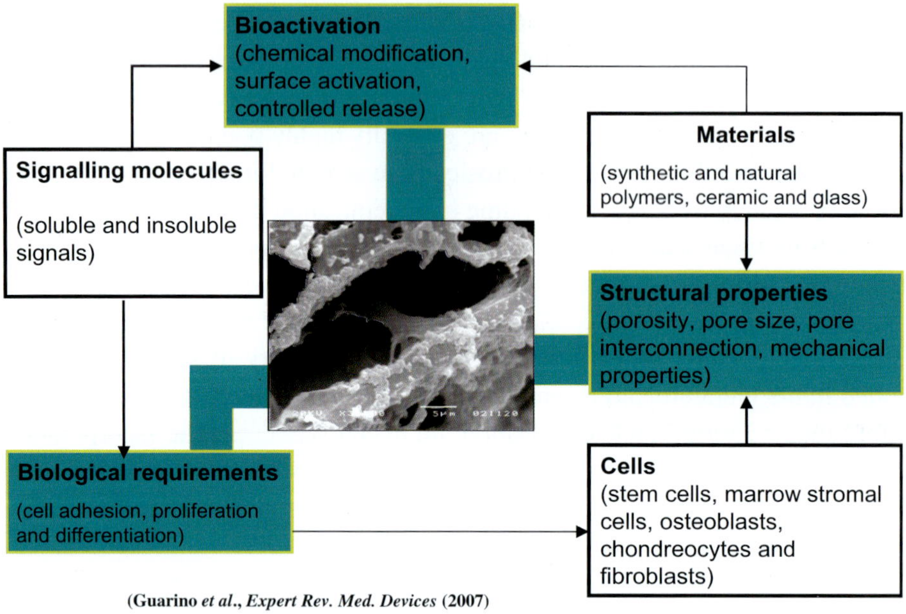

Fig. 1. Most important factors involved in the design of optimal scaffolds for bone tissue engineering (modified after Ref. 3).

substrate material properties. This is related to the property of osteoconductivity, which is important not only to avoid the formation of encapsulating tissue but also to induce a strong bond between the scaffold and host bone.[5] The rate of biodegradation *in vivo* is another criterion for selection of biomaterials for fabricating scaffolds, which should be tailored to match the rate of regeneration of new tissue. When considering biodegradable materials, it is also important to understand the time-dependent variation of their mechanical properties and structural integrity since the scaffolds' mechanical strength should be sufficient to provide mechanical stability in load bearing sites during the period of new tissue formation. Further requirements are related to the scaffold architecture: an ideal bone tissue scaffold should possess an interconnected porous structure with porosity >90% and pore diameters in the range 300–500 μm for cell seeding, tissue ingrowth and vascularisation, and nutrient delivery.[11,12] Moreover, scaffolds should be amenable to fabrication in complex or irregular shapes in order to match specific defects in bone of individual patients. Finally, material synthesis and fabrication of the scaffold should be suitable for commercialisation, i.e. the technology of scaffold production must be scalable and cost-effective.

Since the requirements for optimal scaffolds are manifold, the development of composites comprising biodegradable polymers and bioactive inorganic particles, e.g. HA or Bioglass®, becomes an attractive option to fulfil the requirements of bioactivity, degradability and mechanical competence.[11]

3. Composite Materials Approach for Tissue Engineering Scaffolds

3.1. Advantages of composites materials

A composite material consists of two or more chemically distinct phases (metallic, ceramic, or polymeric) which are separated by an interface.[16] Biodegradable composites for tissue engineering applications must exhibit mechanical competence characterised by suitable initial fracture strength and elastic modulus values, as well as controlled strength and modulus degradation *in vivo* in order for them to provide the necessary support for cell attachment and proliferation. Polymers by themselves are generally flexible and exhibit lack of mechanical strength and stiffness, whereas inorganic materials such as ceramics and glasses are known to be too stiff and brittle. Moreover, polymers can be easily fabricated to form complex shapes and structures yet, in general, they lack bioactive function (e.g. strong bonding to living tissue), being too flexible and weak to meet the mechanical demands in the physiological environment. The reasons behind the development of composites by smart combination of biodegradable polymers and bioactive ceramics or glasses for tissue engineering (Fig. 2) have been discussed extensively.[3,4,7,11] Briefly, they include: (i) the possibility of developing materials with improved mechanical properties due to the inherent higher stiffness and strength of the inorganic material, (ii) the ability to alter the polymer degradation behaviour by buffering the pH of the nearby solution through the dissolution products of the inorganic phase, hence providing a means of controlling the fast acidic degradation of some polymers (e.g. PLA, PDLLA), (iii) the possibility to influence the

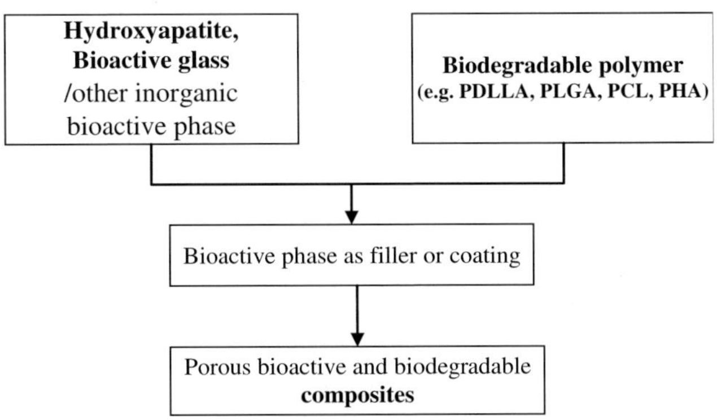

Fig. 2. Schematic diagram showing the types of synthetic bioactive and biodegradable polymer composite scaffolds for tissue engineering applications.

degradation mechanisms of polymers by preventing the autocatalytic effect of the acidic end groups resulting from hydrolysis of the polymer chains, and (iv) the possibility of altering the surface topography of the material by combination of nanosized and microsized inorganic particles as fillers or coatings.[13]

In addition, the incorporation of a bioactive inorganic phase such as HA, Bioglass® or tricalcium phosphate has the extra function of inducing the composite to interact effectively with the surrounding bone tissue by forming a tenacious bond *via* the growth of a carbonate hydroxyapatite layer, as mentioned above.[5]

The development of composite materials for tissue engineering scaffolds is an attractive approach since scaffold properties can be tailored to the particular mechanical and physiologic demands of the host tissue. This can be achieved by controlling the volume fraction, morphology and arrangement of the inorganic particulate phase in the polymer matrix. Also, tailored combinations of bioactive particles both as coatings and fillers can be designed to achieve the required scaffold performance.[11]

3.2. Mechanical properties

The most widely investigated composites for tissue engineering incorporate bioceramic or bioactive glass particles both in micrometre or nanometre size,[3,4,7,11,17] but also composites containing fibres, including carbon nanotubes, have been developed.[18,19] These inorganic inclusions positively affect the mechanical properties leading to reinforcement of the scaffold structure. The enhancement of mechanical properties depends strongly on the inclusion shape and size distribution; as well as on the quality of the inclusion distribution in the matrix and on the strength of the inclusion-matrix interface. Figure 3 shows the elastic modulus and the compressive strength of dense bioactive ceramics, biodegradable polymers, as well as cancellous and cortical bone in comparison with data for porous monophasic scaffolds and composites.[11] It can be seen that the bioactive ceramics region is close to the properties of cortical bone. Porous scaffolds however are at least one order of magnitude weaker than cancellous bone and orders of magnitude weaker than cortical bone. By comparing the mechanical properties of the porous composites to those of porous polymer scaffolds a slight increase of mechanical properties is revealed, as discussed in the literature.[11] However, the increase in stiffness and strength are well below expectations; most probably this result of most investigations can be attributed to the lack of a high interfacial bonding strength between the ceramic phase and the polymer matrix. The increase of bonding at the interface might be achieved by using surfactants chemisorbed on the particle surface prior to composite processing. It has been suggested[11] that using surface

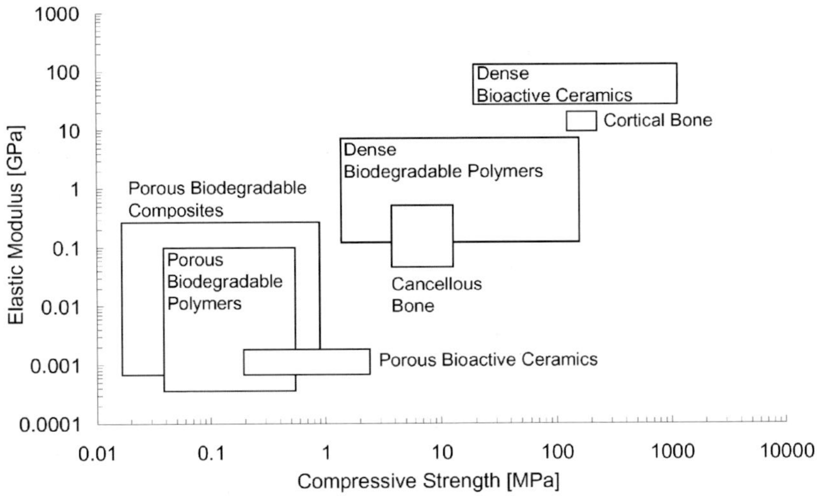

Fig. 3. Elastic modulus versus compressive strength values of biodegradable polymers, bioactive ceramics and composites after Ref. 11. Porosities of the porous scaffolds are >75% and mostly interconnected. Reprinted from Ref. 11, with permission from Elsevier.

functionalised nanoparticles, featuring a higher specific surface and thus a higher interface area, might increase the interfacial bonding strength, and thus the overall mechanical properties of the composite scaffold could be effectively enhanced. However, the increase of interfacial bonding and introduction of surfactants are likely to impact the degradation kinetics and cytotoxicity of the composites. These effects are largely unknown and remain to be investigated.

Nevertheless the inclusion of nanoparticles or carbon nanotubes into the biopolymer matrix with the dual objective of improving the mechanical properties as well as of incorporating nanotopographic features that mimic the nanostructure of natural bone is a currently area of intensive research.[13,18,20] The role of the scaffolds is thus being extended from being simply a mechanical support to include intelligent surfaces capable of providing both chemical and physical signals to guide cell attachment and spreading, possibly influencing also cell differentiation.[14,15,18]

3.3. Fabrication technologies

Among a number of polymer-foaming techniques, solvent casting with and without particle leaching,[21,22] thermally-induced phase separation (TIPS) combined with freeze-drying[8,23,24] and solid freeform fabrication[4,25] have been applied successfully to manufacture synthetic biopolymer-ceramic composite scaffolds. In more recent developments composite scaffolds have been also fabricated by combination of polymer nanofibres, obtained by electrospinning,[26] and inorganic phases developed by biomimetic processes.[27]

Solvent casting for production of composite scaffolds involves the dissolution of the polymer in an organic solvent, mixing with bioactive ceramic or glass granules and casting the solution into a predefined 3D mould.[22] The solvent is subsequently allowed to evaporate. The main advantage of this processing technique is the ease of fabrication without the need of specialised equipment. Potential disadvantages of the method include poor interconnectivity especially at low porosities and difficulty to generate large structures (over 3 mm thick).

Bioactive polymer-ceramic constructs can be fabricated by the combination of solvent casting, particle leaching and microsphere packing methods.[22] Polymer microspheres are firstly formed from traditional water oil/water emulsions. Polymer-bioceramic scaffolds can then be developed by mixing solvent, salt particles (porogens), bioactive glass or ceramic granules and pre-hardened microspheres.[28] A three-dimensional structure of controlled porosity is formed based on this method combined with particle leaching and microsphere packing.

3D resorbable polymer scaffolds with very high porosities (~97%) can be produced using the TIPS technique to give controlled macro- and microstructures.[7,8,23,24] The obtained scaffolds exhibit anisotropic tubular morphology of the pores and extensive pore interconnectivity. Microporosity of TIPS produced foams, their pore morphology, mechanical properties, bioactivity and degradation rates can be controlled by varying the polymer concentration in solution, volume fraction of the secondary phase, quenching temperature and the polymer and solvent used as discussed elsewhere.[29] The TIPS process has been used to produce composite scaffolds based on PLGA and PDLLA foams containing Bioglass® particles,[6,29,30] and transversal sections of PDLLA and PDLLA/ Bioglass® scaffolds developed by this method are shown in Fig. 4.[30] The possibility of coating TIPS produced PDLLA foams with Bioglass® particles has also been investigated.[31]

A number of solid freeform fabrication (SFF) techniques including 3D printing, selective laser sintering, multi-phase jet solidification, and fused deposition modelling (FDM) have been developed to manufacture tissue scaffolds for bone tissue engineering with specific designed properties.[4,32,33] The scaffolds have a high degree of interconnectivity and the porosity can be controlled to a great extent by optimising the processing parameters. The methods can furthermore be used to create scaffolds that both incorporate patient-specific information as well as an explicitly designed micro-environment. Tissue geometry can be extracted from patient's computed tomography (CT) or magnetic resonance imaging (MRI)

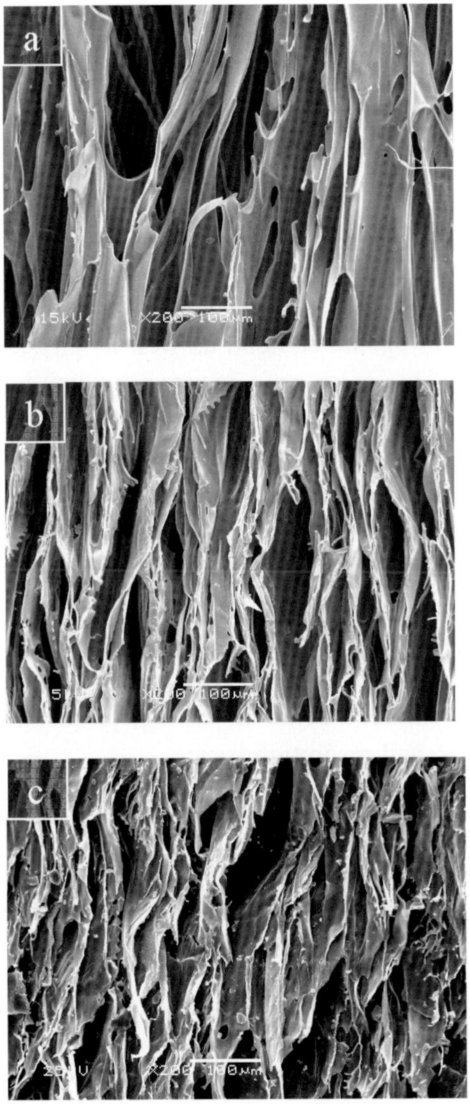

Fig. 4. SEM images of the transversal section showing the typical homogeneous regions of (a) pure PDLLA foam, (b) PDLLA / 2 vol% Bioglass® foam, and (c) PDLLA / 15 vol% Bioglass® foam. Reprinted from Ref. 30, with permission from Elsevier.

data and reconstructed as a 3D model. Additionally, as with most computer-assisted design, analysis of the mechanical and transport properties can aid in the understanding of tissue growth in a scaffold-guided environment. A summary of advantages and disadvantages of the different fabrication process for scaffolds has been given elsewhere.[7]

4. *In Vitro* and *In Vivo* Evaluation

Knowledge concerning interactions between cells and their immediate local environment has deeply improved in the last years[34–36] and efforts have been made in integrating this knowledge in the development of new biomaterials that can be tailored to specific requirements.[37,38] This is particularly true in the development of materials for tissue engineering scaffolds.

As described in previous sections, composite materials are extensively studied due to the considerable versatility and wide range of properties they offer.[11] Polymer composition, inorganic content, as well as physical structure and morphology (e.g. porosity, fibre mesh or foam structure) can be adjusted and tailored in a significant range. As a result, chemical and mechanical properties as well as behaviour regarding degradation *in vitro* and *in vivo* and bio-integration with host tissue vary and can be adapted to specific requirements.

A wide range of composite materials is available for the particular field of bone tissue engineering and the appropriate selection of a particular composite requires a detailed understanding of cells and/or tissue response. This includes cells viability, cell growth and metabolic activity, or inflammatory response, new bone formation or cell recruitment for *in vivo* situations. These *in vitro* and *in vivo* analyses are also important for a better knowledge of the material itself, especially for maturation, stability, and resorption issues.

An overview of the response observed when cells were cultured on several types of composites is given in Table 1, while Table 2 summarises the biological response as well as the biomaterials reaction when placed in an *in vivo* context. In the next sections key findings of previous studies on some of the most widely investigated composite systems are presented and discussed.

4.1. Calcium phosphate-based composites

4.1.1. *HA containing composites*

Hydroxyapatite is being used for hard tissue repair because of its chemical and crystallographic similarities to the natural carbonated apatite of human bone.[82–84]

Back in 1995, Attawia *et al.*[41] prepared porous PLGA/HA scaffolds. Looking at the initial phases of cell/composite interactions, they observed that after 24 hours of rat calvaria osteoblasts culture on these scaffolds, the cells were maintaining their phenotype and were proliferating on the surface as well as migrating inside the pores of the scaffold. Similar results were obtained by Marra *et al.*[42] using a PCL/PLGA (10/90) porous scaffold containing 10% (w/w) HA. Rabbit bone marrow-derived cells (pre-cultured in an osteogenic medium) were cultured over a period of eight weeks. Cells were found on the surface of the materials, but

Table 1. *In vitro* evaluation of artificial composite biomaterials.

Ceramic	Polymer	Cell type	Outcome					
			Viability/ adhesion	Growth	Differentiation/ Kept phenotype	Matrix synthesis	Mineralisation	Remarks
HA	PLGA	Rat calvaria OB[39–41]	+	+	+		+	
		Rabbit-BMSC, bone marrow[42]	+			+		Collagen inside pores
	PLA	Human osteosarcoma cell line[43]	+	+				Effect of HA contents
		Human primary OB (adult/foetal)[44]		+	+/+		+	Colonisation
	PHBV	Mouse-macrophages[45]						Reduced pre-inflammation
		Mouse — calvaria pre-OB[45]			+		+	
		Mouse — OC[45]			+			TRAP+, no resorption
	PHB	Rat — BMSC[46]		+	+			Effect of HA contents
	PE	Human primary OB[47]	+	+	+			Non porous material
	PCL	Human osteosarcoma cell line[48]	+	+	+		+	Effect of HA contents

(*Continued*)

Table 1. (*Continued*).

Ceramic	Polymer	Cell type	Outcome					
			Viability/ adhesion	Growth	Differentiation/ Kept phenotype	Matrix synthesis	Mineralisation	Remarks
Other CaP	PLGA	Rat — BMSC[49]		+			+	
	PHBV	Mouse — calvaria pre-OB[45]			+/−		+/−	
		Mouse — foetal OC[45]			+			TRAP+, no resorption
	PLA	Human primary OB (adult/foetal)[44]		+	+/+		+	
	PCL	Human embryonic MSC[49]	+	+				Pre-coating with ECM proteins
BG	PLGA	Rat-BMSC[50–52]		+	+			
		Human osteosarcoma cell line[53]		+		+	+	
		Mice fibroblasts cell line[54]	+	+				↑ VEGF, effect of BG contents
		Human osteosarcoma cell line[55]		+	+	+	+	

(*Continued*)

Table 1. (*Continued*).

Ceramic	Polymer	Cell type	Outcome					
			Viability/ adhesion	Growth	Differentiation/ Kept phenotype	Matrix synthesis	Mineralisation	Remarks
	PGA	Rat-fibroblasts cell line[56]						↑ VEGF
	PDLLA	Human primary OB[57]	+					Cell spreading, short time study
		Human osteosarcoma cell line[58]	+	+				Colonisation, effect of BG contents
		Human foetal OB[59]			+	+	+	
		Human primary OB[60]	+	+	+			
		Human primary OB[61]	+	+	+			Effect of BG contents
	PLLA	Human osteosarcoma cell line[58]		+	+			Effect of BG contents
	PLA	Mouse pre-OB cell line[62]	+	+/–	–			P glass – PLA based
APC	PGA	Mouse calvaria OB[63]	+	+	+	+		

HA: Hydroxiapatite, CaP: calcium phosphate, BG: bioactive glass, APC: APC carbonated apatite, P: phosphate, PLGA: poly-(D/L-lactic-*co*-glycolic) acid, PGA: polyglycolic acid, PLA: polylactic acid, PLLA: poly-(L-lactic) acid, PDLLA: poly-(D/L-lactic) acid, PHB: polyhydroxybutyrate, PHBV: poly-(3-hydroxybutyrate-*co*-3-hydroxyvalerate), PCL: poly-caprolactone, PE: polyethylene, PPF: poly-(propylene fumarate), BMSC: bone marrow stromal cells, OB: osteoblast, OC: osteoclast, ECM: extracellular matrix, VEGF: vascular endothelial growth factor, TRAP: tartrate resistant acid phosphatase.

Table 2. *In vivo* evaluations of artificial composite biomaterials.

Ceramic	Polymer	Model	Outcome					
			Inflammation/ encapsulation	Bone formation	Bone bridging	Vascularisation	Implant resorption	Remarks
HA	PLGA	Nude mice — subcutaneous[39]		+				Woven/lamellar
	PLA	Rat — femur[64]		+			+	
		Rat — calvaria[65]	+/–	+	+			Lamellar
		Rabbit — lumbar spine[66]		+			+	+ collagen, formation/ resorption
		Sheep — femur[67]	+/–	+				Screw
	PDLLA	Rabbit — femur diverse locations[68]		+			+/++	Rods 4–5 years, effect of location
		Rabbit — distal femur[69]		+			++	5–7 years study, calcined vs. un-HA
	PHBV	Rabbit — tibia[70]		+				Lamellar at interface
	PHB	Rat — subcutaneous +/– BMSC[71]	+/–	+		+		Bone and vessel at periphery
		Rabbit — femur[71]	–	+				Bone at periphery, low HA effect
	PPF	Rat — mandibular defect[72]	+/–	+				Bone at periphery
		Rabbit — spine[73]						autologous HA
		Rat — femur[74]	+/–	+				Porous cement — endochondral bone

(*Continued*)

Table 2. *(Continued)*

Ceramic	Polymer	Model	Outcome					
			Inflammation/ encapsulation	Bone formation	Bone bridging	Vascularisation	Implant resorption	Remarks
Other CaP	PLGA	Nude mice — subcutaneous + BMSC[49]		+				Bone formation in pores
		Rat — cranial[75]	–	+		+	–	Cement — effect porosity, % polymer
		Rat — cranial[76]	+	+/–			+/–	Cement — importance bone/implant contact
		Rat — subcutaneous[75]	–	+		+	+/–	Cement — bone-like structure
		Rat — femur[77]	+/–	+				Coating — Bone on surface
		Rat — subcutaneous[78]	–					Coating — effect on inflammation
	PDLLA	Rat — intramuscular[79]	+				+	TCP — screws
		Sheep — tibia[80]	+	+			+	α-TCP — Bone in pores — osteolysis
	PLA	Rat — calvaria[65]	+/–	+	+			β-TCP — Lamellar bone

(Continued)

Table 2. *(Continued)*

Ceramic	Polymer	Model	Outcome					
			Inflammation/ encapsulation	Bone formation	Bone bridging	Vascularisation	Implant resorption	Remarks
BG	PLGA	Mouse — subcutaneous[54]	+			+		Peripheral granulation tissue
	PGA	Rat — subcutaneous[56]				+		
	PDLLA	Rabbit — subcutaneous, femur[81]	–	+			+	Rods implants
		Nude mice — subcutaneous + BMSC[61]		+				Collagen I synthesis
		Sheep — tibia[80]	+/– ⇒ +	++			+	Bone inside pores — osteolysis

HA: Hydroxiapatite, CaP: calcium phosphate, BG: bioactive glass, APC: APC carbonated apatite, P: phosphate, TCP (α or β): tri-calcium phosphate, PLGA: poly-(D/L-lactic-*co*-glycolic) acid, PGA: polyglycolic acid, PLA: polylactic acid, PLLA: poly-(L-lactic) acid, PDLLA: poly-(D/L-lactic) acid, PHB: polyhydroxybutyrate, PHBV: poly-(3-hydroxybutyrate-*co*-3-hydroxyvalerate), PCL: poly-caprolactone, PE: polyethylene, PPF: poly-(propylene fumarate), BMSC: bone marrow stromal cells, OB: osteoblast, OC: osteoclast, ECM: extracellular matrix, VEGF: vascular endothelial growth factor, TRAP: tartrate resistant acid phosphatase.

they also widely invaded the pores in which they started to produce collagen. Kim and co-workers[39] focused their interest on PLGA/HA composites as well. In their recently published work[39,40,85] they compared the effect of the composite fabrication method on *in vitro* and *in vivo* behaviour of the scaffolds. PLGA/HA scaffolds were fabricated either using particulate leaching method or gas foaming followed by a particulate leaching phase. The resulting materials differed in their porosity; the latter showing two levels of porosity (interconnected macropores from 100 to 200 μm, and closed smaller pores of 10–45 μm). At day 56 of an *in vitro* study using rat calvaria osteoblasts, a significant superiority of the scaffolds containing HA made by gas foaming was observed. However, in the HA free materials, significantly lower cell growth, ALP activity and calcium deposition was obtained. When implanted into the subcutaneous space of Balb-c mice[39] or in rat calvaria,[85] the pre-cellularised scaffolds showed the same efficiency differences as *in vitro*, the HA free PLGA showing more fibrous encapsulation than the two other materials. In agreement with these results, in their study of cell adhesion and cell viability using MG-63 human cell line, Sui *et al.*[43] found enhanced cell adhesion, viability and growth on PLA/HA hybrid nanofibrous membrane compared to PLA alone. The analysis of pH according to the degradation time showed a faster degradation in the case of PLA alone, associated with a strong acidification of the solution. The presence of HA particulates slowed down the composite's degradation, and maintained a constant pH which preserved cell viability and growth. Here, a clear case of the positive effect of inorganic inclusions counteracting the acidic degradation of PLA, discussed in Section 3, was confirmed.

Moreover, if the amount of HA included in the polymer matrix largely determines the mechanical properties of the composite, most studies also show that the resulting biological behaviour strongly depends on HA content as well, with the actual effect of HA amount on cellular response being complex and highly dependent on the model used. Causa *et al.*[48] prepared poly-e-caprolactone (PCL) based composites with different percentages of HA particles (13%, 20% and 32% in weight). Osteoblast-like cells (SaOS-2 cell line) or human trabecular bone osteoblasts were cultured in these different scaffolds. After three weeks, the SaOS-2 cells number was found equivalent on either PCL/HA 13% or PCL/HA 32%. However, dosage of the alkaline phosphatase activity showed higher values when the cells were cultured on PCL/HA 13% when compared to the composite containing more HA. SEM analysis at three weeks showed a higher cell density on the 13% HA containing composites. Cells were well spread on the surface, presenting pseudopodia and globular structures that the authors interpreted as possibly mineralisation nodules. The same trends were obtained with trabecular osteoblasts, for which matrix mineralisation was confirmed by alizarin red uptake. Similar results were obtained by Shishatskaya *et al.*[46] who investigated the influence of HA content in polyhydroxybutyrate (PHB)-based composites on rat bone marrow stromal cells (MSC). PHB containing 0% to 50% (w/w) HA were seeded with rat MSC after their expansion in classical osteogenic medium (dexamethasone). After ten days of culture on the scaffolds, cells were quantified and a larger cell number on materials containing 10% and 20% HA was found, when compared to the composites containing 30%, 40% or 50% HA. Looking at the ALP activity, the same trend was obtained, confirming the enhanced bioactive function of the composites.

In their comparison of several composites using a macrophages assay, Cool *et al.*[45] found that HA/PHBV composite was the most effective in reducing pre-inflammatory response when compared to PHBV alone. In the same study, and looking at osteoclast cells fusion process and activity, they found a lower number of multinucleated cells on the composite materials when compared to bone control, while on the standard PHBV only mononucleated cells were found. This suggests a positive influence of the presence of HA on osteoclast cells formation. However, those cells did not show any resorption activity on the composite material. Even when removed from the PHBV/HA composite and seeded on bovine cortical bone ship (standard resorption assay), osteoclast formed on the composite failed to resorb the bone substrate, indicating that if PHBV/HA materials permitted the osteoclastic cell fusion process to occur, the resulting cells were not active.

4.1.2. *Other calcium phosphate containing composites*

In addition to hydroxyapatite, which is the most studied bioceramic in the field of bone tissue engineering, other calcium phosphates have also been combined with polymer materials to fabricate novel composites. Using a poly-(3-hydroxybutyrate-*co*-3-hydroxyvalate) (PHBV) polymer, Cool *et al.*[45] created various composites by incorporation of, in addition to HA, carbonated HA (caHA) or beta-tri-calcium phosphate (βTCP) particulates. The *in vitro* biological response was analysed for various cell types. Murine calvaria pre-osteoblasts cells (MC3T3) were cultured on the different materials over a period of 21 days. Seven and 21 days after seeding, no significant differences were observed concerning the cell morphology when grown on the different materials, and areas of mineralisation were also observed. However, calcium incorporation analyses showed a clear superiority of mineralisation occurring on cells grown on HA/PHBV when compared to caHA/PHBV and βTCP/PHBV composites.[45] Still, calcium levels were higher when the cells were culture on neat HA or βTCP alone. This result is in contradiction with the observations of Shishatskaya *et al.*[46] and Causa *et al.*[48] who showed that high HA content does not lead necessarily to a better cellular response. Cool and co-workers[45] confirmed their results by von Kossa staining. In the same experiment, they also observed a higher performance of PHBV alone compared to caHA/PHBV or βTCP/PHBV, indicating that while the introduction of HA in the polymer has a positive effect on murine osteoblasts, the presence of carbonated HA or βTCP does not. In parallel studies, Montjovent and colleagues[44,65] compared composites made of PLA reinforced with either 5% HA or 5% βTCP particulates (w/w) through both *in vitro* and *in vivo* experiments. Foetal and adult human osteoblasts were studied in parallel. In this case, they observed a better cellular colonisation of the composite scaffold (either βTCP or HA) compared to the polymer foam alone (Fig. 5). Looking at ALP activity and osteocalcin synthesis, no differences were observed if HA or βTCP was added to the PLA. However, in both cases the performance of the composite materials was better than the one of the polymer alone and favoured differentiation of each cell type. When implanted in the rat calvaria,[65] all tested materials (PLA/HA, PLA/TCP, and βTCP alone) induced similar host tissue reaction including inflammation and fibrous capsule formation (Fig. 6). In each case, encapsulation regressed

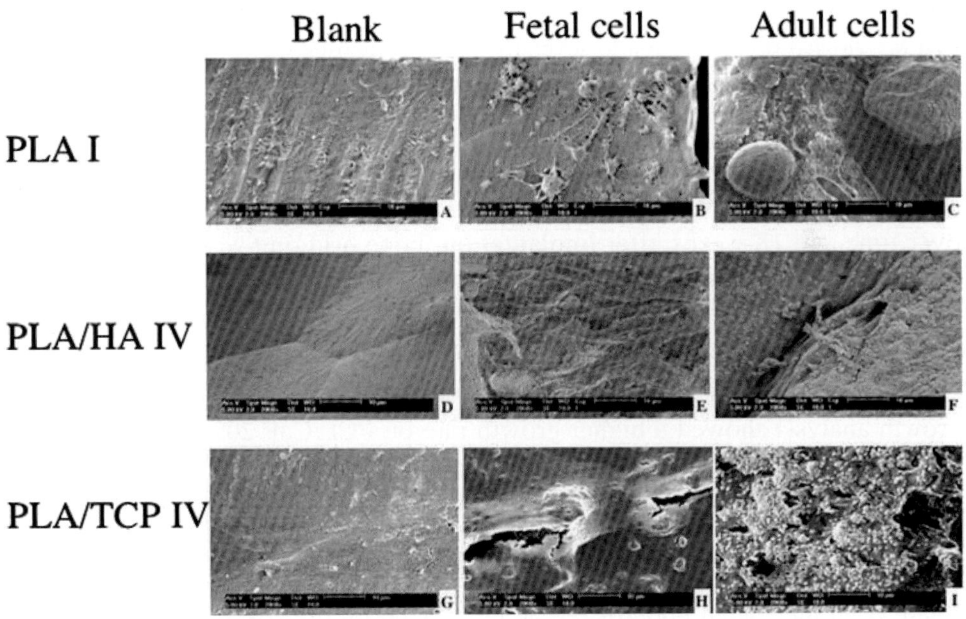

Fig. 5. Different response of human foetal and adult bone cells cultured on neat PLA, HA containing PLA or TCP containing composite scaffolds. Macropores were gradually invaded by cells. Scale bars: 10 μm. Micrographs from Montjovent *et al.*,[44] published with permission of Mary Ann Liebert, Inc. publishers.

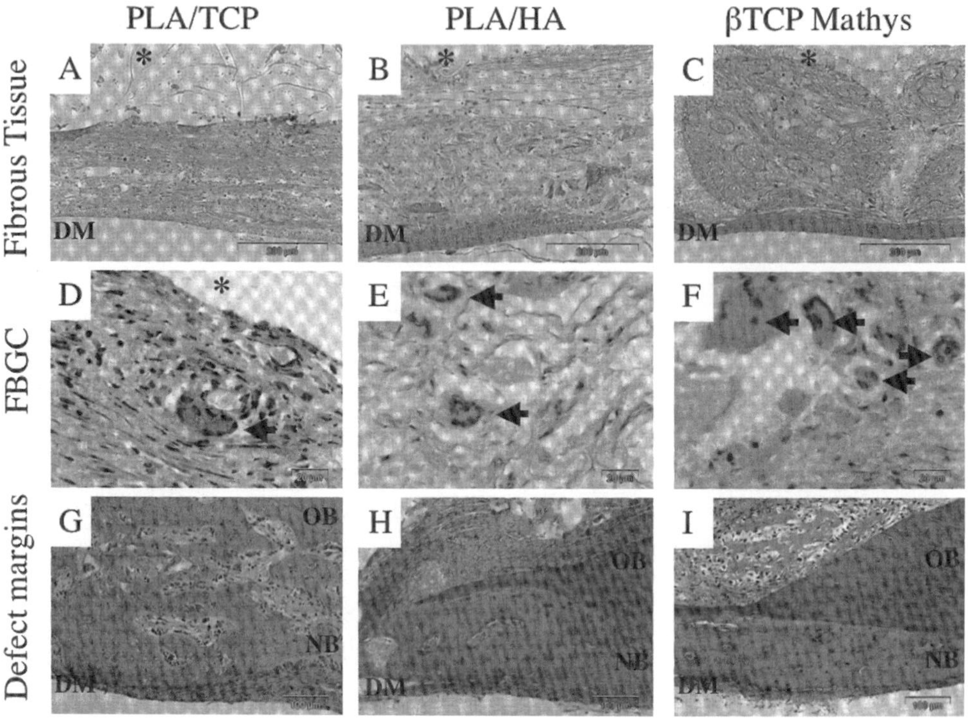

Fig. 6. *In vivo* tissue reaction to several types of composite materials. NB = New bone, OB = old bone, * = implant localisation, and arrows = foreign body giant cells. Micrographs from Montjovent *et al.*,[65] reprinted with permission from John Wiley & Sons, Inc.

over time and 18 weeks after implantation, bone bridging was found in groups receiving PLA composites.

Similar work has been carried out by Mondrinos *et al.*[50] who investigated the influence of ceramic content in composites by incorporating several amounts of calcium phosphate in a PCL scaffold. Human embryonic and mesenchymal cells growth analysis showed a higher growth rate on PCL/CaP 20% (w/w) compared to standard PCL or PCL/CaP 10%.

4.2. Bioactive glass containing composites

There is an increasing body of research focusing on the use of phosphate or silicate bioactive glasses as reinforcing filler for biodegradable polymers. Totally biodegradable composite materials involve phosphate-based glasses,[62] while partially degradable composites include silica-based bioactive glasses (BG), with 45S5 Bioglass® being the most investigated.[7,29–31,51–62] The latter has been extensively studied over the past few years, and, as shown in Tables 1 and 2, the majority of the composites comprising BG are prepared with either PLGA or PDLLA biodegradable polymer matrices.

4.2.1. *Silicate bioactive glass*

Several studies have recently investigated the biological response of different cell types to composite materials made of polylactide-co-glycolyde and bioactive glass. For example, the scaffolds used in the work of Lu *et al.*[53,55] were prepared from PLGA/BG microspheres, using a water-oil emulsion method. The resulting foams presented an average porosity of 43% with a pore size of about 85 μm. In their study, Lu *et al.*[55] compared the SaOS-2 cell growth and mineralisation on PLGA/BG, PLGA only and TCP scaffolds. Extensive cell growth was observed on the outer surface and within the pores of the PLGA/BG scaffold. ALP activity and collagen I synthesis were also found at higher level when the cells were grown on the composite materials when compared to the neat PLGA or TCP materials. In another publication of the same group using PLGA/BG composites prepared in the same way, Lu *et al.*[53] compared the effect of BG content on several physico-chemical parameters, and also the cellular response using SaOS-2. After seven days of culture on neat PLGA, PLGA/BG wt.10% or PLGA/BG wt.50%, they found higher cell proliferation on the wt.10% material compared to the 0 or 50 wt.% BG composites (Fig. 7A). Although similar ALP activity in the 0 and 10 wt.% BG materials was found, mineralisation was higher on the PLGA/BG 10 wt.% scaffold. Alizarin red staining after 28 days in culture showed (Fig. 7B) the positive effect of BG content on the matrix mineralisation. Using the same composite preparation method as in the previous cited study, Yao *et al.*[51] confirmed the ability of PLGA/BG composites to promote rat MSC osteogenic differentiation. The results showed the positive effect of the presence of bioactive glass in developing osteogenic properties of PLGA composite scaffolds. Moreover, a dose dependent effect was determined. Some other groups have also addressed the question of bioactive glass filler amount, not only on the mechanical properties of the materials, but also on the biological response.[58,61]

In a recent study by Tsigkou *et al.*,[59] the effect of bioactive glass (45S5 Bioglass®) content in a PDLLA-based composite on osteoblast cell differentiation and mineralisation was investigated. In this study, using human foetal osteoblasts, the cellular metabolic activity was found to be similar for the different materials, however in terms of cell differentiation an overall superiority of the composites containing either 5 wt.% or 40 wt.% BG was observed, compared to the neat PDLLA. Moreover no significant differences were observed over incubation time between cells on 5 wt.% BG or 40 wt.% BG materials concerning ALP activity,

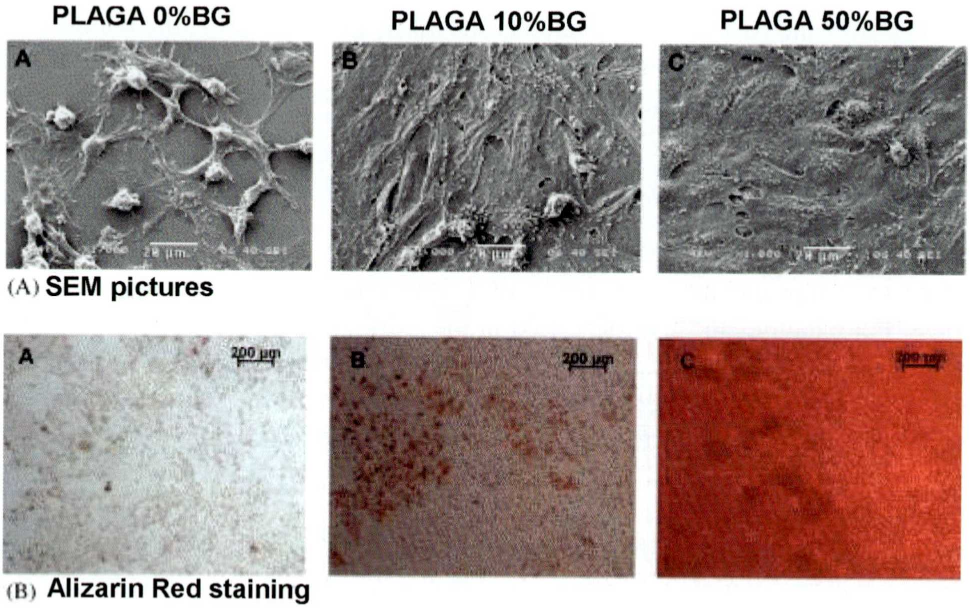

Fig. 7. Effect of BG on SaOS-2 cells. **(A)** shows lower amount of cells on PLAGA 0%BG when compared to the two BG containing composites. Alizarin Red staining after 28 days in culture showed in **(B)** indicates the positive effect of BG content on matrix mineralisation. Images adapted from Lu *et al.*,[53] with permission from Elsevier.

bone-specific gene expression or protein synthesis. However, osteocalcin cell secretion was found to be higher when the cells where cultured on PDLLA/BG 5 wt.% in comparison to the results obtained on PDLLA/BG 40 wt.%. The same observation was made when considering matrix mineralisation: the number and surface area of bone nodules formation was much higher on PDLLA/BG 5 wt.% than on PDLLA/BG 40 wt.% composites, indicating again a BG dose effect. Using comparable scaffolds, (PDLLA alone, PDLLA/BG 5 wt.% and PDLLA/BG 40 wt.%), Yang *et al.*[61] found very similar results. Namely significant increase of ALP activity was noticed for the material containing 5 wt.% Bioglass® compared to the neat scaffold or the 40 wt.% BG scaffold. After eight-week implantation under the skin of nude mice, bone formation was observed throughout the scaffolds. However, no significant differences were found concerning collagen synthesis between the different materials.

In a longer study (two years) performed in a larger animal model (sheep), Ignatius and co-workers[80] compared the tissue response toward composite materials made of PDLLA containing βTCP or bioactive glass GB14N. Six months post-implantation, both material types were well integrated by the surrounding bone, and woven bone and soft tissues were observed inside the scaffold pores. After 24 months, however, both materials showed strong inflammation reactions accompanied with local osteolysis. In this case, no differences were observed between the different materials either.

4.2.2. *BG containing composites: new developments*

As mentioned above, there has been significant research carried out to investigate the biological effect of composite materials for bone tissue engineering scaffolds. Most research has been conducted regarding osteoblast cells/material interactions. The performance of scaffolds is however not only related to their interaction with bone cells, as shown by Cool *et al.*,[45] but also other cell types are meant to interact with the scaffold surface.

In a study published in 2004, Verrier *et al.*[58] worked on PDLLA/Bioglass® composites and compared the effect of Bioglass® content on adhesion and proliferation properties of two different cell types. MG63 osteoblast cells were used as hard tissue cell type, while A549, a lung epithelial-like cell line, was used as soft tissue origin cells. In this set of experiments, MG63 cells were more adherent on PDLLA 40 wt.% BG, but no significant differences were found concerning cell growth between 5 wt.% and 40 wt.% BG. By contrast, A549 cell proliferation rate was found higher on PDLLA/BG 5 wt.% compared to other materials. This study showed that bioactive glass containing composites can significantly influence soft tissue cell behaviour. Beside osteoblast response, inflammatory reaction and osteoclastic resorption activity, another important parameter to take in consideration in the development of an ideal bone substitute is its ability to promote a neovascularisation effect. Day and co-workers[56] specifically investigated the effect of the incorporation of bioactive glass particles into a polyglycolic acid (PGA) scaffold on soft tissue cells. Adhesion and proliferation of rat embryonic fibroblast (208F) was first studied on bioactive glass coatings made using a wide range of 45S5 Bioglass® percentages (0.01% to 10% w/v). The authors observed a strong inhibition of cell proliferation for percentages higher than 0.2% of BG. Likewise, vascular endothelial growth factor (VEGF) secretion increased after 24, 48 and 72 hours for percentages below 0.2%. Moreover a stimulation of VEGF production was obtained for cells cultured on the 0.01 and 0.02% BG composites in comparison to the control (0% BG). For their *in vivo* investigations, the authors prepared a 45S5 Bioglass®/PGA composite mesh that they implanted under the skin of Lewis rats; PGA alone was used as control. Twenty-eight and 42 days after implantation, it was observed that BG containing materials presented a significant higher number of blood vessels than the neat PGA. These results were confirmed in a more recent study using mouse connective tissue fibroblasts (L929),[54] and PLGA /Bioglass® composites which were implanted subcutaneously in mice.

In more recent developments, Helen and Gough[86–88] investigated the response of bovine annulus fibrosus cells according to the BG content in PDLLA matrix composites. A higher glycosaminoglycan production was found on the PDLLA/BG 30 wt.% scaffold, indicating that PDLLA/BG composite films are an appropriate substrate for the culture of annulus fibrous cells *in vitro* and could therefore also be a suitable material for intervertebral disc tissue repair.

5. Discussion

This chapter has discussed the development and application of biodegradable and bioactive polymer/inorganic phase composite scaffolds for bone tissue engineering. General aspects related to material selection, design and processing of highly porous 3D composite structures, mechanical properties and cell/tissue response have been highlighted. The positive influence of the presence of inorganic bioactive particulates in different polymer matrices on cell growth and differentiation has been demonstrated in the reviewed studies. Researchers have shown that the presence of these particulates can stimulate cell proliferation and differentiation. However, differences in the biological responses have been observed according to cell types studied, as well as to the method used for composite preparation which leads to different pore morphologies. Moreover, fabrication processes influence several other parameters, notably specific surface area and surface topography, the latter having a direct effect on cell adhesion.[13–15] Variations in the level of exposure of the ceramic particulates to the biological environment, which influences the composite degradation and the ion exchange mechanism at the interface between the composite and the surrounding environment, are also of importance.

Mineral deposition has been shown to slow down the polymer-based scaffold degradation and concurrently increase its mechanical properties. The presence of a mineral phase within a polymer scaffold always induces an increase of the ALP activity and improves the overall osteo-inductive and osteo-conductive properties of a composite. A dose effect response has been, however, underlined in several investigations, using both HA or BG particulates. Several hypotheses to explain this effect have been suggested, including the effect of inorganic particles inclusions on the material wettability, or micro-/macrosurface topography, both influencing protein adsorption and subsequent cell adhesion mechanisms. Another significant parameter that has also been suggested to affect the biological behaviour of scaffolds is the pH of the culture media and cell environment, which tends to acidify due to the polymer degradation. The inclusion of inorganic particulates within biodegradable polymer scaffolds will not only slow down the scaffold resorption, but also the alkalinity of the calcium phosphate or BG inclusions should neutralise the acidic degradation products of many polymers. In the same way, the formation of a calcium phosphate layer on the surface of BG goes through a series of ion exchange steps and precipitation/dissolution reactions, inducing an elevation of the local pH. Xynos *et al.*[6,89] have investigated the influence of ion release during BG degradation *in vitro*, and found an important direct influence of Bioglass® dissolution products on human osteoblast cell proliferation. Further studies with Bioglass® containing composites have shown encouraging results regarding the potential angiogenic effects of Bioglass® dissolution products, i.e. increase secretion of VEGF *in vitro* and enhancement of vascularisation *in vivo*, suggesting scaffolds containing controlled amounts of Bioglass® might stimulate neovascularisation which is beneficial to large tissue engineered constructs.[56]

6. Conclusions and Future Work

One of the most significant challenges in bone tissue engineering remains the fabrication of scaffolds exhibiting suitable mechanical and biological properties to replace large (critical size) cortical bone defects and capable of load transmission. Although a number of composite scaffolds with favourable properties are available as discussed in this chapter, several issues need to be addressed prior to clinical application, such as mechanical reliability of scaffolds, induction of vascularisation and tailored degradability. One alternative to accelerate bone growth is the incorporation of biomolecules such as growth factors into the scaffold structure. Further improvement of scaffold function is related to surface modification, through the use of protein adsorption or plasma treatment, to provide more cues to cell attachment and response, thus making the scaffold more biocompatible.[3,4,11]

There is limited understanding regarding the long-term *in vitro* and *in vivo* characterisation of porous 3D composite scaffolds, specifically regarding the long-term effect of the incorporation of inorganic bioactive phases on the degradation and ion release kinetics of these highly porous systems. Moreover, more focus on *in vivo* studies is inevitable and there is need for more research on composite scaffolds in realistic biological systems. Engineered composite scaffolds design by suitable combination of biodegradable polymers and bioactive inorganic phase shall continue to be improved and optimised, and due to their high versatility they might represent the "scaffolds of choice" in future developments in combination with stem cells. The use of nanomaterials such as ceramic nanoparticles and carbon nanotubes (CNTs) may also improve the environment to enhance cell attachment and proliferation as well as adding extra functionalities to the base scaffold, however possible toxicity issues associated with nanoparticles and CNTs will have to be comprehensively addressed.[18]

References

1. Williams, D. Benefit and risk in tissue engineering. *Mater. Today* **7**(5), 24–29 (2004).
2. Hench, L.L. and Polak, J.M. Third generation biomedical materials. *Science* **295**(5557), 1014–1017 (2002).
3. Guarino, V., Causa, F. and Ambrosio, L. Bioactive scaffolds for bone and ligament tissue. *Expert Rev. Med. Dev.* **4**(3), 405–418 (2007).
4. Hutmacher, D.W., Schantz, J.T., Lam, C.X.F., Tan, K.C. and Lim, T.C. State of the art and future directions of scaffold-based bone engineering from a biomaterials perspective. *J. Tissue Eng. Regen. Med.* **1**, 245–260 (2007).
5. Hench, L.L. Bioceramics. *J. Am. Ceram. Soc.* **81**, 1705–1728 (1998).
6. Xynos, I.D., Edgar, A.J., Buttery, L.D.K., Hench, L.L. and Polak, M. Gene expression profiling of human osteoblasts following treatment with the ionic products of Bioglass® 45S5 dissolution. *J. Biomed. Mater. Res.* **55**, 151–157 (2001).
7. Boccaccini, A.R. and Blaker, J.J. Bioactive composite materials for tissue engineering scaffolds. *Expert Rev. Med. Dev.* **2**(3), 303–317 (2005).
8. Ma, P.X. and Zhang, R.Y. Microtubular architecture of biodegradable polymer scaffolds. *J. Biomed. Mater. Res.* **56**(4), 469–477 (2001).
9. Agrawal, C.M. and Ray, R.B. Biodegradable polymeric scaffolds for musculoskeletal tissue engineering. *J. Biomed. Mater. Res.* **55**(2), 141–150 (2001).
10. Misra, S.K., Valappil, S.P., Roy, I. and Boccaccini, A.R. Polyhydroxyalkanoate (PHA)/inorganic phase composites for tissue engineering applications. *Biomacromolecules* **7**(8), 2249–2258 (2006).
11. Rezwan, K., Chen, Q.Z., Blaker, J.J. and Boccaccini, A.R. Biodegradable and bioactive porous polymer/inorganic composite scaffolds for bone tissue engineering. *Biomaterials* **27**, 3413–3431 (2006).
12. Yang, S., Leong, K., Du, Z. and Chua, C. The design of scaffolds for use in tissue engineering. Part 1. Traditional factors. *Tissue Eng.* **7**(6), 679–689 (2001).
13. Webster, T.J. and Ahn, E.S. Nanostructured biomaterials for tissue engineering bone in tissue engineering II: basics of tissue engineering and tissue applications. *Adv. Biochem. Eng./Biotechnol.* **103**, 275–308 (2007).
14. Berry, C.C., Dalby, M.J., Oreffo, R.O.C., McCloy, D. and Affrosman, S. The interaction of human bone marrow cells with nanotopographical features in three-dimensional constructs. *J. Biomed. Mater. Res.* **79A**, 431–439 (2006).

15. Stevens, M.M. and George, J.H. Exploring and engineering the cell surface interface. *Science* **310**(5751), 1135–1138 (2005).
16. Matthews, F.L. and Rawlings, R.D. *Composite Materials: Engineering and Science* (Chapman and Hall, London, 1994).
17. Ramakrishna, S., Mayer, J., Wintermantel, E. and Leong, K.W. Biomedical applications of polymer-composite materials: a review. *Comp. Sci. Technol.* **61**(9), 1189–1224 (2001).
18. Harrison, B.S. and Atala, A. Carbon nanotube applications for tissue engineering. *Biomaterials* **28**, 344–353 (2007).
19. Shi, X., Hudson, J.L., Spicer, P.S., Tour, J.M., Krishnamoorti, R.M. and Mikos, A.G. Injectable nanocomposites of single-walled carbon nanotubes and biodegradable polymers for bone tissue engineering. *Biomacromolecules* **7**, 2237–2242 (2006).
20. Misra, S.K., Watts, P.C.P., Valappil, S.P., Silva, S.R.P., Roy, I. and Boccaccini, A.R. Poly(3-hydroxybutyrate)/Bioglass® composite films containing carbon nanotubes. *Nanotechnology* **18**, 075701.1–075701.7 (2007).
21. Guan, L.M. and Davies, J.E. Preparation and characterization of a highly macroporous biodegradable composite tissue engineering scaffold. *J. Biomed. Mater. Res. Part A* **71A**(3), 480–487 (2004).
22. Laurencin, C.T., Lu, H.H. and Khan, Y. Processing of polymer scaffolds: polymer-ceramic composite foams. In: Atala, A. and Lanza, R.P. (eds.), *Methods of Tissue Engineering* (Academic Press, California, 2002), pp. 705–714.
23. Maquet, V., Martin, D., Scholtes, F. *et al.* Poly (D,L-lactide) foams modified by poly(ethylene oxide)-block-poly(D,L-lactide) copolymers and a a-FGF: *in vitro* and *in vivo* evaluation for spinal cord regeneration. *Biomaterials* **22**(10), 1137–1146 (2001).
24. Zhang, R.Y. and Ma, P.X. Poly (alpha-hydroxyl acids) hydroxyapatite porous composites for bone-tissue engineering. I. Preparation and morphology. *J. Biomed. Mater. Res.* **44**(4), 446–455 (1999).
25. Taboas, J.M., Maddox, R.D., Krebsbach, P.H. and Hollister, S.J. Indirect solid free form fabrication of local and global porous, biomimetic and composite 3D polymer-ceramic scaffolds. *Biomaterials* **24**, 181–194 (2003).
26. Chen, H., Chu, B. and Hsiao, B. Mineralization of hydroxyapatite in electrospun nanofibrous poly(L-lactic acid) scaffolds. *J. Biomed. Mater. Res.* **79A**, 307–317 (2006).
27. Oliveira, A.L., Mano, J.F. and Reis, R.L. Nature-inspired calcium phosphate coatings: present status and novel advances in the science of mimicry. *Curr. Opin. Solid State Mater. Sci.* **7**, 309–318 (2003).
28. Ma, P.X. and Choi, J.W. Biodegradable polymer scaffolds with well-defined interconnected spherical pore network. *Tissue Eng.* **7**(1), 23–33 (2001).

29. Boccaccini, A.R. and Maquet, V. Bioresorbable and bioactive polymer/Bioglass® composites with tailored pore structure for tissue engineering applications. *Comp. Sci. Technol.* **63**, 2417–2429 (2003).
30. Blaker, J.J., Maquet, V., Jerome, R., Boccaccini, A.R. and Nazhat, S.N. Mechanical properties of highly porous PDLLA/Bioglass® composite foams as scaffolds for bone tissue engineering. *Acta Biomater.* **1**(6), 643–652 (2005).
31. Roether, J.A., Boccaccini, A.R., Hench, L.L., Maquet, V., Gautier, S. and Jerome, R. Development and *in vitro* characterisation of novel bioresorbable and bioactive composite materials based on polylactide foams and Bioglass® for tissue engineering applications. *Biomaterials* **23**(18), 3871–3878 (2002).
32. Hutmacher, D.W. and Cool, S. Concepts of scaffold-based tissue engineering — the rationale to use solid free-form fabrication techniques. *J. Cell. Mol. Med.* **11**, 654–669 (2007).
33. Xiong, Z., Yan, Y.N., Wang, S.G., Zhang, R.J. and Zhang, C. Fabrication of porous scaffolds for bone tissue engineering via low-temperature deposition. *Scr. Mater.* **46**(11), 771–776 (2002).
34. Saltzman, W.M. Cell interactions with polymers. In: Lanza, R.P., Langer R. and Vacanti, J. (eds.), *Principles of Tissue Engineering*, 2nd Ed. (Academic Press, 2000).
35. Meyer, U., Buchter, A., Wiesmann, H.P., Joos, U. and Jones, D.B. Basic reactions of osteoblasts on structured material surfaces. *Eur. Cell Mater.* **9,** 39–49 (2005).
36. Anselme, K. Osteoblast adhesion on biomaterials. *Biomaterials* **21**(7), 667–681 (2000).
37. Discher, D.E., Janmey, P. and Wang, Y.L. Tissue cells feel and respond to the stiffness of their substrate. *Science* **310**(5751), 1139–1143 (2005).
38. Kirkpatrick, C.J., Krump-Konvalinkova, V., Unger, R.E., Bittinger, F., Otto, M. and Peters, K. Tissue response and biomaterial integration: the efficacy of *in vitro* methods. *Biomol. Eng.* **19**(2–6), 211–217 (2002).
39. Kim, S.S., Sun, P.M., Jeon, O., Yong, C.C. and Kim, B.S. Poly(lactide-co-glycolide)/hydroxyapatite composite scaffolds for bone tissue engineering. *Biomaterials* **27**(8), 1399–1409 (2006).
40. Kim, S.S., Park, M.S., Gwak, S.J., Choi, C.Y. and Kim, B.S. Accelerated bonelike apatite growth on porous polymer/ceramic composite scaffolds *in vitro*. *Tissue Eng.* **12**(10), 2997–3006 (2006).
41. Attawia, M.A., Herbert, K.M. and Laurencin, C.T. Osteoblast-like cell adherence and migration through 3-dimensional porous polymer matrices. *Biochem. Biophys. Res. Commun.* **213**(2), 639–644 (1995).

42. Marra, K.G., Szem, J.W., Kumta, P.N., DiMilla, P.A. and Weiss, L.E. *In vitro* analysis of biodegradable polymer blend/hydroxyapatite composites for bone tissue engineering. *J. Biomed. Mater. Res.* **47**(3), 324–335 (1999).
43. Sui, G., Yang, X., Mei, F., Hu, X., Chen, G., Deng, X. and Ryu, S. Poly-L-lactic acid/hydroxyapatite hybrid membrane for bone tissue regeneration. *J. Biomed. Mater. Res. A* **82**(2), 445–454 (2007).
44. Montjovent, M.O., Mathieu, L., Hinz, B., Applegate, L.L., Bourban, P.E., Zambelli, P.Y., Manson, J.A. and Pioletti, D.P. Biocompatibility of bioresorbable poly(L-lactic acid) composite scaffolds obtained by supercritical gas foaming with human fetal bone cells. *Tissue Eng.* **11**(11–12), 1640–1649 (2005).
45. Cool, S.M., Kenny, B., Wu, A., Nurcombe, V., Trau, M., Cassady, A.I. and Grondahl, L. Poly (3-hydroxybutyrate-co-3-hydroxyvalerate) composite biomaterials for bone tissue regeneration: *in vitro* performance assessed by osteoblast proliferation, osteoclast adhesion and resorption, and macrophage proinflammatory response. *J. Biomed. Mater. Res. A* **82**(3), 599–610 (2007).
46. Shishatskaya, E.I., Khlusov, I.A. and Volova, T.G. A hybrid PHB-hydroxyapatite composite for biomedical application: production, *in vitro* and *in vivo* investigation. *J. Biomater. Sci. Polym. Ed.* **17**(5), 481–498 (2006).
47. Zhang, Y., Tanner, K.E., Gurav, N. and Di Silvio, L. *In vitro* osteoblastic response to 30 vol% hydroxyapatite-polyethylene composite. *J. Biomed. Mater. Res. A* **81**(2), 409–417 (2007).
48. Causa, F., Netti, P.A., Ambrosio, L., Ciapetti, G., Baldini, N., Pagani, S., Martini, D. and Giunti, A. Poly-epsilon-caprolactone/hydroxyapatite composites for bone regeneration: *in vitro* characterization and human osteoblast response. *J. Biomed. Mater. Res. A* **76**(1), 151–162 (2006).
49. Gomi, K., Kanazashi, M., Lickorish, D., Arai, T. and Davies, J.E. Bone marrow genesis after subcutaneous delivery of rat osteogenic cell-seeded biodegradable scaffolds into nude mice. *J. Biomed. Mater. Res. A* **71**(4), 602–607 (2004).
50. Mondrinos, M.J., Dembzynski, R., Lu, L., Byrapogu, V.K., Wootton, D.M., Lelkes, P.I. and Zhou, J. Porogen-based solid freeform fabrication of polycaprolactone-calcium phosphate scaffolds for tissue engineering. *Biomaterials* **27**(25), 4399–4408 (2006).
51. Yao, J., Radin, S., Leboy, S. and Ducheyne, P. The effect of bioactive glass content on synthesis and bioactivity of composite poly(lactic-co-glycolic acid)/bioactive glass substrate for tissue engineering. *Biomaterials* **26**(14), 1935–1943 (2005).
52. Yao, J., Radin, S., Reilly, G., Leboy, P.S. and Ducheyne, P. Solution-mediated effect of bioactive glass in poly (lactic-co-glycolic acid)-bioactive glass

composites on osteogenesis of marrow stromal cells. *J. Biomed. Mater. Res. A* **75**(4), 794–801 (2005).
53. Lu, H.H., Tang, A., Oh, S.C., Spalazzi, J.P. and Dionisio, K. Compositional effects on the formation of a calcium phosphate layer and the response of osteoblast-like cells on polymer-bioactive glass composites. *Biomaterials* **26**(32), 6323–6334 (2005).
54. Day, R.M., Maquet, V., Boccaccini, A.R., Jerome, R. and Forbes, A. *In vitro* and *in vivo* analysis of macroporous biodegradable poly(D,L-lactide-co-glycolide) scaffolds containing bioactive glass. *J. Biomed. Mater. Res. A* **75**(4), 778–787 (2005).
55. Lu, H.H., El Amin, S.F., Scott, K.D. and Laurencin, C.T. Three-dimensional, bioactive, biodegradable, polymer-bioactive glass composite scaffolds with improved mechanical properties support collagen synthesis and mineralization of human osteoblast-like cells *in vitro*. *J. Biomed. Mater. Res. A* **64**(3), 465–474 (2003).
56. Day, R.M., Boccaccini, A.R., Shurey, S., Roether, J.A., Forbes, A., Hench, L.L. and Gabe, S.M. Assessment of polyglycolic acid mesh and bioactive glass for soft-tissue engineering scaffolds. *Biomaterials* **25**(27), 5857–5866 (2004).
57. Roether, J.A., Gough, J.E., Boccaccini, A.R., Hench, L.L., Maquet, V. and Jerome, R. Novel bioresorbable and bioactive composites based on bioactive glass and polylactide foams for bone tissue engineering. *J. Mater. Sci. Mater. Med.* **13**(12), 1207–1214 (2002).
58. Verrier, S., Blaker, J.J., Maquet, V., Hench, L.L. and Boccaccini, A.R. PDLLA/Bioglass composites for soft-tissue and hard-tissue engineering: an *in vitro* cell biology assessment. *Biomaterials* **25**(15), 3013–3021 (2004).
59. Tsigkou, O., Hench, L.L., Boccaccini, A.R., Polak, J.M. and Stevens, M.M. Enhanced differentiation and mineralization of human fetal osteoblasts on PDLLA containing Bioglass composite films in the absence of osteogenic supplements. *J. Biomed. Mater. Res. A* **80**(4), 837–851 (2007).
60. Ruuttila, P., Niiranen, H., Kellomaki, M., Tormala, P., Konttinen, Y.T. and Hukkanen, M. Characterization of human primary osteoblast response on bioactive glass (BaG 13–93) coated poly-L,DL-lactide (SR-PLA70) surface *in vitro*. *J. Biomed. Mater. Res. B Appl. Biomater.* **78**(1), 97–104 (2006).
61. Yang, X.B., Webb, D., Blaker, J., Boccaccini, A.R., Maquet, V., Cooper, C. and Oreffo, R.O. Evaluation of human bone marrow stromal cell growth on biodegradable polymer/bioglass composites. *Biochem. Biophys. Res. Commun.* **342**(4), 1098–1107 (2006).

62. Brauer, D.S., Russel, C., Vogt, S., Weisser, J. and Schnabelrauch, M. Degradable phosphate glass fiber reinforced polymer matrices: mechanical properties and cell response. *J. Mater. Sci. Mater. Med.* **19**(1), 121–127 (2008).
63. Linhart, W., Peters, F., Lehmann, W., Schwarz, K., Schilling, A.F., Amling, M., Rueger, J.M. and Epple, M. Biologically and chemically optimized composites of carbonated apatite and polyglycolide as bone substitution materials. *J. Biomed. Mater. Res.* **54**(2), 162–171 (2001).
64. Higashi, S., Yamamuro, T., Nakamura, T., Ikada, Y., Hyon, S.H. and Jamshidi, K. Polymer-hydroxyapatite composites for biodegradable bone fillers. *Biomaterials* **7**(3), 183–187 (1986).
65. Montjovent, M.O., Mathieu, L., Schmoekel, H., Mark, S., Bourban, P.E., Zambelli, P.Y., Laurent-Applegate, L.A. and Pioletti, D.P. Repair of critical size defects in the rat cranium using ceramic-reinforced PLA scaffolds obtained by supercritical gas foaming. *J. Biomed. Mater. Res. A* **83**(1), 41–51 (2007).
66. Liao, S.S. and Cui, F.Z. *In vitro* and *in vivo* degradation of mineralized collagen-based composite scaffold: nanohydroxyapatite/collagen/poly(L-lactide). *Tissue Eng.* **10**(1–2), 73–80 (2004).
67. Lewandrowski, K.U., Bondre, S.P., Shea, M., Untch, C.M., Hayes, W.C., Hile, D.D., Wise, D.L. and Trantolo, D.J. Composite resorbable polymer/hydroxylapatite composite screws for fixation of osteochondral osteotomies. *Biomed. Mater. Eng.* **12**(4), 423–438 (2002).
68. Shikinami, Y., Matsusue, Y. and Nakamura, T. The complete process of bioresorption and bone replacement using devices made of forged composites of raw hydroxyapatite particles/poly l-lactide (F-u-HA/PLLA). *Biomaterials* **26**(27), 5542–5551 (2005).
69. Hasegawa, S., Ishii, S., Tamura, J., Furukawa, T., Neo, M., Matsusue, Y., Shikinami, Y., Okuno, M. and Nakamura, T. A five- to seven-year *in vivo* study of high-strength hydroxyapatite/poly(L-lactide) composite rods for the internal fixation of bone fractures. *Biomaterials* **27**(8), 1327–1332 (2006).
70. Luklinska, Z.B. and Schluckwerder, H. *In vivo* response to HA-polyhydroxybutyrate/polyhydroxyvalerate composite. *J. Microsc.* **211**(Pt. 2), 121–129 (2003).
71. Doyle, C., Tanner, E.T. and Bonfield, W. *In vitro* and *in vivo* evaluation of polyhydroxybutyrate and of polyhydroxybutyrate reinforced with hydroxyapatite. *Biomaterials* **12**(9), 841–847 (1991).
72. Trantolo, D.J., Sonis, S.T., Thompson, B.M., Wise, D.L., Lewandrowski, K.U. and Hile, D.D. Evaluation of a porous, biodegradable biopolymer scaffold for mandibular reconstruction. *Int. J. Oral Maxillofac. Implants* **18**(2), 182–188 (2003).

73. Hile, D.D., Kandziora, F., Lewandrowski, K.U., Doherty, S.A., Kowaleski, M.P. and Trantolo, D.J. A poly(propylene glycol-co-fumaric acid) based bone graft extender for lumbar spinal fusion: *in vivo* assessment in a rabbit model. *Eur. Spine J.* **15**(6), 936–943 (2006).
74. Hile, D.D., Kowaleski, M.P., Doherty, S.A., Lewandrowski, K.U. and Trantolo, D.J. An injectable porous poly(propylene glycol-co-fumaric acid) bone repair material as an adjunct for intramedullary fixation. *Biomed. Mater. Eng.* **15**(3), 219–227 (2005).
75. Ruhe, P.Q., Hedberg-Dirk, E.L., Padron, N.T., Spauwen, P.H., Jansen, J.A. and Mikos, A.G. Porous poly(DL-lactic-co-glycolic acid)/calcium phosphate cement composite for reconstruction of bone defects. *Tissue Eng.* **12**(4), 789–800 (2006).
76. Link, D.P., van den, D.J., Jurgens, W.J., Wolke, J.G. and Jansen, J.A. Mechanical evaluation of implanted calcium phosphate cement incorporated with PLGA microparticles. *Biomaterials* **27**(28), 4941–4947 (2006).
77. Lickorish, D., Guan, L. and Davies, J.E. A three-phase, fully resorbable, polyester/calcium phosphate scaffold for bone tissue engineering: Evolution of scaffold design. *Biomaterials* **28**(8), 1495–1502 (2007).
78. Lickorish, D., Chan, J., Song, J. and Davies, J.E. An *in-vivo* model to interrogate the transition from acute to chronic inflammation. *Eur. Cell Mater.* **8,** 12–19 (2004).
79. Heidemann, W., Jeschkeit, S., Ruffieux, K., Fischer, J.H., Wagner, M., Kruger, G., Wintermantel, E. and Gerlach, K.L. Degradation of poly(D,L) lactide implants with or without addition of calciumphosphates *in vivo*. *Biomaterials* **22**(17), 2371–2381 (2001).
80. Ignatius, A.A., Betz, O., Augat, P. and Claes, L.E. *In vivo* investigations on composites made of resorbable ceramics and poly(lactide) used as bone graft substitutes. *J. Biomed. Mater. Res.* **58**(6), 701–709 (2001).
81. Pyhalto, T., Lapinsuo, M., Patiala, H., Niiranen, H., Tormala, P. and Rokkanen, P. Fixation of distal femoral osteotomies with self-reinforced poly(L/DL)lactide 70:30 and self-reinforced poly(L/DL)lactide 70:30/bioactive glass composite rods. an experimental study on rabbits. *J. Biomater. Sci. Polym. Ed.* **16**(6), 725–744 (2005).
82. Rizzi, S.C., Heath, D.J., Coombes, A.G., Bock, N., Textor, M. and Downes, S. Biodegradable polymer/hydroxyapatite composites: surface analysis and initial attachment of human osteoblasts. *J. Biomed. Mater. Res.* **55**(4), 475–486 (2001).
83. Auclair-Daigle, C., Bureau, M.N., Legoux, J.G. and Yahia, L. Bioactive hydroxyapatite coatings on polymer composites for orthopedic implants. *J. Biomed. Mater. Res. A* **73**(4), 398–408 (2005).

84. LeGeros, R.Z., Lin, S., Rohanizadeh, R., Mijares, D. and LeGeros, J.P. Biphasic calcium phosphate bioceramics: preparation, properties and applications. *J. Mater. Sci. Mater. Med.* **14**(3), 201–209 (2003).
85. Kim, S.S., Ahn, K.M., Park, M.S., Lee, J.H., Choi, C.Y. and Kim, B.S. A poly(lactide-co-glycolide)/hydroxyapatite composite scaffold with enhanced osteoconductivity. *J. Biomed. Mater. Res. A* **80**(1), 206–215 (2007).
86. Helen, W., Merry, C.L., Blaker, J.J. and Gough, J.E. Three-dimensional culture of annulus fibrosus cells within PDLLA/Bioglass composite foam scaffolds: assessment of cell attachment, proliferation and extracellular matrix production. *Biomaterials* **28**(11), 2010–2020 (2007).
87. Helen, W. and Gough, J.E. Cell viability, proliferation and extracellular matrix production of human annulus fibrosus cells cultured within PDLLA/Bioglass composite foam scaffolds *in vitro*. *Acta Biomater.* **4**(2), 230–243 (2008).
88. Helen, W. and Gough, J.E. *In vitro* studies of annulus fibrosus disc cell attachment, differentiation and matrix production on PDLLA/45S5 Bioglass composite films. *Biomaterials* **27**(30), 5220–5229 (2006).
89. Xynos, I.D., Edgar, A.J., Buttery, L.D.K., Hench, L.L. and Polak, J.M. Ionic products of bioactive glass dissolution increase proliferation of human osteoblasts and induce insulin-like growth factor II mRNA expression and protein synthesis. *Biochem. Biophys. Res. Commun.* **276**, 461–465 (2000).

Chapter 15

Aggregation of Cells Using Biomaterials and Bioreactors

Zahia Bayoussef and Kevin M. Shakesheff

Abstract

Aggregation of cells into clusters is a tissue engineering approach used for rapid and controlled formation of structures that have certain architectural and functional properties of native tissue. Some mammalian cells aggregate spontaneously in suspension, for others, methods have been devised to force them to aggregate. This chapter focuses on approaches based on biomaterials and bioreactors to induce or accelerate cell aggregation. It also describes a recently introduced cell surface engineering method used to induce aggregation of homotypic as well as heterotypic cell types.

Keywords: Cell; Aggregation; Suface Modification; Biomaterials.

Outline

1. Introduction
2. Cell Adhesion and Natural Cell Aggregation
   2.1. Natural cell aggregation
3. Methods of Cell Aggregation
   3.1. Aggregation on low-adherence surfaces
   3.2. Aggregation in rotation culture
   3.3. Microgravity culture (hanging drop method)
4. Synthetic Cell Aggregation
   4.1. Functionalized polymers
      4.1.1. *Chitosan*
      4.1.2. *Modified PEG*
      4.1.3. *Lactone modified eudragit*
   4.2. PLGA nanospheres

4.3. Lectins and derivatives
4.4. Chemical cell surface modification
4.4.1. *Biotinylated cell cross-linking*
4.4.2. *Intercellular polymeric cross-linker*
5. Cell Aggregation on Scaffolds
6. Bioreactors and Cell Aggregation
6.1. Types of bioreactors used in aggregation studies
6.1.1. *Rotating wall vessel*
6.1.2. *Spinner flasks*
7. Summary and Conclusion
References

1. Introduction

Tissue engineering aims to replicate natural tissue arrangement *in vitro*. Various methods have been developed to grow and maintain cells within three-dimensional structures including the use of scaffolds, entrapment of cells in microcapsules and the formation of cell aggregates, the subject of this chapter (see also Chapters 24 and 25 of this book).

Aggregation of cells into clusters is an approach that allows rapid formation of small units of tissue. It allows intimate contacts between cells and binding of cell adhesion molecules, potentially leading to enhancement of cell functionality and maintenance of viability. This advantage of cell aggregation and clustering has raised interest in methods to form aggregates of different cell types and has driven research to accelerate, enhance and control the process of cell aggregation for use in tissue engineering applications. The fact that some cell types fail to aggregate naturally in suspension has also directed interest in external interventions such as use of chemicals and biomaterials to encourage cell adhesion and aggregation. In addition, various types of anchorage dependent cells are physically induced to grow in the form of aggregates in suspension culture for large scale manipulations.

2. Cell Adhesion and Natural Cell Aggregation

Cell–cell interaction or adhesion is one of the important factors that control and determine the state and functionality of a mammalian cell. Cells in higher order organisms are arranged in a 3D network and adhered to neighbouring cells or/and extracellular matrix. Adhesion of cells is usually mediated by one or more transmembrane proteins named cell adhesion molecules (CAMs) including four major families. These include the integrins involved mainly in adhesion to extracellular matrix, selectins which mediate transient adhesion, members of the immunoglobulin (Ig) superfamily and cadherins which mediate stable homophilic binding and contribute to the formation of tight junctions.[1]

In addition to the physical role of CAMs in the organisation and maintenance of tissue integration, these transmembrane proteins have cytoplasmic domains that bind specific proteins that convey a cascade of processes upon activation (binding). This impacts on the overall cell functionality, and in some cases differentiation of the cells. For instance, cadherins (a family of calcium-dependent CAMs) bind to few cytoplasmic proteins such as catenins which bind actin microfilament system within the cell, turning adhered cells into one mechanical unit.

2.1. Natural cell aggregation

Some anchorage-dependent cells assemble into aggregates naturally in suspension culture. This process is the result of interaction of cell adhesion molecules on different cells. In addition, culture medium usually contains serum proteins, some of which are cell adhesive including fibronectin and vibronectin, resulting in cell aggregation. Epithelial cells are the main cells prone to aggregation in suspension such as skin, corneal cells and hepatocytes.[2]

Cell aggregation is recognised as a crucial process during embryonic development[3] and cell aggregate formation (embryoid bodies) is an important step in embryonic stem cell differentiation *in vitro*. In addition, various progenitor cells go through phases of aggregation during their differentiation *in vitro* such as keratinocytes and neuroectodermal progenitors.[4,5]

3. Methods of Cell Aggregation

3.1. Aggregation on low-adherence surfaces

Some surfaces discourage cell adhesion and spreading via surface chemistries or topographies that inhibit protein and cell interactions. On such materials, cells remain in suspension and adhesion to each other becomes kinetically favoured over surface adhesion assuming there is sufficient cell movement to cause regular physical interaction of cells with their neighbours. Surfaces that can prevent cell adhesion include non-treated tissue culture plastic, Petri dishes and $P_{DL}LA$[6] but the precise interaction of cells with these surfaces is dependent on cell type and medium composition. Other surfaces encourage cell aggregation due to their mechanical properties. Surface properties such as rigidity affect the distribution of forces on the membrane of attached cells. This can increase cell contraction and locomotion, causing adhesion to other cells and aggregation.[7] Certainly, some extracellular matrix coated surfaces were found to encourage cell locomotion and cell to cell coupling. These corresponded to surfaces that encourage aggregate formation as confirmed by Powers and co-workers.[2,8] The dominant surface material studied in relation to cell aggregation is Matrigel®, a secretory matrix from Engelbroth-Holm-Swarm sarcoma. It has been particularly used for hepatocyte aggregation.[9] In these examples, the surface provides a 2D environment for cell movement and in turn accelerates the rate of cell-cell interaction.

3.2. Aggregation in rotation culture

Cells may be suspended at a very high density (hundred thousands or millions per millilitre) and placed in rotation conditions to increase probability of cell collision and aggregation. In some cases, fluid stresses and hydrodynamics encourage the expression of cytoskeletal proteins important for aggregation.[7] The set up usually involves agitation of suspended cells in Petri dishes or well plates on a shaker, use of roller bottles and bioreactors. Aggregation in rotation culture has been favoured for its high capacity, ease of scale up and production of aggregates with a more uniform size distribution when compared to other methods.[10] In some studies, high volume bioreactors (up to ten litres) were used to produce aggregates.[11]

3.3. Microgravity culture (hanging drop method)

The hanging drop method for aggregate production was designed for the production of embryoid bodies for stem cell differentiation studies and to model developmental processes. It was later applied for the generation of tumour spheroids and aggregation of various cell lines.[12,13] The method involves placing around 20 μl of cell suspension on a tissue culture dish lid which is then inverted. In the first day of incubation, cells accumulate at the bottom of the drop, followed by gradual compaction of cells to result in a spheroid.[14]

4. Synthetic Cell Aggregation

4.1. Functionalised polymers

The concept is to create a polymer bridge that connects two cells. This polymer can be substituted with a cell adhesive molecule which upon attachment to the cell surface results in cell cross-linking and thus agglomeration (see Fig. 1). The substituting molecule is usually a segment of an extracellular matrix protein that binds to a cell surface receptor. The polymer material used has to be biocompatible, biodegradable, water soluble and preferentially inert to minimise any adverse effects on the cells to be clustered. In the design of the cross-linking unit, length

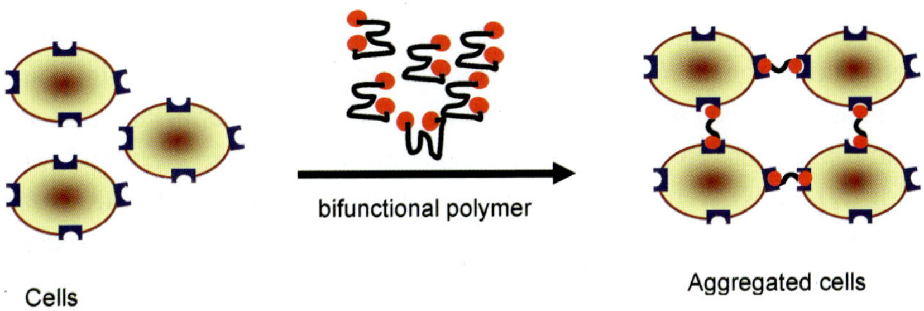

Fig. 1. Schematic for the aggregation of cells using functionalised polymers. The polymer is usually substituted with a ligand that binds a cell surface receptor. Upon suspension with cells, the ligands on polymer couple to the receptor of interest on different cells resulting in aggregation. The polymer acts as a spacer and a bridge to present the ligand and connect the cells.

of polymer (MW) is an important determinant of the efficiency of aggregation as it has to take into account that the receptor to which the binding molecule attaches might not be at the outermost of the cell glycocalix.

4.1.1. *Chitosan*

Chitosan is a natural biodegradable polymer. It has the ability to trap cells in certain solution conditions and has been used previously to aggregate Chinese hamster ovary cells (CHO). Used at a concentration of 0.003% of 810,000 Mw, it speeded the aggregation of recombinant CHO cells in stirred tank bioreactors (spinner flasks).[11] Chitosan coated surfaces were used for growth and aggregation of human melanocytes. This aided the maintenance of cell morphology and function. Donati *et al.* also reported the aggregation of primary pig articular chondrocytes grown on a lactose modified chitosan,[15] which enhanced the secretion of glycosaminoglycans (GAGs). Recently, chitosan films were reported to aggregate and enhance liver function of the hepatocyte cell line HepG2.[16] HepG2 cells aggregated after 24 hours of seeding on chitosan films (prepared using 2% w/v chitosan) while those seeded on control plate remained spread. Urea and albumin secretion were at least twofold higher from aggregates on chitosan films compared to cells on control plates throughout a nine-day study period.

4.1.2. *Modified PEG*

Polyethylene glycol (PEG) modified at both ends with an RGD moiety has been used to aggregate various cell types including rat pheochromocytoma cells (PC12), mouse neuroblastoma cells (neuro-2a), foetal brain cells and 3T3 fibroblasts. Arg-gly-asp

(RGD) is a short peptide sequence found in various extracellular matrix proteins, such as collagen and fibronectin, responsible for binding with integrin receptors on the cell surface. It has been used to encourage cell adhesion to biomaterial surfaces and tissue engineering scaffolds. RGD modified PEG with a molecular weight of 3400 was superior to 20,000 Mw for cell aggregation as reported by Dai and co-workers,[17] which emphasises the importance of molecular weight of the functionalised polymer.

When neuronal cells were placed in rotation culture with various concentrations in serum free nutrient media, aggregates were observed after few hours. These increased in size gradually over time to result in rounded or spherical aggregates.[18] Studies on aggregation of 3T3 cells also showed an increase in aggregates produced with increasing concentration of the polymer conjugate. The aggregates produced were significantly larger than those seen with the control or PEG alone.[17] In addition, aggregation of transfected 3T3 cells increased the production of nerve growth factor (NGF) by two folds, while aggregation of PC12 cells with this polymer peptide conjugate enhanced dopamine and norepinephrine release compared to single cells.[19]

In addition to modification with RGD, PEG has been conjugated with various peptide sequences to test for cell aggregation effect, including YIGSR from laminin.[17,18]

4.1.3. *Lactone modified eudragit*

Some synthetic polymers have the ability to agglomerate specific cells without substitution such as aggregation of primary hepatocytes with eudragit (a co-polymer of methacrylicacid and methylmethacrylate). Hepatocytes in static culture formed smooth spheroids when 0.05%–0.18% eudragit was used and 80% of cells were aggregated.[20] Similarly in rotation culture (spinner flasks), 0.1% eudragit caused rapid aggregation of hepatocytes with a high density of spheroids observed at 48 hours compared to control. The use of eudragit, in addition to enhancement of cell aggregation and spheroid formation, resulted in an increase and maintenance of some liver cell functions for longer period of time such as albumin secretion, ammonia removal and urea synthesis.[21] Yamada and co-workers modified eudragit with a lactone including β-galactose residue, which serves as an asialoglycoprotein receptor ligand.[22] This enhanced cell aggregation and reduced spheroid formation time. This case serves as an example of exploring receptors present at the cell surface to cause agglomeration.

4.2. PLGA nanospheres

The purpose for the use of PLGA nanospheres does not differ from that of bifunctional polymers as the nanospheres act as glue between two cell membranes resulting in aggregation. The nanospheres can be prepared using the oil/water emulsion and solvent evaporation method. They could also be coated with serum proteins to encourage cell adhesion.

PLGA nanospheres of about 0.9 μm have been used by Ryu and colleagues to aggregate human embryonic kidney cells in stirred culture (spinner flask). The nanospheres attached between cells to result in agglomeration aided by the collision promoted with stirring. In the presence of nanospheres, the cells formed aggregates within hours resulting in spherical aggregates of 100 μm diameter at day two compared to single cells and very small clusters in the control samples.[24] In addition, the total cell number increased more than fourfold over seven days, which was significantly higher than control. Attraction of cells to the nanospheres was encouraged with the adsorption of serum proteins from culture medium to the PLGA surface which was confirmed later with the use of fibronectin-coated nanospheres in serum free media.[25] PLGA nanospheres were also used to aggregate rCHO cells where the aggregation rate was faster than control. In this case serum free media was used for culture and the nanospheres were incubated with fetal bovine serum (FBS) overnight prior to use.[23] Cell adhesive proteins such as fibronectin and vibronectin were present on the nanosphere surface and encouraged adhesion and aggregation. This was confirmed with histology (Fig. 2) as the nanospheres were evenly distributed within aggregates and were trapped between cell membranes. In addition to enhancement of aggregation and proliferation, the use of nanospheres in these studies impacted on cell function as confirmed by the significant increase in protein production by the used CHO cells.

PLGA nanospheres were also used to aggregate hepatocytes in spinner flasks.[26] The use of nanospheres accelerated spheroid formation and enhanced efficiency of aggregation. Spheroids of 50–100 μm were formed within 24–36 hours compared to controls (culture without nanospheres). Additional benefits include enhanced viability and liver specific gene expression. In addition, PLGA nanospheres have been used to aggregate human dermal fibroblasts.[27]

The use of PLGA as a biocompatible and biodegradable material (nanospheres degrade within two months) is advantageous in that it could be used with various cell types, and implantation of formed aggregates in the body would have minimal adverse effects.

4.3. Lectins and derivatives

Some proteins possess multiple ligand binding sites rendering them good candidates for agglomeration of the ligand to which they bind. These proteins include antibodies and lectins. Lectins are a large class of polysaccharide-binding proteins, most of which are homo-oligomers assembled from two or four repeat units.

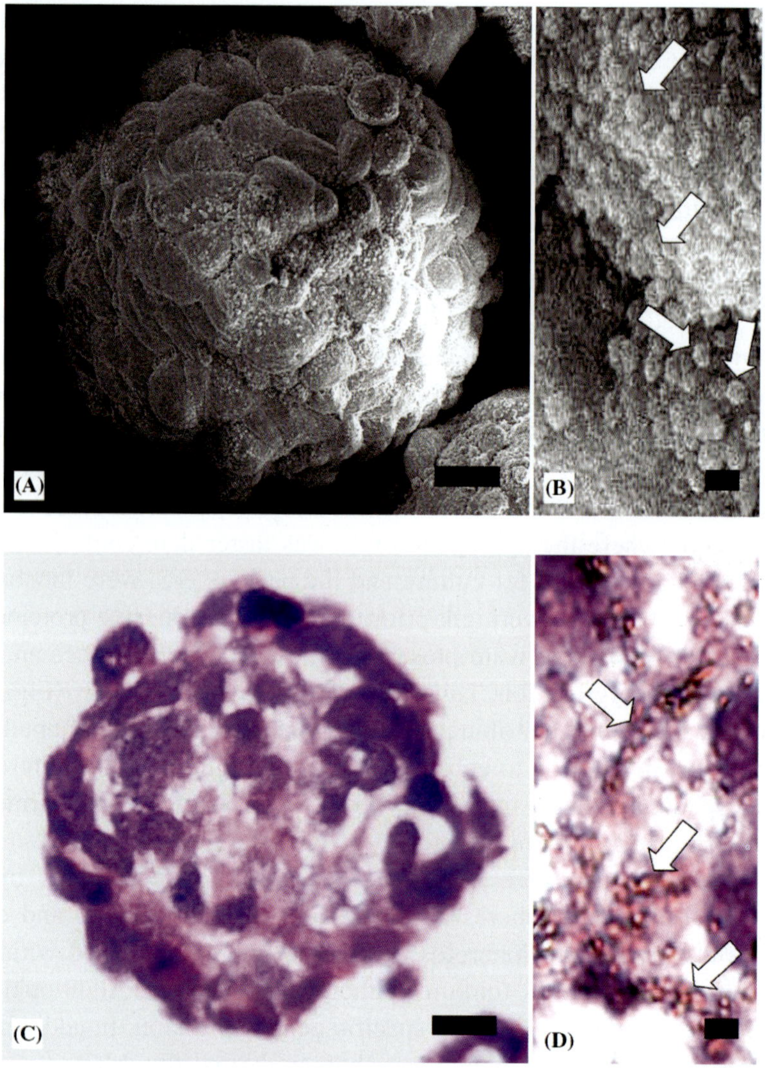

Fig. 2. Aggregates of rCHO cells formed with PLGA nanospheres. **(A)** SEM photographs of cell aggregates on day 12 at a low (scale bar = 10 μm) and **(B)** high magnification (scale bar = 1 μm). **(C)** Histological sections (H&E staining) of cell aggregates on day 5 at a low (scale bar = 10 mm) and **(D)** high magnification (scale bar = 1 μm). The arrows indicate nanospheres.[23] Copyright © 2004 Elsevier Ltd. Reprinted with permission from Elsevier.

They have the ability to cross-link glycoproteins and glycolipids, which enables them to cross-link cells. For instance, concanavalin A has been used to aggregate red blood cells.[28] Gestwicki *et al.* described the aggregation of Jurkat human T-cell leukaemia cell line with concanavalin A and concanavalin A adhered to a mannose displaying polymers.[29] An enhancement of 60% in aggregation was observed with concanavalin A displayed on a polymer compared to concanavalin A alone, which might result from the provision of a high density of binding sites in a close proximity.

4.4. Chemical cell surface modification

4.4.1. *Biotinylated cell cross-linking*

This is a method that enables the aggregation of homotypic as well as heterotypic cell types and was described first by De Bank and co-workers.[30] It relies on the decoration of the cell surface with biotin molecules and the use of the biotin-binding protein avidin as a cross-linker. The cells are first treated with sodium periodate, an oxidising agent that has been used since the 1970s to create aldehydes on the cell surface sialic acids for the purpose of quantification and estimation of glycoproteins, most of which have sialic acid as terminal residue.[31] The aldehydes are sites for coupling biotin hydrazide under mild acidic conditions, which results in the decorated cells. The cells are then suspended in culture media with avidin which results in the agglomeration of cells with gentle agitation. This coagulation is the result of an avidin molecule, which has four biotin binding sites, binding two biotin molecules on different cells. A schematic diagram for the aggregation process with biotinylated cell cross-linking is shown in Fig. 3.

Although this approach starts with an oxidation step, it has no detrimental effects on the cell viability and proliferation. The concentration of sodium periodate and manipulation conditions were set to eliminate such a risk. The periodate

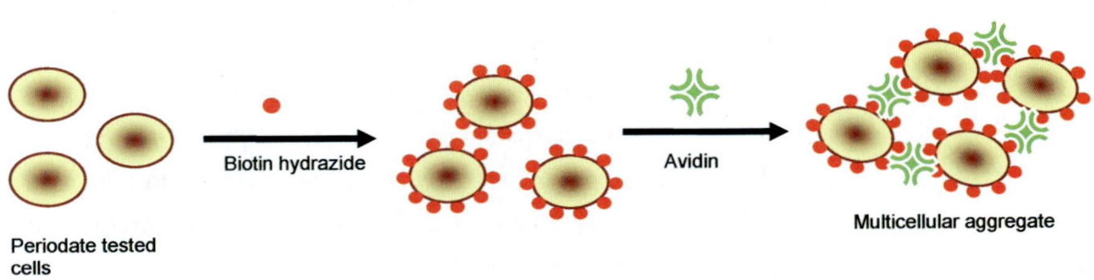

Fig. 3. Schematic diagram of the process of aggregate formation with biotinylated cell cross-linking. Cells of interest are first treated with sodium periodate to create reactive aldehyde handles on the cell surface sialic acids. This is followed by coupling of biotin hydrazide to the new functionality under mild acidic conditions. Biotinylated cells are then suspended in nutrient media containing a sub-saturating concentration of avidin resulting in agglomeration of cells.

concentration used did not exceed 2 mM and the reaction was performed at 4°C for no longer than 15 minutes. The cold temperature for the reaction minimises any cell active transport eliminating access of periodate to the cytoplasm, thus achieving selectivity for the cell surface sialic acids.

This new synthetic cell aggregation approach has been used with various cell lines and primary cells including: the mouse fibroblast cell line 3T3, mouse embryonic stem cells, L6 rat myoblast cell line, C2C12 mouse myoblast cell line, primary human keratinocytes as well as primary rat hepatocytes. In addition to accelerated aggregate formation compared to natural aggregation in mass suspension and bioreactors, as shown in Fig. 4, this new tool allows the control over structure and architecture of the resulting aggregates with regards to the use of heterotypic cell types.[32] The ability to aggregate cells quickly appeared to increase the expression of some developmental regulatory proteins accompanied by an enhancement of embryonic stem cell differentiation.

Different approaches for the decoration of the cell surface with biotin are described in the literature. One method is the use of N-hydroxysuccinimidyl-substituted biotin derivatives (NHS-X-biotin) which react with nitrogen species on the cell surface.[33] An alternative is the insertion of a biotin-cholesteryl derivative to the cell membrane. Cholesterol-tethered molecules are one of the methods

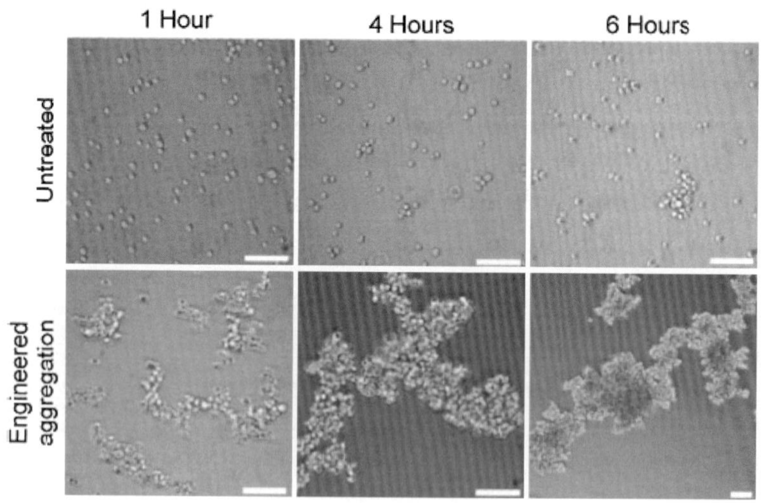

Fig. 4. Representative phase contrast images of aggregated L6 myoblasts with biotinylation and cross-linking with avidin (10 μg/ml). Images also demonstrate the differences with control, where engineered cells agglomerate within an hour of suspension with avidin and grow gradually in size in a spinner flask at 85 rpm. Scale bars = 100 μm.[32] Copyright © 2007 Wiley Periodicals, Inc. Reprinted with permission of Wiley-Liss, Inc., a subsidiary of John Wiley & Sons, Inc.

used for altering cell surface characteristics as the cholesterol moiety is anchored rapidly to membrane lipid bilayer leaving the molecule of interest exposed.[34] Meier[35] used a series of polyethylene oxides (PEO) modified with cholesterol at one end and a biotin molecule at the other (cholesterol-PEO-biotin), where PEO acts as a spacer to reveal the biotin moiety from cell glycocalix.

4.4.2. *Intercellular polymeric cross-linker*

This method also relies on engineering the cell surface by creation of reactive cell surface handles. These handles are sites for reaction with a multi-arm cross-linking polymer (has reactive groups that couple to the cell surface handles) which results in cell agglomeration.[36] Ong and co-workers created reactive aldehyde groups on HepG2 cells with periodate treatment, which were sites of coupling with a hydrazide-substituted non-toxic low molecular weight (1800) polyethyleneimine (PEI). Thiols were first conjugated onto the amines in PEI by reaction with 2–iminothiolanes. These thiols were sites of addition of hydrazide groups using E-maleimidocaproic acid hydrazide. This modified PEI was non-toxic to cells at concentrations lower than 0.1 mM.

HepG2 aggregates were induced by reaction of periodate-treated cells with 0.01 mM hydrazide-PEI for 30 minutes in cold PBS buffer (4°C) on an orbital shaker. This was followed by washing with PBS to remove excess polymer and transfer to growth media for culture. In addition to the hydrazides coupling to aldehydes on the cell surface, the positive charge on PEI backbone was an important factor for cell aggregation as cells failed to aggregate in the presence of a neutral hydrazide-PEI.[36] The interaction between the positively charged PEI and the negative cell surface might contribute to the effect. The charge difference causes the concentration of modified polymer at the cell surface, which facilitates the interaction of the hydrazide groups with the cell surface aldehyde handles.

Aggregated cells maintained high viability and proliferation over a seven-day growth period. The linking polymer disappeared from the cell surface over time as the cells secreted their own extracellular matrix as confirmed by collagen fibre deposition. Other cell types have also been aggregated using this method including primary porcine hepatocytes and rat bone marrow stem cells.

5. Cell Aggregation on Scaffolds

Cell aggregates can be formed on highly porous scaffolds. The confined environment of the pores within scaffolds forces cells to aggregate when seeded at a high cell density. In addition, rotation culture of the scaffold can enhance aggregation and cell proliferation. The aggregation of cells on scaffolds can be used as a method for the control of aggregate size and to ensure cluster diameter uniformity. The scaffold pore size acts as a limit for the size of the clusters to be formed.

Glickilis *et al.*[37] reported the formation of primary rat hepatocyte spheroids after seeding on alginate scaffolds. Alginate scaffolds with high porosity (>90%) have been formed by cross-linking a concentrated solution of a high glucuronic acid content alginate with calcium gluconate. These sponge-like scaffolds had a high pore interconnectivity with a pore diameter of 100–150 μm. Although these scaffolds were hydrophilic, the non-adherent nature of alginate resulted in aggregation of cells when seeded on the scaffold. Over 90% of the seeded cells participated in aggregates after 24 hours of seeding and spheroid diameter reached 100 μm at day 4. The aggregation of hepatocytes on the alginate scaffold promoted and enhanced albumin secretion.[37]

Alginate sponges have been used for the formation of embryoid bodies (EBs) from human embryonic stem cells as shown in Fig. 5.[38] In a study by Gerecht-Nir and colleagues, hESC were dynamically seeded onto a 50–200 μm pore scaffold at a high

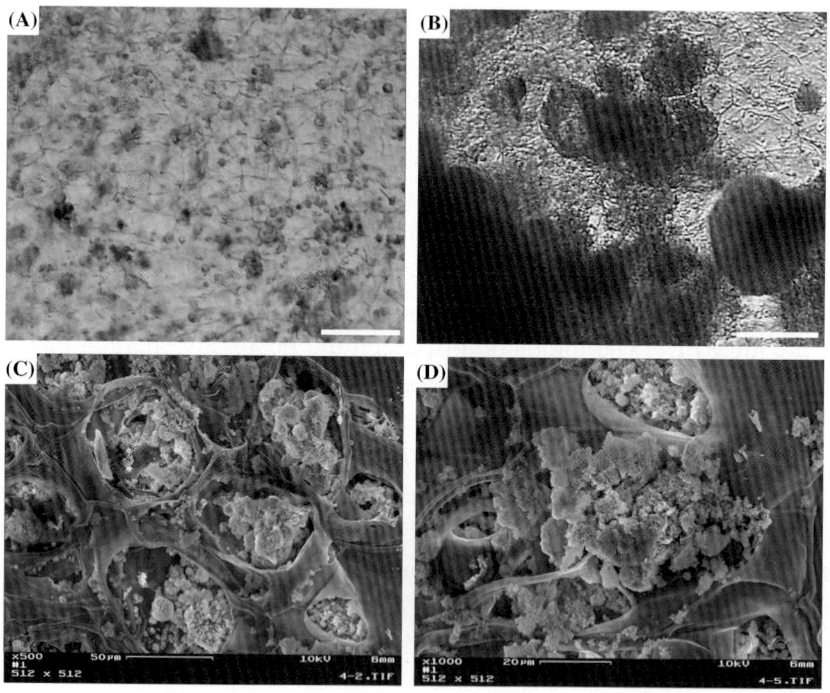

Fig. 5. Human EBs formation within alginate scaffolds. **(A)** and **(B)** are light microscope images of a cell seeded scaffold at days 4 and 20, respectively, showing transition from single cells to spheroids (bar = ×100). **(C)** and **(D)** are scanning electron micrographs of alginate seeded scaffolds after one month of culture showing EBs formed within pores and on the surface of scaffold.[38] Copyright © 2004 Wiley Periodicals, Inc. Reprinted with permission of Wiley-Liss, Inc., a subsidiary of John Wiley & Sons, Inc.

density and transferred for culture in slow turning lateral vessel (STLV) bioreactor. Cell aggregation was enhanced with the use of this bioreactor and was controlled by manipulation of fluid dynamics. The formed EBs were highly proliferative and had a uniform diameter which enabled their differentiation with evidence of vasculogenisis. An additional advantage of the alginate scaffolds described is that they can be dissolved in PBS to release the formed aggregates. Phosphate ions chelate calcium ions (cross-link alginate), which disassembles the scaffold.

Highly porous alginate scaffolds have also been formed by cross-linking alginate with chitosan.[39] By use of galactose modified chitosan, hepatocytes were attracted to the scaffold due to interaction of galactose with the asialoglycoprotein receptor on hepatocytes. This enhanced cell seeding on the scaffold and resulted in spheroid formation.

6. Bioreactors and Cell Aggregation

Some bioreactors provide a good environment for cell aggregation (see also Chapter 25 of this book). The fluid movement within these bioreactors permits an even distribution of cells, therefore, allowing a good nutrient and oxygen exchange as well as increasing the possibility of cell collision and adhesion. Most modelling studies that involved cell aggregation indicated that rate of collision is an important factor for cell aggregation.[40–42] The rate of collision is affected by factors such as hydrodynamic conditions, cell extension and migration speeds. In suspension culture, hydrodynamic condition (media movement) plays an important role. In addition, some bioreactors provide an even suspension of cells and media components, which allows a good distribution of biomaterials used in synthetic cell aggregation. Table 1 summarises some of the studies involving cell aggregation within bioreactors. It also references examples of use of different bioreactors for aggregation of cells with biomaterials.

6.1. Types of bioreactors used in aggregation studies

6.1.1. *Rotating wall vessel*

This bioreactor was developed by NASA to simulate microgravity conditions for cell growth purposes. It maintains cells/aggregates in a suspended state with minimal sheer forces. There are two major types, high aspect rotating vessel (HARV) and slow turning lateral vessel (STLV). The rotating wall vessel bioreactor has been used for growth of tissue and to facilitate cell aggregation. Various cells have been aggregated using this bioreactor including chondrocytes, hepatocytes, embryonic stem cells and various cell lines.[43]

Table 1. Examples of use of bioreactors for cell aggregation.

Bioreactor type	Cell type used	Outcome	Reference(s)
RWV	PC12 HepG2	Aggregation was monitored using a real time system.	43
RWV (STLV)	Human embryonic stem cells	Enhanced efficiency of embryoid body formation and differentiation. Homogenous EBs formed.	44
RWV (STLV)	Human embryonic stem cells	Homogenous EBs formed by cell seeding and aggregation on alginate sponge scaffold.	38
Spinner flask 50 ml	Human dermal fibroblasts	Aggregation with PLGA nanospheres.	27
Spinner flask 50 ml	Human embryonic kidney cells	Aggregation with serum protein coated PLGA nanospheres.	24, 25
Spinner flask 50 ml	L6 myoblasts	Cell aggregation via surface engineering (biotinylated cells cross-linked with avidin).	32
Spinner flask 100 ml	Primary mouse hepatocytes	Cell aggregation using PLGA nanospheres. Enhanced liver functions compared to spheroids formed without nanospheres.	26
Spinner flask 250 ml	Primary rat hepatocytes	Enhanced aggregation and liver function using the polymer eudragit.	20, 21
Spinner flask 250 ml	rCHO	Growth of cells in aggregate form for the production of a recombinant protein.	11
Wavy walled spinner flask	Primary chondrocytes	Accelerated aggregation rate compared to smooth spinner flask.	45

6.1.2. *Spinner flasks*

These are the simplest form of bioreactors, also called stirred tank bioreactors, and exist in different sizes from 50 ml to a few litres. They provide the possibility of scaling up the production of aggregates and long term maintenance of clustered cells.[11] Spinner flasks are also used for culture of aggregates formed via other methods and some examples of their uses for aggregation studies are shown in Table 1.

7. Summary and Conclusion

In this chapter, a few of the approaches designed to aggregate cells were described and discussed. Most of the studies discussed have shown the benefits and possible uses of the cell aggregation approaches in tissue engineering application. The aggregated cells usually had a higher viability and functionality compared to controls (cells on monolayers) and a faster proliferation rate. The basis for these manipulations varied from a simple physical change, such as modification of hydrodynamic conditions with use of bioreactors, to a more structured cell surface chemical modification and engineering. This emphasises the range of options available to achieve cell clustering. Some of these approaches also highlight the importance of understanding the biology and characteristic of the cell type (or tissue) of interest and design of ligands or use of materials that target these cells specifically. Some of the methods described for cell aggregation are also generic and could be applied to a wide variety of cells. This allows the clustering of heterotypic cells which models the state of cells in higher order organisms and could bring benefits for modelling or engineering some tissue types.

References

1. Cooper, G.M. and Hausman, R.E. *The Cell: A Molecular Approach* (ASM Press and Sinauer Associates, 2007).
2. Powers, M.J. and Griffith-Cima, L. Motility behavior of hepatocytes on extracellular matrix substrata during aggregation. *Biotechnol. Bioeng.* **50**, 392–403 (1996).
3. Gilbert, S.F. *Developmental Biology* (Sinauer Associates, Sunderland, Massachusetts, 1997).
4. Schlett, K., Czirok, A., Tarnok, K., Vicsek, T. and Madarasz, E. Dynamics of cell aggregation during *in vitro* neurogenesis by immortalized neuroectodermal progenitors. *J. Neurosci. Res.* **60**, 184–194 (2000).
5. Voroteliak, E.A., Chermnykh, E.S., Vasil'ev, A.V. and Terskikh, V.V. Population structure of *in vitro* cultured keratinocytes. *Izv. Akad. Nauk. Ser. Biol.* **6**, 645–649 (2005).
6. Thomas, R.J., Bennett, A., Thomson, B. and Shakesheff, K.M. Hepatic stellate cells on poly(DL-lactic acid) surfaces control the formation of

3D hepatocyte co-culture aggregates *in vitro*. *Eur. Cell Mater.* **11**, 16–26; discussion 26 (2006).
7. Georges, P.C. and Janmey, P.A. Cell type-specific response to growth on soft materials. *J. Appl. Physiol.* **98**, 1547–1553 (2005).
8. Powers, M.J., Rodriguez, R.E. and Griffith, L.G. Cell-substratum adhesion strength as a determinant of hepatocyte aggregate morphology. *Biotechnol. Bioeng.* **53**, 415–426 (1997).
9. Semler, E.J., Ranucci, C.S. and Moghe, P.V. Mechanochemical manipulation of hepatocyte aggregation can selectively induce or repress liver-specific function. *Biotechnol. Bioeng.* **69**, 359–369 (2000).
10. Glicklis, R., Merchuk, J.C. and Cohen, S. Modeling mass transfer in hepatocyte spheroids via cell viability, spheroid size, and hepatocellular functions. *Biotechnol. Bioeng.* **86**, 672–680 (2004).
11. Han, Y., Liu, X.M., Liu, H., Li, S.C., Wu, B.C., Ye, L.L., Wang, Q.W. and Chen, Z.L. Cultivation of recombinant Chinese hamster ovary cells grown as suspended aggregates in stirred vessels. *J. Biosci. Bioeng.* **102**, 430–435 (2006).
12. Kelm, J.M. and Fussenegger, M. Microscale tissue engineering using gravity-enforced cell assembly. *Trends Biotechnol.* **22**, 195–202 (2004).
13. Kelm, J.M., Timmins, N.E., Brown, C.J., Fussenegger, M. and Nielsen, L.K. Method for generation of homogeneous multicellular tumor spheroids applicable to a wide variety of cell types. *Biotechnol. Bioeng.* **83**, 173–180 (2003).
14. Robinson, E.E., Zazzali, K.M., Corbett, S.A. and Foty, R.A. Alpha 5 beta1 integrin mediates strong tissue cohesion. *J. Cell Sci.* **116**, 377–386 (2003).
15. Donati, I., Stredanska, S., Silvestrini, G., Vetere, A., Marcon, P., Marsich, E., Mozetic, P., Gamini, A., Paoletti, S. and Vittur, F. The aggregation of pig articular chondrocyte and synthesis of extracellular matrix by a lactose-modified chitosan. *Biomaterials* **26**, 987–998 (2005).
16. Verma, P., Verma, V., Ray, P. and Ray, A.R. Formation and characterization of three dimensional human hepatocyte cell line spheroids on chitosan matrix for *in vitro* tissue engineering applications. *In Vitro Cell Dev. Biol. Anim.* **43**, 328–337 (2007).
17. Dai, W. and Saltzman, W.M. Fibroblast aggregation by suspension with conjugates of poly(ethylene glycol) and RGD. *Biotechnol. Bioeng.* **50**, 3349–3356 (1996).
18. Dai, W., Belt, J. and Saltzman, W.M. Cell-binding peptides conjugated to poly(ethylene glycol) promote neural cell aggregation. *Biotechnology (N.Y.)* **12**, 797–801 (1994).

19. Baldwin, S.P. and Saltzman, W.M. Aggregation enhances catecholamine secretion in cultured cells. *Tissue Eng.* **7**, 179–190 (2001).
20. Yamada, K., Kamihira, M., Hamamoto, R. and Iijima, S. Efficient induction of hepatocyte spheroids in a suspension culture using a water-soluble synthetic polymer as an artificial matrix. *J. Biochem. (Tokyo)* **123**, 1017–1023 (1998).
21. Kamihira, M., Yamada, K., Hamamoto, R. and Iijima, S. Spheroid formation of hepatocytes using synthetic polymer. *Ann. N. Y. Acad. Sci.* **831,** 398–407 (1997).
22. Yamada, K., Kamihira, M. and Iijima, S. Enhanced cell aggregation and liver functions using polymers modified with a cell-specific ligand in primary hepatocyte cultures. *J. Biosci. Bioeng.* **88**, 557–562 (1999).
23. Ryu, J.H., Kim, M.S., Lee, G.M., Choi, C.Y. and Kim, B.S. The enhancement of recombinant protein production by polymer nanospheres in cell suspension culture. *Biomaterials* **26**, 2173–2181 (2005).
24. Ryu, J.H., Oh, D.J., Choi, C.Y. and Kim, B.S. Suspension culture of anchorage-dependent animal cells using nanospheres of the biodegradable polymer, poly(lactic-co-glycolic acid). *Biotechnol. Lett.* **25**, 1363–1367 (2003).
25. Ryu, J.H., Kim, S.S., Cho, S.W., Choi, C.Y. and Kim, B.S. HEK 293 cell suspension culture using fibronectin-adsorbed polymer nanospheres in serum-free medium. *J. Biomed. Mater. Res. A* **71**, 128–133 (2004).
26. Gwak, S.J., Choi, D., Paik, S.S., Cho, S.W., Kim, S.S., Choi, C.Y. and Kim, B.S. A method for the effective formation of hepatocyte spheroids using a biodegradable polymer nanosphere. *J. Biomed. Mater. Res. A* **78**, 268–275 (2006).
27. Ryu, J.H. and Kim, B.S. Comparison of mammalian cell suspension culture methods using poly(lactic-co-glycolic acid) nanospheres and dextran microcarriers. *Key Eng. Mater.* **342–343**, 113–116 (2007).
28. Greig, R.G. and Brooks, D.E. Shear-induced concanavalin A agglutination of human erythrocytes. *Nature* **282**, 738–739 (1979).
29. Gestwicki, J.E., Strong, L.E., Cairo, C.W., Boehm, F.J. and Kiessling, L.L. Cell aggregation by scaffolded receptor clusters. *Chem. Biol.* **9**, 163–169 (2002).
30. De Bank, P.A., Kellam, B., Kendall, D.A. and Shakesheff, K.M. Surface engineering of living myoblasts via selective periodate oxidation. *Biotechnol. Bioeng.* **81**, 800–808 (2003).
31. Van Lenten, L. and Ashwell, G. Studies on the chemical and enzymatic modification of glycoproteins. A general method for the tritiation of sialic acid-containing glycoproteins. *J. Biol. Chem.* **246**, 1889–1894 (1971).

32. De Bank, P.A., Hou, Q., Warner, R.M., Wood, I.V., Ali, B.E., Macneil, S., Kendall, D.A., Kellam, B., Shakesheff, K.M. and Buttery, L.D. Accelerated formation of multicellular 3-D structures by cell-to-cell cross-linking. *Biotechnol. Bioeng.* **97**, 1617–1625 (2007).
33. Baqui, M., Botero, D., Gereben, B., Curcio, C., Harney, J.W., Salvatore, D., Sorimachi, K., Larsen, P.R. and Bianco, A.C. Human type 3 iodothyronine selenodeiodinase is located in the plasma membrane and undergoes rapid internalization to endosomes. *J. Biol. Chem.* **278**, 1206–1211 (2003).
34. Kellam, B., De Bank, P.A. and Shakesheff, K.M. Chemical modification of mammalian cell surfaces. *Chem. Soc. Rev.* **32**, 327–337 (2003).
35. Meier, W. Reversible cell aggregation induced by specific ligand-receptor coupling. *Langmuir* **16**, 1457–1459 (2000).
36. Ong, S.M., He, L., Thuy Linh, N.T., Tee, Y.H., Arooz, T., Tang, G., Tan, C.H. and Yu, H. Transient inter-cellular polymeric linker. *Biomaterials* **28**, 3656–3667 (2007).
37. Glicklis, R., Shapiro, L., Agbaria, R., Merchuk, J.C. and Cohen, S. Hepatocyte behavior within three-dimensional porous alginate scaffolds. *Biotechnol. Bioeng.* **67**, 344–353 (2000).
38. Gerecht-Nir, S., Cohen, S., Ziskind, A. and Itskovitz-Eldor, J. Three-dimensional porous alginate scaffolds provide a conducive environment for generation of well-vascularized embryoid bodies from human embryonic stem cells. *Biotechnol. Bioeng.* **88**, 313–320 (2004).
39. Yang, J., Chung, T.W., Nagaoka, N., Goto, M., Cho, C. and Akaike, T. Hepatocyte-specific porous polymer-scaffolds of alginate/galactosylated chitosan sponge for liver-tissue engineering. *Biotechnol. Lett.* **23**, 1385–1389 (2001).
40. Han, B., Akeprathumchai, S., Wickramasinghe, S.R., Qian, X. Flocculation of biological cells: experiment vs. theory. *AIChE J.* **49**, 1687–1701 (2003).
41. Tandon, P. and Diamond, S.L. Hydrodynamic effects and receptor interactions of platelets and their aggregates in linear shear flow. *Biophys. J.* **73**, 2819–2835 (1997).
42. Bell, D.N., Spain, S. and Goldsmith, H.L. Adenosine diphosphate-induced aggregation of human platelets in flow through tubes. I. Measurement of concentration and size of single platelets and aggregates. *Biophys. J.* **56**, 817–828 (1989).
43. Manley, P. and Lelkes, P.I. A novel real-time system to monitor cell aggregation and trajectories in rotating wall vessel bioreactors. *J. Biotechnol.* **125**, 416–424 (2006).

44. Gerecht-Nir, S., Cohen, S. and Itskovitz-Eldor, J. Bioreactor cultivation enhances the efficiency of human embryoid body (hEB) formation and differentiation. *Biotechnol. Bioeng.* **86**, 493–502 (2004).
45. Bueno, E.M., Bilgen, B., Carrier, R.L. and Barabino, G.A. Increased rate of chondrocyte aggregation in a wavy-walled bioreactor. *Biotechnol. Bioeng.* **88**, 767–777 (2004).

Chapter 16

Nanotechnology for Tissue Engineering

Jean S. Stephens-Altus and Jennifer L. West

Abstract

Future advances in tissue engineering are likely to require new tools and materials that provide novel capabilities for the construction of engineered tissue as well as the analysis and monitoring of such products. Nanotechnology research is yielding many new materials with unprecedented control over structure and function, often with highly unique properties. An example of nanotechnology in tissue engineering are nanoparticles, including quantum dots, where control over the size of semiconductor nanocrystals determines the emission spectra from these highly stable fluorescent probes, carbon nanotubes, which offer the potential to make very strong nanocomposites, and polymeric dendrimers, where the highly branched nature of the material offer opportunity to generate multi-functional particles. Nanotechnology also has advanced nanoscale control over material assembly. In this chapter, we will explore some of the ways that these technologies are being used to improve tissue engineering scaffolds, enhance cellular imaging capabilities, and enable monitoring of biological environments and processes.

Keywords: Nanomaterials; Nanostructured Scaffolds; Self-Assembly; Quantum Dots; Imaging.

Outline

1. Introduction
2. Nanostructured Scaffolds
   2.1. Self-assembling scaffold materials
   2.2. Nanocomposites
   2.3. Nanofibers
3. Nanoparticles for Cellular Imaging
   3.1. Optical imaging strategies

3.2. Magnetic resonance imaging
3.3. Bioresponsive probes
4. Conclusions
References

1. Introduction

Increasingly sophisticated advances in chemistry and physics are generating exciting new nanotechnologies, including some that may help to advance the field of tissue engineering. Controlling material structures at the nanoscale has led to a variety of new materials with highly controlled and interesting properties — from exceptionally high strength to unique optical properties and even the ability to mimic native proteins. This control over structures at nanoscale dimensions enables one to control and potentially tailor the properties of nanomaterials to meet the needs of a specific application. A classic example of this control and tenability comes from the so-called "quantum dots." These are semiconductor nanocrystals that are highly luminescent and whose absorbance onset and emission maxima shift to higher energy with decreasing particle size due to quantum confinement effects.[1] This feature makes it possible to easily fabricate a broad range of colors of stable fluorophores, which are candidates for multi-plexed detection and imaging applications.[2] Table 1 provides additional examples of nanomaterials, their unique properties, and some of their potential biomedical applications.

In addition to the unique material properties that can be achieved through nanotechnology, another advantage for the use of nanotechnology in tissue engineering is the biological relevance of this size scale. Nanoparticles are similar in size scale to many common biomolecules, making them interesting for applications such as intracellular tagging, bioconjugate applications such as antibody-targeted molecular imaging, protein-mimetic scaffold materials, or bio-responsive probes. Subcellular access, combined with unique material properties, offers the potential to monitor complex cellular processes such as signaling, gene expression and enzyme activity in living cells and tissues. In this chapter, we will discuss examples of how nanotechnology can improve the scaffolds used in tissue engineering and regenerative medicine, enhance cellular imaging and tracking, provide information about the biological status of engineered tissues, and increase our abilities to manipulate cells. We believe that these advances will increase the sophistication and control possible in tissue engineering applications.

2. Nanostructured Scaffolds

Advances in nanotechnology in the area of 3D tissue engineering scaffolds and the materials to construct them have lead to the development of stronger materials and

Table 1. Properties of nanomaterials for biomedical applications.

Material	Description	Unique properties	Potential applications
Fullerenes (buckyballs and nanotubes)[3–9]	Comprised entirely of carbon atoms joined via hexagonal and pentagonal rings. Can form spheres, ellipsoids or tubes.	Can have high strength, high electrical conductivity, chemically quite inert, infrared fluorescence.	Encapsulation of imaging agents and therapeutics, nanocomposites with strength and conductivity.
Quantum dots[1,2,10,11]	Semiconductor nanocrystals.	Strong, stable and tunable fluorescence.	Imaging and biosensing.
Liposomes[12–15]	Particulates formed via self-assembly of phospholipids. Usually spherical, unilamellar or multilamellar.	Ability to carry either hydrophilic or hydrophobic molecules. Easily functionalized.	Drug and gene delivery, encapsulation of imaging agents.
Dendrimers[16–18]	Highly controlled, highly branched polymer structures.	High carrying capacity. Can incorporate multiple functionalities through composition and derivatization.	Drug and gene delivery, imaging agents, multi-functional platform.
Gold nanoparticles[19–23]	Colloidal gold, gold nanoshells, gold nanorods, and other shapes. Similar structures with other metals also possible.	Strong and tunable optical resonances provide control over absorption and scattering.	Cellular imaging, biosensing, photothermal cell ablation, photothermally responsive composites.
Super-paramagnetic iron oxide particles[24,25]	Iron oxide particles sufficiently small as to be composed of a single crystalline domain.	Large magnetic moment in the presence of an external magnetic field.	MRI contrast agents, MR cellular imaging.

more biomimetic constructs. Many of the material properties of native tissue arise from the highly controlled nanoscale assembly of extracellular matrix molecules, so it is not surprising that numerous studies have found differences in a variety of cellular responses to materials with nanoscale features when compared to homogeneous bulk materials of similar overall composition. This section will focus on nanotechnology and its influence on materials and their use in 3D tissue engineering scaffolds as they relate to self-assembled structures, nanocomposites, and nanofibers.

2.1. Self-assembling scaffold materials

In nanotechnology there are two main approaches to construction; the first is a top-down approach, which starts with a larger structure and breaks it down to generate the desired nanoscale constituent. The second approach is a bottom-up design, which builds the structure molecule by molecule. Biological systems utilize a bottom-up approach, and the establishment of research in the area of self-assembly to form supramolecular architectures is based on the understanding of natural self-assembly, such as phospholipids in the cell membrane, protein folding, fibrillogenesis, and hydrogen bonding between DNA strands.[26,27] Most self-assembly is driven by non-covalent, inter- and intra-molecular interactions. The bonds that hold these structures together are inherently weak hydrogen bonds, ionic bonds (electrostatic forces), hydrophobic interactions, and van der Waals forces. Alone these forces may be weak; however in supramolecular formations they are strong enough to control the formation and stability of these 3D assemblies.

Utilizing the knowledge of naturally occurring assemblies and the knowledge of the interactions of between molecules, new and novel macromolecules are being designed and synthesized to function in a biomimetic manner and enhance bio-interactions. Development of self-assembling polymers for tissue engineering applications draws direct parallels to biology, not only in structure and assembly, but most importantly in function. This is achieved through understanding how amino acid sequences influence peptide and protein folding, i.e. alpha helices or beta sheets, and their amphiphilic properties. This level of control over polymer or macromolecule architecture is an ongoing challenge, but has been attained through conventional organic synthesis, peptide synthesis, or the use of recombinant DNA synthesis in bacterial or yeast host.[28,29] The structures that self-assembled molecules can form are quite numerous, but are most commonly classified as nanofibrils and nanotubules that form a 3D macrostructures called a hydrogel, as well as nanocoatings, micelles, and vesicles.[28,30,31] These 3D assemblies and their functionality are driven by the architecture of the polymer backbone as well environmental cues, such as pH, temperature, and polarity of solution,

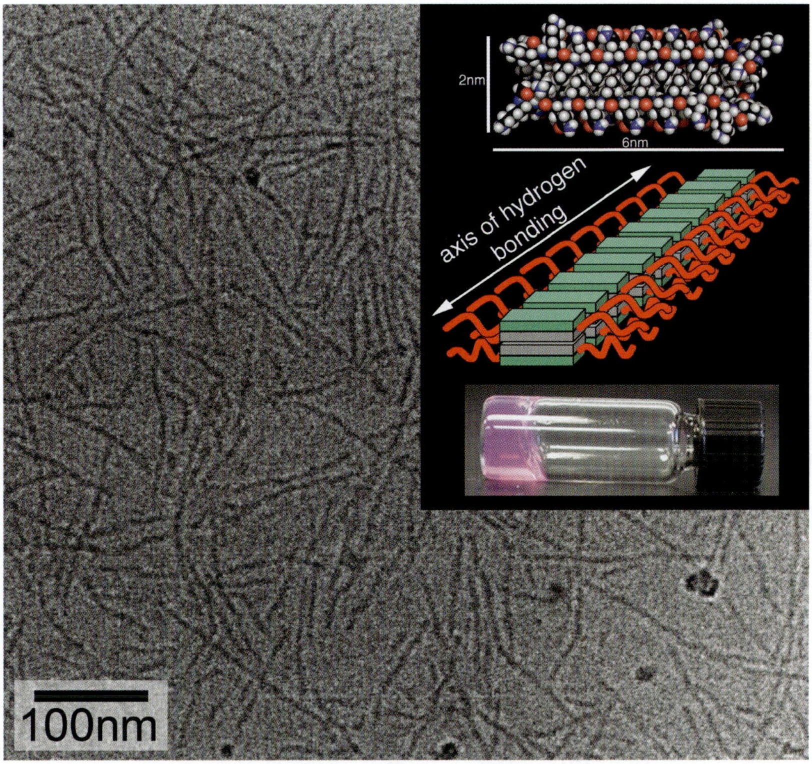

Fig. 1. Self-assembling peptides can organize into nanofibers capable of forming hydrogel scaffolds.

under which they assemble. A hydrogel scaffold formed via self-assembly of peptides is shown in Fig. 1. The use of peptides in this application provides diverse material properties that can drive self-assembly, but also the opportunity to present bioactive sequences such as cell adhesion peptides or enzyme substrate sequences.[29]

2.2. Nanocomposites

Nanocomposites are a combination of nanoparticles embedded within a matrix material, for tissue engineering most commonly a polymer. Nanocomposites are most frequently generated to improve mechanical properties of a scaffold material, though it is also possible to use nanocomposites to achieve other properties such as electrical conductivity or optical absorbance. The use of nanoparticles, versus larger particles, helps to reduce the weight of the scaffold, provide increased durability, and aid in biocompatibility.[32] The ratio of nanoparticle to polymer as well as the physico-chemical nature of both components and their interfacial interaction directly affects the overall properties of the nanocomposite.[32,33] A wide variety of biocompatible natural and man-made polymers are used in tissue engineering nanocomposites, a few examples are poly(lactic-co-glycolic) acid (PLGA), poly(ε-caprolactone) (PCL), polyurethanes, and polyesters, as well as collagen, gelatin, and chitosan, and commonly used nanoparticles are hydroxapatite (HA), titania, silica, carbon nanotubes, and fullerenes.[31] One example of this strategy comes from mechanical reinforcement of biodegradable polymer scaffolds for orthopedic applications, in this case carbon nanotubes in poly(propylene fumarate) (PPF). Carbon nanotubes were functionalized to enhance PPF — carbon nanotube interfacial interactions in order to optimize mechanical properties.[34,35] As shown in Fig. 2, cells retained high viability when cultured on porous scaffolds composed of these nanocomposites.

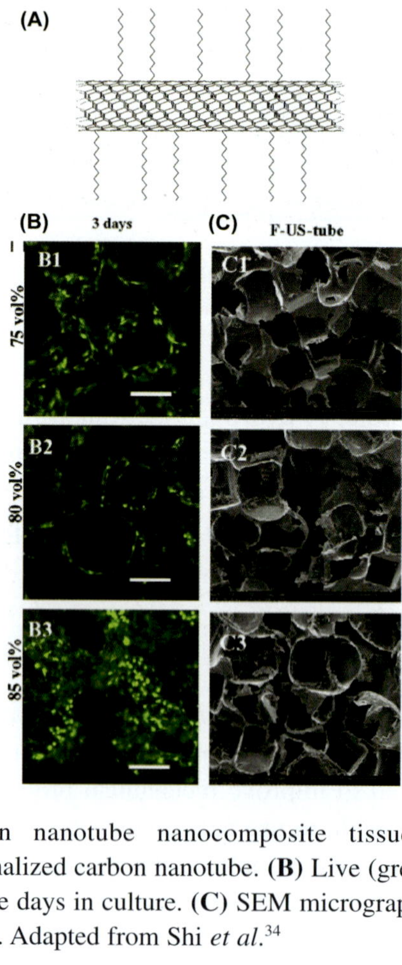

Fig. 2. Porous carbon nanotube nanocomposite tissue engineering scaffold. **(A)** Schematic of functionalized carbon nanotube. **(B)** Live (green)/dead (red) staining of cells in scaffold after three days in culture. **(C)** SEM micrographs to illustrate 3D porous structures of the scaffolds. Adapted from Shi *et al.*[34]

2.3. Nanofibers

There are several approaches to making nanofibers tissue engineering scaffolds. As discussed above self-assembly can be used as well as phase separation effects and a process called electrospinning.[36–39] All of these processes have their unique advantages and disadvantages, but the unifying property is that each can be used to create nanofibrous scaffolds with fiber dimensions that are on a similar size scale to proteins in the extracellular matrix (ECM), for example collagen fibrils that are typically 50 to 500 nm in diameter. An enormous number of polymers have been used to make nanofibers. Figure 3 shows an example of poly-l-lactic acid nanofibrous scaffolds used in orthopedic tissue engineering.[39]

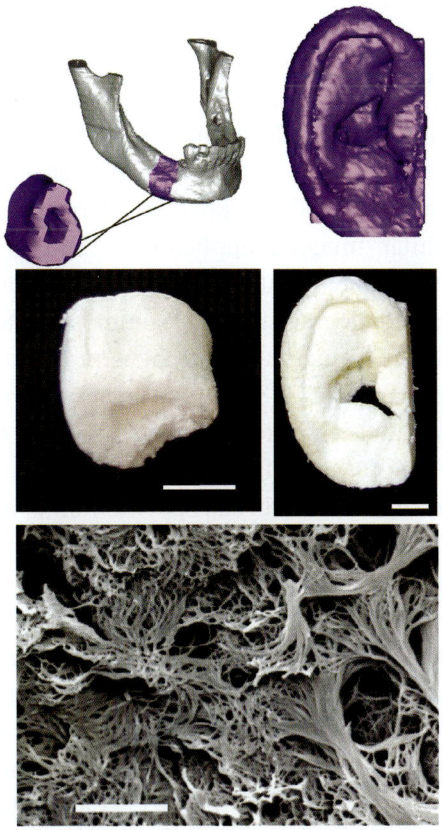

Fig. 3. PLLA nanofibrous scaffolds were formed via phase separation and then fabricated to match anatomical shapes. The top two images are reconstructions of CT scans; the middle two images are corresponding nanofibrous scaffolds to fit the defects shown in the CT scans, both scale bars are 10 mm; the bottom image is an SEM micrograph of the nanofibrous structure of the scaffold scale bar = 5 μm. Adapted from Chen *et al.*[39]

3. Nanoparticles for Cellular Imaging

Tissue engineering research and product development will be substantially advanced by increased availability and power of non-invasive, non-destructive imaging technologies with cellular resolution. These techniques can allow monitoring of tissue formation, as well as long term fate and function of engineered tissues following implantation. It may also be necessary to track cells long term following implantation, ideally with resolution at the single cell level. While the capabilities of imaging technologies are continuously expanding, contrast agents can provide drastic benefits. New contrast agents based on nanoparticles are under development. In addition to increasing signal, nanoparticle imaging agents can generally be targeted to specific cells or cellular structures to increase the information content as well.

3.1. Optical imaging strategies

Optical imaging techniques, particularly fluorescent techniques, can provide high resolution of cellular, and in some cases molecular, structures. Several types of nanoparticles have been developed that provide interesting optical properties that may be of use in cellular imaging applications. Semiconductor nanocrystals, known also as quantum dots, are highly light absorbing, luminescent nanoparticles whose absorbance onset and emission maximum shift to higher energy with decreasing particle size due to quantum confinement effects.[1] These nanocrystals are typically in the size range of 2–8 nm, but generally require passivation and functionalization layers that significantly increase the hydrodynamic radius. Unlike molecular fluorophores, which typically have very narrow excitation spectra, quantum dots absorb light over a broad spectral range, and the excitation spectra are relatively unaffected by changes in particle size. Thus, it is possible to excite a variety of quantum dot materials using a single excitation wavelength, which may enable one to simultaneously image multiple cells or structures, each tagged with a differently colored nanoparticle. Specific binding of bioconjugated quantum dots to cell surfaces, cellular uptake, and nuclear localization have all been demonstrated.[1,2,10,11,40–43] An example of use of quantum dots in cellular imaging is shown in Fig. 4. Recent developments in near-infrared quantum dots can allow deeper imaging of constructs and tissues.[44] Other fluorescent nanoparticle probes are also under development, including fluorescent virus particles[45] and fluorescent dendrimers.[46]

Optical coherence tomography (OCT) is another near-infrared imaging modality that can allow deep imaging in tissue engineered constructs. OCT uses a low coherence light source in a Michelson interferometer to enable high resolution

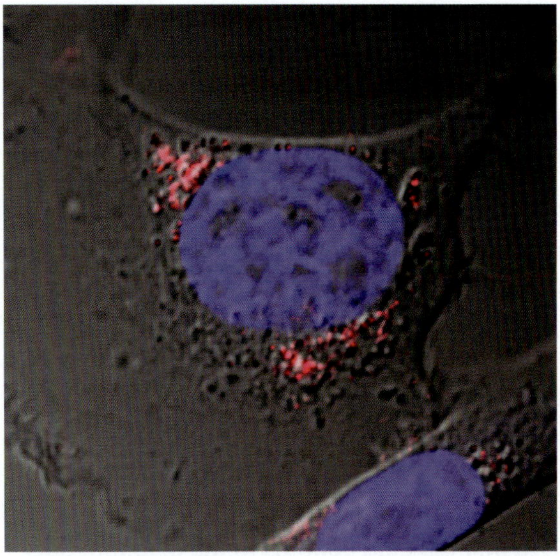

Fig. 4. The ability to easily construct semi-conductor nanocrystals with similar excitation spectra yet very different emission allows for simultaneous imaging of multiple biomarkers.

optical sectioning in tissue. In the system used in this study, low coherence light emitted from the source is directed onto a beamsplitter. One beam is sent toward a reference mirror of known pathlength distance while the other is directed toward the sample. The reflected waves from both the reference mirror and sample are recombined at the beamsplitter and directed to a detector where interference patterns are monitored as the reference mirror is moved. By monitoring these patterns while scanning the reference mirror, it is possible to map depth resolved reflectivity of the tissue.[47] While the inherent optical properties of the tissue can be utilized for OCT imaging, substantially increased image contrast can be generated through introduction of highly scattering agents, such as some types of nanoparticles. For example, gold nanoshells have been used to enhance OCT contrast in tumor imaging, providing at least a fourfold increase in contrast.[21]

3.2. Magnetic resonance imaging

In addition to optical techniques, magnetic resonance imaging (MRI) techniques are advancing to the point that single cell resolution can be achieved.[48] Several nanoparticle-based MRI contrast agents have been developed, including iron oxides[24,25] and gadolinium-containing particulates.[49–51] The introduction of such agents can allow *in vivo* cell tracking or selective imaging of targeted structures. Additionally, several nanoparticle designs have accommodated both optical and MR contrast.[52–54]

These types of advanced, multi-functional nanoparticles can allow both anatomical and molecular imaging strategies simultaneously for the same markers.

3.3. Bioresponsive probes

Complex nanostructures are being developed that have the ability to sense and respond to their biological environment. In general, the approach involves designing a nanostructure such that proximity of components changes in response to a biological event, like the action of an enzyme or expression of a gene, and this leads to a change in a measurable signal from the material. Fluorescence quenching or fluorescence energy resonance transfer (FRET) are common approaches in this type of strategy. An example is shown in Fig. 5, where in order to generate a probe to report cellular protease activity, quantum dot-gold conjugates were constructed such that the quantum dot fluorescence was quenched. Protease-sensitive peptides were used to link the gold colloids to the quantum dots, and as the peptides were cleaved, fluorescence was restored.[55] Such a reporter could be utilized to

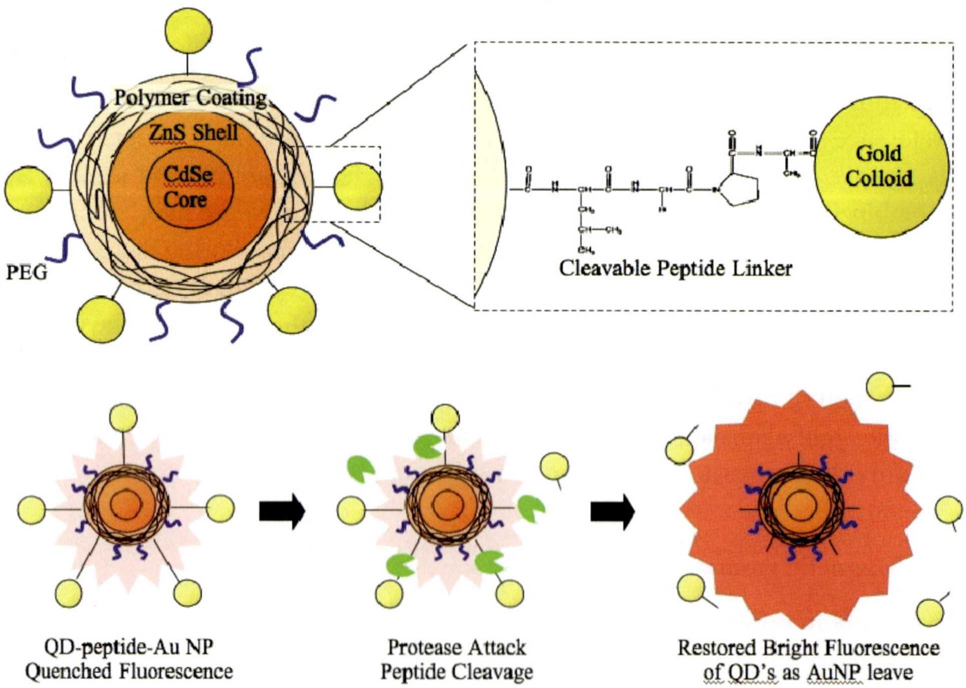

Fig. 5. When a quantum dot is conjugated to gold nanoparticles, its fluorescence is quenched. By using a linker that is an enzyme substrate, though, it is possible to restore fluorescence is response to biological activities such as extracellular proteases.

monitor matrix remodeling activity over time in engineered tissues. Similar probes might also monitor intracellular enzymatic activities.

4. Conclusions

Advances in nanotechnology are rapidly emerging, and may provide novel strategies to improve efforts in tissue engineering, either in terms of tissue construction, or analysis and monitoring of the resultant tissue substitutes. While this is a very exciting and fast-paced field, it is critical to consider that new materials with nanoscale dimensions may elicit adverse biological responses, even when the bulk material is inherently biocompatible. An example is the C60 Buckminsterfullerene, composed entirely of carbon. This material was expected to be bioinert based on experiences with pyrolytic carbon. However, pristine C60 in an aqueous environment assembles into aggregates that have demonstrated marked cytotoxicity.[56] Surprisingly, relatively minor surface modifications of the C60 molecule can essentially eliminate the cytotoxicity.[56] This highlights the importance of rigorous biological characterization of new nanomaterials before they are employed in biological systems, particularly since structure/function relationships related to nanoparticle biocompatibility are currently lacking. Despite these concerns, the promise for nanomedicine is considerable.

References

1. Bruchez, M., Moronne, M., Gin, P., Weiss, S. and Alavisatos, A.P. Semiconductor nanocrystals as fluorescent biological labels. *Science* **281**, 2013–2016 (1998).
2. Weissleder, R. A clearer vision for *in vivo* imaging. *Nat. Biotech.* **19**, 316–317 (2001).
3. Kroto, H.W., Heath, J.R., O'Brien, S.C., Curl, R.F. and Smalley, R.E. C60: Buckminsterfullerene. *Nature* **318**, 162–163 (1985).
4. Jensen, A.W., Wilson, S.R. and Schuster, D.I. Biological applications of fullerenes. *Bioorg. Med. Chem.* **4**, 767–779 (1996).
5. Dugan, L.L., Lovett, E.G., Quick, K.L., Lotharius, J., Lin, T.T. and O'Malley, K.L. Fullerene-based anti-oxidants and neurodegenerative disorders. *Parkinsonism Relat. Disord.* **7**, 243–246 (2001).
6. Tegos, G., Demidova, T., Arcila-Lopez, D., Lee, H., Wharton, T., Gali, H. and Hamblin, M. Cationic fullerenes are effective and selective antimicrobial photosensitizers. *Chem. Biol.* **12**, 1127–1135 (2005).
7. Bekyarova, E., Ni, Y., Malarkey, E.B., Montana, V., McWilliams, J.L., Haddon, R.C. and Parpura, V. Applications of carbon nanotubes in biotechnology and medicine. *J. Biomed. Nanotechnol.* **1**, 3–17 (2005).

8. Cherukuri, P., Gannon, C.J., Leeuw, T.K., Schmidt, H.K., Smalley, R.E., Curley, S.A. and Weisman, R.B. Mammalian pharmacokinetics of carbon nanotubes using intrinsic near-infrared fluorescence. *Proc. Natl. Acad. Sci. USA* **103**, 18882–18886 (2006).
9. O'Connell, M.J., Bachilo, S.M., Huffman, C.B., Moore, V.C., Strano, M.S., Haroz, E.H., Rialon, K.L., Boul, P.J., Noon, W.H., Kittrell, C., Ma, J., Hauge, R.H., Weisman, R.B. and Smalley, R.E. Band gap fluorescence from individual single-walled carbon nanotubes. *Science* **297**, 593–596 (2002).
10. Akerman, M.E., Chan, W.C.W., Laakkonen, P., Bhatia, S.N. and Ruoslahti, E. Nanocrystal targeting *in vivo*. *Proc. Nat. Acad. Sci. USA* **99**, 12617–12621 (2002).
11. Ruan, G., Agrawal, A., Marcus, A.I. and Nie, S. Imaging and tracking of tat peptide-conjugated quantum dots in living cells: new insights into nanoparticle uptake, intracellular trafficking, and vesicle shedding. *J. Am. Chem. Soc.* **129**, 14759–14766 (2007).
12. Barenholz, Y. Liposome application: problems and prospects. *Curr. Opin. Coll. Interface Sci.* **6**, 66–77 (2001).
13. Gabizon, A., Goren, D., Cohen, R. and Barenholz, Y. Development of liposomal anthracylcines: from basics to clinical applications. *J. Control Release* **53**, 275–279 (1998).
14. Samad, A., Sultana, Y. and Aqil, M. Liposomal drug delivery systems: an update review. *Curr. Drug Deliv.* **4**, 297–305 (2007).
15. Immordino, M.L., Dosio, F. and Cattel, L. Stealth liposomes: review of the basic science, rationale, and clinical applications, existing and potential. *Int. J. Nanomed.* **1**, 297–315 (2006).
16. Sadler, K. and Tam, J.P. Peptide dendrimers: applications and synthesis. *Rev. Mol. Biotechnol.* **90**, 195–229 (2002).
17. Najlah, M. and D'Emanuele, A. Synthesis of dendrimers and drug-dendrimer conjugates for drug delivery. *Curr. Opin. Drug Discov. Devel.* **10**, 756–767 (2007).
18. Satija, J., Gupta, U. and Jain, N.K. Pharmaceutical and biomedical potential of surface engineered dendrimers. *Crit. Rev. Ther. Drug Carrier Syst.* **24**, 257–306 (2007).
19. Rao, C.N.R., Kulkarni, G.U., Thomasa, P.J. and Edwards, P.P. Metal nanoparticles and their assemblies. *Chem. Soc. Rev.* **29**, 27–35 (2000).
20. Hirsch, L.R., Gobin, A.M., Lowery, A.R., Tam, F., Drezek, R.A., Halas, N.J. and West, J.L. Metal nanoshells. *Ann. Biomed. Eng.* **34**, 15–22 (2006).
21. Gobin, A.M., Lee, M.H., Halas, N.J., James, W.D., Drezek, R.A. and West, J.L. Near-infrared resonant nanoshells for combined optical imaging and photothermal cancer therapy. *Nano Lett.* **7**, 1929–1934 (2007).

22. Liao, H. and Hafner, J.H. Gold nanorod bioconjugates. *Chem. Mat.* **17**, 4636–4641 (2005).
23. Nehl, C.L., Liao, H. and Hafner, J.H. Optical properties of star-shaped gold nanoparticles. *Nano Lett.* **6**, 683–688 (2006).
24. McCarthy, J.R., Kelly, K.A., Sun, E.Y. and Weissleder, R. Targeted delivery of multifunctional magnetic nanoparticles. *Nanomedicine* **2**, 153–167 (2007).
25. Weissleder, R., Elizondo, G., Wittenberg, J., Rabito, C.A., Bengele, H.H. and Josephson, L. Ultrasmall superparamagnetic iron oxide: characterization of a new class of contrast agents for MR imaging. *Radiology* **175**, 489–493 (1990).
26. Branden, C.I. and Tooze, J. *Introduction to Protein Structure*, 2nd Ed. (Garland, NY, 1999).
27. Ferrari, M. Cancer nanotechnology: opportunities and challenges. *Nat. Rev. Cancer* **5**, 161–171 (2005).
28. Zhang, S. Fabrication of novel biomaterials through molecular self-assembly. *Nat. Biotechnol.* **21**, 1171–1178 (2003).
29. Dong, H., Paramonov, S.E., Aulisa, L., Bakota, E.L. and Harterink, J.D. Self-assembly of multi-domain peptides: Balancing molecular frustration controls conformation and nanostructure. *J. Am. Chem. Soc.* **129**, 12468–12472 (2007).
30. Silva, G.A., Czeisler, C., Niece, K.L., Beniash, E., Harrington, D.A., Kessler, J.A. and Stuppe, S.I. Selective differentiation of neural progenitor cells by high epitope density nanofibers. *Science* **303**, 1352–1355 (2004).
31. Goldberg, M., Langer, R. and Jia, X. Nanostructured materials for applications in drug delivery and tissue engineering. *J. Biomat. Sci. Polym. Ed.* **18**, 241–268 (2007).
32. Hule, R.A. and Pochan, D.J. Polymer nanocomposites for biomedical applications. *MRS Bull.* **32**, 354–358 (2007).
33. Christenson, E.M., Anseth, K.S., van der Beukanm, J.J.J.P., Chen, C.K., Ercan, B., Jansen, J.A., Laurencin, C.T., Murugan, R., Nair, L.S., Ramakrishna, S., Tuan, R.S., Webster, T.J. and Mikos, A.G. Nanobiomaterial applications in orthopedics. *J. Orthop. Res.* **54**, 11–22 (2007).
34. Shi, B., Sitharaman, B., Pham, Q.P., Liang, F., Wu, K., Billups, W.E., Wilson, L.J. and Mikos, A.G. Fabrication of porous ultra-short single-walled carbon nanotube nanocomposite scaffolds for bone tissue engineering. *Biomaterials* **28**, 4078–4090 (2007).
35. Sitharaman, B., Shi, X., Tran, L.A., Spicer, P.P., Rusakova, I., Wilson, L.J. and Mikos, A.G. Injectable *in situ* crosslinkable nanocomposites of biodegradable polymers and carbon nanostructures for bone tissue engineering. *J. Biomater. Sci. Polym. Ed.* **18**, 655–671 (2007).

36. Ma, P.X. and Zhang, R. Synthetic nano-scale fibrous extracellular matrix. *J. Biomed. Mat. Res.* **46**, 60–72 (1999).
37. Murugan, R. and Ramakrishna, S. Design strategies of tissue engineering scaffolds with controlled fiber orientation. *Tissue Eng.* **13**, 1845–1866 (2007).
38. Vasita, R. and Katti, D.S. Nanofibers and their applications in tissue engineering. *Int. J. Nanomed.* **1**, 15–30 (2006).
39. Chen, V.J., Smith, L.A. and Ma, P.X. Bone regeneration on computer-designed nano-fibrous scaffolds. *Biomaterials* **27**, 3973–3979 (2006).
40. Lidke, D.S., Nagy, P., Heintzmann, R., Arndt-Jovin, D.J., Post, J.N., Grecco, H.E., Jares-Erijman, E.A. and Jovin, T.M. Quantum dot ligands provide new insights into erbB/HER receptor-mediated signal transduction. *Nat. Biotechnol.* **22**, 169–170 (2004).
41. Lagerholm, B.C., Wang, M., Ernst, L.A., Ly, D.H., Liu, H., Bruchez, M.P. and Waggoner, A.S. Multi-color coding of cells with cationic peptide coated quantum dots. *Nano Lett.* **10**, 2019–2022 (2004).
42. Xing, Y., Smith, A.M., Agrawal, A., Ruan, G. and Nie, S. Molecular profiling of single cancer cells and clinical tissue specimens with semiconductor quantum dots. *Int. J. Nanomed.* **1**, 473–481 (2006).
43. Koeppel, F., Jaiswal, J.K. and Simon, S.M. Quantum dot-based sensor for improved detection of apoptotic cells. *Nanomedicine* **2**, 71–78 (2007).
44. Sun, J., Zhu, M.Q., Fu, K., Lewinski, N. and Drezek, R.A. Lead sulfide near-infrared quantum dot bioconjugates for targeted molecular imaging. *Int. J. Nanomed.* **2**, 235–240 (2007).
45. Manchester, M. and Singh, P. Virus-based nanoparticles (VNPs): platform technologies for diagnostic imaging. *Adv. Drug Deliv. Rev.* **58**, 1505–1522 (2006).
46. Taite, L.J. and West, J.L. Poly(ethylene glycol)-lysine dendrimers for targeted delivery of nitric oxide. *J. Biomat. Sci. Polym. Ed.* **17**, 1159–1172 (2006).
47. Fujimoto, J.G. Optical biopsy and imaging using optical coherence tomography. *Nat. Med.* **1**, 970–972 (1995).
48. Delikatny, E.J. and Poptani, H. MR techniques for *in vivo* molecular and cellular imaging. *Radiol. Clin. North Am.* **43**, 205–220 (2005).
49. Winter, P.M., Cai, K., Caruthers, S.D., Wickline, S.A. and Lanza, G.M. Emerging nanomedicine opportunities with perfluorocarbon nanoparticles. *Expert Rev. Med. Dev.* **4**, 137–145 (2007).
50. Hifumi, H., Yamaoka, S., Tanimoto, A., Citterio, D. and Suzuki, K. Gadolinium-based hybrid nanoparticles as a positive MR contrast agent. *J. Am. Chem. Soc.* **128**, 15090–15091 (2006).
51. Toth, E., Bolskar, R.D., Borel, A., Gonzalez, G., Helm, L., Merbach, A.E., Sitharaman, B. and Wilson, L.J. Water-soluble gadofullerenes: toward high

relaxivity, pH-responsive MRI contrast agents. *J. Am. Chem. Soc.* **127**, 799–805 (2005).

52. Prinzen, L., Miserus, R.J., Dirksen, A., Hackeng, T.M., Deckers, N., Bitsch, N.J., Megens, R.T.A., Douma, K., Heemskerk, J.W., Kooi, M.E., Frederik, P.M., Slaaf, D.W., van Zandvoort, M.A. and Reutelingsperger, C.P.M. Optical and magnetic resonance imaging of cell death using annexin A5-functionalized quantum dots. *Nano Lett.* **7**, 93–100 (2007).
53. Kim, J., Park, S., Lee, J.E., Jin, S.M., Lee, J.H., Lee, I.S., Yang, I., Kim, J., Kim, S.K., Cho, M. and Hyeon, T. Designed fabrication of multifunctional magnetic gold nanoshells and their application to magnetic resonance imaging and photothermal therapy. *Agnew Chem. Int. Ed.* **45**, 7754–7758 (2006).
54. Rieter, W.J., Kim, J.S., Taylor, K.M., An, H., Lin, W., Tarrant, T. and Lin, W. Hybrid silica nanoparticles for multimodal imaging. *Agnew Chem. Int. Ed.* **46**, 3680–3682 (2007).
55. Chang, E., Miller, J.S., Sun, J., Yu, W.W., Colvin, V.L., Drezek, R. and West, J.L. Protease-activated quantum dot probes. *Biochem. Biophys. Res. Comm.* **334**, 1317–1321 (2005).
56. Sayes, C.M., Fortner, J.D., Lyon, D., Boyd, A.M., Ausman, K.D., Guoh, W., Tao, Y.J., Sitharam, B., Wilson, L.J., West, J.L. and Colvin, V.L. The differential cytotoxicity of water-soluble fullerenes. *Nano Lett.* **4**, 1881–1887 (2004).

Chapter 17

Microscale Technologies for Tissue Engineering

Ali Khademhosseini,* Yanan Du,* Bimal Rajalingam, Joseph P. Vacanti and Robert S. Langer

Abstract

Despite the enormous advances in tissue engineering, several challenges still prevent the widespread clinical application of tissue engineering products, such as how to acquire adequate number of cells, and how to engineer complex vascularized tissues that mimic the complexity and function of native tissues. The merger of biomaterials and microscale technologies offer new opportunities to overcome the challenges in tissue engineering to fabricate scaffolds and direct stem cell differentiation. In this chapter, various applications of microscale technologies have been illustrated in controlling stem cell fate and building complex artificial tissues. It is envisioned that with the rapid growth of this burgeoning research field, microscale technologies will transform the conventional tissue engineering approaches and greatly contribute to the therapeutic potential of tissue engineering.

Keywords: Microfabrication; Soft Lithography; Stem Cells; Microfluidics; Tissue Engineering.

Outline

1. Introduction
2. Microscale Technologies for Controlling Stem Cell Fate
   - 2.1. Regulating stem cell fate by controlling cell shape
   - 2.2. Microwells for uniform embryoid body culture and control of cell-cell contact
   - 2.3. Microarrays for directing stem cell fates
   - 2.4. Microfluidic system for controlling stem cell fate

---

* These authors contributed equally.

3. Microscale Technologies for Engineering Complex Tissues Containing Different Cell Types and Vasculature
   3.1. Microscale technologies for template-based cell assembly into 3D micro-tissues
   3.2. Scaffolds with micro- and nano-topography
   3.3. Microengineered hydrogels for tissue engineering
   3.4. 3D tissue/organ printing
   3.5. Microfluidics for engineering the vasculature
4. Conclusion
   References

1. Introduction

Tissue engineering aims to generate tissues that restore, maintain, or enhance tissue function,[1] with the ultimate goal of providing a permanent solution to the problem of organ failure. Commonly-adopted tissue engineering approaches involve (i) isolated cells or cell substitutes, (ii) biocompatible materials for cellular support and regeneration, or (iii) cell-biomaterial (i.e. scaffold) composites.[2] Cells for tissue engineering have been traditionally derived either as autografts (from the patient), allografts (from a human donor) or xenografts (from a different species). In most tissue engineering approaches, isolated cells are then cultured on biocompatible scaffolds, which provide physical and chemical support and guide the cell growth and organization into three-dimensional (3D) tissues. Despite the enormous advances in tissue engineering which have resulted in clinically viable products such as skin, several challenges still prevent the widespread clinical application of tissue engineering products. These challenges include a number of business, regulatory and ethical issues as well as scientific barriers. These scientific issues include (a) how to acquire adequate source of cells, (b) how to engineer complex vascularized tissues that mimic the complexity of native tissue architecture, and (c) how to generate tissues with the biomechanical and metabolic functions that mimic normal tissues.

Microscale technologies are emerging as a potentially powerful tool for controlling cell microenvironment and generating tissue constructs. Most microscale technologies are adopted from the micro-fabrication processes in semiconductor and microelectronics industries, which can achieve control of microscale features ranging from sub-micron range to greater than 1 cm. Soft lithography is one of the most popular microscale technologies, which can fabricate microscale devices without extensive usage of clean-room and photolithography facilities.[3] In soft lithography, elastomeric stamps (i.e. poly(dimethylsiloxane), PDMS) are fabricated from patterned silicon wafers to print or mold materials, which can control the topography and spatial distribution of biomaterials at sub-micron resolution in a convenient, rapid and inexpensive manner. Since many microscale techniques are compatible with cells, they can be readily used to engineer the cell microenvironment. In particular, microscale technologies have been extensively used to control the cell-cell, cell-extracellular matrix (ECM) and cell soluble factor interactions in tissue culture. Furthermore, microengineering approaches are now being increasingly used to generate tissue scaffolds with enhanced architectural and mechanical properties. Thus microscale technologies, such as soft lithography and photolithography are emerging as promising tools for addressing some of the challenges in tissue engineering. Furthermore, this miniaturization has enabled high-throughput testing of the cell behavior in response to various stimuli in an inexpensive, rapid and reproducible manner.

In this chapter, numerous applications of microscale technologies have been introduced, which can potentially solve the challenges in tissue engineering. Although we provide a broad perspective of the field we focus on the applications of the microscale systems in controlling stem cell fate and engineering complex vascularized tissue constructs.

2. Microscale Technologies for Controlling Stem Cell Fate

One of the major challenges to the clinical feasibility of tissue engineering is how to acquire an adequate number of cells that are immunologically compatible with the patient. Many cell types isolated from the adult tissues are difficult to expand in culture and quickly lose their phenotypes. Recently stem cells from the adult and embryonic sources have generated much excitement that a renewable source of cells may be obtainable for tissue engineering applications. Stem cells are pluripotent cells that have the capacity to self-renew and to differentiate into various lineages. Reproducible and directed regulation of the stem cell fate (i.e. self-renewal and differentiation decisions) is critical for the clinical success of stem cell-based therapies. From years of biological research, it is becoming increasingly evident that both intra- and extracellular cues regulate the resulting cell fate decisions. Most stem cells in the body reside in specific niches that signal the cells to behave in response to physiological conditions. Thus signals in the microenvironment that are regulated in space and time direct the cells and result in their subsequent differentiation.[4] Although standard tissue culture techniques have greatly increased our understanding of the stem cell microenvironment, they lack the spatial and temporal regulation of the microenvironment to which stem cells are exposed. Furthermore, it is difficult to perform high-throughput studies to examine the complexities of the combinations and concentrations of the various signals on stem cell behavior. In contrast, the ability of the microscale technologies to miniaturize experiments and increase experimental control can be used to provide new opportunities in studying and directing stem cell fate responses.

2.1. Regulating stem cell fate by controlling cell shape

Extracellular cues are important in regulating stem cell fate decision. In general, the physical properties of the extracellular matrix (ECM) contribute to cell function and behavior. For example, mechanical cues experienced by the cells in its environment regulates its function including cell shape, DNA synthesis, motility and lineage commitment.[5] Several hypotheses have been put forward to explain how the mechanical cues affect the cell physiology. For example, Ingber and colleagues have suggested that shape-dependent control of cell growth and function appears to be mediated by tension-dependent changes in the actin cytoskeleton. They used computer simulations based on dynamic Boolean networks to show that generalized stimuli (e.g. mechanical forces) and specific molecular cues elicit signals which follow different trajectories, but eventually converge onto one of a small set of common end points (growth, quiescence, differentiation, apoptosis, etc.).[6]

Cell shape has been earlier suggested to play a role in differentiation and apoptosis. For example, Watt *et al.* found that the cell shape influences the terminal differentiation in epidermal keratinocytes.[7] When cell spreading is restricted using micropatterned substrates, more round-shaped cells entered apoptosis compared to cells that were allowed to spread on identically fabricated unpatterned substrates, thus demonstrating a role of cell shape as a critical determinant that switches cells between life and apoptosis.[8] Thus cell shape and the cytoskeletal structure play an important role in cellular functions.[9] While changes in the cell shape appear to regulate several cellular processes, its role in the commitment of the multipotential stem cells is not as clearly understood. Chen and colleagues reported that cell shape and cytoskeletal tension determines the lineage commitment of stem cells[10] (Fig. 1). They observed that human mesenchymal stem cells (MSCs) that were allowed to adhere, flatten, and spread, differentiated into osteogenic cells, while unspread, round cells gave rise to adipocytes. Furthermore, they demonstrated that this behavior was mediated through the modulation of the endogeneous RhoA pathway.

Micropatterning enables the confinement of the individual or group of cells within defined spatially controlled spaces. Cell shape can be easily controlled by using elastomeric stamps.[11] Singhvi *et al.* used an elastomeric stamp to imprint gold surfaces with specific patterns of self-assembled monolayers of alkanethiols to create islands of defined shape and size that support protein adsorption and cell attachment. Using this technique cells were arrayed in

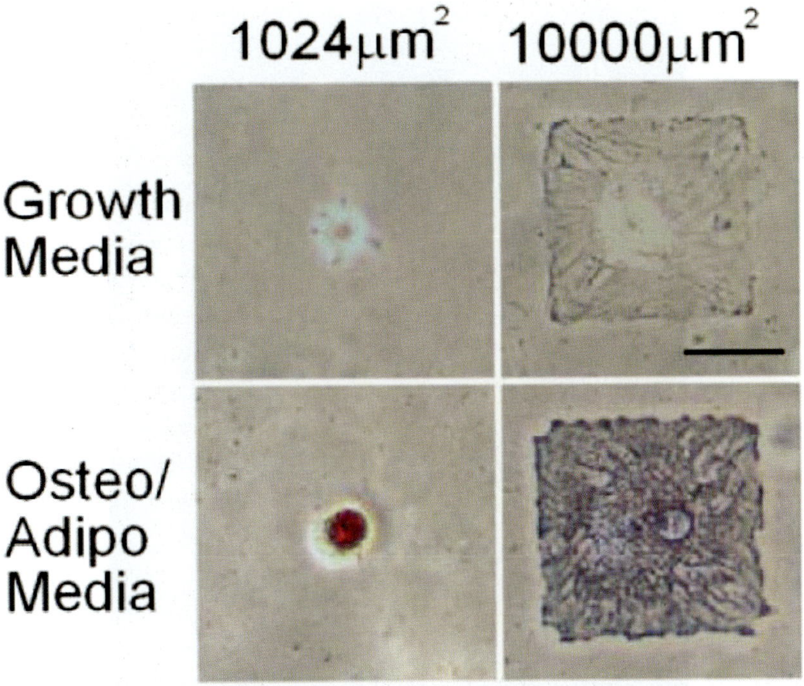

Fig. 1. Cell shape regulates commitment of human mesenchymal stem cells to adipocyte or osteoblast fate.[10] When cells are exposed to a mixture of Osteo/Adipo differentiated media, the cells cultured on the small islands stained for lipids (red), indicating differentiation into adipogenic fates; whereas cells on large islands stained for alkaline phosphatase (blue) indicating differentiation into osteoblasts.

predetermined locations in defined shapes. It was demonstrated that by limiting the degree of cell extension, cell growth and protein secretion could be controlled.[11]

2.2. Microwells for uniform embryoid body culture and control of cell-cell contact

Embryonic stem (ES) cells in suspension culture form multicellular aggregates called embryoid bodies (EBs) that contain the three germ layers. EBs recapitulate the early stages of the embryonic development and give rise to a wide spectrum of cell types. EBs are usually generated by means of the traditional hanging drop method or by suspension cultures in non-tissue culture treated plates.[12] Other methods to fabricate EBs include culture in methylcellulose semisolid media, in round-bottomed 96-well plates, in stirred-suspension cultures using spinner flasks and in

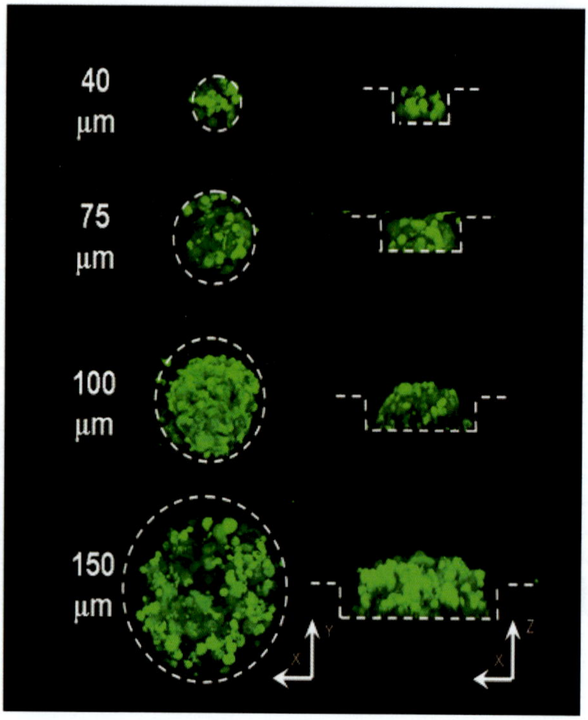

Fig. 2. Microwells for culturing EBs:[14] Confocal laser light microscopy images of CFSE-stained cell aggregates within 40, 75, 100, and 150 µm microwells after five days. Cells formed 3D embryoid bodies with diameters closely corresponding to those of their respective microwells. The first column of images shows aerial views of the cell aggregates whereas the second column shows vertical cross-sections of cell aggregates within PEG microwells.

rotary flasks.[12,13] These methods prevent cell adhesion to the surface of the vessel and promote cell aggregation. The major limitations of the above mentioned methods are the large size and shape distribution of the EBs as well as cumbersome procedures involved in making more homogeneous EBs. To overcome these limitations, Karp *et al.* used non-adhesive microwells made from poly(ethylene glycol) (PEG) to culture homogeneous EBs with controllable sizes and shapes in a rapid and reproducible manner[14] (Fig. 2). PEG inhibits protein adsorption and cell adhesion which results in cell aggregation and the formation of EBs. Moreover they demonstrated that EBs can be retrieved from microwells with a greater than 95% viability. Also it is hypothesized that the size of the EBs influences the subsequent differentiation and hence this technique can be useful for inducing more directed differentiation without addition of exogenous growth factors. The microfabrication-based generation of the EBs is scalable, cost-effective and simple to perform. Furthermore it can be used for performing high-throughput screening and toxicity studies. For example, it is known that microwells can be used to immobilize cells within microfluidic arrays to generate devices that can be used to perform high-throughput experiments.[15]

Direct cell-cell contact is known to affect many stem cell fates decisions. Therefore, methods that can be used to control cell-cell contact between the same cell types (homotypic) and different cell types (heterotypic) will be beneficial for stem cell differentiation studies. To effectively control heterotypic cell-cell interactions, Bhatia and colleagues have developed a number of methods to control the interaction of hepatocytes with non-parenchymal cell types.[16,17] Furthermore, Khademhosseini and colleagues have generated several techniques by using layer-by-layer assembly of electrostatic polymers to engineer surface properties. In their studies they used layer-by-layer deposition of ECM components such as hyaluronic acid (HA), fibronectin and collagen to generate co-cultures by sequentially patterning adhesive and non-adhesive regions.[18]

In addition to static cell-cell contact, dynamic studies of cellular interactions are important in understanding many biological processes such as stem cell homing and embryonic development. To engineer cell-cell contact dynamically, Hui and Bhatia have generated a device based on interdigitating silicon plates that could be seeded with different cell types and brought together in close proximity.[19] Furthermore, a technique based on micropatterned stencils can also be used to generate patterned co-cultures with dynamic control. In this approach stencils made of reversibly sealable parylene-C membranes, a biocompatible material, were sealed and peeled off from a variety of substrates such as PDMS, polystyrene and standard cell culture plates. Using parylene-C and the layer-by-layer technique, a system was developed in which one cell type could be sequentially exposed to various cell types.[20]

2.3. Microarrays for directing stem cell fates

Microarrays have been used extensively in basic biology research, screening diseases, drug discovery and toxicology for molecular profiling of samples at the DNA, RNA, and protein level. Due to the complexity of the factors affecting stem cell differentiation, it is essential to analyze the stem cell microenvironment in a high-throughput manner. The use of microarrays would make such analysis faster and cheaper since miniaturized experiments can be done in a rapid manner without the use of extensive reagents. Microarrays can be used for the identification of lineage specific markers expressed after the differentiation of stem cells into a particular lineage. Microarrays have been applied for identifying the regulatory and cell-fate signaling pathways of stem cell differentiation.[21] Microarrays have also been used to control the ES cell microenvironment. For example, Karp *et al.* used PEG microwells to control the shape and size of EBs by controlling the features of the microwells.[14] They found that the size of the EBs influences the eventual differentiation of the cells. In a similar experiment, Park *et al.* used PDMS stencils to control the size of ES cell aggregates. They found that aggregates with 100 μm diameter showed a greater expression of ectodermal markers while the aggregates with 500 μm diameter displayed increased expression of mesodermal and endodermal markers.[22] Thus initial size of the ES cell aggregate may play a role in ES cell differentiation.

Recently, robotic spotters that can be used to dispense nanoliters of fluid have been used to fabricate microarrays, in which cell–matrix interactions can be screened in a high-throughput manner. For example, synthetic biomaterial arrays have been fabricated to test the interaction of stem cells with various extracellular signals.[23] In this approach, thousands of polymeric materials were synthesized, and their effect on differentiation of human ES cells and human MSC was evaluated.[22] These interactions have led to unexpected cell–material interactions. This technology may be widely applicable in cell–microenvironment studies and in the identification of cues that induce desired cell responses. In addition to analyzing synthetic material libraries, the effect of natural ECM molecules on cell fate can be evaluated in a high-throughput manner. In one example, combinatorial matrices of various natural ECM proteins were tested for their ability to maintain the function of differentiated hepatocytes and to induce hepatic differentiation from murine ES cells.[24]

2.4. Microfluidic system for controlling stem cell fate

Microfluidic systems are becoming increasingly used in biological applications for manipulating small quantities of samples in a fast and low cost manner.[25] Microfluidic systems can provide a powerful tool to investigate the extracellular signals that regulate cell fate, because they can control cell-soluble factor interactions and be merged with high-throughput technologies to test many environmental factors simultaneously. The advantages include reduced consumption of samples and reagents, shorter analysis times, greater sensitivity, portability, and disposability.[26–29] These characteristics make microfluidic systems beneficial for the analysis of the stem cell microenvironment and for directing the stem cell fate. An example of the use of microfluidic systems in studying stem cell behavior was performed by Chung *et al.*[30] In their studies a gradient generating microfluidic device was used to test the effects of various concentrations of growth factors on the response of neural stem cell differentiation. Although the long term culture of cells in microchannels remains a challenge, a number of researchers are working on various problems associated with seeding, culturing and analyzing cells within microfluidic channels. Thus the use of these systems in analyzing stem cell behavior has the potential to contribute in developing optimized conditions for directed differentiation of stem cells.

3. Microscale Technologies for Engineering Complex Tissues Containing Different Cell Types and Vasculature

Living tissues are ensembles of different cell types embedded in complex and well-defined geometries and within a defined matrix that is unique to each tissue type. In tissue engineering, cells are cultured on degradable scaffolds that provide the physical and chemical cues to guide cellular differentiation and assembly into 3D tissues. The assembly of cells into tissues is a highly orchestrated set of events that requires time scales ranging from seconds to weeks and dimensions ranging from 0.0001 to 10 cm.[2] Conventional methods to fabricate artificial tissues rely on cell assembly which in some cases does not proceed to the degree that mimics native tissues, resulting in 3D constructs that lack the complexity associated with the architecture of tissues *in vivo*. Another major challenge of engineering tissues *in vitro* is lack of proper vascularization. Oxygen and other nutrients can only diffuse through a short distance before being consumed (a few hundred micrometers at most), which constraints the size of the engineered tissue. Several approaches resulting from microscale technologies provide new hope to overcome these challenges to build tissues with vascularized structures in a reproducible manner.

3.1. Microscale technologies for template-based cell assembly into 3D micro-tissues

Microscale tissues that mimic the *in vivo* tissue architecture and function can be obtained by inducing controlled cell aggregation. Self-assembled spheroids from single or multi-type of cells have been fabricated by several approaches, such as hanging drop,[31] spinner culture,[32] or by using non-adhesive substrates.[33] These approaches lack the ability to build tissues with the well-defined architectures. Microscale technologies can provide a solution to this challenge by enabling the fabrication of micro-templates for tissue formation. For instance, by using an approach that combines microcontact printing and micromachining, hepatocyte spheroids have been formed.[34] In other examples, microtissues with predefined microscale geometries have been achieved by directed self-assembly of cells on micro-molded non-adhesive agarose gels.[35] Cells were self-assembled into complex structures such as rods and honeycombs (Fig. 3). This study indicates that self-assembled microtissues can be generated from cell suspensions,

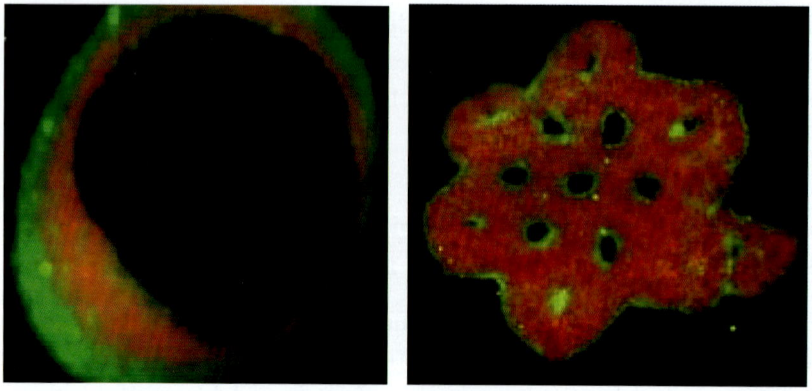

Fig. 3. Template-based self-assembly of cells into microtissues with prescribed structures (i.e. tori and honeycombs). Copyright © 2007 by the Federation of American Societies for Experimental Biology. Reprinted with permission from FASEB.[35]

without being limited to the conventional spheroidal structure, but with more complex and diverse shapes. In addition, template-based assembly of cells could be used to build microtissues containing multiple cell types, which are organized with specific geometries relative to each other. It is envisioned that the integration of template-based microscale technologies with biomaterials and microfluidic will enable the construction of more complex artificial tissues for therapeutic applications.

3.2. Scaffolds with micro- and nano-topography

It is known that topography induces change in morphology and motility of many cell types[36] by a process called "contact guidance." Contact guidance refers to the reactions of cells with the topography of their substratum.[37,38] In natural tissues the ECM provides the nanotopography for the cell. Mimicking the original tissue environment may provide avenues for generating improved tissue engineered constructs. To achieve this aim, scaffolds with micro and nanotopography have been generated by various means[39] including photolithography,[40] electrospinning,[41] and chemical etching.[42] Electrospinning is a relatively inexpensive technique used to generate highly porous nanofibrous scaffolds from natural or synthetic polymers such as collagen,[43] poly(lactic-co-glycolic acid) (PLGA), polycaprolactone (PCL) and poly(L-lactic acid) (PLLA). Cellular functions like adhesion, growth and proliferation are enhanced when cells are cultured on nanoscale electrospun scaffolds.[44] Nanofibrous scaffolds have more surface area and hence offer more binding sites for cells. Electrospinning also enables control of various parameters like fiber diameter, surface topology, porosity, and mechanical properties that can potentially affect the cell behavior.[44–46]

3.3. Microengineered hydrogels for tissue engineering

Hydrogels are networks of hydrophilic polymers that are attractive for tissue engineering since their physical (i.e. mechanical strength and biodegradability) and biological properties (i.e. the biocompatibility and resemblance to the natural ECM matrix) can be tailored to mimic tissues. Commonly-used hydrogels include natural hydrogels (i.e. collagen, hyaluronic acid, alginate), synthetic hydrogels (i.e. Poly(ethylene glycol)-diacrylate (PEGDA), poly(vinyl alcohol) (PVA), poly(lactic acid),[47] poly(lactic-co-glycolic acid) (PLGA)) and hybrid natural-synthetic hydrogels.[48] Photocrosslinkable hydrogels have been used for the encapsulation of various cells[49–51] as scaffolds for tissue engineering.[52–54] The merger of microengineered hydrogels and microfabrication techniques has significant potential to generate tissue constructs. Both "top-down" and "bottom-up" approaches have been used in using microengineered hydrogels for tissue engineering.[55] Top-down tissue engineering approach control the microscale features (i.e. shape and size) of relatively large pieces of hydrogels. Bottom-up approach refers to fabrication of tissue engineered constructs by the assembly of smaller building blocks.[55] In bottom-up approaches, functional units can be assembled in a modular approach to generate larger tissue structures. Small tissue building blocks are usually formed from microgels or cell aggregates. Patterning of the hydrogels in 2D can be done by soft lithography[56] and photopatterning.[57,58] An example of bottom-up tissue engineering is to pack rod-shaped collagen microgels seeded with HepG2 hepatocyte inside and endothelial cells on the surface within a bioreactor. The microgel-packed bioreactor exhibited interconnected channels between the microgel modules, where medium or blood can penetrate without mass transfer limitations.[59] In another example, cell-laden hydrogels with a defined shape were generated by micro-molding photocrosslinkable hydrogels.[49,57] Different cell types were encapsulated within microgels and assembled to generate larger 3D tissue structures with controlled architecture and cell-cell interactions (Fig. 4). Furthermore, to create 3D structures by a bottom-up approach, Bhatia and colleagues used a multilayer photopatterning platform by polymerizing two or more overlapping cell laden PEG structures to form a complex 3D structure.[60]

3.4. 3D tissue/organ printing

3D printing of cells and biomaterials has also been used to generate 3D tissue constructs. Tissue printing offers the ability to deposit cells and other biomaterials in

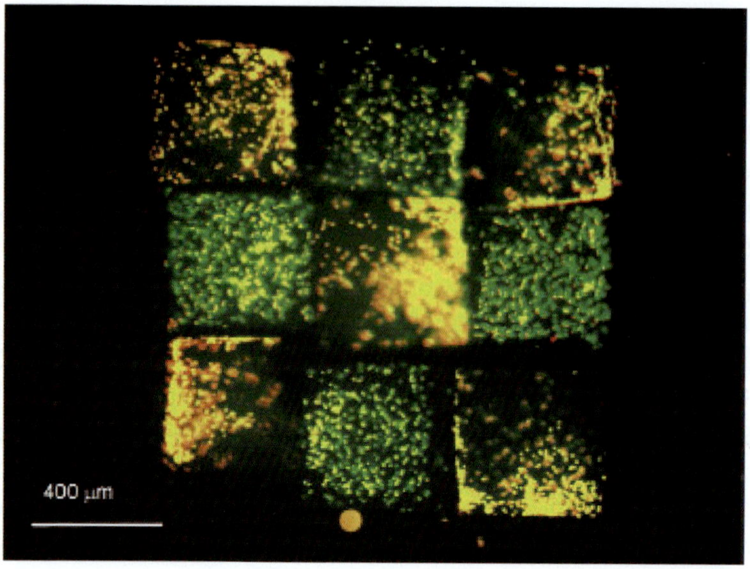

Fig. 4. Microgel arrangement and assembly. Rhodamine (red) and FITC (green) stained cells were encapsulated in separate HA microgels and subsequently arranged in an alternating checkerboard pattern.[57]

a rapid layer-by-layer fashion to create tissue structures. The most common conventional printing methods are laser printing and inkjet printing. Ringeisen *et al.* used laser printing to print pluripotent embryonal carcinoma cells. They found that cells printed onto a layer of hydrogel had greater than 95% viability and minimal single-strand DNA damage. Also the printed cells expressed microtubular associated protein 2 and myosin heavy chain protein after appropriate stimulus indicating successful muscular pathway differentiation.[61] Barron *et al.* used biological laser printing (BioLP) to print human osteosarcoma cells into a biopolymer matrix, and after six days of incubation, the printed cells showed 100% viability.[62] To print complicated structures like the vasculature, laser-guided direct writing (LGDW) has been used. For example, Nahmias *et al.* utilized LGDW to pattern human umbilical vein endothelial cells (HUVEC) in two and three dimensions with micrometer accuracy. Furthermore, they co-cultured the vascular structures with hepatocytes to generate tubular structures similar to hepatic sinusoid.[63,64] The main drawbacks of the laser printing are the heat generated in the process, which may affect the cells as well as the difficulties in scaling up the process for larger structures involving thousands of cells.

Boland and colleagues created complex cellular patterns and structures by automated and direct inkjet printing of primary embryonic hippocampal and cortical neurons. They also generated 3D cellular structures by alternate inkjet printing of NT2 cells and fibrin gels in a layer-by-layer fashion. Various analyses showed that the printed cells were healthy and displayed normal cellular functions. Piezoelectric-based droplet ejectors have also been used to achieve continuous or drop-on-demand ejection of the fluid. The ejector is harmless to sensitive fluids and biological samples can be ejected.[65,66] The non-contact piezoelectric-based ejector technology has been applied to fabricate protein microarrays,[67–69] biosensors, and cell-based assays.[70] Finally, acoustic-based non-contact printing is an emerging technology that can be used for 3D cell printing that overcomes some of the shortcomings of the previous technologies. The advantages of the acoustics-based printing are that it is devoid of heat, pressure and shear. Demirci *et al.* used acoustic-based printers to print picoliter droplets with single to few cells in each droplet. The ejected cells were printed with >90% viability across various cell types.[71]

The scope of the 3D tissue printing has widened to organ printing in recent years due to the advancement in the printing technology. 3D organ printing is an emerging science that tries to overcome the main challenges in tissue engineering, replicating the complexity of the tissues and providing vascular supply. Mironov *et al.* have shown that closely placed cell aggregates and embryonic heart mesenchymal (cushion tissue) fragments could fuse into ring and tube-like structures in a 3D gel[72] (Fig. 5). Despite these impressive results which show the potential of organ printing, several challenges, such as scalability and clogging, need to be addressed.

3.5. Microfluidics for engineering the vasculature

In vivo, cells reside in close proximity to blood vessels that supply tissues with nutrients and oxygen and remove waste products and carbon dioxide. The ability to create thick tissues is a major tissue engineering challenge, requiring the development of a vascular network for a suitable vascular supply. Although conventional techniques for scaffold fabrication, such as solvent casting and particulate leaching, cannot be used to fabricate scaffolds with controllable pore geometry, size, and interconnectivity, the ability to engineer more complex features such as a vasculature in the scaffolds is of interests. Microscale technologies have been used to construct tissue engineering scaffolds with desired microvasculature structures. Engineering vascularized tissue scaffolds have been realized by micromaching on silicon wafer[73] and soft lithography. For soft lithography, biocompatible polymers, such as PDMS, PLGA[74] and PGS[75] have been used respectively in replica molding to fabricate capillary networks. After being coated with fibronectin, these capillary networks can be seeded with endothelial cells to form blood vessels. Microfluidic devices have also been shown as a promising

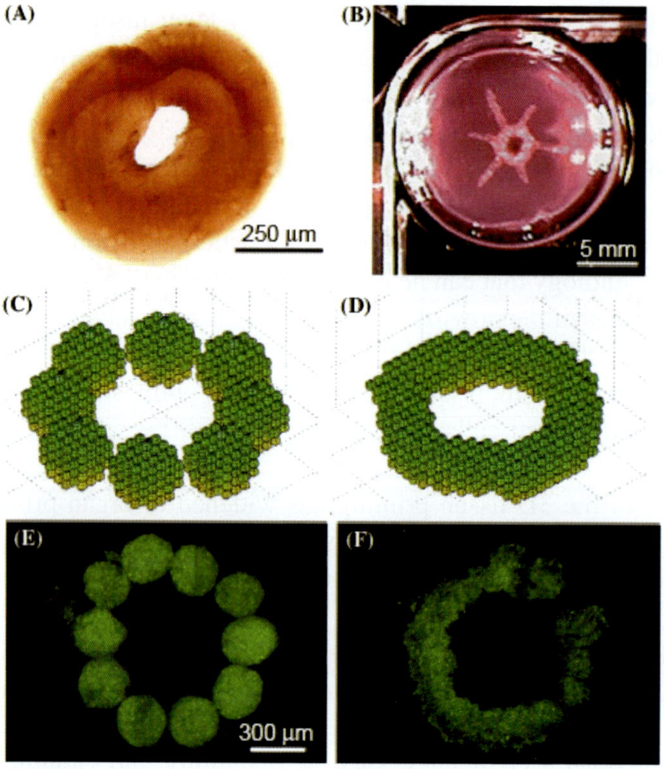

Fig. 5. Organ printing.[72] **(A)** Sequentially printed layers of collagen Type I gel. **(B)** Manually printed living tube with radial branches from the chick 27 stage Hamburger-Hamilton (HH) embryonic heart cushion tissue placed in 3D collagen Type 1 gel. Tube was formed as a result of fusion of three sequential rings. Every ring consists of 16–18 closed placed and fused embryonic cushion tissue explants. **(C** and **D)** Mathematical model of cell aggregate behavior when implanted in a 3D model gel. **(E** and **F)** Fusion of ten aggregates of Chinese Hamster Ovary cells implanted into RGD containing thermo-reversible gel and genetically labeled with green fluorescent protein: **(E)** before fusion and **(F)** final disc-like configuration after fusion.

tool to facilitate the exchange of nutrients and soluble factors in 3D tissue constructs. In microfluidics system, the controlled flow of fluids with minimal reagent consumption can be achieved within microscale channels in high-throughput manner. In a recent example, microfluidic channels from cell-laden hydrogels have been developed by using a soft lithographic technique.[51] Only those cells near the microfluidic channels remained viable after three days, demonstrating the importance of a perfused network of microchannels for delivering nutrients and oxygen to maintain cell viability in large hydrogels (Fig. 6). Cell-laden microfluidic

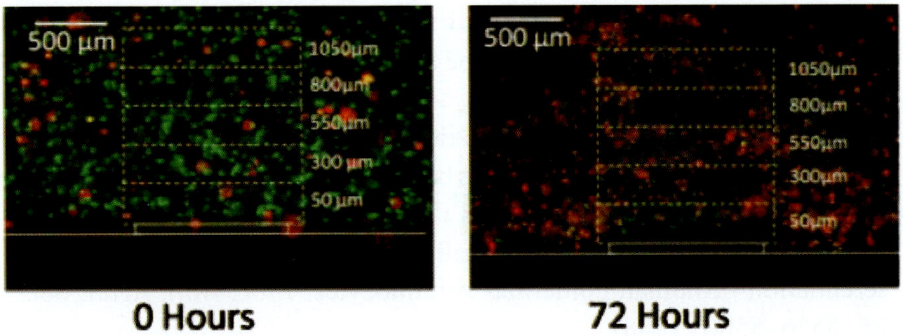

Fig. 6. Microfluidics for supplying nutrients to tissues.[51] The majority of cells were viable (as stained green) upon initial device fabrication; however only those cells near the microfluidic channels remained viable after three days.

hydrogels can also be scaled up by stacking the biomimetic vascular patterns to generate multi-layer vascularization in multiple discrete planes.

4. Conclusion

The merger of biomaterials and microscale technologies and their applications in tissue engineering offer new opportunities to overcome the challenges faced by existing technologies to fabricate scaffolds and direct stem cell differentiation. In this review, the various applications of microscale technologies have been illustrated in controlling the stem cell fate and building complex artificial tissues with well-controlled and vascularized structures. It is believed that with the rapid growth of this burgeoning research field, microscale technologies will transform the conventional tissue engineering approaches and greatly contribute to the therapeutic potential of tissue engineering.

References

1. Langer, R. and Vacanti, J.P. Tissue engineering. *Science* **260**, 920–926 (1993).
2. Griffith, L.G. and Naughton, G. Tissue engineering — current challenges and expanding opportunities. *Science* **295**, 1009–1014 (2002).
3. Whitesides, G.M., Ostuni, E., Takayama, S., Jiang, X. and Ingber, D.E. Soft lithography in biology and biochemistry. *Annu. Rev. Biomed. Eng.* **3**, 335–373 (2001).
4. Czyz, J. and Wobus, A. Embryonic stem cell differentiation: the role of extracellular factors. *Differentiation* **68**, 167–174 (2001).

5. Bhadriraju, K. and Hansen, L.K. Extracellular matrix- and cytoskeleton-dependent changes in cell shape and stiffness. *Exp. Cell. Res.* **278**, 92–100(2002).
6. Huang, S. and Ingber, D.E. Shape-dependent control of cell growth, differentiation, and apoptosis: switching between attractors in cell regulatory networks. *Exp. Cell Res.* **261**, 91–103 (2000).
7. Watt, F.M., Jordan, P.W. and O'Neill, C.H. Cell shape controls terminal differentiation of human epidermal keratinocytes. *Proc. Natl. Acad. Sci. USA* **85**, 5576–5580 (1988).
8. Chen, C.S., Mrksich, M., Huang, S., Whitesides, G.M. and Ingber, D.E. Geometric control of cell life and death. *Science* **276**, 1425–1428 (1997).
9. Le Beyec, J., Xu, R., Lee, S.Y., Nelson, C.M., Rizki, A., Alcaraz, J. and Bissell, M.J. Cell shape regulates global histone acetylation in human mammary epithelial cells. *Exp. Cell Res.* **313**, 3066–3075 (2007).
10. McBeath, R., Pirone, D.M., Nelson, C.M., Bhadriraju, K. and Chen, C.S. Cell shape, cytoskeletal tension, and RhoA regulate stem cell lineage commitment. *Dev. Cell* **6**, 483–495 (2004).
11. Singhvi, R., Kumar, A., Lopez, G.P., Stephanopoulos, G.N., Wang, D.I., Whitesides, G.M. and Ingber, D.E. Engineering cell shape and function. *Science* **264**, 696–698 (1994).
12. Kurosawa, H. Methods for inducing embryoid body formation: *in vitro* differentiation system of embryonic stem cells. *J. Biosci. Bioeng.* **103**, 389–398 (2007).
13. Wang, X.L., Wang, C.Y., Yu, X.J., Zhao, Y.S., Li, J., Duan, C.M. and Guo, X.M. Scalable production of embryoid bodies with the rotay cell culture system. *Sheng Li Xue Bao* **57**, 486–492 (2005).
14. Karp, J.M., Yeh, J., Eng, G., Fukuda, J., Blumling, J., Suh, K.Y., Cheng, J., Mahdavi, A., Borenstein, J., Langer, R. and Khademhosseini, A. Controlling size, shape and homogeneity of embryoid bodies using poly(ethylene glycol) microwells. *Lab Chip* **7**, 786–794 (2007).
15. Khademhosseini, A., Yeh, J., Eng, G., Karp, J., Kaji, H., Borenstein, J., Farokhzad, O.C. and Langer, R. Cell docking inside microwells within reversibly sealed microfluidic channels for fabricating multiphenotype cell arrays. *Lab Chip* **5**, 1380–1386 (2005).
16. Bhatia, S.N., Balis, U.J., Yarmush, M.L. and Toner, M. Effect of cell–cell interactions in preservation of cellular phenotype: co-cultivation of hepatocytes and non-parenchymal cells. *FASEB J.* **13**, 1883–1900 (1999).
17. Bhatia, S.N., Yarmush, M.L. and Toner, M. Controlling cell interactions by micropatterning in co-cultures: hepatocytes and 3T3 fibroblasts. *J. Biomed. Mater. Res.* **34**, 189–199 (1997).

18. Fukuda, J., Khademhosseini, A., Yeh, J., Eng, G., Cheng, J., Farokhzad, O.C. and Langer, R. Micropatterned cell co-cultures using layer-by-layer deposition of extracellular matrix components. *Biomaterials* **27**, 1479–1486 (2006).
19. Hui, E.E. and Bhatia, S.N. Micromechanical control of cell–cell interactions. *Proc. Natl. Acad. Sci. USA* **104**, 5722–5726 (2007).
20. Wright, D., Rajalingam, B., Selvarasah, S., Dokmeci, M.R. and Khademhosseini, A. Generation of static and dynamic patterned co-cultures using microfabricated parylene-C stencils. *Lab Chip* **7**, 1272–1279 (2007).
21. Chang, H.Y., Thomson, J.A. and Chen, X. Microarray analysis of stem cells and differentiation. *Meth. Enzymol.* **420**, 225–254 (2006).
22. Park, J., Cho, C.H., Parashurama, N., Li, Y., Berthiaume, F., Toner, M., Tilles, A.W. and Yarmush, M.L. Microfabrication-based modulation of embryonic stem cell differentiation. *Lab Chip* **7**, 1018–1028 (2007).
23. Anderson, D.G., Levenberg, S. and Langer, R. Nanoliter-scale synthesis of arrayed biomaterials and application to human embryonic stem cells. *Nat. Biotechnol.* **22**, 863–866 (2004).
24. Flaim, C.J., Chien, S. and Bhatia, S.N. An extracellular matrix microarray for probing cellular differentiation. *Nat. Methods* **2**, 119–125 (2005).
25. Haeberle, S. and Zengerle, R. Microfluidic platforms for lab-on-a-chip applications. *Lab Chip* **7**, 1094–1110 (2007).
26. Manz, A., Harrison, D.J., Verpoorte, E.M.J., Fettinger, J.C., Paulus, A., Ludi, H. and Widmer, H.M. Planar chips technology for miniaturization and integration of separation techniques into monitoring systems: capillary electrophoresis on a chip. *J. Chromatogr.* **593**, 253–258 (1992).
27. Atencia, J. and Beebe, D.J. Controlled microfluidic interfaces. *Nature* **437**, 648–655 (2005).
28. Ramsey, J.M. The burgeoning power of the shrinking laboratory. *Nat. Biotechnol.* **17**, 1061–1062 (1999).
29. Whitesides, G.M. and Stroock, A.D. Flexible methods for mircofluidics. *Phys. Today* **54**, 42–48 (2001).
30. Chung, B.G., Flanagan, L.A., Rhee, S.W., Schwartz, P.H., Lee, A.P., Monuki, E.S. and Jeon, N.L. Human neural stem cell growth and differentiation in a gradient-generating microfluidic device. *Lab Chip* **5**, 401–406 (2005).
31. Del Duca, D., Werbowetski, T. and Del Maestro, R.F. Spheroid preparation from hanging drops: characterization of a model of brain tumor invasion. *J. Neurooncol.* **67**, 295–303 (2004).
32. Lee, K.W., Lee, S.K., Joh, J.W., Kim, S.J., Lee, B.B., Kim, K.W. and Lee, K.U. Influence of pancreatic islets on spheroid formation and functions of hepatocytes in hepatocyte-pancreatic islet spheroid culture. *Tissue Eng.* **10**, 965–977 (2004).

33. Du, Y., Chia, S.M., Han, R., Chang, S., Tang, H. and Yu, H. 3D hepatocyte monolayer on hybrid RGD/galactose substratum. *Biomaterials* **27**, 5669–5680 (2006).
34. Fukuda, J., Mizumoto, H., Nakazawa, K., Kajiwara, T. and Funatsu, K. Hepatocyte organoid culture in elliptic hollow fibers to develop a hybrid artificial liver. *Int. J. Artif. Organs* **27**, 1091–1099 (2004).
35. Dean, D.M., Napolitano, A.P., Youssef, J. and Morgan, J.R. Rods, tori, and honeycombs: the directed self-assembly of microtissues with prescribed microscale geometries. *FASEB J.*, **21**, 4005–4012 (2007).
36. Chua, K.N., Chai, C., Lee, P.C., Ramakrishna, S., Leong, K.W. and Mao, H.Q. Functional nanofiber scaffolds with different spacers modulate adhesion and expansion of cryopreserved umbilical cord blood hematopoietic stem/progenitor cells. *Exp. Hematol.* **35**, 771–781 (2007).
37. Oakley, C. and Brunette, D.M. The sequence of alignment of microtubules, focal contacts and actin filaments in fibroblasts spreading on smooth and grooved titanium substrata. *J. Cell Sci.* **106**(Pt 1), 343–354 (1993).
38. Peterson, E.T. and Papautsky, I. Microtextured polydimethylsiloxane substrates for culturing mesenchymal stem cells. *Methods Mol. Biol.* **321**, 179–197 (2006).
39. Borenstein, J.T., Weinberg, E.J., Orrick, B.K., Sundback, C., Kaazempur-Mofrad, M.R. and Vacanti, J.P. Microfabrication of three-dimensional engineered scaffolds. *Tissue Eng.* **13**, 1837–1844 (2007).
40. Wang, G.J., Hsu, Y.F., Hsu, S.H. and Horng, R.H. JSR photolithography based microvessel scaffold fabrication and cell seeding. *Biomed. Microdev.* **8**, 17–23 (2006).
41. Yoshimoto, H., Shin, Y.M., Terai, H. and Vacanti, J.P. A biodegradable nanofiber scaffold by electrospinning and its potential for bone tissue engineering. *Biomaterials* **24**, 2077–2082 (2003).
42. Wang, Z., Hu, H., Wang, Y., Wang, Y., Wu, Q., Liu, L. and Chen, G. Fabrication of poly(3-hydroxybutyrate-co-3-hydroxyhexanoate) (PHBHHx) microstructures using soft lithography for scaffold applications. *Biomaterials* **27**, 2550–2557 (2006).
43. Matthews, J.A., Wnek, G.E., Simpson, D.G. and Bowlin, G.L. Electrospinning of collagen nanofibers. *Biomacromolecules* **3**, 232–238 (2002).
44. Boudriot, U., Dersch, R., Greiner, A. and Wendorff, J.H. Electrospinning approaches toward scaffold engineering — a brief overview. *Artif. Organs* **30**, 785–792 (2006).
45. Nair, L.S., Bhattacharyya, S. and Laurencin, C.T. Development of novel tissue engineering scaffolds via electrospinning. *Expert Opin. Biol. Ther.* **4**, 659–668 (2004).

46. Pham, Q.P., Sharma, U. and Mikos, A.G. Electrospinning of polymeric nanofibers for tissue engineering applications: a review. *Tissue Eng.* **12**, 1197–1211 (2006).
47. Ivanov, I., Schaab, C., Planitzer, S., Teichmann, U., Machl, A., Theml, S., Meier-Ewert, S., Seizinger, B. and Loferer, H. DNA microarray technology and antimicrobial drug discovery. *Pharmacogenomics* **1**, 169–178 (2000).
48. Flores-Ramirez, N., Elizalde-Pena, E.A., Vasquez-Garcia, S.R., Gonzalez-Hernandez, J., Martinez-Ruvalcaba, A., Sanchez, I.C., Luna-Barcenas, G. and Gupta, R.B. Characterization and degradation of functionalized chitosan with glycidyl methacrylate. *J. Biomater. Sci. Polym. Ed.* **16**, 473–488 (2005).
49. Yeh, J., Ling, Y., Karp, J.M., Gantz, J., Chandawarkar, A., Eng, G., Blumling, J., 3rd, Langer, R. and Khademhosseini, A. Micromolding of shape-controlled, harvestable cell-laden hydrogels. *Biomaterials* **27**, 5391–5398 (2006).
50. Elisseeff, J., McIntosh, W., Anseth, K., Riley, S., Ragan, P. and Langer, R. Photoencapsulation of chondrocytes in poly(ethylene oxide)-based semi-interpenetrating networks. *J. Biomed. Mater. Res.* **51**, 164–171 (2000).
51. Ling, Y., Rubin, J., Deng, Y., Huang, C., Demirci, U., Karp, J.M. and Khademhosseini, A. A cell-laden microfluidic hydrogel. *Lab Chip* **7**, 756–762 (2007).
52. Jager, M., Degistirici, O., Knipper, A., Fischer, J., Sager, M. and Krauspe, R. Bone healing and migration of cord blood derived stem cells into a critical size femoral defect after xenotransplantation. *J. Bone Miner Res.* **22**, 1224–1233 (2007).
53. Hahn, M.S., Teply, B.A., Stevens, M.M., Zeitels, S.M. and Langer, R. Collagen composite hydrogels for vocal fold lamina propria restoration. *Biomaterials* **27**, 1104–1109 (2006).
54. Fedorovich, N.E., Alblas, J., de Wijn, J.R., Hennink, W.E., Verbout, A.J. and Dhert, W.J. Hydrogels as extracellular matrices for skeletal tissue engineering: state-of-the-art and novel application in organ printing. *Tissue Eng.* **13**, 1905–1925 (2007).
55. Khademhosseini, A. and Langer, R. Microengineered hydrogels for tissue engineering. *Biomaterials* **28**, 5087–5092 (2007).
56. Suh, K.Y., Seong, J., Khademhosseini, A., Laibinis, P.E. and Langer, R. A simple soft lithographic route to fabrication of poly(ethylene glycol) microstructures for protein and cell patterning. *Biomaterials* **25**, 557–563 (2004).
57. Khademhosseini, A., Eng, G., Yeh, J., Fukuda, J., Blumling, J., 3rd, Langer, R. and Burdick, J.A. Micromolding of photocrosslinkable hyaluronic acid for cell encapsulation and entrapment. *J. Biomed. Mater. Res. A* **79**, 522–532 (2006).

58. Hahn, M.S., Taite, L.J., Moon, J.J., Rowland, M.C., Ruffino, K.A. and West, J.L. Photolithographic patterning of polyethylene glycol hydrogels. *Biomaterials* **27**, 2519–2524 (2006).
59. Uludag, H., De Vos, P. and Tresco, P.A. Technology of mammalian cell encapsulation. *Adv. Drug Deliv. Rev.* **42**, 29–64 (2000).
60. Liu Tsang, V., Chen, A.A., Cho, L.M., Jadin, K.D., Sah, R.L., DeLong, S., West, J.L. and Bhatia, S.N. Fabrication of 3D hepatic tissues by additive photopatterning of cellular hydrogels. *FASEB J.* **21**, 790–801 (2007).
61. Ringeisen, B.R., Kim, H., Barron, J.A., Krizman, D.B., Chrisey, D.B., Jackman, S., Auyeung, R.Y. and Spargo, B.J. Laser printing of pluripotent embryonal carcinoma cells. *Tissue Eng.* **10**, 483–491 (2004).
62. Barron, J.A., Krizman, D.B. and Ringeisen, B.R. Laser printing of single cells: statistical analysis, cell viability, and stress. *Ann. Biomed. Eng.* **33**, 121–130 (2005).
63. Nahmias, Y., Schwartz, R.E., Verfaillie, C.M. and Odde, D.J. Laser-guided direct writing for three-dimensional tissue engineering. *Biotechnol. Bioeng.* **92**, 129–136 (2005).
64. Nahmias, Y. and Odde, D.J. Micropatterning of living cells by laser-guided direct writing: application to fabrication of hepatic-endothelial sinusoid-like structures. *Nat. Protoc.* **1**, 2288–2296 (2006).
65. Percin, G. and Khuri-Yakub, B.T. Piezoelectrically actuated flextensional micromachined ultrasound droplet ejectors. *IEEE Trans. Ultrason. Ferroelectr. Freq. Control* **49**, 756–766 (2002).
66. Percin, G. and Khuri-Yakub, B.T. Piezoelectrically actuated flextensional micromachined ultrasound transducers. *Ultrasonics* **40**, 441–448 (2002).
67. Delehanty, J.B. Printing functional protein microarrays using piezoelectric capillaries. *Methods Mol. Biol.* **264**, 135–143 (2004).
68. Delehanty, J.B. and Ligler, F.S. Method for printing functional protein microarrays. *Biotechniques* **34**, 380–385 (2003).
69. Hsieh, H.B., Fitch, J., White, D., Torres, F., Roy, J., Matusiak, R., Krivacic, B., Kowalski, B., Bruce, R. and Elrod, S. Ultra-high-throughput microarray generation and liquid dispensing using multiple disposable piezoelectric ejectors. *J. Biomol. Screen* **9**, 85–94 (2004).
70. Sumerel, J., Lewis, J., Doraiswamy, A., Deravi, L.F., Sewell, S.L., Gerdon, A.E., Wright, D.W. and Narayan, R.J. Piezoelectric ink jet processing of materials for medical and biological applications. *Biotechnol. J.* **1**, 976–987 (2006).
71. Demirci, U. and Montesano, G. Single cell epitaxy by acoustic picolitre droplets. *Lab Chip* **7**, 1139–1145 (2007).

72. Mironov, V., Boland, T., Trusk, T., Forgacs, G. and Markwald, R.R. Organ printing: computer-aided jet-based 3D tissue engineering. *Trends Biotechnol.* **21**, 157–161 (2003).
73. Kaihara, S., Borenstein, J., Koka, R., Lalan, S., Ochoa, E.R., Ravens, M., Pien, H., Cunningham, B. and Vacanti, J.P. Silicon micromachining to tissue engineer branched vascular channels for liver fabrication. *Tissue Eng.* **6**, 105–117 (2000).
74. Vozzi, G., Flaim, C., Ahluwalia, A. and Bhatia, S. Fabrication of PLGA scaffolds using soft lithography and microsyringe deposition. *Biomaterials* **24**, 2533–2540 (2003).
75. King, K., Wang, C., Kaazempur-Mofrad, M., Vacanti, J. and Borenstein, J. Biodegradable microfluidics. *Adv. Mater.* **16**, 2007–2012 (2004).

Part IV

Non-Invasive Methods to Monitor Tissue Re-Modelling

Chapter 18

Biosensors

Tony Cass

Abstract

Increasingly biological sciences are being built on a quantitative foundation, based upon our ability to accurately determine the temporal and spatial variations in the concentration of key molecules. It is this quantitative analytical data that provides the testable basis for building hypotheses about biological systems. Although many, diverse analytical techniques have been used to collect quantitative data these are often destructive of the system being analysed. Biosensors by contrast promise the ability to measure selected molecules continuously and in real-time with good spatial resolution. They use specific molecular recognition at the surface of a transducer such that an electrical signal is generated in proportion to the concentration of the target analyte. Many different molecular recognition reagents have been exploited in this respect and include enzymes, binding proteins and nucleic acids, whilst electrochemical, optical and mass sensitive signal transduction devices have been used to generate the electrical signal (typically a voltage or current).

Using biosensors *in vivo* presents additional challenges over and above simply relating signal to concentration. Biocompatibility is an ever-present issue and includes both the effect of the biological matrix on the sensor as well as sensor components on the biology. In this chapter a review of the different sensing modes is presented and their potential applicability to the monitoring of cells, tissue and tissue constructs is presented.

Keywords: Electrochemical; Optical; Biocompatibility; Molecular Recognition; Mass Transport.

Outline

1. Introduction to Sensor Technology
2. The Importance of Mass Transport in Sensor Performance

3. Electrochemical Biosensors
   3.1. Potentiometric sensors
   3.2. Amperometric sensors
   3.3. Impedance sensors
4. Optical Biosensors
   4.1. Fluorescence biosensors
   4.2. Integrated optical devices
   4.3. Plasmonic nanomaterials
5. Mass Sensors
6. Cell Sensing Strategies
7. Tissue Sensing Strategies
8. Conclusions and Outlook
   References

1. Introduction to Sensor Technology

There are many bioanalytical tools that can be used to assess the physiology of stem cells. At the level of transcription, DNA microarrays offer a global analysis of gene expression,[1] whilst proteomics has been used to elucidate changes in cell signalling pathways that are linked to the transition from stem to progenitor status.[2] Combining the two "omics" can be even more informative.[3] At a less global level identification and separation of stem cells makes use of surface markers such as $CD34^+$ for haematopoetic stem cells.

Whilst transcriptomics and proteomics offer a huge level of detail they are unsuited to *in situ* real-time analysis. Similarly cell surface markers, although widely used to identify stem cells in mixed populations, do not yield insights into the metabolic state of the cells.

The production of clinically useful quantities of stem cells of high quality and in a way that lends itself to good manufacturing practise requires that they be expanded under controlled conditions and a corollary of this is the need for real-time measurement of critical culture parameters in a spatially resolved manner. Sensors, particularly chemical sensors and biosensors offer an answer to this need. The principle of chemical sensors is familiar from the pH electrode, a device that converts chemical information, the concentration (strictly activity) of protons, into an electrical signal, voltage. The pH electrode achieves this through a proton specific and sensitive surface and is an example of an electrochemical sensor, more precisely a potentiometric sensor. Few other analytes are similarly amenable to sensing in such a straightforward fashion and for most molecules of interest more complex devices are necessary. This complexity is manifest in several ways; the specific recognition of the analyte at the sensor surface, the transduction of this recognition into an

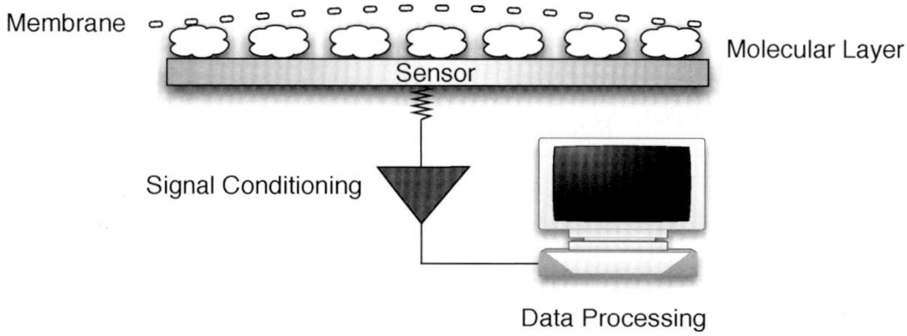

Fig. 1. A schematic diagram of a generic biosensor. The membrane is not present in all biosensor formats.

electrical signal, minimising background (i.e. signals arising from species other than the target analyte resulting in an over-estimate of concentration) and avoiding loss of signal through degradation or fouling of the sensor (resulting in an underestimate of concentration). Figure 1 shows a schematic diagram of a typical biosensor where the molecular recognition is mediated by a biomolecule and for many target analytes it is this reaction that gives the sensor its specificity.

The major issues in applying biosensor technology to monitoring stem cells in culture are several-fold:

Biofouling — Adsorption of proteins and cells to the surface of the sensor can substantially lower its response resulting in drift and at worst failure of the device.

Location — The heterogeneity of stem cells growing in two– and especially three–dimensional cultures means that multiple distributed sensors will be necessary to accurately monitor the total environment.

Selection of analytes — Many factors are known to affect the expansion, differentiation and death of stem cells in culture. Some of these are basic chemical parameters (pH, ammonium ions, pCO_2), others are major components of the medium (glucose, glutamine), whilst others include trace constituents (growth factors). Their concentrations range from mM to nM hence different sensing mechanisms are needed and the dynamics of their concentrations can change on very different timescales.

Minaturisation — The multiplicity of analytes and the desirability of spatially defined measurements, as well as sensor redundancy (to improve reliability) means that micro- or nanosensors will be necessary.

Toxicity — The effect of the sensors or sensor components, especially those leaching from the device that affect cell viability.

In designing biosensors there are many choices in both the recognition chemistry and the signal transduction mechanism. These will determine the overall format as

Table 1. Some examples of combinations of recognition molecules and sensing mechanisms for common analytes.

Analyte	Recognition Molecule	Sensing Mechanism
Oxygen	None	Amperometry
pH	None	Potentiometry
pH	Dyes	Fluorescence
Ammonia	Ionophore	Potentiometry
Glucose	Enzyme	Amperometry
Glutamine	Enzyme	Potentiometry
Glutamine	Binding Protein	Fluorescence
Lactate	Enzyme	Amperometry
DNA	DNA	Impedance
DNA	DNA	Raman Scattering
Protein	Antibody	Surface Plasmon Resonance

well as additional factors such as anti-fouling coatings, control of mass transport and the dynamic concentration range being sensed. Sensors can be broadly grouped into consuming and non-consuming devices, in consuming sensors the signal is generated by the conversion of the target analyte to another molecule whilst non-consuming sensors generate a signal from the analyte binding to the sensor surface. Where biomolecules are involved in the sensing process these two formats correspond to catalytic and affinity sensors respectively. Table 1 is a non-exhaustive collection of sensors grouped by analyte, format and the underlying physical sensor.

Two important parameters for sensors (or indeed any analytical method) are limit of detection (LoD) and dynamic range. The former is defined as the lowest measurable concentration, whilst the latter is the range of concentrations over which measurements can be made with the required degree of accuracy. The LoD is usually determined by the noise floor or by the occurrence of non-specific interactions; it determines the lower bound of the dynamic range. Site saturation effects often determine the upper bound where biomolecules are involved.

2. The Importance of Mass Transport in Sensor Performance

Biosensors by their very nature depend upon reactions occurring at the solid-liquid interface where the molecular recognition reaction occurs. Initially the

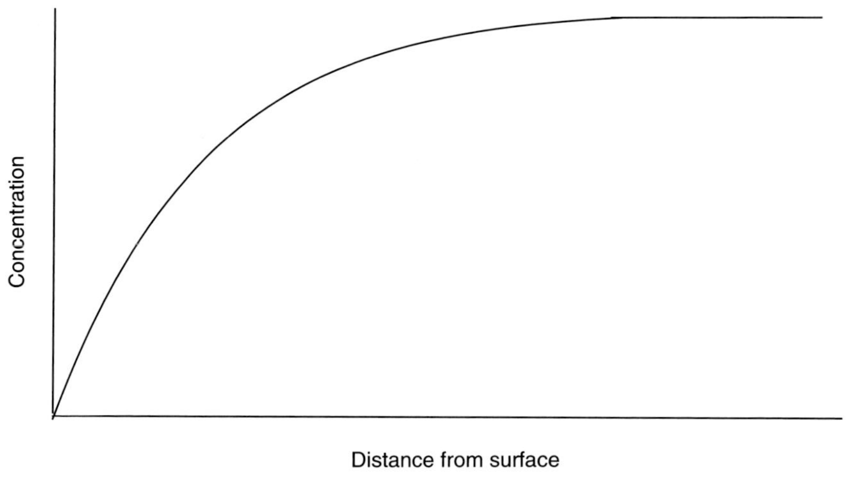

Fig. 2. The concentration profile of analyte close to the surface of a sensor. The steepness of the initial region of the curve depends upon the balance between mass transport and the reaction at the sensor surface.

concentration of analyte at the surface will be the same as in bulk solution, however once the interfacial reaction starts the surface concentration will fall and this sets up a concentration gradient between the bulk and the surface down which molecules now diffuse. There will therefore be a depleted layer (often referred to as the Nernst or diffusion layer) that grows in thickness with time (Fig. 2).

Where the surface reaction is purely the formation of a complex between the analyte and the reagent on the surface then eventually an equilibrium is reached where the fraction of binding sites on the surface that are occupied by the analyte (Γ) is given by the Langmuir Isotherm:

$$\Gamma = \frac{[c]_\infty}{[c]_\infty + K_D} \tag{1}$$

where K_D is the dissociation constant of the complex. Typically it is not Γ that is measured but the signal from the transducer, which is proportional to fractional occupancy. If the surface reaction is allowed to reach equilibrium then the concentration can be determined from Eq. (1). If measurements are made at times shorter than those required to reach equilibrium, i.e. the transient response is measured, then the rate of change in signal (i.e. Γ) can be controlled by either the rate of reaction at the surface or the rate of transport to the surface. These mass transport kinetics will depend on a host of factors including the sensor size and geometry, convection if present and the viscosity of the solution in which the sensor is placed.

Where the surface reaction is catalytic then equilibrium is never reached, rather a steady state is established in which the rates of mass transport and catalysis balance, with one rate controlling. To a degree it is possible to design biosensors through geometry, size, choice of biological material, immobilisation method and the presence of other components such as membranes.

3. Electrochemical Biosensors

The earliest chemical sensors were based upon either the flow of current at an electrified interface that resulted from molecules being oxidised or reduced, or the establishment of an interfacial potential due to different concentrations (strictly activities) of ions either side of the interface. The former are referred to as amperometric sensors whilst the latter are called potentiometric sensors. More recently a third class of electrochemical sensors, impedimetric sensors, measure changes in the interfacial impedance have also been established. Examples of the three classes of sensors include those for dissolved oxygen (amperometric), pH (potentiometric) and cell density (impedimetric). Although all three exploit the properties of an electrified interface their operating principles are quite different and their responses to chemical or biological species show distinct characteristics. In particular potentiometric and impedimetric sensors are equilibrium devices whilst amperometric sensors are inherently kinetic. This means that whilst the first two are directly sensitive to the concentration of analyte, the last measures the flux of material to the surface. A second important distinction is that whilst the first two do not effects a chemical transformation of the analyte, amperometric sensors by their very nature consume the analyte transforming it to a different product.

3.1. Potentiometric sensors

These devices are sensitive to charged species and the Nernst Equation gives the relationship between the potential difference and the analyte concentration in the sample:

$$\left|\Delta E\right| = \frac{RT}{nF} \cdot \ln[analyte]_{sample} \tag{2}$$

where n is the number of charges on the ion, R, T and F are the gas constant, absolute temperature and the Faraday, respectively. Potentials are measured relative to a reference electrode Inserting the relevant values for the constants into Eq. (1) gives potential change of 57 mV per decade change in concentration at 37°C. Membrane electrodes of this type, whilst easy to make and versatile suffer from problems of mechanical fragility and large (mm) sizes so they are being replaced by solid state alternatives.

The most ubiquitous of potentiometric sensors is the pH electrode, often encountered with a glass membrane and an integrated reference electrode, showing typical Nernstian responses of 57 mV/pH. Alternative structures that are more suited to miniaturisation are coated metal wires[4] and the ion sensitive field effect transistors (ISFETs).[5] The former have surface coatings that can be reversibly protonated and so change the potential on the electrode surface, examples include iridium oxide (IrOx)[6] or conducting organic polymers where protonation causes a shift in redox potential.

ISFETs are modified silicon devices that use a sensitised layer to modulate the flow of electrons between a source and drain in a semiconductor structure. The sensitised layer is often a metal oxide such as aluminium oxide and Bergveld, who pioneered them, has reviewed these devices in an article on "Isfetology".[7] Other solid-state potentiometric sensors based on silicon structures include the Electrolyte-Insulator-Semiconductor (EIS) and Light Activated Potentiometric Sensor (LAPS). EIS devices are in essence capacitors where the layer in contact with the electrolyte and the semiconductor act as the two plates and the insulator is the dielectric between them. Changes in the charge on the electrolyte layer alter the capacitance of the device. LAPS devices use visible light to bring about charge separation in the semiconductor layer that senses changes in ionic composition.[5]

Solid-state potentiometric sensors often show sub-Nernstian responses, i.e. potential changes of less than 57 mV/pH unit due to the complex relationship between protonation stoichiometry and voltage. Other biologically relevant analytes that can directly sensed using potentiometric sensors include NH_4^+, Na^+, K^+ and the Nernstian repose of Eq. (2) gives these sensors a large dynamic range, although in practice biological variation is unlikely to be more than one or two decades.

Classically potentiometric sensors have been used to quantify small ions and extending the analytical range of compounds that can be sensed has developed in two ways; either by using an enzyme catalysed reaction to generate of consume ions (commonly but not exclusively protons) or through a binding reaction that changes the surface charge density on a solid-state device. In this latter instance it is macromolecules (usually DNA or proteins) that are sensed.

As an example of the former the hydrolysis of the essential amino acid glutamine by the enzyme glutaminase causes a decrease in pH:

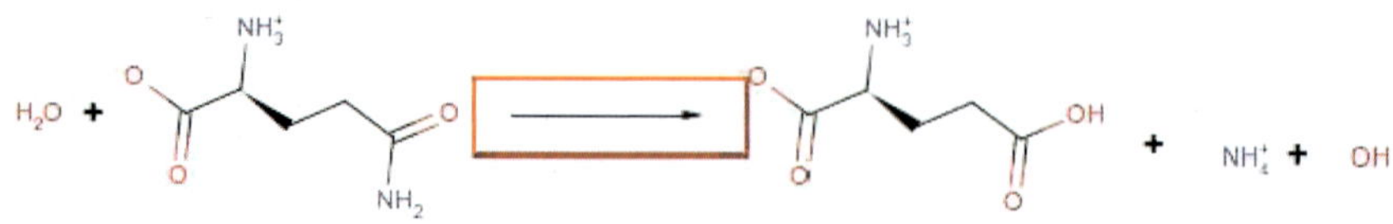

A pH sensor coated with glutaminase will therefore show a response to glutamine with the same 57 mV/decade response as for protons. Where the pH sensor is an ISFET these devices are often referred to as EnzFETs.[8,9] Irrespective of how the pH is sensed there are several circumstances that can complicate the relationship between potential and glutamine concentration; in particular the relationship will depend on the starting pH and the buffer capacity of the matrix (i.e. the solution in which the measurement is being made). If either or both of these are different from the calibration solution then the difference needs to be compensated for. Whilst changes in pH can be corrected for by using a separate, uncoated, pH sensor buffer capacity is more challenging. Although discussed in the context of glutamine sensing, the same would be true for any enzyme-linked system that relies on pH measurement. Table 2 gives some examples of enzyme substrates that have been measured with potentiometric sensors.

Table 2. Examples of analytes that can be measured with potentiometric sensors.

Analyte	Enzyme	Ion sensed
Urea	Urease	H^+ NH_4^+
Acetylcholine	Acetylcholinesterase	H^+
Organophosphorous compounds[a]	Acetylcholinesterase	H^+
Glucose[b]	Glucose oxidase	H^+
Glutamine	Glutaminase	H^+
Creatinine	Creatininase	H^+
DNA	DNA polymerase	H^+
Penicillin	β-lactamase	H^+
Hydrogen peroxide[c]	Peroxidase	F^-
Triglycerides	Lipase	H^+

Notes: [a] These compounds inhibit acetylcholinesterase. [b] Glucose is oxidised to gluconolactone in a reaction that does not produce or consume protons, it is the subsequent spontaneous hydrolysis of gluconolactone to gluconic acid that causes a pH change. [c] Peroxidase reacts with hydrogen peroxide to oxidise a fluorine containing substrate and release fluoride.

Mass transport effects on enzyme potentiometric sensors have been simulated using a coupled reaction-diffusion model assuming planar diffusion to the enzyme surface.[10] Whilst this is appropriate for large sensors, as the size of the sensor is decreased there is a change from linear to hemi-spherical diffusion and this can dramatically shift the balance between mass transport and enzyme catalysis limitation. Although not likely with conventional potentiometric biosensors the effect may be important with the coated wire and EnzFET devices if the sensing surface is less than about 20 μm in diameter and needs to be borne in mind when considering miniaturisation.[11] Ideally the sensor response should be limited by mass transport (under conditions where changes in convection do not occur) as this reduces the sensitivity to factors that affect enzyme loading for example inhibition or inactivation of the enzyme.

Typically ISFETs are operated in the so-called "strongly inverted region" where the current voltage characteristics are linear and power consumption is high. A recent series of papers by Toumazou and co-workers have highlighted the advantages to be gained by working at the on-off threshold (the weakly inverted region)[12–14] and have used this for single nucleotide polymorphism (SNP) detection via a polymerase catalysed reaction.[15,16]

Potentiometric sensing of macromolecules has been achieved using solid-state devices that measure changes in surface charge associated with receptor ligand binding or DNA hybridisation.[4,17,18] The latter has been extended to using allele specific hybridisation for the detection of SNPs.[19]

3.2. Amperometric biosensors

Interfacial charge transfer between a conducting solid phase and molecules in solution lies at the heart of these sensors. Rates of electron transfer are dependent on the electrode, the molecule being oxidised or reduced and the applied potential, the rates of electron transfer therefore determining the current.[20] Amperometric sensors are therefore inherently kinetic devices whereby the analyte is consumed in the electrode reaction and where a sustained current is dependent on mass transfer of analyte to the electrode surface. The relationship between current and concentration arises from the fact that both mass transfer and the redox reaction are concentration dependent. Mass transfer rates can be dependent on a variety of other factors including the size/geometry of the electrode, convection and the presence of membranes or other layers on the electrode. In a complex biological background time dependent changes in the nature of the surface, such as protein adsorption can change the electrode's response and hence a drift in calibration.

The base amperometric sensor can detect oxidisable (or less commonly, reducible) molecules directly and has been used to measure oxygen, nitric oxide, adrenaline, dopamine, serotonin, hydrogen peroxide in cells and tissues, often with microelectrodes that give good spatial and temporal resolution.[21] There are many compounds that have redox potentials in the accessible range in water that are, however, not directly detectable electrochemically. Usually this lack of direct detection can be attributed to one of two factors; slow electrode kinetics or overlapping redox potentials. Slow electrode kinetics means that a much larger voltage than the redox potential must be applied before electron transfer occurs and this over-potential can amount to more than 1 V. A good example of this over-potential is seen with the cell's redox "currency" NADH. The thermodynamic oxidation potential (at pH 7) is around –0.4 V whereas a voltage of nearer to 1 V is needed to oxidise NADH at an unmodified metal electrode such as platinum. A similar situation applies to glucose and such large over-potentials shift the current-voltage curves into regions where other redox active molecules such as ascorbate or glutathione are oxidised. Many of these poorly responding molecules can be catalytically oxidised by the appropriate enzymes that then transfer the electrons to the electrode in a second step. This second step rarely is directly from the active site of the enzyme as these tend to be buried in the middle of the protein and therefore too far for rapid electron transfer to take place. Redox enzymes can be broadly divided into those where there is a protein bound redox centre that undergoes alternate oxidation and reduction reactions and those where two substrates bind to the enzyme and electron transfer takes place with that ternary complex. Protein bound redox centres include haem, flavin or iron/sulphur clusters whilst the ternary complex substrates are exemplified by $NAD(P)^+$/NAD(P)H. The electron flows in these two cases are shown in Fig. 3. The aforementioned large over-potentials seen with NADH means that although these enzymes are by far the common redox enzymes it is the flavin or haem enzymes that have figured most prominently in amperometric biosensors.[22–24] In these enzymes the challenge is the transfer of electrons form the active site of the enzyme to the electrode and a variety of methods have been used including metallic nanoparticles, conducting polymers and diffusing electron transfer molecules (mediators). Examples of substances that can be detected with amperometric biosensors are given in Table 3, and the various kinetic steps are illustrated in Fig. 4.

Glucose biosensors have been by far the most extensively investigated, driven by the therapeutic importance of measuring glucose in diabetes and the relevance of glucose to cellular metabolism.[25,26] Considerable inventiveness and ingenuity has gone into developing new glucose sensing devices (not just amperometric) and the literature is far too large to discuss here and is regularly reviewed.[27,28] The major challenges in all amperometric sensors and for glucose in particular is to ensure that the current reflects the analyte concentration and has neither contributions from

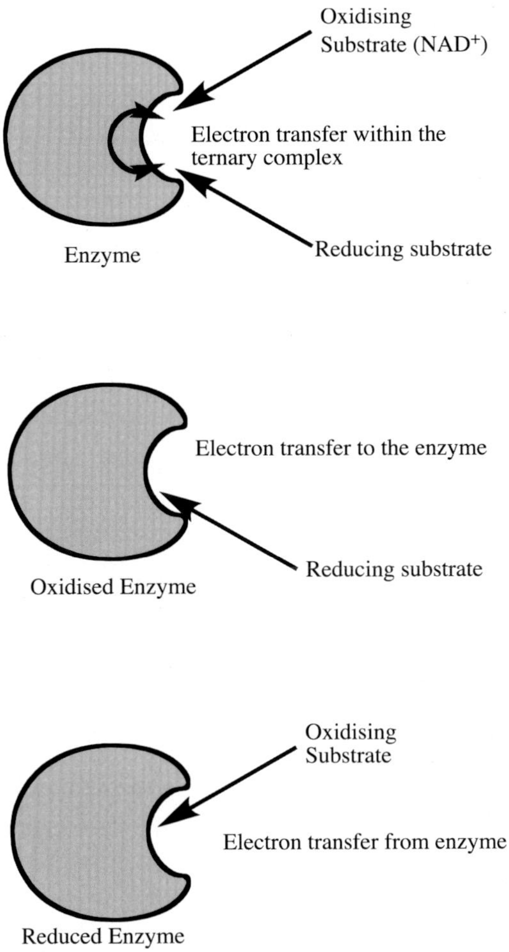

Fig. 3. A comparison of electron flow in dehydrogenase and oxidase enzymes.

Table 3. Examples of relevant analytes that can be sensed amperometrically.

Analyte	Enzyme (if applicable)
Oxygen, NO	
Dopamine, ascorbate	
Glucose	Glucose oxidase/dehydrogenase
Lactate	Lactate oxidase/dehydrogenase
Glutamate	Glutamate oxidase
Other amino acids	L-aminoacid oxidase

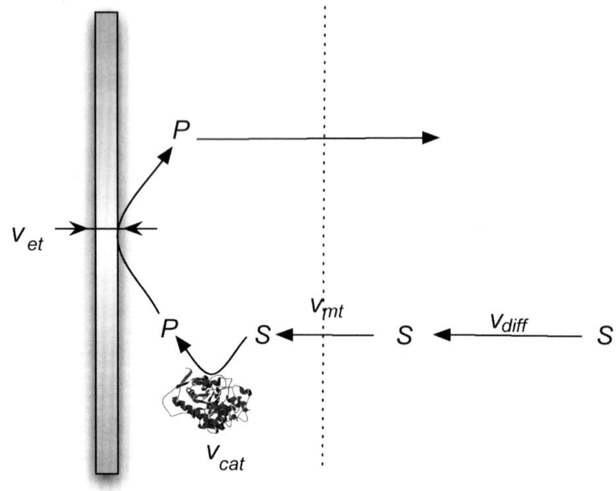

Fig. 4. The kinetic processes that typically take place in amperometric enzyme sensors. V_{diff} is the diffusion rate, V_{mt} the membrane transport rate, V_{cat} the enzyme reaction rate and V_{et} the heterogenous electron transfer rate. Typically the enzyme will have multiple substrates generating multiple products, although only one of the latter undergoes an electrochemical transformation at the electrode surface.

other redox active molecules (i.e. overestimating the analyte) nor that the electron flow from substrate to electrode is intercepted by other molecules (i.e. underestimating analyte concentrations).

3.3. Impedance sensors

These have been much less extensively exploited than the previous two types of electrochemical sensor. In contrast to the amperometric sensors the potential excitation used in impedance sensors is of low amplitude and high frequency (kHz–MHz) applied between a pair of electrodes (often interdigitated). The equivalent circuit of such an arrangement (the "Randles Cell") is shown in Fig. 5. This describes an interface where both kinetics and diffusion are important.

It is the sensitivity of the impedance to the nature of the interface (resistance, capacitance) that leads to its application in biosensors, a reaction at the surface that changes its electrical properties will give rise to a change in impedance.[29–32]

4. Optical Biosensors

The transduction of molecular recognition reactions into changes in the optical properties can be achieved either through the observation of inherent optical properties or

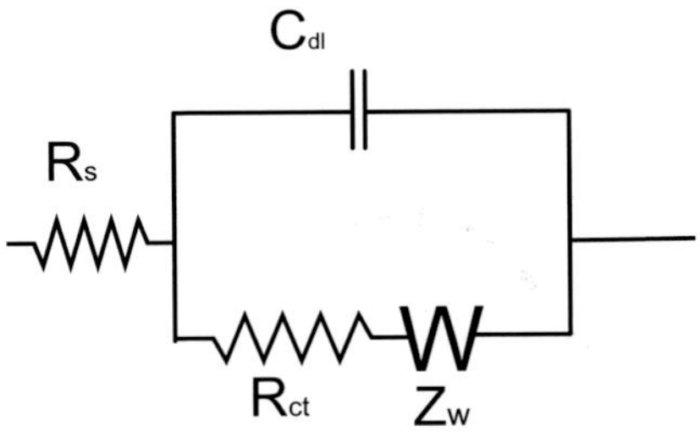

Fig. 5. The Randles equivalent circuit, often used to model the behaviour of an electrode in impedance-based sensors. R_s and R_{ct} are the resistance of the solution and the charge transfer reaction, C_{dl} is the double layer capacitance and Z_w the Warburg impedance.

more commonly through those of an associated label. Labels are many and varied and serve to "tag" one or other components or to report on a binding reaction. Traditionally optical labels have been organic dyes although more recently these have been supplemented by quantum dots. Although many different optical properties could be measured, however in most instances those that have proved most useful are fluorescence, Raman spectroscopy and light scattering. Optical interrogation/readout (i.e. bringing in and out photons) can be carried out in either the near- or far-field.

4.1. Fluorescence biosensors

The most common optical sensing technology is based around fluorescence, partly for reasons of its excellent limits of detection and partly on account of the versatility of measuring modes. A Jablonski diagram[33] (Fig. 6) illustrates the

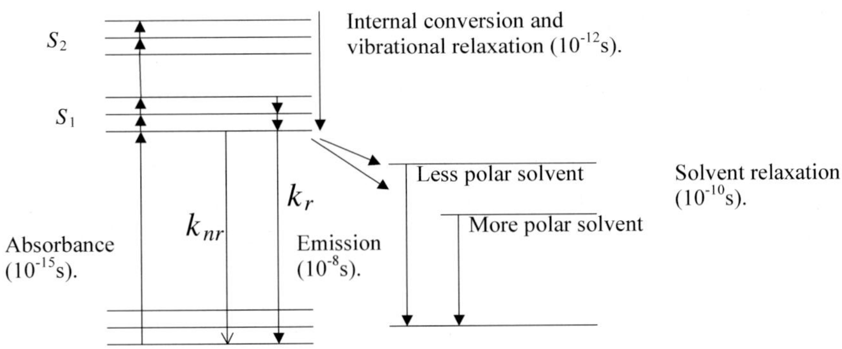

Fig. 6. A Jablonski diagram showing the main photophysical processes associated with fluorescence. k_r is the radiative decay rate and k_{nr} the non-radiative rate.

main photophysical processes involved and the parameters that can be derived there from.

Two broad classes of fluorescence sensor have been described; those where the *quantity* of the fluorescent species is determined based upon its intensity and those where the *environment* of the fluorophore is determined based upon its emission intensity and/or wavelength. In this latter case the change in environment is brought about by the interaction with the analyte. An example of quantity-based fluorescence is the level of green fluorescent protein (GFP) expression, whilst an example of environmental effects is the binding of calcium to a dye such as Fura-2 whose fluorescence is different between bound and unbound states.

Intensity is the most easily measured fluorescence property, but it is not the only one and fluorescence lifetime, polarisation and energy transfer have also been used as a basis for sensing. Traditionally organic dyes have dominated fluorescence sensing although they suffer from disadvantages such as photolability (i.e. decomposition by the excitation light), broad emission peaks and small Stokes shifts. More recently the use of quantum dots[34] and fluorescent proteins[35] has broadened the range of fluorophores that are available with complementary photophysical, chemical and biological properties. Early fluorescence sensors for small molecules included those for pH (based on the different emission spectra of protonated and unprotonated dyes) and for oxygen (which quenches the fluorescence of ruthenium complexes). In the case of pH sensing (as well as for other species) a ratiometric strategy is often employed whereby the emission maxima of the two states are different.[15,36–38] This approach, rather than just observing a change in intensity, has the advantage that factors that affect the latter such as photobleaching and other degradative reactions, or quenching by other molecules in the sample do not distort the results. Oxygen sensing takes advantage of the fact that as a triplet state molecule (two unpaired electrons) it can efficiently relax the excited state of many fluorophores through collisional processes. As with amperometric and potentiometric sensors where the proximal analyte is coupled to other more complex molecules through enzymatic reactions so the simple fluorescence sensing schemes described above can be similarly coupled and glucose has been measured using glucose oxidase, both from the reduction in oxygen quenching and from the acidification resulting from gluconic acid production. In a similar vein cell-based assays have used the measurement of oxygen consumption or the production of acid as indicators of cellular activity. In this respect the oxygen-dependent quenching of ruthenium complexes has the additional advantage that the fluorophore can be trapped behind a hydrophobic barrier, impermeable to most other molecules except oxygen.[38]

One of the major challenges in producing fluorescence sensing reagents is achieving the desired specificity and affinity as well as change in fluorescence. Whilst there have been some noticeable successes the this area (pH, divalent cations, saccharides[39]) the synthetic challenge is considerable and therefore an alternative approach has been to start with a protein that has the desired binding properties and then to engineer in a fluorescent reporter group using a combination of mutagenesis and chemical labelling. Precise placement of the (environment sensitive) reporter group makes it responsive to ligand binding and this approach has been shown to be broadly applicable to a wide range of different proteins including periplasmic binding proteins from *E. coli,*[40] β-lactamase[41] and antibody fragments.[42] Mutagenesis also allows additional functionality to be introduced including altered specificity[43] or affinity for the ligand and immobilisation sequences.[44] Gene technology has also been used to produce wholly protein-based fluorescence sensing entities with applications in intracellular sensing.[45] In this latter case the sensing mode is that of fluorescence resonance energy transfer (FRET) and relies on the distance dependence of the non-radiative transfer of energy from the fluorophore that is excited (the donor) to that which emits (the acceptor). This results in the donor emission being quenched and the acceptor emission being sensitised. The two main criteria for FRET to occur are that the emission band of the donor should overlap that of the acceptor and that the distance between them should be of the order of molecular dimensions.[33] This make FRET sensing particularly amenable where there is either an assembly or disassembly of structures carrying the two fluorophores (Fig. 7), or where there is a (non-covalent) change in structure as a result of ligand binding. In the special case where the acceptor is non-fluorescent then a quenching of the donor is observed but no emission from the acceptor.[46,47] This has been exploited in molecular beacons for nucleic acid analysis and is described later in the cell sensing section.

At its simplest fluorescence measurements can be made on bulk solutions, however for a variety of reasons the use of integrated optics to deliver and collect light has many advantages. The most common of such integrated optical systems are light guides, either as optical fibres[48,49] or planar waveguides.[50,51] They allow the fluorescence to be measured at spatially well-defined locations avoiding bulk fluorescence from the surroundings, they can be made sensitive only to fluorescence at the surface (within a few hundreds of nanometres)[52] and in the case of fibre optic bundles can measure multiple species.[49] Fibre optic sensors can operate in either distal tip mode, where the sensing reagent is attached to the end of the fibre, or in evanescent wave mode whereby the sensing reagent is immobilised along the length of the fibre. In the former case the fibre interrogates a region some distance from the tip, usually millimetres to centimetres, whilst in the later only species a few hundred nanometres from the surface are detected.

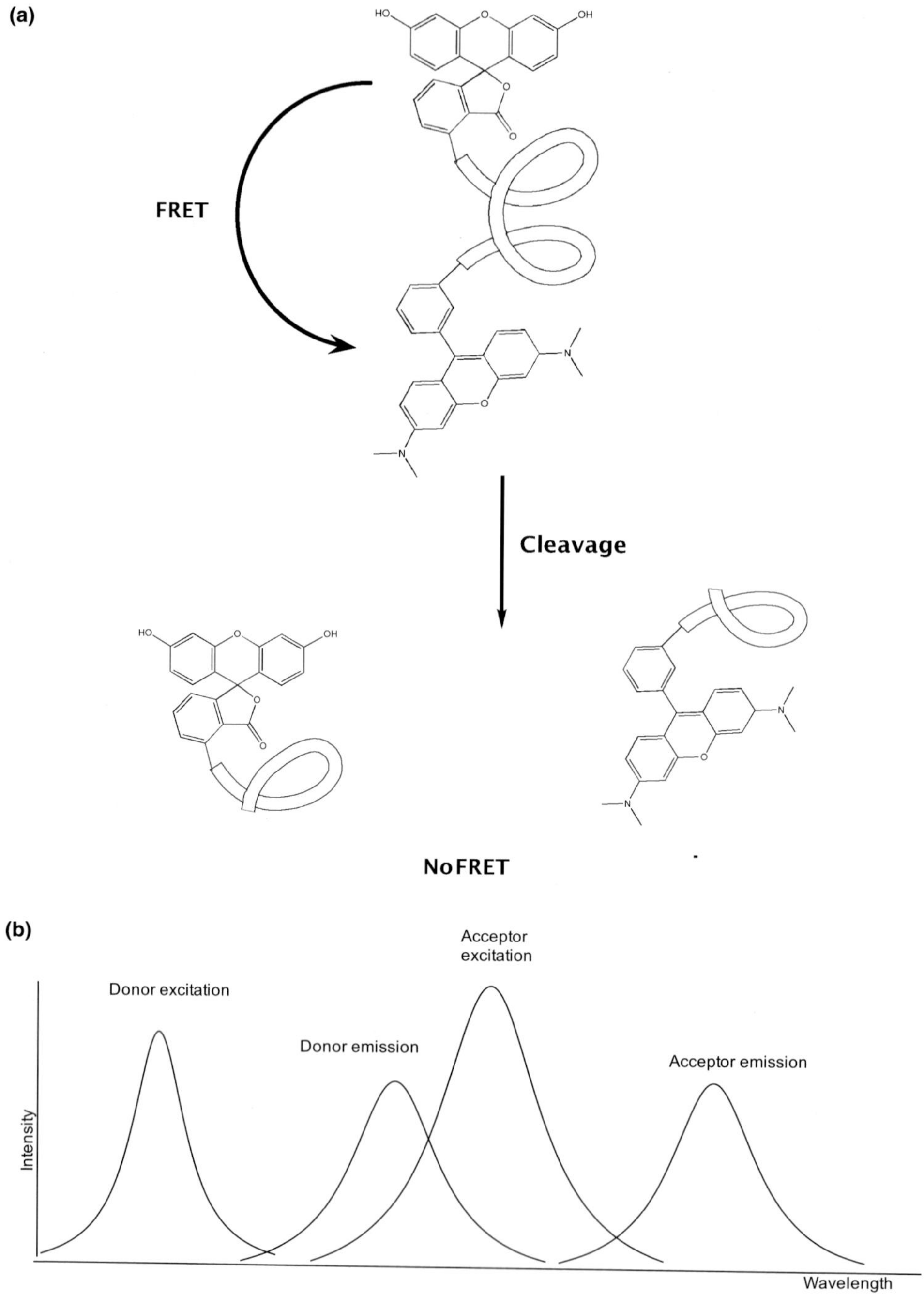

Fig. 7. Fluorescence resonance energy transfer (FRET). **(A)** Cleavage of a doubly labelled substrate (for example a peptide) such that the two fluorophores (donor and acceptor) are close together in the substrate but not in the products and hence the FRET is lost. **(B)** Typical excitation and emission spectra for a FRET pair are shown.

4.2. Integrated optical devices

In addition to their role as passive light guides, more complex optical structures can be employed to measure changes in optical thickness of the order of macromolecular dimensions associated with binding reactions at the surface of the device. The great attraction of such devices is that they require no labelling in order to generate a signal, however because the signal reflects only the surface density of bound molecules the control of non-specific binding is crucial to their reliability. One of the earliest and most widely used of such integrated optical sensors is that based upon surface plasmon resonance (SPR). In this method a thin gold film is coated onto a diffraction element (prism or grating) and monochromatic light shone through the diffraction element to reflect off the underside of the gold film (Fig. 8). At a critical angle of incidence the light excites collective electron waves in the gold film (surface plasmons) and there is a relationship

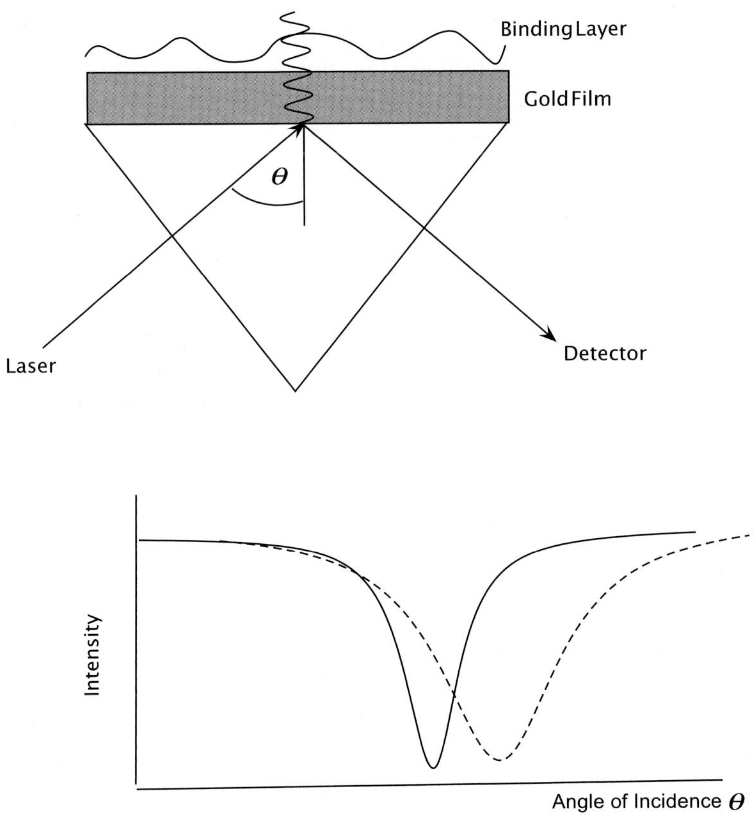

Fig. 8. Label-free analysis of binding through surface plasmon resonance. The binding of a congnate molecule shifts the resonance angle of the SPR the extent of the shift is a measure of the degree of binding.

between this critical angle and the refractive index of any coating on the gold. In the most common SPR format employed, a layer of a specific capture molecule (typically but not exclusively an antibody) is bound to the gold surface. When the capture molecule binds its target there is a change in refractive index and hence a shift in the resonance angle.[53] SPR can also be performed in an imaging mode allowing for multiplex measurements or spatially resolved detection.

Other integrated optical devices offering label free detection of molecular interactions include dual polarisation interferometry,[54] grating couplers[55] and resonant mirror structures.[56]

4.3. Plasmonic nanomaterials

Generation of surface plasmons as described above is a feature of a planar metal-dielectric interface and only occurs with light at a particular angle of incidence. The advent of nanostructured materials has opened the area of plasmonics beyond the confines of physics and two recent books discuss some of the sensing applications of these materials[57,58] Nanomaterials such as particles and rods above a certain size (10 s of nm) will absorb light at a particular (size dependent) wavelength and generate a localised surface plasmon. This size dependence can be manipulated through molecular interactions giving rise to changes in colour associated with the assembly or disassembly of the nanoparticles into differently sized structures. In many ways this is the plasmonic equivalence of FRET.

In the field of Raman spectroscopy the sensitivity gains from surface enhanced Raman scattering (SERS) and in particular surface enhanced resonance Raman scattering (SERRS) offers single molecule detection limits. Moreover the richness of the Raman spectrum (essentially a vibrational spectrum) makes it well suited to multiplex measurements using different resonance Raman active dyes. Typical SERS active surfaces include gold or silver colloids, silver nanostructures on silicon, photonic crystals and gold or silver nanorods and they have been used to sense enzymes,[59] antigens[60,61] and DNA.[62]

5. Mass Sensors

The most fundamental change that can occur in a reaction is a change in mass. The SPR and other integrated optical sensors described above measure the mass change indirectly through changes in refractive index. Sensors that will be described in this section are directly mass sensitive. Unlike the electrochemical and optical sensors covered in earlier sections most mass sensing devices are mechanical. Just like the integrated optical sensors, because they rely on specific molecular interactions to generate the signal, control of non-specific binding is essential to their reliable operation.

The earliest mass sensor was the quartz crystal microbalance (QCM) wherein a particular crystal type (AT-cut) exhibits piezoelectricity, that is a mechanical deformation when an alternating potential is applied across the crystal via a pair of evaporated metal (typically gold) electrodes. The crystal typically forms part of a resonant circuit whose frequency is given by the Sauerbrey equation:

$$\Delta f = -S_t \Delta m$$

where Δf is the frequency shift associated with a change in mass loading of Δm and S_t is the Sauerbrey constant that depends, amongst other factors, upon the square of the fundamental frequency (typically a few 10 s of MHz).[63] The Sauerbrey equation is strictly only valid for thin, rigid films and where the film is soft and viscous there is a damping of the oscillation through coupling to the solvent. To separate the effects of mass loading from viscoeleastic coupling the resonance is induced transiently and then the decay in its amplitude followed with time (QCM-D).[64]

Alternative mass sensors to the QCM include surface acoustic wave (SAW),[65] flexural plate wave (FPW) and acoustic plate mode (APM) devices. Each device has its own characteristic operating frequency, mass sensitivity and viscoelastic coupling characteristics and the reader is referred to the article by Janshoff *et al.*[66] for a comparative discussion of their properties and performance.

Cantilever sensors provide a different approach to mass sensing, being based upon differential surface strain on the two sides of a microfabricated cantilever as a consequence of a binding reaction. The small deflections that are produced can be measured using the same technology as are employed in the atomic force microscope (AFM) whereby a laser beam is reflected from the end of the cantilever onto a position sensitive detector. Typical materials used in cantilever fabrication are silicon and PZT, although polymeric cantilevers fabricated from SU8 have also been described. One of the advantages of silicon based cantilevers is that offer the opportunity for both read out circuit incorporation and cost reduction by using standard CMOS processing methods. It also offers the opportunity to make multi-cantilever arrays. Cantilever sensors can be operated in one of two modes, either through measurement of the static bending or by driving the cantilever into an oscillation and measuring changes in the resonant frequency.[67,68]

6. Cell Sensing Strategies

The role of biosensors in understanding the cellular behaviour that informs tissue engineering stems from their application to real-time minimally invasive analysis of cellular physiology. Cell sensing approaches are typically based on "probe" sensors whereby the device (electrode, optical fibre, cantilever) is placed in close proximity to the cells and the relevant chemical species is sensed. This is particularly relevant to culturing stem cells prior to transplantation where there is a need for close control over the culture conditions to achieve the correct balance between expansion, differentiation and apoptosis. Although such culture sensing approaches are well established in microbial bioprocessing they are not yet routine in stem cell cultures and research is just beginning in this field. Unlike the well-stirred deep cultures characteristic of microbial bioprocesses where a relatively uniform culture environment is achieved, the stem cell cultures, especially where 3D scaffolds are also present, are likely to be spatially heterogeneous. A decision therefore has to be made of not only what to measure but where to measure it and we have developed a design of experiments (DoE) model to both choose sensor locations and also to interpret the results.[69] As to what should be measured, the obvious "low hanging fruits" are simple molecular species such as pH, ammonia, oxygen and glucose. Even with these the sensitivity of the cells to their environment along with the potential for interferents in the complex culture media presents a considerable challenge to the performance and longevity of the biosensors. In fact the literature on stem cells and biosensors is almost exclusively concerned with the use of stem cells as **reagents** rather than samples.[70,71] In the wider perspective of animal cell culture monitoring there are still relatively few examples of *in situ* monitoring as compared with offline (flow injection analysis or micro-reactor devices) measurements. Whilst the latter may be technically simpler to implement they sacrifice both temporal resolution and any knowledge of spatial inhomogeneities in metabolite levels. A further challenge to successful introduction of a multiplicity of sensors into stem cell bioreactors is the sheer amount of connections needed to instrument the sensors. Adopting a modular approach can help in this regard but in the longer term wireless technologies will be needed.

The alternative to sensor probes is to use micro- or nanosensor materials that are optically responsive and measure in an imaging format. Fluorescence is the method of choice here and has employed either engineered biomolecules or encapsulated reagents. Both are however primarily research tools rather than process monitoring devices as their presence in the (cell) products are unlikely to be acceptable. The advantage of both approaches is that they can monitor intracellular analytes as well as culture medium components. PEBBLES (Probes Encapsulated By Biologically Localised Embedding) were introduced by Koopleman and co-workers to avoid both interference of the probe molecule by the cellular milieu and to protect the cells from the possible toxicity of the probes.[38,72–74] Encapsulation of the probes in polymer nanospheres has been used to measure intracellular concentrations of glucose, oxygen and pH by fluorescence imaging. Whilst cell uptake is necessary for these applications the same materials could also be employed to measure culture metabolites.

Fluorescence has also been used to measure intracellular concentrations of maltose in yeast cells by expressing a maltose binding protein fused to green fluorescent protein and yellow fluorescent protein and measuring changes in FRET upon ligand binding.[45] Molecular beacons have been similarly employed to image mRNA levels inside animal cells to assess gene expression.[75]

7. Tissue Sensing Strategies

As we move from cell- to tissue-based sensing the issue of biocompatibility looms large, in cell sensing the primary biocompatibility challenge is protein adsorption on the sensor surface, either reducing the sensors responsiveness or where the sensor is a label-free device giving a spurious signal. In tissue, protein adsorption is the first step in a more complex process that involves frustrated phagocytosis, an inflammatory response and then capsule formation such that the sensor is walled off from the surrounding tissue. Whilst the problems of biocompatibility are not unique to sensing applications and there is a large literature on the biocompatibility of tissue implants and medical devices, the difference is that with a sensor even a few per cent change in performance gives unreliable data as far as measuring changes in tissue biochemistry is concerned. There is not space here to review the literature on biocompatibility but it is clear that prevention of protein adsorption is a key first step. As a general model proteins initially adsorb to surfaces in their native (i.e. folded state), however if the surface is hydrophobic then the protein can lower its surface energy by folding to expose its hydrophobic core which then binds more tightly to the surface so producing a layer of denatured protein that is essentially irreversibly bound. Increasing its hydrophilic nature helps prevent this surface induced unfolding as does making it negatively charged (most human proteins at physiological pH have a net negative charge) and using long flexible polymers such as poly(ethyleneglycol). Of course for sensors all these modifications have also to be compatible with the molecular recognition and signal transduction processes also occurring at the surface.[76–78]

Another strategy that may help is to reduce the size of the sensor such that there is minimal trauma during implantation, although this will in many cases also reduce the signal generated by the device. Miniaturisation also has additional benefits in both enabling multiplexing and allowing for sensor redundancy such that multiple sensors for the same analyte can be "polled" and a sensor showing a response inconsistent with the others is subsequently ignored. As with cell sensing it is likely that in the future low power wireless technologies will be increasingly important as the link between the internal environment of the tissue and the data capture, processing and reporting to the outside world. We have called this fusion of the biological with the electronic "tissue bionics". Again in line with developments in cell sensing it is likely that early targets for monitoring will be glucose, pH and oxygen along with nitric oxide and lactate. In the longer term monitoring of cytokines and other signalling molecules could give early warning of likely problems with engineered tissues.

8. Conclusions and Outlook

The field of biosensing in clinical science is rather like a "funnel" whereby at the research end there is a huge amount of activity in areas as diverse as microfabrication, synthesis, protein engineering, nanomaterials and signal processing. Many new and innovative sensing modalities are described each year with improved selectivity and sensitivity. Much of the work comes from chemistry and physics laboratories and only a small proportion of it is actually used in "real" biological samples. Even less of the work is turned to "challenging" biotechnological applications such as stem cell culture monitoring or tissue monitoring.

Undoubtedly part of this is cultural, physical scientists and engineers finding the complexity, fragility and variability of biological systems very different from the quantitative world they are used to. Similarly biomedical scientists primarily want a "black box" that works and lets them get on and study the interesting biology without needing to understand the physical phenomena underlying (and limiting the device). Part is also the difficulties of translating research into application; many of the sensors are hand-made, possibly in low yield and the transition from these devices to semi-industrialised prototypes is an uncertain and expensive process often requiring resources beyond the research laboratory. Finally there is the sheer difficulty of the problem as demonstrated by the effort and ingenuity that has gone into trying to produce long term implantable glucose sensors for control of diabetes over the past 30 years.

References

1. Tsai, M.S. *et al.* Functional network analysis of the transcriptomes of mesenchymal stem cells derived from amniotic fluid, amniotic membrane, cord blood, and bone marrow. *Stem Cells* **25**(10), 2511–2523 (2007).
2. Maurer, M.H. *et al.* Comprehensive proteome expression profiling of undifferentiated versus differentiated neural stem cells from adult rat hippocampus. *Neurochem. Res.* **29**(6), 1129–1144 (2004).

3. Park, H.W., Shin, J.S. and Kim, C.W. Proteome of mesenchymal stem cells. *Proteomics* **7**(16), 2881–2894 (2007).
4. Koncki, R. Recent developments in potentiometric biosensors for biomedical analysis. *Anal. Chim. Acta* **599**(1), 7–15 (2007).
5. Schoning, M.J. "Playing around" with field-effect sensors on the basis of EIS structures, LAPS and ISFETs. *Sensors* **5**(3), 126–138 (2005).
6. Wang, M., Yao, S. and Madou, M. A long-term stable iridium oxide pH electrode. *Sensor Actuat. B-Chem.* **81**(2–3), 313–315 (2002).
7. Bergveld, P. Thirty years of ISFETOLOGY — What happened in the past 30 years and what may happen in the next 30 years. *Sensor Actuat. B-Chem.* **88**(1), 1–20 (2003).
8. Dzyadevych, S.V. *et al.* Biosensors based on enzyme field-effect transistors for determination of some substrates and inhibitors. *Anal. Bioanal. Chem.* **377**(3), 496–506 (2003).
9. Kullick, T. *et al.* Application of enzyme field-effect transistor sensor arrays as detectors in a flow-injection system for simultaneous monitoring of medium components: preparation and calibration. *Anal. Chim. Acta* **296**(3), 263–269 (1994).
10. Martinoia, S., Massobrio, G. and Lorenzellib, L. Modeling ISFET microsensor and ISFET-based microsystems: a review. *Sensor Actuat. B-Chem.* **105**(1), 14–27 (2005).
11. D'Amico, A. *et al.* Sensors small and numerous: always a winning strategy? *Sensor Actuat. B-Chem.* **106**(1), 144–152 (2005).
12. Toumazou, C. and Shepherd, L. Using transistors to linearise biochemistry. *Electron. Lett.* **43**, 65–67 (2007).
13. Shepherd, L. and Toumazou, C. Weak Inversion ISFETs for ultra-low power biochemical sensing and real-time analysis. *Sensor Actuat. B-Chem.* **107**(1), 468–473 (2005).
14. Shepherd, L. and Toumazou, C. Towards direct biochemical analysis with weak inversion ISFETs. *IEEE Int. Workshop Biomed. Circ. Syst.*, pp. S1/5–S5/8 (2004).
15. Grant, R.L. and Acosta, D. Ratiometric measurement of intracellular pH of cultured cells with BCECF in a fluorescence multi-well plate reader. *In Vitro Cell. Dev. Biol.-AN.* **33**(4), 256–260 (1997).
16. Purushothaman, S., Toumazou, C. and Ou, C.P. Protons and single nucleotide polymorphism detection: a simple use for the ion sensitive field effect transistor. *Sensor Actuat. B-Chem.* **114**(2), 964–968 (2006).
17. Qiang, Z. *et al.* A new potentiometric immunosensor for determination of alpha-fetoprotein based on improved gelatin-silver complex film. *Electrochim. Acta* **51**(18), 3763–3768 (2006).

18. Besselink, G.A.J., Schasfoort, R.B.M. and Bergveld, P. Modification of ISFETs with a monolayer of latex beads for specific detection of proteins. *Biosensor Bioelectron.* **18**(9), 1109–1114 (2003).
19. Sakata, T. and Miyahara, Y. Potentiometric detection of single nucleotide polymorphism by using a genetic field-effect transistor. *Chembiochem* **6**(4), 703–710 (2005).
20. Ramirez-Garcia, S. *et al.* Carbon composite electrodes: surface and electrochemical properties. *Analyst* **127**(11), 1512–1519 (2002).
21. Porterfield, D.M., Measuring metabolism and biophysical flux in the tissue, cellular and sub-cellular domains: recent developments in self-referencing amperometry for physiological sensing. *Biosensor Bioelectron.* **22**(7), 1186–1196 (2007).
22. Willner, I., Basnar, B. and Willner, B. Nanoparticle-enzyme hybrid systems for nanobiotechnology. *FEBS J.* **274**(2), 302–309 (2007).
23. Ahuja, T. *et al.* Biomolecular immobilization on conducting polymers for biosensing applications. *Biomaterials* **28**(5), 791–805 (2007).
24. Zhang, S.X. *et al.* Immobilization of glucose oxidase on gold nanoparticles modified Au electrode for the construction of biosensor. *Sensor Actuat. B-Chem.* **109**(2), 367–374 (2005).
25. Li, C.M. *et al.* Implantable electrochemical sensors for biomedical and clinical applications: progress, problems, and future possibilities. *Curr. Med. Chem.* **14**(8), 937–951 (2007).
26. Koschwanez, H.E. and Reichert, W.M. *In vitro*, *in vivo* and post-explantation testing of glucose-detecting biosensors: current methods and recommendations. *Biomaterials* **28**(25), 3687–3703 (2007).
27. Pickup, J.C. *et al.* *In vivo* glucose monitoring: the clinical reality and the promise. *Biosensor Bioelectron.* **20**(10), 1897–1902 (2005).
28. Pickup, J., Developing glucose sensors for *in vivo* use. *Trends Biotechnol.* **11**(7), 285–291 (1993).
29. Daniels, J.S. and Pourmand, N. Label-free impedance biosensors: opportunities and challenges. *Electroanalysis* **19**(12), 1239–1257 (2007).
30. Pejcic, B. and De Marco, R. Impedance spectroscopy: over 35 years of electrochemical sensor optimization. *Electrochim. Acta* **51**(28), 6217–6229 (2006).
31. Rodriguez, M., Kawde, A. and Wang, J. Aptamer biosensor for label-free impedance spectroscopy detection of proteins based on recognition-induced switching of the surface charge. *Chem. Commun. (Camb.)* **34**, 4267–4269 (2005).
32. Farace, G. *et al.* Reagentless biosensing using electrochemical impedance spectroscopy. *Bioelectrochemistry* **55**(1–2), 1–3 (2002).

33. Lakowicz, J.R., *Principles of Fluorescence Spectroscopy*, 3rd Ed. (Springer, New York, 2006), p. 954.
34. Jaiswal, J.K. *et al.* Long-term multiple color imaging of live cells using quantum dot bioconjugates. *Nat. Biotechnol.* **21**(1), 47–51 (2003).
35. Zimmer, M. Green fluorescent protein (GFP): applications, structure, and related photophysical behavior. *Chem. Rev.* **102**(3), 759–781 (2002).
36. Ge, X.D., Tolosa, L. and Rao, G. Dual-labeled glucose binding protein for ratiometric measurements of glucose. *Anal. Chem.* **76**(5), 1403–1410 (2004).
37. Tolosa, L., Ge, X.D. and Rao, G. Reagentless optical sensing of glutamine using a dual-emitting glutamine-binding protein. *Anal. Biochem.* **314**(2), 199–205 (2003).
38. Xu, H., Aylott, J.W. and Kopelman, R. Fluorescent nano-PEBBLE sensors designed for intracellular glucose imaging. *Analyst* **127**(11), 1471–1477 (2002).
39. Cao, H.S. and Heagy, M.D. Fluorescent chemosensors for carbohydrates: a decade's worth of bright spies for saccharides in review. *J. Fluoresc.* **14**(5), 569–584 (2004).
40. Dwyer, M.A. and Hellinga, H.W. Periplasmic binding proteins: a versatile superfamily for protein engineering. *Curr. Opin. Struct. Biol.* **14**(4), 495–504 (2004).
41. Chan, P.H. *et al.* Rational design of a novel fluorescent biosensor for beta-lactam antibiotics from a class A beta-lactamase. *J. Am. Chem. Soc.* **126**(13), 4074–4075 (2004).
42. Renard, M. *et al.* Knowledge-based design of reagentless fluorescent biosensors from recombinant antibodies. *J. Mol. Biol.* **318**(2), 429–442 (2002).
43. Antikainen, N.M. and Martin, S.F. Altering protein specificity: techniques and applications. *Bioorg. Med. Chem.* **13**(8), 2701–2716 (2005).
44. Wada, A. *et al.* Design and construction of glutamine binding proteins with a self-adhering capability to unmodified hydrophobic surfaces as reagentless fluorescence sensing devices. *J. Am. Chem. Soc.* **125**(52), 16228–16234 (2003).
45. Stitt, M., Imaging of metabolites by using a fusion protein between a periplasmic binding protein and GFP derivatives: from a chimera to a view of reality. *Proc. Natl. Acad. Sci. USA* **99**(15), 9614–9616 (2002).
46. Santangelo, P., Nitin, N. and Bao, G. Nanostructured probes for RNA detection in living cells. *Ann. Biomed. Eng.* **34**(1), 39–50 (2006).
47. Medintz, I. *et al.* Self-assembled nanoscale biosensors based on quantum dot FRET donors. *Nat. Mater.* **2**(9), 630–638 (2003).
48. Leung, A., Shankar, P.M. and Mutharasan, R. A review of fiber-optic biosensors. *Sensor Actuat. B-Chem.* **125**(2), 688–703 (2007).

49. Monk, D.J. and Walt, D.R. Optical fiber-based biosensors. *Anal. Bioanal. Chem.* **379**(7–8), 931–945 (2004).
50. Rowe-Taitt, C.A. *et al.* Array biosensor for detection of biohazards. *Biosens. Bioelectron.* **14**(10–11), 785–794 (2000).
51. Lukosz, W. *et al.* Output grating couplers on planar waveguides as integrated optical chemical sensors. *Sensor Actuat. B* **1**, 585–588 (1990).
52. Taitt, C.R., Anderson, G.P. and Ligler, F.S. Evanescent wave fluorescence biosensors. *Biosensor Bioelectron.* **20**(12), 2470–2487 (2005).
53. Homola, J., Yee, S.S. and Gauglitz, G. Surface plasmon resonance sensors: Review. *Sensor Actuat. B-Chem.* **54**(1–2), 3–15 (1999).
54. Cross, G.H. *et al.* A new quantitative optical biosensor for protein characterisation. *Biosensor Bioelectron.* **19**(4), 383–390 (2003).
55. Gauglitz, G., Direct optical sensors: principles and selected applications. *Anal. Bioanal. Chem.* **381**(1), 141–155 (2005).
56. Hengsakul, M. and Cass, A.E.G. Alkaline phosphatase-strep tag fusion protein binding to streptavidin: resonant mirror studies. *J. Mol. Biol.* **266**(3), 621–632 (1997).
57. Maier, S.A. *Plasmonics: Fundamentals and Applications* (Springer-Verlag, New York, 2007).
58. Prasad, P.N. *Nanophotonics* (John Wiley and Sons, Hoboken, 2004), p. 415.
59. Ingram, A. *et al.* 8-hydroxyquinolinyl azo dyes: a class of surface-enhanced resonance Raman scattering-based probes for ultrasensitive monitoring of enzymatic activity. *Anal. Chem.* **79**(22), 8578–8583 (2007).
60. Manimaran, M. and Jana, N.R. Detection of protein molecules by surface-enhanced Raman spectroscopy-based immunoassay using 2–5 nm gold nanoparticle lables. *J. Raman Spectrosc.* **38**(10), 1326–1331 (2007).
61. Cui, Y. *et al.* Multianalyte immunoassay based on surface-enhanced Raman spectroscopy. *J. Raman Spectrosc.* **38**(7), 896–902 (2007).
62. Liu, F.M. *et al.* A note on distance dependence in surface enhanced Raman spectroscopy. *Chem. Phys. Lett.* **430**(1–3), 173–176 (2006).
63. Rickert, J., Brecht, A. and Gopel, W. Quartz crystal microbalances for quantitative biosensing and characterizing protein multilayers. *Biosensor Bioelectron.* **12**(7), 567–575 (1997).
64. Edvardsson, M. *et al.* A dual-frequency QCM-D setup operating at elevated oscillation amplitudes. *Anal. Chem.* **77**(15), 4918–4926 (2005).
65. Berkenpas, E. *et al.* Pure shear horizontal SAW biosensor on langasite. *IEEE Trans. Ultrason. Ferroelectr. Freq. Control* **51**(11), 1404–1411 (2004).
66. Janshoff, A., Galla, H.-J. and Steinem, C. Piezoelectric mass-sensing devices as biosensors — an alternative to optical biosensors? *Angew. Chem.* **39**(22), 4004–4032 (2000).

67. Hansen, K.M. and Thundat, T. Microcantilever biosensors. *Methods* **37**(1), 57–64 (2005).
68. Ziegler, C. Cantilever-based biosensors. *Anal. Bioanal. Chem.* **379**(7–8), 946–959 (2004).
69. Lim, M. *et al.* Intelligent bioprocessing for haemotopoietic cell cultures using monitoring and design of experiments. *Biotechnol. Adv.* **25**(4), 353–368 (2007).
70. Liu, Q.J. *et al.* Embryonic stem cells as a novel cell source of cell-based biosensors. *Biosensor Bioelectron.* **22**(6), 810–815 (2007).
71. Liu, Q. *et al.* Embryonic stem cells biosensor and its application in drug analysis and toxin detection. *IEEE Sensors J.* **7**(11–12), 1625–1631 (2007).
72. Webster, A., Compton, S.J. and Aylott, J.W. Optical calcium sensors: development of a generic method for their introduction to the cell using conjugated cell penetrating peptides. *Analyst* **130**(2), 163–170 (2005).
73. Koo, Y.E.L. *et al.* Real-time measurements of dissolved oxygen inside live cells by organically modified silicate fluorescent nanosensors. *Anal. Chem.* **76**(9), 2498–2505 (2004).
74. Buck, S.M. *et al.* Optochemical nanosensor PEBBLEs: photonic explorers for bioanalysis with biologically localized embedding. *Curr. Opin. Chem. Biol.* **8**(5), 540–546 (2004).
75. Medley, C.D. *et al.* Simultaneous monitoring of the expression of multiple genes inside of single breast carcinoma cells. *Anal. Chem.* **77**(15), 4713–4718 (2005).
76. Wilson, G.S. and Ammam, M. *In vivo* biosensors. *FEBS J.* **274**(21), 5452–5461 (2007).
77. Wilson, G.S. and Gifford, R. Biosensors for real-time *in vivo* measurements. *Biosensor Bioelectron.* **20**(12), 2388–2403 (2005).
78. Yotter, R.A., Lee, L.A. and Wilson, D.M. Sensor technologies for monitoring metabolic activity in single cells — part I: optical, methods. *IEEE Sensors J.* **4**(4), 395–411 (2004).

Chapter 19

Tissue-Engineering Monitoring Using Microdialysis

Zhaohui Li, Olga Boubriak, Jill Urban and Zhanfeng Cui

Abstract

Monitoring of cell status and tissue formation within a three-dimensional cultured tissue is important to the success of engineered tissue development. Here we describe a method using microdialysis probes for the continuous monitoring of local extracellular environment, cell metabolic activity, cell functions and possible tissue formation. Using 3D culture of chondrocytes and culture of intervertebral disc explant as examples, the effectiveness of this method is demonstrated. Challenges and practical issues in using this method are discussed as well.

Keywords: Monitoring; Microdialysis; Metabolic Activity; Protein Markers; Tissue Formation.

Outline

1. Introduction
2. Methodology of Microdialysis
   2.1. Principle of microdialysis
   2.2. Characteristics of probes
   2.3. Recovery of probes
   2.4. Calibration of probes
      2.4.1. *In vitro recovery*
      2.4.2. *Zero-net-flux recovery*
      2.4.3. *Internal reference recovery*
   2.5. Microdialysis sampling for proteins
   2.6. Assessment of probe membrane fouling
   2.7. Microdialysis with different pumping methods
3. Microdialysis for Tissue Engineering Monitoring — Case Studies
   3.1. Experimental method
   3.2. Internal standard calibration of probes and of membrane fouling

3.3. Monitoring cell metabolic activities in engineered cartilage
3.4. Monitoring tissue metabolism in cultured IVD explants
3.5. Monitoring tissue formation
4. Summary
References

1. Introduction

The process of growing engineered tissues and organs for medical purpose is time consuming and takes weeks or even months to complete.[1–3] It is vital to monitor the cell and tissue status throughout engineered tissue during the culture period, preferably on-line using non-destructive methods. Apart from a few non-destructive methods such as NMR, MRI,[4–6] micro-CT,[7] optical method[8] or measuring the averaged parameters in the effluent of the bioreactor, current methods for assessing tissue culture outcome are largely based on destructive biochemical or histological methods. The cultured bio-constructs are usually taken out of the bioreactor and sectioned for various histological and biochemical assays such as cell counting, viability, extracellular matrix composition and localisation of critical matrix proteins such as aggrecan or collagens.[9,10] These methods can provide the quality assessment of the final products; but they do not allow monitoring the cell growth and tissue formation during the tissue culture, and they are destructive. Also current "on-line" monitoring methods have certain disadvantages: MRI and micro-CT are expensive to be used routinely, and it is difficult for them to be incorporated into the manufacture process. The monitoring of tissue development by analysing the metabolic products in the bioreactor effluent can only provide the data on collective cell activity within the whole construct. It does not allow to assess the cell response to local environment changes or to detect the possible non-uniformity in tissue formation (due to, for example, mass transfer limitation of nutrient and oxygen supply).[11]

One of the problems in engineering 3D constructs is the nutrient supply and waste removal which rely only on diffusion in most engineered tissue cultures. Mass transfer limitation is hence a major problem if bulky tissue is produced for some clinical applications. Cells inside 3D constructs may not proliferate or differentiate properly due to nutrient supply limitations and may even die, as occurs in tumour spheroids.[12] This damage can progress with tissue engineered construct development and growth as cells on construct surfaces survive better and produce and accumulate extracellular matrices more readily then cells in deeper layers. An increase in extracellular matrix density near the construct surface can aggravate mass transfer limitation. In many cases, the cultured tissue would be highly heterogeneous with a high cell density near the surface and a very low one or even lacking the cells in the central area.[11] This is a common problem for any bulky tissues without vascular network.

Therefore non-destructive methods that could monitor changes in cell metabolism, viability or tissue deposition locally within the construct and could point out adverse responses early during culture would be desirable.

Here we propose an alternative method of on-line monitoring of tissue engineered constructs using the principle of microdialysis to determine local production of metabolites and tissue components in the extracellular fluid of the construct. It is a minimally invasive and non-destructive method which should not interfere with the normal tissue formation.

2. Methodology of Microdialysis

2.1. Principle of microdialysis

Microdialysis is performed by perfusing a small semi-permeable hollow fibre membrane probe inserted into the tissue with a physiological fluid (the *perfusate*). Molecules from the probe surrounding will diffuse through the membrane due to the concentration gradient if they can pass the pores. The solution that exits the probe, the *dialysate*, which contains the molecules of interest, can be collected for analysis. The size of the molecules diffusing through the semi-permeable probe membrane depends on pore size (Fig. 1). The first microdialysis experiments were conducted in order to measure dynamic release of substances in the brain and blood plasma.[13] It has been used since to study metabolism in numerous tissues of living animals and human subjects such as brain,[14] muscles,[15] subcutaneous adipose tissue,[16] lungs,[17] kidneys[18] and liver.[19] Microdialysis is widely used for pharmacokinetic research[20] and has also been used to monitor cell metabolites in cell culture media.[21] The most important feature of microdialysis is that it allows non-destructive continuous sampling of the extracellular fluids.

Most tests with microdialysis typically use membranes of a molecular weight cut-off (MWCO) around 20 kDa to exclude bigger molecules and thus to simplify the analytical procedures.[14,15,22] However, many bio-molecules have molecular size bigger than 20 kDa. Recently attempts were hence taken to sample high molecular weight species in muscles, tendons, adipose tissue and dermis using microdialysis probes with 100 — 3000 kDa MWCO.[23–28]

2.2. Characteristics of probes

The choice of microdialysis probe and membrane pore size used for *in vitro* and/or *in vivo* sampling depends on characteristics of the molecules aimed to be collected by microdialysis. Molecules smaller than the membrane pores can diffuse into inner fibre lumen and be carried to the outlet by the continuously flowing perfusion fluid. Larger molecules will be rejected by the membrane

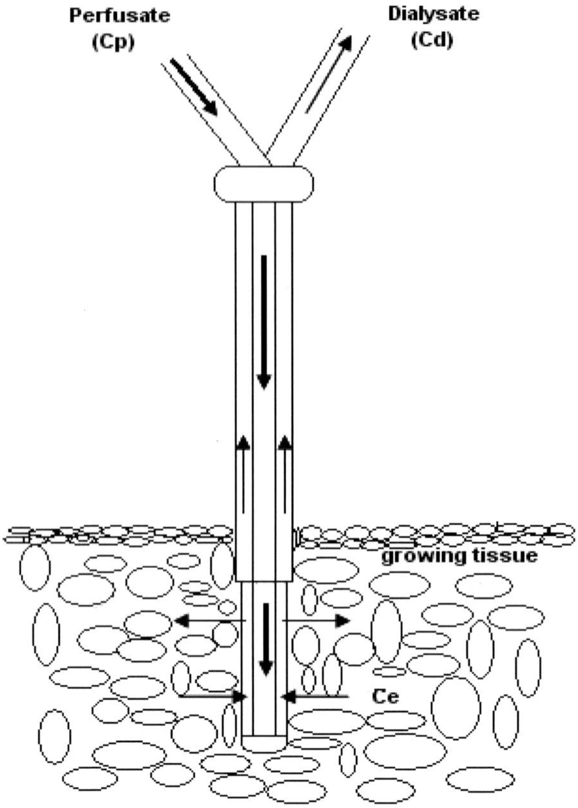

Fig. 1. Principle of microdialysis.

pores. Therefore membrane pore size is the most critical factor for the selection of the probes with non-charged membranes. Theoretically, the MWCO of microdialysis membrane three times greater than the molecular weight of the target molecule is sufficient to permit diffusion across the dialysis membrane.[29] The recovery of a membrane probe (for definition see Section 2.3) is directly proportional to membrane length and diameter, e.g. the membrane area. Practically the selection of the membrane area is the result of optimisation or compromise on the recovery and spatial resolution as well as practicality. Typical commercially available probes are from 1 to 30 mm in lengths with an outer diameter (o.d.) from 0.2 to 0.5 mm, respectively. The membrane probe should have a low diffusion resistance to ensure an overall mass transfer coefficient (for definition see Section 2.3).[30] Membrane materials are important too. Different materials and the way the membranes are manufactured will provide membranes with different surface properties, such as electrostatic charge and hydrophilicity. The surface charge can become a dominant factor to determine which protein goes through the membrane, more important than the pore size.[31] This is particularly important for solution systems at low ionic strength. However, the choices of semi-permeable membrane do not much influence *in vivo* microdialysis results.[32] This is possibly because body fluid has a reasonably high ionic strength to provide electrostatic shielding effect and serum proteins are likely to adsorb on the membrane surface to change the surface properties anyway. A variety of hollow-fibre membranes of different materials (polysulfone, polyamide, polycarbonate-polyether copolymer, cuprophan, and polyethersulfone), different surface morphology and pore size are commercially available.

2.3. Recovery of probes

The dialysing properties of microdialysis membrane are routinely expressed as a relative recovery for a particular solute. The relative recovery (*RR*) of the probe is expressed as a percentage of the solute diffused from the probe surrounding into a perfusate.

$$RR = (C_d/C_e) \times 100\% \qquad (1)$$

where C_e is the concentration of a given solute in probe surrounding, e.g. in extracellular fluid, and C_d is the concentration of a solute in the dialysate.

The concentration of a solute in dialysate is equal to the solute concentration in the probe surrounding if the flow rate of the perfusate equals 0. Otherwise the concentration of the solute in dialysate will always be lower than that in the probe surrounding.

RR is dependent on the membrane area (A) and perfusion flow rate (Q) as

$$C_d/C_e = 1 - e^{-(KA/Q)} \qquad (2)$$

where K is the overall mass transfer coefficient.[33]

K is a complex coefficient of the solute mass transport through the sample, the membrane, and the perfusion fluid. The convective resistances in the dialysate (R_d), the diffusional resistance of the membrane (R_m), and mass transfer resistance outside probe, e.g. external factors (R_{ext}) (such as tissue, solution or particles in a complex extracellular matrix) influence the value of K with a negative relationship.[34,35] Reducing the resistances of any factors mentioned above leads to increase the value of K, hence increase the *RR*. However, how to determine the value of K precisely is not easy. In practice, *RR* is determined by various calibration methods experimentally.

2.4. Calibration of probes

2.4.1. *In vitro recovery*

The simplest microdialysis calibration method is a determination of *RR in vitro*, e.g. immersing the probe into a solution of known concentration. Ungerstedt *et al.* first recognised that the dialysate concentration was governed by probe recovery[36] and the *in vitro* calibrated *RR* values have been routinely used to evaluate extracellular concentrations, based on the assumption that recovery for a given probe is constant regardless of the sampling environment.[37] However, increasing evidence has proved that it is not reliable to apply *RR* measured *in vitro* to *in vivo* studies.[38] *In vivo* recoveries, where the probe is inserted into tissues, are expected to be lower than those *in vitro* as the properties of a tissue surrounding the probe reduce the mass transport of substances to the membrane, e.g. an extra-diffusional resistance exists. Obviously the biological factors, such as the tissue tortuosity, free fluid space between the cells, metabolic processes, as well as intra- and extracellular exchange would affect the *RR* in *in vivo* applications.

2.4.2. *Zero-net-flux recovery*

The most directly experimental method of determining *in vivo* recovery and hence extracellular concentrations of a solute is to use zero-net-flux (ZNF) recovery. This method was described by Lönnroth for human subcutaneous measurement of glucose[39] and adenosine.[40] According to the ZNF method the solute of interest is added at known concentrations to the perfusate. The direction of diffusion of the solute is then either into or out of the probe, depending on the applied concentration. Perfusate concentrations are chosen to balance the expected extracellular concentrations, so there will be a point at which there is no net flux across the membrane where the probe neither loses to nor gains the solute from surrounding the probe. At this equilibrium point of ZNF the concentration present in the perfusate is equal to the *in vivo* extracellular concentration. The extracellular concentration and *in vivo* recovery are determined by plotting the difference between perfusate and dialysate concentrations against perfusate concentrations, zero point on the y-axis indicate the extracellular concentration and the gradient of the line is equal to the *in vivo* probe recovery.

ZNF method has been widely applied in practice for the determination of extracellular concentrations of metabolites for a given solute, such as glucose,[41,42] lactate,[43] ascorbate,[44] dopamine,[45] and glutamate.[46] Unfortunately the ZNF method is time consuming. Also it cannot be used to determine the concentration of unknown molecules others than those identified and added into the dialysate.

2.4.3. *Internal reference recovery*

The internal reference calibration as described by Scheller and Kolb[47] is widely used for *in vivo* probe calibration.[48–53] Since diffusion across the membrane is bi-directional it is possible to add to the perfusate a known amount of an identifiable marker that is chemically and structurally the same or similar to the solute of interest, and monitor its loss during the perfusing through the probe. The relative loss (*RL*) is a measurement of a standard solute diffusing from the probe perfusate into the surrounding interstitial fluid. It can be calculated according to the formula:

$$RL = ((C_p - C_d)/C_p) \times 100\% \tag{3}$$

where C_p is the concentration of an internal reference in the perfusate (e.g. fluorescently labelled or radio-labelled substances) and C_d is the concentration of the internal reference in collected dialysate.

RL is assumed to be equal to *RR* if the solute taken as internal reference has the same or similar molecular size and diffusion properties as the solute of interest.[52,53] Knowing *RR*, C_e can be readily determined from the measured C_d (Eq. (1)).

In our study, we aim to investigate soluble macromolecules released from early stage of cultured cells or tissue explants which are completely unknown from previous research. In this case, determination of *RR* using *in vitro* recovery or the ZNF method is not suitable for our purpose; we then focus on the internal reference method. A fluorescence labelled molecule of molecular weight similar to targeted macromolecule can be selected as internal reference and RL of reference molecule can assume to be equivalent to RR of molecule of interest.

2.5. Microdialysis sampling for proteins

Microdialysis currently represents the best available technique for monitoring the effects on release of bio-molecules and assessing physiological or functional changes in the tissue.[54,55] However, in contrast to determination of small solutes in the tissue, sampling and analysing macromolecules such as proteins is highly challenging due to the following reasons:[56]

(a) low concentration of macromolecules in the extracellular fluids;
(b) low recovery due to the low effective diffusivity of macromolecules;
(c) adsorption of proteins onto the probe membrane which is leading to probe fouling;
(d) fluid loss through large pores of probe membrane caused by the difference in osmotic pressure between interstitial fluid and perfusate;

(e) ultrafiltration due to positive pressure inside the probe created by pumping of perfusate; and
(f) difficulty of analysing low concentration of macromolecules in the dialysate.

Thus it is desirable to minimise the fluid loss or gain and increase the relative recovery of macromolecules in order to assess and quantify them. To minimise fluid loss from the probe due to the osmotic effect, the osmolarity of the perfusate is often adjusted to balance the physiological osmolarity by adding osmotic agents, e.g. dextran-70 [57,58] or a protein such as bovine serum albumin (BSA).[59] Creation of transmembrane pressure (*TMP*) by imposing a negative hydrostatic pressure inside of the probe[27] can increase a net fluid flow through the membrane and enhance protein recovery. However this may cause a non-zero permeate flux and the dialysate will gain fluid, which is not desirable for most *in vivo* applications. Decreasing the flow rate, Q, may achieve higher recovery (see Eq. (2)).[60] However, low perfusion flow rates might be hampered by problems associated with sample evaporation as well as poor temporal resolution.[61] Equation (2) also shows that an increase of membrane area will also increase the recovery. However, large probes (either in length or diameter) will reduce spatial resolution and cause increased invasion which is a disadvantage particularly for *in vivo* application of microdialysis.

2.6. Assessment of probe membrane fouling

Fouling has always been a problem when using membranes in complex systems. Torto and co-workers showed that most membranes used for microdialysis sampling exhibit some degree of interaction with proteins.[29] As mentioned in Section 2.4, diffusion process through saline into a probe *in vitro* differs from the diffusion of molecules through tissue *in vivo*; hence the degree of fouling that occurs *in vitro* and *in vivo* may differ. There are few studies on probe fouling during *in vivo* applications of microdialysis. Boubriak *et al.*[62] has demonstrated *in situ* monitoring membrane fouling by monitoring changes of phenol red (PhR) concentration in dialysate during engineered tissue culture. PhR could be a good candidate for monitoring the possible changes in probe recovery because it is routinely present in culture medium at constant concentration as a pH indicator and it is not consumed by the cells during the experiment. Therefore, the RR of phenol red measured by the probe can be used for assessing the occurrence and extent of membrane fouling.

2.7. Microdialysis with different pumping methods

The perfusion of the microdialysis probe can be achieved by three different methods: push, pull, or push-and-pull. Normally the pump is employed to feed perfusate continuously to the probe. This is termed a push pumping system. When a push pump is used, the hydrodynamic pressure within the membrane probe will be higher than that of the atmosphere (the dialysate collector) and tissue culture environment. As most experiments are carried out at atmospheric pressure, there is a net hydrodynamic pressure difference across the membrane with a positive pressure inside the probe.

The fluid loss (F_L) of the membrane probe can be estimated by:

$$F_L = (TMP - \Delta\pi)/\mu R_m \tag{4}$$

where *TMP* is the averaged transmembrane pressure. $\Delta\pi = \pi_p - \pi_e$ where π_p is the osmotic pressure of the perfusate solution and π_e is the osmotic pressure of the extracellular fluid (outside the membrane probe). R_m is the hydrodynamic resistance of the probe membrane.

It is expected that *TMP* will not be high due to the low flow rate (Q). Nevertheless this *TMP* may cause fluid loss through the membrane probe. In practice, fluid loss can be minimised by increasing the osmotic pressure of the perfusate (π_p). As π_e is usually physiological (i.e. constant), increasing π_p would reduce the driving force and hence fluid loss (see Eq. (4)).

Alternatively the pump may be connected to the probe outlet and used to draw the perfusate through the probe; this is termed a "pull" pumping system. When a pull pump is used, the hydrodynamic pressure within the membrane probe is lower than that of extracellular fluid. Equation (4) is still valid but the probe would gain fluid. This also should be avoided.

Although adjusting the osmotic pressure of perfusion fluid can minimise fluid loss or gain, it is difficult to eliminate this problem completely because the pressure profile within the probe is difficult to predict precisely. The way to eliminate fluid loss or fluid gain is to use two pumps and perfuse the fluid with a "push-and-pull" method[26] in which the perfusate is simultaneously pumped in and withdrawn from the probe at equal rates.

However, it was found that the pumping methods affect not only the fluid loss but also the *RR* of interested molecules. Experimental results indicated the *RR* of small molecules is less affected, but the *RR* of protein molecules varied greatly depending on pumping systems.[56] Hence selection of pumping methods becomes critically important when the target molecules are macromolecular markers, such as those for extracellular matrix turnover.

3. Microdialysis for Tissue Engineering Monitoring — Case Studies

Microdialysis probes have been used to monitor 3D tissue growth[62,63] and tissue explants culture,[55] by collecting soluble molecular markers which indicate the cell viability, type, functions and extracellular matrix turnover. The possibility of monitoring chemical gradients within a tissue engineered construct also attempted to determine local changes in cell metabolism.[62,63] Methods of probe calibration *in situ* and identifying any possible probe fouling have been developed using phenol red (present in the culture media), radioactively labelled methyl-glucose, and fluorescent labelled dextrans.[55,62,63]

3.1. Experimental method

In the engineered cartilage case, a cylindrical construct consisting of bovine articular chondrocytes in a 3D alginate matrix was formed inside a bioreactor. One or more microdialysis probes were inserted into the construct at known locations to monitor nutrient and metabolite gradient on the construct. For tissue explants study, intact bovine intervertebral discs dissected from adjacent vertebral bodies were similarly inserted into a bioreactor with mechanical loading system. The top and bottom surfaces of the engineered construct or disc were perfused with a tissue culture medium to supply the cells with nutrients and to remove metabolic waste. Nutrient and metabolite gradients were formed inside the 3D construct as a result of cell metabolic activity. Concentrations of nutrients and metabolites were dependent on rates of cellular activity and rates of nutrient/metabolite supply or removal. Nutrient and metabolite concentrations at defined locations within the construct were continuously monitored using the microdialysis probes. Cell responses known to affect chondrocyte metabolism such as changes in medium pH, osmolarity, serum concentration as well as tissue explant metabolic changes under mechanical loading stimulation were monitored. Fluorescent dextran and phenol red were used to determine *in situ* the relative recovery of the solute of interest (e.g. bio-molecules), and to detect membrane fouling, respectively. The dialysate was collected daily and the samples were analysed using FPLC, SDS-PAGE and Western blotting. At the end of the culture period (up to 14 days), the construct or disc was analysed histologically stained for live-dead assay, and analysed for spatial variations in cell viability, cell number, glycosaminoglycan (GAG) and collagen content, to validate and correlate to the microdialysis measurements.

For measurement of concentrations of low molecular weight solutes (<1 kD) in alginate gels or alginate-chondrocyte constructs, autoclavable microdialysis probes of standard design with a polyethersulfone (PES) dialysis membrane of 15 kDa cut-off , an effective length of 4 mm, outer diameter (o.d.) of 0.6 mm and a 35 mm flexible polyurethane shaft were used.[62,63] For monitoring macromolecular markers during early stage of cell and tissue culture,[55] CMA/20 probes with 100 kD MWCO membrane or in-house probe with 3000 kDa MWCO membrane and an effective length of 10 mm, outer diameter of 0.4 mm and a 25 mm flexible polyurethane shaft were used.

Probe recovery in our experiments was expressed as a percentage of the concentration of the solute in the probe *dialysate* relative to that in the surrounding external solution (relative recovery). In the experiments, *in situ* probe relative recovery was assessed on the basis of the percentage of phenol red in the dialysate after equilibration of the construct with DMEM containing phenol red.[55,62,63] In this way the relative recovery of the probe was continuously monitored during the experiment to identify any membrane fouling. As the relative recovery depends on molecular weight of the molecules, several tracer molecules were used for the relative loss measurements to determine the relative recovery of molecules of similar size, including fluorescent dextran (40 kD, for protein markers)[55] and radioactively labelled 3-methyl-glucose and lactate (for glucose and lactate).[63]

3.2. Internal standard calibration of probes and of membrane fouling

To determine the value of *RR* or *RL*, and check whether they change with time, we have used three independent approaches.

(a) Addition of known concentration of ^{14}C-labelled lactate (for lactate) or ^{14}C-3-methyl-D-glucose (not metabolised by cells) into the perfusate as the internal standard, to measure *RL*, which would give the *RR* values of lactate and glucose.[63]
(b) Measurement of the concentration of phenol red (PhR) in the dialysate. To determine the total concentration of PhR, the $HPhR^-$ form (yellow) in dialysate was first converted to PhR^{2-} (red) by adding 2 μl of 5 M NaOH to 40–50 μl of dialysate (final pH 12.5). Absorbance of the samples was read at 540 nm on the plate reader (TECAN, GENios). The *RR* of phenol red (coming from the cell culture medium) can then be determined.[62]
(c) Addition of fluorescent dextrans of molecular weight of 40 kD as an internal standard to measure the *RL* for the evaluation of the *RR* of proteins of similar size.[55]

Based on validation of microdialysis calibration before experimental measurement and cross-validation of these *in situ* methods, it is concluded that (i) these calibration methods are reliable, and (ii) the measured *RR* and/or *RL* do not change with time, after an initial period of "adaptation", and hence the membrane probes are not fouled significantly over a period of up to eight weeks.

3.3. Monitoring cell metabolic activities in engineered cartilage

Figure 2 shows the typical results of the monitoring of lactic acid, a main cell metabolite, within 3D constructs seeded with chondrocytes.[63] Concentrations of lactic acid within the constructs reached a steady value within 30 hours of initial seeding. Cells have reacted to the decrease of glucose content and increase of osmolarity in the medium supplied to the construct by alteration of their metabolic activity consequently. The effect of glucose concentration, osmolarity and addition of foetal bovine serum (growth factors) on the cell metabolic activity and glucose and lactic acid gradients within the 3D construct were successfully studied by microdialysis.[63] Further glucose and lactic acid concentration gradients were observed in the 3D constructs as a result of the cell metabolic activity. Low glucose concentrations or high levels of lactic acid in the construct centre were found to be linked to the loss of cell viability and low rate of matrix accumulation.[62] In large constructs (> 4 mm) viable cells were mainly concentrated near the nutrient supply unless the construct was seeded at very low cell densities ($<2 \times 10^6$ cells/ml).[63]

3.4. Monitoring tissue metabolism in cultured IVD explants

Sampling macromolecular markers is necessary for the monitoring of extracellular matrix turnover and other specific functions such as protein synthesis. A commercial

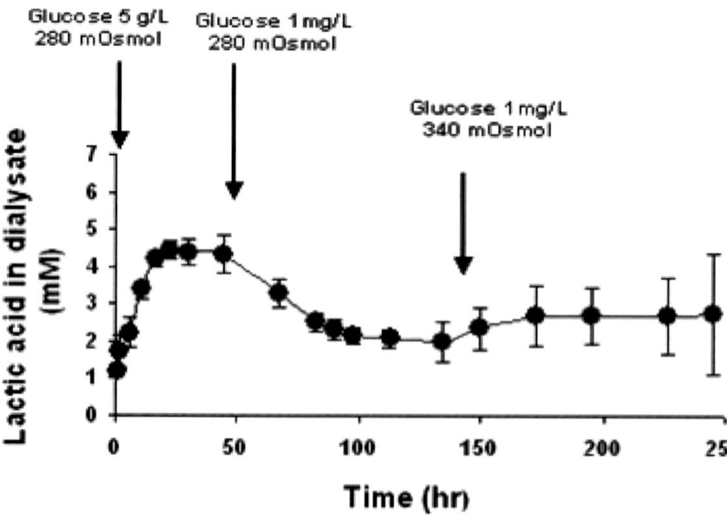

Fig. 2. Glucose and osmolarity affect chondrocyte metabolic activity as measured by micro-membrane probe.[63]

microdialysis probe of 100 kDa MWCO and in-house microdialysis probe of 3000 kDa of MWCO were used for monitoring both small metabolites (MW < 1 kDa) and soluble macromolecules (MW > 20 kDa) in a bovine intervertebral disc (IVD) explant for up to one week. The *in situ* calibration technique (measuring relative loss of fluorescent 40 kDa dextran and phenol red) was proved valid in the experiments and membrane fouling was not significant. The relative recovery of macromolecules of interests remained roughly stable at 9%. The tissue metabolism and soluble proteins released from the extracellular matrix during the seven-day culture were continuously monitored to investigate the effect of different loads. We identified three soluble matrix proteins, which we found were clearly related to cellular activities of chondrocytes and effect on the changes of culture conditions. One of these, which was clearly related to cellular activities of chondrocytes, has now been identified successfully using mass spectrometry and Western blotting (antibodies kindly supplied by Dr. A. Recklies, Shriners Inst, Montreal) as chitinase-3-like protein-1 (CHI3L1). Figure 3 shows the analyses of dialysates collected by a 100 kD microdialysis probe.[55] The effect of physico-chemical and mechanical stimuli (e.g. osmolarity, pH, oxygen tension and mechanical load) on secretion of CHI3L1 by cultured IVD explant and isolated disc cells encapsulated in alginate beads were investigated. The production of CHI3L1 was correlated with the production of lactic acid for both cell types and this could be readily monitored by microdialysis. The concentrations of glucose and lactate within the IVD explant, and their change with mechanical loading, were readily measured.[55]

3.5. Monitoring tissue formation

As neo-tissue is formed during engineered tissue development, the accumulation of extracellular matrix as well as cell proliferation (not much for cartilage) will

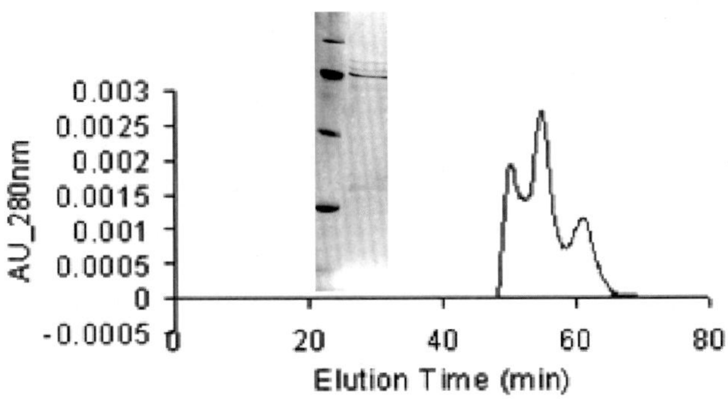

Fig. 3. Protein assays using FPLC and SDS-PAGE with silver stain (left inset). Lane 1: molecular mass standard (Bio-Rad). Lane 2: crude dialysate.[55]

hinder the solute diffusion within the construct. Hence by measuring the effective diffusivity of known solutes and its increase with time, tissue formation can be quantified.

We embedded four microdialysis probes into a cartilage construct (chondrocyte seeded in alginate gel). Probe 2 was used to release a labelled solute; the capture of the labelled solute by the other probes was monitored. The solutes tested to date were ^{14}C-lactate, ^{14}C-methyl-D-glucose, and fluorescently labelled 10 kDa dextran. The profile of appearance of the solutes in the dialysate collected by probes 1, 3 and 4, e.g. concentration variation with time was measured, and was characterised by the time for the concentration of the solute in dialysate reaching to its maximum (t_{max}) and the time at which the concentration of the solute in dialysate decreased to one-half of its maximal value ($t_{50\%}$). Figure 4 shows the

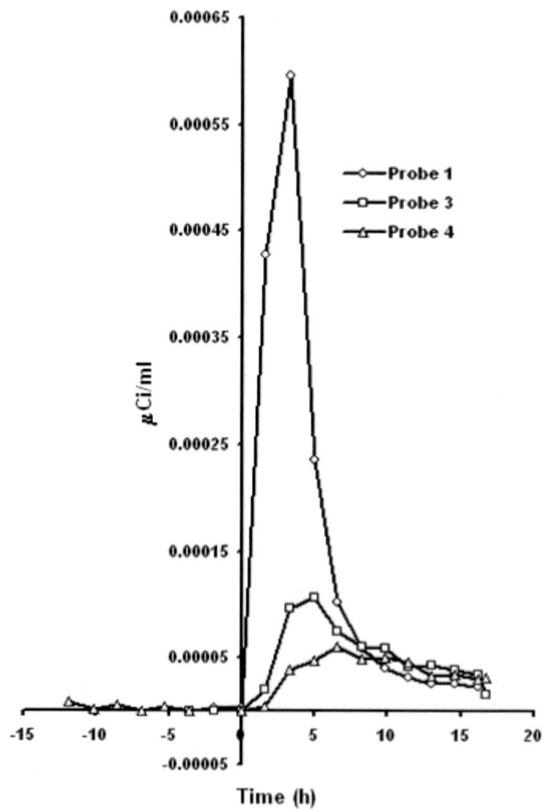

Fig. 4. Typical results showing capture of ^{14}C-methyl-D-glucose probe after a pulse released by probe 2. Probes 1, 3 and 4 were positioned from probe 2 on the distance of 2, 3.5 and 6.4 mm, respectively. t_{max} falls with distance from probe 2 while $t_{50\%}$ increases.

Table 1. Measurement of tissue formation by ^{14}C - methyl-D-glucose release. Change in diffusivity parameters with time in the culture indicates the increase in diffusion resistance and hence increase in tissue accumulated.

Probe no.	Distance (mm)	3 days		12 days	
		t_{max}	$t_{50\%}$	t_{max}	$t_{50\%}$
		h	*h*	*h*	*h*
Probe 1	2	3	3.1	3.3	4.75
Probe 3	3.5	4.5	5.1	4.95	10.4
Probe 4	6.4	—	—	6.6	16.1

typical profile of monitoring of ^{14}C-methyl-D-glucose diffusion into three probes after its pulse release by probe 2 in the construct. Table 1 gives t_{max} and $t_{50\%}$ values at two time intervals. As seen in Table 1, both values increase over culture time due to the increase of diffusion resistance caused by extracellular matrix accumulation. Further analysis to link the increase in diffusion hindrance in the growing tissue with the neo-tissue formation quantitatively is required to characterise the tissue growth.

4. Summary

(1) The concept of using microdialysis probes to monitor cell metabolic activities, tissue functions and their transient changes across 3D tissue engineered constructs and IVD explant was shown. Using this minimally invasive technique, the nutrient concentration distribution, cell metabolic activity, and cell functions in terms of protein release could be readily measured.

(2) Apart from cell metabolic activity, the development of microdialysis helps to analyse the investigation of cell and tissue functions. The methodology of application of microdialysis probes with large pore size membrane for sampling of macromolecular bio-functional markers was established. The effects of pumping methods, affecting the transmembrane pressure and hence the fluid balance, and their influence the relative recovery of small molecules and proteins were experimentally studied. The validity of the internal reference *in situ* calibration was examined in detail. It was concluded that a push-and-pull system was the most effective method for the elimination of fluid loss or gain during the sampling by the probes of high MWCO. Sampling by the probes of low MWCO was not affected by pumping methods. However, the relative recovery of macromolecules was significantly influenced by them.

(3) The *in situ* calibration technique using phenol red as internal reference for small molecules (lactate, glucose) and fluorescent dextrans 10–70 kDa as internal references for the proteins of similar MW range can provide reliable RR values.
(4) Calibration of the probes *in situ* to identify any possible probe fouling was developed using phenol red (present in the culture media), radioactively labelled methyl-glucose, and fluorescent labelled dextrans.
(5) Substantial effort was directed towards the identification of functional biomolecules collected by microdialysis sampling. Amongst several proteins sampled, CHI3L1, a major soluble protein secreted by cultured IVD explants and chondrocytes, was identified and established as a suitable functional bio-molecule for cell metabolism during the early stage of cell and tissue culture. The effect of physico-chemical and mechanical stimuli (e.g. osmolarity, pH, oxygen tension and mechanical load) on secretion of CHI3L1 by cultured IVD cells and IVD explants were investigated. CHI3L1 release was sensitive to physico-chemical stimulation and this could be readily monitored with microdialysis.
(6) Using a multiple probe approach, tissue formation could be monitored through determination of the decrease of effective diffusivity in the extra-cellular matrix by measuring the time delay between tracers released from one probe and captured by other probes.
(7) The developed system has been successfully used to monitor metabolism within cultured tissue explants and the established technique could be applied to monitor tissue repair and tissue integration *in vivo*.

References

1. Seidel, J.O., Pei, M. *et al.* Long-term culture of tissue engineered cartilage in a perfused chamber with mechanical stimulation. *Biorheology* **41**, 445–458 (2004).
2. Freyria, A.M., Cortial, D. *et al.* Influence of medium composition, static and stirred conditions on the proliferation of and matrix protein expression of bovine articular chondrocytes cultured in a 3-D collagen scaffold. *Biomaterials* **25**, 687–697 (2004).
3. Seguin, C.A., Grynpas, M.D. *et al.* Tissue engineered nucleus pulposus tissue formed on a porous calcium polyphosphate substrate. *Spine* **29**, 1299–1306 (2004).
4. Lohmeier-Vogel, E., Skoog, K. *et al.* ^{31}P-nuclear magnetic resonance study of the effect of azide on xylose fermentation by *Candida tropicalis*. *Appl. Env. Microbiol.* **55**, 1974–1980 (1989).
5. Dijkema, C., de Vries, S.C. *et al.* Substrate utilization by suspension cultures and somatic embryos of *Daucus carota* L measured by ^{13}C NMR. *Plant Physiol.* **88**, 1332–1337 (1988).

6. Potter, K., Butler, J.J. *et al.* Response of engineered cartilage tissue to biochemical agents as studied by proton magnetic resonance microscopy. *Arthritis Rheum.* **43**, 1580–1590 (2000).
7. Boyd, S.K., Davison, P. *et al.* Monitoring individual morphological changes over time in ovariectomized rats by *in vivo* micro-computed tomography. *Bone* **39**, 854–862 (2006).
8. Xu, X., Wang, R.K. *et al.* Investigation of changes in optical attenuation of bone and neuronal cells in organ culture or three-dimensional constructs in vitro with optical coherence tomography: relevance to cytochrome oxidase monitoring. *Eur. Biophys. J.* **32**, 355–362 (2003).
9. Martin, I., Obradovic, B. *et al.* Method for quantitative analysis of glycosaminoglycan distribution in cultured natural and engineered cartilage. *Ann. Biomed. Eng.* **27**, 656–662 (1999).
10. Obradovic, B., Meldon, J.H. *et al.* Glycosaminoglycan deposition in engineered cartilage: experiments and mathematical model. *AICHE J.* **46**, 1860–1871(2000).
11. Martin, I., Wendt, D. *et al.* The role of bioreactors in tissue engineering. *Trends Biotechnol.* **22**, 80–86 (2004).
12. Freyer, J.P. and Sutherland, R.M. Proliferative and clonogenic heterogeneity of cells from EMT6/Ro multicellular spheroids induced by the glucose and oxygen supply. *Cancer Res.* **46**, 3513–3520 (1986).
13. Bito, L., Davson, H. *et al.* The concentrations of free amino acids and other electrolytes in cerebrospinal fluid, *in vivo* dialysate of brain, and blood plasma of the dog. *J. Neurochem.* **13**, 1057–1067 (1966).
14. Jones, D.A., Ros, J. *et al.* Dynamic changes in glucose and lactate in the cortex of the freely moving rat monitored using microdialysis. *J. Neurochem.* **75**, 1703–1708 (2000).
15. Henriksson, J. Microdialysis of skeletal muscle at rest. *Proc. Nutr. Soc.* **58**, 919–923 (1999).
16. Wientjes, K.J. and Schoonen, A.J. Determination of time delay between blood and interstitial adipose tissue glucose concentration change by microdialysis in healthy volunteers. *Int. J. Artif. Organs* **24**, 884–889 (2001).
17. Herkner, H., Muller, M.R. *et al.* Closed-chest microdialysis to measure antibiotic penetration into human lung tissue. *Am. J. Respir. Crit. Care Med.* **165**, 273–276 (2002).
18. Baicu, S.C., Simmons, P.M. *et al.* Interstitial fluid analysis for assessment of organ function. *Clin. Transplant.* **18**, 16–21 (2004).
19. Nowak, G., Ungerstedt, J. *et al.* Metabolic changes in the liver graft monitored continuously with microdialysis during liver transplantation in a pig model. *Liver Transplant.* **8**, 424–432 (2002).

20. Tunblad, K., Hammarlund-Udenaes, M. *et al.* An integrated model for the analysis of pharmacokinetic data from microdialysis experiments. *Pharm. Res.* **21**, 1698–1707 (2004).
21. Wu, Y.S., Tsai, T.H. *et al.* Determination of pyruvate and lactate in primary liver cell culture medium during hypoxia by on-line microdialysis-liquid chromatography. *J. Chromatogr.* **913**, 341–347 (2001).
22. de la Peña, A., Liu, P. *et al.* Microdialysis in peripheral tissues. *Adv. Drug Del. Rev.* **45**, 189–216 (2000).
23. Langberg, H., Skovgaard, D. *et al.* Type I collagen synthesis and degradation in peritendinous tissue after exercise determined by microdialysis in humans. *J. Physiol.* **521**, 299–306 (1999).
24. Karamouzis, M., Langberg, H. *et al. In situ* microdialysis of intramuscular prostaglandin and thromboxane in contracting skeletal muscle in humans. *Acta Physiol. Scand.* **171**, 71–76 (2001).
25. Dostálová, I., Pacák, K. *et al.* Application of *in vivo* microdialysis to measure leptin concentration in adipose tissue. *Int. J. Biol. Macromol.* **32**, 205–208 (2003).
26. Sjögren, F., Svensson, C. *et al.* Cutaneous biology: technical prerequisites for *in vivo* microdialysis determination of interleukin-6 in human dermis. *Br. J. Dermatol.* **146**, 375–382 (2002).
27. Rosenbloom, A.J., Sipe, D.M. *et al.* Microdialysis of proteins: performance of the CMA/20 probe. *J. Neurosci. Methods* **148**, 147–153 (2005).
28. Ao, X. and Stenken, J.A. Microdialysis sampling of cytokines. *Methods* **38**, 331–341 (2006).
29. Torto, N., Bâng, J. *et al.* Optimal membrane choice for microdialysis sampling of oligosaccharides. *J. Chromatogr. A* **806**, 265–278 (1998).
30. Torto, N., Mwatseteza, J. *et al.* Microdialysis sampling — challenges and new frontiers. *LCGC* **19**, 462–475 (2001).
31. Wan, Y. H., Ghosh, R. *et al.* Fractionation of bovine serum albumin and monoclonal antibody alemtuzumab using carrier phase ultrafiltration. *Biotechnol. Bioeng.* **90**, 303–315 (2005).
32. Chaurasia, C.S. *In vivo* microdialysis sampling: theory and applications. *Biomed. Chromatogr.* **13**, 317–332 (1999).
33. Jacobson, I., Sandberg, M. *et al.* Mass transfer in brain dialysis devices — a new method for the estimation of extracellular amino acid concentration. *J. Neurosci. Methods* **15**, 263–268 (1985).
34. Hsiao, J.K., Ball, B.A. *et al.* Effects of different semipermeable membranes on *in vitro* and *in vivo* performance of microdialysis probes. *J. Neurochem.* **54**, 1449–1452 (1990).

35. Bungay, P.M., Morrison, P.F. *et al.* Steady-state theory for quantitative microdialysis of solutes and water *in vivo* and *in vitro*. *Life Sci.* **46**, 105–119 (1990).
36. Ungerstedt, U., Herrera-Marschitz, M. *et al.* Dopamine synaptic mechanisms reflected in studies combining behavioural recordings and brain dialysis. In: Kotisaka, M. (ed.), *Advances in Dopamine Research* (Pergamon, New York, 1982), pp. 219–231.
37. Song, Y. and Lunte, C.E. Comparison of calibration by delivery versus no net flux for quantitiative *in vivo* microdialysis sampling. *Anal. Chem. Acta* **379**, 251–262 (1999).
38. Chen, K.C., Höistad, M. *et al.* Quantitative dual-probe microdialysis: mathematical model and analysis. *J. Neurochem.* **81**, 108–121 (2002).
39. Lönnroth, P., Jansson, P.A. *et al.* A microdialysis method allowing characterisation of water space in humans. *Am. J. Physiol.* **253**, E228–231 (1987).
40. Lönnroth, P., Jansson, P.A. *et al.* Microdialysis of intracellular adenosine concentration in subcutaneous tissue in humans. *Am. J. Physiol.* **256**, E250–255 (1989).
41. Bolinder, J., Hagstrom, E. *et al.* Microdialysis of subcutaneous adipose tissue *in vivo* for continuous glucose monitoring in man. *Scand. J. Clin. Lab. Invest.* **49**, 465–474 (1989).
42. Fray, A.E., Boutelle, M. *et al.* Extracellular glucose turnover in the striatum of unanaesthetized rats measured by quantitative microdialysis. *J. Physiol. (Lond.)* **504**, 721–726 (1997).
43. Kurosawa, M., Hallstrom, A. *et al.* Changes in cerebral blood flow do not directly affect *in vivo* recovery of extracellular lactate through microdialysis probe. *Neurosci. Lett.* **126**, 123–126 (1991).
44. Miele, M. and Fillenz, M. *In vivo* determination of extracellular brain ascorbate. *J. Neurosci. Methods* **70**, 15–19 (1996).
45. Justice, J.B., Jr. Quantitative microdialysis of neurotransmitters. *J. Neurosci. Methods* **48**, 263–276 (1993).
46. Miele, M., Berners, M. *et al.* The determination of the extracellular concentration of brain glutamate using quantitative microdialysis. *Brain Res.* **707**, 131–133 (1996).
47. Scheller, D. and Kolb, J. The internal reference technique in microdialysis: a practical approach to monitoring dialysis efficiency and to calculating tissue concentration from dialysate samples. *J. Neurosci. Methods* **40**, 31–38 (1991).
48. Larsson, C.I. The use of an "internal standard" for control of the recovery in microdialysis. *Life Sci.* **49**, 73–78 (1991).
49. Yokel, R.A., Allen, D.D. *et al.* Antipyrine as a dialyzable reference to correct differences in efficiency among and within sampling devices during *in vivo* microdialysis. *J. Pharmacol. Toxicol. Methods* **27**, 135–142 (1992).

50. Lönnroth, P. and Strindberg, L. Validation of the "internal reference technique" for calibrating microdialysis catheters *in situ*. *Acta Physiol. Scand.* **153**, 375–380 (1995).
51. Clement, R., Malinovsky, J.M. *et al. In vitro* and *in vivo* microdialysis calibration using retrodialysis for the study of the cerebrospinal distribution of bupivacaine. *J. Pharm. Biomed. Anal.* **17**, 665–670 (1998).
52. Rizell, M., Naredi, P. *et al.* Validation of the internal reference technique for microdialysis measurement of interstitial histamine in the rat. *Life Sci.* **65**, 1765–1771 (1999).
53. Strindberg, L. and Lonnroth, P. Validation of an endogenous reference technique for the calibration of microdialysis catheters. *Scand. J. Clin. Lab. Invest.* **60**, 205–211 (2000).
54. Kendrick, K.M. Microdialysis measurement of *in vivo* neuropeptide release. *J. Neurosci. Methods* **34**, 35–46 (1990).
55. Li, Z., Boubriak, O.A. *et al.* Microdialysis for monitoring the process of functional tissue culture. *Int. J. Artif. Organs* **29**, 858–865 (2006).
56. Li, Z., Hughes, D. *et al.* Effect of pumping methods on transmembrane pressure, fluid balance and relative recovery in microdialysis. *J. Memb. Sci.* **310**, 237–245 (2008).
57. Rosdahl, H., Ungerstedt, U. *et al.* Interstitial glucose and lactate balance in human skeletal muscle and adipose tissue studied by microdialysis. *J. Physiol.* **471**, 637–657 (1993).
58. Rosdahl, H., Hamrin, K. *et al.* A microdialysis method for the *in situ* investigation of the action of large peptide molecules in human skeletal muscle: detection of local metabolic effects of insulin. *Int. J. Biol. Macromol.* **28**, 69–73 (2000).
59. Trickler, W.J. and Miller, D.W. Use of osmotic agents in microdialysis studies to improve the recovery of macromolecules. *J. Pharm. Sci.* **92**, 1419–1427 (2003).
60. Kjellstrom, S., Appels, N. *et al.* Microdialysis — a membrane based sampling technique for quantitative determination of protein. *Chromatographia* **50**, 539–546 (1999).
61. Menacherry, S., Hubert, W. *et al. In vivo* calibration of microdialysis probes for exogenous compounds. *Anal. Chem.* **64**, 577–583 (1992).
62. Boubriak, O.A., Urban, J.P.G. *et al.* Monitoring of lactate and glucose in tissue engineered constructs by microdialysis. *J. Memb. Sci.* **273**, 77–83 (2006).
63. Boubriak, O.A., Urban, J.P. *et al.* Monitoring of metabolite gradients in tissue-engineered constructs. *J. R. Soc. Interface* **3**, 637–648 (2006).

Chapter 20

Characterisation of Tissue Engineering Constructs by Raman Spectroscopy and X-ray Micro-Computed Tomography (μCT)

Ioan Notingher and Julian R. Jones

Abstract

In regenerative medicine it is important to be able to understand how cells are behaving in response to stimuli. The stimuli can be biological signals or materials. Raman spectroscopy allows the non-invasive real time monitoring of live cells *in vitro* by interpretation of spectra. Materials are being developed to use as templates (scaffolds) for tissue regeneration. The morphology of the pore structure is critical if tissue is to populate the scaffold. X-ray microcomputed tomography is the only method that can obtain 3D images of pore networks. Novel image analysis has been developed that can quantify pore networks. There is potential for this technique to be used to image tissue growth into scaffolds *ex vivo*. The next challenge is to adapt these two promising techniques to monitor the response of cells to porous scaffolds, including that of cells within the porous network.

Keywords: Raman Spectroscopy; Non-Invasive; Tissue Monitoring; X-Ray Micro-Computed Tomography; Scaffolds.

Outline

1. Introduction
   - 1.1. Tissue engineering constructs
   - 1.2. The need for characterisation
   - 1.3. Optical microscopy and Raman spectroscopy
   - 1.4. Characterising porous materials in 3D

2. Principles and Instrumentation
   2.1. Raman spectroscopy
   2.2. X-ray micro-computed tomography (μCT)
3. Applications of Raman Micro-Spectroscopy to Cells
   3.1. Raman spectra of biomolecules and cells
   3.2. Differentiation of stem cells
   3.3. Phenotypic characterisation of cells
   3.4. Bone nodule formation and mineralisation *in vitro*
4. Application of μCT to the Quantification of Scaffolds
   4.1. Quantification of open pore networks
   4.2. Monitoring of cells within scaffolds
5. Conclusions
   References

1. Introduction

1.1. Tissue engineering constructs

There are two main strategies for the regeneration of diseased or damaged tissues using these materials; tissue engineering or *in situ* tissue regeneration. Each of these strategies can use cells (usually stem cells) or biomaterials or combinations of cells and materials.

Tissue engineering involves the growth (engineering) of a tissue or tissue construct in a laboratory and implanting it. A tissue construct can be created by seeding cells on a template (scaffold). The scaffold should degrade over time into safe degradation products. The construct can either be implanted as a tissue/material composite, or the scaffold can be degraded in the laboratory so that only the tissue is implanted.

For *in situ* tissue regeneration stem cells can be treated in the laboratory with biological signals and then injected to the host site. Alternatively, a biologically active material can be directly implanted into the defect site, where it must recruit and stimulate cells in the body to repair the tissue. The implant must then degrade at the same rate the tissue re-grows.

1.2. The need for characterisation

Tissue regeneration strategies are relatively new and largely unproven. In many cases, the mechanisms of the positive effects that certain materials and signals have on cells are unkown. This will need co-operation between engineers and biologists. Biologists traditionally have used optical microscopy, including fluorescence and confocal, to image and observe cell responses to biological factors. There are two main problems with these techniques. First, immunohistochemical techniques that allow the use of fluorescence are destructive as the cells must be fixed. Ideally characterisation of the cells would be in real time. Second, cells are often grown on tissue culture plastic, a thin, two-dimensional (2D), optically transparent material. However, scaffold materials are often three-dimensional (3D), non-transparent materials, many with complex macroporous networks. This means that conventional optical microscopy cannot be used to image the materials or the cellular response to them. Confocal microscopy can be used to look at cells on the surface of flat surfaces of materials or within shallow pores of a surface, but not *within* the material.

One of the criteria for an ideal scaffold for bone regeneration is that it has an interconnected porous structure, similar to cancellous bone. The interconnects should be at least 100 μm in diameter, otherwise blood vessels will not be able to penetrate the scaffold and any new bone generated will die. Scaffolds are being designed to optimise the material properties (pore network and nanotopography) for optimised cellular response. It is imperative therefore that techniques are developed that can monitor cell behaviour, quantify scaffold morphology and ideally, monitor cell behaviour when they are growing on or in the scaffolds.

This chapter will first discuss how the technique of Raman spectroscopy can be used to obtain real time data of cells in culture, and its application to cells on materials. It will then describe advances in 3D imaging of materials using X-ray micro-computed tomography (μCT) and quantification of the images and discuss the potential for this technique to be used in monitoring cellular response to 3D materials.

1.3. Optical microscopy and Raman spectroscopy

Raman spectroscopy uses light to probe live cells and provide information regarding the intrinsic molecular properties of cells. This technique offers unique advantages for studying live cells and monitoring growing tissue engineering grafts as it does not require invasive procedures and can achieve the high-spatial resolution needed for studying single cells. Conventional optical microscopy allows cell biologists to observe the morphology of live cells but the information obtained is usually limited to qualitative evaluations. Additional information can be obtained by measuring the fluorescence emission of certain cellular components. However, fluorescence experiments have limited applications when applied to single living cell studies because the number of fluorescent cellular components is small and the broad bands present in fluorescence spectra limit the amount of information which can be obtained. Techniques based on transfection of cells with plasmids designed to ligate proteins and peptides to fluorescent labels, such as green fluorescent protein, have become popular in the last two decades. However, genetic manipulation of cells is expensive and labour intensive as selection protocols need to be developed for particular genes.

Compared to fluorescence spectroscopy, which involves absorption and re-emission of light following the excitation of electrons in biomolecules, vibrational spectroscopy uses light to probe the vibrations of molecules. This method has a higher chemical specificity since the frequencies of molecular vibrations depend on the atomic species found in molecules but also on bond strength and angles, molecular structure and environment. Optical techniques based on vibrational spectroscopy are well suited for non-invasive biochemical analysis of samples at a micrometre scale. In particular, Raman micro-spectroscopy has a clear advantage over infrared spectroscopy due to reduced background signals from the culture medium and intracellular water. Additionally, well-established optical instrumentation developed to work in the visible range of the electromagnetic range, such as optical microscopes, can be attached to Raman spectrometers in order to obtain the micro-scale spatial resolution required to study individual cells. In Raman spectroscopy, both excitation and detection is performed optically without the need of cell staining or labelling. The information obtained from Raman spectra represents a fingerprint of molecular vibrations, which have high chemical specificity.

Since the development of lasers in 1960s, Raman micro-spectroscopy has been extensively used in biology and medicine, from studies of isolated biopolymers to complex tissues.[1–4] The potential of Raman micro-spectroscopy arises from its ability to detect biochemical changes in cells at a molecular level, and therefore, can be used for diagnosis, or as a tool for developing therapies, as well as testing and evaluation of drugs. Additionally, Raman spectroscopy is able to provide information regarding the extracellular matrix and mineral phase present in certain tissues such as bone. During tissue development, the time evolution of the mineral phase and extracellular matrix are affected by various environmental conditions. Raman spectroscopy has been proven to provide useful information regarding transformation of the initial mineral phase and its maturation.

Due to its ability to study both live cells and properties of biomaterials, Raman spectroscopy may be further developed for non-invasive monitoring the overall development of engineered tissues and follow in real-time the interactions between cells and scaffold materials. It is well-known that biomaterials have an active role in the development of tissues *in vitro* as they are designed to influence cell adhesion, differentiation and proliferation, and modulate gene expression.[5] Considering the high-chemical specificity of Raman spectroscopy, it may be possible to sample individual cells and the biomaterial for hours, days or weeks, in order to monitor spectral changes that can be correlated with changes in the cell phenotype, cell growth as well as biodegradability of the scaffold, release of bio- and chemical stimuli, and other parameters.

1.4. Characterising porous materials in 3D

One of the most important criteria of an ideal scaffold is that it must act as a 3D template for tissue growth. That means it must contain an interconnected macroporous network. For bone regeneration the interconnects must have a diameter of at least 100 μm to stimulate bone ingrowth and vitally blood vessel ingrowth, without which new bone will die.[6] It is therefore important to be able to quantify both the pores and interconnects to optimise tissue scaffolds. The interconnectivity of the pores also dominates the flow properties, which will ensure adequate delivery of cells during seeding and nutrients during subsequent culture.[7] Therefore it would also be important to be able to predict the permeability of the scaffolds and the likelihood of cells and blood vessels penetrating the constructs.

Traditionally, porous materials have been characterised with scanning electron microscopy (SEM) and mercury intrusion porosimetry (MIP).[8] However these characterisation techniques have several disadvantages. SEM micrographs can be misleading as they are images of fracture surfaces; therefore the real 3D shape and connectivity of the pores cannot be imaged or quantified without using stereology. Unless a field emission gun (FEG) SEM is available polymeric and ceramic scaffolds have to be coated, as they are non-conductive, making the technique destructive. Mercury intrusion porosimetry works by forcing mercury into a sample. It is therefore destructive. It measures volume of mercury intruded as a function of pressure applied, and uses mathematical models to calculate a pore size distribution[9] and only measures a distribution of constrictions in a pore network, i.e. it provides only an indirect indication of the interconnect distribution in a pore network and no information on shape or pore size.

For clinical trials, or product quality assurance, it is vital to know the exact morphology of the pore network before implantation, therefore a non-destructive characterisation technique is required. X-ray micro-computed tomography (μCT) is one such technique. Figure 1 shows μCT images of human trabecular bone and a typical bioactive glass scaffold. The scaffold was produced by foaming sol-gel derived bioactive glass.[8] The bone was sectioned from a femur of an adult female, 1 cm from the centre of the medial condyle. The macroporous network of the bioactive glass scaffold is highly interconnected and qualitatively very similar to the μCT image of trabecular bone. However, what is needed is a quantitative comparison.

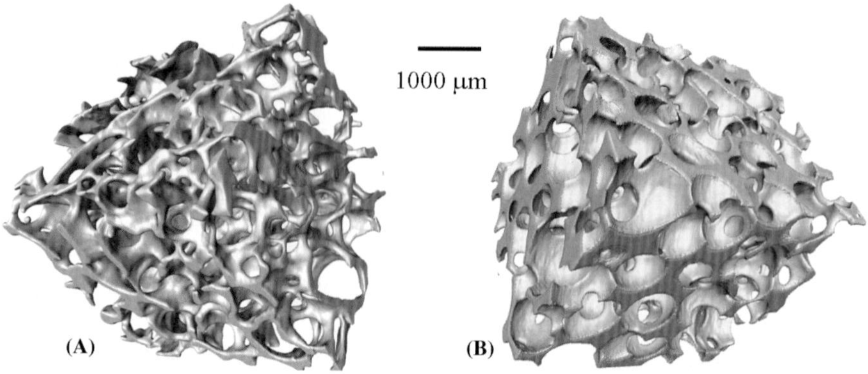

Fig. 1. μCT reconstructions of **(A)** human trabecular bone, and **(B)** a bioactive glass foam scaffold.

Reliable methods to quantify the pore networks of scaffolds are required to ensure scaffolds are suitable for tissue regeneration and to enable standards for their regulatory approval to be developed.

2. Principles and Instrumentation

2.1. Raman spectroscopy

Raman spectroscopy is based on inelastic scattering (Raman scattering) of photons following their interaction with vibrating molecules of the sample. This effect was discovered in 1928 by C. V. Raman,[10] who was awarded the Nobel Prize. The inelastic interaction between photons and molecules of the sample leads to frequency shifts of the incident photons as they transfer/receive energy to/from the sample molecules. Therefore, the energy change of the scattered photons is determined by the vibrational energy levels of the molecules (Fig. 2). Since the vibrational energy spectrum depends on the physical and chemical properties of the sample (type of atoms, bond strength, bond angles, symmetry, etc.), a Raman spectrum represents a physico-chemical fingerprint of the sample. (For more detailed description of the physics of the Raman effect, see Refs. 11 to 13.)

The intensity of the Raman scattering is proportional with the fourth power of the frequency of the incident laser photons, but it is still typically 15 orders of magnitude lower than fluorescence emission of dye molecules. Certain molecules present in cells, such as carotenoids and haeme proteins, can produce enhanced Raman signals when specific laser wavelengths are used for excitation.[14,15] If the laser wavelength corresponds to a strong electronic absorption band of a molecule, the intensity of some Raman-active vibrations of the molecule can be enhanced by

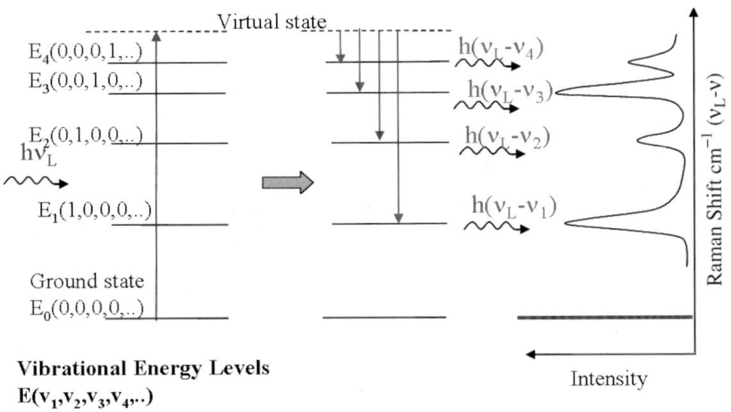

Fig. 2. Schematic description of the Raman scattering (incident photon transfers energy to the molecule, i.e. Stokes scattering). Raman spectrum represents a fingerprint of the vibrational energy level of molecules (frequency shifts ν_L-ν of incident photons, ν_L is the laser frequency and ν represents a molecular vibrational frequency).

a factor of 10^2–10^4. Higher enhancements of Raman spectra (typically 10^7 or even higher) can be achieved when molecules are adsorbed on rough metallic substrates or added to metal colloids (surface enhanced Raman spectra, SERS).[16,17] The most popular metals for SERS when excitation is realised in the visible or near-infrared regions are silver and gold. Recent studies proved that SERS is able to study even single molecules, suggesting Raman scattering enhancements as large as 10^{14}–10^{15}.[18,19]

Although these enhancing mechanisms produce stronger Raman signals, their applications in tissue engineering has been yet rather limited compared to non-resonant Raman spectroscopy.[20]

To achieve the high spatial resolution required for studying living cells, a Raman spectrometer is coupled to an optical microscope. Thus, spatial resolution limited by optical diffraction to approximately half wavelength of the excitation laser (~250 nm if a 514 nm laser is used for excitation) can be achieved in the horizontal plane. The axial resolution can be improved by using a pinhole of approximately 100 μm (confocal set-up) to reject the out-of-focus photons. Certain Raman micro-spectrometers can achieve improved spatial resolution by using the actual detector pixels or the aperture of the collection optical fibre as a confocal pinhole. Confocal Raman instruments have reported sampling volumes as small as 1.4 fL from inside single cells.[21] The high spatial resolution achievable in Raman micro-spectroscopy make this technique attractive to studying single cells and cellular organelles.

Both upright and inverted optical microscopes have become common in Raman micro-spectrometers and are used both for focusing the laser on a cell/biomaterial

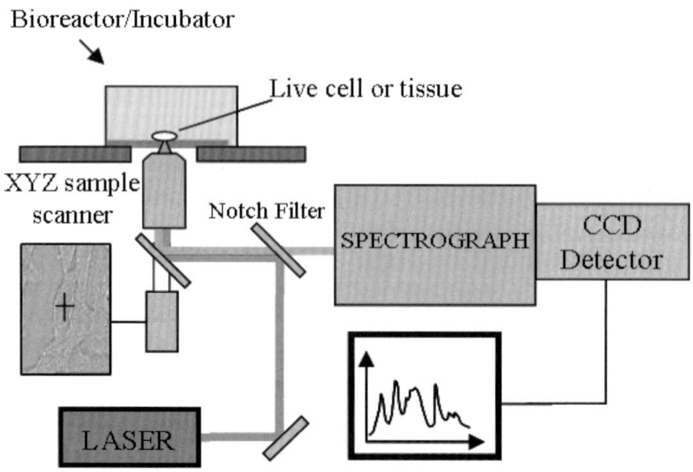

Fig. 3. Schematic diagram of a Raman micro-spectrometer based on inverted microscope. Cells or growing tissue can be enclosed in a sterile temperature controlled chamber to ensure cell viability during measurements.

and at the same time collects the Raman scattered photons (see Fig. 3). A laser in the visible or near-infrared regions is used as the monochromatic excitation source. The frequency spectrum (the shift from the laser frequency) of the scattered photons is usually analysed using a dispersion spectrometer. The collected Raman photons are collimated on a diffraction grating which spatially disperses the Raman scattered beam into multiple beams corresponding to specific frequencies which are subsequently focused on an array of detectors. Modern Raman spectrometers are equipped with high sensitivity charge coupled devices (CCD). A notch or edge filter is commonly used to reject the elastically scattered photons (Rayleigh photons), which have the same energy as the laser photons and would otherwise produce an intense background covering the Raman photons.

The recent developments in laser technology and CCDs during the last decades have had a high impact on Raman micro-spectroscopy. Raman spectroscopy was previously known to be a weak signal technique that required long integration time. Additionally, the fluorescence background commonly observed when exciting biological molecules with visible lasers limited the applications of Raman spectroscopy to biomedical applications. The use of modern near-infrared lasers, cooled silicon CCD, laser-rejection filters and high-throughput dispersive spectrographs makes it possible to obtain high-quality Raman spectra of cells within tens of seconds.

The fluorescence emission by biological materials can be significantly reduced by using near-infrared lasers (700–850 nm). New detectors, such as back-illuminated deep-depletion silicon CCD quantum are optimised to have improved quantum efficiency in this spectral range required to acquire the Stokes spectrum. It was also shown that near-infrared lasers induce less photodamage compared to UV or visible lasers.[20,22,23] This finding allowed the use of higher laser power for the excitation of Raman photons, increasing the signal strength and subsequently reducing the measurement time.

Raman micro-spectroscopy can also be used to build Raman spectral maps in two and three dimensions. Such maps can be achieved by representing the intensity of a certain spectral peak,[21,24] score of a principal component[25] or weight obtained in a least square fitting analysis[26] for each individual location in the 3D region where Raman spectra were acquired.

2.2. X-ray micro-computed tomography (μCT)

If the equipment is available, it is not difficult to obtain 3D images of single phase porous materials. Figure 4 shows a schematic of a sample being scanned. During a μCT scan, a series of 2D transmission X-ray images of a rotating sample are taken and reconstructed to form a 3D image. Geometric enlargement is used to magnify the image by placing the object close to a micron-sized spot source, producing a magnified image which is projected onto a solid-state detector a large distance from the object (relative to the source-object distance). Reconstruction software is then used. The intensity of the image is dependent on the integrated density and atomic number of the matter in each voxel (volume pixel). The only difficulties come in the way of resolution. The highest specification laboratory based sources can now give 3D resolution of 2 μm, however this requires the sample to have dimensions of approximately 2 mm.

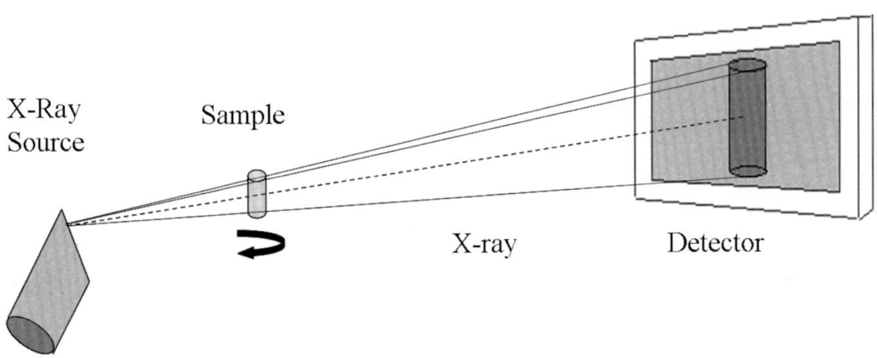

Fig. 4. A schematic of a μCT instrument.

This is not ideal if the macropore size of a scaffold is 600 μm. The sample may not be respresentative. Lower resolutions are therefore used to allow imaging of larger samples, but these may conceal fine components of the material.

A problem is also seen for composite materials, such as bioactive ceramics dispersed in biodegradable polymer matrix. Although both materials have good X-ray attenuation (and therefore good image contrast) compared to air, the difference in attenuation of the ceramic phases is extremely large compared to the difference between polymer and air, therefore it is difficult to obtain enough constrast between polymer and air to image the polymer in these systems.

3. Applications of Raman Micro-Spectroscopy to Cells

3.1. Raman spectra of biomolecules and cells

A typical Raman spectrum of a cell is shown in Fig. 5 along with the spectra of main cellular biopolymers. The Raman spectrum of cells contains peaks

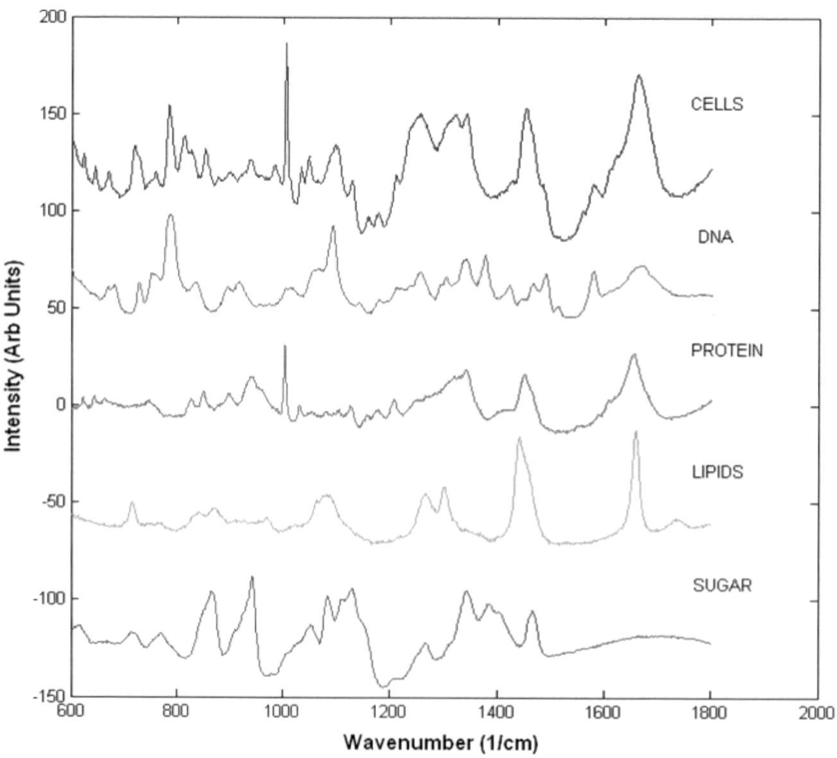

Fig. 5. Raman spectra of cells and biomolecules find in cells (spectra have been offset for clarity).

corresponding to all biomolecules found in cells, nucleic acids, proteins, lipids and carbohydrates.

The vibrations of the sugar-phosphate backbone of nucleic acids can provide useful information regarding the secondary structure of DNA, and can also be used to discriminate between RNA and DNA: the phosphodiester bond is found at 788 cm^{-1} for DNA and at 813 cm^{-1} for RNA. The vibration of the phosphodioxy group produces a relatively strong Raman peak at 1095 cm^{-1}, but which is rather insensitive to conformational changes of the nucleic acids. Additionally, Raman peaks corresponding to nucleotides can also be identified at 782 cm^{-1} (thymine, cytosine and uracil) and 1578 cm^{-1} (guanine and adenine). The peaks corresponding to proteins in the ranges 1660–1670 cm^{-1} (Amide I) respectively 1200–1300 cm^{-1} (Amide III) can be used for determining the secondary structure of the proteins. Certain amino acids produce strong Raman peaks, such as phenylalanine (1005 cm^{-1}), tyrosine (854 cm^{-1}) and tryptophan (760 cm^{-1}). The large number of C-H bonds in proteins also leads to a strong Raman band around 1449 cm^{-1}. Lipids are characterised by intense Raman peaks at 1449 cm^{-1} and 1301 cm^{-1} corresponding to C–H vibrations. Peak assigned to the stretching vibrations of unsaturated C=C bonds may also appear around 1660 cm^{-1}. C–O–C vibrations of the glycosidic bonds and ring vibrations in carbohydrates produce specific Raman peaks in the 800–1100 cm^{-1} range.

3.2. Differentiation of stem cells

One of the main challenges in tissue engineering is to maintain the appropriate phenotype of cells during the growth of the tissue *in vitro*. It has been recently recognised that stem cells have a great potential in tissue engineering, as given the right environment and stimuli, they can differentiate into the required mature cell type. Cell differentiation involves many intracellular biochemical and biophysical changes, including production of specific proteins which allow them to perform their functions.

Raman micro-spectroscopy was used to identify spectral markers for monitoring the differentiation status of embryonic stem (ES) as function of differentiation time (Fig. 6).[27] The main biochemical changes observed were related to higher RNA concentration in undifferentiated ES cells than in differentiated cells. The massive down-regulation is consistent with the hypothesis that ES cells "keep their options open" by maintaining many genes at intermediate levels and selecting only a few for continuous expression that are needed for differentiation to a specific phenotype. The rest will be down-regulated after commitment to a cell fate for which they are not needed.

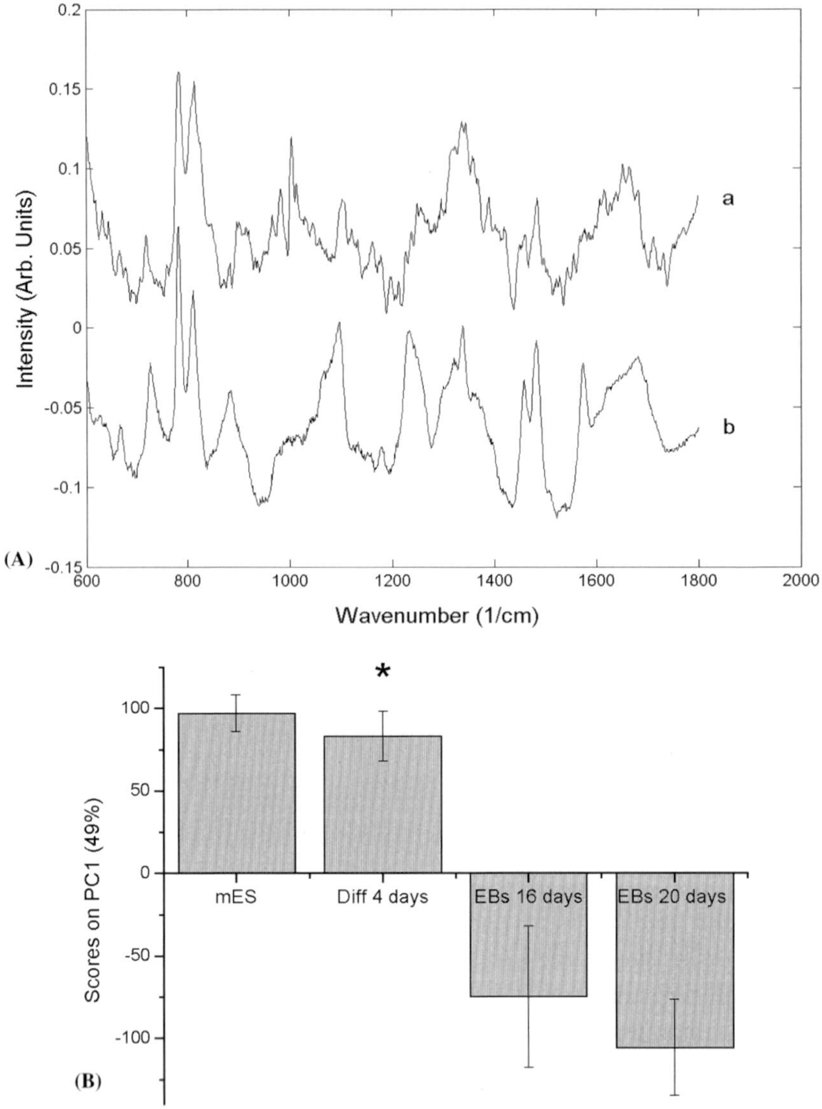

Fig. 6. Monitoring differentiation of mouse ES cells by Raman spectroscopy and principal component analysis. **(A)** Comparison between PC1 loading (a) and Raman spectrum of reference RNA (b). **(B)** PCA analysis: scores of PC1.

3.3. Phenotypic characterisation of cells

Primary osteoblasts obtained directly from patients are the ideal cells for bone tissue engineering. However, the use of primary osteoblasts *in vitro* is sometimes limited because of difficulty in obtaining sufficient bone fragments for harvesting cells. Furthermore, primary osteoblasts are difficult to culture *in vitro*, have a finite life span and also can lose their phenotype as the number of passages increases. To overcome the limitations in using primary osteoblasts, osteosarcoma and immortalised bone cell lines have been established and have higher proliferation rate and longer life span. However, the major disadvantage of using these cell lines relates to whether the phenotype of these cells reflect the true phenotype of normal osteoblasts.[28,29]

Raman spectroscopy can be used to determine biochemical differences between various cell-lines. Three types of bone cells commonly used in tissue engineering were compared: human primary osteoblasts, retroviral transfected human alveolar bone cells with SV40 large T-antigen and osteoblast-like human osteosarcoma-derived cells (MG63 cell line). A LDA cell classification model was built to determine the accuracy of Raman spectroscopy to discriminate between the primary osteoblasts, alveolar bone cells and MG63 cells. The LDA scores were computed to produce maximum separation between the cell groups (Fig. 7). The tumour-derived and non-tumour cells were well separated along the LD1 direction, while only a slight non-significant separation between primary osteoblasts and alveolar bone cells was observed along the

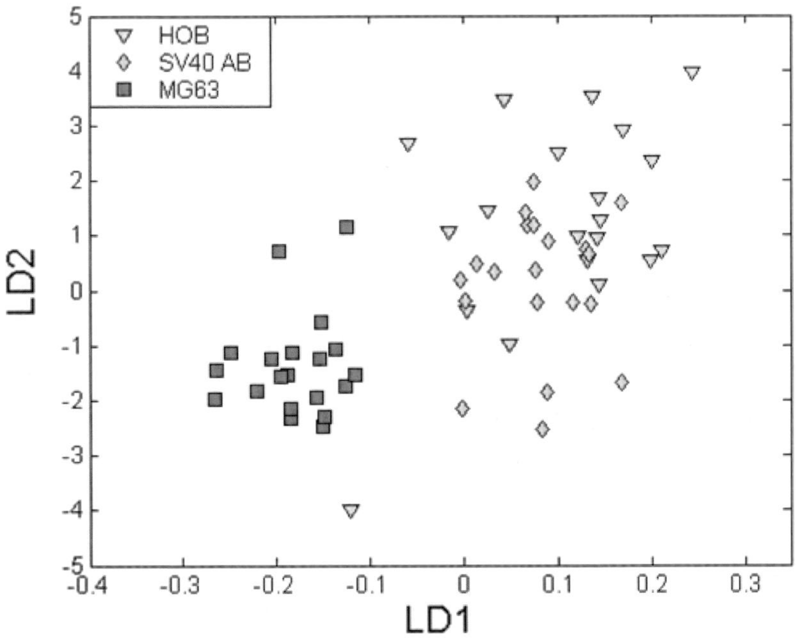

Fig. 7. LDA scores for the human primary osteoblasts (HOB), immortalised bone cells (SV40 AB) and osteosarcoma cell-line (MG63).

Table 1. Accuracy of the LDA model for phenotypic identification of bone cells.

Cross-validation prediction accuracy of the LDA model				
True classification	Predicted phenotype by Raman HOB	AB	MG63	Total
HOB	**11**	7	2	20
AB	9	**11**	0	20
MG63	0	0	**20**	20
Sensitivity	55%	55%	100%	
Specificity	77.5%	82.5%	95%	

LD2 direction. The accuracy of a prediction model based on PCA-LDA analysis of Raman spectra was tested using the cross-validation method (Table 1). The LDA model showed that the biochemical differences between the primary osteoblasts and the alveolar bone cells were very small, confirming the biochemical similarity between the transfected alveolar bone cell-line and primary osteoblasts.

3.4. Bone nodule formation and mineralisation *in vitro*

Raman micro-spectroscopy can be used to study mineralisation of bone and bone nodules produced in cultures *in vitro*. The mineral environment in bone tissue and the CaP species can be determined by measuring the vibration frequency and intensity of the symmetric stretching vibration (ν_1) of the PO_4^{3-} group.[30,31] The frequency of ν_1 vibration changes with ionic incorporation and crystallinity of the apatite: 955–959 cm^{-1} in B-type apatite (carbonate substituted phosphate in apatite lattice), 962–964 cm^{-1} in non-substituted apatite, and 945–950 cm^{-1} in disordered HA lattice (probably A-type carbonate substitution, i.e. carbonate for hydroxide or amorphous calcium phosphate).[30,31] The effect of surface roughness on the formation and mineralisation of bone nodules by human primary osteoblast can be investigated using Raman spectroscopy.[32] Raman spectroscopic measurement showed higher concentration of CaP in the bone nodules formed by osteoblasts on rough bioactive glass (45S5 Bioglass®) compared to similar but smooth substrates.

In principle, these techniques can be applied to cells growing *in vitro* on porous scaffolds so that the cell response to a macroporous surface can be monitored in real time. However limitations in the depth of field of optical microscopy mean that it is difficult to locate individual cells on a material surface.

Therefore more advanced imaging techniques must be combined with the Raman spectrometer.

4. Application of μCT to the Quantification of Scaffolds

4.1. Quantification of open pore networks

Many scaffold engineers have used μCT to image scaffolds. However, there are few reports of quantitative measurements of the pore size in scaffolds with connected pores[20–23] and only recently has an image analysis technique been developed that will quantify the size of the interconnects (regions of open space connecting adjacent pores).

Moore *et al.*[33] quantified some aspects of the porosity and interconnectivity of a polymer scaffold by analysing μCT images. They calculated the volume fraction of porosity that had a pathway of air to the outside of the material. However, they did not obtained quantitative data for either pore or interconnect sizes. Otsuki *et al.*[34] mapped the path length from the surface of a scaffold to connected interior pores using a well-established algorithm developed for percolation studies in rock.[35] Combining this with the image analysis operation of dilation, they categorised the pore throats into rough groupings based on the rate of closure of the pathways. However, this technique did not allow the individual quantification of pore and interconnect size and shape.

The first pore size distributions of both macropores and their interconnects from 3D μCT images were obtained by Atwood *et al.*,[36] which combined established algorithms, together with a novel method of generating a distance map for the watershed identification of interconnected pores. The method was then recently updated by Jones *et al.*[37]

Any 3D image analysis of μCT images involves the conversion of the images from grayscale images (e.g. Fig. 1B) to quantified descriptors of the structures by the development and use of appropriate mathematical morphological operators. Figure 8 shows a step-by-step demonstration of the quantification of the 3D pore network. In Figs. 8A to 8C, 2D images are shown for clarity, but all operations were performed on the 3D data. A filter is applied to the reconstructed raw image data to remove noise and then a threshold was applied, classifying each voxel as either scaffold or empty space (Fig. 8A). This ensures the algorithm that will be applied next knows what is material and what is air.

A dilation algorithm is then applied, which starts from the pore edge and fills the pores in equal steps (Fig. 8B). The point at which the rings converge is the centroid of the pore. The number of steps taken to fill each pore is noted. This is then

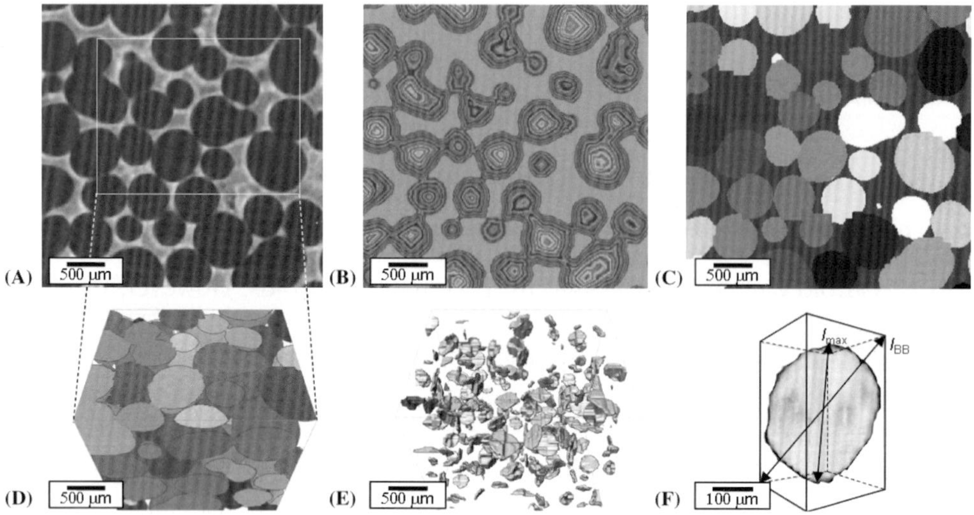

Fig. 8. Step-by-step process of the quantification of 3D pore networks from μCT data. **(A)** 2D slice of raw data, **(B)** 2D slice showing application of the dilation algorithm, **(C)** 2D representation of pores derived from the watershed algorithm, **(D)** 3D image of identified pores, **(E)** 3D image of the interconnects obtained from the top down algorithm, and **(F)** demonstration of the bounding box method of measuring the interconnect length. Reprinted with permission from Jones *et al.*[37]

converted into a distance map. This is possible because each step was equal and known.

Using the centroids, together with the distance map, a 3D watershed algorithm is applied on the distance map. Watershed algorithms find the set of points in a function, considered as a height map, that divide regions in which virtual water would flow to the same final point. In this case the highest points were taken as being the centroids of the pores. The watershed therefore divides the pore network into individual pores (Fig. 8C), identifying the pores (Fig. 8D). The algorithm used was developed by Mangan and Whitaker[38] and implemented by Ibanez *et al.*[39] in open-source form (www.itk.org).

The third algorithm applied is a top down algorithm that groups voxels with neighbours on the same two pores, which are then defined as interconnects (Fig. 8E).

The individual pores and interconnects can then be quantified to determine their volume/area and maximum diameter. An equivalent spherical diameter was measured for the pores. For each numbered pore, the number of voxels assigned to that pore was taken as a measure of its volume, and the equivalent spherical diameter was calculated from this volume.

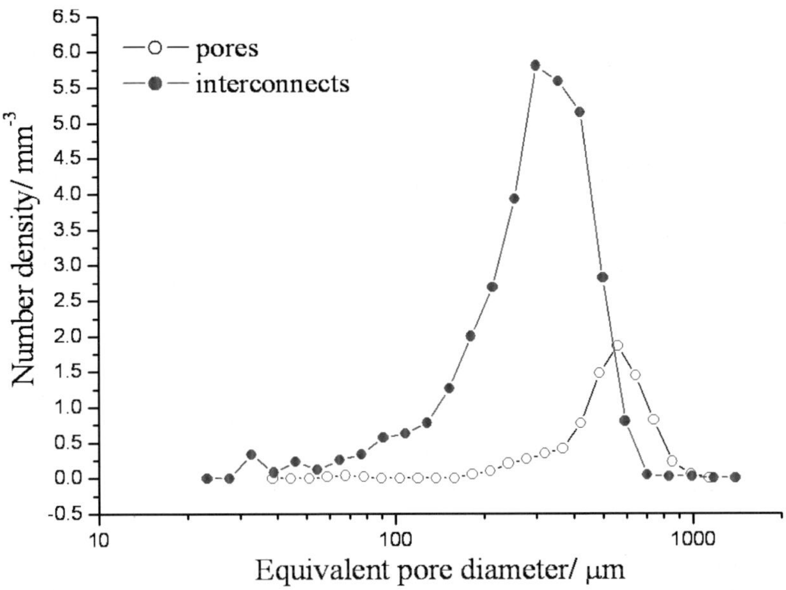

Fig. 9. Size distributions of the pore diameter and interconnect diameter of sol-gel derived bioactive glass foam scaffolds, determined by image analysis of 3D μCT images.

The diameters of the interconnects were measured as the diagonal of a bounding box of the interconnect (Fig. 8F). The bounding box method provides an upper limit on maximum dimension (I_{BB} in Fig. 8B). I_{BB} must then be divided by a correction factor of $\sqrt{2}$, which approximately corresponds to the difference in diagonal length between a bounding circle and square, to obtain I_{MAX}.[37] Figure 9 shows size distributions of macropores and their interconnects obtained by the 3D image analysis process. Figure 9 shows that quantified data can be obtained non-destructively using this technique. These techniques are applicable to many types of porous scaffolds.

4.2. Monitoring of cells within scaffolds

It is vital to be able to understand what pore networks are optimal to stimulate cell migration into a scaffold. At present, the only method for investigating tissue ingrowth into 3D pore networks is histology, which is a labourious process that requires the embedding and sectioning of an implant *ex vivo* into micron thick samples. Each slice is then stained and optical imaged. As μCT is a technique that can image a porous scaffold in 3D, there is potential that it could be used to non-destructively monitor cell migration and behaviour inside the pore network of the scaffold. However, it is very difficult to resolve cells growing on materials.

Therefore tagging technology must be developed where the tags given to cells can be resolved in the μCT image.

5. Conclusions

Raman spectroscopy is a novel technique that can be used to monitor cellular response non-invasively in real time. Single cells or groups of cells can be monitored. Models can be used to interpret the spectra. Future developments of the optical interface will allow the observation of cell responses to materials. μCT is a useful technique for the 3D imaging of porous scaffolds for tissue regeneration applications, but the images have limited use if the networks cannot be quantified. Recent developments have used combinations of algorithms to quantify images of bioactive glass scaffolds. These techniques should be developed further so that standard specifications for optimised scaffolds can be set for each tissue application. Techniques must be developed that will allow imaging of cell behaviour in the centre of scaffolds.

References

1. Gremlich, H.U. and Yan, B. *Infrared and Raman Spectroscopy of Biological Materials* (Marcel Dekker, 2001).
2. Hanlon, E.B., Manoharan, R., Koo, T.W., Shafer, K.E., Motz, J.T., Fitzmaurice, M., Kramer, J.R., Itzkan, I., Dasari, R.R. and Feld, M.S. Prospects for *in vivo* Raman spectroscopy. *Phys. Med. Biol.* **45**, R1–R59 (2000).
3. Manoharan, R., Wang, Y. and Feld, M.S. Histochemical analysis of biological tissues using Raman spectroscopy. *Spectrochim. Acta Part A Mol. Biomol. Spectrosc.* **52**, 215–249 (1996).
4. Tu, A.T. *Raman Spectroscopy in Biology: Principles and Applications* (John Wiley and Sons, 1982).
5. Hench, L.L. and Polak, J.L. Third-generation biomedical materials. *Science* **295**, 1014–1017 (2002).
6. Hulbert, S.F., Morrison, S.J. and Klawitte, J.J. Tissue reaction to three ceramics of porous and non-porous structures. *J. Biomed. Mater. Res.* **6**, 347–374 (1972).
7. Gomes, M.E., Sikavitsas, V.I., Behravesh, E., Reis, R.L. and Mikos, A.G. Effect of flow perfusion on the osteogenic differentiation of bone marrow stromal cells cultured on starch-based three-dimensional scaffolds. *J. Biomed. Mater. Res. Part A* **67A**, 87–95 (2003).

8. Jones, J.R., Ehrenfried, L.M. and Hench, L.L. Optimising bioactive glass scaffolds for bone tissue engineering. *Biomaterials* **27**, 964–973 (2006).
9. Lowell, S. and Shields, J.E. Theory of mercury porosimetry hysteresis. *Powder Technol.* **38**, 121–124 (1984).
10. Raman, C.V. and Krishnan, K.S. A new type of secondary radiation. *Nature* **121**, 501–502 (1928).
11. Long, D.A. *Raman Spectroscopy* (McGraw-Hill, 1977).
12. Lewis, I. and Edwards, H. *Handbook of Raman Spectroscopy* (CRC Press, 2001).
13. Ferraro, J.R., Nakamoto, K. and Brown, C.W. *Introductory Raman Spectroscopy* (Academic Press, 2003).
14. Myers, A.B. and Mathies, R.A. Resonance Raman spectra of polyenes and aromatics. In: Spiro, T.G. (ed.), *Biological Applications of Raman Spectroscopy* (John Wiley and Sons, 1987).
15. Spiro, T.G. *Biological Applications of Raman Spectroscopy, Vol. 1, Raman Spectra and the Conformations of Biological Macromolecules* (Wiley, 1987).
16. Moskovits, M. Surface-enhanced spectroscopy. *Rev. Mod. Phys.* **57**, 783–828 (1985).
17. Otto, A., Mrozek, I., Grabhorn, H. and Akemann, W. Surface-enhanced Raman scattering. *J. Phys. Condens. Matter* **4**, 1143–1212 (1992).
18. Nie, S. and Emory, S.R. Probing single molecules and single nanoparticles by surface-enhanced Raman spectroscopy. *Science* **275**, 1102–1106 (1997).
19. Kneipp, K., Kneipp, H., Corio, P., Brown, S.D., Shafer, K., Motz, J., Perelman, L.T., Hanlon, E.B., Marucci, A., Dresselhaus, G. and Dresselhaus, M.S. Surface-enhanced and normal stokes and anti-stokes Raman spectroscopy of single-walled carbon nanotubes. *Phys. Rev. Lett.* **84**, 3470–3473 (2000).
20. Notingher, I. and Hench, L.L. Raman microspectroscopy: a non-invasive tool for studies of individual living cells *in vitro*. *Expert Rev. Med. Dev.* **3**, 215–234 (2006).
21. Uzunbajakava, N., Lenferink, A., Kraan, Y., Willekens, B., Vrensen, G., Greve, J. and Otto, C. Non-resonant Raman imaging of protein distribution in single human cells. *Biopolymers* **72**, 1–9 (2003).
22. Neuman, K.C., Chadd, E.H., Liou, G.F., Bergman, K. and Block, S.M. Characterization of photodamage to *Escherichia coli* in optical traps. *Biophys. J.* **77**, 2856–2863 (1999).
23. Puppels, G.J., Olminkhof, J.H.F., Segersnolten, G.M.J., Otto, C., Demul, F.F.M. and Greve, J. Laser irradiation and Raman-spectroscopy of single living cells and chromosomes — sample degradation occurs with 514.5nm but not with 660 nm laser-light. *Exp. Cell Res.* **195**, 361–367 (1991).

24. Uzunbajakava, N., Lenferink, A., Kraan, Y., Volokhina, E., Vrensen, G., Greve, J. and Otto, C. Non-resonant confocal Raman imaging of DNA and protein distribution in apoptotic cells. *Biophys. J.* **84**, 3968–3981 (2003).
25. Krafft, C., Knetschke, T., Siegner, A., Funk, R.H.W. and Salzer, R. Mapping of single cells by near infrared Raman microspectroscopy. *Vibrat. Spectrosc.* **32**, 75–83 (2003).
26. Shafer-Peltier, K.E., Haka, A.S., Motz, J.T., Fitzmaurice, M., Dasari, R.R. and Feld, M.S. Model-based biological Raman spectral imaging. *J. Cell Biochem. Suppl.* **39**, 125–137 (2002).
27. Notingher, I., Bisson, I., Bishop, A.E., Randle, W.L., Polak, J.M.P. and Hench, L.L. *In situ* spectral monitoring of mRNA translation in embryonic stem cells during differentiation *in vitro*. *Anal. Chem.* **76**, 3185–3193 (2004).
28. Bilbe, G., Roberts, E., Birch, M. and Evans, D.B. PCR phenotyping of cytokines, growth factors and their receptors and bone matrix proteins in human osteoblast-like cell lines. *Bone* **19**, 437–445 (1996).
29. Clover, J. and Gowen, M. Are Mg-63 and Hos Te85 human osteosarcoma cell-lines representative models of the osteoblastic phenotype. *Bone* **15**, 585–591 (1994).
30. Tarnowski, C.P., Ignelzi, M.A., Jr. and Morris, M.D. Mineralization of developing mouse calvaria as revealed by Raman microspectroscopy. *J. Bone Miner. Res.* **17**, 1118–1126 (2002).
31. Timlin, J.A., Carden, A., Morris, M.D., Rajachar, R.M. and Kohn, D.H. Raman spectroscopic imaging markers for fatigue-related microdamage in bovine bone. *Anal. Chem.* **72**, 2229–2236 (2000).
32. Gough, J.E., Notingher, I. and Hench, L.L. Osteoblast attachment and mineralized nodule formation on rough and smooth 45S5 bioactive glass monoliths. *J. Biomed. Mater. Res. A* **68**, 640–650 (2004).
33. Moore, M.J., Jabbari, E., Ritman, E.L., Lu, L.C., Currier, B.L., Windebank, A.J. and Yaszemski, M.J. Quantitative analysis of interconnectivity of porous biodegradable scaffolds with micro-computed tomography. *J. Biomed. Mater. Res. Part A* **71A**, 258–267 (2004).
34. Otsuki, B., Takemoto, M., Fujibayashi, S., Neo, M., Kokubo, T. and Nakamura, T. Pore throat size and connectivity determine bone and tissue ingrowth into porous implants: three-dimensional micro-CT based structural analyses of porous bioactive titanium implants. *Biomaterials* **27**, 5871–5966 (2006).
35. Stauffer, D. *Percolation Theory* (Taylor and Francis, 1985).
36. Atwood, R.C., Jones, J.R., Lee, P.D. and Hench, L.L. Analysis of pore interconnectivity in bioactive glass foams using X-ray microtomography. *Scr. Mater.* **51**, 1029–1033 (2004).

37. Jones, J.R., Poologasundarampillai, G., Atwood, R.C., Bernard, D. and Lee, P.D., Non-destructive quantitative 3D analysis for the optimisation of tissue scaffolds. *Biomaterials* **28**, 1404–1413 (2007).
38. Mangan, A.P. and Whitaker, R.T. Partitioning 3D surface meshes using watershed segmentation. *IEEE Trans. Vis. Comput. Graph.* **5**, 308–321 (1999).
39. Ibanez, L., Schroeder, W., Ng, L. and Cates, J. *The ITK Software Guide* (Kitware Inc., 2003).

Chapter 21

Role of Stem Cell Imaging in Regenerative Medicine

Gabriella Passacquale and Kishore Bhakoo

Abstract

Stem cell research is undergoing a critical transition from being a discipline of the basic sciences to being recognised as a potential component of medical practice. Cell transplants to replace cells lost due to injury or degenerative diseases, for which there are currently no cures, are being pursued in a wide range of experimental models.

The monitoring of cellular grafts, non-invasively, is an important aspect of the ongoing efficiency and safety assessment of cell-based therapies. Magnetic resonance imaging (MRI) methods are potentially well suited for such an application as they produce non-invasive "images" of opaque tissues. For transplanted stem cells to be visualised and tracked by MRI, they need to be tagged so that they are "MR visible". We are developing and implementing a programme of molecular imaging in pre-clinical models that is directed towards improving our understanding of stem cell migration in the context of the whole organism.

In order to achieve these goals we are engineering novel MRI contrast agents and developing specific tagging molecules to deliver efficient amounts of contrast agents into stem cells. The intracellular contrast agents are based on either paramagnetic nanoparticles, such as dextran-coated iron oxide, or other MR contrast agents. Methods for monitoring implanted stem cells non-invasively *in vivo* will greatly facilitate the clinical realisation and optimisation of the opportunities of stem cell-based therapies.

Keywords: Cell Tracking; MRI Contrast Agents; PET; Optical; Nuclear Imaging.

Outline

1. Introduction
2. Ideal Imaging Technology for Non-Invasive Stem Cell Tracking

2.1. X-ray-based imaging
2.2. Radionuclide imaging
2.3. Optical imaging
2.4. Ultrasound
3. Non-Invasive Tracking of Stem Cells Using MRI
3.1. Intracellular MRI contrast agents
3.2. MRI tracking of stem cells in the heart
4. Role of Imaging in Stem Cell-Based Therapy for the Central Nervous System
5. Multimodality
6. Conclusions
References

1. Introduction

Stem cell research is undergoing a significant evolution from being a discipline of the basic sciences to being recognised as a potential component of medical practice. Cell transplants to replace cells lost due to injury or degenerative diseases, for which there are currently no cures, are being pursued in a wide range of experimental models. Both cardiovascular and neurological disorders have a high clinical impact in terms of mortality, morbidity, quality of life as well as being a financial burden on society. Stem cell transplantation is emerging as potential therapy to treat these disorders since their demonstrated clonogenic and renewal capabilities, which could support a programme of cell replacement in organs and tissues leading to functional recovery.

The encouraging results obtained from numerous recent experimental studies in regenerative medicine, originally born as an academic curiosity, have allowed this field to move towards clinical translation. The growing number of early phase human studies aims to demonstrate the feasibility and potential efficacy of stem cell therapy in a clinical environment. The development of these new stem cell-based therapies requires both quantitative and qualitative assessments of stem cell migration towards target organs (homing), differentiation outcome and engraftment. In this context, stem cell imaging represents a crucial aspect in the evaluation of stem cell therapy. It provides both a useful method of studying stem cell behaviour following transplantation, in terms of migration and integration into damaged tissues, and an evaluation of functional recovery following cell-based therapy. In this chapter, we have attempted to summarise the available evidence on the role of imaging in stem cell therapy with principal attention on regeneration of cardiac and nervous system lesions.

Once implanted, it is clear that the migration dynamics of stem cells will determine the extent of tissue regeneration at the site of implantation and surrounding tissue. Methods for monitoring implanted stem cells non-invasively *in vivo* will greatly facilitate the clinical realisation and optimisation of the opportunities for cell-based therapies. Due to the seamless integration into the host parenchyma, and migration over long distances, cell grafts cannot be detected based on their mass morphology. To monitor cell migration and positional fate after transplantation, current methods use either reporter genes or chimeric animals. These methods are cumbersome, involve sacrifice of the animal and removal of tissue for histological procedures, and cannot be translated to human studies.[1,2] However, this approach lacks the temporal analysis of the donor cells, so, in practices its uses are limited. The monitoring of cellular grafts, non-invasively, is an important aspect of the ongoing efficiency and safety assessment of cell-based therapies. Molecular imaging is potentially well suited for such an application. For transplanted stem cells to be visualised and tracked by imaging technologies, they need to be tagged so that they are "visible". Moreover, imaging and biosensor technologies are moving from diagnostic towards therapeutic and interventional roles.

2. Ideal Imaging Technology for Non-Invasive Stem Cell Tracking

The ideal imaging modality should provide integrated information relating to the entire process of cell engraftment, survival, and functional outcome.The *in vivo* monitoring of stem cells after grafting is essential to our understanding the first steps in cellular replacement process, i.e. cellular migratory potential towards a lesion and subsequent incorporation into the damaged tissue, especially when stem cells are administrated systemically.

The imaging technologies of MRI, radionuclide and optical imaging are emerging as key modalities for *in vivo* molecular imaging because of their ability to detect molecular events. However, in order to exploit sensitivity, specificity, temporal resolution and spatial resolution offered by these modalities they require an interaction with respective contrast agents that exerts an "effect size" enough to be visualised by complementary imaging hardware. Effective use of the tools of molecular imaging requires knowledge of the basis of detection of the imaging modality, mechanism of contrast agent interactions and the biological environment.[3] Obviously, the use of a contrast agent for *in vivo* studies needs to be biocompatible, safe and non-toxic. Moreover, it is necessary to avoid any kind of genetic modification of the stem cell that could perturb its genetic programme. Sensitivity and specificity of imaging stem cells is another important aspect that influences its suitability for clinical applications. Indeed, the possibility of quantifying the exact cell number at any anatomical localisation is particularly important for the study of cellular migration, but also reports on the resident time of stem cells implanted locally. Finally, the ideal imaging should permit tracking of stem cells for months to years allowing a long-term follow-up of tissue function and donor survival. So far, no imaging technologies satisfy all these criteria, although some came close.

2.1. X-ray-based imaging

X-ray-based methods, including plain films and computed tomography (CT) are the most readily available clinical imaging modalities. Unfortunately, generation of contrast requires high concentrations of high-density/high-atomic number materials, rendering them unsuitable for stem cell tracking at the present time. Moreover, the exposure of the stem cell recipient to ionising radiation limits the suitability of this technique for a long-term and frequent follow-up.

2.2. Radionuclide imaging

Nuclear medicine is the main form of molecular and cellular imaging in current clinical use. Because of their exquisite (10^{-11} to 10^{-12} mol/l) sensitivity, single-photon emission CT (SPECT) and positron emission tomography (PET) imaging modalities are able to detect tracer quantity of radioisotopes for studying biological processes in living subjects. Technological developments of both PET and SPECT have led to the implementation of specialised systems for small animal imaging with much greater spatial resolution (1–2 mm),[4–6] which has dramatically advanced the field of cell tracking in animal models *in vivo*.

A number of strategies for stem cells labelling have been described with the use of radioactive atoms in SPECT. They include the direct loading with a radiometal,[7–9] the enzymatic conversion and retention of a radioactive substrate[10] and the receptor-mediated binding.[10,11] All these strategies of cell labelling have inherent disadvantages limiting the use of radionuclide technique for stem cell tracking. Enzymatic conversion and retention of a radioactive substrate requires genetic manipulation of stem cells *ex vivo* and the administration of a substrate intravenously for each imaging session. Receptor-mediated binding requires stable expression of a receptor not found elsewhere in the body and intravenous injection of a radioactive receptor ligand. Direct labelling of cells with radionuclides appears to be the only choice in radionuclide technique for stem cell tracking. However, the major limitations are the exposure to ionising radiation and the short half-life time of radioisotopes leading to loss of the imaging signal within few days. A more advanced technique for stem cell imaging with radionuclides over the course of months has been described with PET. It consists of the stable integration of a mutant herpes simplex type 1 thymidine kinase (TK) into stem cells and periodic intravenous injection of TK substrate 9-(4-[18F]-fluoro-3hydroxymethylbutyl)guanine (^{18}F-HBG).[12]

Unfortunately, the need for genetic manipulation of stem cells and exposure to radiation of the recipient limits the use of this technique in clinical practice. This concern was demonstrated in a study, in which hematopoietic progenitor cells were labeled with In-111-labelled-oxine and injected into the cavity of the left ventricle heart in a rat model of myocardial infarct; although gamma imaging revealed homing of the progenitor cells to infarcted myocardium, significant impairment of proliferation and function of labelled cells was observed. Moreover, rapid efflux of labels out of cells over the course of time leads to label loss from viable cells. So far, PET and SPECT techniques, already used in diagnostic clinical practice for specific diseases, represent important method to evaluate the functional effects of stem cell-based therapy.

2.3. Optical imaging

This imaging modality measures the intensity of the emitted light with a detector system (generally CCD cameras in a black box). Two distinct contrast mechanisms can be used for molecular and genomic imaging studies, one involving fluorescence and the other bioluminescence. Both can be used for stem cell tracking, but several methodological limitations preclude the use of these techniques in a clinical setting. In fluorescence imaging an external organic light source,[13] or organic/inorganic hybrids (quantum dots) as exogenous contrast agents,[14] is used to excite fluorescent molecules inside the subject. Green fluorescent protein (GFP) or small molecule polymethines are the most commonly used optical reporters. Conversely in bioluminescence there is no need for external light stimulation. In this case *in vivo* imaging involves introducing reporter genes that encode for enzymes (known as luciferases) that can catalyse a light-producing chemical reaction using exogenous substrates.[15] This method requires genetic modification of stem cells to stably express non-human genes, and the injection of strong immunogenic non-human substances such as luciferine and coelenterazine. The method also has limited tissue penetration. Thus, optical imaging, however, is generally limited to use in small rodents.[16–18]

In spite of the opaque nature of tissue, optical imaging is an increasingly important technique for molecular imaging in small animals. It is also being developed for clinical use, in particular using near infrared (NIR) wavelengths (~650–900 nm), which can theoretically travel 5–15 cm through tissue.[19]

A notable theoretical advantage of optical techniques is the fact that multiple probes with different spectral characteristics can be used for multi-channel imaging, similar to *in vivo* karyotyping.[20]

Optical imaging is increasingly used in animal models of disease. In humans the poor penetration will limit its applications, although the development of NIR intravascular imaging catheters and endoscopes may allow this approach to delve deeper into the body! The other drawback of optical imaging is its resolution, which is good near the surface, but poorer for deeper structures. New approaches such as fluorescence molecular tomography, which allows volumetric reconstruction of the source of fluorescent light, can improve spatial resolution to 1 mm at the surface to 3 mm in the centre of small animals, as well as increasing sensitivity (as low as 200 femtomoles).[19,21] Hence, attachment of a fluorophore to a stem cell would allow detection of stem cells by histology using conventional microscopes. Both luminescence and fluorescence have been used for stem cell tracking *in vivo.*[22,23]

2.4. Ultrasound

Ultrasound (US) allows real time imaging with a high resolution (~50 microns) in the absence of ionising radiation. Ultrasound imaging relies on the acoustic properties of the tissue causing reflection of a high frequency ultrasound (40 MHz) beam. Contrast agents are being developed that promise to give US a major role in molecular imaging and therapeutics. The main contrast agent used is microbubbles formed from albumin, polymers or lipids and containing perfluorocarbon or air. These bubbles adsorb and scatter the sound waves, and in turn the bubbles can resonate and emit secondary sound waves at differing multiples of the original input frequency. This "non-linear" property of microbubbles makes them a powerful contrast agent as only the microbubbles are detected at these bubble specific frequencies, making the contrast such than single bubbles can be detected. Targeting is possible by decorating the surface of bubble with targeting molecules such as antibodies or other ligands.

Ultrasound has a number of advantages, including safety and portability. The short half-life and the benign composition of the bubbles allow rapid repeated scans. However, their large size (10–300 microns) limits its application to intravascular investigations.

3. Non-Invasive Tracking of Stem Cells Using MRI

Magnetic resonance imaging (MRI) is well suited for stem cell tracking because it can provide both whole-body and detailed information of host organs with near microscopic anatomical resolution and excellent soft-tissue contrast. MRI provides good anatomical information (resolution <100 microns). Conventionally MRI relies on differences in proton density and the local magnetic environment of the hydrogen atoms to detect differences between and within tissues. Images can be enhanced using contrast agents. MR only indirectly detects most contrast agents;

they are "seen" as a result of their affect on water molecules. These characteristics have led to this imaging modality to become the method of choice for following anatomical changes in soft tissue. Furthermore MRI has the ability to measure physiological parameters such as water diffusion and blood oxygenation. Current developments of contrast agents are based on the use of gadolinium-analogues (Gd^{3+}-DTPA) or iron oxide particles as contrastophores. They are designed to alter either T_1 or T_2 relaxation time and, in turn, providing either hyper- or hypo-intense imaging, respectively. Despite the inherent reduced sensitivity of MR compared to techniques using radionuclides (SPECT, PET), MR is more sensitive to cell detection due to the higher concentration of achievable intracellular contrastophores.

There are numerous ways of increasing the signal-to-noise ratio in micro-MRI when imaging small animals, and thus achieving near microscopic resolution. These include working at relatively high magnetic fields (4.7–14T), using hardware and software customised to the small size of animals of interest, and the relative flexibility of much longer acquisition times during imaging. Magnetic resonance imaging can provide detailed morphologic and functional information and, therefore, seems ideally suited to integrate efficacy assessments with the capability for cell tracking. Yet studies show that the lowest detectable number of cells is 10^5 with the use of conventional MRI scanners without any sequence modification. This threshold of detection can be lowered using high-field magnets (11.7 T) such that single cells containing a single iron particle can be detected and tracked as well as other approaches appeared in recent literature.[24,25]

MRI is now a rapidly evolving molecular and cellular imaging strategy. The era of modern molecular imaging began with the first successful demonstration of a functionalised Gd-complex to visualise gene expression in 2000.[26] Since then successful linking of receptor molecules to contrast agents for a broad variety of applications has been published.

For stem cells to be visualised and tracked by MRI, they need to be tagged so that they are "MR visible". At present there are two types of MR contrast agent used clinically. These are gadolinium-analogues (e.g. Gd^{3+}-DTPA) or iron oxide nanoparticles. However, these reagents were designed as blood-pool contrast agents and are impermeable to cells. Several approaches have been deployed to enhance cell labelling to allow *in vivo* cell tracking by conjugating MRI contrast agents to a range of ancillary molecules to enhance their uptake. With the growing array of cell labeling techniques, cells tagged with various monocrystalline MR probes have been evaluated both *in vitro* and *in vivo*.[27–29]

The development of methods for monitoring stem cell grafts non-invasively, with sufficiently high sensitivity and specificity to identify and map the fate of transplanted cells, is an important aspect of application and safety assessment of stem cell therapy. MRI methods are potentially well suited for such applications as this produces non-invasive "images" of opaque tissues or structures inside the body.

3.1. Intracellular MRI contrast agents

Recent work in the design of MRI contrast agents has opened the possibility of combining the spatial resolution available in MRI for anatomic imaging with the ability to "tag" cells, and thus enable non-invasive detection and study of cell migration from the site of implantation. *In vivo* monitoring of stem cells after grafting is essential for understanding their migrational dynamics, which is an important aspect in determining the overall therapeutic index in cell therapies. Despite recent advances in both the synthesis of paramagnetic molecules and the basic cell biology of stem cells, methods for achieving effective cell labelling using molecular MR-tags are still in their infancy.

MRI contrast agents are used extensively in the clinic to improve sensitivity/selectivity in order to facilitate diagnosis of an underlying pathology. MRI contrast agents either alter the T1 and/or T2 relaxation time, making the local tissue hyper- or hypo-intense respectively. The contrast agents used most extensively can be classed as either paramagnetic or superparamagnetic.

Examples of paramagnetic agents are Fe^{3+}, Mn^{2+}, and Gd^{3+}. The effect of the magnetic moment in solution results in a dipolar magnetic interaction between the paramagnetic ion and neighbouring water molecules. Fluctuations in this magnetic interaction causes the decrease in T1/T2 relaxation time.[30] Paramagnetic compounds produce, predominantly, T1 effect, giving a hyperintense region.

The other class of contrast agents is superparamagnetic reagents. These consist of an iron oxide core, typically 4–10 nm in diameter, where several thousand iron atoms are present. A biocompatible polymer surrounds the core to provide steric and/or electrostatic stabilisation. This is required due to the large surface area to volume of the nanoparticles. If no stabilisation is present, the particles spontaneously precipitate out of solution due to colloidal instability. Many polymeric coating materials have been proposed, such as dextran, carboxymethylated dextran, carboxy-dextran, starch, PEG, arabinogalactan, glycosaminoglycan, organic siloxane, and sulphonated styrene-divinylbenzene.[31]

There are two types of superparamagnetic contrast agents, SPIO (superparamagnetic iron oxide) and USPIO (ultrasmall superparamagnetic iron oxide). The difference between the two is that the SPIOs consist of several magnetic cores surrounded by a polymer matrix whereas USPIOs are individual cores surrounded by a polymer. Superparamagentic contrast agents provide predominantly a T_2 effect, but smaller particles have shown to act as a T_1 agent.[30] A new class of USPIO has been produced known as cross-linked iron oxide (CLIO), whereby the dextran coat of the USPIO is cross-linked in the presence of epichlorohydrin, and then subjected to amination in the presence of ammonia to produce amine terminated nanoparticles suitable for conjugation.[32] Thus, due to their biocompatibility and strong effect on T_2 relaxation time, iron oxide nanoparticles are the MR contrast agent of choice for cell labelling.

However, none of the contrast agents used in the clinic was designed for cellular internalisation. To cross the cell membrane, contrast agents must either be used in conjunction with a transfection agent or conjugated to a biological entity, such as a peptide transduction domain. For example, incubation with non-derivatised, dextran-coated iron oxide particles,[33–38] incubation with liposome-encapsulated iron oxide particles,[39] and lectin-mediated uptake.[40] In general, uptake is low, and further improvements in magnetic nanoparticle uptake are needed. Several approaches have been described to optimise the internalisation process,[41,42] including the link between nanoparticles and the highly cationic HIV-tat peptide[27] or the use of an anti-transferrin receptor monocloncal antibody covalently linked to nanoparticles (MION-46L).[43] Unmodified SPIOs and USPIOs have been also successfully used at high concentration without transfection agents to label cells (fluid phase-mediated endocytosis).[41,43] However, the need for transfecting agent to improve cell labelling is dependent on the cell type.

Uptake of nanoparticles is time dependent, but cell viability decreases with increasing incubation time. An incubation of two to four hours seems to be an optimal compromise for haematopoietic stem cells.[44] Iron oxide-labelled cells showed a gradual decline of intracellular iron particles, due to cell division and exocytosis or release of iron from non-viable cells. Moreover, the gradual disappearance of magnetic labelling may also be attributed to the incorporation of iron into metabolic pathways.

Some superparamagnetic contrast agents have been approved by the US Food and Drug Administration (FDA) for non-stem cell applications. The same FDA-approved iron oxide have been shown to affect neither haematopoietic nor mesenchymal stem cell function or differentiation capacity.[45] The problem with the use of these substances for stem cell tracking is the dilution of contrast with cell division, difficulty in quantification because of susceptibility artefact, the potential transfer to non stem cells, such macrophages after stem cell death. Furthermore, a significant clinical problem common to all MR methods is the contraindication of scanning when implantable devices, such pacemakers or defibrillators, or prosthesis are present. However, due to its safety profile and non-invasive property MR represents, so far, the major imaging method of tracking stem cells *in vivo*.[41,42]

3.2. MRI tracking of stem cells in the heart

Myocardial infarction is by nature an irreversible injury. The extent of the infarction depends on the duration and severity of the perfusion defect.[46] Beyond contraction and fibrosis of myocardial scar, progressive ventricular remodelling of non-ischaemic myocardium can further reduce cardiac function in the weeks to months after initial event.[47]

Many of the therapies available to clinicians today can significantly improve the prognosis of patients following an acute myocardial infarction.[47,48] However, no pharmacological or interventional procedure used clinically has shown efficacy in replacing myocardial scar with functioning contractile tissue.

Stem cell transplantation is being widely assessed as potential therapy for cell death-related heart disease.[49] Several cell types, including embryonic stem cells,[50] skeletal myoblasts,[51,52] mesenchymal stem cells (MSCs),[53] bone marrow-derived cells,[2,54,55] cardiac resident stem cells[56,57] and endothelial progenitor cells (EPCs)[58] have demonstrated their capacity to reduce scar formation and fibrosis and preserving cardiac function, when administered after a myocardial infarction. Moreover, different subsets of progenitor cells were shown to augment perfusion. All these effects may be related to a direct effect of differentiation into endothelial cells, smooth muscle cells and cardiomyocytes and/or to the release of paracrine factors by progenitor cells affecting resident mature tissue.[49] Imaging is crucial to *in vivo* investigations and understanding the effect of cell therapy. Initial animal studies using either micron scale particles[59,60] or nanoparticles of iron oxide,[61,62] showed their potential for non-toxic labelling of haematopoietic bone marrow-derived and mesenchymal stem cell populations. In the cardiovascular setting, this cell labelling technology was coupled with direct delivery methods using endomyocardial injections, demonstrating that transplanted stem cells could be imaged shortly after delivery with high degree of spatial resolution MRI.[59] Iron oxide labelling was used also to track a smaller number of stem cells migrated to infarcted myocardium after intravenous injection.[60] Yet, in all these studies, the lowest detectable number of stem cells was 10^5 with the use of conventional MRI scanners without any sequence modification. Moreover, it was demonstrated that magnetically labelled MSCs could be detected by MRI for up to three weeks.[61] Studies with MSCs were performed in rodents and large animals, and the fate and movement of stem cells was followed with various imaging techniques. Recently, human embryonic stem cells (hES) were labelled with MRI contrast agents and injected in animal models of myocardial infarction demonstrating the feasibility to magnetically label and *in vivo* visualisation of hES.[63] Similar results were obtained with mouse cardiac embryonic stem cells transplanted in animal models of myocardial infarction and detected by MRI that simultaneously permits the monitoring of cardiac function.[64]

A series of studies investigating stem cell-based therapy for cardiac diseases have been performed in humans. A number of studies were conducted in the setting of acute myocardial infarction,[64–76] and several other studies in the setting of chronic heart disease.[77–88] Stem cell therapy was performed by using bone marrow-derived stem cells injected either intracoronary or intramyocardially. In all these studies the imaging was particularly helpful in evaluating the functional effect of stem cell therapy. The most accurate assessment of left ventricular (LV) function and LV volumes was carried out using MRI where Fernandez-Avilès *et al.*[71] demonstrated an increase of 5.8% in LVEF, whereas no improvement was seen in control patients. A number of other studies evaluated the effect of stem cell therapy on myocardial perfusion by using nuclear imaging with PET or SPECT. SPECT was used in nine of the 11 studies available to date, and only Janssens *et al.*[74] used PET to evaluate effect of cell therapy on perfusion. It is worth noting, however, that only PET permits absolute quantification of myocardial perfusion, whereas SPECT offers information on relative tracer uptake. In the clinical environment, PET and SPECT, contrast-enhanced MRI and myocardial contrast echocardiography were used to investigate the infarct size following stem cell therapy. The most accurate method was contrast-enhanced MRI, allowing precise detection of scar tissue and is currently the only technique allowing the differentiation between subendocardial and transmural infarction.[89] Finally, imaging technique offered an important method of studying the myocardial viability, basically with nuclear imaging or low dose dobutamine echocardiography.

It is interesting to note from recent reports, that imaging played a key role in evaluating the effect of cardiac stem cell therapy in a clinical setting. Tracking of injected stem cells remains, thus far, an experimental model for the study of cell engraftment into the heart. Moreover, all the reports in the literature provide evidence for the utility of using a non-invasive imaging technique to combine anatomical/functional evaluation of the heart with transplanted stem cell mapping.

4. Role of Imaging in Stem Cell-Based Therapy for the Central Nervous System

Neuronal replacement therapy is based on the concept that neurological function lost due to injury or degenerative disease can be improved by implanting new cells that can differentiate and integrate appropriately into the brain. The literature provides significant evidence that the adult brain environment is capable of supporting production of new functional neural cell types from undifferentiated stem cells. This therapeutic strategy offers a new approach to treat a number of degenerative diseases such as Parkinson's disease and demyelinating disorders, as well stroke and neuronal traumatic injuries. The first pioneering studies were performed in 1970s when cells derived from foetal tissue from the mouse *Substantia nigra* was transplanted into the adult rat and found to develop into mature dopamine neurons.[90] In the 1980s, several groups showed that transplantation of this type of tissue could reverse Parkinson's-like symptoms in rats and monkeys when implanted into the damaged areas. The success of the animal studies led to several human trials beginning in the mid-1980s[91,92] showing, in some cases, a lessening of symptoms in treated patients suffering from Parkinson's disease. The clinical efficacy of stem cell-based therapy observed in treated patients was confirmed and investigated further using PET imaging of the brain, which demonstrated an increase of dopamine function after surgery. Later, in the 1990s a study of engraftment of human mesencephalic tissue into the brain of a patient with Parkinson's further demonstrated functional recovery, and L-DOPA withdrawal, followed by an increase in dopamine release.[93] Since the ethical and political problems related to the use of foetal or embryonic tissue, the possibility to obtain dopaminergic neurons from stem cell types derived from alternative sources was investigated.

The most obvious kind of stem cells for neuronal replacement is the cell population derived from the brain. That includes neural stem cells localised in the *subventricular zone* (SVZ) that have been shown to differentiate into dopaminergic neurons,[94] astrocytes and oligodendrocytes,[95] as well as spinal cord motor neurons.[96] Bone marrow-derived mesenchymal stem cells also have been shown to be capable of generating astrocytes, oligodendrocytes and neurons,[97] and more specifically to develop GABAergic, dopaminergic and serotoninergic neurons.[98] Similarly, haematopoietic stem cells also show promise with their ability to express neural markers, thus as a potential source for neural replacement.[99,100]

Independently of the kind of stem cells investigated, the imaging technique is particularly helpful to investigate the phases of colonisation and integration of transplanted cells into the damaged tissue. In particular, MRI was shown to be particularly useful in tracking migration of stem cells toward a neurological lesion together with the utility of identifying the different types of brain lesions such as those of ischaemic or demyelinating disorders. Several studies have reported that it is possible to visualise magnetically labelled cells in the brain after transplantation. This includes rat foetal neural cells labelled with SPIO conjugated to the lecithin wheat germ agglutinin,[101] rat foetal brain tissue labelled with SPIO containing reconstituted *Sendai* virus envelopes,[102] and CG4 oligodendrocyte progenitor cells labelled with plain dextran-coated SPIO.[103]

Local migration of transplanted stem cells was studied with MRI tracking of SPIO-labelled rat foetal brain cells[101,102] and CG4 oligodendrocyte progenitor cells.[103] Similarly, ferromagnetic particle-labelled SVZ cells were followed with MRI within the brain after transplantation into the healthy rat striatum.[104] Animal models of stroke were used to investigate more extensive migration of stem cells within the brain. Particularly, neural stem cells derived from the Maudsley hippocampal clone 36 (MHP36) cell line were grafted unilaterally in the hemisphere contralateral to an ischaemic lesion induced by middle cerebral artery occlusion. Gadolinium rhodamine dextran (GRID) labelled cells migrated to the ipsilateral hemisphere along the *corpus callosum* to colonise the injured area.[105] A similar model of stroke was used to investigate the migration of USPIO-labelled embryonic stem cells.[106] In all these studies immunohistochemical analysis of brain supported the imaging study and demonstrated the expression of neural, astrocytic and oligodendrocyte markers in transplanted stem cells suggesting their terminal differentiation at the site of incorporation. Finally, an MRI study investigated the incorporation of adipose-derived adult stem (ADAS) cells intravenously injected in animal model of stroke.[107] Although the scanning was performed on post-mortem brains, the results of this study support all the previous published reports demonstrating the suitability of MRI in tracking stem cell migration after transplantation. *In vivo* analysis of stem cell migration toward brain ischaemic lesions after an intravenous injection was performed with magnetically labelled ES cells and MSCs. In a different set of animal, ES cells and MSCs were transplanted intracerebrally. The MRI showed stem cell migration towards the lesion within two weeks after either intravenous or intracerebral injection of stem cells.[108]

MRI imaging was also used to track glial progenitor in demyelinating diseases. The oligodendrocytes progenitor cells (OPCs), found in the embryonic spinal cord, has both greater migratory and myelinating potential into mature oligodendrocytes. Experimental transplantation used two approaches. The first one was to transplant cells into persistently demyelinated lesions in the spinal cord that are created by focal irradiation and injection of gliotoxic chemicals.[109] In this model, it was found that transplanted glial cells migrate over a greater distance and remyelinated the demyelinated lesions more rapidly than endogenous cells.[110] In the second approach, mutant animal models of dysmyelination have been used.[111–114]

New myelination in the CNS by transplanted cells will be of relevance to human demyelinating disease, but only if it results in restoration of nerve function. Transplantation of adult human neural precursors cells,[115] or pig olfactory ensheathing cells[116] in rat models of chemical demyelination or transection has resulted in restoration of nerve conduction. These data suggest that cell transplantation in demyelinated and dysmyelinated human disease may have significant potential for functional restoration. The use of stem and progenitor cells in human clinical studies will require a technique that can monitor their migration and tissue biodistribution non-invasively. MR imaging has the capabilities of non-invasive whole-body *in vivo* imaging, with a resolution of 25 to 50 microns, approaching the resolution of single cells. Bulte *et al.*[43] first demonstrated that the migration and distribution of rat oligodendrocyte progenitor cells labelled with MION-46L can be efficiently monitored by MRI following transplantation into the spinal cord of myelin-deficient rats.

5. Multimodality

The challenge for understanding the very complex phenomena behind the stem cell therapy is multimodality. In fact no unique technique is capable of addressing all the critical issues and their complementary role has to be fully exploited. So a multidisciplinary approach is needed to validate the therapy.

Most available molecular imaging techniques are applicable not only in animals but also in humans, but some problems arise. Optical techniques are not applicable to humans (and large animals); so an intelligent combination of it with MRI and radionuclide (SPECT or PET) is needed to validate the human technologies and protocols to close existing gaps between basic science and clinical trials of cardiac cell therapy. The goal of imaging is to monitor the stem cells trafficking, homing and fate, viability and differentiation providing also quantifiable information. Imaging-based cell-tracking methods can potentially evaluate the short-term distribution of infused cells or their long-term survival and differentiation status. Therefore, these methods would play an indispensable role in detailed pre-clinical studies to optimise the cell type, delivery methods, and strategies for enhancing cell survival.

6. Conclusions

Molecular and cellular imaging has enormous potential for transplantation. In the first place this will largely be as a research tool in both animal and clinical studies — providing early response markers for trials and allowing the development of new therapies. Eventually these applications may translate into more routine clinical applications, allowing monitoring of individual patients. While this review can provide some idea of the potential, it cannot show the difficulties of developing this technology for a particular application. As well as the input of clinicians and biologists there is a need for chemists to help develop the contrast agents, physicists for refining the imaging process itself and mathematicians to improve the data analysis and handling. None of these elements is trivial!

However, in the next decade some of the approaches outlined in this review may become essential tools for research and management of transplants.

Safety and ethical issues imposed on a human study would limit the clinical translation of those methods that require extensive manipulation of cells; however, most direct labelling methods have already applied in patients or would be suitable for the clinic.

These promising experiments demonstrate the need for future studies to delineate the fate of injected stem cells by incorporating non-invasive tagging methods to monitor myocardial function following cells engraftment in the myocardial infarction. Consequently, MRI may lead to a better understanding of the myocardial pathophysiology as well as the assessing of a proper implantation and the effects of stem cell therapy by allowing a multimodal approach to evaluating anatomy, function, perfusion, and regional contractile parameters in a single non-invasive examination.

In conclusion, the use of stem cells therapy to treat neurodegenerative diseases is a realistic possibility in the near future. However, the need for non-invasive imaging techniques is a prerequisite in order to monitor these transplants to determine clinical efficacy. Examination by MRI ensures that the stem cells are not only injected to the lesion site, but it also allows the monitoring of inappropriate cellular migration, and furthermore identify damage to surrounding tissues.

Methods for monitoring implanted stem cells non-invasively *in vivo* will greatly facilitate the clinical realisation and optimisation of the opportunities for stem cell-based therapies. In writing this short review we have concentrated on the application of MRI to tracking of stem cells in only two types of tissue repair, namely cardiac and nervous system. There are however, numerous examples where similar methodologies of cell tracking can aid in clinical diagnosis or can be used to trace other cells types, such as those from a tumour or following an inflammatory response.[38,117]

References

1. Lagasse, E., Connors, H., Al-Dhalimy, M. *et al.* Purified hematopoietic stem cells can differentiate into hepatocytes *in vivo*. *Nat. Med.* **6**, 1229–1234 (2000).
2. Orlic, D., Kajstura, J., Chimenti, S. *et al.* Bone marrow cells regenerate infarcted myocardium. *Nature* **410**, 701–705 (2001).
3. Bengel, F.M., Schachinger, V. and Dimmeler, S. Cell-based therapies and imaging in cardiology. *Eur. J. Nucl. Med. Mol. Imag.* **32**(Suppl. 2), S404–S416 (2005).

4. Massoud, T.F. and Gambhir, S.S. Molecular imaging in living subjects: seeing fundamental biological processes in a new light. *Genes Dev.* **17**, 545–580 (2003).
5. Weber, D.A. and Ivanovic, M. Ultra-high-resolution imaging of small animals: implications for preclinical and research studies. *J. Nucl. Cardiol.* **6**, 332–344 (1999).
6. Acton, P.D. and Kung, H.F. Small animal imaging with high resolution single photon emission tomography. *Nucl. Med. Biol.* **30**, 889–895 (2003).
7. Gao, J., Dennis, J., Muzic, R. *et al.* The dynamic *in vivo* distribution of bone marrow-derived mesenchymal stem cells after infusion. *Cells Tissues Organs* **169**, 12–20 (2001).
8. Chin, B., Nakamoto, Y., Bulte, J. *et al.* ^{111}In oxine labelled mesenchymal stem cell SPECT after intravenous administration in myocardial infarction. *Nucl. Med. Commun.* **24**, 1149–1154 (2003).
9. Barbash, I., Chouraqui, P., Baron, J. *et al.* Systemic delivery of bone marrow-derived mesenchymal stem cells to the infarcted myocardium: feasibility, cell migration, and body distribution. *Circulation* **108**, 863–868 (2003).
10. Gambhir, S., Herschman, H., Cherry, S. *et al.* Imaging transgene expression with radionuclide imaging technologies. *Neoplasia* **2**, 118–138 (2000).
11. Simonova, M., Shtanko, O., Sergeyev, N. *et al.* Engineering of technetium-99m-binding artificial receptors for imaging gene expression. *J. Gene Med.* **5**, 1056–1066 (2003).
12. Wu, J., Chen, I., Sundaresan, G. *et al.* Molecular imaging of cardiac cell transplantation in living animals using optical bioluminescence and positron emission tomography. *Circulation* **108**, 1302–1305 (2003).
13. Becker, A., Hessenius, C., Licha K. *et al.* Receptor-targeted optical imaging of tumors with near-infrared fluorescent ligands. *Nat. Biotechnol.* **19**, 327–331 (2001).
14. Frangioni, J.V. *In vivo* near-infrared fluorescence imaging. *Curr. Opin. Chem. Biol.* **7**, 626–634 (2003).
15. Sato, A., Klaunberg, B. and Tolwani, R. *In vivo* bioluminescence imaging. *Comput. Med.* **54**, 631–634 (2004).
16. Wang, X., Rosol, M., Ge, S. *et al.* Dynamic tracking of human hematopoietic stem cell engraftment using *in vivo* bioluminescence imaging. *Blood* **102**, 3478–3482 (2003).
17. Tang, Y., Shah, K., Messerli, S.M. *et al. In vivo* tracking of neural progenitor cell migration to glioblastomas. *Hum. Gene Ther.* **14**, 1247–1254 (2003).
18. Cao, Y.A., Wagers, A.J., Beilhack, A. *et al.* Shifting foci of hematopoiesis during reconstitution from single stem cells. *Proc. Natl. Acad. Sci. USA* **101**, 221–226 (2004).

19. Weissleder, R. and Ntziachristos, V. Shedding light onto live molecular targets. *Nat. Med.* **9**, 123–128 (2003).
20. Zacharakis, G., Kambara, H., Shih, H. *et al.* Volumetric tomography of fluorescent proteins through small animals *in vivo*. *Proc. Natl. Acad. Sci. USA* **102**, 18252–18257 (2005).
21. Ntziachristos, V., Tung, C.H., Bremer, C. *et al.* Fluorescence molecular tomography resolves protease activity *in vivo*. *Nat. Med.* **8**, 757–760 (2002).
22. Wu, J.C., Inubushi, M., Sundaresan, G. *et al.* Optical imaging of cardiac reporter gene expression in living rats. *Circulation* **105**, 1631–1634 (2002).
23. Niyibizi, C., Wang, S., Mi, Z. *et al.* The fate of mesenchymal stem cells transplanted into immunocompetent neonatal mice: implications for skeletal gene therapy via stem cells. *Mol. Ther.* **9**, 955–963 (2004).
24. Heyn, C., Bowen, C.V., Rutt, B.K. *et al.* Detection threshold of single SPIO-labeled cells with FIESTA. *Magn. Reson. Med.* **53**, 312–320 (2005).
25. Shapiro, M.G., Atanasijevic, T., Faas, H. *et al.* Dynamic imaging with MRI contrast agents: quantitative considerations. *Magn. Reson. Imag.* **24**, 449–462 (2006).
26. Louie, A.Y., Huber, M.M., Ahrens, E.T. *et al.* *In vivo* visualization of gene expression using magnetic resonance imaging. *Nat. Biotechnol.* **18**, 321–325 (2000).
27. Josephson, L., Tung, C.H., Moore, A. *et al.* High-efficiency intracellular magnetic labeling with novel superparamagnetic-Tat peptide conjugates. *Bioconjug. Chem.* **10**, 186–191 (1999).
28. Bhorade, R., Weissleder, R., Nakakoshi, T. *et al.* Macrocyclic chelators with paramagnetic cations are internalized into mammalian cells via a HIV-tat derived membrane translocation peptide. *Bioconjug. Chem.* **11**, 301–305 (2000).
29. Lewin, M., Carlesso, N., Tung, C.H. *et al.* Tat peptide-derivatized magnetic nanoparticles allow *in vivo* tracking and recovery of progenitor cells. *Nat. Biotechnol.* **18**, 410–414 (2000).
30. Merbach, A.E. and Toth, E. *The Chemistry of Contrast Agents in Medical Magentic Resonance Imaging* (John Wiley and Sons, 2001).
31. Mornet, S., Vasesur, F., Grasest, F. *et al.* Magnetic nanoparicle design for medical diagnosis and therapy. *J. Mater. Chem.* **14**, 2161–2175 (2004).
32. Wunderbaldinger, P., Josephson, L. and Weissleder, R. Crosslinked iron oxides (CLIO): a new platform for the development of targeted MR contrast agents. *Acad. Radiol.* **9**(Suppl. 2), S304–S306 (2002).
33. Dodd, S.J., Williams, M., Suhan, J.P. *et al.* Detection of single mammalian cells by high-resolution magnetic resonance imaging. *Biophys. J.* **76**, 103–109 (1999).

34. Schoepf, U., Marecos, E.M., Melder, R.J. *et al.* Intracellular magnetic labeling of lymphocytes for *in vivo* trafficking studies. *Biotechniques* **24**, 642–646 (1998).
35. Sipe, J.C., Filippi, M., Martino, G. *et al.* Method for intracellular magnetic labeling of human mononuclear cells using approved iron contrast agents. *Magn. Reson. Imag.* **17**, 1521–1523 (1999).
36. Weissleder, R., Cheng, H.C., Bogdanova, A. *et al.* Magnetically labeled cells can be detected by MR imaging. *J. Magn. Reson. Imag.* **7**, 258–263 (1997).
37. Yeh, T.C., Zhang, W., Ildstad, S.T. *et al.* Intracellular labeling of T-cells with superparamagnetic contrast agents. *Magn. Reson. Med.* **30**, 617–625 (1993).
38. Yeh, T.C., Zhang, W., Ildstad, S.T. *et al.* *In vivo* dynamic MRI tracking of rat T-cells labeled with superparamagnetic iron-oxide particles. *Magn. Reson. Med.* **33**, 200–208 (1995).
39. Bulte, J.W., Ma, L.D., Magin, R.L. *et al.* Selective MR imaging of labeled human peripheral blood mononuclear cells by liposome mediated incorporation of dextran-magnetite particles. *Magn. Reson. Med.* **29**, 32–37 (1993).
40. Bulte, J.W., Laughlin, P.G., Jordan, E.K. *et al.* Tagging of T cells with superparamagnetic iron oxide: uptake kinetics and relaxometry. *Acad. Radiol.* **3**(Suppl. 2), S301–S303 (1996).
41. Daldrup-Link, H.E., Rudelius, M., Oostendorp, R.A. *et al.* Targeting of hematopoietic progenitor cells with MR contrast agents. *Radiology* **228**, 760–767 (2003).
42. Bulte, J.W., Arbab, A.S., Douglas, T. *et al.* Preparation of magnetically labeled cells for cell tracking by magnetic resonance imaging. *Meth. Enzymol.* **386**, 275–299 (2004).
43. Bulte, J.W., Zhang, S., van Gelderen, P. *et al.* Neurotransplantation of magnetically labeled oligodendrocyte progenitors: magnetic resonance tracking of cell migration and myelination. *Proc. Natl. Acad. Sci. USA* **96**, 15256–15261 (1999).
44. Dunning, M.D., Lakatos, A., Loizou, L. *et al.* Superparamagnetic iron oxide-labeled Schwann cells and olfactory ensheating cells can be traced *in vivo* by magnetic resonance imaging and retain functional properties after transplantation into the CNS. *J. Neurosci.* **24**, 9799–9810 (2004).
45. Arbab, A.S., Yocum, G.T., Rad, A.M. *et al.* Labeling of cells with ferumoxides-protamine sulfate complexes does not inhibit function or differentiation capacity of hematopoietic or mesenchymal stem cells. *NMR Biomed.* **18**, 553–555 (2005).
46. Reimer, K.A., Lowe, J.E., Rasmussen, M.M. *et al.* The wavefront phenomenon of ischemic cell death. 1. Myocardial infarct size vs. duration of coronary occlusion in dogs. *Circulation* **56**, 786–794 (1977).

47. Pfeffer, M.A. Left ventricular remodeling after acute myocardial infarction. *Annu. Rev. Med.* **46**, 455–466 (1995).
48. Ryan, T.J., Antman, E.M., Brooks, N.H. *et al.* 1999 update: ACC/AHA guidelines for the management of patients with acute myocardial infarction. A report of the American College of Cardiology/American Heart Association Task Force on Practice Guidelines (Committee on Management of Acute Myocardial Infarction). *J. Am. Coll. Cardiol.* **34**, 890–911 (1999).
49. Woller, K.C. and Drexler, H. Clinical application of stem cells for the heart. *Circ. Res.* **96**, 151–163 (2005).
50. Lev, S., Kehat, I. and Gepstein, L. Differentiation pathways in human embryonic stem cell-derived cardiomyocytes. *Ann. N. Y. Acad. Sci.* **1047**, 50–65 (2005).
51. Menasche, P., Hagege, A.A., Scorsin, M. *et al.* Myoblast transplantation for heart failure. *Lancet* **357**, 279–280 (2001).
52. Smits, P.C., van Geuns, R.J., Poldermans, D. *et al.* Catheter-based intramyocardial injection of autologous skeletal myoblasts as a primary treatment of ischemic heart failure: clinical experience with six-month follow-up. *J. Am. Coll. Cardiol.* **42**, 2063–2069 (2003).
53. Shake, J.G., Gruber, P.J., Baumgartner, W.A. *et al.* Mesenchymal stem cell implantation in a swine myocardial infarct model: engraftment and functional effects. *Ann. Thorac. Surg.* **73**, 1919–1925 (2002).
54. Kocher, A.A., Schuster, M.D., Szabolcs, M.J. *et al.* Neovascularization of ischemic myocardium by human bone-marrow-derived angioblasts prevents cardiomyocyte apoptosis, reduces remodeling and improves cardiac function. *Nat. Med.* **7**, 430–436 (2001).
55. Silva, G.V., Litovsky, S., Assad, J.A. *et al.* Mesenchymal stem cells differentiate into an endothelial phenotype, enhance vascular density, and improve heart function in a canine chronic ischemia model. *Circulation* **111**, 150–156 (2005).
56. Dawn, B., Stein, A.B., Urbanek, K. *et al.* Cardiac stem cells delivered intravascularly traverse the vessel barrier, regenerate infarcted myocardium, and improve cardiac function. *Proc. Natl. Acad. Sci. USA* **102**, 3766–3771 (2005).
57. Messina, E., De Angelis, L., Frati, G. *et al.* Isolation and expansion of adult cardiac stem cells from human and murine heart. *Circ. Res.* **95**, 911–921 (2004).
58. Kawamoto, A., Gwon, H.C., Iwaguro, H. *et al.* Therapeutic potential of *ex vivo* expanded endothelial progenitor cells for myocardial ischemia. *Circulation* **103**, 634–637 (2001).
59. Hinds, K.A., Hill, J.M., Shapiro, E.M. *et al.* Highly efficient endosomal labeling of progenitor and stem cells with large magnetic particles allows magnetic resonance imaging of single cells. *Blood* **102**, 867–872 (2003).

60. Hill, J.M., Dick, A.J., Raman, V.K. *et al.* Serial cardiac magnetic resonance imaging of injected mesenchymal stem cells. *Circulation* **108**, 1009–1014 (2003).
61. Kraitchman, D.L., Heldman, A.W., Atalar, E. *et al.* *In vivo* magnetic resonance imaging of mesenchymal stem cells in myocardial infarction. *Circulation* **107**, 2290–2293 (2003).
62. Shapiro, E.M., Skrtic, S., Sharer, K. *et al.* MRI detection of single particles for cellular imaging. *Proc. Natl. Acad. Sci. USA* **101**, 10901–10906 (2004).
63. Tallheden, T., Nannmark, U., Lorentzon, M. *et al.* *In vivo* MR imaging of magnetically labeled human embryonic stem cells. *Life Sci.* **79**, 999–1006 (2006).
64. Ebert, S.N., Taylor, D.G., Nguyen, H.L. *et al.* Non-invasive tracking of cardiac embryonic stem cells *in vivo* using magnetic resonance imaging techniques. *Stem Cells* **25**, 2936–2944 (2007).
65. Strauer, B.E., Brehm, M., Zeus, T. *et al.* Repair of infarcted myocardium by autologous intracoronary mononuclear bone marrow cell transplantation in humans. *Circulation* **106**, 1913–1918 (2002).
66. Assmus, B., Schachinger, V., Teupe, C. *et al.* Transplantation of progenitor cells and regeneration enhancement in acute myocardial infarction (TOPCARE-AMI). *Circulation* **106**, 3009–3017 (2002).
67. Britten, M.B., Abolmaali, N.D., Assmus, B. *et al.* Infarct remodeling after intracoronary progenitor cell treatment in patients with acute myocardial infarction (TOPCARE-AMI): mechanistic insights from serial contrast-enhanced magnetic resonance imaging. *Circulation* **108**, 2212–2218 (2003).
68. Schachinger, V., Assmus, B., Britten, M.B. *et al.* Transplantation of progenitor cells and regeneration enhancement in acute myocardial infarction: final one-year results of the TOPCARE-AMI trial. *J. Am. Coll. Cardiol.* **44**, 1690–1699 (2004).
69. Wollert, K.C., Meyer, G.P., Lotz, J. *et al.* Intracoronary autologous bone-marrow cell transfer after myocardial infarction: the BOOST randomised controlled clinical trial. *Lancet* **364**, 141–148 (2004).
70. Meyer, G.P., Wollert, K.C., Lotz, J. *et al.* Intracoronary bone marrow cell transfer after myocardial infarction: eighteen months' follow-up data from the randomized, controlled BOOST (BOne marrOw transfer to enhance ST-elevation infarct regeneration) trial. *Circulation* **113**, 1287–1294 (2006).
71. Fernandez-Avilès, F., San Roman, J.A., Garcia-Frade, J. *et al.* Experimental and clinical regenerative capability of human bone marrow cells after myocardial infarction. *Circ. Res.* **95**, 742–748 (2004).
72. Chen, S.L., Fang, W.W., Ye, F. *et al.* Effect on left ventricular function of intracoronary transplantation of autologous bone marrow mesenchymal stem

cell in patients with acute myocardial infarction. *Am. J. Cardiol.* **94**, 92–95 (2004).

73. Bartunek, J., Vanderheyden, M., Vandekerckhove, B. *et al.* Intracoronary injection of CD133-positive enriched bone marrow progenitor cells promotes cardiac recovery after recent myocardial infarction: feasibility and safety. *Circulation* **112**, I178–183 (2005).
74. Janssens, S., Dubois, C., Bogaert, J. *et al.* Autologous bone marrow-derived stem-cell transfer in patients with ST-segment elevation myocardial infarction: double-blind, randomised controlled trial. *Lancet* **367**, 113–121 (2006).
75. Schachinger, V., Erbs, S., Elsasser, A. *et al.* Intracoronary bone marrow-derived progenitor cells in acute myocardial infarction. *N. Engl. J. Med.* **355**, 1210–1221 (2006).
76. Lunde, K., Solheim, S., Aakhus, S. *et al.* Intracoronary injection of mononuclear bone marrow cells in acute myocardial infarction. *N. Engl. J. Med.* **355**, 1199–1209 (2006).
77. Assmus, B., Walter, D.H., Lehmann, R. *et al.* Intracoronary infusion of progenitor cells is not associated with aggravated restenosis development or atherosclerotic disease progression in patients with acute myocardial infarction. *Eur. Heart J.* **27**, 2989–2995 (2006).
78. Strauer, B.E., Brehm, M., Zeus, T. *et al.* Regeneration of human infarcted heart muscle by intracoronary autologous bone marrow cell transplantation in chronic coronary artery disease: the IACT study. *J. Am. Coll. Cardiol.* **46**, 1651–1658 (2005).
79. Katritsis, D.G., Sotiropoulou, P.A., Karvouni, E. *et al.* Transcoronary transplantation of autologous mesenchymal stem cells and endothelial progenitors into infarcted human myocardium. *Catheter Cardiovasc. Interv.* **65**, 321–329 (2005).
80. Fuchs, S., Satler, L.F., Kornowski, R. *et al.* Catheter-based autologous bone marrow myocardial injection in no-option patients with advanced coronary artery disease: a feasibility study. *J. Am. Coll. Cardiol.* **41**, 1721–1724 (2003).
81. Fuchs, S., Kornowski, R., Weisz, G. *et al.* Safety and feasibility of transendocardial autologous bone marrow cell transplantation in patients with advanced heart disease. *Am. J. Cardiol.* **97**, 823–829 (2006).
82. Beeres, S.L., Bax, J.J., Dibbets, P. *et al.* Effect of intramyocardial injection of autologous bone marrow-derived mononuclear cells on perfusion, function, and viability in patients with drug-refractory chronic ischemia. *J. Nucl. Med.* **47**, 574–580 (2006).
83. Beeres, S.L., Bax, J.J., Kaandorp, T.A. *et al.* Usefulness of intramyocardial injection of autologous bone marrow-derived mononuclear cells in patients

with severe angina pectoris and stress-induced myocardial ischemia. *Am. J. Cardiol.* **97**, 1326–1331 (2006).

84. Beeres, S.L., Bax, J.J., Dibbets-Schneider, P. *et al.* Sustained effect of autologous bone marrow mononuclear cell injection in patients with refractory angina pectoris and chronic myocardial ischemia: twelve-month follow-up results. *Am. Heart. J.* **152**, 684.el1–684.e16 (2006).
85. Perin, E.C., Dohmann, H.F., Borojevic, R. *et al.* Transendocardial, autologous bone marrow cell transplantation for severe, chronic ischemic heart failure. *Circulation* **107**, 2294–2302 (2003).
86. Perin, E.C., Dohmann, H.F., Borojevic, R. *et al.* Improved exercise capacity and ischemia 6 and 12 months after transendocardial injection of autologous bone marrow mononuclear cells for ischemic cardiomyopathy. *Circulation* **110**, II213–218 (2004).
87. Erbs, S., Linke, A., Adams, V. *et al.* Transplantation of blood-derived progenitor cells after recanalization of chronic coronary artery occlusion: first randomized and placebo-controlled study. *Circ. Res.* **97**, 756–762 (2005).
88. Kang, H.J., Kim, H.S., Zhang, S.Y. *et al.* Effects of intracoronary infusion of peripheral blood stem-cells mobilised with granulocyte-colony stimulating factor on left ventricular systolic function and restenosis after coronary stenting in myocardial infarction: the MAGIC cell randomised clinical trial. *Lancet* **363**, 751–756 (2004).
89. Cao, F., Lin, S., Xie, X. *et al.* *In vivo* visualization of embryonic stem cell survival, proliferation, and migration after cardiac delivery. *Circulation* **113**, 1005–1014 (2006).
90. Olson, L. and Malmfors, T. Growth characteristics of adrenergic nerves in the adult rat. Fluorescence histochemical and 3H-noradrenaline uptake studies using tissue transplantations to the anterior chamber of the eye. *Acta Physiol. Scand.* **348**, 1–11 (1970).
91. Madrazo, I., Leon, V., Torres, C. *et al.* Transplantation of fetal *substantia nigra* and adrenal medulla to the caudate nucleus in two patients with Parkinson's disease. *N. Engl. J. Med.* **318**, 51 (1988).
92. Lindvall, O., Rehncrona, S., Brundin, P. *et al.* Human fetal dopamine neurons grafted into the striatum in two patients with severe Parkinson's disease. A detailed account of methodology and a 6-month follow-up. *Arch Neurol.* **46**, 615–631 (1989).
93. Piccini, P., Brooks, D.J., Bjorklund, A. *et al.* Dopamine release from nigral transplants visualized *in vivo* in a Parkinson's patient. *Nat. Neurosci.* **2**, 1137–1140 (1999).

94. Ourednik, J., Ourednik, V., Lynch, W.P. *et al.* Neural stem cells display an inherent mechanism for rescuing dysfunctional neurons. *Nat. Biotechnol.* **20**, 1103–1110 (2002).
95. Yamamoto, S., Yamamoto, N., Kitamura, T. *et al.* Proliferation of parenchymal neural progenitors in response to injury in the adult rat spinal cord. *Exp. Neurol.* **172**, 115–127 (2001).
96. Shibuya, S., Miyamoto, O., Auer, R.N. *et al.* Embryonic intermediate filament, nestin, expression following traumatic spinal cord injury in adult rats. *Neuroscience* **114**, 905–916 (2002).
97. Weimann, J.M., Charlton, C.A., Brazelton, T.R. *et al.* Contribution of transplanted bone marrow cells to Purkinje neurons in human adult brains. *Proc. Natl. Acad. Sci. USA* **100**, 2088–2093 (2003).
98. Jiang, Y., Jahagirdar, B.N., Reinhardt, R.L. *et al.* Pluripotency of mesenchymal stem cells derived from adult marrow. *Nature* **418**, 41–49 (2002).
99. Kabos, P., Ehtesham, M., Kabosova, A. *et al.* Generation of neural progenitor cells from whole adult bone marrow. *Exp. Neurol.* **178**, 288–293 (2002).
100. Goolsby, J., Marty, M.C., Heletz, D. *et al.* Hematopoietic progenitors express neural genes. *Proc. Natl. Acad. Sci. USA* **100**, 14926–14931 (2003).
101. Norman, A.B., Thomas, S.R., Pratt, R.G. *et al.* Magnetic resonance imaging of neural transplants in rat brain using a superparamagnetic contrast agent. *Brain Res.* **594**, 279–283 (1992).
102. Hawrylak, N., Ghosh, P., Broadus, J. *et al.* Nuclear magnetic resonance (NMR) imaging of iron oxide-labeled neural transplants. *Exp. Neurol.* **121**, 181–192 (1993).
103. Franklin, R.J., Blaschuk, K.L., Bearchell, M.C. *et al.* Magnetic resonance imaging of transplanted oligodendrocyte precursors in the rat brain. *Neuroreport* **10**, 3961–3965 (1999).
104. Zhang, R.L., Zhang, L., Zhang, Z.G. *et al.* Migration and differentiation of adult rat subventricular zone progenitor cells transplanted into the adult rat striatum. *Neuroscience* **116**, 373–382 (2003).
105. Modo, M., Mellodew, K., Cash, D. *et al.* Mapping transplanted stem cell migration after a stroke: a serial, *in vivo* magnetic resonance imaging study. *Neuroimage* **21**, 311–317 (2004).
106. Hoehn, M., Kustermann, E., Blunk J. *et al.* Monitoring of implanted stem cell migration *in vivo*: a highly resolved *in vivo* magnetic resonance imaging investigation of experimental stroke in rat. *Proc. Natl. Acad. Sci. USA* **99**, 16267–16272 (2002).

107. Rice, H.E., Hsu, E.W., Sheng, H. *et al.* Superparamagnetic iron oxide labeling and transplantation of adipose-derived stem cells in middle cerebral artery occlusion-injured mice. *Am. J. Roentgenol.* **188**, 1101–1108 (2007).
108. Jendelova, P., Herynek, V., Urdzikova, L. *et al.* Magnetic resonance tracking of transplanted bone marrow and embryonic stem cells labeled by iron oxide nanoparticles in rat brain and spinal cord. *J. Neurosci. Res.* **76**, 232–243 (2004).
109. Blakemore, W.F. and Franklin, R.J. Transplantation of glial cells into the CNS. *Trends Neurosci.* **14**, 323–327 (1991).
110. Blakemore, W.F., Gilson, J.M. and Crang, A.J. Transplanted glial cells migrate over a greater distance and remyelinate demyelinated lesions more rapidly than endogenous remyelinating cells. *J. Neurosci. Res.* **61**, 288–294 (2000).
111. Duncan, I.D., Hammang, J.P., Jackson, K.F. *et al.* Transplantation of oligodendrocytes and Schwann cells into the spinal cord of the myelin-deficient rat. *J. Neurocytol.* **17**, 351–360 (1988).
112. Utzschneider, D.A., Archer, D.R., Kocsis, J.D. *et al.* Transplantation of glial cells enhances action potential conduction of amyelinated spinal cord axons in the myelin-deficient rat. *Proc. Natl. Acad. Sci. USA* **91**, 53–57 (1994).
113. Tontsch, U., Archer, D.R., Dubois Dalcq, M. *et al.* Transplantation of an oligodendrocyte cell line leading to extensive myelination. *Proc. Natl. Acad. Sci. USA* **91**, 11616–11620 (1994).
114. Zhang, S.C., Ge, B. and Duncan, I.D. Adult brain retains the potential to generate oligodendroglial progenitors with extensive myelination capacity. *Proc. Natl. Acad. Sci. USA* **96**, 4089–4094 (1999).
115. Akiyama, Y., Honmou, O., Kato, T. *et al.* Transplantation of clonal neural precursor cells derived from adult human brain establishes functional peripheral myelin in the rat spinal cord. *Exp. Neurol.* **167**, 27–39 (2001).
116. Imaizumi, T., Lankford, K.L., Burton, W.V. *et al.* Xenotransplantation of transgenic pig olfactory ensheathing cells promotes axonal regeneration in rat spinal cord. *Nat. Biotechnol.* **18**, 949–953 (2000).
117. Taupitz, M., Schmitz, S. and Hamm, B. Superparamagnetic iron oxide particles: current state and future development. *Rofo Fortschr. Geb. Rontgenstr. Neuen. Bildgeb. Verfahr.* **175**, 752–765 (2003).

PART V

BIOTECHNOLOGY SECTOR

Chapter 22

Lessons Learnt

Nancy L. Parenteau, Susan J. Sullivan,
Kelvin G. M. Brockbank and Janet Hardin Young

Abstract

Tissue engineering can be broadly defined as the combination of biology and engineering to repair or replace lost tissue function. From an industry perspective, the field encompasses implanted biomaterials, cell and tissue transplants and therapies, and even extracorporeal cellular devices. To achieve its goals, tissue engineering must effectively utilize not only multiple aspects of engineering but also several aspects of biology that govern mechanisms of organ development, repair and regeneration. The field has always had a strong focus on application yet the challenge of integrating biological science, engineering and medicine has kept many past efforts from reaching their therapeutic and commercial potential. This chapter covers the evolution of tissue engineering, looking at the change in emphasis from bioengineering to stem cell biology and the potential impact of this shift in focus from an industrial perspective. In addition, we have analyzed four major commercial thrusts from past to present: vascular tissue engineering, cartilage repair, liver-assist devices and skin constructs, paying particular attention to how the biomedical disciplines must be integrated to achieve commercial feasibility and therapeutic success. Each example yields one or more important and practical lessons learnt that could be instructive for most future medical and commercial efforts in tissue engineering.

Keywords: Commercialization; Vascular; Liver; Skin; Cartilage.

Outline

1. Background: The Marriage of Biology and Engineering
2. The Evolving Emphasis from Engineering to Biology
3. Vascular Tissue Engineering
4. Cartilage

5. Extracorporeal Devices
6. Skin
7. Conclusions
References

1. Background: The Marriage of Biology and Engineering

From the late 1970s, the concept of guided tissue regeneration was being pursued as a clinical concept.[1] By the end of the 1980s the term "tissue engineering" was coined to describe the combination of engineering and biology to repair and replace lost tissues.[2] Work in cell therapy and artificial organs was often conducted separately until the unifying concept of regenerative medicine emerged in the 1990s. Today, tissue engineering incorporates many strategies and technologies including the use of adult and embryonic stem cells.

Tissue engineering has always had a strong focus on application, but few applications have made it to the clinic, and even fewer to the marketplace. In this chapter, we examine four major commercial thrusts from the past to find lessons for the way forward.

Scaffolds dominated much of tissue engineering in the US during the first ten to 15 years.[3] The bioengineering of scaffolds was seen as a way to direct cell migration, growth and behavior to achieve three-dimensional structure. However, in practice, the engineering of scaffolds suffered from lack of a true multidisciplinary partnership with biologists. There is some evidence that this is changing, but the change has been slow and advances in polymer science and extracellular matrices are often left without the enabling cell biology to accompany and guide their efforts. The advent of stem cell biology has created a greater interest in cell biology, but it is unclear whether this has yet translated into better partnerships.

The most widely used biomaterials in tissue engineering are resorbable polymers.[4] Since removal is through chemical dissolution rather than an enzymatic reaction, they can serve as a temporary support for three-dimensional tissue formation. Biological polymers like collagen provide the benefits of natural cell interaction and adhesion but require biological remodeling. Today, there are approved products from tissue engineering containing both synthetic and natural polymers.

Current research has moved biomimetic[5] and instructional[6,7] scaffolds to better direct tissue formation. This strategy is derived principally from an engineering perspective. From a biological perspective, the scaffold still faces the challenge of enabling the biological interaction and response of cell populations without getting in the way, both physically and biologically.

Biologically, a scaffold might not act at the level of tissue organization *per se* but rather at the level of a biological process that modulates the regenerative response. For example, recent work highlights the new, but growing, appreciation that inflammatory cytokines and inflammatory cells are influential in determining the tissue response to an implanted scaffold. A scaffold's impact on the innate immune response may be more influential in fostering functional tissue formation than any other property. In addition, inflammatory cell signaling will be an important factor contributing to the death or survival of implanted cells and tissues almost irrespective of the scaffold used. Therefore, the design of any cellular implant should take into account not only the tissue reaction to the implant but also the implant's reaction to the recipient environment, perhaps making an immunologist one of the most valuable members to include on today's multidisciplinary team.

Another consideration is the dynamic biological regulation inherent in, and controlled by, the cell population itself. This regulation not only results in the generation of different cell phenotypes but also matrix biosynthesis, cytokine cross-talk, cell organization and ultimately formation of an organized structure.[8] Stem cell niches may be important for maintenance of the stem cell population in tissues[9] but, not surprisingly, the key to generation of that niche may lie in the natural development and differentiation of the stem cell population itself.[10] A scaffold that clears the way for, and enables, that response to occur as part of the natural development of functional tissue is likely to be more biologically and physiologically robust than one that attempts to force it artificially. The closer one can recapitulate an enabling environment in the midst of host response, the better. So another valuable lesson is to be sure to have the input of a developmental biologist.

2. The Evolving Emphasis from Engineering to Biology

The growing interest in embryonic and adult stem cells has fuelled an interest in the cell biology of growth, differentiation and regeneration, bringing cell biology to the forefront in tissue engineering. But it is unclear whether the biology needed for tissue engineering has advanced because of this, or simply changed. It is questionable whether the interest in incorporating stem cells will lead to faster solutions near term or whether it has caused biological efforts to step back, drawing attention away from other limiting issues, thus delaying their solution.

One of the primary hurdles is finding a safe source of cells that will not only yield enough cells but, as importantly, ultimately yield tissues with the proper functions. There have been various strategies ranging from *in situ* promotion of tissue regeneration to the use of autologous, allogeneic and xenogeneic cells and tissues as well as genetically engineered cells. The ability to generate human embryonic stem cells presents the possibility of a highly expandable cell source capable of generating a wide variety of tissues. However, the ultimate goal is still the same — formation of functional tissue.

In reality, pluripotent stem cells have real, practical hurdles of their own and also share some of the same hurdles that limit their more lineage-committed relatives — hurdles that must be surmounted for their clinical potential to be realized. The functional benefit of stem cells over other sources is still largely implied. Biological differences and benefits must be established, not assumed.

In general, tissue engineering must be committed to going beyond anecdotal biological evidence. Biological mechanisms cannot remain in a black box if one hopes to move a tissue engineering application into the marketplace and broad clinical use. The more one understands the cells and engineered tissues and their relationship to native situations, the better chance one has to apply technology in the best way possible.

It is important to be clear on the principal factors that have enabled or thwarted progress in the past and to address them. In the following sections, we examine how both cell biology and engineering have been used to tackle some of major commercial applications in tissue engineering and regenerative medicine and what remains to be conquered in each area.

3. Vascular Tissue Engineering

Autogenous saphenous veins are the "gold standard" for coronary grafting after internal mammary arteries. However, many patients do not have veins suitable for grafting. This translates to a clear clinical need for an engineered, small-diameter vascular graft.

Experimental and clinical vein grafting[11] indicate that it is not necessary to implant a fully developed artery as veins undergo arterialization upon transplantation. Therefore, a tissue or material that can be reliably sutured, is strong enough to hold arterial pressures, compliant enough to form non-hyperplastic anastomoses, and cell compatible enough to allow arterialization, while maintaining structure and patency, is a reasonable clinical and commercial target for vascular tissue engineering.

Decellularized tissues are strong, naturally compatible options for both blood vessels and heart valves. The decellularized matrices for blood vessels range from xenogeneic blood vessels to ureters, and intestinal submucosa. Decellularized heart valve allografts (CryoLife, Inc.) perform well clinically.[12–14] Regrettably, the first clinical experience with decellularized porcine valves manufactured without cross-linking resulted in early failures and patient deaths due to residual tissue antigenicity. Also, both human and "processed" xenograft scaffolds appear more proinflammatory when cellular debris, cytokines, and other inflammatory moieties are not thoroughly removed.[15,16]

The compact, collagenous submucosa of the small intestine (SIS) was first studied as a large diameter vascular substitute in dogs.[17–20] Lantz *et al.*[21] reported 75% patency using a small-diameter sewn SIS graft in the canine carotid and femoral positions. Explant histopathology indicated promising arterialization.[22] However, small diameter SIS grafts were thrombogenic[23] and the studies of Lantz *et al.* employed anti-coagulant protocols that would be problematic in humans. In addition, the native material would also require additional processing to make the material acceptable for human use.

A cleaned intestinal collagen layer intubated using a wrapping technique and varying degrees of carbodiimide cross-linking was developed (Organogenesis Inc.). The luminal surface was modified by treatment with benzalkonium-chloride heparin complex and collagen[24] and the construct was sterilized with peracetic acid.[25] Implanted as interpositional grafts in rabbit carotid arteries, the grafts were remodeled by host cells, showed some endothelialization of the luminal surface and became physiologically responsive.[26] However, explant pathology suggested a chronic inflammatory response and the cellularity in the intima and along the luminal surface was variable. In subsequent studies,[27] process modifications, including decreased cross-linking, led to improved cellular response and accelerated tissue remodeling.

While an off-the-shelf acellular graft remains an attractive concept, acellular grafts must rely on patient response, which could be variable and ultimately limiting in aged, diseased patients. Could stem cell recruitment or the addition of a cellular component remove this limitation in a feasible and reliable way? Recently, human pulmonary valve allografts decellularized and reseeded with peripheral blood mononuclear cells as a source of endothelial progenitor cells performed well in two paediatric patients after more than three years.[28]

The concept of a man-made cellular tissue-engineered graft began with the work of Weinberg and Bell[29] where a collagen/fibrin mixture was combined with cells in concentric layers of cell-contracted matrices around a mandril. In later work, the constructs were mechanically conditioned prior to implantation. Many of the issues faced by Weinberg and colleagues in the 1980s and early 1990s (Organogenesis Inc.) are still hurdles today. While these first tissue-engineered vessels faced significant biological and process challenges, their primary limitation was their lack of suturability.

Despite the challenges, work on cellular constructs has continued.[30] Shin'oka *et al.*,[31] were able to reconstruct a low-pressure pulmonary outflow tract in paediatric patients using autologous bone marrow cells seeded into 12 to 24 mm diameter tubes of the lactic acid-caprolactone copolymer reinforced with a polyglycolic acid sleeve with good success. Although the implants were of a large diameter in the relatively low-pressure pulmonary circulation, the study demonstrates the clinical use of an engineered cellularized graft.

An alternative strategy is *in vivo* creation of a tissue engineered blood vessel (TEBV). Campbell and co-workers surgically implanted a mandril of foreign material in the peritoneal cavity of the recipient to induce the growth of tubular tissues.[32,33] This has shown some promise, although reproducibility appears to be an issue,[34] illustrating the challenge of *in vivo* variability.

A completely cell-produced TEBV was first described in 1998.[35] A cellular/acellular, layered modification of this concept is currently being developed for clinical use (Cytograft, Inc.). The tube is constructed from multiple layers of extracellular matrix produced by cultured autologous fibroblasts.[36] The primary purpose of the tube is the delivery of a biocompatible, structurally sound extracellular matrix, as the intimal layer is dehydrated and decellularized. But the current clinical construct incorporates an outer layer containing viable cells.[37] The tubes are lined with autologous endothelium prior to use. So far, the grafts have handled and functioned well in early safety testing.[38] Theoretically, since cell-derived matrices can be produced in a controlled, sterile environment, this approach has potential processing advantages as well as the potential for additional biological modification, provided manufacturing times and costs are reasonable. An alternative construct made from allogeneic fibroblasts in which both layers are dehydrated is being investigated as an off-the-shelf modification (personal communication, L' Heureux), however production time and cost will still be a hurdle.

A lesson to be learned is that every component and process step should be there for a reason that relates to function or safety. If viable cells are not necessary, then why include them in the final product? If viable cells can contribute, then can a better product be made by working to keep them? The goal is to achieve clinically relevant function within a realistic framework. Proceed to the clinic only when methods that will lead to both are reasonably in place. More often than not, there will be no second chance.

Can a better TEBV be achieved by including cells? Adult stem cells from bone marrow, cord blood or adipose tissue are of interest as accessible cell sources for smooth muscle and stromal cells as well as autologous endothelial cells and their precursors. Neither arterialized veins nor most TEBV models used to date show evidence of *de novo* elastin biosynthesis unless conditioned by cyclic distension.[39,40] Literature suggests that less mature[41,42] cells may be more likely to produce elastin. Bone marrow-derived smooth muscle progenitor cells appear superior to vascular smooth muscle cells in elastin biosynthesis, matrix organization and physical performance.[43] Other studies have shown that vessels established from infant cells exhibit higher levels of proliferation, extracellular matrix deposition and enhanced physical properties when compared with vessels established from adult cells.[44] TEBV constructed with cells originating from older donors exhibited inferior mechanical properties that prohibited implantation, although both cell proliferation and mechanical properties were improved by the expansion of cellular lifespan through telomerase expression.[45,46] The degree to which reported differences represent true biological limitations of adult cells or represent technical challenges is unclear. Perhaps the most important concern is whether cellularization will be ultimately required to assure adequate, reliable development and remodeling of vascular implants in a majority of patients.

It is clear that the physical character and biological response to the biomaterial component are among the most critical aspects of TEBV design. Following that, refinements in how one uses and/or incorporates cells will be the most enabling aspect to delivering a safe, effective construct that is a true clinical substitute for an autologous vein graft and perhaps even providing some superior function.

4. Cartilage

Articular cartilage provides its own particular challenges for tissue engineering. Cartilage forms readily from mesenchymal stem cells *in vivo* as a callous during bone repair. Its structure appears simple and it only contains one cell type. But cartilage is avascular, is not innervated and normal mechanisms of tissue repair, involving the recruitment of cells to the site of damage, do not occur. The culture of cartilage cells and Type II collagen deposition has been studied for decades. However, the challenge of cartilage repair still lies in achieving regeneration of its complex highly organized extracellular matrix (ECM).

The physical properties of articular cartilage depend on the structure and organization of the macromolecules in the ECM. They can largely be understood in terms of the contribution made by fibrillar and nonfibrillar components.[47] The predominantly type II collagen provides the tensile strength. Proteoglycans, such as aggrecan, fill the interfibrillar matrix to produce a tissue that is not only strong in tension but also resistant to compression. Articular cartilage thus forms a tough but compliant load-bearing surface. Part of the challenge of tissue engineering cartilage is to provide or encourage the critical cells and signals that will permit appropriate cartilage differentiation, i.e. establish a cartilage ECM and recapitulate this molecular organization that forms the basis for the essential mechanical properties of the tissue.

Autologous chrondrocyte implantation (ACI) involves culturing the patient's own cells (collected from a knee biopsy). Once the cells have expanded to sufficient

Table 1. Companies commercializing ACI/MACI products.

Company	Location	Product(s)
BioTissue Technologies	Germany	BioSeed®
co.don Ag	Germany	Chondrotransplant®
Fidia Advance Biomaterials	Italy	HYALOGRAFT®
Genzyme BioSurgery	USA	Carticel®
Pro Biotech LTD	USA	BioCart™/ BioCart™-II
Tigenix	Belgium	CHONDROCELECT
Verigen	Germany	ACI/MACI/MACI-A

numbers, they are transferred to the damaged knee and a periosteal flap is sutured over the lesion. This procedure was first commercialized in 1997 (Genzyme). Since then, a number of companies have developed variations of ACI (Table 1). Today biomaterials such as collagen are used to replace the periosteal flap in an effort to simplify the surgical procedure.

The second generation ACI involves cell-seeded membranes or matrix-induced ACI (MACI). Chondrocytes are seeded on biological matrices such as hyaluronic acid,[48,49] collagen membranes[50] or atelocollagen gel.[51] One of the advantages of the MACI procedure over ACI is that it is arthroscopic whereas the ACI procedure requires a more invasive open knee procedure. There are a number of companies globally which either have a commercial product or are in late-stage clinical testing (Table 1).

Several randomized prospective clinical trials have looked at the efficacy of ACI and MACI. Three studies compared ACI to standard surgical procedures such as microfracture or mosaicplasty;[52–54] while two studies compared the results of patients treated with ACI or MACI. Results showed the formation of nearly normal cartilage in a good number of ACI and MACI treated patients compared to surgical controls.[55] However in the two side-by-side studies, no significant difference was seen between ACI and MACI patients.[56] In addition, neither ACI nor MACI is devoid of issues related to demanding surgical technique, and problems of cell retention in the site.[57,58]

The search for cell sources superior to mature chondrocytes is driven in part by the concerns associated with donor site morbidity, cell dedifferentiation, and the limited life span of mature chondrocytes. A number of research groups[59,60] and companies[61] are trying to identify appropriate genes which can be used to screen and monitor cultures for the correct chondrocyte phenotype. The SOX transcription factors may be one such set of markers.[59,60]

Alternative cell sources such as mesenchymal progenitor cells (MPCs) or mesenchymal stem cells (MSCs) are being explored for use in MACI procedures[62–64] although implants of MSC for meniscal repair have met with disappointment (data from Osiris Inc.). MPCs are resident within a host of musculoskeletal and connective tissues, and the multipotent nature of MPCs makes them a potential candidate for cartilage repair. Regardless of the cell source, a key hurdle is the ultimate differentiation into hyalin cartilage rather than fibrocartilage or scar tissue. Since differentiation of cartilage is at least partially modulated by biomechanics, scaffolds have been seen as a possible way to achieve both adequate cell delivery and differentiation.

Three-dimensional scaffolds provide better mechanical properties and facilitate cell adhesion and retention. Clinical and preclinical studies have shown the ability of cell/scaffold constructs to promote chondrocyte proliferation, maturation, and differentiation. Table 2 summarizes the major approaches being pursued both in industry and academics to improve the scaffold for engineered cartilage. However after at least 20 years, the differentiation of articular cartilage still appears difficult to direct, regardless of the approach.

Further advancement in cartilage repair will depend on additional advances in understanding the regulation of cartilage biology. Additional knowledge of mesenchymal cell regulation should contribute to this understanding. But this knowledge must be met with innovative biomaterials, engineering and surgical techniques.

The difficulty of clinical testing has often been blamed for the comparatively slow advance of innovation in commercial cartilage technologies. However, the ability to sell an autologous cell product without first proving effectiveness in a prospective clinical trial is as likely to be responsible for steering resources away from more rapid innovation in both the biology and the engineering.

5. Extracorporeal Devices

As the field of hybrid artificial organs emerged in the 1970s, an extracorporeal liver assist device (LAD) was quickly identified as one of the most compelling applications for this new technology (See also Chapters 34 and 40). In 1980, it was considered a "holy grail"[65] and in 2008, the clinical message is essentially the same. Yearly, about 17,000 patients are on the waiting list for a liver transplant in the US and 2000 will die awaiting a donor. A clinically effective liver assist device has the potential to save lives and use donor organs more efficiently by (i) bridging to transplant, (ii) providing time for liver recovery for fulminant liver failure, (iii) supporting the 5% of liver recipients that experience primary non-function of the transplanted liver, and (iv) offering an treatment option that could delay the onset of acute-on-chronic liver failure. Unfortunately, the clinical reality of a liver

Table 2. Approaches to improving scaffolds for tissue engineered cartilage.

Scaffold type	Approach	Example
Injectable	Naturally derived polysaccharide gels to deliver cells and matrix to the defect site. *Goal: To improve cell retention and the surgical procedure.*	Amalgamation of a biodegradable polymer with alginate as a scaffold to support cells and significantly improve mechanical strength of the injectable scaffold.
Heterogeneous	Scaffolds to structurally organize zone-specific cells and encourage heterotypic cell interactions. *Goal: To improve biological and functional properties.*	Chondrocytes, from different articular cartilage zones, placed in bilayered photopolymerized hydrogels organize into a stratified framework which demonstrate greater shear and compressive strength than homogenous cell constructs.
Nanoscopic	Nanoscopic biodegradable scaffolds mimic the cells' natural environment while providing structural stability. *Goal: Structural and morphological properties of native extracellular matrix.*	Electrospun poly-caprolactone-based nanofibrous scaffolds have been produced which support the chondrocytic phenotype of fetal bovine chondrocytes.
Self-assembled	This approach relies entirely on creating the appropriate biological environment to allow cells to produce the appropriate 3D cartilagous structure. *Goal: To develop a cell-produced scaffold.*	In preclinical studies, chondrocytes produced dime-sized disks of cartilage. Disks displayed mechanical and biochemical properties very similar to the native cartilage.

assist device is no closer in 2008 than a decade ago when four companies had liver assist devices in clinical trials (Table 3).

Since the first description of rodent hepatoma cells seeded onto the outer surface of semi-permeable hollow fibers,[66] there has been a vast amount of *in vitro* data suggesting that various LAD configurations could support liver cell function and survival. Furthermore, in contrast to renal dialysis, non-biological therapies such as hemodialysis and plasma exchange were not very effective, leading to the widely held belief that incorporating living cells into extracorporeal perfusion circuit was key. Employing hollow fiber cartridges already in use for dialysis

Table 3. Liver assist devices.

Company	Device design	Cell source	Clinical outcome	Current status
Circe Biomedical HepatAssist 2000[86]	Plasma perfusion through a hollow cartridge	Porcine hepatocytes	Phase II/III trial completed in 2002 did not achieve Circe clinical endpoint	Unable to raise funds for a full Phase III study, dissolved in 2003. Assets transferred to Arbios Systems in 2004
Vitagen Inc. ELAD[87]	Plasma perfusion through hollow-fiber cartridge	C3A immortalized human cell line	Phase I/II clinical trials initiated in 1999, completed in 2002	Assets acquired by Vital Therapies in 2003
Excorp Medical Mark II[88]	Blood perfusion through cartridge containing hollow fibers with cells within the luminal space	Porcine hepatocytes	Phase I/II trials initiated in 1998	Ongoing
HybridOrgan GmbH BELS	Tripartite hollow fiber system designed for plasma inflow, outflow and oxygenation	Porcine and human hepatocytes	Completed Phase I/II trials in 2004	Being developed by HepaLife Inc. for use with PICM-19, a porcine liver cell line

applications along with porcine primary hepatocytes presented a strategy for getting to the clinic quickly. Refining device design and optimizing cell function were seen as follow on activities. However, the path to the clinic has been much more difficult than anticipated and illustrates the need for specific science and innovation. While progress has been made, the struggle to show efficacy in what was to be a "rapid" first generation model has hampered the evolution of this technology and the challenges identified early in the development of a bioartificial liver, cell source, device design, manufacturability and clinical evaluation, remain.

LAD's extracorporeal, limited use nature makes it one of the most amenable cell-based therapeutics for xenogeneic cells. Devices employing primary porcine hepatocytes continue in clinical development; Arbios Systems HepatAssist, Excorp Medical (Mark II). Phase I/II trials demonstrated safety and concerns about retroviral transmission have been minimized.[67] Preclinical studies have shown that the primary cells can maintain differentiated hepatocyte function for at least a week in the device.[68] However, the current interest in human stem cells as a hepatocyte source now makes it more difficult to advance a xenogeneic cell therapy to market.

An alternative cell source strategy has been to use human hepatocyte-like cell lines. But the human C3A cell line used in the ELAD system (Vitagen) has lower levels of activity than primary hepatocytes so its availability must be balanced against its practical and physiological limitations. Further, recent studies indicate these cells are not capable of ammonia detoxification because the urea cycle is non-functional. Since ammonia detoxification is likely to be critical to clinical efficacy,[69] this questions its clinical usefulness in most LAD applications. The cell line, PICM-19, which is related to bile duct progenitor cells, appears able to maintain activity more comparable to primary cells.[70–72] The lessons to be learned in the use of cell lines and any hepatocyte-like cells emerging from stem cell biology is that achieving a high level of function is essential. Without being able to deliver meaningful function, other features of the cell source provide little benefit.

The challenge of defining the key attributes of hepatocyte function that must be optimized and understanding what levels of activity are likely to be clinically relevant remains. The correlation between ammonia and hepatic encephalopathy is clear. However, biosynthetic capacity, often used as an indicator of differentiated function, is likely to be the least clinically relevant although many worked on optimizing metabolic function based largely on the assumption it would correlate with clinical utility. This lesson has been learned largely from the progress in non-cellular detoxification systems. It should also be appreciated that the barrier to entry also moves with progress. Future biologic liver assist devices will be likely to incorporate state of the art adsorbent pheresis systems like MARS (Teraklin) and Hemocleanse DT (Hemocleanse Technologies) into the plasma treatment circuit. Cellular devices will have to demonstrate efficacy in this combination to be most clinically and commercially competitive.

While not well established, about 2×10^{10} hepatocytes or roughly 10% of the hepatocytes present in the adult liver has been a device goal. Adequately supporting the viability of this number of cells is a challenge. Supporting differentiated function at a level comparable to that of a normal hepatocyte *in situ* is even more daunting.

First generation devices were based on commercially available dialysis cartridges. A second-generation device has been designed to support more cells (Arbios, LIVERAID™), but it is still a traditional fiber-based approach. This design is difficult to optimize for high mass transfer and small concentration gradients.[73] Adequate oxygenation is particularly important for highly metabolic cells. To address this issue, some have incorporated in-series plasma oxygenators (HepatAssist, ELAD and ExCorp) while the tripartite fiber design developed by Gerlach includes integral oxygenation.[74]

Alternative configurations include non-woven polyester matrices,[75] packed bed bioreactors,[76] immobilization in polyurethane foams[77] and matrices[78] and flat plate bioreactors.[79,80] Better support of differentiated hepatocyte function is achieved by promoting morphology and cell organization.[81,82] Unfortunately, the lack of clinical first generation proof-of-concept has severely limited commercial support of innovative device designs again highlighting the important need to develop an effective first product. A device that supports hepatocyte function, maximizes mass transfer to effectively treat the necessary plasma volume, and uses cost-effective manufacturing is still needed.[83]

The first controlled clinical experience should also help to better define the clinical trial designs going forward. HepatAssist, tested by Circe in 171 liver failure patients in a Phase II/III prospective, randomized, controlled trial in 20 centers, highlighted the challenges associated with a heterogeneous patient population, varying local standards of care and the severity of illness in the study population.

The LAD experience exemplifies the ongoing need to understand key underlying biological principles of hepatocyte function as well as clearly defined clinical targets. Having a better grasp of these factors can then be used to inspire innovative engineering that will deliver a physiologically effective extracorporeal therapy.

6. Skin

The engineering of skin tissue is the most successful area of tissue engineering and cell therapy (See also Chapter 28). This is due in part to the early vision of scientists like Burke, Yannas, Green and Bell and the extensive biological research on the epidermal keratinocyte.

The delivery of a manufactured epidermis is enabled first, by the ability to significantly expand a keratinocyte progenitor cell population and second, by the ability to permit keratinocyte differentiation to proceed unimpeded — both enabled by assimilation of a body of sound biological research. Many areas of tissue engineering do not yet have such a foundation to work from and the biology must not be side-tracked by stem cell phenomena but rather, stem cell biology must be used to understand how one can generate enabling organ progenitor cells for other tissues, irrespective of the source. The heterogeneous and differentiating nature of keratinocyte cell culture still holds many lessons for the stem cell biologist.

The idea that skin was first because it was simple is naively too narrowly focused on its dimension. Skin does have an advantage of thinness; however, it has a distinct disadvantage in having to respond to both the host and the environment. This leads to many additional considerations which have been discussed elsewhere.[84,85] Response to inflammation, and the construct's ability to withstand it, are key factors for future generation products, particularly in burns. Skin constructs must also contend with wounds that are highly variable and indications that have different underlying etiologies. Perhaps the most encouraging fact during the development of the skin constructs was that skin grafts worked and could serve as a gold standard and a design goal. However, since Apligraf®,[a] many commercial strategies have taken a risky "less is more" approach. Expediency for the product developer and convenience for the manufacturer or customer are only of value if the product works. Alternative designs must firmly establish mechanisms to be sure they have not thrown out the baby with the bath water.

We learned from skin that the costs of production often had little to do with the science but were rather associated with "non-science" issues such as the utilization of clean rooms, packaging and delivery. The most enabling aspects to skin's commercialization comes from an engineering contribution, not only in process development, but also in ways to deliver it to the patient. Today, the engineering component is as important as ever, but one cannot lose sight of the fact that the biologist must ensure that the team is delivering something worthwhile. The first-generation Apligraf was a successful clinical product that could be produced and delivered reasonably, even though there was ample room for improvement. Organogenesis has since had the opportunity to implement further process improvements continually developed over time. The lesson to be learned from this experience is that while the science is first, supporting technology makes it feasible and improvement is an ongoing process.

---

[a] Apligaf is a registered trademark of Novartis Pharmaceuticals.

Although work continues on new scaffold designs to form the dermal layer, the next generation skin products are more likely to be entirely cell-produced. At least two companies (Intercytex and Organogenesis) are in, or near, clinical trials with cell-derived constructs. Our current understanding of epidermal stem cell biology and mesenchymal cell biology also strongly suggest that more complete skin organogenesis, including hair and sweat glands, is biologically feasible and not far in the future.[8] The question will be whether other areas of tissue engineering can learn from the skin experience and finally clear the hurdles to reach commercialization and broad clinical use.

7. Conclusion

In conclusion, some common themes emerge. One is that a true multidisciplinary partnership is needed to deliver the science and technology with the required level of sophistication. Biologists trained in developmental biology, cell biology and immunology along with biomedical, mechanical and chemical engineers, are needed not only to work on tissues, but to provide pivotal process and delivery innovations like closed systems, cryopreservation, scale-up of complex tissues, and more. Despite recent efforts, academic multidisciplinary teams continue to look more like a conglomerate of individual studies rather than a coordinated effort. Industry is still the ideal place to foster a multidisciplinary effort focused on a common goal — if there is support for it in the marketplace.

From these examples, it should be clear that there is no substitute for exacting science used in a rational way, that the product created from a technology must be effective, and that it must be accompanied by medical and commercial feasibility if the promise of tissue engineering is to be realized.

References

1. Yannas, I.V. and Burke, J.F. Design of an artificial skin. I. Basic design principles. *J. Biomed. Mater. Res.* **14**(1), 65–81 (1980).
2. Skalak, R. and Fox, C.F. (eds.). Introduction. In: *Tissue Engineering* (Alan R. Liss, New York, 1988).
3. Langer, R. and Vacanti, J.P. Tissue engineering. *Science* **260**(5110), 920–926 (1993).
4. Hubbell, J.A. Biomaterials in tissue engineering. *Biotechnology (N. Y.)* **13**(6), 565–576 (1995).
5. Lutolf, M.P., Weber, F.E., Schmoekel, H.G. *et al.* Repair of bone defects using synthetic mimetics of collagenous extracellular matrices. *Nat. Biotechnol.* **21**(5), 513–518 (2003).

6. Lutolf, M.P. and Hubbell, J.A. Synthetic biomaterials as instructive extracellular microenvironments for morphogenesis in tissue engineering. *Nat. Biotechnol.* **23**(1), 47–55 (2005).
7. Furth, M.E., Atala, A. and van dyke, M.E. Smart biomaterials design for tissue engineering and regenerative medicine. *Biomaterials* **28**(34), 5068–5073 (2007).
8. Parenteau, N.L., Rosenberg, L. and Hardin Young, J. The engineering of tissues using progenitor cells. *Curr. Top. Dev. Biol.* **64**, 101–139 (2004).
9. Scadden, D.T. The stem-cell niche as an entity of action. *Nature* **441**(7097), 1075–1079 (2006).
10. Bendall, S.C., Stewart, M.H., Menendez, P. *et al.* IGF and FGF cooperatively establish the regulatory stem cell niche of pluripotent human cells *in vitro*. *Nature* **448**(7157), 1015–1021 (2007).
11. Davies, M.G. and Hagen, P.-O. The pathophysiological consequences of autologous vein implantation. In: Brockbank, K.G. (ed.), *Principles of Autologous, Allogeneic and Cryopreserved Venous Transplantation* (RG Landes Company, Austin, TX, 1995), pp. 31–39.
12. Clarke, D.R., Elkins, R.C., Doty, D.B. and Tweddell, J.S. Decellularized allograft heart valves (CryoValve SG) — early clinical results from a multicenter study. In: *Advances in Tissue Engineering and Biology of Heart Valves Meeting* (2004).
13. Tavakkol, Z., Gelehrter, S., Goldberg, C.S., Bove, E.L., Devaney, E.J. and Ohye, R.G. *Superior Durability of SynerGraft Decellularized Pulmonary Allografts Compared to Standard Cryopreserved Allografts*, Society of Thoracic Surgeons (2005).
14. Zehr, K.J., Yagubyan, M., Connolly, H.M. *et al.* Aortic root replacement with a novel decellularized cryopreserved aortic homograft: postoperative immunoreactivity and early results. *J. Thorac. Cardiovasc. Surg.* **130**(4), 1010–1015 (2005).
15. Rieder, E., Seebacher, G., Kasimir, M.T. *et al.* Tissue engineering of heart valves: decellularized porcine and human valve scaffolds differ importantly in residual potential to attract monocytic cells. *Circulation* **111**(21), 2792–2797 (2005).
16. Ketchedjian, A., Kreuger, P., Lukoff, H. *et al.* Ovine panel reactive antibody assay of HLA responsivity to allograft bioengineered vascular scaffolds. *J. Thorac. Cardiovasc. Surg.* **129**(1), 159–166 (2005).
17. Matsumoto, T., Holmes, R.H., Burdick, C.O. *et al.* The fate of the inverted segment of small bowel used for the replacement of major veins. *Surgery* **60**(3), 739–743 (1966).

18. Egusa, S. Replacement of inferior vena cava and abdominal aorta with the autogenous segment of small intestine. *Acta Med.* **22**, 153–165 (1968).
19. Lawler, Jr., M.R., Foster, J.H. and Scott, Jr., H.W. Evaluation of canine intestinal submucosa as a vascular substitute. *Am. J. Surg.* **122**(4), 517–519 (1971).
20. Badylak, S.F., Lantz, G.C., Coffey, A. and Geddes, L.A. Small intestinal submucosa as a large diameter vascular graft in the dog. *J. Surg. Res.* **47**(1), 74–80 (1989).
21. Lantz, G.C., Badylak, S.F., Hiles, M.C. *et al.* Small intestinal submucosa as a vascular graft: a review. *J. Invest. Surg.* **6**(3), 297–310 (1993).
22. Badylak, S. Small intestinal submucosa: a biomaterial conducive to smart tissue remodeling. In: Bell, S. (ed.), *Tissue Engineering: Current Perspectives* (Burkhauser, Cambridge, 1993), pp. 179–189.
23. Termin, P.L., Carr, R.M, O'Neil, K.D. and Connolly, R.J. Thrombogenicity of intestinal submucosa: results of canine *ex vivo* shunt and acute rabbit implant studies. In: *AAMI Cardiovascular Science and Technology Conference* (1993).
24. Termin, P.L., Carr, R.M. and O'Neil, K.D. Remodeling of an absorbable synthetic collagen graft. In: *The 20th Annual Meeting of the Society for Biomaterials* (1994).
25. Hardin Young, J., Carr R.M., Downing, G.J., Condon, K.D. and Termin, P.L. Modification of native collagen reduces antigenicity but preserves cell compatibility. *Biotechnol. Bioeng.* **49**, 675–682 (1996).
26. Huynh, T., Abraham, G., Murray, J. *et al.* Remodeling of an acellular collagen graft into a physiologically responsive neovessel. *Nat. Biotechnol.* **17**(11), 1083–1086 (1999).
27. Murray, J., Brockbank, K., Bachrach, N. and Sullivan, S. Development of an acellular collagen construct as a small-diameter vascular graft. In: *8th Biennial Meeting of the International Society for Applied Cardiovascular Biology* (2002). Abstract published in the *J. Cardiovasc. Pathol.*, St. Gallen, Switzerland.
28. Cebotari, S., Lichtenberg, A., Tudorache, I. *et al.* Clinical application of tissue engineered human heart valves using autologous progenitor cells. *Circulation* **114**(1 Suppl.), I132–I137 (2006).
29. Weinberg, C.B. and Bell, E.A. Blood vessel model constructed from collagen and cultured vascular cells. *Science* **231**(4736), 397–400 (1986).
30. Isenberg, B.C., Williams, C. and Tranquillo, R.T. Small-diameter artificial arteries engineered *in vitro*. *Circ. Res.* **98**(1), 25–35 (2006).
31. Shin'oka, T., Imai, Y. and Ikada, Y. Transplantation of a tissue-engineered pulmonary artery. *N. Engl. J. Med.* **344**(7), 532–533 (2001).

32. Campbell, J.H., Efendy, J.L. and Campbell, G.R. Novel vascular graft grown within recipient's own peritoneal cavity. *Circ. Res.* **85**, 1173–1178 (1999).
33. Chue, W.L., Campbell, G.R., Caplice, N. *et al.* Dog peritoneal and pleural cavities as bioreactors to grow autologous vascular grafts. *J. Vasc. Surg.* **39**(4), 859–867 (2004).
34. Cebotari, S., Walles, T., Sorrentino, S. *et al.* Guided tissue regeneration of vascular grafts in the peritoneal cavity. *Circ. Res.* **90**(8), e71 (2002).
35. L'Heureux, N., Paquet, S., Labbe, R. *et al.* A completely biological tissue-engineered human blood vessel. *FASEB J.* **12**(1), 47–56 (1998).
36. L'Heureux, N., Dusserre, N., Marini, A. *et al.* Technology insight: the evolution of tissue-engineered vascular grafts — from research to clinical practice. *Nat. Clin. Pract. Cardiovasc. Med.* **4**(7), 389–395 (2007).
37. L'Heureux, N., Dusserre, N., Konig, G. *et al.* Human tissue-engineered blood vessels for adult arterial revascularization. *Nat. Med.* **12**(3), 361–365 (2006).
38. L'Heureux, N., McAllister, T.N. and de la Fuente, L.M. Tissue-engineered blood vessel for adult arterial revascularization. *N. Engl. J. Med.* **357**(14), 1451–1453 (2007).
39. Isenberg, B.C. and Tranquillo, R.T. Long-term cyclic distention enhances the mechanical properties of collagen-based media-equivalents. *Ann. Biomed. Eng.* **31**(8), 937–949 (2003).
40. Swartz, D.D., Russell, J.A. and Andreadis, S.T. Engineering of fibrin-based functional and implantable small-diameter blood vessels. *Am. J. Physiol. Heart Circ. Physiol.* **288**(3), H1451–H1460 (2005).
41. McMahon, M.P., Faris, B., Wolfe, B.L. *et al.* Aging effects on the elastin composition in the extracellular matrix of cultured rat aortic smooth muscle cells. *In Vitro Cell Dev. Biol.* **21**(12), 674–680 (1985).
42. Johnson, D.J., Robson, P., Hew, Y. and Keeley, F.W. Decreased elastin synthesis in normal development and in long-term aortic organ and cell cultures is related to rapid and selective destabilization of mRNA for elastin. *Circ. Res.* **77**(6), 1107–1113 (1995).
43. Liu, J.Y., Swartz, D.D., Peng, H.F. *et al.* Functional tissue-engineered blood vessels from bone marrow progenitor cells. *Cardiovasc. Res.* **75**(3), 618–628 (2007).
44. Solan, A., Prabhakar, V. and Niklason, L. Engineered vessels: importance of the extracellular matrix. *Transplant Proc.* **33**(1–2), 66–68 (2001).
45. Poh, M., Boyer, M., Solan, A. *et al.* Blood vessels engineered from human cells. *Lancet* **365**(9477), 2122–2124 (2005).
46. Gong, Z. and Niklason, L.E. Blood vessels engineered from human cells. *Trends Cardiovasc. Med.* **16**(5), 153–156 (2006).

47. Hardingham, T. Articular cartilage. In: Maddison, P.J., Isenberg, D.A., Woo, P. and Glass, D.N. (eds.), *Oxford Textbook of Rheumatology* (Oxford Medical Publications, Oxford, 1998), pp. 405–420.
48. Gobbi, A., Kon, E., Berruto, M. *et al.* Patellofemoral full-thickness chondral defects treated with Hyalograft-C: a clinical, arthroscopic, and histologic review. *Am. J. Sports Med.* **34**(11), 1763–1773 (2006).
49. Marcacci, M., Zaffagnini, S., Kon, E. *et al.* Arthroscopic autologous chondrocyte transplantation: technical note. *Knee Surg. Sports Traumatol. Arthrosc.* **10**(3), 154–159 (2002).
50. Lee, C.R., Grodzinsky, A.J., Hsu, H.P. and Spector, M. Effects of a cultured autologous chondrocyte-seeded type II collagen scaffold on the healing of a chondral defect in a canine model. *J. Orthop. Res.* **21**(2), 272–281 (2003).
51. Ochi, M., Uchio, Y., Kawasaki, K. *et al.* Transplantation of cartilage-like tissue made by tissue engineering in the treatment of cartilage defects of the knee. *J. Bone Joint Surg. Br.* **84**(4), 571–578 (2002).
52. Bentley, G., Biant, L.C., Carrington, R.W. *et al.* A prospective, randomised comparison of autologous chondrocyte implantation versus mosaicplasty for osteochondral defects in the knee. *J. Bone Joint Surg. Br.* **85**(2), 223–230 (2003).
53. Dozin, B., Malpeli, M., Cancedda, R. *et al.* Comparative evaluation of autologous chondrocyte implantation and mosaicplasty: a multicentered randomized clinical trial. *Clin. J. Sport Med.* **15**(4), 220–226 (2005).
54. Knutsen, G., Drogset, J.O., Engebretsen, L. *et al.* A randomized trial comparing autologous chondrocyte implantation with microfracture. Findings at five years. *J. Bone Joint Surg. Am.* **89**(10), 2105–2112 (2007).
55. Clar, C., Cummins, E., McIntyre, L. *et al.* Clinical and cost-effectiveness of autologous chondrocyte implantation for cartilage defects in knee joints: systematic review and economic evaluation. *Health Technol. Assess.* **9**(47), iii–iv, ix–x, 1–82 (2005).
56. Bartlett, W., Skinner, J.A., Gooding, C.R. *et al.* Autologous chondrocyte implantation versus matrix-induced autologous chondrocyte implantation for osteochondral defects of the knee: a prospective, randomised study. *J. Bone Joint Surg. Br.* **87**(5), 640–645 (2005).
57. Driesang, I.M. and Hunziker, E.B. Delamination rates of tissue flaps used in articular cartilage repair. *J. Orthop. Res.* **18**(6), 909–911 (2000).
58. Micheli, L.J., Browne, J.E., Erggelet, C. *et al.* Autologous chondrocyte implantation of the knee: multicenter experience and minimum 3-year follow-up. *Clin. J. Sport Med.* **11**(4), 223–228 (2001).

59. Babister, J.C., Tare, R.S., Green, D.W. *et al.* Genetic manipulation of human mesenchymal progenitors to promote chondrogenesis using "bead-in-bead" polysaccharide capsules. *Biomaterials* **29**(1), 58–65 (2008).
60. Lee, J.W., Kim, Y.H., Kim, S.H. *et al.* Chondrogenic differentiation of mesenchymal stem cells and its clinical applications. *Yonsei Med. J.* **45**(Suppl.), 41–47 (2004).
61. Gonzalez, R., Maki, C.B., Pacchiarotti, J. *et al.* Pluripotent marker expression and differentiation of human second trimester mesenchymal stem cells. *Biochem. Biophys. Res. Commun.* **362**(2), 491–497 (2007).
62. Gao, J., Dennis, J.E., Solchaga, L.A. *et al.* Repair of osteochondral defect with tissue-engineered two-phase composite material of injectable calcium phosphate and hyaluronan sponge. *Tissue Eng.* **8**(5), 827–837 (2002).
63. Solchaga, L.A., Gao, J., Dennis, J.E. *et al.* Treatment of osteochondral defects with autologous bone marrow in a hyaluronan-based delivery vehicle. *Tissue Eng.* **8**(2), 333–347 (2002).
64. Tuan, R.S., Boland, G. and Tuli, R. Adult mesenchymal stem cells and cell-based tissue engineering. *Arthritis Res. Ther.* **5**(1), 32–45 (2003).
65. Friedman, A.L. Why bioartificial liver support remains the Holy Grail. *ASAIO J.* **44**(4), 241–243 (1998).
66. Wolf, C.F. and Munkelt, B.E. Bilirubin conjugation by an artificial liver composed of cultured cells and synthetic capillaries. *Trans. Am. Soc. Artif. Int. Organs* **21**, 16–27 (1975).
67. Paradis, K., Langford, G., Long, Z. *et al.* Search for cross-species transmission of porcine endogenous retrovirus in patients treated with living pig tissue. The XEN 111 Study Group. *Science* **285**(5431), 1236–1241 (1999).
68. Jauregui, H.O., Chowdhury, N.R. and Chowdhury, J.R. Use of mammalian liver cells for artificial liver support. *Cell Transplant* **5**(3), 353–367 (1996).
69. Mavri-Damelin, D., Damelin, L.H., Eaton, S. *et al.* Cells for bioartificial liver devices: the human hepatoma-derived cell line C3A produces urea but does not detoxify ammonia. *Biotechnol. Bioeng.* **99**(3), 644–651 (2007).
70. Talbot, N.C., Caperna, T.J., Lebow, L.T. *et al.* Ultrastructure, enzymatic, and transport properties of the PICM-19 bipotent liver cell line. *Exp. Cell Res.* **225**(1), 22–34 (1996).
71. Talbot, N.C., Caperna, T.J. and Wells, K.D. The PICM-19 cell line as an *in vitro* model of liver bile ductules: effects of cAMP inducers, biopeptides and pH. *Cells Tissues Organs* **171**(2–3), 99–116 (2002).
72. Talbot, N.C., Rexroad, Jr., C.E., Powell, A.M. *et al.* A continuous culture of pluripotent fetal hepatocytes derived from the 8-day epiblast of the pig. *In Vitro Cell Dev. Biol. Anim.* **30A**(12), 843–850 (1994).

73. Catapano, G. Mass transfer limitations to the performance of membrane bioartificial liver support devices. *Int. J. Artif. Organs* **19**(1), 18–35 (1996).
74. Gerlach, J.C., Schnoy, N., Encke, J. *et al.* Improved hepatocyte *in vitro* maintenance in a culture model with woven multicompartment capillary systems: electron microscopy studies. *Hepatology* **22**(2), 546–552 (1995).
75. Flendrig, L.M., Chamuleau, R.A., Maas, M.A. *et al.* Evaluation of a novel bioartificial liver in rats with complete liver ischemia: treatment efficacy and species-specific alpha-GST detection to monitor hepatocyte viability. *J. Hepatol.* **30**(2), 311–320 (1999).
76. Yanagi, K., Miyoshi, H. and Ohshima, N. Improvement of metabolic performance of hepatocytes cultured *in vitro* in a packed-bed reactor for use as a bioartificial liver. *ASAIO J.* **44**(5), M436–M440 (1998).
77. Ijima, H., Nakazawa, K., Mizumoto, H. *et al.* Formation of a spherical multicellular aggregate (spheroid) of animal cells in the pores of polyurethane foam as a cell culture substratum and its application to a hybrid artificial liver. *J. Biomater. Sci. Polym. Ed.* **9**(7), 765–778 (1998).
78. Pahernik, S.A., Thasler, W.E., Doser, M. *et al.* High density culturing of porcine hepatocytes immobilized on nonwoven polyurethane-based biomatrices. *Cells Tissues Organs* **168**(3), 170–177 (2001).
79. Bader, A., De Bartolo, L. and Haverich, A. High level benzodiazepine and ammonia clearance by flat membrane bioreactors with porcine liver cells. *J. Biotechnol.* **81**(2–3), 95–105 (2000).
80. Shito, M., Kim, N.H., Baskaran, H. *et al. In vitro* and *in vivo* evaluation of albumin synthesis rate of porcine hepatocytes in a flat-plate bioreactor. *Artif. Organs* **25**(7), 571–578 (2001).
81. Berthiaume, F., Moghe, P.V., Toner, M. and Yarmush, M.L. Effect of extracellular matrix topology on cell structure, function, and physiological responsiveness: hepatocytes cultured in a sandwich configuration. *FASEB J.* **10**(13), 1471–1484 (1996).
82. Bhatia, S.N., Balis, U.J., Yarmush, M.L. and Toner, M. Effect of cell–cell interactions in preservation of cellular phenotype: cocultivation of hepatocytes and nonparenchymal cells. *FASEB J.* **13**(14), 1883–1900 (1999).
83. Hilal, A.A. and Gaylor, J.D.S. Bioartificial liver: review of science requirements and technology. *World Rev. Sci. Technol. Sustain. Dev.* **3**(1), 80–97 (2006).
84. Parenteau, N.L. and Hardin Young, J. Cutaneous healing. In: *Achieving Medical and Commercial Success in Wound Repair and Regeneration, Vol. 1.* (Parenteau BioConsultants, LLC, Fair Haven, 2006), pp. 1–396.
85. Parenteau, N.L. and Hardin Young, J. The biological mechanisms behind injury and inflammation: how they can affect treatment strategy, product

performance, and healing wounds. *Compend. Clin. Res. Pract.* **19**(4), 87–96 (2007).

86. Demetriou, A.A., Brown, Jr., R.S., Busuttil, R.W. *et al.* Prospective, randomized, multicenter, controlled trial of a bioartificial liver in treating acute liver failure. *Ann. Surg.* **239**(5), 660–667 (2004); discussion 667–670.
87. Ellis, A.J., Hughes, R.D., Wendon, J.A. *et al.* Pilot-controlled trial of the extracorporeal liver assist device in acute liver failure. *Hepatology* **24**(6), 1446–1451 (1996).
88. Patzer, 2nd, J.F., Block, G.D., Khanna, A. *et al.* D-galactosamine based canine acute liver failure model. *Hepatobiliary Pancreat. Dis. Int.* **1**(3), 354–367 (2002).

Chapter 23

The Promise of Stem Cells: A Venture Capital Perspective

Cathy Prescott

Abstract

The routine delivery of therapeutics to patients requires a translational infrastructure that includes a substantial level of financial support that is typically provided by the pharmaceutical industry and venture capitalists. However, there is little evidence that the stem cell arena has attracted any significant levels of translational support despite the anticipated revolutionary impact of stem cell-based therapeutics. The pharmaceutical industry has adopted a "watch and wait" stance for stem cell-related products and this attitude together with a complex and evolving patent landscape have to date deterred any major commitment by private investors to support the stem cell sector. However, timing is key and 2007 has been notable for a series of events that will positively influence the stem cell sector. The European Union accepted a harmonised regulatory system for advanced therapy medicinal products that include stem cells; major players within the pharmaceutical industry publicly announced that they would work with human embryonic stem cells and furthermore, stem cell therapy companies were able to successfully attract corporate investors. These events and their impact on the business of stem cells are the subject of discussion within this chapter.

Keywords: Venture Capital; Investment Model; Business Model; Regulatory Framework; Public Private Partnership; Patent Landscape.

Outline

1. Introduction
2. Venture Capital — Balancing Risk
3. The Value Proposition
4. "Watch and Wait"
5. The European Regulatory Environment

6. An Evolving Patent Landscape
7. Future Prospects?
References

1. Introduction

The advance of stem cell-based therapeutics into the clinic is a major achievement; however it is by no means the final step. The delivery of therapeutics to patients on a routine basis requires both science and a translational infrastructure that includes substantial financial support, regulatory processes and manufacturing capability. For conventional therapeutics, this infrastructure is typically provided by the biotechnology and pharmaceutical industries with support from the venture capital community.

To date, however, there has been little appetite by either the venture capital community or the pharmaceutical industry to support stem cell research, especially that focused on the development of therapeutics. Whilst the pharmaceutical industry is beginning to explore the potential of stem cells to generate clinically relevant cell lines for drug and toxicity screening, thus far they have not elected to pursue cell-based therapeutics on any significant scale. The pharmaceutical industry's "watch and wait" stance for stem cell-related products has several consequences including a reticence by venture capitalists to invest in a company that has a diminished prospect of partnering with a major player within the pharmaceutical sector and the need to identify an alternative source of funding and expertise to support products as they advance towards the market.

However, timing is key and 2007 has been notable for three significant events that will have a positive impact on the stem cell sector: (i) a unified regulatory framework was accepted across the European Union; (ii) the UK government established the first ever public-private partnership with industry to work on human embryonic stem cells; and (iii) three corporate venture capital funds invested in two stem cell therapy companies. The stem cell sector also saw a major change in the patent landscape with a successful challenge to a series of dominating patents. These events and their impact on the business of stem cells are the subject of discussion within this chapter.

2. Venture Capital — Balancing Risk

From 1994 to 2004, private venture capital invested approximately US$300 million in US biotechnology companies exploiting embryonic and adult stem cells, representing a mere 1% of the overall venture capital investment activity.[1] The majority of the funding was invested in adult rather than embryonic stem cell companies. Within the UK there are at least 17 regenerative medicine companies, a number of which are focussed on stem cells. Only about half of the companies in the sector have been successful in raising venture capital and even fewer have raised any substantial level (over £5 million) of venture funding. Why is there so little venture capital investment in this industry? In an attempt to answer this question, it is important to understand how venture capitalists operate.

Venture capitalists have two remits, raising and investing funds, with a single objective to maximise shareholder value. A private fund is typically raised from a range of "limited partners" that include for example banks, pension funds, pharmaceutical companies and high net-worth individuals. The venture capitalists invest at their own discretion subject to a mandate agreed with the limited partners. The mandate relates to the industry focus (e.g. biotechnology, high-technology), location (e.g. Europe, UK, worldwide) and stage (e.g. early, mid and/or late stage). Many funds have a fixed duration, typically ten years and must be fully committed by for example, year five. This means that no new investments can be made after the active investment period. At the end of the fund's lifetime, the portfolio companies must have reached an "exit", either via a trade sale or floatation on a public stock market. The stage of the life cycle of the fund will influence the type of investment that the venture capitalist will consider. For example, if the fund is relatively new, then there is more time to grow the company relative to a fund that is coming to the end of its active investment period.

The track record of a venture capitalist is a direct reflection of the fund's performance. A good track record facilitates the venture capitalist in raising a further fund. A successful fund is made up of a portfolio of investments that will include one or more ventures that generate multiple returns on the original investment. However, not all the investments are successful and although some of the companies maybe financially secure, they may not generate a significant return on the investment. Furthermore, other companies may fail and the investment is lost. Therefore, venture capitalists need to take risks in order to generate the high returns to compensate for those companies that do not yield significant returns. The consequence of this is that a fund cannot afford to invest in low value opportunities. Furthermore, the venture capitalist will seek to mitigate risk at every opportunity, including staging an investment against an agreed set of milestones. There is a need to distinguish venture from adventure capital!

During the initial appraisal, the venture capitalist will typically establish what problem the technology will solve, whether a product can be made from the technology and assess the balance between technology-push and market-pull. Consideration will also be given to how much funding the company will likely require over what period of time, in order to reach an exit. In other words, does the investment opportunity suit the size and stage of the venture fund.

3. The Value Proposition

Stem cells and stem cell-related technologies have the potential to deliver multiple commercial products. The products can be broadly ascribed to one of three categories: therapeutics, enabling technologies and thirdly, research reagents and tools.

The cell-based therapeutics for degenerative diseases, organs and replacement tissues are the highest value proposition, targeting markets that are both enormous and growing. For example, according to the International Diabetes Federation, in 2000 over 150 million people worldwide were estimated to have diabetes and by 2025, this figure is expected to rise to 380 million.[2] According to World Health Report, 2003 cardiovascular diseases made up 16.7 million, or 29.2% of total global deaths.[3] This figure is set to rise to an estimated 20 million deaths by 2015.

Products such as culture media, growth factors and manufacturing related technologies represent the enabling technologies sector. These products command a higher value relative to those of the tools and reagents markets. The latter include the differentiated cell lines as tools for drug discovery, disease modelling, screening assays and toxicity screening.

Within the UK, there is a significant number of companies developing products that encompass all of the aforementioned commercial opportunities within the regenerative medicines sector. For example, Intercytex, ReNeuron, NovaThera and Axordia are focused on therapeutics; Reinnervate and Plasticell are developing enabling technologies and companies such as Roslin Cells and Stem Cell Sciences are generating tools and reagents. Some companies such as Stem Cell Sciences, have adopted a hybrid model, generating revenue from the rapid-to-market reagents in order to re-invest in the development of the longer term, high value therapeutic products.

4. "Watch and Wait"

There are a variety of reasons why venture capitalists are reticent to invest in stem cell opportunities. One obvious reason is that the length of time that will be required to develop a product will be longer than the life of the venture fund; although this cannot be the sole reason as many biotechnology companies successfully reach an exit prior to ever developing a product that is ready for market launch.

A second reason for the cautious investor is a consequence of the current stance of the pharmaceutical industry sector to "watch and wait". The pharmaceutical industry is exploring the use of stem cells to generate clinically relevant cell lines for drug and toxicity screening.[4] From an investment perspective, these tools and reagents represent a relatively low value product opportunity by comparison to the enabling technologies and cell-based therapeutics. However, the pharmaceutical industry has not elected to pursue cell-based therapeutics on any significant scale and their stance has several consequences including a need to revise the traditional investment model.

The traditional biotechnology investment model usually starts with a seed investment into an early stage company. As the company progresses, it will typically seek some form of partnership within the pharmaceutical sector to help further develop the company's product portfolio. A successful deal provides third party endorsement for the value of the underlying technology and product, as well as to facilitate progression of the product, by enabling the biotechnology company to access clinical trails expertise, infrastructure and funding. A successful partnership can lead to the acquisition of the biotechnology company by the pharmaceutical partner representing an exit event for the investor. Alternatively, the company may be in a position to exit via floatation on the public market. A successful exit event, by either route, should result in a financial uplift to the investors and their limited partners. A good track record facilitates the venture capitalist to raise another fund, and in this way, the investment cycle is maintained.

Companies developing enabling technologies, tools and reagents, are more aligned with the traditional investment model as pharmaceutical companies become more actively engaged in the development of clinically relevant cell lines derived from stem cells. The Stem Cells for Safer Medicines (SC4SM) consortium, launched in October 2007, is the first public private partnership on human embryonic stem (hES) cells. The consortium has gained the commitment of three of the major pharmaceutical companies: GlaxoSmithKline, AstraZeneca and Roche, each investing £100,000 alongside a £750,000 commitment from the UK government. The initial focus of the consortium is to explore the potential of generating hepatocytes and cardiomyocytes, to use in high throughput toxicology screening of potential new medicines. The consortium will not investigate the therapeutic potential of stem cells.

The pharmaceutical industry stance also reflects the fact that the commercialisation of cell-based products presents quite a different business model relative to the more conventional therapeutics. The "blockbuster" pharmaceuticals are manufactured in high volumes, are prescribed to large patient populations and are usually self-administered. This is in contrast to cell-based products that will likely be tailored to individuals or restricted patient cohorts and require administration by a specialist in a clinical setting. This divergence from the traditional business model creates an additional hurdle that stem cell companies must surmount in order to establish some form of collaboration or partnership deal with the pharmaceutical industry. From an investor's perspective, a diminished prospect of forming an alliance may also reduce the prospect of a trade sale and therefore the major exit route is dependent on the sentiment of the public market, which is often unpredictable. For example in September 2005 ViaCell Inc.'s market value dropped by 22% after the company announced the suspension of enrolment in a Phase I trial for CB001 for use of haematopoietic stem cells to treat cancer.[5] In December 2005 the company was successful in convincing the US Food and Drug Administration to lift suspension on the CB001 Phase I trial.[6] Despite this good news, the confidence of the public investor had apparently been eroded as the company's market value did not significantly increase. Public health care markets are volatile, overly reactive and a clinical failure can seriously undermine the confidence in an emerging sector. Entry onto the public market is often more successful if the company has secured a deal with a major player as this provides a form of independent and qualified endorsement for the underlying technology and product portfolio. Furthermore, the stem cell company and its shareholders face the additional pressure of securing an alternative source of funding and expertise to support the stem cell company as its products advance into the clinic.

There are however, signs that the pharmaceutical industry is becoming more actively engaged with the cell-based therapy sector. In July 2007 Novocell Inc., raised US$25 million in an investment round led by Johnson and Johnson Development Corporation.[7] Two months later, in September 2007, Cellerix successfully raised US$38 million from an investor consortium that included both the Roche and Novartis corporate venture funds.[8] Both Novocell and Cellerix are exploiting stem cells to develop novel therapeutics, including cell-based products.

The pharmaceutical industry will no doubt eventually become more fully engaged in exploiting stem cells for the development of therapeutics. The timing will be key and likely reflect a consolidation of the intellectual property within the industry, through a combination of in-licensing, mergers and acquisition activities. Full integration within the pharmaceutical company is unlikely, and therefore the stem cell "business unit" may be run as an autonomous division within the company. This is analogous to the rapid development of AstraZeneca's biologics capability following the acquisition of Cambridge Antibody Technology in 2006, which has continued to function essentially as an autonomous unit.[9]

There remain many questions relating to the cost of goods, whether these cell-based products will be eligible for reimbursement and therefore generally available to all patients. There are also questions about the logistics of manufacturing for industrial-scale production, transport from the laboratory to the clinical setting and where and how these products will be stored, tracked and administered. However, there may be lessons to be learned from the blood transfusion medicines sector that can be applied to cell-based products.

5. The European Regulatory Environment

A clear regulatory environment is crucial for the successful launch of a product into a defined market. However, the rapid development of complex tissue engineered products has challenged the capacity of existing regulatory frameworks to cope with such novel products. In October 2005 Smith and Nephew, a global leader in advanced wound management announced, "On a global basis the lack of clear regulatory frameworks for tissue engineered products has resulted in delays that have become commercially unacceptable".[10] This statement followed their decision to exit DERMAGRAFT, a skin product, after failing to gain approval in the US for the treatment of venous leg ulcers.

In November 2005 the European Commission published a proposal to establish a single, integrated and tailored Europe-wide framework for advanced therapies. The proposal set out a centralised review and market approval process for all gene, cell and tissue-based therapies within the European Union. The proposal was approved in May 2007 and will enter in to force one year from when the text has been translated into the official European languages and published in the EU Official Journal[11] (see also Chapter 26).

The single, centralised authorisation replaces the existing 27 differing processes. Until now, although a company might have gained approval in one State, it did not automatically result in approval in another territory. Multiple submissions inevitably resulted in delays and in some instances, the requirement for additional data, all of which resulted in increased costs and therefore lower margins and time delays. Such a fragmented market was unattractive to the industry and difficult to secure partnerships with those companies that, for example, had the necessary marketing expertise and distribution infrastructure.

The Regulation means that all advanced therapies are subject to a single EU Marketing Authorisation, valid throughout Europe, following a scientific evaluation by the new Committee of Advanced Therapy (CAT) at the EMEA (the European Agency for the Evaluation of Medical Products). The committee will advise on the authorisation of the products and the creation of product-specific guidance. CAT will work as a sub-committee of the Committee for Human Medicinal Products to which it will report its recommendations for acceptance or review by that committee. The overall impact as that market authorisation will be more cost effective, time saving and ensure that the products meet a Europe-wide common safety and efficacy standard.

Although centrally approved, the decision for the use of products derived from hES cells, will be left to the individual member states. Therefore the market for these products remains fragmented and is subject to the afore-mentioned repercussions. However, it remains to be seen which member states will enforce such a ban and whether such a stance remains in force in the face of an expanding medical tourism industry.

6. An Evolving Patent Landscape

The stem cell patent landscape is complex, with existing patents to claims to the processes of isolating stem cells, their expansion, differentiation, selection, purification, and administration (see also Chapter 3). Whilst a patent may be granted for a particular invention based on its novelty, non-obviousness and industrial application, this does not necessarily mean that the inventor has the freedom to operate. Patents represent a form of "biotech currency" and therefore investors will focus attention on the integrity of the patent portfolio. Having to obtain a license to operate is not necessarily an issue, if a license can be obtained under reasonable terms. Navigating through a complex landscape takes time and requires specialist support and therefore investors may elect to focus their attention on a less complex investment prospect.

The stem cell sector is dominated by the WARF (Wisconsin Alumni Research Foundation) patents (US patents 6,200,806 and 5,843,780), which will expire in 2015. The patents include broad claims to all human embryonic stem cells and the method by which they are made. Commercial licences are available for a US$100,000 upfront fee and a US$25,000 annual fee,[12] a significant cost for an early stage company. By May 2005 it was reported that only seven licences had been granted by comparison to the 231 research groups that had received cell lines under non-commercial terms. In January 2007, WARF relaxed its licensing strategy by permitting companies to sponsor research by academic or non-profit scientists without a licence. Geron has negotiated exclusive rights to develop therapeutics and diagnostics from hES cell-derived neural, cardiomyocyte and pancreatic islet cells, representing three of the major therapeutic market sectors.

In October 2006 the Santa Monica-based Foundation for Taxpayer and Consumer Rights requested that the United States Patent and Trademark Office (PTO) re-examine the "over-reaching" stem cell patents owned by WARF. In April 2007, the US PTO ruled in favour of the Foundation and rejected every WARF claim regarding the method of deriving stem cells, and the claim to stem cells themselves. This is not the final step as WARF can appeal the decision, a process that may take many years. In the meantime WARF continue to assert that the patents remain in force.

WARF applied to the European Patent Office (EPO) for a European patent (published by the EPO as EP0770125). The examining division of the EPO rejected the application on moral grounds under Article 53(a) of the European Patent Convention. WARF appealed the decision to the Technical Board of

Appeal, who subsequently referred four questions (relating to the technology, morality and patent law) to the Enlarged Board of Appeal.

The response of the Enlarged Board could set an important precedent for European patent applications that relate to hES cells. If the EPO refused to grant patents on moral grounds then this would deny patent protection in those European states where hES cell technologies are morally acceptable and therefore permitted.

Whilst scientists and consumer groups are generally enthusiastic about the prospect of the WARF patents being rejected or at least restricted to a narrower set of claims, the evolving landscape represents a double-edged sword for investors. On the one hand, the decision by the US PTO would open the field to companies seeking to exploit hES cells. On the other hand, an evolving patent landscape means that there is a degree of uncertainty in terms of the long-term legitimacy and therefore value of the patent portfolio.

7. Future Prospects?

In 1996 Hillary Clinton famously said "it takes a village to raise a child". A similar sentiment can be applied to the emerging field of stem cells. The routine delivery of stem cell-based therapeutics to patients will require an integrated and coordinated multidisciplinary approach that includes scientists, clinicians, engineers, regulators, and financiers. The approach is necessary to address significant challenges such as scale-up manufacturing, which will impact on the cost of goods and therefore the eligibility of these products for reimbursement by the public and private health insurance industries and therefore their marketability. Within the UK, several research centres have been established to support an integrated approach to regenerative medicine including the Scottish Centre for Regenerative Medicine (SCRM) (University of Edinburgh) and the Wolfson Centre for Stem Cells, Tissue Engineering and Modelling (STEM) (University of Nottingham) as well as initiatives such as remedi an EPSRC Innovative Manufacturing Grand Challenge. It is the anticipation and resolution of these challenges that will underpin the success of the stem cell industry.

Although the markets for stem cell-related products are attractive in terms of size and value, for many of these novel and often complex products their route to market is unclear by comparison to the more conventional products and technologies developed by the biotechnology and pharmaceutical industries. The lack of clarity has, at least for the present, deterred any substantial level of venture capital investment. Perhaps it is time to discard the traditional investment and business models and explore new approaches to support the business of stem cells.

2007 has proven to be an interesting year for the industry as denoted by the acceptance of a harmonised European regulatory system for advanced therapy medicinal products and the positive change in attitude by major representatives of the pharmaceutical industry towards investment in stem cells. The rate of this change is key to the success of the sector and no doubt we will see the pace accelerate as the stem cell-related products deliver on their promise.

References

1. Gottlieb, S. *California's Stem-Cell Follies*, Forbes.com (2004).
2. International Diabetes Federation, www.idf.org.
3. *Shaping the Future*, The World Health Report (2003).
4. Stem Cells for Safer Medicine a public private partnership, www.sc4sm.org.
5. *ViaCell Suspends Enrollment in CB001 Phase 1 Clinical Trial*, News Release of ViaCell Inc. (2005).
6. *FDA Lifts Clinical Hold on ViaCell CB001 Trial*, News Release of ViaCell Inc. (2005).
7. *Novocell Raises $25 Million*, Press Release of Novocell Inc. (2007).
8. *Cellerix Closes €27.2M Financing Round to Bring Product Cx401 to Market*, Press Release of Cellerix Inc. (2007).
9. *Cambridge Antibody Technology Appoints New Chief Executive*, News Release of Cambridge Antibody Technology (2006).
10. *Positive Outlook After a Slower Quarter*, News Release of Smith and Nephew (2005).
11. *Advanced Therapies: Tissue Engineering, Cell Therapy and Gene Therapy* (European Commission, 2007).
12. Waldman, M. Licensing fees slow advance of stem cells. *Nature* **435**, 272–273 (2005).

PART VI

TISSUE ENGINEERING PRODUCTS

Chapter 24

Cell Expansion, Cell Encapsulation, 3D Cultures

Julia M. Polak and Athanasios Mantalaris

Abstract

Regenerative medicine is reaching maturity. There are numerous examples of translation of research efforts into initial clinical practices. Stem cells have played a pivotal role, as exemplified by the 2007 Nobel Prize accolade to one of the discoverers of embryonic stem cells (ESCs). In spite of this, numerous challenges lay ahead if the conversion from a few clinical successes to treatment to a vast number of people with incurable illnesses is to succeed.

Within these challenges the ability to obtain large and uniform number of specific cell lineages is of paramount importance. This short chapter reviews the continuous efforts that are being made towards this goal, by learning how to grow the cells in a more natural 3D environment to sophisticated and automated ways of stem cell bioprocessing.

Keywords: Bioreactors; Stem Cell Expansion; Alginate Encapsulation; Bone Tissue Engineering; Clinical Applications.

Outline

1. Introduction
2. Controlled Differentiation
3. Generation of Clinically Relevant Number of Cells in 3D Cultures as an Integrated and Scalable Process
4. Encapsulation
5. Bioprocessing for Regenerative Medicine
   5.1. Bioreactors
   5.2. Stem cell bioprocessing
6. Discussion

   References

1. Introduction

The field of regenerative medicine/tissue engineering/cell therapy has witnessed lately an unprecedented resurgence parallel to that experienced by the appearance of a better, more mature World Wide Web back in 2004.[1] When the dot.com bubble burst back in 2001, it appeared almost impossible that a more mature Web, more integrated with the users needs, was going to surge. The Web 2, as it was termed then, to distinguish it from the early internet attempts, began to appear (Google, You Tube, others). A similar transformation is being witnessed by the field of regenerative medicine whereby the early and pioneer work of the 1985–2002 mostly centred in scientific discoveries with little understanding/regard for its applications to health care is being gradually replaced by a much more mature field where the emphasis is on translation of scientific discoveries to cures for large number of patients and on public involvement (e.g. public awareness, large number of public debates, California Proposition 71, the UK Stem Cell Initiative, to name a few) (see Table 1).

In spite of this exponential increase in translational activity, on public participation and on investment interest, numerous challenges are still present (see Table 2).

Table 1. Regenerative medicine changes analogies to the World Wide Web.

Web 1.0 = commerce	Regenerative medicine 1.0 = science and research (little benefit to patients and shareholders)
Web 2.0 = people and active participation	Regenerative medicine 2.0 = pragmatic translation of science into routine clinical practice

Table 2. Regenerative medicine (current challenges to translational research).

(a) Best cell source
(b) Efficient lineage differentiation
(c) Robust expansion of specific cell lineages
(d) Cultures:
3D
Automation
GMP
(e) Effect of freezing/thawing of samples
(f) Immunological barriers
(g) Vascularisation
(h) Mechanical strength of constructs

In this short review we will discuss the need to develop robust technology for cell expansion, 3D cell cultivation methods, what have we learnt so far and what barriers lay ahead.

Of paramount importance for the success of stem cell therapy is to find the appropriate cells, which should be non-immunogenic, highly proliferative, easy to harvest, and should have the ability to differentiate into cell types with specialised functions. Several cell sources have been explored, including cells taken from adult tissues. In most cases, fully differentiated cells taken from adult tissues exhibit a limited proliferation capacity.[2,3] This places limitations on their expansion capabilities in culture and their subsequent use in health care. For this reason, more rapidly proliferating cell sources such as foetal,[4,5] neonatal,[6] genetically modified[7] (see also chapters in Section 2 of the book), and stem cells (adult and embryonic) have been used and are an active area of investigation.[8] Embryonic stem cells (ESCs), are expected to become a powerful tool for regenerative medicine due to their capacity for self-renewal, pluripotency and lack of immunogenecity.[9] To date, bone marrow stem cells are the source most frequently used for clinical applications. However, a number of limitations exist with bone marrow cells: (1) the limited numbers of mesenchymal stem cells (MSCs) in the bone marrow (BM) and invasive harvest procedures; (2) the variability in the expansion capability of the MSCs due to donor age. Stem cells taken from the umbilical cord are also coming to the fore (see Chapter 27). This is evidenced by the increasing number of newly opened cord blood banks both private and public. But again, these stem cells have gone down the developmental path and may show less plasticity than ESCs. To date umbilical cord cells have been used in Phase I–II of clinical trials for haematological disorders only.[10] In order for stem cells to be utilised clinically certain important challenges need to be overcome, such as (a) their controlled differentiation into a homogeneous population of the desired cells and (b) generation of clinically relevant numbers through integrated and scaleable processes.

2. Controlled Differentiation

Attempts to differentiate stem cells into specific lineages are being energetically pursued by numerous researchers worldwide (e.g. Chapter 5 of Section 2 and Refs. 11–15).

Specific cell lineages from ES cells are predominantly obtained via embryoid body formation. Recently efforts are being focussed in the development of lineage specific cells from ES cells avoiding embryoid body formation. This involves the immediate separation of ESC colonies into single cells, which are then plated directly into a cell adhesive culture dish.[16–18] Using this method Karp *et al.* for instance[16] obtained a seven-fold greater number of osteogenic cells derived from hES cells and spontaneous bone nodule formation after ten to 12 days in culture.

An alternative strategy for enhancing lineage specific differentiation is the use of conditioned culture media to enhance the development of a specific germinal layer. A previous study was able to generate a cell population similar to primitive streak/nascent mesoderm from stem cells following treatment with condition medium from a human hepatocarcinoma cell line (HepG2).[17] The authors based their investigations on previous observations that treatment of stem cells with HepG2-conditioned medium directed differentiation to a cell type with a gene expression profile resembling that of primitive streak/nascent mesoderm-stage cells of the mouse embryos.[19–22] Interestingly, it was shown that by reducing the length of embryoid body culture time, to one to two days (currently embryoid bodies are cultured for at least five days) using the conditioned medium, it was possible to enhance mesoderm formation and increase (five- to ten-fold) osteogenic lineage differentiation while suppressing cardiac differentiation (Figs. 1A–1D).

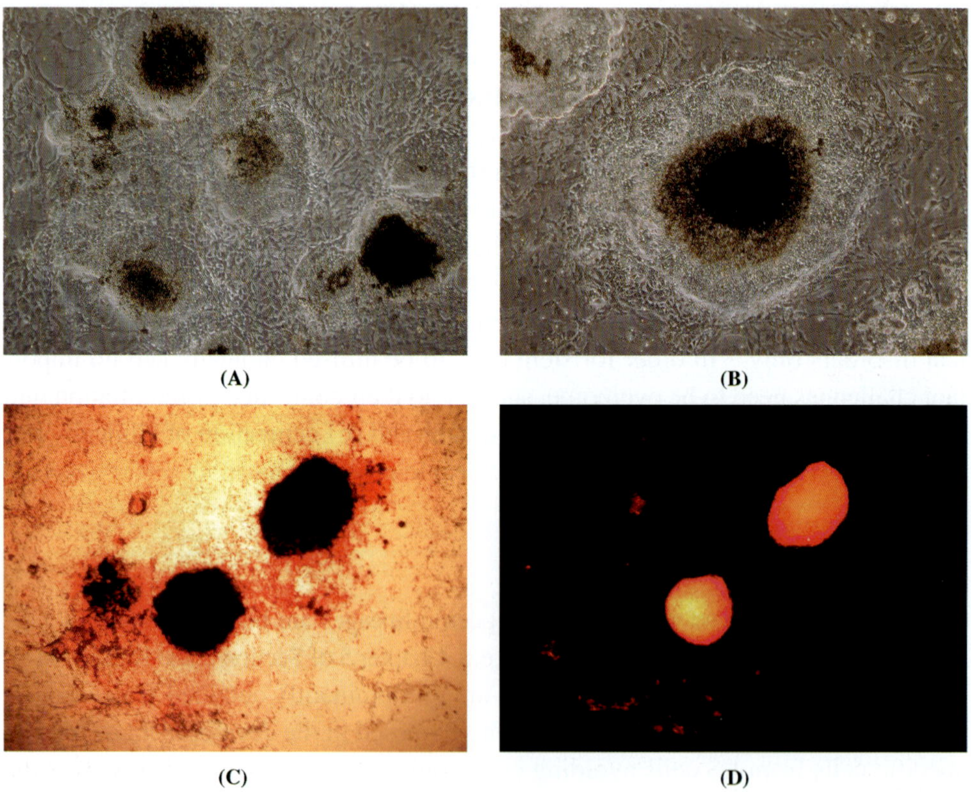

Fig. 1. **(A)** and **(B)** Bone nodules seen under inverted light microscope; **(C)** mineralised bone nodules under light microscopy; and **(D)** mineralised bone nodules stained with Alizarin Red. Reprinted with permission from *Tissue Engineering*.[46]

By contrast, if embryoid bodies were cultured for a longer period of time in a comparable conditioned medium it was then possible to obtain, preferentially, cardiac cells.[23] This is important, since it is well known that lineage differentiation via embryoid body formation produces unwanted lineages of different germinal layer derivatives with potential inhibition for the development of specific lineages and suggests that it may be possible to customise the production of lineage specific cells.

3. Generation of Clinically Relevant Number of Cells in 3D Cultures as an Integrated and Scalable Process

In order to move forward and be of clinical use, cell therapy needs to advance by developing methodologies to expand the required number of cells. Cell expansion methodology has been used for a long time (bone marrow stem cell transplantation for haematological disorders, see the contribution in Section 8 of this book, or expansion of adult cells: human bladder urothelia — from 1 cm^2 to 4.200 m^2) (quoted in Ref. 24) but refinements are clearly needed. Recently, Shafritz managed to expand hepatocytes *in vivo* using an animal model of liver failure.[25] Traditional stem cell culture methodologies are labourious and arbitrary in nature, requiring the employment of highly trained operators to conduct routine culture work. Two-dimensional cultures in which normal 3D relationships with the extracellular matrix and other cells are distorted alter cellular behaviour. Analysing cell/cell interactions in more natural 3D settings is likely to provide conditions that are closer to what actually occurs *in vivo*.[26–29]

Since the pioneering work of Peter Zandstra,[30] much attention has recently been paid to the development of technologies to obtain sufficient number of pure and differentiated functioning stem cells. These include a single-step bioprocess that enables both medium perfusion and direct monitoring of cell viability and metabolite production[28]; prevention of ES cell aggregation (e.g. hydrogel encapsulation)[31,32]; and purification of specific cell types (e.g. genetic manipulation).[7]

De Bank *et al.*[32] were able to gain an accelerated formation of multicellular 3D structures by developing a surface modifying method on the cell and matrix. Mondrinos *et al.* have recently published the production of 3D structures using foetal lung explants and especially designed matrigel and polymer scaffolds.[33] Przyborski and colleagues have provided an ingenious way of culturing the cells in 3D by allowing the cells to grow in a ten-pence sized highly porous polystyrene scaffold riddled with holes like a sponge.[34]

4. Encapsulation

The use of encapsulated cells is a fairly simple concept. The principle is to develop a capsule with sufficient permeability that nutrients and oxygen can reach the transplanted cells, and appropriate cellular products can be released into the bloodstream or to adjacent tissues. At the same time, the capsular material must be restrictive enough to exclude immune cells and antibodies that would cause rejection and destroy the implant.

Alginate encapsulation has been carried out frequently with adult cells[35,36] and more recently, Magyar *et al.* encapsulated ES cells and researchers in Australia encapsulated both mouse and human ES cells to treat diabetic patients.[37] Chondrogenic differentiation from MSCs encapsulated in alginate beads has been reported and a combination of alginate with gelatin has been considered to provide a biodegradable delivery vehicle for tissue engineering applications.[38] Normally, alginate hydrogels lose Ca^{2+} cations after prolonged culture, but the incorporation of gelatin enables cell-mediated contraction and packing of the scaffold material.[39] The use of alginate to enhance chondrogenesis from encapsulated EBs derived from mESCs has been attempted, albeit with limited success.[40]

Recently, Barnett *et al.*[41] used microencapsulation technology in combination with magnetic resonance imaging (MRI)[41] to simultaneously immunoprotect pancreatic β-cells and monitor, non-invasively and in real-time hepatic delivery and engraftment.

Zur Nieden *et al.*[42] were able to maintain cells in a pluripotent state using cell aggregates and bioreactors and we have successfully demonstrated that hESCs (H1) can be encapsulated within the hydrogels and remain in an undifferentiated state for up to 260 days.[43]

5. Bioprocessing for Regenerative Medicine

This is an entirely new field, which can only be partially compared to that of mass production of molecular medicines (see Table 3 and Ref. 28). It is a challenging field, whereby the cells, unlike in molecular medicines, are important at all stages

Table 3. Regenerative medicine bioprocess and molecular medicines.

Molecular medicines	Cell bioprocessing
Clear cut-off point in the process: cells only important during culture.	Cells important at **every** step (harvesting, expansion, tissue formation, implantation).

of the process. Currently much of the cultivation methods are done manually, with skilled, personal workers and at a great cost. If regenerative medicine is to succeed in providing therapies to the masses, novel methodologies will have to be devised to ensure that the end product is consistent and reliable.

5.1. Bioreactors

Bioreactors have been defined as devices in which environmental and operating conditions can be closely monitored and tightly controlled to permit or induce the desired biological and/or biochemical process[44,45] (see for details Chapter 25 of Section 6). The ultimate bioreactors, namely the organs/tissues within the body, share common operational characteristics. Specifically, mass transport in the form of circulation and diffusion to the cells is excellent — no cell is located more than 400 μm from a blood supply. In order for regenerative medicine to come of age there is the need to create optimal bioreactors which will achieve production not by embracing traditional scale-up principles (larger bioreactors) but through process integration (see Chapter 25 of Section 6).

5.2. Stem cell bioprocessing

Stem cell bioprocessing can be defined as the development of bioprocessing technology for the successful transfer of laboratory-based practice of stem cells and tissue culture to the clinic as therapeutics through the application of engineering principles and practices, in order to achieve reproducibility, control and automation.[30]

Static cultures suffer from several limitations including: (a) lack of mixing results in heterogeneity in dissolved oxygen, pH, cytokines and nutrients/metabolites and (b) frequent media exchange is impractical. Bioreactors provide a dynamic cultivation system within a controlled environment and enhance the expansion of cells, including in a 3D environment (see Chapter 25 of Section 6).

Randle *et al.*[46] have been able to develop a simplified, integrated, and reproducible bioprocess for the production of osteogenic cells from ES cells that could be amenable to automation and scale-up for the generation of clinically relevant numbers of high-quality bone cells and mineralised tissue. Specifically, the authors encapsulated mESCs in alginate hydrogels and cultured them in high aspect ratio vessels (HARV). In this one-step, integrated process, Randle *et al.* differentiated ES cells into osteogenic cells capable of producing 3D mineralised tissue identified by demonstration of stained mineralised aggregates and expression of osteogenic markers using micro-CT, FTIR and elemental SEM (Figs. 2 and 3).

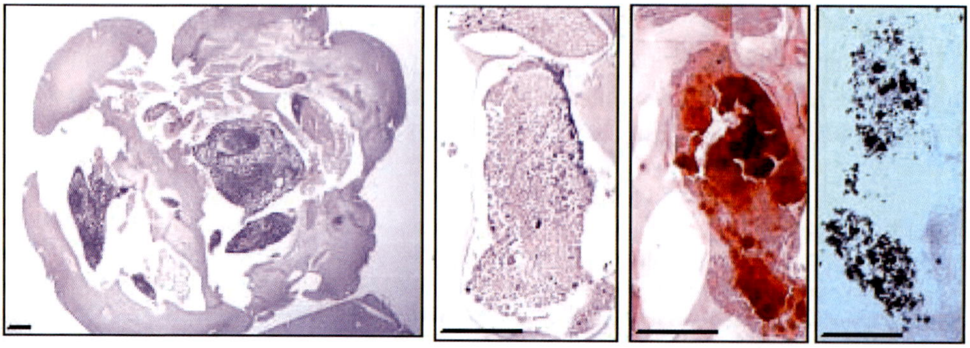

Fig. 2. Mineralised tissue formation characterised by using Alizarin Red staining. Reprinted with permission from *Tissue Engineering*.[46]

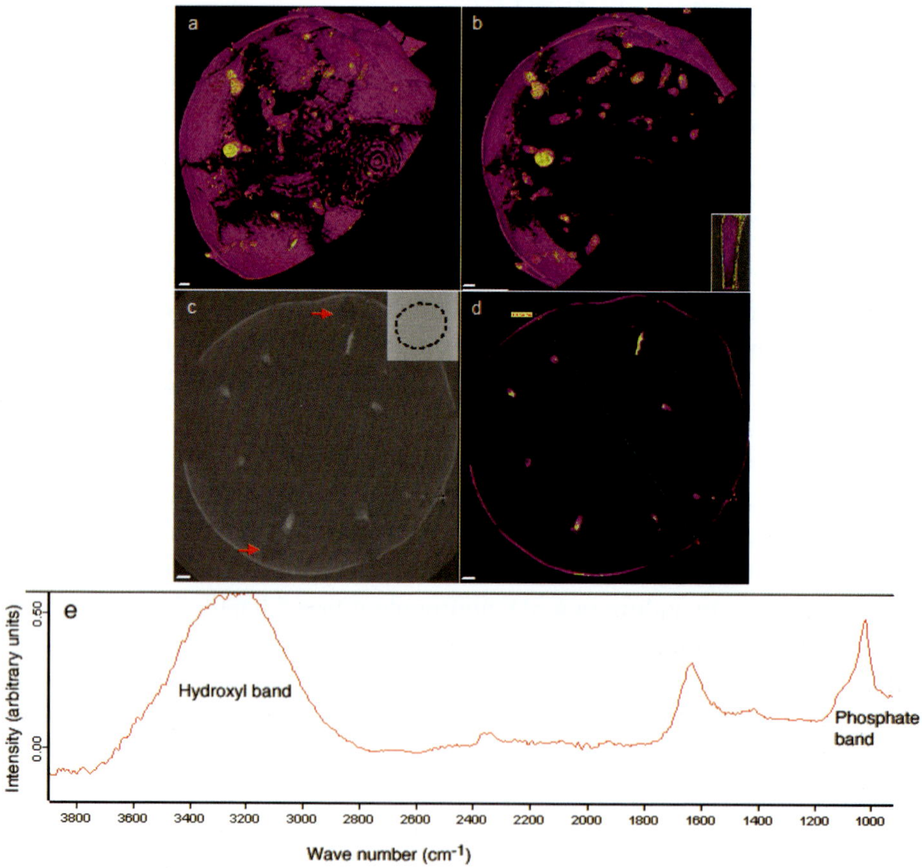

Fig. 3. Evaluation of tissue mineralisation using micro-computed tomography (micro-CT). The alginate beads were evaluated at day 29. Reprinted with permission from *Tissue Engineering*.[46]

Specifically, an 80-fold increase in the cell number from a starting density of 10,000 cells/alginate bead was observed after 29 days in culture.[46,47] Using this integrated procedure it is potentially possible to generate the equivalent of 20–30 T75 flask cultures in a single 50 ml bioreactor.

6. Discussion

Regenerative medicine stands ready to create the next major wave of the biotechnology revolution, following in the footsteps of recombinant proteins and monoclonal antibodies. The field has the potential to impact the whole spectrum of health and change current medical practices.

Regenerative medicine is a new way of treating injuries and diseases, by using combinations of specially grown cells, tissues and laboratory-made compounds to replace or amplify the natural healing process. This is an entirely new field that brings together expertise in biology, chemistry, engineering, genetics, medicine and other specialities to find solutions to some of the most challenging medical problems faced by human kind.

Advances in medicine over the past 30 years have resulted in an increased life span in the industrialised world. However, increased longevity should be accompanied by a good quality of life. For instance, even though kidney failure can be managed by regular blood dialysis or transplantation, the disturbance to normal life may be significant even when a suitable organ is secured.

Tissue engineering and regenerative medicine may offer a solution to such challenges. To date, several engineered skin-equivalents and synthetic bone and cartilage composites are commercially available. Continued efforts are being made in advancing bone and cartilage tissue engineering, cardiac, venous and arterial replacements, urological structures, and haematopoietic stem cell products (see chapters under Section 8). Unquestionably, the development of bioprocessing technologies for the successful transfer of the current laboratory-based practice of stem cell and tissue culture to the clinic as therapeutics necessitates radical changes in tissue culture practices in order for the cellular product to be reproducible and safe. The successful translation will require contributions from fundamental research and from current industrial practice (biologics), especially on automation, quality assurance and regulation.

In general, bioprocessing for the production of cellular products (biologics) and cells/tissues (cellular therapeutics) is impeded by the lack of real-time, online, *in situ*, quantitative information with respect to cellular behaviour in culture. This bottleneck results in the culture and processing of the cells being essentially manual (empirical), resulting in sub-optimal productivity and product quality.

In order to develop optimal bioprocesses, the need exists to analyse the culture parameters that influence the process and identify the critical ones. The existing approaches, however, are limited because they focus on a few parameters and do not examine the interaction of a multitude of parameters, thus missing important interactions.

Though many of these tasks cannot be readily addressed and may require long-term commitment, some of the current challenges must remain the primary focus of our research and development. Many successful stories of initial clinical applications of stem cells and tissue engineering are currently in the literature.[48–51] (see also chapters under section 8 of this book). The major challenges that face the field include the need of obtaining "off the shelf" cells that are GMP compliant and immunologically tolerant. There is no, as yet, consensus as to which will be the best cell type for clinical applications and it is likely that not one single cell type will be the cell to use.[52] As always in medicine, cell therapy will have to adapt to specific clinical problems and be patient specific. One fact remains clear. There is an urgent need to develop reliable and robust culture procedures that will produce vast quantities of identical batches of cells for a given clinical application. There is no question that scale-up of differentiated cells will come at the forefront of the challenges faced by clinical application of regenerative medicine. Reliable and reproducible stem cell bioprocessing offers a potential solution.

References

1. Mason, C. Regenerative medicine 2.0 editorial. *Regenerative Med.* **2**, 11–18 (2007).
2. Tsuji, T., Hughes, F.J. and McCulloch, C.A. The effects of donor aged on osteogenic cells of rat marrow *in vitro*. *Mech. Ageing Dev.* **51**, 121–132 (1990).
3. Dodson, S.A., Bernard, G.W., Kenney, E.B. and Carranza, F.A. *In vitro* comparison of aged young osteogenic and hemopoietic bone marrow stem cells and their derivative colonies. *J. Periodontol.* **67**, 184–196 (1996).
4. Christodoulou, I., Buttery, L.D., Tai, G., Hench, L.L. and Polak, J.M. Characterization of human fetal osteoblasts by microarray analysis following stimulation with 58S bioactive gel-glass ionic dissolution products. *J. Biomed. Mater. Res. B Appl. Biomater.* **77**, 431–446 (2006).
5. Christodoulou, I. *et al.* Dose- and time-dependent effect of bioactive gel-glass ionic-dissolution products on human fetal osteoblast-specific gene expression. *J. Biomed. Mater. Res. B Appl. Biomater.* **74**, 529–537 (2005).
6. De Coppi, P. *et al.* Isolation of amniotic stem cell lines with potential for therapy. *Nat. Biotechnol.* **25**, 62–63 (2007).

7. Tai, G., Polak, J.M., Bishop, A.E., Christodoulou, I. and Buttery, L. Differentiation of osteoblasts from murine embryonic stem cells by overexpression of the transcriptional factor osterix. *Tissue Eng.* **10**, 1456–1466 (2004).
8. Lanza, R. *et al. Essentials of Stem Cell Biology* (Academic Press, 2006), pp. 1–531.
9. Daniels, J.T. Stem cells: moving the biology towards the clinic. *Regenerative Med.* **2**, 313–315 (2007).
10. Rocha, V. *et al.* Transplant of umbilical-cord blood or bone marrow from unrelated donors in adults with acute leukemia. *N. Engl. J. Med.* **352**, 2276–2285 (2004).
11. Buttery, L.D. *et al.* Differentiation of osteoblasts and *in vitro* bone formation from murine embryonic stem cells. *Tissue Eng.* **7**, 89–99 (2001).
12. Ali, N.N. *et al.* Derivation of type II alveolar epithelial cells from murine embryonic stem cells. *Tissue Eng.* **8**, 541–450 (2002).
13. Bielby, R.C., Boccaccini, A.R., Polak, J.M. and Buttery, L.D.K. *In vitro* differentiation and *in vivo* mineralization of osteogenic cells derived from human embryonic stem cells. *Tissue Eng.* **10**, 18–25 (2004).
14. Shimko, D.A., Burks, C.A., Dee, K.C. and Nauman, E.A. Comparison of *in vitro* mineralization by murine embryonic and adult stem cells cultured in an osteogenic medium. *Tissue Eng.* **10**, 1386–1398 (2004).
15. Sottile, V., Thomson, A. and McWhir, J. *In vitro* osteogenic differentiation of human ES cells. *Cloning Stem Cells* **5**, 149–155 (2003).
16. Karp, J.M. *et al.* Cultivation of human embryonic stem cells without the embryoid body step enhances osteogenesis *in vitro*. *Stem Cells* **24**, 835–843 (2006).
17. Hwang, Y.-S., Randle, W.L., Bielby, R.C., Polak, J.M. and Mantalaris, A. Enhanced derivation of osteogenic cells from murine embryonic stem cells after treatment with HepG2-conditioned medium and modulation of the embryoid body formation period: application to skeletal tissue engineering. *Tissue Eng.* **12**, 1381–1392 (2006).
18. Wang, D., Haviland, D.L., Burns, A.R., Zsigmond, E. and Wetsel, R.A. A pure population of lung alveolar epithelial type II cells derived from human embryonic stem cells. *Proc. Natl. Acad. Sci. USA* **104**, 4449–4454 (2007).
19. Calhoun, J.D. *et al.* Transcriptional profiling of initial differentiation events in human embryonic stem cells. *Biochem. Biophys. Res. Commun.* **323**, 453–464 (2004).
20. Mummery, C. *et al.* Cardiomyocyte differentiation of mouse and human embryonic stem cells. *J. Anat.* **200**, 233–242 (2002).

21. Lake, J., Rathjen, J., Reimiszewski, J. and Rathjen, P.D. Reversible programming of pluripotent cell differentiation. *J. Cell. Sci.* **113**, 555–566 (2000).
22. Rathjen, J. *et al.* Formation of a primitive ectoderm like cell population, EPL, cells, from ES cells in response to biologically derived factors. *J. Cell. Sci.* **112**, 601–612 (1999).
23. Mummery, C. *et al.* Cardiomyocyte differentiation of mouse and human embryonic stem cells. *J. Anat.* **200**, 233–242 (2002).
24. Atala, A. Engineering tissues, organs and cells. *J. Tissue Eng. Regen. Med.* **1**, 83–96 (2007).
25. Shafritz, D.A. A human hepatocyte factory. *Nat. Biotechnol.* **25**, 871–872 (2007).
26. Yamada, K.M. and Clark, K. Cell biology: survival in three dimensions. *Nature* **419**, 790–791 (2002).
27. Abbott, A. Biology's new dimension. *Nature* **424**, 870–872 (2003).
28. Mason, C. and Hoare, M. Regenerative medicine bioprocessing: building a conceptual framework based on early studies. *Tissue Eng.* **13**, 301–311 (2007).
29. Napolitano, A., Dean, D. and Morgan, J. Dynamics of the self-assembly of complex cellular aggregates on micromolded nonadhesive hydrogels. *Tissue Eng.* **13**, 2087–2094 (2007).
30. Zandstra, P.W. and Nagy, A. Stem cell bioengineering. *Annu. Rev. Biomed. Eng.* **3**, 275–305 (2001).
31. Magyar, J.P. *et al.* Mass production of embryoid bodies in microbeads. *Ann. N. Y. Acad. Sci.* **944**, 135–143 (2001).
32. De Bank, P.A. *et al.* Accelerated formation of multicellular 3D structures by cell-to-cell cross linking. *Biotechnol. Bioeng.* **97**, 1617–1625 (2007).
33. Mondrinos, M.J. *et al.* Engineering three-dimensional pulmonary tissue constructs. *Tissue Eng.* **12**, 1–11 (2007).
34. Bokhari, M., Carnachan, R.J., Cameron, N.R. and Przyborski, S.A. Culture of HepG2 liver cells on three dimensional polystyrene scaffolds enhances cell structure and function during toxicological challenge. *J. Anat.* **211**, 567–576 (2007).
35. Sun, A., Parisius, W., Macmorine, H., Sefton, M. and Stone, R. An artificial endocrine pancreas containing cultured islets of Langerhans. *Artif. Organs* **4**, 275–278 (1980).
36. Sefton, M.V. and Broughton, R.L. Microencapsulation of erythrocytes. *Biochim. Biophys. Acta* **717**, 473–477 (1982).
37. Foster, J.L., Williams, G., Willimas, L.H. and Tuch, B.E. Differentiation of transplanted microencapsulated fetal pancreatic cells. *Transplantation* **83**, 1440–1448 (2007).

38. Ma, H.L., Hung, S.C., Lin, S.Y., Chen, Y.L. and Lo, W.H. Chondrogenesis of human mesenchymal stem cells encapsulated in alginate beads. *J. Biomed. Mater. Res.* **64A**, 273–281 (2003).
39. Awad, H.A., Quinn, W.M., Leddy, H.A., Gimble, J.M. and Guilak, F. Chondrogenic differentiation of adipose-derived adult stem cells in agarose, alginate, and gelatin scaffolds. *Biomaterials* **25**, 11–22 (2004).
40. Tanaka, H. *et al.* Chondrogenic differentiation of murine embryonic stem cells: effects of culture conditions and dexamethasone. *J. Cell Biochem.* **93**, 454–462 (2004).
41. Barnett, B.P. *et al.* Magnetic resonance-guided, real-time targeted delivery and imaging of magnetocapsules immunoprotecting pancreatic islet cells. *Nat. Med.* **13**, 986–991 (2007).
42. Zur Nieden, N.I., Cormier, J.T., Rancourt, D.E. and Kallos, M.S. Embryonic stem cells remain highly pluripotent following long term expansion as aggregates in suspension bioreactors. *J. Biotechnol.* **129**, 421–432 (2007).
43. Ismail, S., Bishop, A.E., Polak, J.M. and Mantalaris, A. Encapsulation of human embryonic stem cells supports prolonged feeder-free maintenance. *Biomaterials*, submitted (2007).
44. Martin, I., Wendt, D. and Heberer, M. The role of bioreactors in tissue engineering. *Trends Biotechnol.* **22**, 80–86 (2004).
45. Liu, J.Y., Hafner, J., Dragieva, G. and Burg, G. Bioreactor microcarrier cell culture system (Bio-MCCS) for large-scale production of autologous melanocytes. *Cell Transplant.* **13**, 809–816 (2004).
46. Randle, W.L. *et al.* Integrated 3D expansion and osteogenic differentiation of murine embryonic stem cells. *Tissue Eng.* **13**, 2957–2970 (2007).
47. Mantalaris, A., Randle, W. and Polak, J.M. Stem cell bioprocessing for clinical applications of regenerative medicine. In: *3rd Stem Cell Symposium Proceedings on Repair and Regeneration*, review submitted (Imperial College Press, 2007).
48. Assmus, B.J. and Honold, J. Transcoronary transplantation of progenitor cells after myocardial infarcation. *N. Engl. J. Med.* **355**, 1222–1232 (2006).
49. Lunde, K. *et al.* Intracoronary injection of mononuclear bone-marrow cells in acute myocardial infarction. *N. Engl. J. Med.* **355**, 1199–1209 (2006).
50. Schachinger, V. *et al.* Intracoronary bone marrow-derived progenitor cells in acute myocardial infarction. *N. Engl. J. Med.* **355**, 1210–1221 (2006).
51. Atala, A., Bauer, S.B., Soker, S., Yoo, J.J. and Retik, A.B. Tissue-engineered autologous bladders for patients needing cystoplasty. *Lancet* **367**, 1241–1246 (2006).
52. Ghen, M.J. *et al.* Potential clinical applications using stem cells derived from human umbilical cord blood. *Reprod. Biomed.* **13**, 562–572 (2006).

Chapter 25

Bioreactor Engineering: Regenerating the Dynamic Cell Microenvironment

Tal Dvir and Smadar Cohen

Abstract

The replacement of the "flat biology" of the Petri dish with three-dimensional (3D) cell cultures has shown to narrow the gap between cell behaviours and function *in vitro* and at the physiological settings. A fundamental challenge to realise the potential of the 3D cell culture is the design and application of "smart" bioreactor systems. These systems should provide homogenous mass transport into the internal volume of the cultured cell constructs as well as to efficiently propagate physical and mechanical stimuli. Herein, we describe the design principles of various bioreactors, starting with the conventional spinner flasks, the rotary wall vessels and up to the latest perfusion vessels. In particular, the key role of perfusion bioreactors in regenerating the dynamic 3D cell microenvironment is demonstrated by providing a few successful examples of engineering thick functional tissues, such as the cardiac muscle tissue. In closing this chapter, we envision future innovations in bioreactors.

Keywords: 3D Microenvironment; Cardiac Tissue Engineering; Interstitial Fluid Flow; Perfusion Bioreactor; Mass Transport; Mechanical Stress.

Outline

1. Tissue Engineering — The Introduction of 3D Cell Cultures
2. Mass Transport Challenges in 3D Cell Cultures
3. First Generation of Tissue Engineering Bioreactors
4. Perfusion Bioreactors — Theory and Practice
5. Examples of Perfusion Bioreactors in TE

6. Bioreactors Providing Physical Signals
   6.1. Mechanical stimuli
   6.2. Electrical stimuli
7. Microfabricated Bioreactors
8. Concluding Remarks and Future Aspects
   References

1. Tissue Engineering — The Introduction of 3D Cell Cultures

Tissue engineering (TE) combines the principles of engineering, biology and medicine for the development of biological tissue substitutes to restore, maintain or improve body functions, and for *in vitro* applications, such as screening for drug toxicity.[1] In one of the proposed approaches to TE, isolated cells are seeded in three-dimensional (3D) temporary polymeric scaffolds until the cells secrete their own extracellular matrix (ECM) and organised into a functional tissue.[2] Ideally, when this process is completed, the synthetic scaffolds are resorbed to allow the integration of the engineered tissue with the host.

The science of TE has generated a wealth of information on cellular behaviour and function in a physiological-relevant microenvironment. Growing cells within 3D scaffolds has shown to reduce the gap between cell cultures and physiological tissues. Thus, the 3D culture has been offered as an appropriate *ex vivo* approach to replace the "flat biology" of the Petri dish. In parallel, the use of 3D cell cultures has raised a few challenges, such as mass transport in avascular environment.

In this chapter, we will describe the development of bioreactors designed for *ex vivo* TE, from being a supplementary device to a novel science niche standing on its own. We will discuss the challenges in 3D cell cultivation and the theory and technical aspects of bioreactor design. In addition, we will present examples of innovative bioreactors supplying perfused medium, mechanical and electrical signals and microfabricated bioreactors to create the optimal 3D cell microenvironment.

2. Mass Transport Challenges in 3D Cell Cultures

In vivo, the transport of dissolved oxygen and nutrients through the interior regions of tissues is aided by tangled vascular networks. These networks serve as transport channels for convective blood flow, thus enabling the diffusive nutrients to transport more efficiently through the cellular space and ECM. *Ex vivo*, in static 3D cell cultures (no mixing in the culture medium), with the lack of vasculature, two main engineering issues challenge the development of a functional tissue inside the scaffold. First, a boundary layer around the cell constructs decreases the transport of nutrients and dissolved oxygen from the bulk medium to the surface of the constructs. The second is the limited internal mass transfer rate from the surface of the construct into its core. The regular diffusion coefficient is reduced to an effectiveness diffusion coefficient, which is influenced by the cell construct porosity and tortuosity, as described in Eq. (1).

$$D_{eff} = D_N \frac{\varepsilon_p}{\tau} \tag{1}$$

where D_{eff} is the effectiveness diffusion coefficient, D_N is the nutrient diffusion coefficient and ε_p and τ are the cell construct porosity and tortuosity, respectively.

All together, these external and internal diffusion limitations affect the tissue growth and development and restrict the tissue thickness. In practice, the internal cells in the developing tissue do not survive due to the evolving anoxic conditions and poor nutrient transfer and waste removal (Fig. 1). Therefore, one of the major hurdles in *ex vivo* tissue engineering is attaining functional tissues that constitute more than a few layers of viable cells.

Among the nutrients required for tissue growth and survival, the dissolved oxygen diffusion distance is considered to be the main limiting factor restricting tissue thickness. Mathematically, this can be described by the oxygen balance equation [Eq. (2)], wherein oxygen concentration is calculated as the balance of the input oxygen diffusion into the cell construct as described by Fick's second law, and the output oxygen consumption by the cells, assuming Monod-type kinetics.

$$\frac{\partial C_{O_2}}{\partial t} = \underbrace{D_{eff} \frac{\partial^2 C_{O_2}}{\partial x^2}}_{\text{Diffusion}} - \underbrace{\frac{\mu_{max} \times C_{O_2}}{K_S + C_{O_2}}}_{\text{Consumption}} \tag{2}$$

where C_{O_2} is the dissolved oxygen concentration, x is half of the tissue thickness, μ_{max} is the maximum oxygen consumption rate and K_S is the oxygen saturation constant.

As seen in Eq. (2), when the cell construct thickness increases, the dissolved oxygen concentration decreases and oxygen is deprived from the internal cells. The ratio output to input, that is consumption to diffusion, can be expressed in terms of the Thiele modulus: $\phi = x(\mu_{max} / K_S D_{eff})^{1/2}$. Usually, under static conditions, the x is large, namely $\phi > 1$. Ideally, in an efficient cultivation system, where the dissolved oxygen and nutrients do not limit tissue thickness, ϕ must be ≤ 1.

Our group has recently combined a theoretical and experimental approach to define the optimal hepatic tissue thickness, which enables maximal cell viability

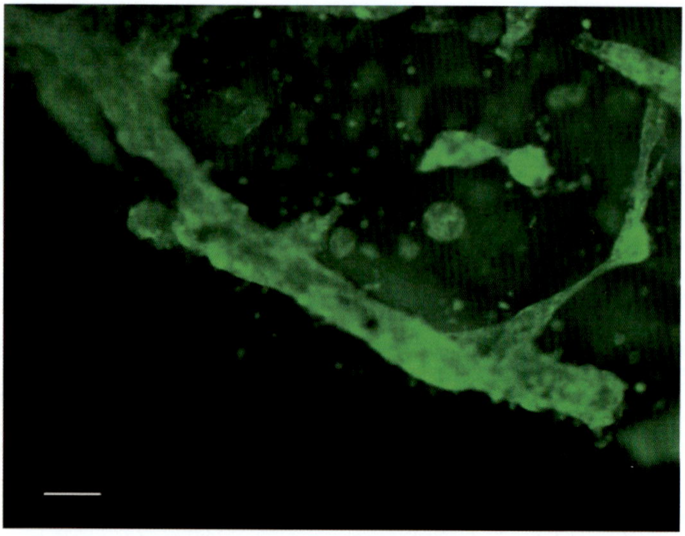

Fig. 1. Fluorescence microscope imaging of cardiac cell construct, cultivated under static conditions for 13 days. The viable cells are stained green with fluorescein diacetate. The external and internal mass transfer in the static culture restrict cell survival to the construct periphery and limit the tissue thickness to 100 mm.[7] Bar = 100 μm.

under static 3D environment conditions, assuming that the dissolved oxygen is the main limiting factor. The resulting model described that no oxygen limitations is taking place in tissue thickness up to 100 μm, under static conditions.[3] Since the 3D engineered constructs should be at least a few millimetres in thickness to serve as a graft for tissue replacement, mass transfer limitations represent one of the greatest challenges to be addressed. Bioreactors have been implemented for this task.

3. First Generation of Tissue Engineering Bioreactors

The first bioreactors employed in TE were adapted from mammalian cell culture facilities, designed to cultivate large volumes of concentrated cell suspensions. The bioreactors represented different patterns of fluid dynamics and vessel geometry, aiming to promote the efficient mixing of culture medium and obtain a homogeneous environment with no nutrient gradients. The simplest bioreactor in use has been the spinner flask (Fig. 2A), a cylindrical glass container with the cell constructs suspended in the medium and an impeller ensuring medium mixing and homogeneous distribution of the dissolved oxygen and other necessary nutrients in the culture medium.[4] With mixing, the stagnant layer around the cell constructs has been significantly reduced, enabling higher oxygen and nutrient concentrations at the surface of the cell constructs. Thus, the 3D cultivation in spinner flasks

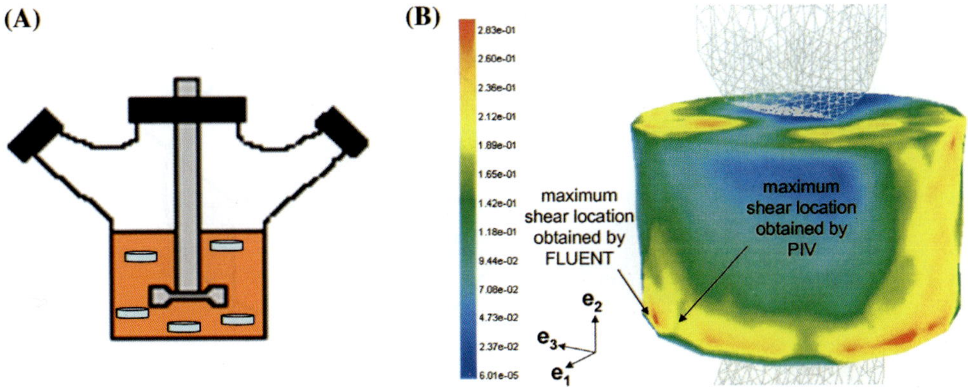

Fig. 2. 3D cell cultivation in spinner flask. **(A)** A schematic description of spinner flask. It consists of a cylindrical container and an impeller for medium mixing. **(B)** Mathematical modelling of the fluid flow around cell construct cultivated in spinner flask. The colour scale represents different levels of shear stress. The arrows indicate the locations on the cell construct surface subjected to a maximal shear stress, according to model simulation and experiments.[4]

improved, to some extent, tissue homogeneity and viability[5] and enhanced cell proliferation and differentiation.[6] Yet, in cell constructs thicker than 200 μm, the inner cells were still dying. Furthermore, cultivation within the spinner flask subjected the cells to high shear stress, due to turbulence and eddies within the vessel. A mathematical model to elucidate the shear stress acting on the cell constructs inside a spinner flask revealed a maximal shear stress of more than 2.8 dynes/cm^2 (Fig. 2B),[4] known to be harmful for most mammalian cells.[7,8]

The introduction of the rotary wall vessel (RWV), originally developed by NASA for microgravity simulations, enabled 3D cell cultivation under medium mixing with a minimal shear stress on the cultivated cells. The bioreactor rotates about its horizontal axis while maintaining the cell constructs in a free-fall state (Fig. 3). In the vessel, fluid flow mixing is generated by settling of the cell constructs, which is associated with oscillation, tumbling, wake formation and vortex shedding.[9] In addition, the RWV has a coaxial tubular silicone membrane for efficient oxygenation of the medium. Cultivation of cardiac cell constructs within the RWV produced an engineered tissue with an improved cellularity and high expression of cardiac muscle-specific markers.[10] Rotating bioreactors also encouraged chondrogenesis and yielded larger cartilaginous constructs with better structural, functional, and molecular properties.[11]

In addition to cultivation of differentiated cells, the RWV has become a useful tool in stem cell culture. Recently, we have shown that cultivation of human

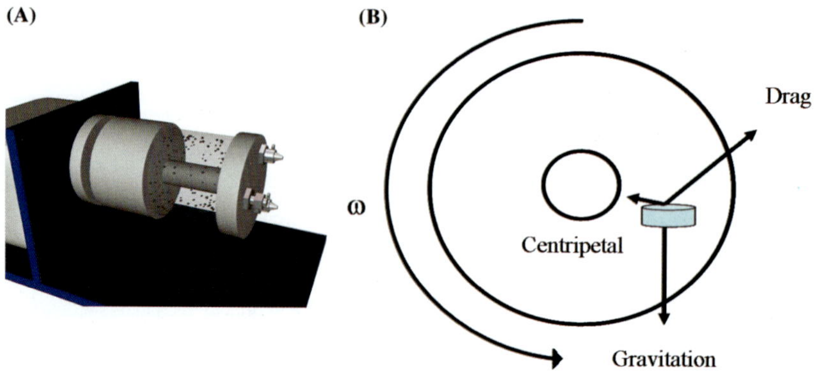

Fig. 3. Cultivation within the rotary wall vessel (RWV) bioreactor. **(A)** Illustration of the bioreactor configuration.[12] **(B)** The accumulative forces acting on the cell construct, maintaining it in a free-fall state.[9]

embryonic stem cells (hESCs) within a rotating bioreactor increased the cell proliferation rate and maintained cell viability in the culture. Furthermore, efficient medium mixing within the vessel prevented embryoid body (EB) aggregation and the formation of anoxic areas within the EBs.[12]

Although the RWV bioreactor has been efficient in reducing the boundary layer surrounding the construct surface, it did not solve the problem of the limiting internal mass transfer into the core of the cell constructs. To address this challenge, perfusion bioreactors, designed to force the culture medium into the cell constructs, were developed.

4. Perfusion Bioreactors — Theory and Practice

In vivo, perfusion is the process of nutritive delivery of arterial blood into a tissue via capillaries. *Ex vivo*, perfusion describes the transfer of culture medium directly through a porous cell-seeded matrix (Fig. 4). With the fluid flow directly forced through the cell construct, the convection element (V-flow velocity) is added to the oxygen balance [Eq. (3)].

$$\frac{\partial C_{O_2}}{\partial t} = \underbrace{V\frac{\partial C_{O_2}}{\partial x}}_{\text{Convection}} + \underbrace{D_{eff}\frac{\partial^2 C_{O_2}}{\partial x^2}}_{\text{Diffusion}} - \underbrace{\frac{\mu_{max} \times C_{O_2}}{K_S + C_{O_2}}}_{\text{Consumption}}. \tag{3}$$

According to this, oxygen transport in perfused systems is by both diffusion and convection, thus substantially increasing its transport distance in a tissue.

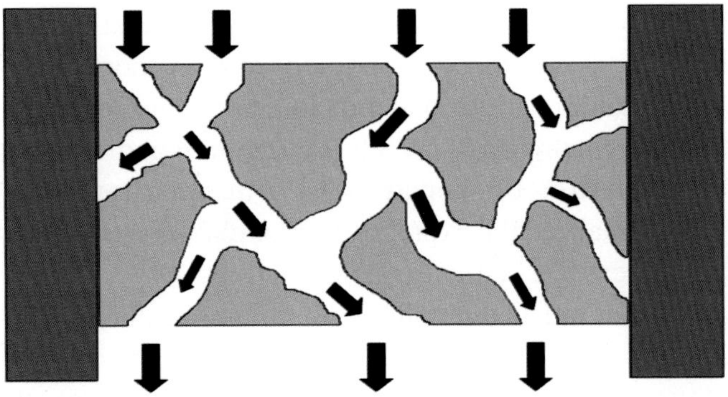

Fig. 4. Medium perfusion through macroporous matrix. The culture medium is forced through the cell construct by high pressure. In this system, the dissolved oxygen transport is via both convection and diffusion, thus increasing the oxygen balance, according to Eq. (3).

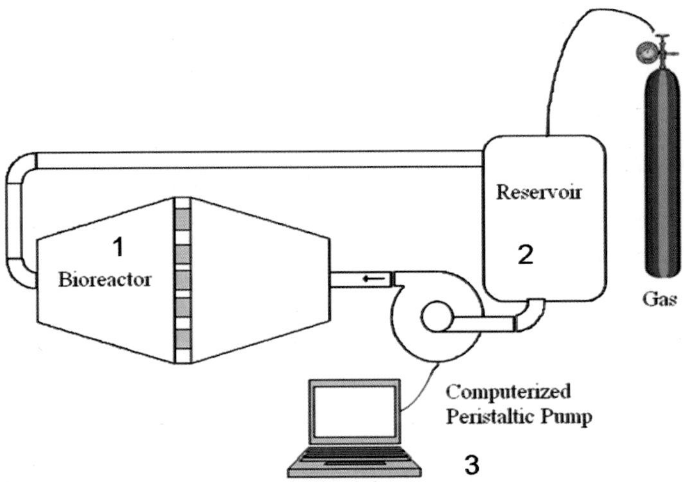

Fig. 5. A typical perfusion system. It consists of a bioreactor body, wherein the constructs are fixed between nets (1), a reservoir, wherein the medium is heated and oxygenated (2), and a peristaltic pump that fosters the medium through the cell constructs (3).

Perfusion systems usually consist of a few elements (Fig. 5): (1) the bioreactor body, wherein the cell constructs are fixed in space to allow medium perfusion, (2) a reservoir, where the medium is oxygenated and heated, and (3) a peristaltic pump to circulate the medium. The system may be placed inside a humidified incubator or serve as an independent system; either way, various essential parameters, such as pH, pO_2, pCO_2 and temperature, must be monitored and rigorously controlled.

With a proper control of these parameters, 3D cell cultivation within perfusion bioreactors has been shown to be superior over conventional bioreactors. For example, 3D cultivation in perfusion bioreactors has better maintained cell viability,[7,13] encouraged mesenchymal stem cell proliferation and differentiation,[14] promoted albumin secretion by hepatocytes,[15] increased GAG synthesis by chondrocytes[16] and enhanced the expression of cardiac-specific markers and ultrastructural features in cardiac cell-cultivated constructs.[8,17]

The ability to foster the medium directly through the porous scaffold has been also applied to increase the efficiency of cell seeding into a matrix, to obtain high cell density cultures. Cell seeding in porous scaffolds represents the first step in establishing the 3D culture and thus it plays a critical role in determining the progression of tissue formation. The complex and tortuous architecture of most of the macroporous scaffolds cause the static seeding to be inefficient and the seeded cells are largely confined to the top region of the scaffold (Fig. 6A). In perfusion systems, the cells are forced together with the medium into the depth of the tortuous porous scaffold (Fig. 6B). Thus, on-line cell seeding, using perfusion systems, has promoted an efficient and high-density cell-seeding and the development of homogeneous tissues, *ex vivo*.[18–20]

5. Examples of Perfusion Bioreactors in TE

Different groups have designed various types of perfusion bioreactors with one common feature, namely the culture medium had to perfuse the cell constructs. Carrier and colleagues utilised perfusion cartridges consisted of 13 mm filter holders. A stainless steel screen was placed at the cartridge inlet to disperse the fluid flow over the construct surface and a nylon mesh to fix the downstream side

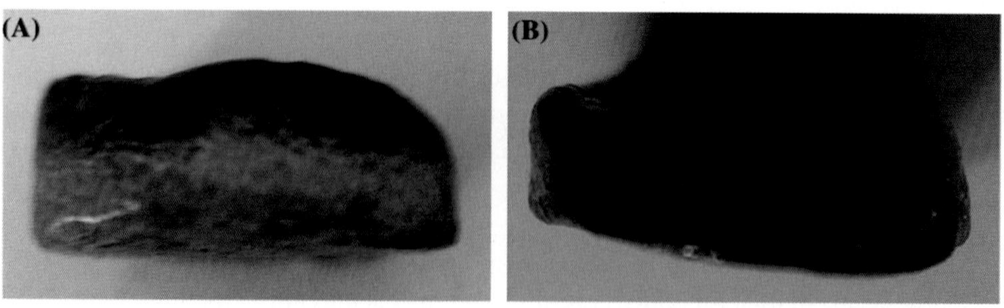

Fig. 6. Cardiac cell distribution in macroporous alginate scaffolds post-seeding. **(A)** In static seeding, the cells are confined to the top region of the scaffold. **(B)** Dynamic seeding, the cells are distributed in the entire matrix. The cardiac cells are stained purple with MTT cell viability assay, for visualisation (Shachar, M., unpublished results).

of the construct. The cartridges were arranged in a series of four, and each series was connected to a four-channel peristaltic pump, perfusing the medium through the constructs. During cultivation, the dissolved oxygen concentration decreased due to cell respiration, in a manner depending on the construct position in the series. Using this system, the researchers were able to investigate the influence of different pO_2 levels on a developing cardiac tissue.[21]

A different bioreactor for 3D osteoblast culture in porous scaffolds was designed by Bancroft and colleagues.[22] Their system consisted of six individual medium flow chambers, operated and controlled in parallel by a multichannel pump. In each chamber, one cell construct was held in a cassette, sandwiched between two O-rings in a manner which prevented undesirable non-perfusing flow. To avoid bubble entrapment within the cell constructs, the medium entered the column through a top hole, passed through the cell-seeded construct and exited through the bottom. The system had dual-reservoir system, allowing an easy access for medium supplementation and removal, and the complete purging of the medium during its replacement.

In most perfusion bioreactors, the cell constructs are held in place by fixing nets, which block, in part, medium perfusion into the cell constructs. This usually results in the formation of inconsistent tissues within the construct. In addition, the character of fluid flow in a tube also promotes non-homogeneous flow in perfusion bioreactors. The velocity profile of developed laminar flow is unequal along the cross-section area due to frictions of the fluid flow with the bioreactor walls. Thus, the level of shear stress acting on the cell constructs is different in various locations in the bioreactor and the consequence is non-homogeneous tissue.

To address the non-homogeneous milieu within perfusion systems, we designed two unique cell construct-fixing nets for our bioreactor, with an open area of 95.8%.[7] The net architecture (Fig. 7A) is an assembly of equally spaced, 380-square-based micropyramidal structures, with their heads pointing toward and holding the multiple cell constructs (Fig. 7B). Four round openings are located at the corners of the square pyramid, enabling medium perfusion from the small holes toward the pyramid head, at an angle of 60°, thus maintaining the fluid flow direction. Using this net architecture, 99.88% of the cell construct volume has been perfused by the culture medium (Fig. 7C). In addition, to impose equal shear stress at the bioreactor cross-section, a micromesh designed to disrupt the perfusing media, for a split of a second, before entering the cell constructs has been inserted to the bioreactor. It transforms the developed laminar fluid flow velocity profile into an undeveloped one, having equal velocity vectors along the bioreactor cross-section. Locating the micromesh in proximity to the cell construct compartment ensures that the velocity profile will not develop before the medium reaches the cell constructs, thus subjecting them to an equal shear stress (Fig. 8).

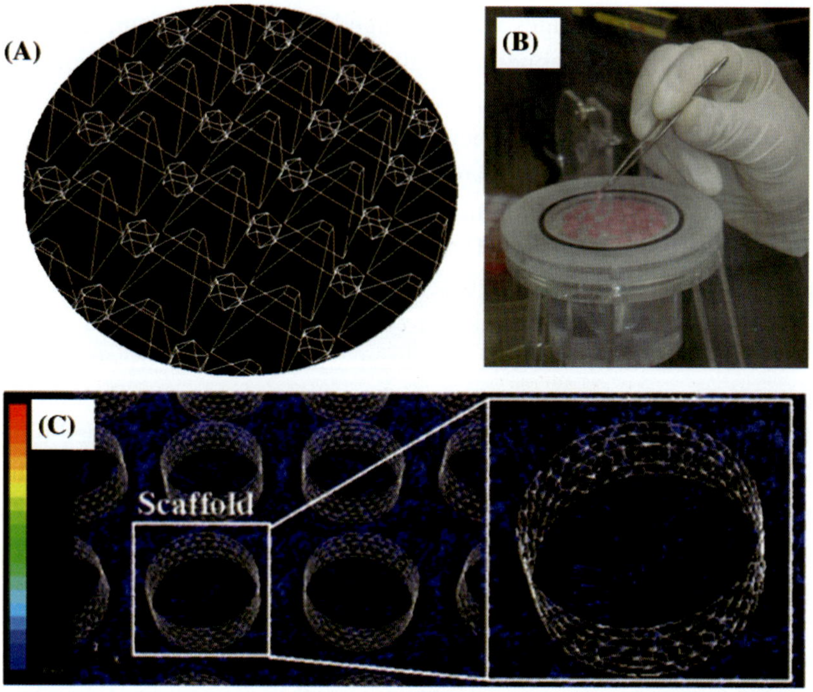

Fig. 7. Our perfusion bioreactor with its unique construct fixing nets. **(A)** Architecture of the fixing nets. The nets are comprised of 380-square-based pyramidal structures with their heads pointing towards and holding the cell constructs. **(B)** Multiple cell constructs are placed on the fixing net. **(C)** The perfusing medium (blue) flows through the constructs (grey), according to mathematical modelling.[7]

By employing this advanced perfusion bioreactor, the cardiac cell constructs maintained the viability of almost 100% of the seeded cells, while less than 60% of the cells in static cultures were viable after seven-day cultivation.[7] Moreover, a thick (> 500 μm) cardiac tissue was generated, composed of elongated and aligned cells with a massive striation. Ultrastructural morphology analyses revealed organised sarcomeres, defined Z-lines and intercalated disks, resembling the native adult heart.[8]

6. Bioreactors Providing Physical Signals

6.1. Mechanical stimuli

It has long been recognised that cultured cells can sense and transduce a broad range of mechanical stimuli into distinct sets of biochemical signals that ultimately regulate cellular processes, including cell adhesion, proliferation and differentiation.[23]

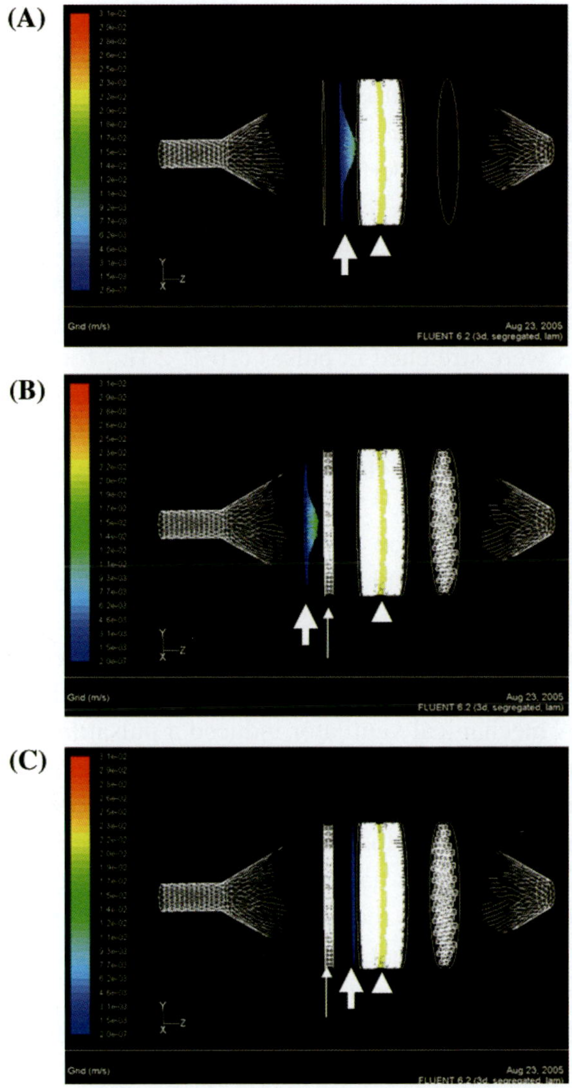

Fig. 8. Profile of the fluid flow velocity in the perfusion bioreactor, constructed via the computerised fluid dynamics software Fluent after solving Navier-Stock equations.[7] The velocity profiles in a perfusion bioreactor with no fluid-distributing mesh **(A)** or when approaching it in a bioreactor with the flow-distributing mesh **(B)** reveal a well-developed flow with large velocity vectors (green) at the centre of the bioreactor and smaller ones on the sides (blue). Once past the mesh **(C)**, the flow is interrupted and transforms to undeveloped flow, with equal velocity vectors (blue line) along the bioreactor cross-section. Arrowhead indicates the location of the cellular constructs; thin, long arrows indicate the location of the flow-distributing mesh 1.5 cm upstream from the construct compartment, and thick, short arrows indicate the flow velocity profile. Velocity scale-bar is presented on the left. Dark blue represents the smallest vectors.

In 3D cell cultures, the homogeneous distribution of these mechanical stimuli represents another challenge for bioreactor engineers. Different approaches to induce mechanical stimuli in 3D cell constructs have been employed; among them are cyclic stretching of the matrix, direct or indirect compression of the construct and via the fluid flow regime in perfusion systems. The signals were administrated to the cell constructs as a means to induce *ex vivo* tissue regeneration.

Zimmermann *et al.* applied unidirectional cyclic stretch on ring-shaped circular molds of cardiomyocytes, mixed with collagen and matrigel. The stretching device was composed of two poles on which the ring-shaped constructs were hung. The poles were stretched in pulses (10%, 2 Hz) for seven days. In addition to the matrix stretching, this activity was most likely accompanied with medium mixing. Compared with static systems, the stretched constructs exhibited a better cardiac tissue/matrix ratio, high degree of cardiomyocyte differentiation and an improved contractile function.[24] While the exact mechanism responsible for inducing cardiac tissue regeneration in the stretching system is not yet clear, it may involve stimulation and activation of integrins, receptors and other adhesion molecules connecting the cells to the stretched matrix.

In a different study, Thompson and colleagues utilised a different system to expose the 3D cultivated cells to mechanical stimuli and thus engineer blood vessels *in vitro*.[25] A mechanical ventilator induced a pulsatile air flow, followed by a laminar medium flow in a column, wherein a vascular construct was housed. This set up closely mimics the mammalian physiology of blood flow through vessels, allowing a tight control over the pulse pressure and radial expansion/compliance of the tissue. The researchers have shown that exposure of the vascular cell construct to this flow regime has led to the development of engineered blood vessels with increased collagen content, suture retention and burst pressure.[25]

Compression is another form of inducing mechanical stimuli in a developing tissue. Compression may be induced indirectly by compressing the medium against the construct or directly by compressing the construct with a piston. When designing a compression bioreactor, the main parameters to consider are the frequency of the applied load, the strain or force to be used, and the duration of the exposure.[26] Direct compression has been extensively used to apply external load on cartilage constructs, resulting in massive production of matrix.[27–29]

Our group has recently investigated the effects of a homogeneous pulsatile, interstitial fluid flow, provided by the perfusion bioreactor described above, on cardiac tissue assembly. For the first time, the various molecular and cellular events taking place in the cell construct, from the cell signalling level up to the ultrastructural phenotype of the regenerated cardiac muscle tissue, have been elucidated.[8] We demonstrated that immediately after operating the bioreactor at a shear stress of 0.6 dynes/cm^2, the extracellular signal-regulated kinase (ERK)

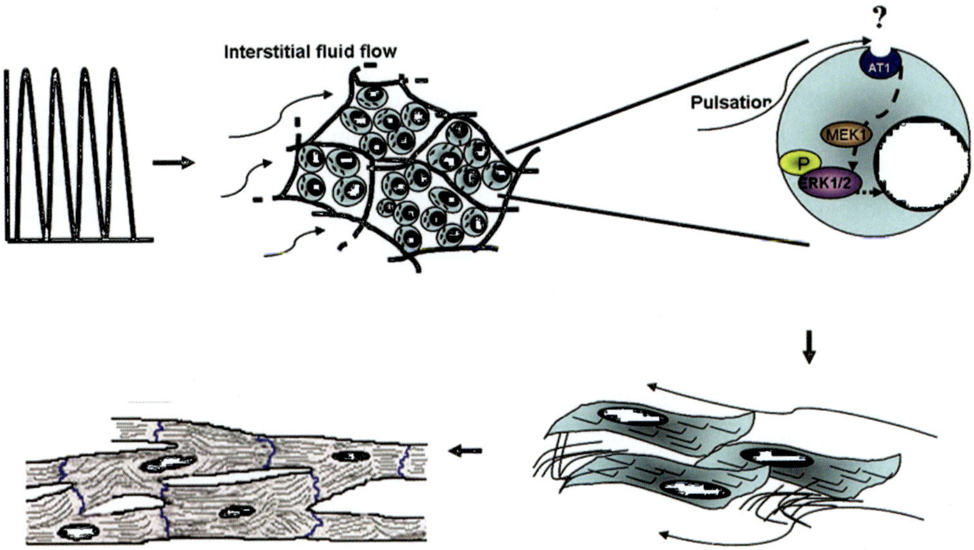

Fig. 9. Our proposed model to describe the mechanistic effects of the pulsatile interstitial fluid flow/shear stress on cardiac tissue regeneration. Effective medium pulsation in the entire cell construct promotes ERK1/2 activation, the synthesis of contractile and cell-cell interaction proteins, cardiac cell alignment and regeneration of striated cardiac muscle.[8]

1/2 signal transduction pathway has been activated, inducing the synthesis of cardiac contractile proteins (troponin T and sarcomeric actinin) and cell-cell interaction proteins (connexin-43 and N cadherin). Furthermore, we showed that prolonged cultivation under this stimulation regime promoted the assembly of a thick cardiac tissue, resembling in many respects to the native adult heart. A model describing the proposed mechanism, translating mechanical cues to tissue regeneration, is presented in Fig. 9.

6.2. Electrical stimuli

In vivo, many physiological processes, such as embryonic and heart development as well as wound healing, are influenced by electrical activities. In the myocardium, contraction is driven by waves of electrical excitation generated by a number of pacing cells. In 2D cardiomyocyte cultures, a short-term rapid electrical stimulus caused an upregulation of connexin-43 via activation of ERK and p38 pathways, and was accompanied by an increase in the conduction velocity.[30]

In cardiac TE, an electrical stimulation of the cell constructs has been tested as a means to induce cardiac cell alignment with the electrical field lines and consequently to *ex vivo* tissue regeneration.[31] The constructs were placed in a glass chamber, fitted with two carbon rods placed 1 cm apart and connected to a cardiac stimulator with platinum wires. The stimulation applied by this system (rectangular, 2 ms, 5 V/cm, 1 Hz) promoted the establishment of numerous gap junctions between neighbouring cells, an efficient propagation of the pacing signals and generation of action potentials, inducing synchronous macroscopic contractions. The regenerated tissue revealed the ultrastructural morphology of a native cardiac tissue, with multiple sarcomeres that have increased the construct responsiveness to pacing.[31]

Due to static culture conditions, the functional tissue thickness has been limited to 100 μm. We foresee that by combining perfusion with electrical stimulation, it would be possible to engineer an aligned thick cardiac tissue suitable for replacing a full-thickness scar tissue post-myocardial infarction.

7. Microfabricated Bioreactors

Recent progress in micro- and nanotechnology enabled microfabrication of bioreactors, designed to mimic the 3D cell microenvironment with built-in microchannels similar to the blood vessels that nourish physiological tissues. The microfabricated bioreactors are manufactured by processes adapted from the microelectronics industry, using the flexible silicone rubber poly(dimethyl siloxane) (PDMS).[32] Due to its biocompatibility and high gas permeability, PDMS is a suitable material for fabricating cell-based systems. One of the most important advantages of microfluidic bioreactors over the conventional systems is their large surface to volume ratio, which allows high mass transfer rates to maintain cell and tissue viability.[33]

Microfabricated bioreactors have additional added values; they combine the advantages of microarray technology, such as small working volume, high throughput and independent culture wells, with the high mass transfer and biophysical signals that bioreactors can offer[34] (Fig. 10). With these features, microbioreactors provide a means to study the growth and differentiation of cells, under controlled conditions and multi-array settings. Furthermore, due to the simplicity of the microchannel fabrication process, microbioreactors can be designed to encourage different tissue morphogenesis, such as hepatic[15] and cardiac tissues.[35]

Innovations in bioreactor design are continuing to be made. In a recent study, Christen and Andreou cleverly combined principles from microfluidics, electrical engineering and life sciences to create a thumb-size microincubator for cell culture. In this device, cell seeding was accomplished by injecting the cells into the incubator's microchannels, fabricated from PDMS. Following their gravitation towards the microchip surface, the cells adhered to the surface. Computer-controlled

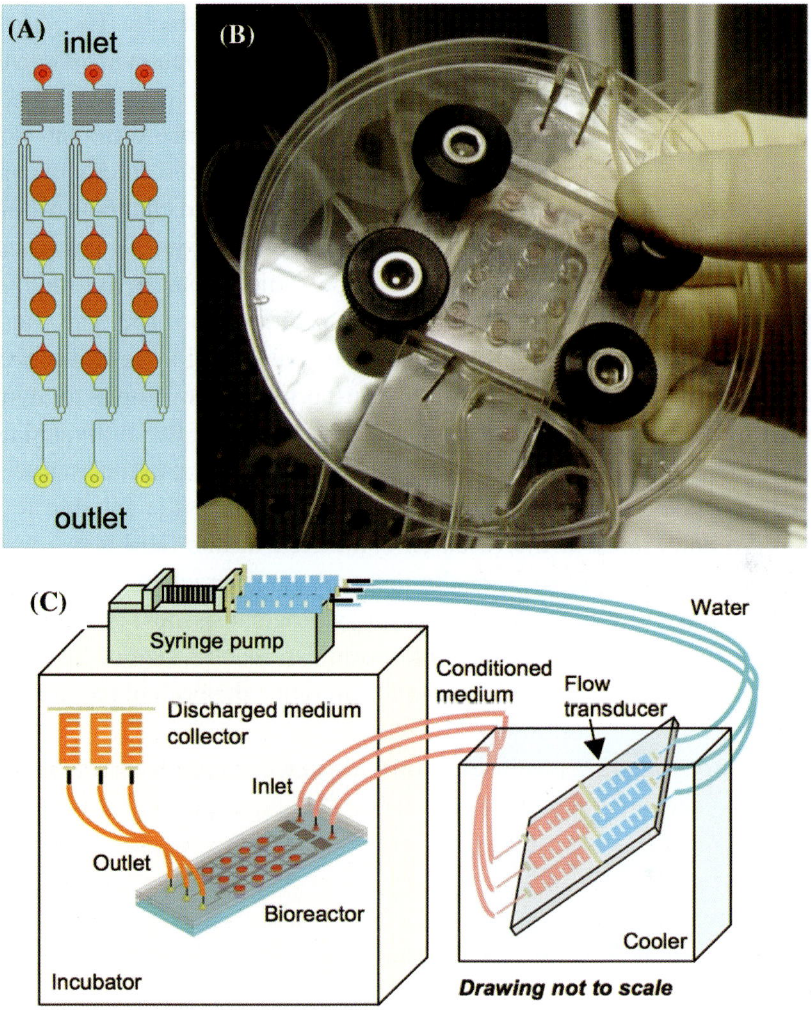

Fig. 10. The microfabricated bioreactor by Figallo *et al.*[34] **(A)** Schematic of the microbioreactor with its 4 × 3 multi-well array. **(B)** Image of the microbioreactor array with the fluidic connectors. **(C)** Overall view of the microfabricated bioreactor array.

electronics autonomously maintained the media under physiological conditions and the BHK-21 hamster cells, cultivated in the incubator, survived in this environment over a three-day period.[36]

8. Concluding Remarks and Future Aspects

In this chapter, we describe the evolution of bioreactors for tissue engineering, from adapting devices originally designed for cell suspensions to the fabrication of custom-made bioreactors for cultivating 3D cell constructs. The bioreactors were designed to supply a homogeneous medium environment, efficient mass transport and physical stimuli during tissue development.

In the future, we envision a bioreactor that combines the different technologies described herein, to promote the ideal physiological cell microenvironment. It may be fabricated from an assembly of multi-microfluidic channels for efficient medium perfusion and to enable high mass transfer rate to the 3D cultivated cells located between the channels. The cells may be subjected to an assorted of electrical and mechanical stimuli, according to the specific tissue type. The conditions, such as temperature, pH and dissolved gases in this "ideal" bioreactor would be self-maintained under a tight regulation. Furthermore, the bioreactor would be transparent to allow *in situ* visualisation and monitoring of the cultivated cells.

Another foreseen innovation in the future bioreactor for tissue engineering is its scaling-up for the purpose of engineering a full-size tissue suitable for transplantation. To date, most of the research conducted in this field was confined to the laboratory research and so were the bioreactors. In practice, the *ex vivo* engineered tissues have to replace large thick physiological tissues. For example, the area of a scar tissue post-myocardial infarction can reach a size of 20 cm^2. Thus, scaling-up the bioreactor size and function is amongst the critical parameters to be addressed.

Bioreactor engineering promises to provide new adventures for biotechnology engineers.

References

1. Langer, R. and Vacanti, J.P. Tissue engineering. *Science* **260**, 920–926 (1993).
2. Leor, J., Aboulafia-Etzion, S., Dar, A., Shapiro, L., Barbash, I.M., Battler, A., Granot, Y. and Cohen, S. Bioengineered cardiac grafts: a new approach to repair the infarcted myocardium? *Circulation* **102**, III56–61 (2000).
3. Glicklis, R., Merchuk, J.C. and Cohen, S. Modeling mass transfer in hepatocyte spheroids via cell viability, spheroid size, and hepatocellular functions. *Biotechnol. Bioeng.* **86**, 672–680 (2004).
4. Sucosky, P., Osorio, D.F., Brown, J.B. and Neitzel, G.P. Fluid mechanics of a spinner-flask bioreactor. *Biotechnol. Bioeng.* **85**, 34–46 (2004).
5. Meinel, L., Karageorgiou, V., Fajardo, R., Snyder, B., Shinde-Patil, V., Zichner, L., Kaplan, D., Langer, R. and Vunjak-Novakovic, G. Bone tissue engineering using human mesenchymal stem cells: effects of scaffold material and medium flow. *Ann. Biomed. Eng.* **32**, 112–122 (2004).

6. Sikavitsas, V.I., Bancroft, G.N. and Mikos, A.G. Formation of three-dimensional cell/polymer constructs for bone tissue engineering in a spinner flask and a rotating wall vessel bioreactor. *J. Biomed. Mater. Res.* **62**, 136–148 (2002).
7. Dvir, T., Benishti, N., Shachar, M. and Cohen, S. A novel perfusion bioreactor providing a homogenous milieu for tissue regeneration. *Tissue Eng.* **12**, 2843–2852 (2006).
8. Dvir, T., Levy, O., Shachar, M., Granot, Y. and Cohen, S. Activation of the ERK1/2 cascade via pulsatile interstitial fluid flow promotes cardiac tissue assembly. *Tissue Eng.* **13**, 2185–2193 (2007).
9. Shachar, M. and Cohen, S. Cardiac tissue engineering, *ex-vivo*: design principles in biomaterials and bioreactors. *Heart Fail Rev.* **8**, 271–276 (2003).
10. Carrier, R.L., Papadaki, M., Rupnick, M., Schoen, F.J., Bursac, N., Langer, R., Freed, L.E. and Vunjak-Novakovic, G. Cardiac tissue engineering: cell seeding, cultivation parameters, and tissue construct characterization. *Biotechnol. Bioeng.* **64**, 580–589 (1999).
11. Pei, M., Solchaga, L.A., Seidel, J., Zeng, L., Vunjak-Novakovic, G., Caplan, A.I. and Freed, L.E. Bioreactors mediate the effectiveness of tissue engineering scaffolds. *FASEB J.* **16**, 1691–1694 (2002).
12. Gerecht-Nir, S., Cohen, S. and Itskovitz-Eldor, J. Bioreactor cultivation enhances the efficiency of human embryoid body (hEB) formation and differentiation. *Biotechnol. Bioeng.* **86**, 493–502 (2004).
13. Radisic, M., Euloth, M., Yang, L., Langer, R., Freed, L.E. and Vunjak-Novakovic, G. High-density seeding of myocyte cells for cardiac tissue engineering. *Biotechnol. Bioeng.* **82**, 403–414 (2003).
14. Braccini, A., Wendt, D., Jaquiery, C., Jakob, M., Heberer, M., Kenins, L., Wodnar-Filipowicz, A., Quarto, R. and Martin, I. Three-dimensional perfusion culture of human bone marrow cells and generation of osteoinductive grafts. *Stem Cells* **23**, 1066–1072 (2005).
15. Powers, M.J., Janigian, D.M., Wack, K.E., Baker, C.S., Beer Stolz, D. and Griffith, L.G. Functional behavior of primary rat liver cells in a three-dimensional perfused microarray bioreactor. *Tissue Eng.* **8**, 499–513 (2002).
16. Seidel, J.O., Pei, M., Gray, M.L., Langer, R., Freed, L.E. and Vunjak-Novakovic, G. Long-term culture of tissue engineered cartilage in a perfused chamber with mechanical stimulation. *Biorheology* **41**, 445–458 (2004).
17. Carrier, R.L., Rupnick, M., Langer, R., Schoen, F.J., Freed, L.E. and Vunjak-Novakovic, G. Perfusion improves tissue architecture of engineered cardiac muscle. *Tissue Eng.* **8**, 175–188 (2002).

18. Kim, S.S., Sundback, C.A., Kaihara, S., Benvenuto, M.S., Kim, B.S., Mooney, D.J. and Vacanti, J.P. Dynamic seeding and *in vitro* culture of hepatocytes in a flow perfusion system. *Tissue Eng.* **6**, 39–44 (2000).
19. Wendt, D., Marsano, A., Jakob, M., Heberer, M. and Martin, I. Oscillating perfusion of cell suspensions through three-dimensional scaffolds enhances cell seeding efficiency and uniformity. *Biotechnol. Bioeng.* **84**, 205–214 (2003).
20. Zhao, F. and Ma, T. Perfusion bioreactor system for human mesenchymal stem cell tissue engineering: dynamic cell seeding and construct development. *Biotechnol. Bioeng.* **91**, 482–493 (2005).
21. Carrier, R.L., Rupnick, M., Langer, R., Schoen, F.J., Freed, L.E. and Vunjak-Novakovic, G. Effects of oxygen on engineered cardiac muscle. *Biotechnol. Bioeng.* **78**, 617–625 (2002).
22. Bancroft, G.N., Sikavitsas, V.I. and Mikos, A.G. Design of a flow perfusion bioreactor system for bone tissue-engineering applications. *Tissue Eng.* **9**, 549–554 (2003).
23. Vogel, V. Mechanotransduction involving multimodular proteins: converting force into biochemical signals. *Annu. Rev. Biophys. Biomol. Struct.* **35**, 459–488 (2006).
24. Zimmermann, W.H., Schneiderbanger, K., Schubert, P., Didie, M., Munzel, F., Heubach, J.F., Kostin, S., Neuhuber, W.L. and Eschenhagen, T. Tissue engineering of a differentiated cardiac muscle construct. *Circ. Res.* **90**, 223–230 (2002).
25. Thompson, C.A., Colon-Hernandez, P., Pomerantseva, I., MacNeil, B.D., Nasseri, B., Vacanti, J.P. and Oesterle, S.N. A novel pulsatile, laminar flow bioreactor for the development of tissue-engineered vascular structures. *Tissue Eng.* **8**, 1083–1088 (2002).
26. Darling, E.M. and Athanasiou, K.A. Articular cartilage bioreactors and bioprocesses. *Tissue Eng.* **9**, 9–26 (2003).
27. Schulz, R.M. and Bader, A. Cartilage tissue engineering and bioreactor systems for the cultivation and stimulation of chondrocytes. *Eur. Biophys. J.* **36**, 539–568 (2007).
28. Larsson, T., Aspden, R.M. and Heinegard, D. Effects of mechanical load on cartilage matrix biosynthesis *in vitro*. *Matrix* **11**, 388–394 (1991).
29. Kim, Y.J., Sah, R.L., Grodzinsky, A.J., Plaas, A.H. and Sandy, J.D. Mechanical regulation of cartilage biosynthetic behavior: physical stimuli. *Arch. Biochem. Biophys.* **311**, 1–12 (1994).
30. Inoue, N., Ohkusa, T., Nao, T., Lee, J.K., Matsumoto, T., Hisamatsu, Y., Satoh, T., Yano, M., Yasui, K., Kodama, I. and Matsuzaki, M. Rapid electrical stimulation of contraction modulates gap junction protein in neonatal rat

cultured cardiomyocytes: involvement of mitogen-activated protein kinases and effects of angiotensin II-receptor antagonist. *J. Am. Coll. Cardiol.* **44**, 914–922 (2004).

31. Radisic, M., Park, H., Shing, H., Consi, T., Schoen, F.J., Langer, R., Freed, L.E. and Vunjak-Novakovic, G. Functional assembly of engineered myocardium by electrical stimulation of cardiac myocytes cultured on scaffolds. *Proc. Natl. Acad. Sci. USA* **101**, 18129–18134 (2004).
32. Stroock, A.D., Dertinger, S.K., Ajdari, A., Mezic, I., Stone, H.A. and Whitesides, G.M. Chaotic mixer for microchannels. *Science* **295**, 647–651 (2002).
33. Leclerc, E., Sakai, Y. and Fujii, T. Microfluidic PDMS (polydimethylsiloxane) bioreactor for large-scale culture of hepatocytes. *Biotechnol. Prog.* **20**, 750–755 (2004).
34. Figallo, E., Cannizzaro, C., Gerecht, S., Burdick, J.A., Langer, R., Elvassore, N. and Vunjak-Novakovic, G. Micro-bioreactor array for controlling cellular microenvironments. *Lab Chip* **7**, 710–719 (2007).
35. Khademhosseini, A., Eng, G., Yeh, J., Kucharczyk, P.A., Langer, R., Vunjak-Novakovic, G. and Radisic, M. Microfluidic patterning for fabrication of contractile cardiac organoids. *Biomed. Microdevices* **9**, 149–157 (2007).
36. Christen, J. and Andreou, A. Design, fabrication, and testing of a hybrid CMOS/PDMS microsystem for cell culture and incubation. *IEEE Trans. Biomed. Circ. Syst.* **1**, 2–14 (2007).

Chapter 26

UK Regulatory Issues: The View from the Researcher

Caroline Munro and Neil Harris

Abstract

The United Kingdom regulatory landscape as it applies to cell-based therapies is rapidly evolving and constantly produces new information for researchers. This chapter brings together the plethora of information in the form of a process map of the key stages in the life cycle of a cell-based product, from cell/tissue procurement, processing and manufacture, through pre-clinical trials, clinical trials and on to commercialisation and post-launch activities. The critical components of each stage are described, and key issues which are pertinent to the UK researcher are discussed, for example, use of pre-clinical models, documentation requirements for clinical trials. The text goes on to identify which regulations, codes of practice and standards are already available for use in the UK and links them to the life cycle stages. The most recent regulation to be agreed in 2007 in Europe is also discussed. EC 1394/2007 is an amendment of EU Directive 2001/83/EC, and describes overarching regulations of advanced therapy medicinal products (which encompasses cell-based therapeutics).

This information is taken from a Publicly Available Specification (No. 83) which was written by the authors of this chapter and published by the British Standards Institution (BSI) in 2006. It is intended that this PAS acts as a quick reference source to increase clarity for users on the requirements needed for exploitation of a cell-based therapy in the UK, rather than an in-depth examination of the supporting literature.

Keywords: Regulation(s); Therapy; Cell; Regenerative; Process; Guidance.

Outline

1. Introduction
2. The Product Life Cycle of a Cell-Based Therapeutic

3. Stage 1: Procurement — The Obtaining of Cells or Tissue Components from Donors Under cGCP
4. Stage 2: Analysis — Initial Isolation, Screening, Characterisation and Manipulation of Cells/Other Components and Storage
5. Stage 3: Confirmed Proof of Product and Process — Initial to Final Screening for Potential Use
6. Stage 4: Product Manufacturing — Production of Clinical Grade Material Under cGMP
7. Stage 5: Pre-Clinical Trials — Assessment of Safety and Performance for Regulatory Submission
   7.1. Regulatory studies
   7.2. *In vivo* pre-clinical models
8. Stage 6: Clinical Trials — Clinical Assessments of Product Safety and Performance
   8.1. Manufacturing
   8.2. Pre-trial documents
   8.3. Trial designs
9. Stage 7: Launch — Commercially Available Product
10. Stage 8: Post-Market — Ongoing Processes Following Commercialisation of Product
11. Regulations, Guidelines and Codes of Practice

References

1. Introduction

Regulatory issues impact on the field of cell-based therapeutics not only at the late stages in a product life cycle, but also on the basic fundamental science being performed at the bench by researchers. The use of primary human cells immediately brings the scientist into areas that are closely regulated, for example tissue procurement, tracking and disposal, even before there may be any concept of a product. Thus, a basic science researcher in cell-based therapeutics, unlike in some fields of research, needs to have an awareness of the regulations which govern the use of donated cells and potential cell products. As there are likely to be circumstances in which the regulatory support for such workers is probably less than in a large industrial group, it is interesting to postulate how much regulatory awareness some of the less experienced scientists, or those from smaller groups have in this field. The difficulties for the researcher are compounded in some respects by the regulatory bodies as they attempt to develop and put in place the regulations for cell-based therapeutics. The plethora of new regulations and guidelines, consultation documents and presentations mean that the researcher, who is not directly involved in the regulatory field, may well find it a challenge to keep up to date.

In 2006 we were co-authors of a Publicly Available Specification No. 83 (PAS 83[1]); a concise guidance document for the UK published by the British Standards Institute (BSI) and commissioned by the Department of Trade and Industry. This document was written as a response to the views expressed by those within the UK regenerative medicine industry who felt that in the UK there needed to be greater overall clarity and structure to the currently available legislation, guidance and standards documents that are used in deriving a cell-based therapeutic product. It was also felt that there is a need for ongoing standardisation (in cohesion with the regulatory framework) in technical areas as they become more mature.

We, together with the assistance of a steering group (for participants see PAS 83), wrote the document from the perspective of UK research scientists who required a concise, straightforward, "at the bench or desk" quick reference guide. The PAS contains information on the key steps (as a process map) in the life cycle of a cell-based therapy product, starting with the procurement of the cells/tissue through to commercialisation and post-launch activities. Over-arching the process map we identified the key regulations, codes of practice and standards which are applicable within the UK and that should be used during the product life cycle.

The information in this chapter is based on the definition of a cell-based therapeutic as: "a product in which human somatic cells are administered to the body, for the purpose of replacing, repairing, regenerating, or enhancing function of tissues". The regulatory documents to which we refer are for the UK scenario; however, we aim to raise topics which are more widely applicable. Many of the key points are contained in the PAS document, but we have added our own thoughts and concerns, in order to prompt the reader to consider regulatory issues as they develop their particular cell-based therapeutics.

2. The Product Life Cycle of a Cell-Based Therapeutic

Figure 1 defines a generic map of the process of development of a cell-based therapeutic and condenses the product life cycle into eight stages; beginning with procurement of the tissue and ending with the work that takes place once the product is marketed. Within each stage it is anticipated that there are likely to be cyclical, iterative processes for the purposes of optimisation and

[1] PAS 83 Guidance on codes of practice, standardised methods and regulations for cell-based therapeutics is available from BSI British Standards who can be contacted at www.bsi-global.com.

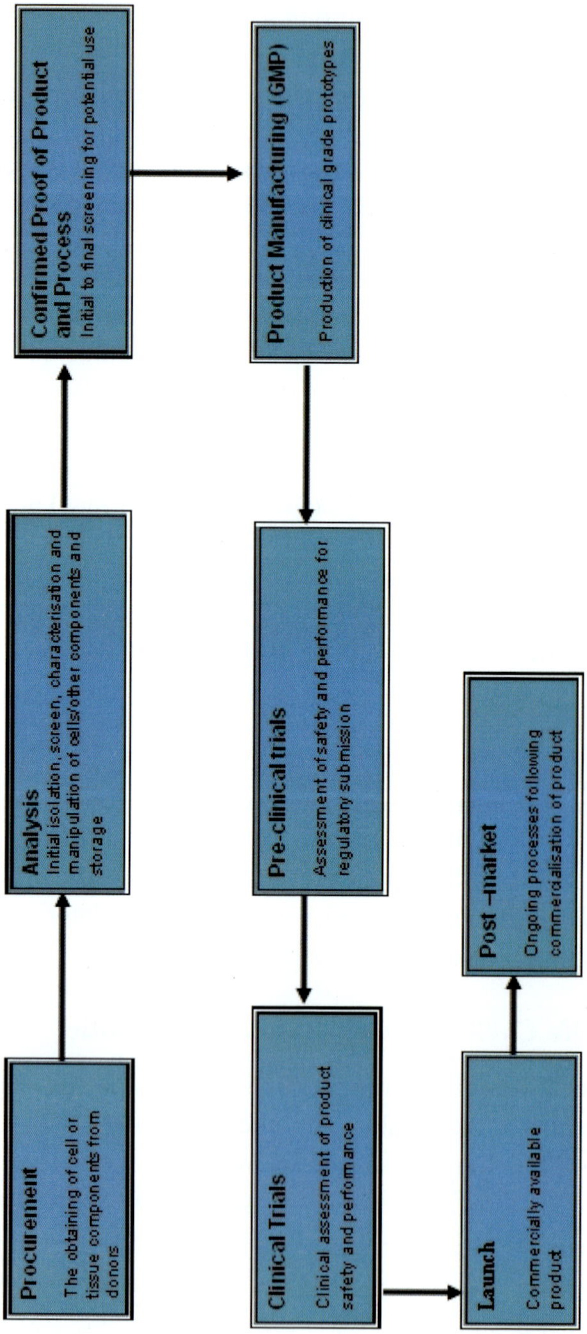

Fig. 1. The eight stages of the product life cycle of a cell-based therapeutic.

also that there may be additional information that can be added to this core diagram.

While key UK regulatory information is discussed where it is relevant to the process stages, it is important to note that the stages of the product life cycle from Stage 3 (confirmed proof of product and process) onwards will be subject to a new regulation from December 2008 on advanced therapy medicinal products (EC 1394/2007), published in December 2007. This is an amendment of EU Directive 2001/83/EC, and has been specifically written in order to bridge a regulatory gap in Europe which developed as the new biotechnology area of tissue engineering emerged. The content of the amendment has been widely debated across Europe, with the key issues being embryonic stem cells, and the relationship between advanced therapy medicinal products and medical devices. The regulation defines an advanced therapy medicinal product, and lays down principles for the evaluation and authorisation of the products, marketing authorisation procedures and post-authorisation vigilance.

EC 1394/2007 presents advanced therapy products as biological medicinal products within the meaning of annex 1 to Directive 2001/83/EC as they are:

— *having properties for treating or preventing disease in human beings;*
— *or may be used in or administered to human beings with a view to restoring, correcting of modifying physiological functions by exerting principally a pharmacological, immunological or metabolic action.*

The regulations go on to give a legal definition specifically for tissue engineered products in Chapter 1, Article 2 of the document.

"tissue engineered product" means a product that:

— *contains or consists of engineered cells or tissues and*
— *is presented as having properties for, or is used and administered to a human being with a view to regenerating, repairing or replacing a human tissue.*

A tissue engineered product may contain cells or tissues of human or animal origin or both. The cells or tissues may be viable or non-viable. It may also contain additional substances, such as cellular products, bio-molecules, biomaterials, chemical substances, scaffolds or matrices.

The sections go on to describe a number of additional conditions including an explanation of what is meant by "engineered".

Many of the Articles described in EC 1394/2007 are highlighted below in our descriptions of each of the stages of the product life cycle where we have summarised the key points and identified additional UK regulations.

3. Stage 1: Procurement — The Obtaining of Cells or Tissue Components from Donors Under cGCP

The primary stage in the development of a cell-based therapeutic is that of the procurement of cells or tissue components from donors. As this is the initial stage in the therapeutic product process cycle, it is critical that the appropriate procedures are followed. Two important considerations at this stage are: (i) donation: that the donors are appropriately screened and selected in order to obtain material of the specification required for further development, and also that the consent and rights of the donor are appropriately respected and maintained during the donation process; (ii) post-donation: that all systems are in place for the manipulation of the samples. In order to ensure that appropriate measures are adopted, the EU Tissue and Cells Directive 2004 (2004/23/EC) requires that all aspects of Good Clinical Practice (GCP) be followed where they relate to donation, including: ethics and full protocols, inclusion/exclusion criteria, mandatory screening and patient history. Specifically in the UK, Directive 2004/23/EC has been enacted by the Human Tissue Act (2004). Under this legislation, the Human Tissue Authority acts as the regulator and has issued eight codes of practice to inform researchers of the necessary requirements for various aspects of this stage from consent through to the import and export of material. In the case of embryonic stem cell donation there are further requirements, with this area being covered by the Human Fertilisation and Embryology (HFE) Act (1990) and more specifically the HFE (Research Purposes) Regulations, 2001.

Patient consent must be explicit in order for an informed choice to be made. Individuals are required to be given both a clear explanation of the future potential use of the cells harvested and also the nature of any analytical tests that are likely to be performed. The recently developed Human Tissue Authority Codes of Practice (2006) contain detailed guidance for users on both patient consent and donation. A procedure for the notification of adverse events and reactions related to procurement should be in place and a robust quality system must be in operation at all facilities where donation procedures are carried out.

Post-donation procedures are used to enable proper transfer of the material into the next stages of the process and should include: manipulation/handling methodologies for donated material; adequate storage facilities for material following removal; appropriate labelling to enable traceability; suitable transportation methods to enable transfer of the material to other establishments; systems for dealing with the disposal of discarded or "waste" tissue and for tracking donated samples. Although the intended final use of the material or potential "product" may not be well defined it is important that appropriate traceability and record keeping is in place, especially if the material may initially be used in a research or academic environment but then take on a more "commercial" aspect later in the process, potentially via a third party.

The origin of the cell population should be defined as autologous or allogeneic and its tissue type confirmed by appropriate analysis (see Analysis stage). For products intended for autologous application it is important that the identity chain of the patient is clearly maintained with an appropriate traceability system to avoid sample mix up. By contrast, for materials intended for allogeneic use, an anonymised link to the donor is appropriate. In both cases, at tissue sourcing, quality criteria should be the same for microbiological/virological safety of the material and the serological status of the patient. In the case of cells destined for autologous use, these may be positive for transmissible agents but will still be useable as they are going to be administered to the same individual. Therefore, care must be taken to avoid cross-contamination and protect staff handling the material.

4. Stage 2: Analysis — Initial Isolation, Screening, Characterisation and Manipulation of Cells/Other Components and Storage

Once tissue or cells have been donated, initial analysis and characterisation of the material is required. This is essential to identify, define and monitor the performance characteristics of the tissues or cells. It is important that the origin and homogeneity of the material are tested to avoid cross-contamination and maintain quality procedures going forward. Such processes should be considered both when deriving a new cell line, and when obtaining cells and/or tissues from external suppliers to confirm assumptions about the characteristics and nature of the material being used. Progressing into the next stages of the process pathway, it is important to incorporate the characterisation methods into any cell quality monitoring regime by regular re-checks throughout the production process. This will ensure that the integrity of the product material is maintained and also that the relevant information is traceable and passed on to all parties involved in the process pathway. This will be of particular importance in ensuring data integrity if a small research group is producing or researching a potential product that they hope to develop with or pass on to a commercial developer or partner for the latter stages of development.

Relevant information and guidance for to the researcher is contained in a number of Directives. The EU Tissue and Cells Directive 2004 (2004/23/EC) used for donation information (see above) also describes information relevant to testing and this directive is supported by 2006/17/EC which describes selection and testing criteria. In addition, the OECD principles of Good Laboratory Practice (GLP) regulated in the EU as EU 99/11/EEC and EU99/12/EEC and transposed in the UK as SI 3106 and SI 994 describe the quality regime that should be adopted. If embryonic stem cells are being studied at this stage, two other pieces of legislation that apply are the Human Fertilisation and Embryology (HFE) Act (1990) and the HFE (Research Purposes) Regulations, 2001. In relation to the last two pieces of legislation it is also important to note that it is a condition of an HFEA licence that a sample of all human embryonic stem cell lines derived in the UK is deposited in the UK Stem Cell Bank.

Key to this stage is the high level of control and monitoring of the conditions in which the cells are held. Cross-contamination, and the absence of any microbial or viral contamination will need to be verified by appropriate testing procedures. Culture conditions should be tracked and monitored to ensure intended performance and quality and cells or tissues handled under controlled conditions to avoid cross-contamination. It is important, at this stage, that master and working cell banks are established. This enables proper storage, retrieval and supplies going forward, while preserving the functionality and integrity of the cells or tissues under development and maintains an archive of material.

The specific tests to be performed by the researcher are not currently specified in any documentation and will, to some extent, be determined by the material and process specifics. However, the following should be considered: microbiology; cell viability and yield; genotypic/phenotypic stability and potential epigenetic changes induced by the culture environment. Several of the guidelines and codes of practice (see Table 2) contain suggestions on appropriate regimes to be adopted by researchers.

5. Stage 3: Confirmed Proof of Product and Process — Initial to Final Screening for Potential Use

Once the cell line or tissue characteristics have been established these have to be translated and potentially assembled into a concept product. It is at this stage that other components of the concept product may be brought into the equation, potentially adding a further tier of complexity and developmental effort to the analysis required by the researcher. The purity, sterility and absence of contaminating agents in additional components should be ensured either by testing or by assurance from the supplier or source. Examples of the additional products that may be used, along with the cells, to form a product, include such items as matrix components and scaffolds or other supports. At this stage, as well as bringing the required materials together, the methodology used for the collection, selection and manipulation of cells will be further defined and should be documented in detail. This information will then form the basis of defined Standard Operating Procedures (SOPs) to enable repeated testing to be carried out under standardised conditions. It is at this point that the process essentially becomes "locked" and starts to move away from the "basic science" of bench-based research into "production". The OECD principles of GLP discussed in the previous section are also used at this stage in combination with directive 2003/94/EC laying down the principles and guidelines of Good Manufacturing Practice (GMP) in respect of medicinal products for human use and EC 1394/2007, the amendment to 2001/83/EC Medicinal Products for Human Use.

As a consequence of this change in emphasis of the research at this stage, business planning and cost/benefit studies should be designed and conducted to demonstrate proof of concept of the product and the economic rationale for proceeding. This stage of the process and the analysis conducted bridge the research side of the product cycle with further development, and are critical in helping to determine whether a potential product has enough merit to warrant proceeding into the most expensive stages of the pathway, or whether research and development efforts should be terminated at this stage. In drafting PAS 83, a common theme from industrial commentators was that many investigators fail to consider this aspect or to carry out enough evaluation when developing a potentially promising cell line or concept product.

This evaluation should consider such aspects as stability, reproducibility, potency and dose definition. It is important to note that the definition of dose may be open to some interpretation, dependent upon the product in question. This is because, in contrast to traditional medicines, and by their very nature, cell-based therapies could have their dose defined by such criteria as the number of cells in a product or their biological activity. The new Committee for Human Medicinal Products (CHMP) Guideline on Human Cell-Based Medicinal Products (2007) contains suggestions on how to assess such characteristics as dosage and potency for cell-based therapies.

An evaluation of increased production, either via scale-out or scale-up for further stages of the process will also be needed and production conditions should be optimised with respect to the intended final function of the cells. Evaluation will require an analysis to be conducted that identifies the basic requirements and the impact on the final cost of a product of various items such as raw materials and production equipment. This will need to be looked at for each of the various process steps that are likely to be required for a particular product, going right through to quality assurance/quality control procedures for the finished product and the requirements for post-market monitoring.

6. Stage 4: Product Manufacturing — Production of Clinical Grade Material Under cGMP

The aim of this stage of the process is to conduct pilot scale manufacture in order to produce sufficient quantities of clinical grade material to be taken forward into pre-clinical and clinical trials (next stages). It is essentially a "tightening up" of the outputs of the previous stage, in that prototypes developed will now need to be repeatedly produced for use in clinical trials to an established level of quality and reproducibility. Simultaneously, accompanying SOPs and other procedures will need to be in place. As cell-based therapeutic products are to be regulated as medicinal products within the EU (see following process steps and further accompanying discussion on the Advanced Therapy Medicinal Product regulations) they fall under Directive 2003/94/EC which requires that all of the manipulation steps for manufacture, including the equipment used, be validated to GMP rules. These cover such aspects as microbiological control, cell viability, growth, identity, activity or function, purity and the procedures and containers for shipment.

The protocols and success criteria that have to be developed will be specific to the particular product and proposed final clinical application and therefore have to be defined by the researcher/manufacturer. However, for all products these should include batch testing of cell characteristics to investigate the effects of scale-out or scale-up procedures and that of longer term production on the stability and efficacy of the prototype product. At this stage, criteria for product release will need to be generated and appropriate tests to monitor these selected or defined. These will also be very product and application specific and will be dependent both on the outcomes of the earlier stage characterisation of the product and definition by the researcher/manufacturer.

7. Stage 5: Pre-Clinical Trials — Assessment of Safety and Performance for Regulatory Submission

In the EC 1394/2007 amendment to Directive 2001/83/EC the European Council agreed that advanced therapy medicinal products should be subject to the same regulatory principles as other types of biotechnology medicinal products. However the technical requirements in particular the type and amount of quality, pre-clinical and clinical data necessary to demonstrate their quality, safety and efficacy may be highly specific. While the new amendment will eventually have guidelines pertinent to advanced therapy medicinal products, in addition to those in 2001/83/EC, researchers still need to debate some of the fundamental issues concerning the design and criteria for pre-clinical trials.

Pre-clinical *in vivo* studies can provide data which is indicative of efficacy and/or safety of a cell-based product or therapy. However, there are important issues with *in vivo* assessments which the researcher should be aware of. The first is the difference between an *in vivo* study assessing proof of principle and efficacy, and a study designed to investigate product safety, and how this fits into the regulatory framework, and secondly and perhaps more crucially, the understanding both of the limitations of animal models, and the need for robust success criteria and study designs.

In vivo studies are tightly regulated in the UK. Directives, Regulations and Acts which are applicable to pre-clinical testing are those relating to the requirements for animal testing; 86/609/EEC transposed in UK to Animals (Scientific Procedures) Act 1986 and the OECD principles of GLP (regulated in EU as EU 99/11/EEC and EU99/12/EEC and transposed in UK as SI 3106 and SI 994). Relevant information is also contained in 2003/94/EC laying down the principles and guidelines of GMP in respect of medicinal products for human use and in 2001/83/EC Medicinal Products for Human Use.

7.1. Regulatory studies

In the UK a non-clinical "regulatory study" is defined as:

(a) A study in which an item is examined under laboratory conditions or in the environment in order to obtain data on its properties or its safety (or both) with respect to human health.
(b) The results of which are intended for submission to the appropriate regulatory authorities.
(c) An experiment which is compliant with the principles of GLP (whether or not compliance with GLP is also a legislative requirement).

In vivo studies which are an early stage proof of principle and/or efficacy are likely to be performed during the Research and Development stage of the product life cycle. Such studies are not required to be GLP compliant. However, this lack of compliance means that the data generated at this stage will be deemed to be from a non-regulatory study, and therefore cannot be submitted as part of a regulatory package (although it may be included as an additional information annex).

With a cell-based therapy it is probably worth a producer suggesting that, if possible, a final prototype used for *in vivo* testing is prepared from clinical grade material and that any cells are taken from substantially the same cell bank as the prototypes planned for clinical trials. Designed in this way, the pre-clinical material will be essentially the same as that used in the clinical trials.

Traditionally, GLP regulatory *in vivo* safety studies have tended to be toxicological in nature and/or determine biocompatibility. For example, studies of efficacy and toxicology of pharmaceuticals can be designed as separate experiments with different outcomes as the pharmaceutical is not often designed to be very long lasting but requires long-term toxicological assessment. As cell-based therapeutics are designed to remain within the human body for considerable lengths of time or indeed indefinitely, the design of safety and efficacy studies may need to be different from those traditionally used, and combine testing of efficacy and toxicology within the same study.

7.2. *In vivo* pre-clinical models

It is widely accepted that there are limitations to the use of animal models, due to their suitability to assess efficacy or compatibility, or to predict the clinical response in humans. Animal models are just that; models that can give some indication of how a potential product may behave in an *in vivo* environment. Many examples exist where a model has some deficiency, either the clinical condition cannot be well mimicked (for example chronic wounds, osteoarthritis), or the animal tissues have some difference to the human counterpart. Additionally, there is often difficulty with the design of control groups, where either the ailment under investigation does not proceed in a manner comparable to that seen in humans, or the response to "gold standard" treatment is likewise difficult to compare. Compounding these difficulties is the issue of xenogenicity when implanting a cell based therapy which may also need to be overcome. These many issues make it necessary for the researcher to put significant thought into study design and outcomes, and there should also be some consideration of how valuable the data obtained from large efficacy *in vivo* studies will actually be, and a weighing up of the cost benefits as part of the justification that is required within the UK regulatory framework for performing *in vivo* studies.

No technical guidelines currently exist that describe pre-clinical testing for cell-based therapies. Tests are described for medicinal products (see 2001/83/EC Medicinal Products for Human Use Annex 1 which describes pre-clinical testing for medicinal products).

As the testing of cell-based therapies is relatively new, the requirements for pre-clinical testing are, quite rightly, judged by the regulatory authorities on a product-by-product basis. An additional factor is that regulatory bodies can and do differ in their requirements for pre-clinical testing and it should not be assumed that requirements stipulated by one regulatory body will match those of another.

It is also good to note that the EMEA are aiming to assist small and medium-sized enterprises (SMEs) by making available a system of early evaluation and certification of quality and non-clinical safety data by the Agency, independent of any marketing authorisation application. This is designed to assist SMEs that focus on the early development aspects, but do not conduct the subsequent clinical trials themselves. The certification of "early-development" data by the Agency should provide an important selling argument to those companies who wish to license out their technology to bigger undertakings.

8. Stage 6: Clinical Trials — Clinical Assessments of Product Safety and Performance

Clinical trials are often some of the final stages of product development, where the products are tested in clinical experiments on humans. Trials are most commonly split into three phases — I, II and III (although some combining of phases can be seen) and the following definitions can be applied:

Phase I — Effects of treatment — *basic information on safety, adverse and beneficial effects. Patient group — healthy volunteers/patients whose illness/disease has still progressed after other treatments.*

Phase II — Efficacy of treatment — *does the product (or range of formulations) produce therapeutic effects and additional information on side-effects and how they could be managed? Patient group — patients with illness/disease who are exhibiting clinically relevant symptoms.*

Phase III — Efficacy of treatment — *may be compared to established best practice therapies — longer-term effects and outcomes. Patient cohort — patients with illness/disease who are exhibiting clinically relevant symptoms.*

Clinical trials are extensively regulated with the main Directives being the 2001/20/EC (Clinical Trials Directive) and 2005/28/EC (Laying down principles and detailed guidelines for GCP as regards investigational medicinal products for human use), as well as the requirements for authorisation of the manufacturing or importation of such products. Information is also contained in 2001/83/EC and an update described in EC 1394/2007 states that clinical trials should be conducted in accordance with the over-arching principles and the ethical requirements laid down in the 2001/20/EC Clinical Trials Directive. In addition detailed guidelines on GCP specific to advanced therapy medicinal products will be drawn up by the

European Commission. 2001/20/EC was transposed in the UK as SI2004/1031 Medicines for Human Use (clinical trials) Regulations 2004.

Authorisation for a clinical trial in the UK is gained from the UK competent authority which is the Medicines and Healthcare products Regulatory Agency (MHRA). Some potential cell-based products may not easily fall within the new regulations and grey areas still exist with, for example, the medical devices directives. Therefore, early contact with the MHRA is very important for the industry to verify where a product will sit with respect to current legislation. In order to address any anomalies it will also be possible for applicants to request a scientific recommendation from the EMEA on the classification of any product based on cells and tissues with a view to resolving borderline issues.

8.1. Manufacturing

Cell-based products intended for human use in clinical trials will need to be manufactured in compliance with 2003/94/EC laying down the principles and guidelines of GMP in respect of medicinal products for human use. Additional information is given in 2001/83/EC and 2001/20/EC. EC 1394/2007 also states that guidelines in line with the principles of GMP and specific to advanced medicinal products will be drawn up by the European Commission.

A manufacturing authorisation licence is required for the production of a cell-based therapy where it is classed as an Investigational Medicinal Product (IMP) and includes all components used in the manufacture of the product. Products should be manufactured to GMP and therefore require a number of processes in place, for example, for storage, transport and labelling. The manufacturing process will require suitable scaling for production of trial products and completed process validation within the timescale of phase I and II clinical trials. Release of products for trials will involve ongoing QC checks.

8.2. Pre-trial documents

The MHRA requires that clinical trial authorisation (CTA) be obtained for all clinical trials, including those on healthy volunteers prior to commencing the study. Cell-based therapeutics studies would currently fall within the definition of "medicinal products with special characteristics". It is a requirement that before submitting a CTA form to MHRA a EudraCT number be obtained from the European Medicines Agency (EMEA), where ethical approvals procedures are laid down.

8.3. Trial designs

GCP is necessary for all clinical trials. Clearly defined clinical end point measures and success criteria should be established, and statistical considerations made for phase II and III trials. This is especially pertinent when establishing the efficacy of a cell-based therapeutic, as the gold standard treatment used to compare treatment against may well be via a different mode of action, or may be measurable over a different time-course. Additionally, the cell-based therapeutic may remain *in situ* indefinitely, potentially requiring long-term follow-up or investigation of possible cell migration to other sites and also long-term follow-up for any possible side effects. Prior to implantation there will need to be considerable consideration regarding transport, storage and manipulation of a cell-based product in the operating theatre. Looking through this short list indicates the difficulties and hurdles to overcome in performing a clinical trial on such products.

Special attention should be paid to informed patient consent. There are many additional items to be taken into account which include, but are not limited to: patient awareness of the outcome of phase I trials — safety and effect but not necessarily direct therapeutic benefit; a need to track patients long-term post-treatment (a robust secure database is required); indefinite link of donor and recipients and to manufacturer-company records.

9. Stage 7: Launch — Commercially Available Product

Products require marketing authorisation prior to their being commercially available. The regulations governing marketing authorisation are contained in 2001/83/EC Medicinal Products for Human Use but more specific information is contained within the Advanced Therapy Medicinal Products Directive amendment EC 1394/2007.

The procedure for the marketing authorisation of cell-based therapeutics is also outlined within this amendment. This principle is already established for gene therapy medicinal products and somatic cell therapy medicinal products resulting from any biotechnology process referred to in Regulation (EC) 726/2004 and it is proposed that the same principles be applied to all advanced therapy medicinal products. The scientific evaluation of such products would also be carried out by Member states experts within the network coordinated by the European Medicines Agency (EMEA). Within the EMEA, a Committee for Advanced Therapies (CAT) will be set up within the Committee for Medicinal Products for Human Use (CHMP).

Packaging, distribution, wholesale selling, advertising and promotion of medicines in the UK are governed by a number of regulations and this is summarised in The Blue Guide, 2005. However, the new amendment to 2001/83/EC Articles 11 to 14 contains information on product characteristics, packaging and leaflets which will also need to be adhered to, including the use of unique donation and product codes and unique patient identifiers for autologous products.

10. Stage 8: Post-Market — Ongoing Processes Following Commercialisation of Product

The essential components of the processes following commercialisation are: (i) patient follow-up, monitoring and pharmacovigilance, (ii) complete traceability, (iii) risk management and (iv) the systems for the acquisition and storage of this data.

The very definition of a cell-based therapeutic is suggestive that, once implanted, it is likely to remain in the human body for a long period of time. Therefore, monitoring patients for long-term follow-up is essential. Additional Pharmacovigilance data collected in parallel will cover such items as adverse reactions and Periodic Safety Update Reports and also any post-authorisation safety studies.

Traceability of the both the patient and the product (together with starting materials and all substances coming into contact with the tissues or cells it may contain) and the relationship between the two are vital to monitor the safety and efficacy of the product from a long-term perspective.

Data obtained from follow-up, monitoring and traceability should be collected as directed in a prewritten and agreed protocol. A data storage system will be needed, together with appropriate back-up systems and should be compatible with requirements laid down in Directive 2004/23/EC with regards to donation, procurement and testing, and include aspects of data protection, confidentiality and the anonymity of both donor and recipient. A suitable risk management system over-arching the data collection will also be necessary. Data should be kept for a minimum of 30 years after placing the product on the market or longer if required as a term of the marketing authorisation

There should be a continual QA/QC aspect of manufacture related to the ongoing GMP for product production and a renewal of Marketing Authorisation will be required at a time given by the regulating authority.

This information is also outlined in a number of Directives, Regulations and Acts. The main directives are 2001/83/EC Medicinal Products for Human Use together with the amendment EC 1394/2007 on advanced therapy medicinal products and regulation (EC) 726/2004 concerning the centralised marketing authorisation procedure. SI 1994 3144 The Medicines for Human Use (Marketing authorisation) Regulation 1994 Amended by SI 2005 2759. Useful Guidelines and

Table 1. Illustration of which key Directives and Acts are applicable at each stage of the process map.

Directives and Acts	Stages of development of the process map							
	Procurement	Analysis	Proof of product and process	Product Manufacturing	Pre-clinical	Clinical	Launch	Post Market
EU 2004/23/EC Tissue Directive	■	■						
EU 2006/17/EC implementing 2004/23/EC	■	■						
UK Human Tissue Act 2004	■							
UK Human Fertilisation and Embryology (HFE) Act 1990 and HFE (Research) Regulations 2001	■	■						
Animals (Scientific Procedures) Act 1986					■			
OECD Principles of Good Laboratory Practice (GLP)		■	■		■			
UK SI3106 1999 GLP Regulations		■	■		■			
EU 2003/94/EC GMP for Medicinal products			■	■	■	■	■	■
EU 2001/20/EC Clinical trials directive						■		
EU 2005/28/EC GCP for Medicinal products						■		
UK SI 1031 2004 Medicines for Human Use Clinical trials Regulations						■		
EU 2001/83/EC Medicinal Products for Human use. Includes 2003/63/EC, 2004/27/EC			■	■	■	■	■	■
EC 1394/2007 on advanced therapy medicinal products amending 2001/83/EC			■	■	■	■	■	■

Source: Adapted from Figure 1 of PAS 83.

Table 2. Illustration of which key Guidance documents are applicable at each stage of the process map.

GUIDANCE and CODES OF PRACTICE	Stages of process map							
	1	2	3	4	5	6	7	8
Code of Practice for the Use of Human Stem Cell lines 2005 (UK Stem Cell Bank)	■	■	■					
MHRA Codes of Practice Production of Human-Derived Therapeutic products 2002, Tissue Banks 2001, and Guidance on the Microbiological safety of Human Organs, Tissues and cells used in transplantation 2000	■	■	■	■	■			
Committee for Proprietary Medicinal Products (CPMP) points to consider on the manufacture and quality control of human somatic cell therapy medicinal products (2001). Replaced in 2007 by a (CHMP) Guideline on Human Cell-Based Medicinal Products	■	■	■	■				
ICH E6 Guidance for industry -Good Clinical Practice (GCP) Consolidation Guidance (2002)	■					■		
Safe Disposal of Clinical Waste (HSE, 1999)	■	■	■					
Human Tissue Authority Codes of Practice 1-8 (2006-2007)	■							
Application guidance for a licence to store human tissue intended for human application 2006 (HTA)	■							
ECVAM Guidance on Good Cell Culture Practice (GCCP) (2003, 2005)		■						
Guide to UK GLP Regulations (1999)			■		■			
Rules and Guidance for Pharmaceutical Manufacturers and Distributors "The Orange Guide" 2002 (MHRA)			■	■	■	■	■	■
Eudralex. The Rules Governing Medicinal Products in the European Union Volume 4 EU Guidelines to Good Manufacturing Practice Medicinal Products for Human and Veterinary Use (2005)			■	■				
Guidance on the Operation of the Animal (SP) Act 1986					■			
Description of SI2004/1031 Medicines for human use (clinical trials) regulations 2004 (MHRA)						■		
European commission EudraCT - Guidance documents on Clinical trial application, Ethics, Adverse event reporting						■		
MHRA 5 Notes for applicants and holders of a manufacturer's licence						■	■	■
Eudralex Volume 2. Pharmaceutical legislation: notice to Applicants – Procedures for Marketing authorization Volumes 2A, 2B 2C							■	■
MHRA 2005 "The Blue Guide" Advertising and Promotion of Medicines in the UK							■	■

Stages; 1 = Procurement, 2 = Analysis, 3 = Proof of product and process, 4 = Product manufacturing, 5 = Pre-clinical, 6 = Clinical, 7 = Launch, and 8 = Post-Market.

codes of practice are volumes 2 (A, B and C) and 9 of the Eudralex rules governing Medicinal Products in the European Union

11. Regulations, Guidelines and Codes of Practice

Key relevant regulations have been referred to in the stages of the product life cycle described in the previous sections but in order to provide the researcher with an easy guide to the documents, two tables have been produced showing the key documents in relation to the process cycle stages. Table 1 illustrates which Directives and acts are used at which stage and Table 2 demonstrates the guidance and codes of practice documents that are available to support the researcher with additional information that often translates the directives into more readable prose.

References

European legislation

86/609/EC: The approximation of laws, regulations and administrative provisions of the Member States regarding the protection of animals used for experimental and other scientific purposes.

99/11/EC and 99/12/EC: Adopting the OECD principles of Good Laboratory Practice (GLP).

2001/20/EC: The approximation of the laws, regulations and administrative provisions of the Member States relating to the implementation of good clinical practice (GCP) in the conduct of clinical trials on medicinal products for human use.

2001/83/EC: The community code relating to medicinal products for human use. Includes 2003/63/EC, 2004/27/EC and draft Advanced Therapy Medicinal Products Directive.

2002/98/EC: Setting standards of quality and safety for the collection, testing, processing, storage and distribution of human blood and blood components and amending Directive 2001/83/EC.

2003/94/EC: Laying down the principles and guidelines of Good Manufacturing Practice (GMP) in respect of medicinal products for human use.

2004/23/EC: Setting standards of quality and safety for the donation, procurement, testing, processing, preservation, storage and distribution of human tissues and cells.

2005/28/EC: Laying down principles and detailed guidelines for good clinical practice (GCP) as regards investigational medicinal products for human use,

as well as the requirements for authorisation of the manufacturing or importation of such products.
EC 1394/2007: On advanced therapy medicinal products and amending Directive 2001/83/EC and Regulation (EC) No 726/2004. UK Legislation.
Animals (Scientific Procedures) Act, HMSO, 1986.
Data Protection Act, HMSO, 1998.
Good Laboratory Practice Regulations 1999 (SI 3106).
Good Laboratory Practice (Codification Amendments Etc.), Regulations 2004 (SI 994).
Human Tissue Act, HMSO, 2004.
Human Fertilisation and Embryology (HFE) Act, HMSO, 1990.
HFE (Research Purposes) Regulations, HMSO, 2001.
Medicines for Human Use (Marketing Authorisation) Regulation 1994 (SI 1994 3144) (Amended by SI 2005 2759).
Medicines for Human Use (Clinical Trials) Regulations 2004 (SI 2004/1031).
Safe Disposal of Clinical Waste, HSE, 1999.

Guidelines and codes of practice

Advertising and Promotion of Medicines in the UK "The Blue Guide", 2005 (MHRA).
Application Guidance for a Licence to Store Human Tissue Intended for Human Application, 2006 (HTA).
Code of Practice for the Production of Human-Derived Therapeutic Products, 2002 (MHRA).
Code of Practice for Tissue Banks, 2001 (MHRA).
Code of Practice for the Use of Human Stem Cell Lines, 2005 (UK Stem Cell Bank).
Description of SI2004/1031 Medicines for Human Use (clinical trials) Regulations 2004 (MHRA).
Eudralex: Volume 2. Pharmaceutical Legislation: Notice to Applicants — Procedures for Marketing Authorisation Volumes 2A, 2B 2C, 2005.
Eudralex: Volume 4. Medicinal Products for Human and Veterinary Use: Good Manufacturing Practice, 2005.
Guidance for Blood Transfusion Services in the UK, 2001 (MHRA).
Guidance on the Operation of the Animal (SP) Act, 1986.
Guidance on the Microbiological Safety of Human Organs, Tissues and Cells Used in Transplantation, 2000 (MHRA).
Guide to UK GLP Regulations, 1999 (DH).

Human Tissue Authority Codes of Practice 1–8, 2006–2007 (HTA).
ICH E6 Guidance for Industry — Good Clinical Practice (GCP) Consolidation Guidance (2002) FDA.
MHRA Guidance Note 5, "Notes for applicants and holders of a manufacturer's licence", 2004 (MHRA).
Points to consider on the manufacture and quality control of human somatic cell therapy medicinal products. Committee for Proprietary Medicinal Products, 2001 (CPMP).
Rules and Guidance for Pharmaceutical Manufacturers and Distributors "The Orange Guide", 2002 (MHRA).

PART VII

TISSUE REPAIR

Chapter 27

Stem Cell Therapy: Past, Present, and Future

Frédéric Baron and Rainer Storb

Abstract

The observations that mice exposed to otherwise lethal irradiation could survive if their spleens or marrows were shielded, or if they received an infusion of bone marrow, led to the first attemps of bone marrow transplantation in humans in the mid-1950s by E. D. Thomas and J. Ferrebee. Thanks to Thomas' persistence despite criticism and initial clinical failures, and thanks to the development of a canine model of bone marrow transplantation by Thomas and Storb, the role of allogeneic hematopoietic cell transplantation (HCT) changed during the last 50 years from a desperate therapeutic maneouver plagued by apparently insurmountable complications to a curative treatment modality for thousands of patients with hematologic diseases. Further, it was recognized that allogeneic immunocompetent cells contained in the graft mediated therapeutic antitumor effects independent of the action of the high-dose therapy. These were termed graft-versus-tumor (GVT) effects. This prompted the recent development of nonmyeloablative conditioning regimens for allogeneic HCT that have allowed offering this treatment modality in elderly patients and those with comorbid conditions. While hematopoietic stem cells were identified in the early 1960s, identification of other types of stem cells such as mesenchymal stem cells or embryonic stem cells might pave the way for stem cell therapy in regenerative medicine in the future.

Keywords: Transplantation; Hematopoietic Stem Cell; Hematopoietic Cell Transplatation; Allogeneic; Dog; Graft-versus-Host Disease; Graft-versus-Tumor Effects; Bone Marrow; Peripheral Blood Mononuclear Cells; Cord Blood; Mesenchymal Stem Cells; Plasticity; Duchenne Muscular Dystrophy.

Outline

1. Hematopoietic Stem Cells and Hematopoietic Stem Cell Transplantation
   1.1. Discovery of hematopoietic stem cells
   1.2. The beginnings of hematopoietic cell transplantation in humans
   1.3. Allogeneic HCT after high-dose conditioning
      1.3.1. *Non-malignant disease*
      1.3.2. *Hematological malignancies*
   1.4. Graft-versus-tumor effects
   1.5. Alternative donors
   1.6. Sources of hematopoietic stem cell for clinical use
   1.7. Allogeneic HCT after low-dose conditioning
2. Plasticity of Adult Hematopoietic Stem Cells: Lessons Learnt from a Canine Model of Duchenne Muscular Dystrophy
3. Mesenchymal Stromal Cells
4. Embryonic Stem Cells
5. Future of Stem Cell Therapy
   5.1. HCT: minimizing pain, maximizing gain
   5.2. Stem cells in regenerative medicine
   Acknowledgments
   References

1. Hematopoietic Stem Cells and Hematopoietic Stem Cell Transplantation

1.1. Discovery of hematopoietic stem cells

After World War II, given concerns about atomic warfare and newly available radioactive isotopes, scientists investigated the biological effects of irradiation.[1–3] It was quickly recognized that bone marrow was most sensitive to radiation, and death associated with low-lethal radiation exposure was due to marrow failure. Twenty-five years earlier, Fabricius-Moeller had observed that postirradiation hemorrhagic diathesis could be prevented by shielding the legs of guinea pigs during total body irradiation (TBI).[3] However, this important observation was largely forgotten until 1949, when Jacobson *et al.* reported that mice could survive otherwise lethal irradiation if their spleens (a hematopoietic organ in the mouse) or femora were protected with lead shields.[4] The field of hematopoietic cell transplantation (HCT) began with the landmark observations by Jacobson *et al.* and by Lorenz *et al.*, both in 1951, that protection against low lethal irradiation exposure

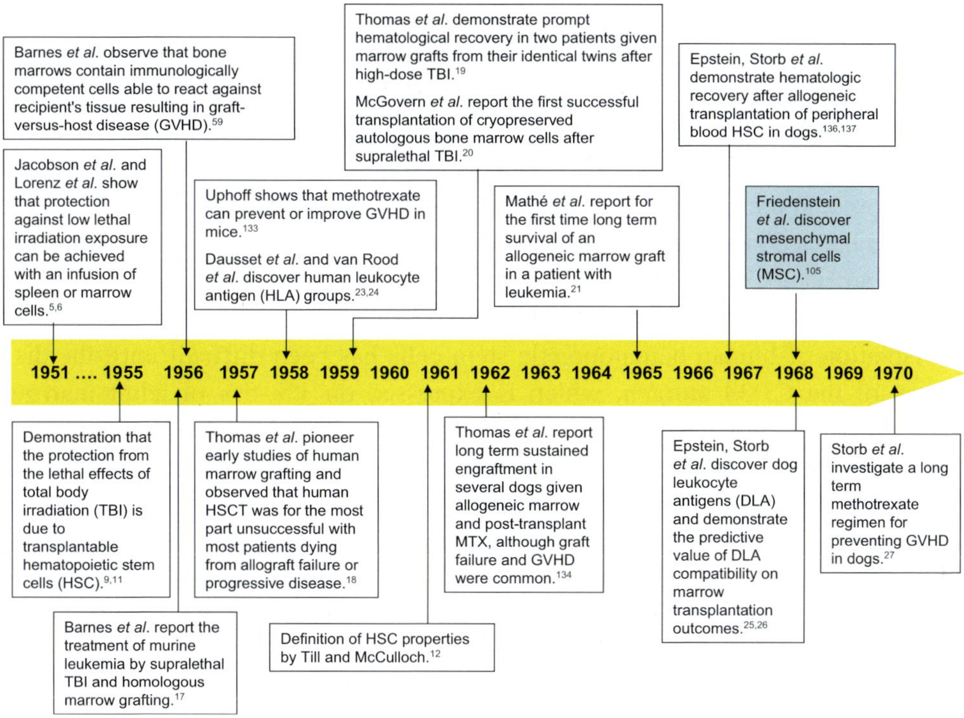

Fig. 1. Timeline. History of stem cell therapy: from 1951 to 1970.[5,6,9–12,17–21,23–27,59,105,133,134,136,137]

could also be achieved with infusions of spleen or marrow cells[5,6] (Fig. 1). Although controversial at the time, many investigators thought that the marrow protective effect was due to "a substance of non-cellular nature" in the spleen or bone marrow that stimulated recovery of marrow, and thus allowed mice to survive (humoral hypothesis), rather than to stem cells (cellular hypothesis) that migrated from the shielded tissue to the irradiated marrow spaces, or home to those spaces after marrow or spleen infusions.[7] However, in the mid-1950s, a number of reports using various histochemical[8] and genetic[9,10] markers demonstrated that the protection from the lethal effects of TBI was due to transplantable hematopoietic stem cells. Animals which carried a foreign hematopoietic system after allogeneic marrow transplantation were termed "chimera" by Ford and colleagues in 1956.[10] The demonstration that transplantation of a limited number of marrow cells allowed transplanting the entire lymphohematopoietic system from one person to another prompted investigations of identifying hematopoietic stem cells and factors that control their proliferation and differentiation.

Hematopoietic stem cells are defined by at least two properties. First, they can divide indefinitely producing a population of identical offspring. Second, they can undergo asymmetric division giving birth to two different daughters, one identical to the parent cell, and the other having more restricted developmental potential. The concepts of hematopoietic stem cells were defined by landmark studies by Till and McCulloch beginning in 1961.[11,12] By analyzing the number and nature of cells developing into spleen colonies, the authors demonstrated that each spleen colony came from a single precursor which was later termed stem cell, that each spleen colony contained erythroid, megakaryocytic, monocytic and granulocytic colonies, and that stem cells were able to self-renew.[11,12] Sixteen years later, Abramson *et al.* demonstrated that T- and B-lymphocytes also derived from stem cells.[13]

Isolation of human hematopoietic stem cells has been difficult until the discovery of the CD34 antigen, which is expressed on 1%–5% of adult marrow cells.[14] It is generally accepted that only a fraction of CD34+ cells represents true stem cells. Studies in humans and in dogs showed that transplantation of allogeneic CD34-selected cells after myeloablative conditioning resulted in prompt hematopoietic reconstitution with mixed or full donor chimerism.[15,16] The phenotype of human stem cells has been further characterized by the expression of not only CD34, but also c-kit, Thy-1, and CD133, being lineage negative, and lacking expression of CD38 and CD45.

1.2. The beginnings of hematopoietic cell transplantation in humans

The discovery of hematopoietic stem cells suggested the possibility of treating patients with life-threatening hematological diseases such as acquired lack of marrow function, e.g. severe aplastic anemia, inborn errors, e.g. hemoglobinopathies and immunodeficiency diseases, and hematological malignancies. In the latter case, the intensity of cytotoxic anticancer drugs could be increased beyond the range that was toxic to the bone marrow, potentially increasing drug efficacy. This hypothesis was supported by Barnes *et al.*, who reported in 1956 on the treatment of murine leukemia by supralethal TBI and homologous marrow grafting.[17]

In 1955, E. D. Thomas and colleagues pioneered early studies of human marrow grafting and observed in 1957 that human HCT was for the most part unsuccessful with only one patient engrafting transiently, and most patients dying from allograft failure or progressive disease.[18] However, in 1959, two landmark papers demonstrated the feasibility of HCT in humans. Thomas *et al.* described two patients with advanced leukemia who were given bone marrow from their identical twins after supralethal TBI.[19] Although both patients relapsed after a few months, their prompt hematologic recovery supported the principle of HCT in humans. In a second paper, McGovern *et al.* reported the first successful transplantation of cryopreserved autologous bone marrow cells after supralethal TBI in a patient with relapsed acute lymphoblastic leukemia.[20] Unfortunately, the leukemia recurred, possibly due to malignant leukemic clones harbored in the stored graft and/or a lack of graft-versus-leukemia effects (see below). In 1965, Mathé *et al.* were the first to report the "long term" survival of an allogeneic bone marrow graft in a patient with leukemia.[21] The patient experienced acute and chronic graft-versus-host disease (GVHD, a disease caused by the destruction of host tissue through donor immune cells contained in the graft) and eventually died from varicella encephalitis 20 months after transplantation, while still in complete remission.

In 1970, Bortin reviewed data from 203 human allogeneic human marrow grafts reported during the 1950s and 1960s.[22] Only three patients were alive, while most remaining patients failed to engraft, developed fatal GVHD, or relapsed. The feasibility of crossing the "allogeneic barrier" in humans was challenged, since it had become clear that graft-versus-host reactions in humans were incomparably more violent than in inbred rodents. While many researchers abandoned the concept that allogeneic HCT could be used to treat human patients, Thomas, Storb *et al.* in Seattle started conducting experiments using a canine model. The outbred nature and the wide genetic diversity of dogs made them a particularly suitable animal model for preclinical studies.[2] Indeed, most dogs given marrows from littermate after lethal TBI experienced the same complications as humans, including graft rejection, GVHD, and death from opportunistic infections, although a few dogs became healthy long-term survivors with hematopoietic cells of donor origin. Thomas and Storb hypothesized that appropriate donor selection was needed to prevent graft rejection and GVHD. Based on the discovery of the human leukocyte antigen (HLA) system by Dausset *et al.*[23] and van Rood *et al.*,[24] Epstein, Storb, and Thomas developed canine histocompatibility typing. In 1968, they demonstrated that dog leukocyte antigen (DLA) compatibility between donors and recipients improved remarkably the outcome of allogeneic HCT in dogs.[25,26] However, they also observed that GVHD occurred even after DLA-identical HCT because of minor histocompatibility differences between donors and recipients. They then developed drug regimens able to prevent GVHD, for example using the antimetabolite drug methotrexate alone or in combination with the T-cell activation inhibitors cyclosporine or tacrolimus,[27,28] allowing most dogs given marrows from DLA-identical littermates to become long term survivors. Among the many other important studies performed in the canine model of marrow transplantation were the description of treating GVHD with antithymocyte serum (ATG; a strategy that was later successfully used to cure many patients with GVHD);[29] documenting the detrimental influence of sensitization by blood transfusions on subsequent allogeneic marrow grafts;[30] developing conditioning regimens capable of overcoming transfusion-induced sensitization;[31] and demonstrating the role of suppressor cells (now called regulatory T-cells) in maintaining long term tolerance

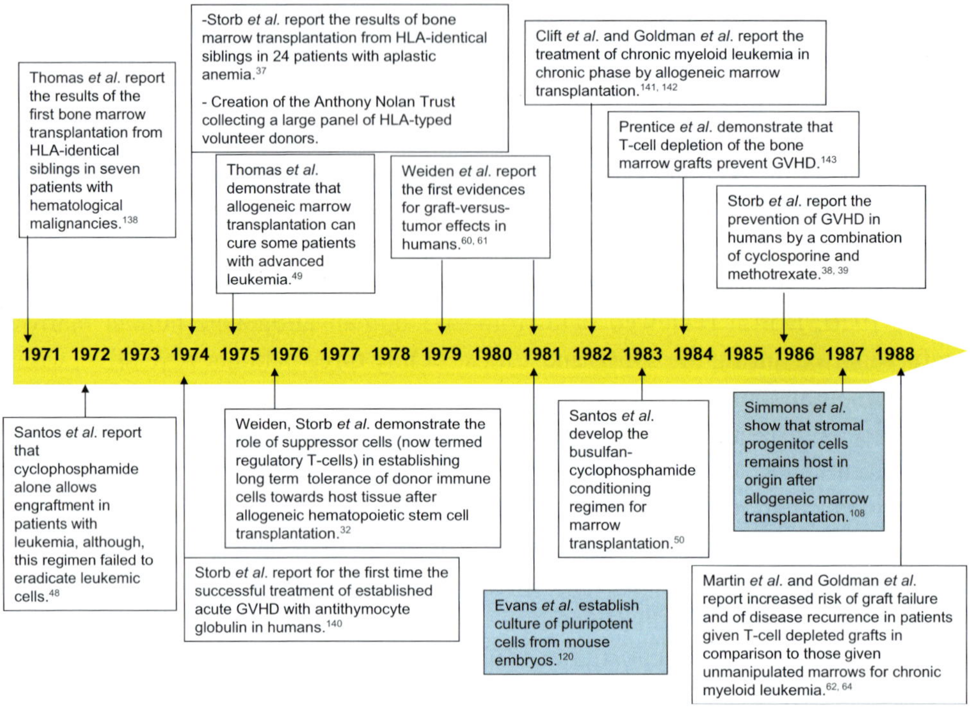

Fig. 2. Timeline. History of stem cell therapy: from 1971 to 1988.[32,37–39,48–50,60–62,64,108,120,138,140–143]

of donor immune cells towards host tissues[32] (Figs. 1 and 2). These observations triggered decades of research in immunology and immunogenetics of transplantation that have markedly improved our knowledge of the immune system. Further, they encouraged trials of allogeneic HCT between HLA-matched human siblings that started in the late 1960s in Seattle. These studies led to the development of effective supportive care of patients without marrow function including blood component transfusions (platelets were often donated by staff members, and granulocytes were obtained using patients with chronic myeloid leukemia as donors), parenteral nutrition, vascular access as well as therapies to prevent or treat bacterial, fungal, and viral infections.

1.3. Allogeneic HCT after high-dose conditioning

1.3.1. *Non-malignant disease*

Gatti *et al.*, Bach *et al.*, and deKoning *et al.* were the first, in 1968–1969, to perform successful marrow transplantations from HLA-matched siblings in three infants with immunological deficiency diseases.[33–35] Because of their underlying diseases, these patients did not require cytotoxic agents to destroy host immunity. These three patients were still leading normal lives more than 25 years after transplantation.[1]

The first successful allogeneic marrow transplantations for aplastic anemia using HLA-identical sibling donors were reported in 1972.[36] Two years later, the Seattle group reported on 24 cases.[37] The conditioning regimen consisted of cyclophosphamide, 200 mg/kg over four days, and postgrafting immunosuppression of intermittent methotrexate. Eleven of 24 patients were alive with donor cells and without GVHD at the time of the report. Graft rejection occurred in four patients, while the nine remaining patients died from transplantation complications including four who died from GVHD. Failure to achieve sustained engraftment was recognized as a major problem in aplastic anemia patients, and could be prevented by limiting the number of transfusions given before transplantation, and by adding ATG to the conditioning regimen. GVHD prevention was improved by post-transplant administration of cyclosporine combined with methotrexate instead of either drug alone. The combination of cyclosporine (or tacrolimus) and methotrexate has remained the most widely used method of GVHD prevention.[38–41] Using cyclophosphamide plus ATG as conditioning regimen, and postgrafting immunosuppression with cyclosporine plus methotrexate, 96% of aplastic anemia patients given marrows from HLA-identical siblings achieved sustained engraftment, and long term survival was 88%.[42]

Following the reports on successful transplants for aplastic anemia, allogeneic marrow transplantation from HLA-identical siblings was carried out in patients with a number of genetic disorders including thalassaemia, and sickle cell disease.[43–47]

1.3.2. *Hematological malignancies*

Two principal conditioning regimens for allogeneic transplantation in patients with hematological malignancies were developed in the 1960s–1970s. Regimens consisted of maximally tolerated doses of either cyclophosphamide or TBI alone.[48,49] Subsequently cyclophosphamide was combined either with TBI or high-dose busulfan.[2,3,49,50] In the early 1970s, most leukemia patients were transplanted in the terminal phase of their diseases with heavy body burdens of leukemic cells. Although some of them survived in remission demonstrating that allogeneic marrow transplantation could cure some patients with refractory leukemia, overall survival remained poor.[49] In 1975, Thomas *et al.* reported that leukemic patients in fair general clinical condition (based on a scale including disease status, transfusion requirements and/or infection at the time of transplantation, and on general condition) at the time of HCT had better survivals than those in poor clinical condition.[49] Also, patients with advanced refractory leukemia/lymphoma had disease recurrence rates of approximately 75% despite the high-dose conditioning regimens.[51]

These observations suggested that transplantation should be performed earlier in the disease course, while patients were still in good medical condition and had lower tumor burdens. Indeed, two studies reported in 1979 showed 50%–60% survival at two years when acute leukemia patients were transplanted in first complete remission.[52,53] These encouraging results prompted investigators to perform transplantation from HLA-matched siblings in patients with chronic myeloid leukemia in chronic phase.[54,55] Long term survival ranged from 50% to 72%.[54,55] Further improvements of conditioning regimens included the use of fractionated instead of single dose TBI,[56] the adjustment of busulfan dosing according to pharmacokinetics ("targeted busulfan"),[57] and the introduction of radiolabeled monoclonal antibodies specifically targeting cancer cells.[58]

1.4. Graft-versus-tumor effects

Barnes and Loutit were the first to postulate that allogeneic HCT given to tumor-bearing mice after lethal TBI destroyed residual leukemia cells by action of immunity.[59] They observed that leukemic cells were eradicated when mice were given allogeneic but not syngeneic grafts.[59] Initial evidence for allogeneic graft-versus tumor effects in humans came from studies published in the late 1970s/ early 1980s by the Seattle group describing reduced leukemic relapse rates in patients who developed acute and/or chronic GVHD compared with those who did not.[60,61] The demonstration that graft-versus-tumor (GVT) effects could eradicate leukemia prompted efforts to develop effective immunotherapy approaches outside of the HCT setting, with T-cell infusion or vaccines. GVT effects were subsequently confirmed by other investigators who also observed increased risk of relapse in patients receiving T-cell depleted[62–64] and syngeneic transplants.[65] Further, direct evidence for antitumor effects of allogeneic cells came from observations that infusion of donor lymphocytes induced complete remissions in a number of patients with hematological malignancies who had relapsed after allogeneic HCT.[66–68]

1.5. Alternative donors

Only approximately 30% of patients in need of HCT have HLA-identical sibling donors. Although a few HLA-matched unrelated transplants were carried out in the 1970s,[69] the complexity of the HLA system underscored the need for large panels of HLA-typed volunteer donors to be enrolled in registries. The Anthony Nolan Trust in England was the first such registry; this was followed by the establishment of similar programs in more than 40 countries.[70] The HLA phenotypes available in registries worldwide have been assembled in the database of Bone Marrow Donor Worldwide (http://www.bmdw.org), and include data from more than 11.5 millions donors and cord blood units — for the latter see below. Advances in the immunogenetics of HLA, especially typing by molecular techniques have made it possible to select unrelated donors who are matched at the allele level for HLA-A, -B, -C, -DRB1 and -DQB1 with their respective recipients. Results with such unrelated transplants have begun resembling those obtained with HLA-identical sibling grafts.[71]

Progress has also been made with regard to HLA haplo-identical HCT (i.e. grafts from donors who share one HLA haplotype with the recipient but are disparate for the other). During the 1980s, bone marrow transplantation from HLA haplo-identical donors was often unsuccessful because of very high incidences of graft failure and severe GVHD. In 1998, Aversa *et al.* reported that transplantation of mega doses of T-cell depleted hematopoietic stem cells from HLA haploidentically related donors after high-dose conditioning resulted in a high rate of sustained engraftment, and little GVHD.[72] While T-cell reconstitution was very slow leading to a high incidence of infectious complications, the relapse rate was relatively low in patients with acute myeloid leukemia transplanted in remission owing to GVT effects mediated by donor NK- and T-cells.[73] Transfer of functional immune responses to pathogens after HLA haploidentical HCT might improve the efficacy of this approach.[74]

1.6. Sources of hematopoietic stem cell for clinical use

In the past, bone marrow was the source of stem cells for most allogeneic transplantations. However, it was known since the 1960s that transplantable stem cells were present in the circulating blood.[75] The development of recombinant human hematopoietic growth factors such as human granulocyte colony stimulating factor (G-CSF) that could "mobilize" stem cells from the marrow into the circulating blood, has allowed using peripheral blood as a stem cell source. Stem cells have also been successfully mobilized with AMD3100, a CXCR4 antagonist.[76] Prospective randomized studies suggested that using allogeneic HLA-matched stem cells derived from G-CSF primed peripheral blood mononuclear cells (G-PBMC) instead of marrow was accompanied by faster hematopoietic recovery, and increased survival in a subgroup of patients with advanced hematological malignancies without increase in acute GVHD, but at the price of more chronic GVHD[77,78] (Fig. 3).

Since the late 1980s, cord blood has been used as an alternative source of transplantable stem cells.[79] Potential advantages include more rapid availability and, because cord blood contains immature naive T-cells, the possibility that a greater degree of HLA mismatching might be tolerable. Small cell doses have been a major limitation affecting cord blood transplantation outcomes.[80] This prompted Wagner *et al.* to investigate simultaneous transplantation of two umbilical cord blood units.[81] Preliminary results supported the feasibility of this

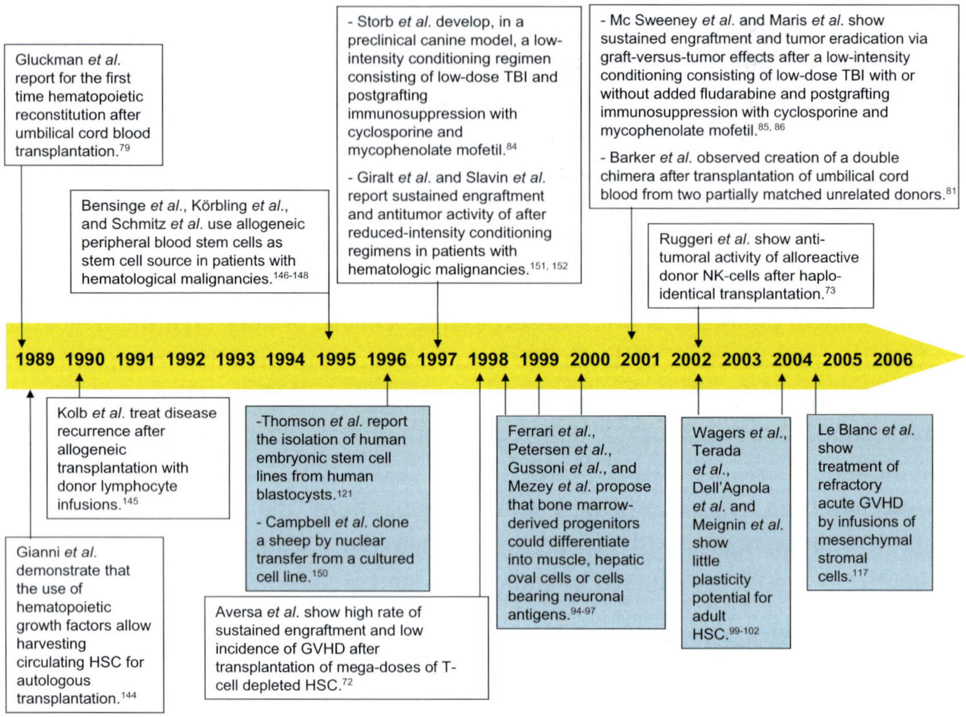

Fig. 3. Timeline. History of stem cell therapy: from 1989 to 2007.[72,73,79,81,84–86,94–97,99–102,117,144–148, 150–152]

approach. Eventually, one of the two cord blood units established "permanent" hematopoiesis in the recipients.[82] Another approach to increase the number of hematopoietic progenitors infused with cord blood included *in vitro* expansion of cord blood precursors (CD34+CD38−) on immobilized Delta-1 (an activator of Notch signaling).[83]

1.7. Allogeneic HCT after low-dose conditioning

Due to toxicities associated with conventional high-dose chemo-radiotherapy conditioning, its use has been limited to younger patients without medical comorbidities. Given that median ages at diagnosis for patients with most hematologic malignancies range from 65–70 years, the majority of such patients could not benefit from potentially curative HCT. Further, some patients have medical comorbidities, which excludes them from high-dose HCT. Since both graft rejection and GVHD are mediated by T-cells after HLA-identical HCT, we hypothesized that optimizing post-transplant immunosuppression could not only control GVHD but also graft rejection, thereby reducing the need for high-intensity pretransplant conditioning.[84] Based on results of preclinical studies in a canine model of HCT,[84] we have developed a very-low intensity regimen of 2 Gy TBI given either alone or combined with fludarabine (90 mg/m²); postgrafting immunosuppression consisted of mycophenolate mofetil and cyclosporine (or tacrolimus).[85–87] This regimen set the stage for successful engraftment of donor hematopoiesis and lymphopoiesis,[88] and relied on tumor eradication via GVT effects.[89] To date, more than 1200 patients have received allogeneic grafts with this regimen in clinical trials carried out in the US and Europe. Data from 834 consecutive patients have recently been analyzed.[90] Patients were given G-PBMC ($n = 816$) or marrow grafts ($n = 18$) from HLA-matched related ($n = 498$) or unrelated ($n = 336$) donors. Median patient age was 55 (range, 5–74) years. The conditioning regimen was remarkably well tolerated, and the transplant procedure could be carried out in the outpatient setting for a majority of patients.[91] Overall, 93% of patients had sustained engraftment. Acute GVHD, grades II–IV, was seen in 47% of related and 59% of unrelated recipients; most of this was grade II (33% and 44%, respectively). Chronic GVHD was seen in 50% of patients; the median time to complete response of chronic GVHD to immunosuppressive therapy was 2.4 years. Among patients with active malignancy at HCT, 60% achieved remissions a median of six months after HCT. Three-year survival ranged from 60% to 21%, depending on disease and disease stage at HCT[90] (Fig. 4). The addition of targeted therapy using radiolabeled monoclonal antibodies to the low-dose conditioning regimen might improve results in patients with bulky disease or relatively fast proliferative tumors (i.e. acute leukemias or aggressive lymphomas in relapse).[92,93]

2. Plasticity of Adult Hematopoietic Stem Cells: Lessons Learnt from a Canine Model of Duchenne Muscular Dystrophy

A series of studies in the late 1990s and early 2000s suggested that human hematopoietic stem cells could contribute to non-hematopoietic organ regeneration.[94–98] However, subsequent reports raised doubts about the plasticity of hematopoietic stem cells;[99–101] it appeared that many of the initial findings were the results of fusion between hematopoietic stem cells and differentiated cells rather than of trans-differentiation.[100]

Duchenne muscular dystrophy (DMD) is the most frequent and severe form of muscular dystrophy. DMD is an X-linked genetic disease caused by a complete or partial lack of dystrophin which plays a pivotal role in linking the cytoskeleton of the muscle fiber to the extracellular matrix. While a mild muscular dystrophy phenotype is present in the mdx mouse, the c-xmd dog has a phenotype closely mimicking DMD in humans. One study in mdx mice reported that bone marrow-derived

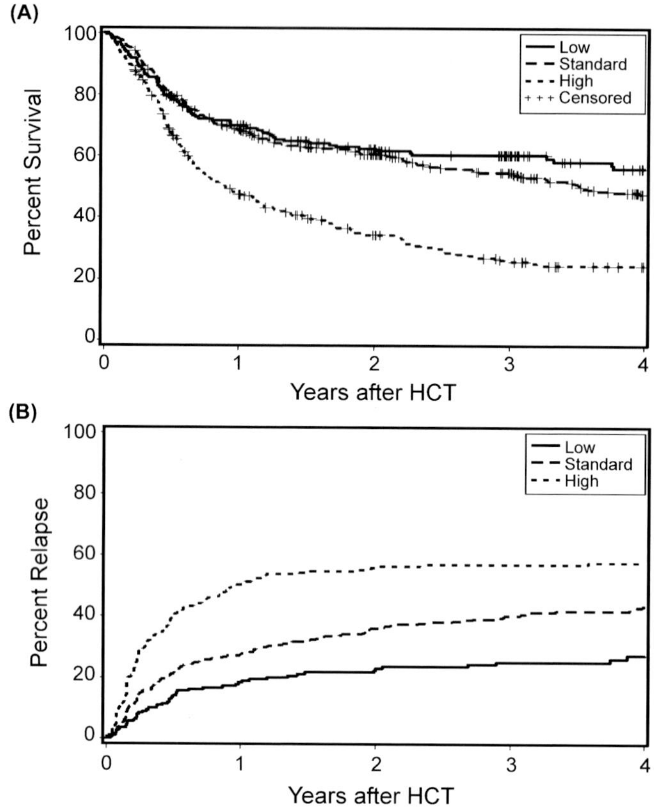

Fig. 4. Relapse risk in patients with malignant diseases given allogeneic hematopoietic cell transplantation after non-myeloablative conditioning. Overall survivals (**A**) and cumulative relapse rates (**B**) according to relapse risk groups. Reprinted with permission from Ref. 90, copyright American Society of Hematology.

cells contributed to skeletal muscle fibers and restored dystrophin expression.[96] Based on these observations, we evaluated whether HCT from healthy DLA-identical littermates could lead to skeletal muscle repair at clinically significant levels in seven c-xmd dogs.[102] Despite successful allogeneic HCT resulting in near complete or complete donor hematopoietic chimerism, dystrophy progressed in all transplanted dogs. Further, there were no significant increases in the percentages of dystrophin-positive fibers after HCT in comparison with baseline levels (Fig. 5). The contribution of donor cells to the diseased skeletal muscle was minimal, if at all. Similar negative data were obtained after the transplanted DMD dogs were treated with a course of G-CSF to mobilize hematopoietic stem cells into their circulation. The lack of donor-derived muscle cells could not be attributed to immune-rejection of dystrophin-positive cells since the allogeneic HCT transferred tolerance for

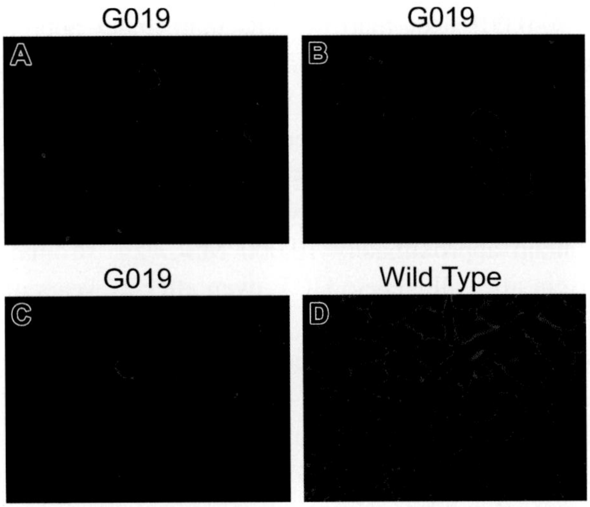

Fig. 5. Dystrophin-positive fibers before and after HCT and at autopsy. Immunofluorescence staining for dystrophin expression (using a mouse anti-human dystrophin monoclonal antibody) was performed on muscle tissue sections from pre- and post-transplantation biopsies; necropsy and wild-type dogs were used as positive controls. Sartorius from **(A)** before HCT and **(B)** bicep from necropsy; **(C)** diaphragm from necropsy; **(D)** sartorius muscle from wild-type dog. Normal mouse immunoglobulin G (IgG; Invitrogen) was used as negative control (not shown). Reprinted with permission from Ref. 102, copyright American Society of Hematology.

dystrophin into the recipients, as evidenced by the survival of muscle cells from the stem cell donors which were transplanted into recipient muscle, and of vascularized donor muscle flaps transplanted into recipient dogs. Finally, direct injection of donor marrow cells, into a recipient's dystrophic muscle failed to reconstitute satellite cells and myofibers.[103] Taken together, these data suggested that trans-differentiation of marrow stem cells into muscle was, at best, an infrequent event, and that occasional fusions between hematopoietic stem cells and muscle cells did not lead to significant restoration of dystrophin expression in a large animal model of DMD. Studies investigating direct infusions of muscle stem cells into immunologically tolerant c-xmd dogs have generated more encouraging results.[104]

3. Mesenchymal Stromal Cells

The bone marrow microenvironment is composed of different elements that support hematopoiesis and bone homeostasis (see also other chapters in this book). Among them, bone marrow stroma contains an adherent fibroblast-like population, representing 0.01%–0.001% of marrow cells that, under appropriate conditions, retains the ability to differentiate into a wide variety of cell lineages, including bone, cartilage, tendon, muscle, or adipose tissue. These cells, first identified by Friedenstein *et al.* in 1968,[105] have been termed bone marrow "stromal cells," "mesenchymal stem cells," or "mesenchymal stromal cells" (MSCs). MSCs have been shown to support hematopoiesis *in vitro* and *in vivo.*[106,107] However, although marrow grafts contain approximately 10,000 MSCs/kg, stromal progenitor cells remain host in origin after allogeneic HCT, even after 27 years of complete donor hematopoietic chimerism,[108,109] suggesting that donor MSC have a limited role in reconstituting the marrow microenvironment after HCT.

In recent years, interest in MSC has been raised by the observation that they exhibit profound immunosuppressive abilities *in vitro* and *in vivo*. *In vitro*, MSCs inhibited T-cell proliferation in mixed lymphocyte reactions or induced by mitogens.[110] The degrees of inhibition were dose-dependent, independent of HLA-matching, and did not require cell-to-cell contact. MSCs also inhibited IL-2 (or IL-15) induced NK-cell proliferation, although the inhibition was only partial.[111] In baboons, infusion of *ex vivo* expanded donor or third-party MSCs prolonged survival of histo-incompatible skin grafts.[112] Further, MSC infusion ameliorated experimental autoimmune encephalomyelitis in mice, by inducing T-cell anergy.[113]

Human pilot studies suggested that infusion of *ex vivo* expanded MSCs together with autologous or allogeneic HCT was safe, and possibly promoted engraftment.[114,115] More recently, Ball *et al.* reported uniform, sustained engraftment in 14 consecutive children co-transplanted with *ex vivo* expanded MSCs and HLA-haploidentical CD34+ cells, while 15% graft failure was seen in 47 historical controls not given MSCs ($p = 0.14$).[116] The authors concluded that co-transplantation of MSCs might reduce the risk of graft failure after HLA-haplo-identical HCT, possibly by inhibiting alloreactive host T-lymphocytes that had escaped the preparative regimen. Clearly, further studies are needed to support this hypothesis.

Based on the immunosuppressive abilities of MSCs, and on the apparent safety of infusion of *ex vivo* expanded MSCs, Le Blanc *et al.* infused such cells in a nine-year-old boy who had refractory gut and liver grade IV acute GVHD after HLA-matched unrelated HCT.[117] Infusion of 2×10^6 MSCs/kg from his HLA-haplo-identical mother resulted in remarkable improvement of his GVHD, with normalization of stools and decline in bilirubin levels. However, long term tolerance was not induced by MSC infusion since GVHD recurred after discontinuation of postgrafting immunosuppression. More recently, Le Blanc *et al.* reported the results among 40 patients given *ex vivo* expanded MSCs [median dose 1.0 (range 0.4–9) $\times$ 10^6 MSC/kg body weight of the recipient] as treatment for steroid-refractory grade III–IV acute GVHD.[118] No side effects were seen after MSC infusions. Nineteen patients received one infusion, 19 two infusions, and two $\geq$ three infusions. MSC donors were either HLA-identical siblings (*n* = 5), HLA-haploidentical relatives (*n* = 19), or third-party HLA-mismatched volunteers (*n* = 41). Some patients received MSCs from more than one donor. Remarkably, 21 of 40 patients achieved full resolution of GVHD, eight showed improvement, and two had stable GVHD. At the time of the report, 20 patients were alive six weeks to 3.5 years after MSC infusion, nine of them with extensive chronic GVHD. Further, MSC donor DNA was detected in lymph nodes and colon of one of the patients at autopsy. Taken together, these results suggested that infusion of *ex vivo* expanded MSC could be a useful tool in the treatment of patients with steroid refractory GVHD. Data from prospective randomized studies are needed to confirm these encouraging results.

Over the last several years, MSC infusions have also been explored both in preclinical and clinical trials to treat osteogenesis imperfecta, bone fractures, and several neurological or cardiac disorders.[119]

4. Embryonic Stem Cells

A landmark paper in 1981 described isolation of murine embryonic stem cells (ESC) from the inner cell mass of a murine blastocyst (see also other chapters in this book).[120] This was followed, 14 and 18 years later, respectively, by isolation of ESC from non-human primate and human blastocysts,[121,122] and from human primordial germ cells.[123] ESCs are unique in the sense that they can proliferate indefinitely in tissue culture, and that they are pluripotential, they have the potential to differentiate into germ cells and all somatic cell types. These observations prompted several investigators to study the role of ESC in preclinical murine models of Parkinson disease, spinal cord diseases, diabetes, or muscle diseases. However, besides the ethical debate on human ESC use, a number of issues have to be resolved before ESC can become useful tools in medicine, including how to avoid immune rejection and formation of teratomas, and how to best differientiate them into the various cell precursors.

5. Future of Stem Cell Therapy

5.1. HCT: minimizing pain, maximizing gain

The development of new transplantation strategies allowing allogeneic HCT after low-dose conditioning has dramatically decreased regimen-related toxicities.[91] In the future, further decreases in the intensity of the conditioning regimen might be achievable by inducing host-anergy towards donor hematopoietic cells using T-cell costimulation blockade.[124,125] This would be particularly desirable in the setting of HCT for non-malignant diseases so that long term adverse effects would be minimized. Further, progress in the identification of tumor and tissue-specific polymorphic minor histocompatibility antigens might allow increasing antitumor efficacy of HCT without inducing GVHD.[126] Better understanding of current and development of new immunosuppressive drugs[127] or the use of regulatory T-cells[128] and MSCs[118] might result in better control of graft-versus-host reactions. Finally, combination of HCT with disease-targeted therapy such as imatinib, thalidomide, lenalidomide, rituximab or radiolabeled monoclonal antibodies, are likely to make the transplant procedure even more effective.

5.2. Stem cells in regenerative medicine

While the initial enthusiasm about plasticity of hematopoietic stem cells and their potential use for tissue repair has been tempered by a number of negative studies,[129,130] better understanding of their physiology and that of other tissue-specific stem cells might make adult stem cell therapy an important therapeutic tool in the future.[104,131] Advances in somatic cell nuclear transfer might allow generating ESC that are genetically similar to the cells of the individual who donated the nucleus, thereby decreasing the risk of immune response in the host.[132] This, along with progress in the understanding of ESC growth and differentiation might bring the use of ESC into the clinic.

Acknowledgments

We thank Bonnie Larson, Helen Crawford, and Sue Carbonneau for help with manuscript preparation.

References

1. Thomas, E.D. and Blume, K.G. Historical markers in the development of allogeneic hematopoietic cell transplantation (Review). *Biol. Blood Marrow Transplant.* **5**, 341–346 (1999).
2. Little, M.-T. and Storb, R. History of haematopoietic stem-cell transplantation. *Nat. Rev. Cancer* **2**, 231–238 (2002).
3. Santos, G.W. Historical background. In: Atkinson, K. (ed.), *Clinical Bone Marrow Transplantation: A Reference Textbook* (Cambridge University Press, Cambridge, UK, 1994), pp, 1–9.
4. Jacobson, L.O., Marks, E.K., Robson, M.J., Gaston, E.O. and Zirkle, R.E. Effect of spleen protection on mortality following X-irradiation. *J. Lab. Clin. Med.* **34**, 1538–1543 (1949).

5. Jacobson, L.O., Simmons, E.L., Marks, E.K. and Eldredge, J.H. Recovery from radiation injury. *Science* **113**, 510–511 (1951).
6. Lorenz, E., Uphoff, D., Reid, T.R. and Shelton, E. Modification of irradiation injury in mice and guinea pigs by bone marrow injections. *J. Natl. Cancer Inst.* **12**, 197–201 (1951).
7. van Bekkum, D.W. and de Vries, M.J. *Radiation Chimaeras* (Logos Press Limited, London, 1967).
8. Nowell, P.C., Cole, L.J., Habermeyer, J.G. and Roan, P.L. Growth and continued function of rat marrow cells in X-radiated mice. *Cancer Res.* **16**, 258–261 (1956).
9. Lindsley, D.L., Odell, T.T. and Tausche, F.G. Implantation of functional erythropoietic elements following total-body irradiation. *Proc. Soc. Exp. Biol. Med.* **90**, 512–515 (1955).
10. Ford, C.E., Hamerton, J.L., Barnes, D.W.H. and Loutit, J.F. Cytological identification of radiation-chimaeras. *Nature* **177**, 452–454 (1956).
11. Becker, A.J., McCulloch, E.A. and Till, J.E. Cytological demonstration of the clonal nature of spleen colonies derived from transplanted mouse marrow cells. *Nature* **197**, 452–454 (1963).
12. Till, J.E. and McCulloch, E.A. A direct measurement of the radiation sensitivity of normal mouse bone marrow cells. *Radiat. Res.* **14**, 213–222 (1961).
13. Abramson, S., Miller, R.G. and Phillips, R.A. The identification in adult bone marrow of pluripotent and restricted stem cells of the myeloid and lymphoid systems. *J. Exp. Med.* **145**, 1567–1579 (1977).
14. Andrews, R.G., Singer, J.W. and Bernstein, I.D. Monoclonal antibody 12.8 recognizes a 115-kd molecule present on both unipotent and multipotent hematopoietic colony-forming cells and their precursors. *Blood* **67**, 842–845 (1986).
15. Bruno, B., Nash, R.A., Wallace, P.M., Gass, M.J., Thompson, J., Storb, R. and McSweeney, P.A. CD34+ selected bone marrow grafts are radioprotective and establish mixed chimerism in dogs given high dose total body irradiation. *Transplantation* **68**, 338–344 (1999).
16. Bensinger, W.I., Buckner, C.D., Shannon-Dorcy, K., Rowley, S., Appelbaum, F.R., Benyunes, M., Clift, R., Martin, P., Demirer, T., Storb, R., Lee, M. and Schiller, G. Transplantation of allogeneic CD34+ peripheral blood stem cells in patients with advanced hematologic malignancy. *Blood* **88**, 4132–4138 (1996).
17. Barnes, D.W.H., Corp, M.J., Loutit, J.F. and Neal, F.E. Treatment of murine leukaemia with X-rays and homologous bone marrow. Preliminary communication. *Br. Med. J.* **2,** 626–627 (1956).

18. Thomas, E.D., Lochte, H.L., Jr., Lu, W.C. and Ferrebee, J.W. Intravenous infusion of bone marrow in patients receiving radiation and chemotherapy. *N. Engl. J. Med.* **257**, 491–496 (1957).
19. Thomas, E.D., Lochte, H.L., Jr., Cannon, J.H., Sahler, O.D. and Ferrebee, J.W. Supralethal whole body irradiation and isologous marrow transplantation in man. *J. Clin. Invest.* **38**, 1709–1716 (1959).
20. McGovern, J.J., Jr., Russel, P.S., Atkins, L. and Webster, E.W. Treatment of terminal leukemic relapse by total-body irradiation and intravenous infusion of stored autologous bone marrow obtained during remission. *N. Engl. J. Med.* **260**, 675–683 (1959).
21. Mathé, G., Amiel, J.L., Schwarzenberg, L., Cattan, A. and Schneider, M. Haematopoietic chimera in man after allogeneic (homologous) bone-marrow transplantation. *Br. Med. J.* **5373**, 1633–1635 (1963).
22. Bortin, M.M. A compendium of reported human bone marrow transplants. *Transplantation* **9**, 571–587 (1970).
23. Dausset, J. Iso-leuco-anticorps. *Acta Haematol.* **20**, 156–166 (1958).
24. van Rood, J.J., Eernisse, J.G. and van Leeuwen, A. Leukocyte antibodies in sera from pregnant women. *Nature* **181**, 1735–1736 (1958).
25. Epstein, R.B., Storb, R., Ragde, H. and Thomas, E.D. Cytotoxic typing antisera for marrow grafting in littermate dogs. *Transplantation* **6**, 45–58 (1968).
26. Storb, R., Rudolph, R.H. and Thomas, E.D. Marrow grafts between canine siblings matched by serotyping and mixed leukocyte culture. *J. Clin. Invest.* **50**, 1272–1275 (1971).
27. Storb, R., Epstein, R.B., Graham, T.C. and Thomas, E.D. Methotrexate regimens for control of graft-versus-host disease in dogs with allogeneic marrow grafts. *Transplantation* **9**, 240–246 (1970).
28. Storb, R. and Thomas, E.D. Graft-versus-host disease in dog and man: the Seattle experience (Review). *Immunol. Rev.* **88**, 215–238 (1985).
29. Storb, R., Kolb, H.J., Graham, T.C., Kolb, H., Weiden, P.L. and Thomas, E.D. Treatment of established graft-versus-host disease in dogs by antithymocyte serum or prednisone. *Blood* **42**, 601–609 (1973).
30. Storb, R., Rudolph, R.H., Graham, T.C. and Thomas, E.D. The influence of transfusions from unrelated donors upon marrow grafts between histocompatible canine siblings. *J. Immunol.* **107**, 409–413 (1971).
31. Storb, R., Floersheim, G.L., Weiden, P.L., Graham, T.C., Kolb, H.-J., Lerner, K.G., Schroeder, M.-L. and Thomas, E.D. Effect of prior blood transfusions on marrow grafts: abrogation of sensitization by procarbazine and antithymocyte serum. *J. Immunol.* **112**, 1508–1516 (1974).
32. Weiden, P.L., Storb, R., Tsoi, M.-S., Graham, T.C., Lerner, K.G. and Thomas, E.D. Infusion of donor lymphocytes into stable canine radiation

chimeras: implications for mechanism of transplantation tolerance. *J. Immunol.* **116**, 1212–1219 (1976).

33. Gatti, R.A., Meuwissen, H.J., Allen, H.D., Hong, R. and Good, R.A. Immunological reconstitution of sex-linked lymphopenic immunological deficiency. *Lancet* **2**, 1366–1369 (1968).
34. Bach, F.H., Albertini, R.J., Joo, P., Anderson, J.L. and Bortin, M.M. Bone-marrow transplantation in a patient with the Wiskott-Aldrich syndrome. *Lancet* **2**, 1364–1366 (1968).
35. deKoning, J., van Bekkum, D.W., Dicke, K.A., Dooren, L.J., Radl, J. and van Rood, J.J. Transplantation of bone-marrow cells and fetal thymus in an infant with lymphopenic immunological deficiency. *Lancet* **1**, 1223–1227 (1969).
36. Thomas, E.D., Buckner, C.D., Storb, R., Neiman, P.E., Fefer, A., Clift, R.A., Slichter, S.J., Funk, D.D., Bryant, J.I. and Lerner, K.G. Aplastic anaemia treated by marrow transplantation. *Lancet* **1**, 284–289 (1972).
37. Storb, R., Thomas, E.D., Buckner, C.D., Clift, R.A., Johnson, F.L., Fefer, A., Glucksberg, H., Giblett, E.R., Lerner, K.G. and Neiman, P. Allogeneic marrow grafting for treatment of aplastic anemia. *Blood* **43**, 157–180 (1974).
38. Storb, R., Deeg, H.J., Farewell, V., Doney, K., Appelbaum, F., Beatty, P., Bensinger, W., Buckner, C.D., Clift, R., Hansen, J., Hill, R., Longton, G., Lum, L., Martin, P., McGuffin, R., Sanders, J., Singer, J., Stewart, P., Sullivan, K., Witherspoon, R. and Thomas, E.D. Marrow transplantation for severe aplastic anemia: methotrexate alone compared with a combination of methotrexate and cyclosporine for prevention of acute graft-versus-host disease. *Blood* **68**, 119–125 (1986).
39. Storb, R., Deeg, H.J., Whitehead, J., Appelbaum, F., Beatty, P., Bensinger, W., Buckner, C.D., Clift, R., Doney, K., Farewell, V., Hansen, J., Hill, R., Lum, L., Martin, P., McGuffin, R., Sanders, J., Stewart, P., Sullivan, K., Witherspoon, R., Yee, G. and Thomas, E.D. Methotrexate and cyclosporine compared with cyclosporine alone for prophylaxis of acute graft versus host disease after marrow transplantation for leukemia. *N. Engl. J. Med.* **314**, 729–735 (1986).
40. Sorror, M.L., Leisenring, W., Deeg, H.J., Martin, P.J. and Storb, R. Twenty-year follow-up in patients with aplastic anemia given marrow grafts from HLA-identical siblings and randomized to receive methotrexate/cyclosporine or methotrexate alone for prevention of graft-versus-host disease (Letter to the Editor). *Biol. Blood Marrow Transplant.* **11**, 567–568 (2005).
41. Sorror, M.L., Leisenring, W., Deeg, H.J., Martin, P.J. and Storb, R. Twenty-year follow-up of a controlled trial comparing a combination of methotrexate plus cyclosporine with cyclosporine alone for prophylaxis of graft-versus-host disease in patients administered HLA-identical marrow

grafts for leukemia (Letter to the Editor). *Biol. Blood Marrow Transplant.* **11**, 814–815 (2005).

42. Kahl, C., Leisenring, W., Deeg, H.J., Chauncey, T.R., Flowers, M.E.D., Martin, P.J., Sanders, J.E. and Storb, R. Cyclophosphamide and antithymocyte globulin as a conditioning regimen for allogeneic marrow transplantation in patients with aplastic anaemia: a long-term follow-up. *Br. J. Haematol.* **130**, 747–751 (2005).
43. Thomas, E.D., Buckner, C.D., Sanders, J.E., Papayannopoulou, T., Borgna-Pignatti, C., De Stefano, P., Sullivan, K.M., Clift, R.A. and Storb, R. Marrow transplantation for thalassaemia. *Lancet* **2,** 227–229 (1982).
44. Lucarelli, G., Galimberti, M., Polchi, P., Giardini, C., Politi, P., Baronciani, D., Angelucci, E., Manenti, F., Delfini, C., Aureli, G. and Moretto, P. Marrow transplantation in patients with advanced thalassemia. *N. Engl. J. Med.* **316**, 1050–1055 (1987).
45. Vermylen, C., Fernandez-Robles, E., Ninone, J. and Cornu, G. Bone marrow transplantation in five children with sickle cell anaemia. *Lancet* **1**, 1427–1428 (1988).
46. Walters, M.C., Patience, M., Leisenring, W., Eckman, J.R., Scott, J.P., Mentzer, W.C., Davies, S.C., Ohene-Frempong, K., Bernaudin, F., Matthews, D.C., Storb, R. and Sullivan, K.M. Bone marrow transplantation for sickle cell disease. *N. Engl. J. Med.* **335**, 369–376 (1996).
47. Johnson, F.L., Look, A.T., Gockerman, J., Ruggiero, M.R., Dalla-Pozza, L. and Billings, F.T., III. Bone-marrow transplantation in a patient with sickle-cell anemia. *N. Engl. J. Med.* **311**, 780–783 (1984).
48. Santos, G.W., Sensenbrenner, L.L., Burke, P.J., Mullins, G.M., Blas, W.B., Tutschka, P.J. and Slavin, R.E. The use of cyclophosphamide for clinical marrow transplantation. *Transplant. Proc.* **4**, 559–564 (1972).
49. Thomas, E.D., Storb, R., Clift, R.A., Fefer, A., Johnson, F.L., Neiman, P.E., Lerner, K.G., Glucksberg, H. and Buckner, C.D. Bone-marrow transplantation. *N. Engl. J. Med.* **292**, 832–843, 895–902 (1975).
50. Santos, G.W., Tutschka, P.J., Brookmeyer, R., Saral, R., Beschorner, W.E., Bias, W.B., Braine, H.G., Burns, W.H., Elfenbein, G.J., Kaizer, H., Mellits, D., Sensenbrenner, L.L., Stuart, R.K. and Yeager, A.M. Marrow transplantation for acute nonlymphocytic leukemia after treatment with busulfan and cyclophosphamide. *N. Engl. J. Med.* **309**, 1347–1353 (1983).
51. Thomas, E.D., Buckner, C.D., Banaji, M., Clift, R.A., Fefer, A., Flournoy, N., Goodell, B.W., Hickman, R.O., Lerner, K.G., Neiman, P.E., Sale, G.E., Sanders, J.E., Singer, J., Stevens, M., Storb, R. and Weiden, P.L. One hundred patients with acute leukemia treated by chemotherapy, total body irradiation, and allogeneic marrow transplantation. *Blood* **49**, 511–533 (1977).

52. Thomas, E.D., Buckner, C.D., Clift, R.A., Fefer, A., Johnson, F.L., Neiman, P.E., Sale, G.E., Sanders, J.E., Singer, J.W., Shulman, H., Storb, R. and Weiden, P.L. Marrow transplantation for acute nonlymphoblastic leukemia in first remission. *N. Engl. J. Med.* **301**, 597–599 (1979).
53. Thomas, E.D., Sanders, J.E., Flournoy, N., Johnson, F.L., Buckner, C.D., Clift, R.A., Fefer, A., Goodell, B.W., Storb, R. and Weiden, P.L. Marrow transplantation for patients with acute lymphoblastic leukemia in remission. *Blood* **54**, 468–476 (1979).
54. Thomas, E.D., Clift, R.A., Fefer, A., Appelbaum, F.R., Beatty, P., Bensinger, W.I., Buckner, C.D., Cheever, M.A., Deeg, H.J., Doney, K., Flournoy, N., Greenberg, P., Hansen, J.A., Martin, P., McGuffin, R., Ramberg, R., Sanders, J.E., Singer, J., Stewart, P., Storb, R., Sullivan, K., Weiden, P.L. and Witherspoon, R. Marrow transplantation for the treatment of chronic myelogenous leukemia. *Ann. Int. Med.* **104**, 155–163 (1986).
55. Goldman, J.M., Apperley, J.F., Jones, L., Marcus, R., Goolden, A.W.G., Batchelor, R., Hale, G., Waldmann, H., Reid, C.D., Hows, J., Gordon-Smith, E., Catovsky, D. and Galton, D.A.G. Bone marrow transplantation for patients with chronic myeloid leukemia. *N. Engl. J. Med.* **314**, 202–207 (1986).
56. Deeg, H.J., Sullivan, K.M., Buckner, C.D., Storb, R., Appelbaum, F.R., Clift, R.A., Doney, K., Sanders, J.E., Witherspoon, R.P. and Thomas, E.D. Marrow transplantation for acute nonlymphoblastic leukemia in first remission: toxicity and long-term follow-up of patients conditioned with single dose or fractionated total body irradiation. *Bone Marrow Transplant.* **1**, 151–157 (1986).
57. Radich, J.P., Gooley, T., Bensinger, W., Chauncey, T., Clift, R., Flowers, M., Martin, P., Slattery, J., Sultan, D. and Appelbaum, F.R. HLA-matched related hematopoietic cell transplantation for chronic-phase CML using a targeted busulfan and cyclophosphamide preparative regimen. *Blood* **102**, 31–35 (2003).
58. Pagel, J.M., Appelbaum, F.R., Eary, J.F., Rajendran, J., Fisher, D.R., Gooley, T., Ruffner, K., Nemecek, E., Sickle, E., Durack, L., Carreras, J., Horowitz, M.M., Press, O.W., Gopal, A.K., Martin, P.J., Bernstein, I.D. and Matthews, D.C. ^{131}I-anti-CD45 antibody plus busulfan and cyclophosphamide before allogeneic hematopoietic cell transplantation for treatment of acute myeloid leukemia in first remission. *Blood* **107**, 2184–2191 (2006).
59. Barnes, D.W.H. and Loutit, J.F. Treatment of murine leukaemia with X-rays and homologous bone marrow: II. *Br. J. Haematol.* **3**, 241–252 (1957).
60. Weiden, P.L., Flournoy, N., Thomas, E.D., Prentice, R., Fefer, A., Buckner, C.D. and Storb, R. Antileukemic effect of graft-versus-host disease in human

recipients of allogeneic-marrow grafts. *N. Engl. J. Med.* **300**, 1068–1073 (1979).

61. Weiden, P.L., Sullivan, K.M., Flournoy, N., Storb, R., Thomas, E.D. and the Seattle Marrow Transplant Team. Antileukemic effect of chronic graft-versus-host disease. Contribution to improved survival after allogeneic marrow transplantation. *N. Engl. J. Med.* **304**, 1529–1533 (1981).
62. Martin, P.J., Hansen, J.A., Torok-Storb, B., Durnam, D., Przepiorka, D., O'Quigley, J., Sanders, J., Sullivan, K.M., Witherspoon, R.P., Deeg, H.J., Appelbaum, F.R., Stewart, P., Weiden, P., Doney, K., Buckner, C.D., Clift, R., Storb, R. and Thomas, E.D. Graft failure in patients receiving T cell-depleted HLA-identical allogeneic marrow transplants. *Bone Marrow Transplant.* **3,** 445–456 (1988).
63. Maraninchi, D., Gluckman, E., Blaise, D., Guyotat, D., Rio, B., Pico, J.L., Leblond, V., Michallet, M., Dreyfus, F., Ifrah, N. and Bordigoni, A. Impact of T-cell depletion on outcome of allogeneic bone-marrow transplantation for standard-risk leukaemias. *Lancet* **2**, 175–178 (1987).
64. Goldman, J.M., Gale, R.P., Horowitz, M.M., Biggs, J.C., Champlin, R.E., Gluckman, E., Hoffmann, R.G., Jacobsen, S.J., Marmont, A.M., McGlave, P.B., Messner, H.A., Rimm, A.A., Rozman, C., Speck, B., Tura, S., Weiner, R.S. and Bortin, M.M. Bone marrow transplantation for chronic myelogenous leukemia in chronic phase: increased risk of relapse associated with T-cell depletion. *Ann. Int. Med.* **108**, 806–814 (1988).
65. Horowitz, M.M., Gale, R.P., Sondel, P.M., Goldman, J.M., Kersey, J., Kolb, H.-J., Rimm, A.A., Ringden, O., Rozman, C., Speck, B., Truitt, R.L., Zwaan, F.E. and Bortin, M.M. Graft-versus-leukemia reactions after bone marrow transplantation. *Blood* **75**, 555–562 (1990).
66. Kolb, H.J., Schattenberg, A., Goldman, J.M., Hertenstein, B., Jacobsen, N., Arcese, W., Ljungman, P., Ferrant, A., Verdonck, L., Niederwieser, D., van Rhee, F., Mittermueller, J., de Witte, T., Holler, E. and Ansari, H. Graft-versus-leukemia effect of donor lymphocyte transfusions in marrow grafted patients. European Group for Blood and Marrow Transplantation Working Party Chronic Leukemia. *Blood* **86**, 2041–2050 (1995).
67. Collins, R.H., Jr., Shpilberg, O., Drobyski, W.R., Porter, D.L., Giralt, S., Champlin, R., Goodman, S.A., Wolff, S.N., Hu, W., Verfaillie, C., List, A., Dalton, W., Ognoskie, N., Chetrit, A., Antin, J.H. and Nemunaitis, J. Donor leukocyte infusions in 140 patients with relapsed malignancy after allogeneic bone marrow transplantation. *J. Clin. Oncol.* **15**, 433–444 (1997).
68. Kolb, H.J., Schmidt, C., Barrett, A.J. and Schendel, D.J. Graft-versus-leukemia reactions in allogeneic chimeras. *Blood* **103**, 767–776 (2004).

69. Speck, B., Zwaan, F.E., van Rood, J.J. and Eernisse, J.G. Allogeneic bone marrow transplantation in a patient with aplastic anemia using a phenotypically HLA-A-identical unrelated donor. *Transplantation* **16**, 24–28 (1973).
70. Hansen, J.A. and Petersdorf, E.W. Unrelated donor hematopoietic cell transplantation. In: Thomas, E.D., Blume, K.G. and Forman, S.J. (eds.), *Hematopoietic Cell Transplantation* (Blackwell Science, Boston, 1999), pp. 915–928.
71. Mielcarek, M., Storer, B., Kahl, C., Sandmaier, B.M., Maloney, D., Maris, M.B. and Storb, R. Nonmyeloablative and myeloablative hematopoietic cell transplantation for hematologic malignancies: outcomes with HLA-matched unrelated donors compared to HLA-identical sibling donors. *Blood* **106**(Part 1), 194a (2005) (#658 Abstract).
72. Aversa, F., Tabilio, A., Velardi, A., Cunningham, I., Terenzi, A., Falzetti, F., Ruggeri, L., Barbabietola, G., Aristei, C., Latini, P., Reisner, Y. and Martelli, M.F. Treatment of high-risk acute leukemia with T-cell-depleted stem cells from related donors with one fully mismatched HLA haplotype. *N. Engl. J. Med.* **339**, 1186–1193 (1998).
73. Ruggeri, L., Capanni, M., Urbani, E., Perruccio, K., Shlomchik, W.D., Tosti, A., Posati, S., Rogaia, D., Frassoni, F., Aversa, F., Martelli, M.F. and Velardi, A. Effectiveness of donor natural killer cell alloreactivity in mismatched hematopoietic transplants. *Science* **295**, 2097–2100 (2002).
74. Perruccio, K., Tosti, A., Burchielli, E., Topini, F., Ruggeri, L., Carotti, A., Capanni, M., Urbani, E., Mancusi, A., Aversa, F., Martelli, M.F., Romani, L. and Velardi, A. Transferring functional immune responses to pathogens after haploidentical hematopoietic transplantation. *Blood* **106**, 4397–4406 (2005).
75. Storb, R., Epstein, R.B. and Thomas, E.D. Marrow repopulating ability of peripheral blood cells compared to thoracic duct cells. *Blood* **32**, 662–667 (1968).
76. Burroughs, L., Mielcarek, M., Little, M.-T., Bridger, G., MacFarland, R., Fricker, S., LaBrecque, J., Sandmaier, B.M. and Storb, R. Durable engraftment of AMD3100-mobilized autologous and allogeneic peripheral blood mononuclear cells in a canine transplantation model. *Blood* **106**, 4002–4008 (2005).
77. Bensinger, W.I., Martin, P.J., Storer, B., Clift, R., Forman, S.J., Negrin, R., Kashyap, A., Flowers, M.E.D., Lilleby, K., Chauncey, T.R., Storb, R. and Appelbaum, F.R. Transplantation of bone marrow as compared with peripheral-blood cells from HLA-identical relatives in patients with hematologic cancers. *N. Engl. J. Med.* **344**, 175–181 (2001).
78. Stem Cell Trialists' Collaborative Group. Allogeneic peripheral blood stem-cell compared with bone marrow transplantation in the management of

hematologic malignancies: an individual patient data meta-analysis of nine randomized trials. *J. Clin. Oncol.* **23**, 5074–5087 (2005).

79. Gluckman, E., Broxmeyer, H.E., Auerbach, A.D., Friedman, H.S., Douglas, G.W., Devergie, A., Esperou, H., Thierry, D., Socié, G., Lehn, P., Cooper, S., English, D., Kurtzberg, J., Bard, J. and Boyse, E.A. Hematopoietic reconstitution in a patient with Fanconi's anemia by means of umbilical-cord blood from an HLA-identical sibling. *N. Engl. J. Med.* **321**, 1174–1178 (1989).
80. Gluckman, E. and Rocha, V. Donor selection for unrelated cord blood transplants (Review). *Curr. Opin. Immunol.* **18**, 565–570 (2006).
81. Barker, J.N., Weisdorf, D.J. and Wagner, J.E. Creation of a double chimera after the transplantation of umbilical-cord blood from two partially matched unrelated donors. *N. Engl. J. Med.* **344**, 1870–1871 (2001).
82. Barker, J.N., Weisdorf, D.J., Defor, T.E., Blazar, B.R., McGlave, P.B., Miller, J.S., Verfaillie, C.M. and Wagner, J.E. Transplantation of two partially HLA-matched umbilical cord blood units to enhance engraftment in adults with hematologic malignancy. *Blood* **105**, 1343–1347 (2005).
83. Ohishi, K., Varnum-Finney, B., Serda, R.E., Anasetti, C. and Bernstein, I.D. The notch ligand, delta-1, inhibits the differentiation of monocytes into macrophages but permits their differentiation into dendritic cells. *Blood* **98**, 1402–1407 (2001).
84. Storb, R., Yu, C., Wagner, J.L., Deeg, H.J., Nash, R.A., Kiem, H.-P., Leisenring, W. and Shulman, H. Stable mixed hematopoietic chimerism in DLA-identical littermate dogs given sublethal total body irradiation before and pharmacological immunosuppression after marrow transplantation. *Blood* **89,** 3048–3054 (1997).
85. McSweeney, P.A., Niederwieser, D., Shizuru, J.A., Sandmaier, B.M., Molina, A.J., Maloney, D.G., Chauncey, T.R., Gooley, T.A., Hegenbart, U., Nash, R.A., Radich, J., Wagner, J.L., Minor, S., Appelbaum, F.R., Bensinger, W.I., Bryant, E., Flowers, M.E.D., Georges, G.E., Grumet, F.C., Kiem, H.-P., Torok-Storb, B., Yu, C., Blume, K.G. and Storb, R.F. Hematopoietic cell transplantation in older patients with hematologic malignancies: replacing high-dose cytotoxic therapy with graft-versus-tumor effects. *Blood* **97**, 3390–3400 (2001).
86. Maris, M.B., Niederwieser, D., Sandmaier, B.M., Storer, B., Stuart, M., Maloney, D., Petersdorf, E., McSweeney, P., Pulsipher, M., Woolfrey, A., Chauncey, T., Agura, E., Heimfeld, S., Slattery, J., Hegenbart, U., Anasetti, C., Blume, K. and Storb, R. HLA-matched unrelated donor hematopoietic cell transplantation after nonmyeloablative conditioning for patients with hematologic malignancies. *Blood* **102**, 2021–2030 (2003).

87. Baron, F., Sandmaier, B.M., Storer, B.E., Maris, M.B., Langston, A.A., Lange, T., Petersdorf, E., Bethge, W., Maziarz, R.T., McSweeney, P.A., Pulsipher, M.A., Wade, J.C., Chauncey, T.R., Shizuru, J.A., Sorror, M.L., Woolfrey, A.E., Maloney, D.G. and Storb, R. Extended mycophenolate mofetil and shortened cyclosporine failed to reduce graft-versus-host disease after unrelated hematopoietic cell transplantation with nonmyeloablative conditioning. *Biol. Blood Marrow Transplant.* **13**, 1041–1048 (2007).
88. Baron, F., Baker, J.E., Storb, R., Gooley, T.A., Sandmaier, B.M., Maris, M.B., Maloney, D.G., Heimfeld, S., Oparin, D., Zellmer, E., Radich, J.P., Grumet, F.C., Blume, K.G., Chauncey, T.R. and Little, M.-T. Kinetics of engraftment in patients with hematologic malignancies given allogeneic hematopoietic cell transplantation after nonmyeloablative conditioning. *Blood* **104**, 2254–2262 (2004).
89. Baron, F., Maris, M.B., Sandmaier, B.M., Storer, B.E., Sorror, M., Diaconescu, R., Woolfrey, A.E., Chauncey, T.R., Flowers, M.E.D., Mielcarek, M., Maloney, D.G. and Storb, R. Graft-versus-tumor effects after allogeneic hematopoietic cell transplantation with nonmyeloablative conditioning. *J. Clin. Oncol.* **23**, 1993–2003 (2005).
90. Kahl, C., Storer, B.E., Sandmaier, B.M., Mielcarek, M., Maris, M.B., Blume, K.G., Niederwieser, D., Chauncey, T.R., Forman, S.J., Agura, E., Leis, J.F., Bruno, B., Langston, A., Pulsipher, M.A., McSweeney, P.A., Wade, J.C., Epner, E., Petersen, F.B., Bethge, W.A., Maloney, D.G. and Storb, R. Relapse risk among patients with malignant diseases given allogeneic hematopoietic cell transplantation after nonmyeloablative conditioning. *Blood* **110**, 2744–2748 (2007).
91. Sorror, M.L., Maris, M.B., Storer, B., Sandmaier, B.M., Diaconescu, R., Flowers, C., Maloney, D.G. and Storb, R. Comparing morbidity and mortality of HLA-matched unrelated donor hematopoietic cell transplantation after nonmyeloablative and myeloablative conditioning: influence of pretransplant comorbidities. *Blood* **104**, 961–968 (2004).
92. Gopal, A.K., Pagel, J.M., Rajendran, J.G., Maloney, D.G., Appelbaum, F.R., Sorror, M.L., Sandmaier, B.M., Storb, R. and Press, O.W. Improving the efficacy of reduced intensity allogeneic transplantation for lymphoma using radiotherapy. *Biol. Blood Marrow Transplant.* **12**, 697–702 (2006).
93. Pagel, J.M., Appelbaum, F.R., Sandmaier, B.M., Rajendran, J.G., Gooley, T., Fisher, D.R., Nemecek, E., Gopal, A.K., Maris, M.B., Ruffner, K., Matthews, D.C., Deeg, H.J., Maloney, D.G., Martin, P.J. and Storb, R. ^{131}I-anti-CD45 antibody plus fludarabine, low-dose total body irradiation and peripheral blood stem cell infusion for elderly patients with advanced acute

myeloid leukemia (AML) or high-risk myelodysplastic syndrome (MDS). *Blood* **106**(Part 1), 119a (2005) (#397 Abstract).

94. Ferrari, G., Cusella-De, A.G., Coletta, M., Paolucci, E., Stornaiuolo, A., Cossu, G. and Mavilio, F. Muscle regeneration by bone marrow-derived myogenic progenitors. *Science* **279**, 1528–1530 (1998).
95. Petersen, B.E., Bowen, W.C., Patrene, K.D., Mars, W.M., Sullivan, A.K., Murase, N., Boggs, S.S., Greenberger, J.S. and Goff, J.P. Bone marrow as a potential source of hepatic oval cells. *Science* **284**, 1168–1170 (1999).
96. Gussoni, E., Soneoka, Y., Strickland, C.D., Buzney, E.A., Khan, M.K., Flint, A.F., Kunkel, L.M. and Mulligan, R.C. Dystrophin expression in the mdx mouse restored by stem cell transplantation. *Nature* **401**, 390–394 (1999).
97. Mezey, E., Chandross, K.J., Harta, G., Maki, R.A. and McKercher, S.R. Turning blood into brain: cells bearing neuronal antigens generated *in vivo* from bone marrow. *Science* **290**, 779–1782 (2000).
98. Orlic, D., Kajstura, J., Chimenti, S., Jakoniuk, I., Anderson, S.M., Li, B., Pickel, J., McKay, R., Nadal-Ginard, B., Bodine, D.M., Leri, A. and Anversa, P. Bone marrow cells regenerate infarcted myocardium (Letter). *Nature* **410**, 701–705 (2001).
99. Wagers, A.J., Sherwood, R.I., Christensen, J.L. and Weissman, I.L. Little evidence for developmental plasticity of adult hematopoietic stem cells. *Science* **297**, 2256–2259 (2002).
100. Terada, N., Hamazaki, T., Oka, M., Hoki, M., Mastalerz, D.M., Nakano, Y., Meyer, E.M., Morel, L., Petersen, B.E. and Scott, E.W. Bone marrow cells adopt the phenotype of other cells by spontaneous cell fusion. *Nature* **416**, 542–545 (2002).
101. Meignin, V., Soulier, J., Brau, F., Lemann, M., Gluckman, E., Janin, A. and Socie, G. Little evidence of donor-derived epithelial cells in early digestive acute graft-versus-host disease. *Blood* **103**, 360–362 (2004).
102. Dell'Agnola, C., Wang, Z., Storb, R., Tapscott, S.J., Kuhr, C.S., Hauschka, S.D., Lee, R.S., Sale, G.E., Zellmer, E., Gisburne, S., Bogan, J., Kornegay, J.N., Cooper, B.J., Gooley, T.A. and Little, M.-T. Hematopoietic stem cell transplantation does not restore dystrophin expression in Duchenne muscular dystrophy dogs. *Blood* **104**, 4311–4318 (2004).
103. Kuhr, C.S., Lupu, M. and Storb, R. Hematopoietic cell transplantation directly into dystrophic muscle fails to reconstitute satellite cells and myofibers. *Biol. Blood Marrow Transplant.* **13**, 886–888 (2007).
104. Sampaolesi, M., Blot, S., D'Antona, G., Granger, N., Tonlorenzi, R., Innocenzi, A., Mognol, P., Thibaud, J.L., Galvez, B.G., Barthelemy, I., Perani, L., Mantero, S., Guttinger, M., Pansarasa, O., Rinaldi, C., Cusella De Angelis, M.G., Torrente, Y., Bordignon, C., Bottinelli, R. and Cossu, G.

Mesoangioblast stem cells ameliorate muscle function in dystrophic dogs. *Nature* **444**, 574–579 (2006).
105. Friedenstein, A.J., Petrakova, K.V., Kurolesova, A.I. and Frolova, G.P. Heterotopic of bone marrow. Analysis of precursor cells for osteogenic and hematopoietic tissues. *Transplantation* **6**, 230–247 (1968).
106. Roecklein, B.A. and Torok-Storb, B. Functionally distinct human marrow stromal cell lines immortalized by transduction with the human papilloma virus E6/E7 genes. *Blood* **85**, 997–1005 (1995).
107. Maitra, B., Szekely, E., Gjini, K., Laughlin, M.J., Dennis, J., Haynesworth, S.E. and Koc, O.N. Human mesenchymal stem cells support unrelated donor hematopoietic stem cells and suppress T-cell activation. *Bone Marrow Transplant.* **33**, 597–604 (2004).
108. Simmons, P.J., Przepiorka, D., Thomas, E.D. and Torok-Storb, B. Host origin of marrow stromal cells following allogeneic bone marrow transplantation. *Nature* **328**, 429–432 (1987).
109. Awaya, N., Rupert, K., Bryant, E. and Torok-Storb, B. Failure of adult marrow-derived stem cells to generate marrow stroma after successful hematopoietic stem cell transplantation. *Exp. Hematol.* **30**, 937–942 (2002).
110. Ringden, O. and Le Blanc, K. Mesenchymal stem cells combined with cyclosporine inhibits cytotoxic T cells. *Biol. Blood Marrow Transplant.* **12**, 693–694 (2006).
111. Spaggiari, G.M., Capobianco, A., Becchetti, S., Mingari, M.C. and Moretta, L. Mesenchymal stem cell-natural killer cell interactions: evidence that activated NK cells are capable of killing MSCs, whereas MSCs can inhibit IL-2-induced NK-cell proliferation. *Blood* **107**, 1484–1490 (2006).
112. Bartholomew, A., Sturgeon, C., Siatskas, M., Ferrer, K., McIntosh, K., Patil, S., Hardy, W., Devine, S., Ucker, D., Deans, R., Moseley, A. and Hoffman, R. Mesenchymal stem cells suppress lymphocyte proliferation *in vitro* and prolong skin graft survival *in vivo*. *Exp. Hematol.* **30**, 42–48 (2002).
113. Zappia, E., Casazza, S., Pedemonte, E., Benvenuto, F., Bonanni, I., Gerdoni, E., Giunti, D., Ceravolo, A., Cazzanti, F., Frassoni, F., Mancardi, G. and Uccelli, A. Mesenchymal stem cells ameliorate experimental autoimmune encephalomyelitis inducing T-cell anergy. *Blood* **106**, 1755–1761 (2005).
114. Koc, O.N., Gerson, S.L., Cooper, B.W., Dyhouse, S.M., Haynesworth, S.E., Caplan, A.I. and Lazarus, H.M. Rapid hematopoietic recovery after coinfusion of autologous-blood stem cells and culture-expanded marrow mesenchymal stem cells in advanced breast cancer patients receiving high-dose chemotherapy. *J. Clin. Oncol.* **18**, 307–316 (2000).

115. Lazarus, H.M., Koc, O.N., Devine, S.M., Curtin, P., Maziarz, R.T., Holland, H.K., Shpall, E.J., McCarthy, P., Atkinson, K., Cooper, B.W., Gerson, S.L., Laughlin, M.J., Loberiza, F.R., Jr., Moseley, A.B. and Bacigalupo, A. Cotransplantation of HLA-identical sibling culture-expanded mesenchymal stem cells and hematopoietic stem cells in hematologic malignancy patients. *Biol. Blood Marrow Transplant.* **11**, 389–398 (2005).
116. Ball, L.M., Bernardo, M.E., Roelofs, H., Lankester, A., Cometa, A., Egeler, R.M., Locatelli, F. and Fibbe, W.E. Cotransplantation of *ex vivo* expanded mesenchymal stem cells accelerates lymphocyte recovery and may reduce the risk of graft failure in haploidentical hematopoietic stem-cell transplantation. *Blood* **110,** 2764–2767 (2007).
117. Le Blanc, K., Rasmusson, I., Sundberg, B., Gotherstrom, C., Hassan, M., Uzunel, M. and Ringden, O. Treatment of severe acute graft-versus-host disease with third party haploidentical mesenchymal stem cells. *Lancet* **363**, 1439–1441 (2004).
118. Le Blanc, K., Frassoni, F., Ball, L., Lanino, E., Sundberg, B., Lonnquist, B., Roelofs, H., Dini, G., Bacigalupo, A., Locatelli, F., Fibbi, G. and Ringden, O. Mesenchymal stem cells for treatment of severe acute graft-versus-host disease. *Blood* **108**, 226a (2006) (#753 Abstract).
119. Rao. M. Meeting report: adult mesenchymal stem cells in regenerative medicine conference, August 2007. *Stem Cells* **25**, 2973–2974 (2007).
120. Evans, M.J. and Kaufman, M.H. Establishment in culture of pluripotential cells from mouse embryos. *Nature* **292**, 154–156 (1981).
121. Thomson, J.A., Kalishman, J., Golos, T.G., Durning, M., Harris, C.P., Becker, R.A. and Hearn, J.P. Isolation of a primate embryonic stem cell line. *Proc. Natl. Acad. Sci. USA* **92**, 7844–7848 (1995).
122. Thomson, J.A., Itskovitz-Eldor, J., Shapiro, S.S., Waknitz, M.A., Swiergiel, J.J., Marshall, V.S. and Jones, J.M. Embryonic stem cell lines derived from human blastocysts [Erratum appears in *Science* **282**(5395), 1827 (1998)]. *Science* **282** (5391), 1145–1147 (1998).
123. Shamblott, M.J., Axelman, J., Wang, S., Bugg, E.M., Littlefield, J.W., Donovan, P.J., Blumenthal, P.D., Huggins, G.R. and Gearhart, J.D. Derivation of pluripotent stem cells from cultured human primordial germ cells [Erratum appears in *Proc. Natl. Acad. Sci. USA* **96**(3), 1162 (1999)]. *Proc. Natl. Acad. Sci. USA* **95**, 13726–13731 (1998).
124. Storb, R., Yu, C., Zaucha, J.M., Deeg, H.J., Georges, G., Kiem, H.-P., Nash, R.A., McSweeney, P.A. and Wagner, J.L. Stable mixed hematopoietic chimerism in dogs given donor antigen, CTLA4Ig, and 100 cGy total body irradiation before and pharmacologic immunosuppression after marrow transplant. *Blood* **94**, 2523–2529 (1999).

125. Jochum, C., Beste, M., Zellmer, E., Graves, S.S. and Storb, R. CD154 blockade and donor-specific transfusions in DLA-identical marrow transplantation in dogs conditioned with 1-Gy total body irradiation. *Biol. Blood Marrow Transplant.* **13**, 164–171 (2007).
126. Bleakley, M. and Riddell, S.R. Molecules and mechanisms of the graft-versus-leukaemia effect. *Nat. Rev. Cancer* **4**, 371–380 (2004).
127. Hogan, W.J., Sorror, M.L., Diaconescu, R., Zellmer, E., Little, M.-T. and Storb, R. Marrow engraftment after nonmyeloablative conditioning followed by postgrafting immunosuppression with mycophenolate mofetil, sirolimus and cyclosporine in a canine model. *Blood* **102**(Part 2), 396b (2003) (#5311 Abstract).
128. Lesnikova, M., Nikitine, A., Pogosov, L., Nash, R.A. and Georges, G.E. Peripheral $CD4^+CD25^+$ regulatory T cells (Treg) block alloreactive host antidonor T cells in canine mixed hematopoietic chimeras. *Blood* **106**(Part 1), 867a (2005) (#3101 Abstract).
129. Janssens, S., Dubois, C., Bogaert, J., Theunissen, K., Deroose, C., Desmet, W., Kalantzi, M., Herbots, L., Sinnaeve, P., Dens, J., Maertens, J., Rademakers, F., Dymarkowski, S., Gheysens, O., Van Cleemput, J., Bormans, G., Nuyts, J., Belmans, A., Mortelmans, L., Boogaerts, M. and Van de Werf, F. Autologous bone marrow-derived stem-cell transfer in patients with ST-segment elevation myocardial infarction: double-blind, randomised controlled trial. *Lancet* **367**, 113–121 (2006).
130. Lunde, K., Solheim, S., Aakhus, S., Arnesen, H., Abdelnoor, M., Egeland, T., Endresen, K., Ilebekk, A., Mangschau, A., Fjeld, J.G., Smith, H.J., Taraldsrud, E., Grogaard, H.K., Bjornerheim, R., Brekke, M., Muller, C., Hopp, E., Ragnarsson, A., Brinchmann, J.E. and Forfang, K. Intracoronary injection of mononuclear bone marrow cells in acute myocardial infarction. *N. Engl. J. Med.* **355**, 1199–1209 (2006).
131. Guzman, R., Uchida, N., Bliss, T.M., He, D., Christopherson, K.K., Stellwagen, D., Capela, A., Greve, J., Malenka, R.C., Moseley, M.E., Palmer, T.D. and Steinberg, G.K. Long-term monitoring of transplanted human neural stem cells in developmental and pathological contexts with MRI. *Proc. Natl. Acad. Sci. USA* **104**, 10211–10216 (2007).
132. Wilmut, I., Beaujean, N., de Sousa, P.A., Dinnyes, A., King, T.J., Paterson, L.A., Wells, D.N. and Young, L.E. Somatic cell nuclear transfer. *Nature* **419**, 583–586 (2002).
133. Uphoff, D.E. Genetic factors influencing irradiation protection by bone marrow. I. The F1 hybrid effect. *J. Natl. Cancer Inst.* **19**, 123–125 (1957).
134. Thomas, E.D., Collins, J.A., Herman, E.C., Jr. and Ferrebee, J.W. Marrow transplants in lethally irradiated dogs given methotrexate. *Blood* **19**, 217–228 (1962).

135. Mathe, G., Amiel, J.L., Schwarzenberg, L., Cattan, A. and Schneider, M. Adoptive immunotherapy of acute leukemia: experimental and clinical results. *Cancer Res.* **25**, 1525–1531 (1965).
136. Epstein, R.B., Graham, T.C., Buckner, C.D., Bryant, J. and Thomas, E.D. Allogeneic marrow engraftment by cross circulation in lethally irradiated dogs. *Blood* **28**, 692–707 (1966).
137. Storb, R., Epstein, R.B., Ragde, H. and Thomas, E.D. Marrow engraftment by allogeneic leukocytes in lethally irradiated dogs. *Blood* **30**, 805–811 (1967).
138. Thomas, E.D., Buckner, C.D., Rudolph, R.H., Fefer, A., Storb, R., Neiman, P.E., Bryant, J.I., Chard, R.L., Clift, R.A., Epstein, R.B., Fialkow, P.J., Funk, D.D., Giblett, E.R., Lerner, K.G., Reynolds, F.A. and Slichter, S. Allogeneic marrow grafting for hematologic malignancy using HL-A matched donor-recipient sibling pairs. *Blood* **38**, 267–287 (1971).
139. Storb, R., Thomas, E.D., Buckner, C., Clift, R.A., Fefer, A., Glucksberg, H. and Neiman, P.E. Transplantation of bone marrow in refractory marrow failure and neoplastic diseases. *Am. J. Clin. Pathol.* **62**, 212–217 (1974).
140. Storb, R., Gluckman, E., Thomas, E.D., Buckner, C.D., Clift, R.A., Fefer, A., Glucksberg, H., Graham, T.C., Johnson, F.L., Lerner, K.G., Neiman, P.E. and Ochs, H. Treatment of established human graft-versus-host disease by antithymocyte globulin. *Blood* **44**, 57–75 (1974).
141. Clift, R.A., Buckner, C.D., Thomas, E.D., Doney, K., Fefer, A., Neiman, P.E., Singer, J., Sanders, J., Stewart, P., Sullivan, K.M., Deeg, J. and Storb, R. Treatment of chronic granulocytic leukaemia in chronic phase by allogeneic marrow transplantation. *Lancet* **2**, 621–623 (1982).
142. Goldman, J.M., Baughan, A.S.J., McCarthy, D.M., Worsley, A.M., Hows, J.M., Gordon-Smith, E.C., Catovsky, D., Batchelor, J.R., Goolden, A.W.G. and Galton, D.A.G. Marrow transplantation for patients in the chronic phase of chronic granulocytic leukaemia. *Lancet* **2**, 623–625 (1982).
143. Prentice, H.G., Blacklock, H.A., Janossy, G., Gilmore, M.J., Price-Jones, L., Tidman, N., Trejdosiewicz, L.K., Skeggs, D.B.L., Panjwani, D., Ball, S., Graphakos, S., Patterson, J., Ivory, K. and Hoffbrand, A.V. Depletion of T lymphocytes in donor marrow prevents significant graft-versus-host disease in matched allogeneic leukaemic marrow transplant recipients. *Lancet* **1**, 472–476 (1984).
144. Gianni, A.M., Siena, S., Bregni, M., Tarella, C., Stern, A.C., Pileri, A. and Bonadonna, G. Granulocyte-macrophage colony-stimulating factor to harvest circulating haemopoietic stem cells for autotransplantation. *Lancet* **2**, 580–585 (1989).

145. Kolb, H.J., Mittermüller, J., Clemm, Ch., Holler, G., Ledderose, G., Brehm, G., Heim, M. and Wilmanns, W. Donor leukocyte transfusions for treatment of recurrent chronic myelogenous leukemia in marrow transplant patients. *Blood* **76**, 2462–2465 (1990).
146. Bensinger, W.I., Weaver, C.H., Appelbaum, F.R., Rowley, S., Demirer. T., Sanders, J., Storb, R. and Buckner, C.D. Transplantation of allogeneic peripheral blood stem cells mobilized by recombinant human granulocyte colony-stimulating factor. *Blood* **85**, 1655–1658 (1995).
147. Korbling, M., Huh, Y.O., Durett, A., Mirza, N., Miller, P., Engel, H., Anderlini, P., van Besien, K., Andreeff, M., Przepiorka, D. and Champlin, R.E. Allogeneic blood stem cell transplantation: peripheralization and yield of donor-derived primitive hematopoietic progenitor cells (CD34+ Thy-1dim) and lymphoid subsets, and possible predictors of engraftment and graft-versus-host disease. *Blood* **86**, 2842–2848 (1995).
148. Schmitz, N., Dreger, P., Suttorp, M., Rohwedder, E.B., Haferlach, T., Löffler, H., Hunter, A. and Russell, N.H. Primary transplantation of allogeneic peripheral blood progenitor cells mobilized by filgrastim (granulocyte colony-stimulating factor). *Blood* **85**, 1666–1672 (1995).
149. Thomson, J.A., Kalishman, J., Golos, T.G., Durning, M., Harris, C.P. and Hearn, J.P. Pluripotent cell lines derived from common marmoset (*Callithrix jacchus*) blastocysts. *Biol. Reprod.* **55**, 254–259 (1996).
150. Campbell, K.H., McWhir, J., Ritchie, W.A. and Wilmut, I. Sheep cloned by nuclear transfer from a cultured cell line. *Nature* **380**, 64–66 (1996).
151. Giralt, S., Estey, E., Albitar, M., van Besien, K., Rondón, G., Anderlini, P., O'Brien, S., Khouri, I., Gajewski, J., Mehra, R., Claxton, D., Andersson, B., Beran, M., Przepiorka, D., Koller, C., Kornblau, S., Körbling, M., Keating, M., Kantarjian, H. and Champlin, R. Engraftment of allogeneic hematopoietic progenitor cells with purine analog-containing chemotherapy: harnessing graft-versus-leukemia without myeloablative therapy. *Blood* **89**, 4531–4536 (1997).
152. Slavin, S., Nagler, A., Naparstek, E., Kapelushnik, Y., Aker, M., Cividalli, G., Varadi, G., Kirschbaum, M., Ackerstein, A., Samuel, S., Amar, A., Brautbar, C., Ben-Tal, O., Eldor, A. and Or, R. Nonmyeloablative stem cell transplantation and cell therapy as an alternative to conventional bone marrow transplantation with lethal cytoreduction for the treatment of malignant and nonmalignant hematologic diseases. *Blood* **91**, 756–763 (1998).

Chapter 28

Tissue Engineered Skin Comes of Age?

Sheila MacNeil

Abstract

Tissue engineered skin was the first out of the stable of tissues that could be made in the laboratory from biopsies of patients skin expanded and then delivered back to them. As patients have been benefiting from cultured skin cells since 1981,[1] at 25 years old this is far from being a new area. In this article the question of to what extent tissue engineered skin has finally come of age will be reviewed.

There are currently three clinical areas where it can benefit man — for the treatment of patients with extensive skin loss due to burns injuries, to accelerate or initiate healing in patients with chronic non-healing ulcers and for reconstructive surgery purposes (an area which is still in its infancy but can encompass the treatment of pigmentation defects and diseases such as vitiligo and scarring and hopefully blistering diseases). There are also many *in vitro* applications where having a physiologically relevant model of skin can teach us more about normal and pathological skin biology than working with monolayers of skin cells.

This chapter looks at why tissue engineered skin was initially developed for burns injuries and how it is now also used for chronic wounds and beginning to be used for reconstructive surgery. The challenges that remain are then considered followed by a summary of some of the *in vitro* applications for tissue engineered skin.

Keywords: Skin; Burns; Chronic Wounds; Reconstructive Surgery.

Outline

1. To What Extent can Tissue Engineered Skin Deliver Normal Skin Structure and Function for Clinical Use?
2. Which Patients can Benefit from Tissue Engineered Products?
3. History of the Development of Tissue Engineered Skin for Burns Injuries
4. Development of Tissue Engineered Skin for Chronic Wounds

5. Development of Tissue Engineered Skin for Reconstructive Surgery
6. The Design Process for Tissue Engineered Products
7. Keratinocyte Stem Cells — Where are We?
8. Clinical and Future Development Issues
9. Laboratory Uses of Tissue Engineered Skin
10. Conclusion
References

1. To What Extent can Tissue Engineered Skin Deliver Normal Skin Structure and Function for Clinical Use?

The immediate thing to note is that tissue engineered skin currently is used only to restore the all essential physical barrier properties of skin. The major function of skin is to act as a barrier to the outside world and when this function is lost through superficial or extensive burns or chronic non-healing ulcers then patients can be susceptible to bacterial infection. In burns patients with extensive skin loss septicaemia is a major cause of patients failing to survive.

Skin barrier function is achieved by layers of differentiated keratinocytes fusing together and forming a barrier impermeable to water, electrolytes and bacteria.[2] The presence of melanocytes also helps skin cope with UV, oxidative and inflammatory stress.[3] Keratinocytes in the epidermis are programmed to be replaced continuously from a population of basal keratinocytes which are located on the basement membrane zone. (The latter is composed of specialised collagen fibrils which attach the epidermis to the underlying dermis.) Cells within the basal layer have the capability of dividing and giving rise to daughter cells which are pushed upwards. This population of cells contains a small percentage of cells with stem cell properties[4] in that they are able to renew continuously throughout a lifetime while the majority of epidermal keratinocytes become stratified, lose their nuclei and become part of an integrated sheet of keratin which is eventually shed from the skin's surface.

At present tissue engineered skin does not contribute to immune surveillance, temperature regulation or assist us in interacting with our environment. Also despite excellent work on understanding the biology of hair follicles[4,5] and culturing these *in vitro*[6] the prospect of receiving skin with hair has not yet translated into clinical use. Similarly, problems of introducing vasculature into skin remain a major challenge (discussed later in this article) and there are no attempts to introduce sweat glands into skin or to tackle problems of innervation to the best of my knowledge. (For a recent article looking at how skin cells interact with cutaneous sensory neurones please see Ref. 7.)

While our normal wound repair process is excellent, problems commonly occur when the blood supply is compromised because of, for example, venous stasis ulcers or diabetes or when the extent of skin loss is so great (as in burns) that the body cannot produce sufficient barrier cover quickly enough to prevent bacterial entry and fluid loss. In the Western world while management of extensive acute burns remains a major challenge[8] and is very expensive the numbers of patients suffering from extensive burns are relatively small. (These remain a major health care problem for developing countries.) Chronic wounds however are on the increase associated with ageing and with diabetes. These are economically expensive to health care systems and have a major impact on the patient's quality of life.[9,10] This is the "volume market" for tissue engineered skin products.

2. Which Patients can Benefit from Tissue Engineered Products?

Table 1 lists which groups of patients can benefit clinically from tissue engineered skin and Table 2 gives examples of products currently available on a commercial basis.

How these tissue engineered products are used will be discussed by looking: (a) at the development of tissue engineered skin for burns injuries, (b) at the very different needs of patients with chronic non-healing ulcers, and (c) reconstructive surgery needs will be described.

3. History of the Development of Tissue Engineered Skin for Burns Injuries

How does tissue engineered skin fit into the management of burns injuries, chronic wounds and reconstructive surgery? Of these, burns are the most extreme acute emergencies. The needs of patients with extensive full thickness skin loss led to the development of cultured skin cells and more complex tissue engineered products.[1,28–31] As most tissue engineered products and research was initiated in response to major burns, the management of these will be discussed in some detail.

In a partial thickness burn in which the upper epidermal layer is completely lost but part of the dermis remains then there are epidermal keratinocytes lining the dermal inclusions (the sweat glands and hair follicles) and it is these that give rise to rapid migration and proliferation of keratinocytes to form a new epithelium in these circumstances. However, problems occur when the epidermis and all of the dermis has been lost. Any loss of full thickness skin of more than 4 cm in diameter will not heal well without surgical intervention.[8,32] The gold standard approach is to take a split thickness skin graft from elsewhere on the body that contains all of the epidermis but only part of the dermis and graft this onto the damaged area. Such skin grafts work well for burn injuries which are not too extensive but when the patient has suffered more than 30% or 40% full thickness body surface area then burns surgeons welcome having cultured skin products.

Table 1. Which patients can benefit from tissue engineered products?

Clinical conditions	Clinical needs	Examples of tissue engineered products being used
Burns — extensive full thickness	Restore barrier function as soon as possible	A combination of Integra and split thickness grafts[11] or Integra and Permaderm[12] Cultured keratinocytes over wide meshed autograft[13]
Burns — superficial burns and scalds	Accelerate re-epithelialisation to reduce pain and reduce the incidence of scars and contractures	SprayCell[14]
Chronic non-healing ulcers	To restore barrier function and mobility where possible	Apligraf[15,16] Dermagraft[17,18] Myskin[19]
Reconstructive surgery following surgical trauma		
• Contracture releases	To restore full mobility and a normal appearance	Ref. 20
Reconstructive surgery following congenital defects or acquired diseases		
• Giant nevi	To remove abnormal pigmentation and restore normal pigmentation	Refs. 21, 22
• Vitiligo	To restore normal pigmentation to depigmented areas	Refs. 23–25
• Blistering diseases	To correct genetic defects in the ability of epidermal cells to attach to the underlying dermis	Refs. 26, 27

If surgeons take sufficient skin graft to treat, e.g. a patient with 50% full thickness burns, they could run the risk of removing too much barrier protection from elsewhere on the body, leading *in extremis* to a patient with very little effective barrier anywhere on the body. In the UK, burns surgeons will only routinely graft

Table 2. Skin replacement materials available commercially for clinical use in 2008.

Epidermal cover	Cultured autologous keratinocytes • Cultured epidermal sheets — Epicell™ (Genzyme)[40] • Cultured epidermal sheets derived from hair biopsies — Epidex™ (Modex Therapeutics)[50] • Subconfluent cells on a synthetic carrier — Myskin™ (CellTran)[51] • Subconfluent cells delivered in a spray (C3)[41] • Cultured allogeneic keratinocytes (no commercially available products as yet although they were available in Europe as Cryoskin until recently)
Dermal replacement materials	• Integra® DRM (Integra)[48] • Donor skin — Alloderm® (Lifecell)[52] • Permacol® (Tissue Science Laboratories)[47] • Dermagraft® (Advanced BioHealing)[18] • Transcyte® (Advanced BioHealing)[53]
Epidermal/dermal replacement materials	• Autologous tissue engineered skin — split thickness skin grafts — Permaderm™ (Cambrex)[45] • Allogeneic TE skin — Apligraf® (Novartis)[16] • Orcel™ (Ortec International)[54]

something like 15%–20% of a patient's body surface area in one go to try to minimise this risk. This in turn necessitates the patient undergoing several operations, often over several weeks. Commonly a patient with extensive burns is managed by surgeons taking further skin grafts from existing donor sites as soon as these have healed. Against this situation burns surgeons from the early 1980s were interested in having laboratory expanded keratinocytes from the patient available as an adjunct to conventional skin grafting. The method now used throughout the world for fast and reliable culture of keratinocytes from mature adult skin was initially developed by Rheinwald and Green in 1975.[28,29]

This was then developed further so that cells grew into small sheets of cells two to three layers thick (known as cultured epithelial autografts, CEAs[31]). These were produced rapidly by growing keratinocytes on a layer of irradiated murine fibroblasts in the presence of a mitogen-rich medium including bovine foetal calf serum. To get the cells to the patient it is then necessary to enzymically detach the CEA sheets from the tissue culture plastic at which point they are very difficult to handle. Accordingly these fragile sheets of cells are wrapped around an inert backing dressing and then delivered to the patients — this is illustrated in Figs. 1A to 1C.

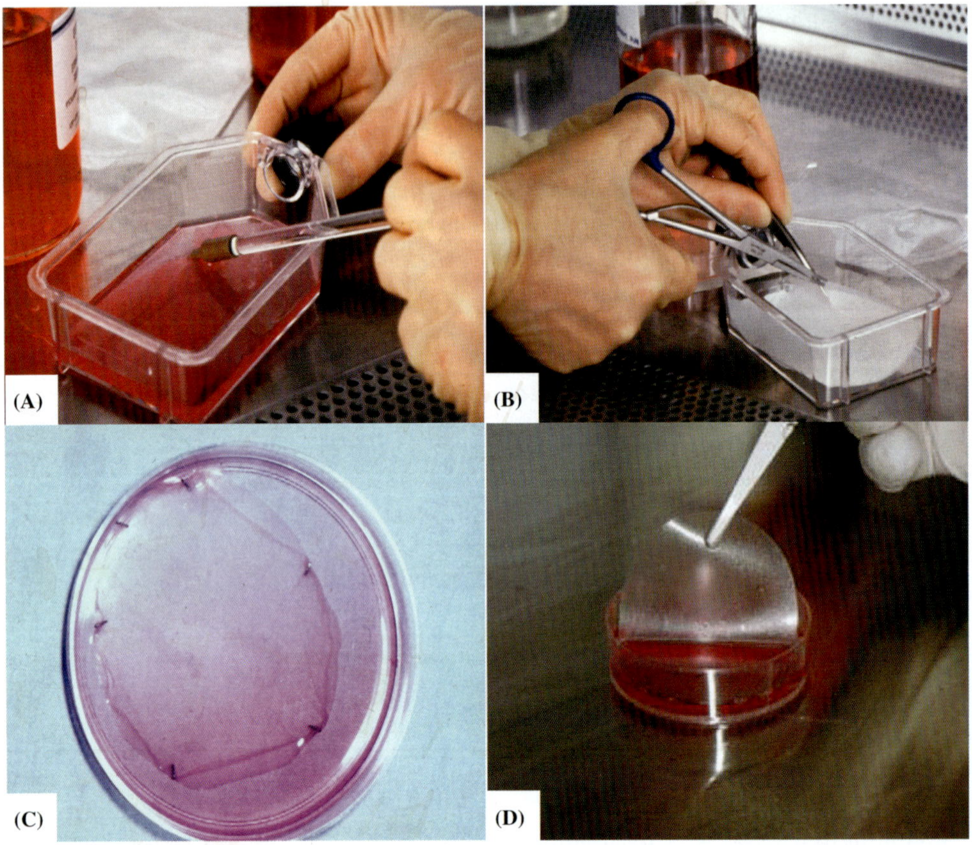

Fig. 1. Preparation of cultured keratinocytes for transfer to the clinic. Cells are initially isolated from a biopsy by enzymic separation of the epidermis and the dermis following an overnight culture in trypsin and then keratinocytes are expanded on a feeder layer of lethally irradiated murine fibroblasts in Greens media plus 10% bovine foetal calf serum. The methodology for delivering cells to the clinic as a sheet cultured epithelial autograft (CEA) as described in[30] is shown in **(A)**, **(B)** and **(C)**. Cells are grown until they reach a confluent sheet approximately two to three layers deep. At this point the whole sheet of cells is enzymically detached from the underlying plastic using a neutral protease, dispase. A rubber-tipped glass rod is used to gently dislodge the whole sheet of cells without tearing it from the tissue culture plastic. In **(B)** the sheet of cells is wrapped around a backing dressing (in this case Tegapore but many different materials can be used) and held in place with surgical staples (liga clips). **(C)** shows a CEA sheet on a backing dressing ready to be delivered to the patient. Cells are delivered cell surface down, backing dressing side uppermost to the patient's wound bed and kept in place for generally five to seven days at which point the backing dressing with whatever cells have not transferred onto the wound bed are taken off and discarded. This methodology requires that cells within the sheet leave the sheet and enter the wound bed. In contrast in **(D)** an alternative methodology for delivering cells to the clinic is shown. Here cells are expanded by conventional culture but then passaged onto this carrier

There is a substantive body of evidence dating from the early 1980s[31,33] to show that cultured skin cells returned to the patient retain the capacity to proliferate throughout the patient's lifetime. There have not been any cases reported of patients treated with cultured cells experiencing a breakdown in their skin (unless subjected to trauma or infection, etc.). Patients treated from the early 1980s onwards will now be wearing cultured skin cells that originated from cultures that are at least 25 years old. Indeed current thinking[4] suggests that skin contains very useful adult stem cell populations. This will be further discussed later in this article.

However, in the treatment of extensive full thickness burns injuries it then became apparent that cultured cells alone were of limited value as they often failed to adhere well to the underlying wound bed. This led to the development of dermal replacement materials using either donor skin to provide a dermis[34] or a dermal substitute material such as Integra.[35] From our own experience of using CEA as illustrated in Fig. 2 in badly burned patients within Sheffield,[36] survival of patients where donor skin was used to provide a well vascularised dermal wound bed prior to treatment with autologous keratinocytes was better in this cohort of patients than for patients who received CEA alone. However as all of those involved in using cultured cells for major burns patients would acknowledge this is a complex area. Survival of burns patients is influenced by several factors of which age, the extent of the burn and the presence of inhalation injuries play a major role in predicting survival.[37]

For these reasons it is not possible to conduct controlled clinical trials of tissue engineered products on badly burned patients where each patient's condition is uniquely different. It is also difficult to see how potentially life saving products can be trialled ethically in major burns patients. Thus in practice burns surgeons treat patients with conventional split thickness skin grafts as their first line of defence and then they will metaphorically reach for tissue engineered products only for those patients with extreme burns injuries where they are concerned that they will not be able to get wound cover rapidly enough.

Donor (cadaveric) skin is extremely beneficial in managing full thickness burns injuries.[34] If sourced from screened consented donors using well regulated tissue banks it can be made to be as low risk for the patient as a blood transfusion. Used as a temporary dressing it provides immediate barrier protection and keeps

---

surface — a medical grade silicone coated with a pin-hole free layer of acrylic acid as described in.[19,51] In this form the cells are then transported to the patient. Discs of cells are placed on the patient's wound bed cell side down, disc uppermost, and kept in place for again usually five to seven days. At this point the silicone disc is discarded, by which time cells will have transferred from the disc onto the wound bed.

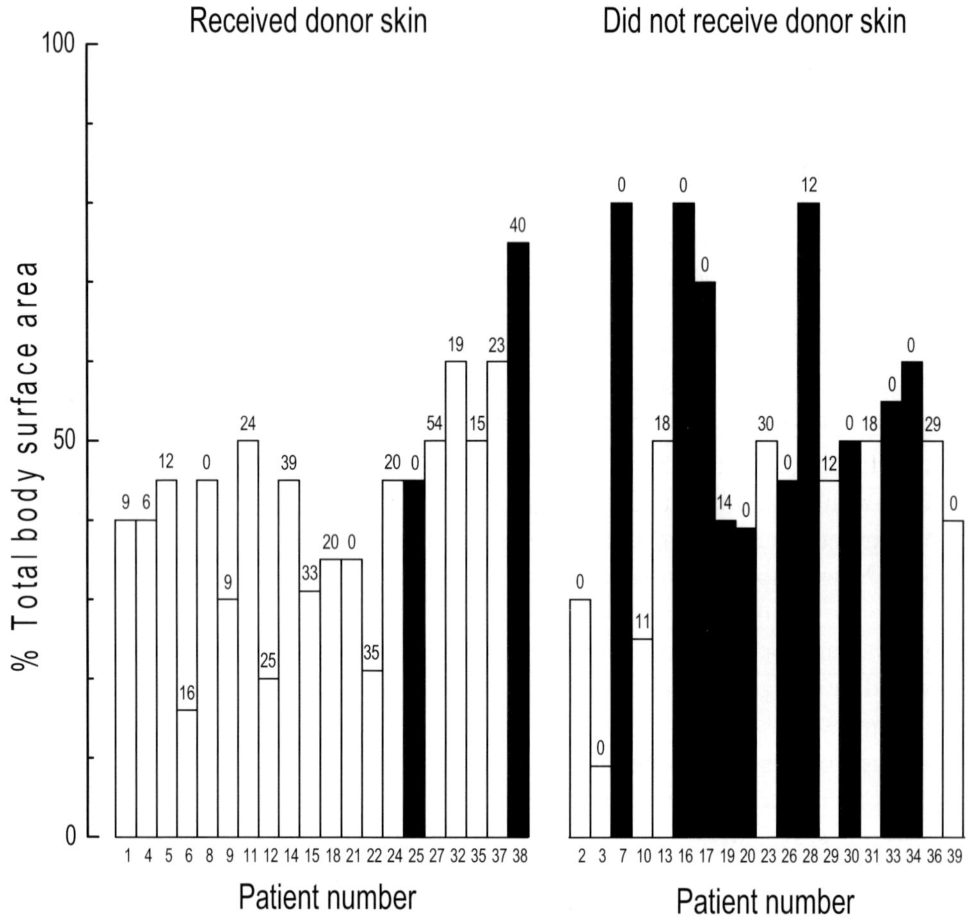

Fig. 2. This figure is taken from an audit of the use of cultured epithelial autografts (CEA) either with or without donor allodermis.[36] The histograms show the extent of the burns injuries of two cohorts of patients, the first group received donor dermis prior to treatment with CEA, the second group did not receive donor dermis and CEAs were added directly to the debrided wound bed. The figures above each bar show how many epithelial autografts they received. Open histograms indicate patients who survived and filled histograms indicate patients who died as a result of their burns injuries. Where patients did not receive CEAs (denoted as a 0) this was either because the patient died before the CEA sheets were ready or that we were unable to match the timing of the CEA production to the clinical needs of the patient as discussed in Ref. 36. (Reproduced from *Regenerative Medicine* **1**(6), 809–821 (2006) with permission of Future Medicine Ltd.)

the underlying wound bed in a good state for future grafting. It can also be used[34,36] to provide a well vascularised donor dermis. Here the split thickness donor skin is grafted onto the excised burned areas of the patient and allowed to engraft. After about three weeks the donor dermis is well vascularised and the upper epidermal layer (which is not immune system compatible) can be detached with little effort and then cultured skin cells are placed onto the vascularised donor dermis.

The tissue engineered products placed on the dermis may vary from cultured keratinocytes alone (delivered as a CEA[1,28,38–40] or a spray[41] or in a fibrin carrier,[42] or on collagen coated[43] or chemically defined membranes[13,19,44] — as illustrated in Fig. 1D and Figs. 3 and 4) or the wound may be grafted with tissue engineered skin of which the best example currently available is that now referred to as Permaderm[45] previously Cincinnati Skin Substitute.[12,46,47] This comprises autologous keratinocytes and fibroblasts in bovine collagen which can provide a permanent skin substitute for burns patients.

Burns surgeons then developed the dermal substitute Integra as an alternative to the human dermis. This was designed specifically for the management of major burns and it is composed largely of bovine collagen with shark chondroitin sulphate. Onto this a silicone membrane has been sealed which acts as a temporary barrier.[35,48]

This material is grafted onto the wound bed and over a period of several weeks vasculogenesis into the neodermis occurs. Only once adequate vascularisation has occurred is the silicone barrier membrane removed and at that point a new and permanent skin barrier must be achieved. It is recommended that surgeons use an autologous split thickness skin graft grafted onto the vascularised dermis at this stage. These can take well on well vascularised Integra.[35] In contrast direct application of cultured cells onto Integra has so far been found to be problematic because cultured cells are unable to attach well enough and quickly enough onto this neodermis which lacks a natural basement membrane. The one product which does work well is the tissue engineered skin previously mentioned[45] as in this case the epidermal cells are already well attached to the fibroblast containing collagen substrate.[12,45]

From the above it will be clear that treatment of major burns injuries is far from simple. For most patients where there is insufficient undamaged skin for autologous skin grafting the patient will require a number of operations and often the damaged skin is replaced in two stages — the first stage achieving a dermal replacement and the second an epidermal replacement (as in Fig. 2). With burns patients the race is to achieve barrier function as quickly as possible, rapid

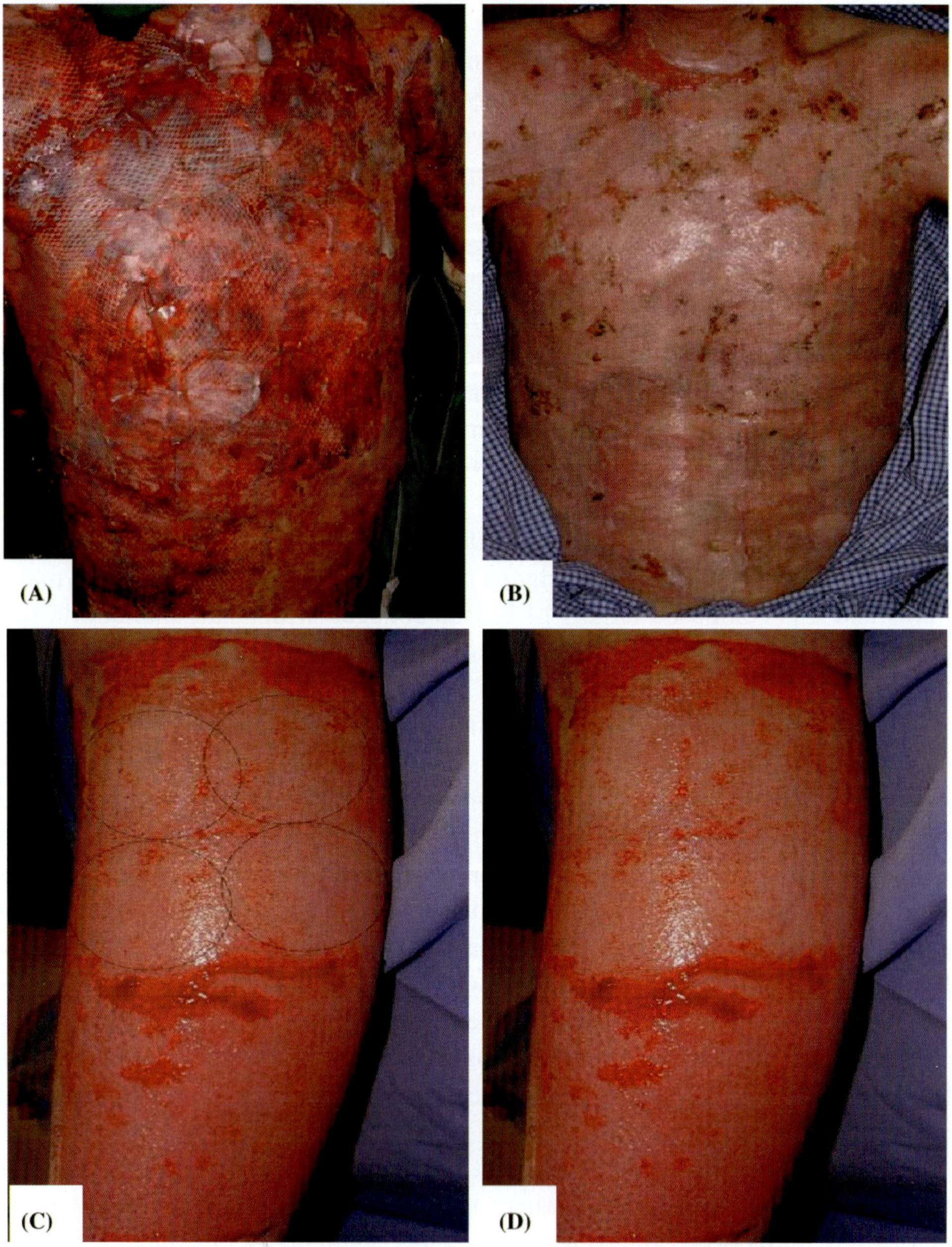

Fig. 3. Clinical uses of tissue engineered skin in two patients who received cultured autologous keratinocytes delivered on the Myskin carrier.[13] In patient **(A)** extensive full thickness injuries were treated with wide meshed autograft over which 20 Myskin discs of autologous keratinocytes were placed. In **(B)** this is the appearance of the patient ten days later after the Myskin carrier discs were removed. The contours of cells transferring from the discs to the wounds can be clearly seen and the degree of epithelialisation at ten days

re-epithelialisation is known to be associated with few complications for the patient. Unfortunately the converse is true. The longer it takes to achieve good barrier function, the higher the chances of the patient experiencing contraction and scarring and often requiring further reconstructive surgery to allow the patient to re-achieve full joint mobility.[49]

4. Development of Tissue Engineered Skin for Chronic Wounds

Chronic wounds often become static through a combination of poor vasculature, infection and chronic inflammation. The current management of chronic wounds can be summarised as debridement, optimisation and then closure. Wounds can be debrided mechanically by scraping or by larval therapy to remove necrotic tissue. Offloading of weight from the affected area and treatment of infection (usually using antimicrobials rather than antibiotics) is also routine for such ulcers. To optimise the underlying wound bed vacuum assisted therapy is sometimes used to stimulate angiogenesis and in many cases a combination of these approaches will enable the wound to then heal without further intervention.

However, despite best wound care practice, there remain large numbers of patients for whom such wounds still refuse to heal. These patients can benefit from the application of tissue engineered skin products which generally act to initiate migration and proliferation of keratinocytes from the edges of the wound. Autologous (as in Fig. 4)[44] or allogeneic keratinocytes and fibroblast-based products (e.g. Dermagraft which consists of donor fibroblasts in a synthetic scaffold[17]) can all prompt wound healing. Repeated applications of cultured skin cells or skin cell-related products have all been shown to prompt chronic non-healing wounds to start healing. In the main, cultured cells are used here as biological factories to assist the body's own wound repair mechanisms and in general the longer the wound has failed to heal the more applications of products and more aggressive treatment will be required to initiate healing.

Chronic wounds represent a different challenge for healing to burns injuries. It is unlikely that one application of any tissue engineered product will result in prompt healing of an ulcer that has been static for months or even years. Patients

---

is very good. In **(C)** this shows the location of four discs of cells placed on a donor site for a patient who required extensive grafting for burns injuries. Here cells were used to accelerate donor site healing. **(D)** shows the appearance of this donor site only three days after removal of the carrier disks. The degree of re-epithelialisation achieved at three days is very much greater than would ordinarily be seen and one can see the contours of the cells transferring from the discs to the donor site. (Reproduced with permission from *The European Journal of Plastic Surgery*.)

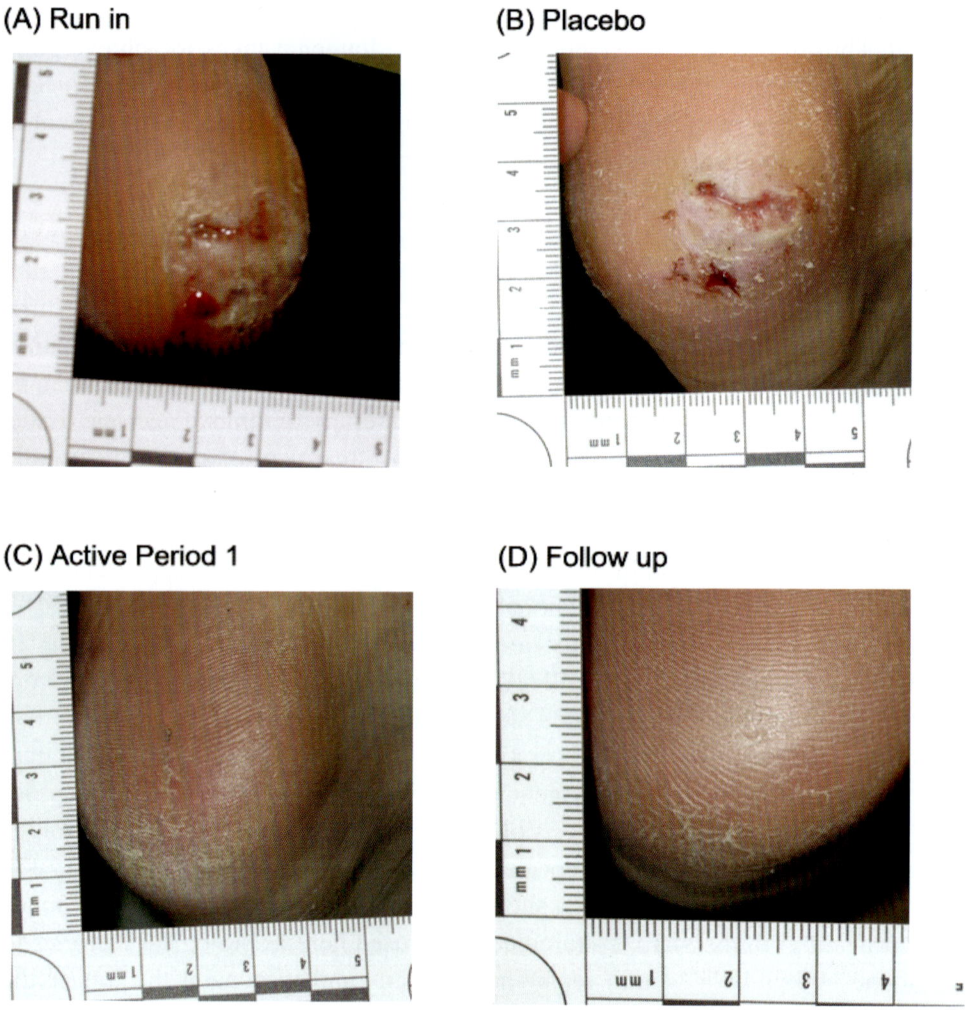

Fig. 4. The use of autologous keratinocytes delivered on the Myskin carrier to treat diabetic foot ulcers. This figure shows the healing progress of an ulcer on the heel of a diabetic patient. This ulcer had been present for nine months without showing any signs of healing despite conventional best practice offloading and wound management care. This patient had three other ulcers which had also persisted for nine months non-healing as described in detail in Ref. 44. This figure shows the appearance of the ulcer at the end of a four-week initial lead in period **(A)**, after six applications of a placebo (an inert carrier dressing without cultured cells **(B)** after treatment with six applications (delivered once per week) of autologous cultured keratinocytes delivered on the Myskin carrier **(C)** and finally **(D)** at the end of the three-month follow-up period. (Reproduced from *Regenerative Medicine* **2**(6), 887–902 (2007) with permission of Future Medicine Ltd.)

are generally treated on an outpatient basis, travelling to specialist clinics (at least in the UK) and the product must be designed to be convenient to be applied in the clinic and easy to be managed by the patient or district nurse. There is also the realisation that, as is the case with diabetic ulcers, the underlying pathology is unchanged hence patients are likely to present with reoccurrence of ulcers at the healed sites or new sites. This is discussed further in a single blind study of the use of cultured autologous keratinocytes delivered on an inert carrier for treatment of diabetic foot ulcers[44] — see also Fig. 4 from this study. Thus having a convenient to use tissue engineered product that can be used rapidly as soon as an ulcer develops rather than when it has become a long-standing problem should be a development goal for tissue engineered products for chronic wounds.

5. Development of Tissue Engineered Skin for Reconstructive Surgery

Finally, reconstructive surgery — where the surgery requires full thickness skin excision then it is usually necessary to take a split thickness skin graft from elsewhere on the body to treat the defect. There are several conditions where tissue engineered skin can benefit patients, e.g. obtaining a better result following excision of cutaneous tumours, scar revision and correction of pigmentation defects ranging from congenital nevi,[21,22] through to vitiligo[23] — see Table 1.

In reconstructive surgery these are usually relatively small defects where there is a good chance of creating a well vascularised wound bed onto which to graft the tissue engineered skin product (even if this must be done in a two-stage operation). These are generally not acute medical emergencies but can be undertaken in a planned manner. For all of these reasons this is an area predicted to grow as the use of a tissue engineered approach will lessen the morbidity of donor sites for the patients.

For vitiligo there is good clinical evidence[24,25] that surgical treatment with cultured melanocytes either on their own or with keratinocytes can offer repigmentation rates of 70%–80% for patients with stable forms of vitiligo. This has the potential to benefit many thousands of patients as vitiligo is an extremely common disease and most dermatologists would agree there is very little to offer these patients.

6. The Design Process for Tissue Engineered Products

The design and evaluation process for tissue engineered products (not just skin) goes through several stages — designing the tissue engineered skin product to work *in vitro*, then designing it to work on an animal model (when required), then designing it to work in a small scale proof of concept clinical study (e.g. such as Moustafa *et al.*[19]) then designing it to work in a multicentre clinical study (e.g. such as Moustafa *et al.*[44]), then finally having it work well for large numbers of patients with clinicians not involved in controlled studies. It may prove necessary to revisit aspects of the product design at each stage.

For example, in developing tissue engineered skin in our laboratory, as shown in Fig. 5, before going to the clinic we evaluated the material in nude mice as shown in Figs. 5E and 5F.[55] With animal experimentation there may then be a need to adjust the design of the tissue engineered product again.[56] Thus our reason for taking our tissue engineered skin to nude mice was to see whether it would become well vascularised. If it had failed to vascularise then this would have been a strong contraindication to moving it forward to the clinic. At this point we would have redesigned the tissue engineered skin.

Looked at this way its easy in hindsight to see why so relatively few tissue engineered skin products have made it through to general clinical use after even 25 years of research.

7. Keratinocyte Stem Cells — Where are We?

It is a time of rapid change in our understanding of what constitutes a stem cell. At present there is good agreement that tissue specific stem cells may not be so committed and that transdifferentiation may occur, e.g. Refs. 57 to 59. There are acknowledged to be two compartments for multipotent stem cells in skin and hair. One is in a niche environment in the rete ridges of the skin and the other is in the bulge region of the hair follicle.

For a recent review of how the epidermis and its appendages develop from a single layer of multipotent embryonic progenitor keratinocytes, please see Ref. 4. Adult skin epithelia maintain populations of stem cells which are involved in re-epithelialisation in response to wounding and in maintaining the natural cycle of hair follicle regeneration. As discussed in Ref. 4, a consistent finding that developmental biologists have long been aware of is that epithelial cells in embryo are sensitive to the mesenchyme to which they are exposed. Thus in birds embryonic epithelial cells confronted with mesenchyme from the wing produce feathers, but if confronted with mesenchyme from the leg produce scales. This responsiveness seems to be lost as development proceeds but the tantalising prospect offered at present is that multipotent stem cells of the skin may retain this potential. This may open doors for the future particularly as skin is an accessible tissue.

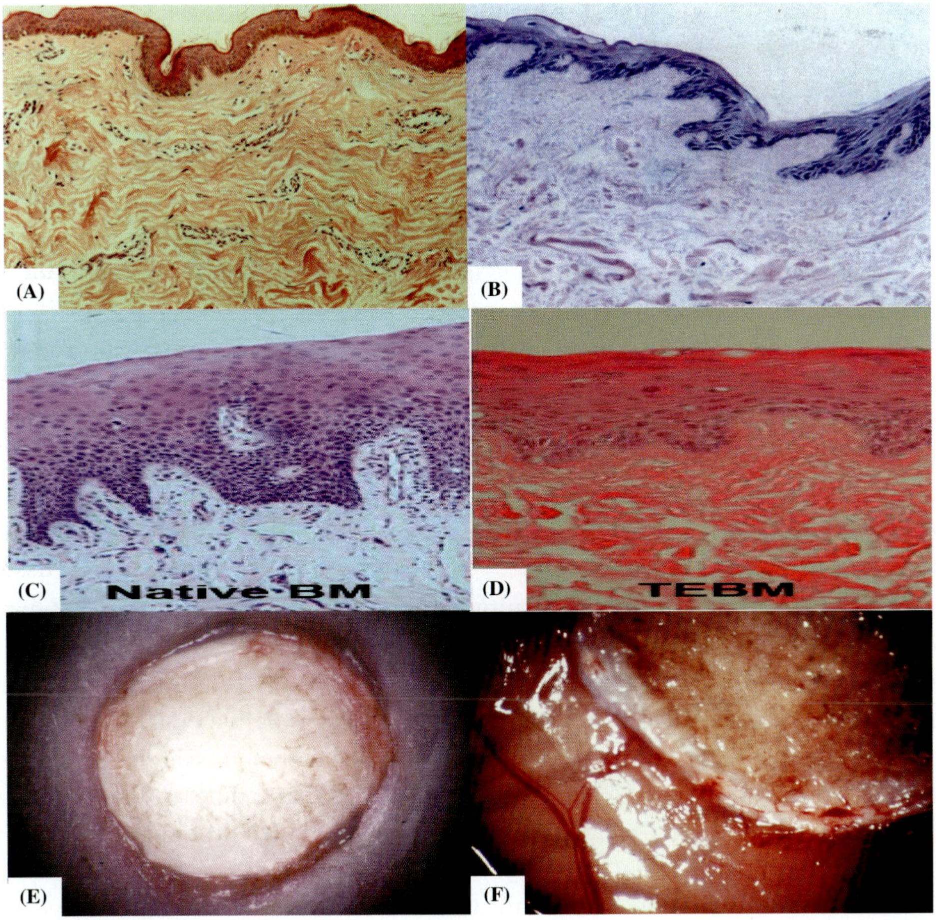

Fig. 5. Development of tissue engineered skin. **(A)** shows the histology of normal skin and **(B)** the histology of tissue engineered skin produced using sterilised de-epidermised acellular dermis to which laboratory expanded keratinocytes and fibroblasts are added. The development of this tissue engineered skin is described in Ref. 55. **(C)** shows a section through normal buccal mucosa and **(D)** shows tissue engineered buccal mucosa developed as above using sterilised de-epidermised acellular dermis to which oral epithelial cells and fibroblasts are added. Details as described in Ref. 63. **(E)** shows the appearance of tissue engineered human skin on the back of nude mice at two weeks, details as described in Ref. 55. **(F)** shows a close-up of tissue engineered skin on the back of a nude mouse showing mouse vasculature entering human skin at one week.

8. Clinical and Future Development Issues

The main methodology for expanding keratinocytes is based on a feeder layer of irradiated murine cells in a bovine serum medium. Neither murine cells nor bovine sera were perceived as clinical risks in 1975 when the methodology was created. The risk of transferring murine viruses can be reduced by lethal gamma irradiation and good cell banking practices whilst the risk of transmission of bovine spongiform encephalitis (BSE) is managed by sourcing serum from Australasia where herds are free from BSE. Ideally, a completely xenobiotic-free media is needed, and although considerable progress has been made, none of the current candidates have regulatory approval. Donor skin (human allodermis) needs to be from screened donors via accredited skin banks. Risks can be further reduced by sterilisation of human skin.[60]

Safety and risk management are key issues when developing skin products for clinical use. Reducing risks of tissue engineered skin will increase its use for less life-threatening conditions such as reconstructive surgery, vitiligo and paediatric scalds (where cultured cells can accelerate healing reducing the risk of contractures and scarring — currently 30% of children with superficial scalds have complications requiring further surgical interventions as the child grows[61]).

A major challenge is getting cultured cells to attach well to the patient's wound beds which are often poorly vascularised, infected and degradative. For burns patients, placing cultured cells over wide meshed autograft (to fill in the interstices) works well and this can be achieved for cells delivered as CEAs[1] or on a carrier dressing such as Myskin[13] or sprayed.[14] Attachment research is progressing (e.g. avidin biotin[62]) but a methodology is needed that can be readily translated to the clinic. A spray-on basement membrane substitute would richly reward further research.

There are several clinical problems that can occur after grafting. Tissue engineered skin, unlike normal skin, has no residual vasculature. Consequently, tissue engineered grafts will only succeed if they are very thin or placed over well-vascularised wound beds. Expensively constructed tissue placed over poorly vascularised wound beds or fat will be lost.[20]

Experience in Sheffield, UK, showed that take was very much dependent on wound bed condition. Tissue engineered oral mucosa[63] has been used to correct fibrosis of the urethra, which in contrast to skin has an excellent blood supply and vasculogenesis of this reconstructed material was good in five out of five patients[64]). Angiogenesis is a challenging multifactorial research area requiring combinations of cells and stimuli. Pioneering work on blood vessel formation in tissue engineered constructs *in vitro* is being undertaken at present[65] but much more is needed.

Other problems are contracture and abnormal pigmentation (see below). Tissue engineered skin based on normal dermis contracts significantly in the laboratory[49,66,67] and also on the patient.[56] Abnormal pigmentation may seem less of an issue when considering life-saving surgery but patients find this to be a major barrier to recovering their self-esteem.[68] Both these areas can be studied *in vitro*.

Other barriers are regulatory and commercial issues. Good research leading to therapies that work clinically does not guarantee commercial success. The field of tissue engineering has suffered from unrealistic commercial expectations. Innovative tissue engineered products are often developed within small research groups or small companies with insufficient funds to progress beyond clinical proof of concept. Larger companies acquiring such technology do not always succeed. As these products are new, they do not fit readily into the established regulatory models of medical devices or drugs and the consequent uncertainty of regulation leads to different solutions in different countries, radically increasing costs. The slow evolution of the regulatory systems' responses to tissue engineering has been a major brake on development. There are however positive cost benefit equations for tissue engineering. For example, healing a chronic diabetic ulcer using autologous cultured skin cells is very much cheaper than the alternative, as well as delivering significant improvements to patients' quality of life. In the UK it was recently estimated that annual expenditure attributable to diabetic foot complications, excluding amputation costs, was 152 million pounds.[69] How health care organisations view these new therapies is as important as the development of the therapies themselves. Other medical advances have similarly struggled to establish a foothold. For example, kidney and heart transplantation, where there is undoubted benefit to the patient but development costs were very high, were initially viewed as unacceptable. Costs remain high, but they are now viewed as acceptable within most Western health care frameworks. The conclusion is that clear clinical benefit and persistence eventually succeed. Tissue engineering today needs more controlled studies showing clinical benefits. These will make the case for commercial sustainability.

9. Laboratory Uses of Tissue Engineered Skin

This is a rapidly growing area driven by the need to reduce or replace animal experimentation in the development of products that are used on human skin and hair[70] and by the realisation that many of the more interesting things that happen in skin happen through a dialogue between different skin cells and are rarely seen when individual cells are studied in monolayers. Table 3 lists some of the *in vitro* uses of tissue engineered skin.

Figure 5 shows examples of 3D tissue engineered skin and oral mucosa based on natural dermis. These mimic the normal physiology of these epithelial tissues but lack any vasculature, immune system or innervation (and these remain big challenges). Although they cannot fully replace *in vivo* models, they are ideal models to examine cell/cell dialogues and cell/extracellular matrix interactions.

Table 3. Laboratory uses of tissue engineered skin.

- Models to replace some animal testing
  - Dermatotoxicity
  - Mucotoxicity
  - Skin irritancy
- Skin barrier/skin penetration studies
  - Wound healing models
  - Angiogenesis models
  - Skin contraction
- Pigmentation research
  - Photoprotection studies
  - Depigmentation studies
- Models to investigate disease processes
  - Melanoma models
  - Aetiology of vitiligo
  - Blistering diseases
  - Psoriasis

The main impetus for developing tissue engineered skin is the need to find alternatives to animal testing for the many thousands of chemical additives used in human skin products. Companies such as L'Oréal[71] and Skin Ethic[72] have developed skin models for these purposes and models are also being assessed for their ability to detect agents that might induce contact dermatitis. As yet most commercial models contain only keratinocytes (on a membrane or collagen) and do not include fibroblasts.

The presence of fibroblasts in 3D skin models helps keratinocytes cope with cytotoxic agents such as silver and nickel[73] and also affects the pigmentary response of melanocytes *in vitro*. Thus melanocytes from pale skin donors, when introduced to reconstructed skin gave barely pigmented skin but spontaneously pigmented if fibroblasts were omitted from the models.[74] Further work in 2D co-culture models demonstrated that while normal fibroblasts will suppress melanocyte pigmentation stressed fibroblasts (stressed by freezing or gamma irradiation) produced soluble factors which induced pigmentation in melanocytes. These results were confirmed for melanocytes derived from skin, hair and eye indicating that this may be a generic biology.[75] All of the former suggest that the fibroblast may play a hitherto undetected role in helping epidermal cells cope with UV, oxidative and inflammatory stress.

Figure 6 shows that normal human melanocytes seem to know where to go, in 3D skin, orientating in the basal layer of keratinocytes, tightly regulated by

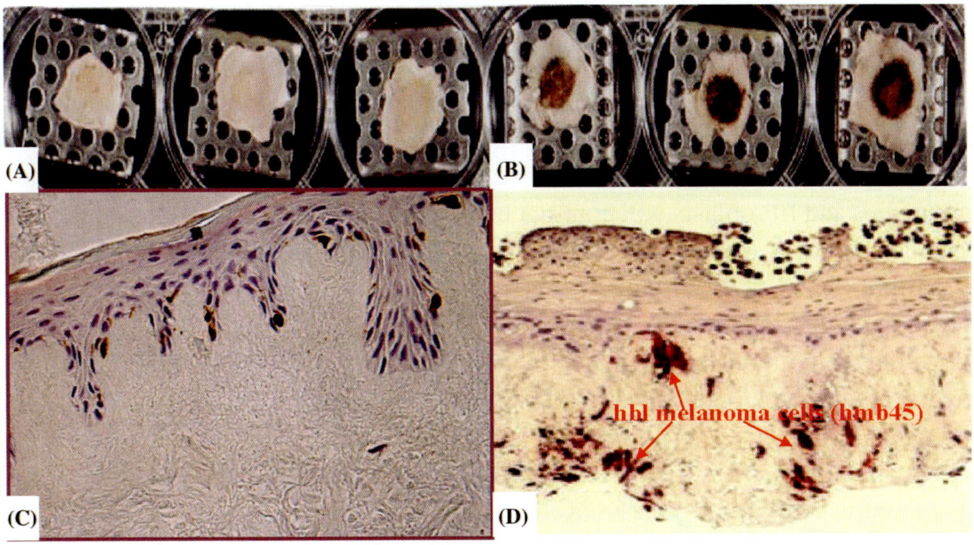

Fig. 6. The behaviour of melanocytes and melanoma cells added to tissue engineered skin. Tissue engineered skin was constructed based on sterilised de-epidermised acellular dermis to which laboratory keratinocytes and fibroblasts and melanocytes (**A** and **C**) or melanoma cells (**B** and **D**) were added. (**A**) and (**B**) show the gross morphology of this reconstructed tissue engineered skin after ten days' culture at an air–liquid interface. In (**A**) the pale morphology is typical of the skin type of the type I skin donors from which the melanocytes were cultured.[74] In (**B**) the pigmentation seen is due to the presence of the human HBL melanoma cell line.[76] (**C**) and (**D**) show H & E histology of these tissue engineered skins. In (**C**) the melanocytes have been stained with S100 and it can be seen that they are distributed in the basal layer of keratinocytes in the deep rete ridges as would be the case in normal skin. In (**D**) the melanoma cells have been stained with HMB45 and the majority of the melanoma cells are on the upper superficial layers of the epidermis but it can be seen that a substantive number have penetrated through the epidermis to the dermis. Full details as described in Ref. 76.

both the keratinocytes, and fibroblasts.[74,76] In contrast, melanoma cells added to the same model re-organise differently. Many cells remain in the upper superficial layers of the epidermis but some invade through the basement membrane into the dermis. Some melanoma cells were found to be only able to invade the dermis when skin cells were added[75,76] while others invaded irrespective of the presence of skin cells.[77] The results suggested that the melanoma cells largely dictated the outcome of their interaction with the adjacent skin cells but also that some cells appeared to use the activated degradative enzymes (metalloproteinases MMP-2 and MMP-9) of the adjacent skin cells for invasion rather than any intrinsic enzyme activity.[77]

Several authors now suggest that inflammation plays a major role in the progression of a range of solid tumours, e.g. see Ref. 78. Studies in both 3D[79] and 2D[80,81] models have suggested that inflammation accelerates melanoma migration.

The 3D skin model also can be used to study skin graft contraction. Both keratinocytes and fibroblasts will contract bovine collagen gels but only keratinocytes seem capable of contracting mature cross-linked human dermal collagen.[66,67] Keratinocytes were then found to contract allodermis through lysyl oxidase, a collagen cross-linking enzyme. Inhibitors of lysyl oxidase blocked contraction to a large degree without having adverse effects on the skin morphology.[49] Currently patients who experience severe skin graft contraction are treated with pressure garments which they may need to wear for up to 12 or even 18 months.[61] Alternative approaches to preventing initial contraction are long overdue.

Finally, skin models can be used to gain further understanding of the process of major skin diseases such as vitiligo, blistering diseases (epidermolysis bullosa) and psoriasis.

One laboratory studying vitiligo with 3D models reports that melanocytes in 3D culture spontaneously produce pigment reflecting their donor phototype origin.[82] Recent work from these and others suggests that both keratinocytes and melanocytes from patients with vitiligo are less able to cope with oxidative stress than cells from unaffected individuals.[83]

The study of the genetic causes of the various blistering diseases is relatively mature and defective genes have been identified. The question now is whether these can be corrected in skin cells such that epidermal cells regain the ability to attach well to the underlying dermis. Studies combining tissue engineering and genetically altered keratinocytes with the aim of improving their adhesion to the patient's dermis are now holding out hope for this condition.[23,24]

With respect to psoriasis recent work has shown that transglutaminase inhibitors induce a phenotype in epidermal keratinocytes which shares many of the features of psoriasis and there is evidence for transglutaminase abnormalities in psoriasis.[84] Psoriasis is a particularly difficult disease in which to do research as there are no good animal or experimental models.

10. Conclusion

In clinical terms tissue engineered skin is now 25 years old. Major problems concerning the hype associated with tissue engineering, in that it would revolutionise medicine, be amazingly fast and generate vast profits for companies have, not surprisingly, proven to be overambitious. Like all products designed for medicine it has been calculated that the average time between conception of the idea and clinical take up into practical use is 28 years. Against this timetable, tissue engineered skin is on schedule to become mainstream for conditions that cannot be readily treated using conventional approaches. It has proven it can deliver clinical benefit. Where it is also delivering value is with the production of a broad array of tissue engineered models of skin. These allow us to gain a better understanding of skin biology, and eliminate some animal experimentation and allow us to undertake experiments that could not be readily tackled using patients, animal models or monolayers of cells. The use of these models to investigate challenging skin diseases has only just begun. Thus in research terms tissue engineered skin is coming of age, while in terms of clinical benefit the argument now is not whether it benefit patients — it clearly can — but of it translating into routine clinical practice.

References

1. O'Connor, N.E., Mulliken, J.B., Banks-Schlegel, S., Kehinde, O. and Green, H. Grafting of burns with cultured epithelium prepared from autologous epidermal cells. *Lancet* **1**, 75–78 (1981).
2. MacNeil, S. Progress and opportunities in tissue engineering of skin. *Nature Insights. Nature* **445**, 874–880 (2007).
3. Lin, J.Y. and Fisher, D.E. Melanocyte biology and skin pigmentation. *Nature Insights. Nature* **445**, 843–850 (2007).
4. Fuchs, E. Scratching the surface of skin development. *Nature Insights. Nature* **445**, 834–842 (2007).
5. Tobin, D.J., Gunin, A., Magerl, M., Handijski, B. and Paus, R. Plasticity and cytokinetic dynamics of the hair follicle mesenchyme: implications for hair growth control. *J. Invest. Dermatol.* **120**, 895–904 (2003).
6. Philpott, M.P., Green, M.R. and Kealey, T. Human hair growth *in vitro*. *J. Cell Sci.* **97,** 463–471 (1990).
7. Lumpkin, E.A. and Caterina, M.J. Mechanisms of sensory transduction in the skin. *Nature Insights. Nature* **445**, 858–865 (2007).
8. Herndon, D.N. *et al.* A comparison of conservative versus early excision therapies in severely burned patients. *Ann. Surg.* **209**, 547–552 (1989).
9. Falanga, V. Chronic wounds: pathophysical and experimental considerations. *J. Invest. Dermatol.* **100**, 721–725 (1993).
10. Philips, T., Santon, B., Provan, A. and Lew, R. A study of the impact of leg ulcers on quality of life: financial, social, and psychologic implications. *J. Am. Acad. Dermatol.* **31**, 49–53 (1994).
11. Stern, R., McPherson, M. and Longaker, M.T. Histologic study of artificial skin used in the treatment of full thickness thermal injury. *J. Burn. Rehabil.* **11**, 7–13 (1990).

12. Boyce, S.T. *et al.* The 1999 Clinical Research Award. Cultured skin substitutes combined with Integra to replace native skin autograft and allograft for closure of full-thickness burns. *J. Burn. Care Rehabil.* **20**, 453–461 (1999).
13. Zhu, N. *et al.* Treatment of burns and chronic wounds using a new cell transfer dressing for delivery of autologous keratinocytes. *Eur. J. Plast. Surg.* **28**, 319–330 (2005).
14. Navarro, F.A. *et al.* Sprayed keratinocyte suspensions accelerate epidermal coverage in a porcine microwound model. *J. Burn Care Rehabil.* **21**(6), 513–518 (2000).
15. Bello, Y.M. and Falabella, A.F. The role of Graftskin (Apligraf®) in difficult-to-heal venous leg ulcers. *J. Wound Care* **11**(5), 182–183 (2003).
16. Apligraf® from Organogenesis: www.organogenesis.com.
17. Marsten, W.A., Hanft, J., Norwood, P. and Pollak, R. The efficacy and safety of Dermagraft in improving the healing of chronic diabetic foot ulcers: results of a prospective randomized trial. *Diabetes Care* **26**(6), 1701–1705 (2003).
18. Dermagraft® from Advanced BioHealing: www.AdvancedBioHealing.com.
19. Moustafa, M. *et al.* A new autologous keratinocyte dressing treatment for non-healing diabetic neuropathic foot ulcers. *Diabetic Med.* **21**, 786–789 (2004).
20. Sahota, P.S. *et al.* Development of a reconstructed human skin model for angiogenesis. *J. Wound Repair Regen.* **11**, 275–284 (2003).
21. Bittencourt, F.V. *et al.* Large congenital melanocytic nevi and the risk of development of malignant melanoma and neurocutaneous melanocytosis. *Pediatrics* **106**, 736–741 (2000).
22. Gallico, G.G. *et al.* Cultured epithelial auto grafts for giant congenital nevi. *Plast. Reconstr. Surg.* **84**, 1–9 (1989).
23. Olsson, M.J. and Juhlin, L. Long-term follow-up of leucoderma patients treated with transplants of autologous cultured melanocytes, ultrathin epidermal sheets and basal cell layer suspension. *Br. J. Dermatol.* **147**, 893–904 (2002).
24. Falabella, R., Escobar, C. and Borerro, I. Treatment of refractory and stable vitiligo by transplantation of *in vitro* cultured epidermal autografts bearing melanocytes. *J. Am. Acad. Dermatol.* **26**, 230–236 (1992).
25. Gupta, S., Mats, O., Kanwar, A. and Ortonne J.-P. (eds.). *Surgical Management of Vitiligo* (Oxford, Blackwell Publishing, 2006), 289 pp.
26. Ferrari, S., Pellegrini, G., Matsui, T., Mavilio, F. and De Luca, M. Gene therapy in combination with tissue engineering to treat epidermolysis bullosa. *Expert Opin. Biol. Ther.* **6**(4), 367–378 (2006).

27. Mavilio, F. *et al.* Correction of junctional epidermolysis bullosa by transplantation of genetically modified epidermal stem cells. *Nat. Med.* **12**(12), 1397–1402 (2006).
28. Rheinwald, J.G. and Green, H. Serial cultivation of strains of human epidermal keratinocytes: the formation of keratinizing colonies from single cells. *Cell* **6**, 331–344 (1975).
29. Rheinwald, J.G. and Green, H. Epidermal growth factor and the multiplication of cultured human epidermal keratinocytes. *Nature* **265**, 421–424 (1977).
30. Green, H., Kehinde, O. and Thomas, I. Growth of cultured human epidermal cells into multiple epithelia suitable for grafting. *Proc. Natl. Acad. Sci. USA* **76**, 5665–5668 (1979).
31. Gallico, G.G., O'Connor, N.E., Compton, C.C., Kehinde, Q. and Green, H. Permanent coverage of large burn wounds with autologous cultured human epithelium. *New Engl. J. Med.* **311**, 448–451 (1984).
32. Herndon, D.N. and Parks, D.H. Comparison of serial debridement and auto grafting and early massive excision with cadaver skin overlay in the treatment of large burns in children. *J. Trauma* **26**, 149–152 (1986).
33. Odessey, R. Addendum: multicenter experience with cultured epidermal autograft for treatment of burns. *J. Burn Care Rehabil.* **13**, 174–180 (1992).
34. Cuono, C.B., Langdon, R. and McGuire, J. Use of cultured epidermal autografts and dermal allografts as skin replacement after burn injury. *Lancet* **1**, 1123–1124 (1986).
35. Burke, J., Yannas, I., Quinby, W.C., Bondoc, C.C. and Jung, W.K. Successful use of a physiologically acceptable artificial skin in the treatment of extensive burn injury. *Ann. Surg.* **194**, 413–427 (1981).
36. Hernon, C.A., Dawson, R.A., Freedlander. E., Short, R., Haddow, D.B., Brotherston, M. and MacNeil, S. Clinical experience using cultured epithelial autografts leads to an alternative methodology for transferring skin cells from the laboratory to the patient. *Regen. Med.* **1**(6), 809–821 (2006).
37. Rose, J.K. and Herndon, D.N. Advances in the treatment of burn patients. *Burns* **23**, 721–725 (1997).
38. Compton, C.C. Current concepts in pediatric burn care: the biology of cultured epithelial autografts: an eight-year study in pediatric burn patients. *Eur. J. Pediatr. Surg.* **2**, 216–222 (1992).
39. Carsin, H.P. *et al.* Cultured epithelial autografts in extensive burn coverage of severely traumatized patients: a five-year single-center experience with 30 patients. *Burns* **26**, 379–387 (2000).
40. Epicel™ from Genzyme: www.genzymebiosurgery.com.

41. CellSpray® from C3: www.clinicalcellculture.com.
42. Horch, R.E., Bannasch, H. and Stark, G.B. Transplantation of cultured autologous keratinocytes in fibrin sealant biomatrix to resurface chronic wounds. *Transpl. Proc.* **33**, 642–644 (2001).
43. Horch, R.E., Debus, M., Wagner, G. and Stark, G.B. Cultured human keratinocytes on type I collagen membranes to reconstitute the epidermis. *Tissue Eng.* **6**, 53–67 (2000).
44. Moustafa, M. *et al.* Randomized, controlled, single-blind study on use of autologous keratinocytes on a transfer dressing to treat non-healing diabetic ulcers. *Regen. Med.* **2**(6), 887–902 (2007).
45. Cincinnati Skin Substitute — now referred to as Permaderm™, www.cambrex.com.
46. Boyce, S.T. *et al.* Cultured skin substitutes reduce donor skin harvesting for closure of excised, full-thickness burns. *Ann. Surg.* **235**, 269–279 (2002).
47. Permacol®, a natural matrix derived from decellularized porcine dermis.
48. Integra® DRT (dermal replacement template) from Integra: www.integra-ls.com.
49. Harrison, C.A. *et al.* Use of an *in vitro* model of tissue-engineered skin to investigate the mechanism of skin graft contraction. *Tissue Eng.* **12**(11), 3119–3133 (2006).
50. EpiDex ™ from Modex Therapeutics: www.modex-t3r.com.
51. Myskin® from Celltran Ltd: www.celltran.co.uk.
52. AlloDerm® regenerative dermis from www.lifecell.com.
53. TransCyte® from Advanced BioHealing: www.AdvancedBioHealing.com.
54. Orcel™: www.ortecinternational.com.
55. Chakrabarty, K.H. *et al.* Development of autologous human epidermal/dermal composites based on sterilised human allodermis for clinical use. *Br. J. Dermatol.* **141**, 811–823 (1999).
56. Sahota, P.S., Burn, J.L., Brown, N.J. and MacNeil, S. Approaches to improve angiogenesis in tissue-engineered skin. *Wound Repair Regen.* **12**, 635–642 (2004).
57. Reynolds, A.J. and Jahoda, C.A.B. Cultured dermal papilla cells induce follicle formation and hair growth by transdifferentiation of an adult epidermis. *Development* **115**, 587–593 (1992).
58. Gharzi, A., Reynolds, A.J. and Jahoda, C.A.B. Plasticity of hair follicle dermal cells in wound healing and induction. *Exp. Dermatol.* **12**, 126–136 (2003).
59. Fliniaux, I., Viallet, J.P., Dhouailly, D. and Jahoda, C.A.B. Transformation of amnion epithelium into skin. *Differentiation* **72**, 558–565 (2004).

60. Huang, Q., Dawson, R., Pegg, D.E., Kearney, J.N., and MacNeil, S. Use of peracetic acid to sterilise human donor skin for production of acellular matrices for clinical use *J. Wound Repair Regen.* **12**, 276–287 (2004).
61. Harrison, C. The mechanism of skin graft contraction: an update on current research and potential future therapies. *Burns* **34**(2), 153–163 (2008).
62. De Bank, P.A. *et al.* Accelerated formation of multicellular 3D structures by cell-to-cell cross-linking. *Biotechnol. Bioeng.* **97**(6), 1617–1625 (2007).
63. Bhargarva, S., Chappell, C., Bullock, A.J., Layton, C. and MacNeil, S. Tissue-engineered buccal mucosa for substitution urethroplasty. *Br. J. Urol.* **93**, 807–811 (2004).
64. Bhargava, S., Patterson, J., Inman, R.D., MacNeil, S. and Chapple, C.R. Tissue engineered buccal mucosa urethroplasty — clinical outcomes. *Eur. Urol.* (available online).
65. Mertsching, H., Walles, T., Hofmann, M., Schanz, J. and Woffram, H.K. Engineering of a vascuarised scaffold for artificial tissue and organ regeneration. *Biomaterials* **26**, 6610–6617 (2005).
66. Ralston, D.R. *et al.* Keratinocytes contract normal human dermal extracellular matrix and reduce soluble fibronectin production by fibroblasts in a skin composite model. *Br. J. Plast. Surg.* **50**, 408–415 (1997).
67. Chakrabarty, K.H. *et al.* Keratinocyte-driven contraction of reconstructed human skin. *Wound Repair Regen.* **9**, 95–106 (2001).
68. Medalie, D.A. *et al.* Differences in dermal analogs influence subsequent pigmentation, epidermal differentiation, basement membrane and rete ridge formation of transplanted composite skin grafts. *Transplantation* **64**, 454–465 (1997).
69. Gordois, A., Scuffman, P., Shearer, A. and Oglesby, A. The health care costs of diabetic peripheral neuropathy in the United States and the United Kingdom. *J. Diabetes Complicat.* **18**, 18–26 (2004).
70. Sauer, U.G., Spielmann, H. and Rusche, B. Fourth EU report on the statistics on the number of animals used for scientific purposes in 2002 — trends, problems, conclusions. *ALTEX* **22**, 59–67 (2002), www.altex.ch/hauptseite_e.htm.
71. Facy, V., Flouret, V., Regnier, M. and Schmidt, R. Reactivity of Langerhans cells in human reconstructed epidermis to known allergens and UV radiation. *Toxicol. In Vitro* **19**, 787–795 (2005).
72. Welss, T., Basketter, D.A. and Schroder, K.R. *In vitro* skin irritation: facts and future. State of the art review of mechanisms and models. *Toxicol. In Vitro* **18**(3), 231–243 (2004).
73. Sun, T., Jackson, S., Haycock, J.W. and MacNeil, S. Culture of skin cells in 3D rather than 2D improves their ability to survive exposure to cytotoxic agents. *J. Biotechnol.* **122**, 372–381 (2006).

74. Hedley, S.J. *et al.* Fibroblasts play a regulatory role in the control of pigmentation in reconstructed human skin from skin types I and II. *Pigment Cell Res.* **15**, 49–56 (2002).
75. Balafa, C. *et al.* Dopa oxidase activity in the hair, skin and ocular melanocytes is increased in the presence of stressed fibroblasts. *Exp. Dermatol.* **14**, 363–372 (2005).
76. Eves, P. *et al.* Characterisation of an *in vitro* model of human melanoma invasion based on reconstructed human skin. *Br. J. Dermatol.* **142**, 210–222 (2003).
77. Eves, P. *et al.* Melanoma invasion in reconstructed human skin is influenced by skin cells — investigation of the role of proteolytic enzymes. *Clin. Exp. Metastasis* **20**, 685–700 (2003).
78. Balkwill, F. and Mantovani, A. Inflammation and cancer: Back to Virchow? *Lancet* **357**, 539–545 (2001).
79. Eves, P. *et al.* Anti-inflammatory and anti-invasive effects of alpha-melanocyte stimulating hormone in human melanoma cells. *Br. J. Cancer* **89**, 2004–2015 (2003).
80. Zhu, N., Lalla, R., Eves, P., Brown, T.L., King, A., Kemp, E.H., Haycock, J.W. and MacNeil, S. Melanoma cell migration is upregulated by tumour necrosis factor-alpha and suppressed by alpha-melanocyte stimulating hormone. *Br. J. Cancer* **90**, 1457–1463 (2004).
81. Katerinaki, E. *et al.* Sodium salicylate inhibits TNF-α induced NF-κB activation, cell migration, invasion and ICAM-1 expression in human melanoma cells. *Melanoma Res.* **16**(1), 11–22 (2006).
82. Harrison, C.A., Layton, C.M., Hau, Z., Bullock, A.J., Johnson, T.S. and MacNeil, S. Transglutaminase inhibitors induce hyperproliferation and parakeratosis in tissue-engineered skin. *Br. J. Dermatol.* **156**(2), 247–257 (2007).
83. Bessou, S. *et al.* *Ex vivo* study of skin phototypes *J. Invest. Dermatol.* **107**, 684–688 (1996).
84. Cario-Andre, M., Pain, C., Gauthier, Y. and Taieb, A. The melanocytorrhagic hypothesis of vitiligo tested on pigmented, stressed, reconstructed epidermis. *Pigment Cell Res.* **20**, 385–393 (2007).

Chapter 29

Liver Repair

Nataša Levičar, Madhava Pai and Nagy A. Habib

Abstract

Stem cell therapy is currently one of the most exciting areas of biomedical research with hopes of providing therapeutic treatments for a myriad of diseases, including liver diseases. Several liver diseases fall under this category, including fibrosis of the liver, and hepatitis B and C viral infection. At the cirrhotic stage, liver disease is considered irreversible and the only alternative is orthotopic liver transplantation. While orthotopic liver transplantation cures chronic liver disease and a variety of metabolic and genetic deficiency disorders, the increased shortage of donor organs restricts liver transplantation. Therefore novel therapeutic options are in demand.

Adult stem cells with their multilineage differentiation potential and self-renewal are possible candidate cells and it is believed that novel cellular therapeutics can perform better than any medical device, recombinant proteins or therapeutic agents. In this chapter we have presented our own experience using adult stem cells for therapeutics for liver disease as well as reviewed the latest literature on this topic.

Keywords: Liver Disease; Cell Therapy; Adult Stem Cells.

Outline

1. Introduction
2. Haematopoietic Stem Cells and Liver Regeneration
3. Clinical Studies
4. Conclusions

References

1. Introduction

Liver plays a major role in metabolism and has a number of functions in the body, including the regulation of energy homeostasis, glycogen storage, plasma protein synthesis and detoxification.[1] It has the remarkable capacity to regenerate itself in response to injury. However, in severe cases of liver injury, its regenerative capacity may prove insufficient and the liver injury may progress to end-stage liver disease and subsequent liver failure. At this stage the only therapeutic option is orthotopic liver transplantation. However, due to the worldwide shortage of available organs, alternative therapies, such as cell therapy are being investigated. Cell transplantation is less invasive than whole-organ transplantation and can be performed repeatedly. In many diseases requiring liver transplantation, correction of hepatocyte functional deficiency is a prime goal of the procedure. The major limiting factor is the inability to produce large quantity of hepatocytes and to keep them ready for use on demand. In search of generating a clinically viable and sustainable source of functional hepatocytes, various strategies are currently being explored. Possibilities include the expansion of existing hepatocytes, differentiation of progenitor/stem cells in the liver, differentiation of embryonic stem cells and extrahepatic adult stem cells, particularly those from the bone marrow (BM). There is a great interest in adult stem cells because of their high plasticity and self-renewal capacity. It has been recently shown that bone marrow stem cells can circulate in peripheral blood and migrate to distant organs and tissues and promote tissue repair at injured sites.[2–4]

The best characterised and most widely understood adult stem cells are haematopoietic stem cells (HSC), which sustain the formation of the blood and immune systems throughout life and were first identified in 1961.[5] The BM compartment is largely made up of HSC and committed progenitor cells, non-circulating stromal cells (called mesenchymal stem cells (MSC)) that have the ability to develop into mesenchymal lineages.[6,7] It was previously thought that adult stem cells were lineage restricted, but recent studies demonstrated that BM-derived progenitors in addition to haematopoiesis also participate in regeneration of ischaemic myocardium,[8] damaged skeletal muscle[9] and neurogenesis.[10] MSC can be isolated as a growing adherent cell population and can differentiate into osteoblasts, adipocytes and chondrocytes.[7]

2. Haematopoietic Stem Cells and Liver Regeneration

Several independent reports have demonstrated that adult BM cells can give rise to different hepatic epithelial cells types, including oval cells, hepatocytes and duct epithelium.[11–16] These observations have resulted in the hypothesis that BM resident stem cells, specifically HSC, could be an important source for liver epithelial cell replacement,[17] particularly during chronic injury.

It has been shown that rat BM contains a subpopulation of hepatocyte-like cells expressing α-fetoprotein (αFP), c-met, CD34 and c-kit.[18,19] Similar observations were made by Okumuto *et al.*,[20] where cells cultured in the presence of hepatocyte growth factor (HGF) expressed liver-enriched transcription factor HNF1α and cytokeratin CK8. A subpopulation of murine mononuclear BM cells isolated by chemotaxis to stromal-derived factor-1 (SDF-1) expressed mRNA for αFP and population enriched for Sca-1 expressed mRNA for αFP, c-met and CK19.[21] Purified murine HSC differentiated into liver cells expressing early αFP, GATA4, HNF4, HNF3β, HNF1α and mature hepatocyte markers: CK18, albumin, transferrin, when co-cultured with injured liver tissue.[22] Studies on human BM cells showed that they are also able to differentiate into cells with liver-like characteristics. These cells cultured on a collagen matrix and in the presence of HGF express liver-specific genes, such as albumin and cytokeratin CK19.[23] When selected by SDF-1 chemotaxis they appear to be multipotent and express αFP.[21]

Subpopulation of human MSC, known as multipotent adult progenitor cells (MAPC), can differentiate into hepatocyte-like cells in the presence of HGF and fibroblast growth factor-4 (FGF-4).[24] However, there is a substantial delay between the time MAPC are isolated and the time they can be shown to differentiate into hepatocyte-like cells and hence their use in the clinic has been questioned. Lee *et al.*[25] showed hepatic differentiation of MSC, in the presence of HGF and oncostatin M. After a long differentiation period of four weeks, cells expressing marker genes specific of liver cells demonstrated *in vitro* liver functions, including albumin production, glycogen storage, urea secretion, uptake of low-density lipoprotein, and phenobarbital-inducible cytochrome P450 activity.

In an animal model, Petersen *et al.*[11] transplanted male BM stem cells into injured livers of female mice. They demonstrated that regenerated hepatic cells were of BM origin by using Y chromosome, dipeptidyl peptidase IV enzyme, and L21-6 antigen as markers to identify donor-derived cells. Grompe *et al.*[26] used $FAH^{(-/-)}$ mice, an inducible animal model of tyrosinemia type I, a lethal hereditary liver disease and showed repopulation of injured liver by donor-derived BM cells. Twenty-two weeks after transplantation one-third of the liver comprised BM-derived cells, suggesting that BM stem cells contribute to hepatocyte generation, when regenerative potential of hepatocytes is impaired. Sakaida *et al.*[27] showed that BM stem cells can reduce liver fibrosis, probably by expressing matrix metalloproteases, which enable degradation of hepatic scars. Mallet *et al.*[28] induced hepatic apoptosis in mice by JO_2 antibody, the murine anti-Fas agonist and transplanted unfractionated BM cells expressing Bcl-2 under the control of a liver-specific promoter. BM-derived hepatocytes expressing Bcl-2 were only seen in the liver of the mice, which had received JO_2 antibodies. Moreover, in mice with induced liver cirrhosis, 25% of the recipient liver was repopulated in four weeks by BM-derived hepatocytes.[29] Similarly, murine HSC converted into viable hepatocytes with increasing liver injury and restored liver function two to seven days after transplantation, suggesting that HSC contribute to liver regeneration by differentiating into functional hepatocytes.[22] However, some studies have shown that BM stem cells can repopulate the liver even in the absence of liver injury. Theise *et al.*[14] identified up to 2.2% donor-derived hepatocytes when they transplanted BM or $CD34^+lin^-$ cells into irradiated mice without acute liver injury. HSC transplanted into irradiated mice engrafted in several organs, including liver, gastrointestinal tract, bronchus and skin of recipient animals and generated albumin-expressing hepatocyte-like cells.[30]

In contrast, several other studies failed to show the contribution of BM stem cells to liver regeneration. HSC reconstituted blood leukocytes of irradiated mice, but did not contribute to non-haematopoietic tissues, including liver, brain, kidney, gut, and muscle.[31] Some groups failed to show a significant contribution of BM stem cells to liver regeneration using various liver injury models.[32] Although the contribution of HSC to hepatocyte lineages *in vivo* still remains divisive, the differences between the studies may in part reflect the types of cells used, different injury models used and the method used to detect engrafted stem cells.

Human studies have also shown the presence of cells of BM origin in the human liver. Alison *et al.*[12] examined livers from female patients that received BM transplantation and found donor derived hepatocytes, suggesting that extrahepatic stem cells can colonise the liver. Theise *et al.*[13] identified hepatocytes (4%–43%) and cholangiocytes (4%–38%) of BM origin in archival autopsy and biopsy liver specimens, suggesting BM stem cells can replenish large numbers of hepatic parenchymal cells. However, in a similar study Korbling *et al.*[33] found only 4%–7% BM-derived hepatocytes. Ng *et al.*[34] identified only a small proportion of donor-derived hepatocytes (1.6%) in liver allografts; most donor-derived cells were macrophages/Kupffer cells. Two other studies did not detect any BM-derived hepatocytes at all.[35,36] The differences in the published studies could be due to use of different techniques and markers to identify recipient-derived hepatocytes in transplanted patients.

3. Clinical Studies

Several studies (Table 1) have shown that stem cell therapy could be the future of treatment for patients with liver diseases. A few clinical trials have been conducted using BM-derived stem cells and have assessed the safety and tolerability of adult stem cell therapy.

Table 1. Stem cell therapy and liver disease.

Type of liver disease	Study group	Route of injection	Cell type	Outcome	Ref. No.
Liver cancer	$n = 3$	Portal vein	$CD133^+$	Well-tolerated 2.5-fold increase in left lobe	39
Hepatitis B or C Alcoholic cirrhosis Primary sclerosing cholangitis	$n = 5$	Portal vein Hepatic artery	$CD34^+$	Well-tolerated Improvement in albumin and bilirubin	37
Hepatitis B or C Cirrhosis	$n = 9$	Peripheral vein	Mononuclear cells	Well-tolerated Improvement in Child's Pugh score and albumin	40
Alcoholic cirrhosis	$n = 2$	Peripheral vein	$CD34^+$	Well-tolerated Improvement in Child's Pugh and MELD score	41
Alcoholic, hepatitis C, cholestatic and cryptogenic cirrhosis	$n = 10$	Hepatic artery	Mononuclear cells	Well-tolerated Improvement in Child's Pugh score, bilirubin and albumin	43
Drug-induced acute liver failure	$n = 1$	Portal vein	$CD34^+$	Well-tolerated Improvement in clotting, AST and ALT Patient died of sepsis at day 60	44
Decompensated liver cirrhosis	$n = 4$	Hepatic artery	$CD34^+$	Not tolerated well, side effects Radiocontrast nephropathy and hepatorenal syndrome	45
Decompensated liver cirrhosis	$n = 4$	Hepatic artery	Mesenchymal stem cells	Well-tolerated MELD score improvement	46

Our group conducted a phase I clinical trial for the transplantation of adult HSC cells into patients with chronic liver disease.[37] We treated five patients with liver insufficiency and mobilised their stem cells by G-CSF. Between 1×10^6 and 2×10^8 autologous $CD34^+$ cells were injected into the either the portal vein or the hepatic artery. The majority of patients showed improvement in serum albumin and bilirubin, which lasted for more than 18 months.[38] Clinically, the procedure was well tolerated with no observed procedure-related complications. Our experience is in keeping with the observation of am Esch *et al.*,[39] who demonstrated increased liver regeneration following portal application of autologous BM cells to the contralateral side during portal vein embolisation. Three patients with large central hepatobiliary malignancies were treated with autologous $CD133^+$ cells, which were selectively implanted to the left-lateral portal branches subsequent to selective portal vein embolisation of right liver segments. Mean daily hepatic growth determined by CT scan volumetry rates was 2.5-fold higher compared to the patients that had been subjected to portal vein embolisation without $CD133^+$ application. These data suggested that stem cell therapy enhances and accelerates liver regeneration and may bear the potential for augmentation of liver regeneration before extensive hepatectomy. Terai *et al.*[40] have shown improvement in liver function following peripheral infusion of autologous BM cells in patients with liver cirrhosis. Nine patients with cirrhosis received portal vein infusion of 5.2×10^9 mononuclear cells ($CD34^+$, $CD45^+$, c-kit$^+$). They reported significant improvement in serum albumin levels, total protein and improved Child-Pugh score. Liver biopsies were taken in some patients and showed increases in proliferating cell nuclear antigen staining, an indirect marker of hepatocyte turnover. Yannaki *et al.*[41] reported two cases where they treated patients with liver cirrhosis due to alcohol abuse with autologous mobilised HSC. Each patient underwent three rounds of G-CSF mobilisation and infusion of $CD34^+$ cells into the peripheral vein. Both patients improved their baseline Child-Turcotte-Pugh (CTP) and model for end-stage liver disease (MELD) scores during the 30 months' follow-up. The results suggested that the procedure may be considered as a bridging therapy until organ transplantation becomes available or to reverse decompensated cirrhosis to a compensated one. Similar findings were reported by Yan *et al.*,[42] who treated two patients with hepatitis B-related decompensated liver cirrhosis. Patients were transplanted with mobilised autologous peripheral blood monocytes (PBMC) and showed improvement in serum albumin, billirubin, alanine aminotransferase (ALT), aspartate aminotransferase (ASP) and CTP score for more than one year after transplantation. Similarly, a study on ten patients with end-stage chronic liver disease, who received 10^8 mononuclear-enriched BM cells showed improvement in serum bilirubin and albumin levels.[43] Gasbarrini *et al.*[44] reported successful use of autologous $CD34^+$ cells as a rescue treatment for a patient with drug-induced acute liver failure, who was not suitable for liver transplantation. A liver biopsy performed at 20 days post-infusion showed increased hepatocyte replication around necrotic foci and the improvement in the synthetic liver function in the first 30 days. However, the patient died due to multi-organ failure related to bacterial infection.

In contrast, a study on four patients showed that infusion of $CD34^+$ stem cells through the hepatic artery is not safe in decompensated cirrhosis. Radiocontrast nephropathy and hepatorenal syndrome were found to be the major side effects.[45] The same group showed that MSC transplantation seems to be feasible and safe in the treatment of decompensated liver cirrhosis.[46]

The mechanics that govern the functional improvement observed in human studies remain a debatable issue. The inability for routine pathological examination of specimens in the clinical trial setting necessitates data collection from experimental models to investigate this. Although both transdifferentiation and fusion provide an acceptable theoretical platform at the cellular level for the documented functional recovery, it is by now widely accepted that their collective clinical effect is probably negligible due to their rare occurrence. There is growing evidence that transplanted cells, being multipotent, not just simply replace missing tissue but also trigger local mechanism to initiate a repair response. Paracrine effects and immune regulation of the transplanted cells play a role in functional restoration of the tissue.[47,48] Another possible improvement in function would be facilitating the release of vascular endothelial growth factor (VEGF) by stem cells, thereby increasing the blood supply to the cells and thus helping to repair the damaged tissue.[49,50] Stem cells may act by up-regulating the Bcl-2 gene and therefore suppressing apoptosis.[49,51] and by suppressing inflammation in the diseased organ via the interleukin-6 (IL-6) pathway.[52] Both these processes may help in the regeneration of normal cells in the damaged organ.

4. Conclusions

Despite the great amount of interest and development in stem cell biology and therapy in the past few years many questions need to be answered before stem cell therapy can be applied to its fullest potential in the clinic. Which type of cells should be used for maximizing potential benefits? What is the ideal route of administration? The optimum timing and method of delivery need to be determined as they may have a significant influence on the outcome of cell transplantation. The long-term consequences of treatment remain largely unknown.

It also remains to be determined how cell survival can be optimised and the optimal number of cells necessary to achieve this therapeutic effect. Ideally, cells for liver therapies should expand extensively *in vitro*, differentiate into mature liver cells, have minimal immunogenicity and be able to reconstitute liver tissue when transplanted *in vivo*.

References

1. Arias, I.M., Boyer, J., Fausto, N., Chisari, F.V., Schachter, D. and Shafritz, D.A. *The Liver: Biology and Pathobiology*, 4th Ed. (Lippincott Williams & Wilkins, Philadelphia, PA, 2001).
2. Kucia, M., Wojakowski, W., Reca, R., Machalinski, B., Gozdzik, J., Majka, M., Baran, J., Ratajczak, J. and Ratajczak, M.Z. The migration of bone marrow-derived non-hematopoietic tissue-committed stem cells is regulated in an SDF-1-, HGF-, and LIF-dependent manner. *Arch. Immunol. Ther. Exp. (Warsz)* **54**, 121–135 (2006).
3. Wojakowski, W., Kucia, M., Kazmierski, M., Ratajczak, M.Z. and Tendera, M. Circulating stem/progenitor cells in stable ischemic heart disease and acute coronary syndromes — relevant reparatory mechanism? *Heart* **94**, 27–33 (2008).
4. Wright, D.E., Wagers, A.J., Gulati, A.P., Johnson, F.L. and Weissman, I.L. Physiological migration of hematopoietic stem and progenitor cells. *Science* **294**, 1933–1936 (2001).
5. Till, J.E. and McCulloch, E.A. direct measurement of the radiation sensitivity of normal mouse bone marrow cells. *Radiat. Res.* **14**, 213–222 (1961).
6. Bianco, P., Riminucci, M., Gronthos, S. and Robey, P.G. Bone marrow stromal stem cells: nature, biology, and potential applications. *Stem Cells* **19**, 180–192 (2001).
7. Pittenger, M.F., Mackay, A.M., Beck, S.C., Jaiswal, R.K., Douglas, R., Mosca, J.D., Moorman, M.A., Simonetti, D.W., Craig, S. and Marshak, D.R. Multilineage potential of adult human mesenchymal stem cells. *Science* **284**, 143–147 (1999).
8. Orlic, D., Kajstura, J., Chimenti, S., Jakoniuk, I., Anderson, S.M., Li, B., Pickel, J., McKay, R., Nadal-Ginard, B., Bodine, D.M., Leri, A. and Anversa, P. Bone marrow cells regenerate infarcted myocardium. *Nature* **410**, 701–705 (2001).
9. Gussoni, E., Soneoka, Y., Strickland, C.D., Buzney, E.A., Khan, M.K., Flint, A.F., Kunkel, L.M. and Mulligan, R.C. Dystrophin expression in the mdx mouse restored by stem cell transplantation. *Nature* **401**, 390–394 (1999).
10. Mezey, E., Chandross, K.J., Harta, G., Maki, R.A. and McKercher, S.R. Turning blood into brain: cells bearing neuronal antigens generated *in vivo* from bone marrow. *Science* **290**, 1779–1782 (2000).

11. Petersen, B.E., Bowen, W.C., Patrene, K.D., Mars, W.M., Sullivan, A.K., Murase, N., Boggs, S.S., Greenberger, J.S. and Goff, J.P. Bone marrow as a potential source of hepatic oval cells. *Science* **284**, 1168–1170 (1999).
12. Alison, M.R., Poulsom, R., Jeffery, R., Dhillon, A.P., Quaglia, A., Jacob, J., Novelli, M., Prentice, G., Williamson, J. and Wright, N.A. Hepatocytes from non-hepatic adult stem cells. *Nature* **406**, 257 (2000).
13. Theise, N.D., Nimmakayalu, M., Gardner, R., Illei, P.B., Morgan, G., Teperman, L., Henegariu, O. and Krause, D.S. Liver from bone marrow in humans. *Hepatology* **32**, 11–16 (2000).
14. Theise, N.D., Badve, S., Saxena, R., Henegariu, O., Sell, S., Crawford, J.M. and Krause, D.S. Derivation of hepatocytes from bone marrow cells in mice after radiation-induced myeloablation. *Hepatology* **31**, 235–240 (2000).
15. Avital, I., Inderbitzin, D., Aoki, T., Tyan, D.B., Cohen, A.H., Ferraresso, C., Rozga, J., Arnaout, W.S. and Demetriou, A.A. Isolation, characterization, and transplantation of bone marrow-derived hepatocyte stem cells. *Biochem. Biophys. Res. Commun.* **288**, 156–164 (2001).
16. Masson, S., Harrison, D.J., Plevris, J.N. and Newsome, P.N. Potential of hematopoietic stem cell therapy in hepatology: a critical review. *Stem Cells* **22**, 897–907 (2004).
17. Austin, T.W. and Lagasse, E. Hepatic regeneration from hematopoietic stem cells. *Mech. Dev.* **120**, 131–135 (2003).
18. Oh, S.H., Miyazaki, M., Kouchi, H., Inoue, Y., Sakaguchi, M., Tsuji, T., Shima, N., Higashio, K. and Namba, M. Hepatocyte growth factor induces differentiation of adult rat bone marrow cells into a hepatocyte lineage *in vitro*. *Biochem. Biophys. Res. Commun.* **279**, 500–504 (2000).
19. Miyazaki, M., Akiyama, I., Sakaguchi, M., Nakashima, E., Okada, M., Kataoka, K. and Huh, N.H. Improved conditions to induce hepatocytes from rat bone marrow cells in culture. *Biochem. Biophys. Res. Commun.* **298**, 24–30 (2002).
20. Okumoto, K., Saito, T., Hattori, E., Ito, J.I., Adachi, T., Takeda, T., Sugahara, K., Watanabe, H., Saito, K., Togashi, H. and Kawata, S. Differentiation of bone marrow cells into cells that express liver-specific genes *in vitro*: implication of the Notch signals in differentiation. *Biochem. Biophys. Res. Commun.* **304**, 691–695 (2003).
21. Kucia, M., Ratajczak, J., Reca, R., Janowska-Wieczorek, A. and Ratajczak, M.Z. Tissue-specific muscle, neural and liver stem/progenitor cells reside in the bone marrow, respond to an SDF-1 gradient and are mobilized into peripheral blood during stress and tissue injury. *Blood Cells Mol. Dis.* **32**, 52–57 (2004).

22. Jang, Y.Y., Collector, M.I., Baylin, S.B., Diehl, A.M. and Sharkis, S.J. Hematopoietic stem cells convert into liver cells within days without fusion. *Nat. Cell Biol.* **6**, 532–539 (2004).
23. Fiegel, H.C., Lioznov, M.V., Cortes-Dericks, L., Lange, C., Kluth, D., Fehse, B. and Zander, A.R. Liver-specific gene expression in cultured human hematopoietic stem cells. *Stem Cells* **21**, 98–104 (2003).
24. Schwartz, R.E., Reyes, M., Koodie, L., Jiang, Y., Blackstad, M., Lund, T., Lenvik, T., Johnson, S., Hu, W.S. and Verfaillie, C.M. Multipotent adult progenitor cells from bone marrow differentiate into functional hepatocyte-like cells. *J. Clin. Invest.* **109**, 1291–1302 (2002).
25. Lee, K.D., Kuo, T.K., Whang-Peng, J., Chung, Y.F., Lin, C.T., Chou, S.H., Chen, J.R., Chen, Y.P. and Lee, O.K. *In vitro* hepatic differentiation of human mesenchymal stem cells. *Hepatology* **40**, 1275–1284 (2004).
26. Grompe, M., Lindstedt, S., al-Dhalimy, M., Kennaway, N.G., Papaconstantinou, J., Torres-Ramos, C.A., Ou, C.N. and Finegold, M. Pharmacological correction of neonatal lethal hepatic dysfunction in a murine model of hereditary tyrosinaemia type I. *Nat. Genet.* **10**, 453–460 (1995).
27. Sakaida, I., Terai, S., Yamamoto, N., Aoyama, K., Ishikawa, T., Nishina, H. and Okita, K. Transplantation of bone marrow cells reduces CCl_4-induced liver fibrosis in mice. *Hepatology* **40**, 1304–1311 (2004).
28. Mallet, V.O., Mitchell, C., Mezey, E., Fabre, M., Guidotti, J.E., Renia, L., Coulombel, L., Kahn, A. and Gilgenkrantz, H. Bone marrow transplantation in mice leads to a minor population of hepatocytes that can be selectively amplified *in vivo*. *Hepatology* **35**, 799–804 (2002).
29. Terai, S., Sakaida, I., Yamamoto, N., Omori, K., Watanabe, T., Ohata, S., Katada, T., Miyamoto, K., Shinoda, K., Nishina, H. and Okita, K. An *in vivo* model for monitoring trans-differentiation of bone marrow cells into functional hepatocytes. *J. Biochem. (Tokyo)* **134**, 551–558 (2003).
30. Wang, X., Ge, S., McNamara, G., Hao, Q.L., Crooks, G.M. and Nolta, J.A. Albumin-expressing hepatocyte-like cells develop in the livers of immune-deficient mice that received transplants of highly purified human hematopoietic stem cells. *Blood* **101**, 4201–4208 (2003).
31. Wagers, A.J., Sherwood, R.I., Christensen, J.L. and Weissman, I.L. Little evidence for developmental plasticity of adult hematopoietic stem cells. *Science* **297**, 2256–2259 (2002).
32. Kanazawa, Y., and Verma, I.M. Little evidence of bone marrow-derived hepatocytes in the replacement of injured liver. *Proc. Natl. Acad. Sci. USA* **100**(Suppl. 1), 11850–11853 (2003).
33. Korbling, M., Katz, R.L., Khanna, A., Ruifrok, A.C., Rondon, G., Albitar, M., Champlin, R.E. and Estrov, Z. Hepatocytes and epithelial cells of donor

origin in recipients of peripheral-blood stem cells. *N. Engl. J. Med.* **346**, 738–746 (2002).
34. Ng, I.O., Chan, K.L., Shek, W.H., Lee, J.M., Fong, D.Y., Lo, C.M. and Fan, S.T. High frequency of chimerism in transplanted livers. *Hepatology* **38**, 989–998 (2003).
35. Fogt, F., Beyser, K.H., Poremba, C., Zimmerman, R.L., Khettry, U. and Ruschoff, J. Recipient-derived hepatocytes in liver transplants: a rare event in sex-mismatched transplants. *Hepatology* **36**, 173–176 (2002).
36. Wu, T., Cieply, K., Nalesnik, M.A., Randhawa, P.S., Sonzogni, A., Bellamy, C., Abu-Elmagd, K., Michalopolous, G.K., Jaffe, R., Kormos, R.L., Gridelli, B., Fung, J.J. and Demetris, A.J. Minimal evidence of transdifferentiation from recipient bone marrow to parenchymal cells in regenerating and long-surviving human allografts. *Am. J. Transplant.* **3**, 1173–1181 (2003).
37. Gordon, M.Y., Levicar, N., Pai, M., Bachellier, P., Dimarakis, I., Al-Allaf, F., M'Hamdi, H., Thalji, T., Welsh, J.P., Marley, S.B., Davis, J., Dazzi, F., Marelli-Berg, F., Tait, P., Playford, R., Jiao, L., Jensen, S., Nicholls, J.P., Ayav, A., Nohandani, M., Farzaneh, F., Gaken, J., Dodge, R., Alison, M., Apperley, J.F., Lechler, R. and Habib, N.A. Characterisation and clinical application of human Cd34+ stem/progenitor cell populations mobilised into the blood by G-Csf. *Stem Cells* **24**, 1822–1830 (2006).
38. Levicar, N.P.M., Habib, N.A., Tait, P., Jiao, L.R., Marley, S.B., Davis, S., Dazzi, F., Smadja, C., Jensen, S.L., Nicholls, J.P., Apperley, J.F. and Gordon, M.Y. Long-term clinical results of autologous infusion of mobilised adult bone marrow derived CD34+ cells in patients with chronic liver disease. *Cell Prolif.* **41**(Suppl. 1), 115–125 (2008).
39. am Esch, 2nd, J.S., Knoefel, W.T., Klein, M., Ghodsizad, A., Fuerst, G., Poll, L.W., Piechaczek, C., Burchardt, E.R., Feifel, N., Stoldt, V., Stockschlader, M., Stoecklein, N., Tustas, R.Y., Eisenberger, C.F., Peiper, M., Haussinger, D. and Hosch, S.B. Portal application of autologous CD133+ bone marrow cells to the liver: a novel concept to support hepatic regeneration. *Stem Cells* **23**, 463–470 (2005).
40. Terai, S., Ishikawa, T., Omori, K., Aoyama, K., Marumoto, Y., Urata, Y., Yokoyama, Y., Uchida, K., Yamasaki, T., Fujii, Y., Okita, K. and Sakaida, I. Improved liver function in patients with liver cirrhosis after autologous bone marrow cell infusion therapy. *Stem Cells* **24**, 2292–2298 (2006).
41. Yannaki, E., Anagnostopoulos, A., Kapetanos, D., Xagorari, A., Iordanidis, F., Batsis, I., Kaloyannidis, P., Athanasiou, E., Dourvas, G., Kitis, G. and Fassas, A. Lasting amelioration in the clinical course of decompensated alcoholic cirrhosis with boost infusions of mobilized peripheral blood stem cells. *Exp. Hematol.* **34**, 1583–1587 (2006).

42. Yan, L., Han, Y., Wang, J., Liu, J., Hong, L. and Fan, D. Peripheral blood monocytes from patients with HBV related decompensated liver cirrhosis can differentiate into functional hepatocytes. *Am. J. Hematol.* **82**, 949–954 (2007).
43. Lyra, A.C., Soares, M.B., da Silva, L.F., Fortes, M.F., Silva, A.G., Mota, A.C., Oliveira, S.A., Braga, E.L., de Carvalho, W.A., Genser, B., dos Santos, R.R. and Lyra, L.G. Feasibility and safety of autologous bone marrow mononuclear cell transplantation in patients with advanced chronic liver disease. *World J. Gastroenterol.* **13**, 1067–1073 (2007).
44. Gasbarrini, A., Rapaccini, G.L., Rutella, S., Zocco, M.A., Tittoto, P., Leone, G., Pola, P., Gasbarrini, G. and Di Campli, C. Rescue therapy by portal infusion of autologous stem cells in a case of drug-induced hepatitis. *Dig. Liver Dis.* **39**, 878–882 (2007).
45. Mohamadnejad, M., Namiri, M., Bagheri, M., Hashemi, S.M., Ghanaati, H., Zare Mehrjardi, N., Kazemi Ashtiani, S., Malekzadeh, R. and Baharvand, H. Phase 1 human trial of autologous bone marrow-hematopoietic stem cell transplantation in patients with decompensated cirrhosis. *World J. Gastroenterol.* **13**, 3359–3363 (2007).
46. Mohamadnejad, M., Alimoghaddam, K., Mohyeddin-Bonab, M., Bagheri, M., Bashtar, M., Ghanaati, H., Baharvand, H., Ghavamzadeh, A. and Malekzadeh, R. Phase 1 trial of autologous bone marrow mesenchymal stem cell transplantation in patients with decompensated liver cirrhosis. *Arch. Iran Med.* **10**, 459–466 (2007).
47. Hess, D., Li, L., Martin, M., Sakano, S., Hill, D., Strutt, B., Thyssen, S., Gray, D.A. and Bhatia, M. Bone marrow-derived stem cells initiate pancreatic regeneration. *Nat. Biotechnol.* **21**, 763–770 (2003).
48. Pluchino, S., Quattrini, A., Brambilla, E., Gritti, A., Salani, G., Dina, G., Galli, R., Del Carro, U., Amadio, S., Bergami, A., Furlan, R., Comi, G., Vescovi, A.L. and Martino, G. Injection of adult neurospheres induces recovery in a chronic model of multiple sclerosis. *Nature* **422**, 688–694 (2003).
49. Rehman, J., Traktuev, D., Li, J., Merfeld-Clauss, S., Temm-Grove, C. J., Bovenkerk, J.E., Pell, C.L., Johnstone, B.H., Considine, R.V. and March, K.L. Secretion of angiogenic and antiapoptotic factors by human adipose stromal cells. *Circulation* **109**, 1292–1298 (2004).
50. Tang, J., Xie, Q., Pan, G., Wang, J. and Wang, M. Mesenchymal stem cells participate in angiogenesis and improve heart function in rat model of myocardial ischemia with reperfusion. *Eur. J. Cardiothorac. Surg.* **30**, 353–361 (2006).

51. Chen, Z., Chua, C.C., Ho, Y.S., Hamdy, R.C. and Chua, B.H. Overexpression of Bcl-2 attenuates apoptosis and protects against myocardial I/R injury in transgenic mice. *Am. J. Physiol. Heart Circ. Physiol.* **280**, H2313–H2320 (2001).
52. Wang, M., Crisostomo, P.R., Herring, C., Meldrum, K.K. and Meldrum, D.R. Human progenitor cells from bone marrow or adipose tissue produce VEGF, HGF, and IGF-I in response to TNF by a p38 MAPK-dependent mechanism. *Am. J. Physiol. Regul. Integr. Comp. Physiol.* **291**, R880–R884 (2006).

Chapter 30

Tissue Engineering for Tooth Regeneration

Ivan. A. Diakonov and Paul Sharpe

Abstract

Teeth are clinically important, easily accessible, non-essential organs whose development, cell biology and physiology are well-understood. Missing or damaged teeth are currently replaced by non-cellular structures such as metal implants. Replacement of teeth with cell-based implants that will form whole teeth or specific tooth structures is a realistic goal of regenerative dentistry. Putative stem cell populations have been identified in several different tooth tissues and the ease with which these cells can be obtained, for example from naturally lost deciduous (milk) teeth makes them an attractive source of mesenchymal stem cells.

Keywords: Dental Stem Cells; Tooth Regeneration; Tooth Repair.

Outline

1. Tooth Development and Morphology
2. Stem Cells in the Tooth
    - 2.1. Dental pulp stem cells
    - 2.2. Periodontal ligament and dental follicle stem cells
    - 2.3. Dental epithelial stem cells
3. Two Strategies for Tissue Engineering Tooth Germs *De Novo*
    - 3.1. Scaffold-based roots
    - 3.2. Reproducing embryonic tooth germs for implantation
4. Challenges
5. Conclusions

References

1. Tooth Development and Morphology

The dentition is an important part of feeding in vertebrates. Tooth shapes often played an instructive role in evolution of vertebrates. In humans, aberrations in dentition as well as poor dental care are not life-threatening conditions in the modern world but they are still challenging problems for many. There is no doubt that tissue engineering of teeth can help a large number of people to improve their quality of life. Here we discuss methods of generating bioartificial tooth implants that can be beneficial for future clinical use as replacement teeth.

In common with other ectodermal organs, teeth develop through a series of reciprocal signalling events between epithelium and underlying mesenchyme.[1] The instructive capacity switches between these tissues. Tooth development can be divided into three major stages: initiation, morphogenesis and terminal differentiation. The first morphological sign during the development of all ectodermal organs is an epithelial thickening at the initiation stage resulting in the formation of an epithelial placode. The dental placode or dental lamina is an ectodermal ridge that spans the length of the developing maxilla and mandible. Tooth development begins at certain locations along dental lamina when the epithelium buds into the mesenchyme while mesenchyme condenses around the epithelium, which is referred as bud stage.[2]

The mesenchyme cells that give rise to the tissues of the face, bone, cartilage and teeth, are derived from cranial neural crest.[3,4] The neural crest cells migrate as specific streams from the anterior neural tube to populate different regions of the embryo. Thus the cells that contribute to tooth formation originate in the caudal midbrain and rostral hindbrain.[3,4] The individual fates of crest cells within a stream are specified by external signals they receive *en route* or at their destinations. The cranial neural crest cells are directed/committed to participate in tooth development by signals they receive from the oral/dental epithelium. The oral/dental epithelium is thus the source of the inductive signals for the odontogenesis. Candidate molecules for these signals include BMPs, FGFs, Wnts and Shh.[1,2,5]

The instructive capacity to determine morphogenesis switches from epithelium to mesenchyme at the bud stage.[2] For many of these instructive signals emanating from either epithelium or mesenchyme timing and intensity are of paramount importance, and many of the signals are only transiently produced by one tissue before being produced by the other.[6,7]

Signalling proteins that belong to all the major gene families of growth and differentiation factors take part in regulating tooth development. The list of such factors includes members of fibroblast growth factor (FGFs), hedgehog (Hh), Notch, transforming growth factor (TGFs, BMPs, activins), tumour necrosis factor (TNF) and Wnt families.[7] The molecular cascades involved in morphogenesis include not only the effector molecules but also corresponding receptors and intracellular signalling molecules. Further complexity to the regulation is added by the large number of specific inhibitors and modulators of signalling. Other major macromolecules that can influence morphogenesis of teeth are extracellular matrices such as laminin and tenascin as well as proteinases such as matrix metalloproteinases (MMPs) and their regulators, for example TIMPs.[8,9]

Rodents, and in particular the mouse, have been model animals for biomedical research for many years in part due to the ease of genetic manipulation. Unfortunately for the dental researchers mouse and human dentitions differ significantly. Mice not only form a single set of teeth but their continuously growing incisors are quite special teeth that cannot be related to human incisors. Mouse tooth buds form at embryonic day 10.5 (E10.5) to E11.5, one incisor and one molar primordia being initiated in each quadrant. Incisors and molars are separated by a toothless region called the diastema in which rudimentary tooth buds first form but fail to develop into teeth.[2] In humans two sets of teeth develop during lifetime. Twenty primary tooth germs start to develop at six weeks after gestation, with 32 permanent tooth germs developing from the primary tooth germs thereafter. Very little is known about the mechanisms governing the generation of human tooth germs. In particular, it is not clear how 32 secondary tooth germs are generated from only 20 primary tooth germs.

Following the bud stage, the epithelium invaginates even more to enclose part of the mesenchyme that becomes the dental papilla to form a "cup". The enamel knot, a signalling centre, created in the epithelium at the early cap stage is a transient structure that directs the growth of the dental tissues and determines the shape of the tooth crown.[1]

After the mesenchyme receives the early odontogenic signals from the epithelium, it expresses master molecules such as homeoproteins, that control morphogenesis.[2] Reciprocity of signalling between epithelium and mesenchyme is required for progressing through later stages. At the next stage, named the bell stage, a recognisable "tooth" is formed that consists of an enamel organ, dental papilla and dental follicle. The inner enamel epithelium of the epithelial enamel organ is the precursor of ameloblasts, cells that produce enamel, which is composed of more than 90% hydroxyapatite and is known to be the hardest tissue in the body.[10] During bell stage the cervical loops keep extending and the epithelium near the oral side folds to an intricate pattern that corresponds to the later cusps of the tooth. At this point in time only the foundations for the crown are established, but later during postnatal tooth maturation the deposition of the enamel finishes and the epithelium degenerates.

During postnatal development the roots are formed as a direct effect of structural changes that are taking place in the cervical loop area. The epithelium of the root does not differentiate into ameloblasts. Rather the root epithelium forms a double layer of basal epithelium known as Hertwig's epithelial root sheath (HERS).[11] The epithelium above HERS fragments and forms a fenestrated network of epithelial cells known as the epithelial cell rests of Malassez (ERM). With the formation of the HERS no ameloblast differentiation occurs in the cervical loop area. Through the network of the ERM, mesenchymal cells from dental follicles can migrate and form the periodontal ligament which attaches the root surface to the jaw bone. FGFs such as FGF10 have been implicated in root formation.[12] A transgenic mouse lacking NFI-C/CTF transcription/replication factor has molars whose crowns grow normally, but lack root formation.[12]

The dental papilla gives rise to pulp tissue, which is a living connective tissue composed of fibroblasts, blood vessels, nerves, lymphatic ducts and odontoblasts. Odontoblasts are the cells derived from the mesenchymal cells in the dental papilla adjacent to the inner enamel epithelium.[10] Functional odontoblasts show polarised columnar morphology that shift into a resting state and become small and flat after primary dentin formation. However, odontoblasts remain functional throughout their life and can produce secondary dentin if trauma is mild.[13]

The dental follicle appears as a transient structure when teeth undergo morphogenesis. It is the origin of three major types of cells: cementoblasts, osteoblasts and fibroblasts.[10] Cementoblasts secrete cementum, which is attached to the root surface. Osteoblasts produce bone around the roots of teeth. Fibroblasts produce collagen giving rise to periodontal ligaments (PDLs), which connect roots to the alveolar bone via the cementum. The PDL functions as a cushion when force is applied, as a source of sensation, and it is regarded as the main impetus for the tooth eruption process. The complex structure that includes the PDL, adjacent cementum and alveolar bone is called the periodontium.

2. Stem Cells in the Tooth

2.1. Dental pulp stem cells

Although dental neural crest-derived cells are considered to retain inherent pluripotency,[14] it is still unclear whether teeth contain true multipotent stem cells, or the potentiality of dental precursor cells is somewhat limited. The notion of "stem cell" as it applies to adult tissues such as tooth can be described as "narrow" and it is often more appropriate to name these cell populations precursor cells, but by popular convention the term "stem cell" remains in circulation; therefore we will describe these multipotent cells as stem cells. It has long been established that regeneration of dentin following tooth injury is achieved by new odontoblasts that emerge close to the injury site. Labelling of proliferating cells with tritiated thymidine following damage has revealed differences in labelling depending on the location related to trauma site.[15,16] The existing odontoblast layer and deeper pulp layers did not show any label, whereas some perivascular labelling suggests that progenitor cells are located around the vessels.[16] The accumulation of labelled cells over time suggests the existence of a continuous source of cells for replacement. Pulse labelling suggested that there is constant migration of cells from deep tissue in the pulp to the periphery. These data support the idea that undifferentiated mesenchymal cells in the pulp have the ability to differentiate into odontoblast-like cells, which are responsible for new dentin formation following dental injury.[15,17] It is still a matter of debate whether these cells are the reminder of earlier neural crest cells or they are somehow associated with blood vessels formed during organogenesis, or even become attracted to the injury site from the circulating precursors presumably derived from bone marrow. One of the methods to address these questions is to isolate precursor cell populations from the pulp and characterise them. Indeed, several populations of dental cells with high proliferative potential have been identified both in animal models and in humans. Human dental pulp cells derived from developing third molars have been cultured under mineralisation-enhancing conditions and shown to form odontoblast-like cells that produce dentin-like structure *in vitro* and express nestin.[18] Other studies have demonstrated that human dental pulp from adult teeth (DPSC) and exfoliated deciduous teeth contains dental pulp stem cells (SHED).[19–22] These cells are capable of multipotent differentiation, which relates them to bone marrow mesenchymal stem cells (BMSC). Both cell types express mesenchymal stem cell markers STRO-1 and CD146. Indeed, dental pulp stem cells form dentin when transplanted into host animals, whereas BMSC do not participate in dentin regeneration *in vivo*. In addition to mineralisation, DPSCs and stem cells from human exfoliated deciduous teeth (SHED) can express neural markers and have the potential to differentiate into adipocytes.[20,22] Compared with DPSCs, SHED show higher proliferation rates and increased population doublings. SHED cells can also form spherical aggregations, which relates them to neural precursors. Based on these observations is has been proposed that SHED are distinct from DPSCs. When DPSCs are implanted subcutaneously into immunocompromised mice, dentin-pulp-like complex but not lamellar bone is formed, although BMSC form bone. In turn, SHED cells produced both dentin and bone but not mature dentin-pulp complexes when transplanted *in vivo*. It has been suggested that both DPSCs and SHED may contain a fraction of cells that can be described as bone fide stem cells, but generally they constitute a heterogeneous population of cells.[20,22] Another unique population of stem cells has been recently isolated from the root apex of the developing human third molar (SCAP, stem cells from root apical papilla). It has been demonstrated that SCAP are able to differentiate into odontoblasts and adipocytes.

Although this cell population has not been characterised in detail, it is apparent that its higher proliferative potential compared with DPSCs makes this population of cells suitable for cell-based regeneration.[22] Less has been reported on the characterisation of mouse dental pulp cells. Studies of transgenic mice carrying lineage trace markers showed that the majority of dental pulp originates from neural crest, therefore ectomesenchymal by origin.[23] It had been established that mesenchymal cells from postnatal mice incisors can give rise to odontoblasts, osteoblasts and chondrocytes in culture. Transplanting mesodermal-derived cells into kidney capsules produces dentin matrix and bone.[24–27] The osteogenic potential of mouse dental pulp is consistent with early studies from adult rodent pulp[28] and human DPSCs and SHED. From investigations of bone marrow it emerged that endothelium, pericytes and surrounding smooth muscle are good candidates for a perivascular niche of stem cells.[29,30] Double immunodetection of STRO-1 and markers of smooth muscle cells, endothelial cells and pericytes has been performed to identify dental mesenchymal cells further and DPSCs are considered to express endothelial and smooth muscle markers.[31] The appearance of STRO-1 positive staining in the pulp around blood vessels that also were positive for endothelial markers CD-146 and 3G5 suggests that pulp stem cells may originate from pericytes.[22] It is feasible to think that pericytes either provide a niche for DPSC or the two types are related. A cell population from DPSCs/SHED has been sorted for their positivity for mesenchymal markers and negativity for haematopoietic markers (ckit+/ STRO-1+/CD34+[/CD45–]). This subpopulation can efficiently produce woven bone even without external osteoinduction *in vitro*. These cells are able to grow into remodelling lamellar bone when implanted into immunocompromised rats.[32,33] The multipotent nature of this population was confirmed by showing that these cells can also differentiate into adipocytes endothelium and myotubes.[34] A subpopulation has been selected from DPSC on the basis of their expression of an ABCG2 transporter (side population).[35] These so-called "side population" cells can self-renew and differentiate into odontoblasts, chondrocytes, adipocytes and glia cells. The functional differentiation of human DPSC has been demonstrated by culturing cells on dentin slices.[36] Rat adult DPSC were subjected to culturing in the conditioned medium of developing tooth germ cells. Subsequent transplantation showed that only treated cells formed mature dentin-pulp complexes.[27]

Expression profiling using microarrays has been attempted comparing DPSC, SHED and BMSC.[37] The major differences were observed between BMSC and the other two cell types. It is a matter of controversy as to what characterises a multipotent mesenchymal stem cell. The list of markers is produced, but the degree of variation between groups in defining that set of markers makes it difficult to draw a proper comparison. Indeed DPSC and SHED fit into a broad MSC definition.

However, they clearly represent a distinct population of cells with specific traits that are not common among other MSCs. The apparent heterogeneity of DPSC and SHED is not a unique feature of these cells. For example, clonal strains of bone marrow MSC exhibiting the ability to regenerate bone *in vivo* can still demonstrate non-identical marker expression.[38] The usefulness of DPSCs is reinforced by the fact that they can be cryopreserved retaining their multipotential differentiation ability.[39,40] In addition, it has been discovered that under appropriate conditions, DPSCs have superior qualities to bone marrow MSC in inducing immune tolerance towards T-cells.[41] One additional interesting feature of DPSC is their ability to attract and maintain growth of neurons.[42] This is probably due to the ability of dental pulp to act as a sensory organ.[10] A close relationship with neurons is perhaps determined by the neural-crest origins of dental pulp. Immortalisation of mouse DPSC has been achieved by limiting dilution passaging.[43] The cells maintained several transcripts of genes such as dental syaloprotein, dental matrix protein-1 and others.

2.2. Periodontal ligament and dental follicle stem cells

As discussed earlier, in common with dental pulp, the auxiliary tissues of the tooth can regenerate to a certain degree after mild trauma.[15] Earlier studies have identified dividing cells emerging from wounded PDL as fibroblast-like heterogeneous cells, derived from the vicinity of blood vessels, but not of haematopoietic origin. As in the case of dental pulp, undifferentiated perivascular progenitors were suggested as the source of cells for repair.[44] The isolation of periodontal ligament stem cells has attracted considerable effort. Many groups reported such cell populations being isolated.[45–48] Multipotent progenitors from human PDL have been characterised after clonal selection and magnetic activated cell sorting with STRO-1.[49] Periodontal stem cells (PDLSC) are characteristically STRO-1 and CD146/MUC18 positive. Under defined culture conditions, PDLSCs are multipotent and show a differentiation ability into cementoblast-like cells, adipocytes and fibroblasts.[49] When PDLSCs are transplanted into host mice cementum/PDL-like structures are formed.[50] Interestingly PDLSC can be isolated from the solid-frozen human primary tissue.[51] Cells isolated from frozen samples have been verified as maintaining their stem cell properties and tissue regeneration capacity. Gene expression profiling of PDLSC has been attempted.[52] The elevated expression of early growth response-1 (EGD-1), osteoprotegerin, elastin and IGF binding protein-3 (IGFBP-3) was found. A primitive subpopulation of PDLSC has been isolated from adult rat PDL by applying the technique of neurosphere formation, suggesting that PDL contains neural-crest-derived precursors.[48] These cells express neural crest markers Nestin, Twist, Slug, Sox9 and Sox2. Under differentiation conditions these cells were found to make myotubes, neuron-like cells and glia suggesting they are true neural-crest stem cells. Another subpopulation that has been characterised in PDLSC is the side population selected by the expression of the ABCG-2 transporter following dye exclusion and fluorescence sorting.[53] Periodontal stem cells have been shown to regenerate PDL in rats when cultured cell sheets were transplanted into previously generated periodontal defects of molar teeth.[50]

The dental follicle shares with periodontal ligament the ability to regenerate.[45] Human dental follicle progenitor cells can be obtained from human third molars and are characterised by their plastic attachment in culture and expression of makers such as Nestin and Notch-1. These cells are considered to be able to differentiate into PDL-like structures, bone and cementum.[54] Further analysis has shown the presence of heterogeneous cell populations in developing dental follicles after analysis of their mineralisation characteristics *in vitro* and the growth factor and matrix protein gene expression patterns from several cloned cell lines under the same culture conditions.[47] Human DFSC acquire cementoblast characteristics upon stimulation with BMP-7 and enamel matrix derivatives *in vitro*.[55,56] Bovine dental follicle cells show a differentiation ability to form cementoblasts when transplanted into immunodeficient mice.[45,54] Numerous successes have been reported in generating immortalised PDLSC and DFSC. Strains of genetically modified mice expressing SV40 large T-antigen have been used to generate cementoblasts and PDLSC.[54,58] Bovine cementoblast progenitors have been transformed with Bmi-1 and TERT to generate a continuous cell line.[59] Mouse PDLSC were immortalised by introducing human papilloma virus type 16 E6 gene lacking the PDZ-domain.[60] Human PDLSC were produced after gene transfer of TERT and SV40 large T-antigen.[61]

2.3. Dental epithelial stem cells

In humans, ectoderm-derived epithelial cells take part in the formation of the tooth crown during tooth morphogenesis and maturation, but once differentiated into ameloblasts they cease to proliferate. Generally it is accepted that epithelium cannot regenerate after it reaches the mature stage of development. Nevertheless in other animals, notably in rodents, continuously growing teeth present an ideal model for studying renewal of dental epithelium. Mouse incisors and continuously growing molars in some mammalian species show replenishing populations of enamel organs composed of a core of stellate reticulum, stratum intermedium and surrounding enamel epithelial cells.[62] An epithelial stem cell niche in a mouse incisor is located at its labial apical end, known as the cervical loop.[63] This is the junction of the inner enamel epithelium (IEE) and the outer enamel epithelium (OEE) at the apex of the enamel organ. Early morphological observations have found most of the mitoses to be located in the distal part of the cervical loop. It may indicate the occurrence of transient population of quickly dividing cells in IEE and stratum intermediate.[64] Being highly proliferative, these cells may not necessarily be stem cells but rather amplifying cells. To date they have only been shown to be capable of forming one cell type — ameloblasts. In a series of experiments, Harada *et al.* determined that FGF10, Notch and Sprouty are essential for the continuous growth of the mouse incisors and maintenance of the stem cell niche.[64–66] The authors established epithelial progenitor cell lines from mouse incisors that can be differentiated in culture in co-cultures with dental papilla cells.[67] Ameloblast cell lines could also be established from newborn mouse molars, suggesting that epithelial stem cell niche is still functional.[68] The addition of Shh stimulated differentiation of these cell both *in vitro* and following subcutaneous injection with Matrigel.[69,70] Spontaneously immortalised ameloblast-like cells have been generated from mouse newborn molars.[68] Generation of epithelial cell lines from embryonic mouse molar tooth germs has been greatly enhanced by using a mouse strain deficient for anti-oncogenic gene p53.[71] Several lines have been established at various stages of ameloblast differentiation. By recombining these cells with DPSC various degrees of efficiency in generating tooth shape were achieved. The degree of success varied from 12% to 56% of recombinations giving functional teeth. Recombination of the apical bud cells from two-day-old rat incisors with DPSC of adult rats led to the development of correct incisor shape in transplantation experiments.[72] Epithelial cells are present in adult teeth in the form of epithelial rests of Mallassez, a loose mesh of interconnected cells that are attached to the root of human teeth.[73] Little is known about their function, although the general idea exists that these cells participate in regulating cementoblast function.[73,74]

3. Two Strategies for Tissue Engineering Tooth Germs *De Novo*

Strategies of the regenerative engineering of new replacement teeth fall into two categories. This reflects two basic conceptual approaches to engineering tooth structures.

3.1. Scaffold-based roots

One strategy aims to create a bio-artificial root implant that can support an artificial crown, in the same way as the standard artificial tooth implant does. Essentially, the construct consists of a scaffold in the shape of a root that is seeded with appropriate cells. This allows growth of cells inside the scaffold and formation of proper anchor to the jaw bone.[72] Remarkably, the use of two stem cell populations allowed the building of proper connections between the implant and the bone. A hydroxyapatite/tricalcium phosphate (HA/TCP) scaffold in a shape of incisor root was seeded with porcine apical root papilla cells (SCAP cells). A central channel space was made for subsequent installation of a porcelain crown, and the channel closed with a filling. The implant was placed in a tooth socket; the outside of the implant was coated with gelatine-based gel (Gelfoam, Pharmacia) that had been seeded with periodontal ligament stem cells (PDLSC). The implant was then sutured into place and kept closed for three months. At the end of this period CT scans revealed the formation of an electron-dense root-shaped structure. Having opened the sutured implant, a prefabricated crown prosthesis was cemented onto the bioengineered implant. Subsequently the implant remained in the jaw for another four weeks to allow for the normal tooth use. Further histological analysis revealed that proper root and periodontal ligament structures were formed and remained intact after continuous use of the tooth for four weeks.

3.2. Reproducing embryonic tooth germs for implantation

An alternative strategy is based upon the idea of making a fully functional tooth by recreating the condition in which stem cells recapitulate morphogenetic events that occur during tooth development and maturation. This approach aims to create both root and crown from stem cells. In principle, it is assumed that when the development process is initiated, it can continue naturally if conditions are maintained. Experiments on rodents have shown that not only embryonic tissues, but the cells as late in development as newborn epithelial and adult dental pulp can produce a tooth after recombination.[75,76] Such artificially formed tooth germs can be later implanted in place of a missing tooth.

As discussed above, tooth initiation requires the cross-talk between the two competent cell populations, epithelial and mesenchymal. The challenge that tooth bioengineering faces is how to put epithelial and mesenchymal counterparts in a correct position and hold them together for a period long enough for them to communicate and arrange themselves. This arrangement needs to be accompanied by the deposition of basement membrane between the two cell populations, which will serve as a depot for the growth factors and morphogenic factors. Hence, various strategies have been proposed to recombine dissociated cells into tooth-initiating structures. Broadly, in a traditional way for hard tissue engineering, scaffolding is used to retain cells in place during culturing and transplantation steps.[77–79] Alternatively, cell retention and close contact can be achieved by making a cell pellet that can be later put into organ culture or transplanted into recipients for maturation.[76,80–82] Lastly, cells can be injected into soluble collagen that is able to retain cells in close apposition when it becomes a gel.[71,83] All three strategies produce teeth, but with various degrees of efficiency.

Entrapment of cells in PGA and PGLA scaffolds was the first method of producing bioengineered teeth.[77,79] Tooth germs of newborn rats were dissected, dental cells dissociated and put in culture for six days without separating epithelium and mesenchyme. Cells from these mixed cultures were harvested and seeded as a single cell suspension into scaffolds, which were subsequently transplanted into the omentum of recipient rats. Twelve weeks later some implants showed development of tooth structures with dentin and enamel. However the shapes of artificially generated teeth were random. Similar experiments have been repeated with pigs, and the only difference was that pig teeth took longer to form, up to 30 weeks.[77] An attempt has been made to generate complete teeth and bone structure that could be transplanted into a site of missing tooth.[78] First, a tooth germ was created by seeding adult pig third molar tooth germ cells into grafted PGA/PLLA scaffolds. The structures were transplanted into rat omentum for four weeks. At the same time adult pig bone marrow stem cells were seeded into bagel-shaped PGLA scaffolds. The scaffolds were grown in a rotating bioreactor for ten days. Next, tooth structures were recovered from the omentum and sutured into the middle of the bone structures. The resulting grafts were reimplanted into rat omentum for a minimum of eight weeks. Roots and periodontal ligaments developed; however, the orientation of bioengineered tooth periodontal structures was often incorrect in respect to the bioengineered bone. Some roots were oriented toward the tooth crown instead of apically toward the bone construct.

Attempts to recreate tooth germs as pellets of epithelial and mesenchymal cells gave more satisfying results. Hu *et al.* combined molar cap-stage embryonic dissociated epithelium with dissociated mesenchyme as well as with non-dissociated mesenchyme to form pellets that were put in organ culture for six to eight days in semi-solid medium.[80,81] The best success rate for tooth initiation was seen when epithelium was dissociated but mesenchyme was intact. Pellets of dissociated mesenchyme with intact epithelium gave poor results, as did pellets when both types of cells were dissociated and randomly mixed. This is in agreement with earlier observations that at the cap stage, mesenchyme constitutes the inducer of morphogenesis. When the cultured recombined tooth germs were transplanted under the skin of immunocompromised mouse only combinations where mesenchyme was not dissociated gave regular teeth with appropriate shape and size.[81] In associations where either epithelium and mesenchyme or only mesenchyme had been dissociated, the overall shape of the tooth was not maintained, crown shape was distorted and cusp number and position were altered. These difficulties seem to be at least partially overcome by Nakao *et al.*, who injected dissociated epithelium and mesenchyme cells from cap-stage mouse incisors and molars into collagen gels.[83] They could observe progressive formation of tissue-specific cell organisation over a period of 14 days in culture. The key element in their technique was the density of the cells. The cell density was found to greatly affect the outcome of recombination. Tooth-specific cell associations were observed only when high densities of cells were achieved (5×10^8/ml). The success of tooth morphogenesis in such associations depended also on the position of two cell types with respect to each other. The close contact of two highly dense pellets of cells is required. When the two cell types were mixed randomly, such pellets did not show morphogenesis. When the artificially formed tooth germs were transplanted under renal capsules only the high density recombinations gave teeth after two weeks. The low density cell aggregates ($0.5–1 \times 10^8$) gave oral epithelium differentiation in the form of keratin cysts and also mesenchymal-derived bone.

Not only cap-stage embryonic dental epithelium and mesenchyme can be associated to produce a tooth shape. Tooth germs have been isolated from young (six-month-old) pigs and the mesenchymal cells have been seeded into collagen sponges at high density, followed by epithelium.[75] The grafts so formed have been transplanted into the omentum of immunocompromised rats for six to 20 weeks. Tooth structures were observed when epithelium and mesenchyme cells were seeded randomly and in grafts with epithelium seeded on top of mesenchyme. In the latter case the grafts seemed to contain only one tooth germ per graft, whereas in the former situation multiple teeth formed.

A considerable challenge for tissue engineering is how to find suitable source of precursor/stem cells. For bioartificial teeth one obvious cell source in adult humans would be from extracted third molars. Cell immortalisation is the way of obtaining cells suitable for allogenic transplantation. Indeed, as discussed earlier, many attempts have been made to get continuous cell lines of both tooth mesenchyme and epithelium.[43,57–59,61,68] However, so far it has not been shown that these cell lines are capable of producing teeth. Nevertheless, one study clearly showed that spontaneously immortalised molar cap-stage epithelial cells can produce a tooth when combined with mesenchyme in a collagen gel.[71]

In the effort to find an acceptable cell source several primitive stem cell populations were examined in associations with the early tooth epithelium, which has been shown to possess inducing capability.[82] Combinations of mouse embryonic stem cells, neural stem cells and bone marrow stem cells (BMSC) with early oral epithelium were performed. It has been assumed that some primitive stem cells may respond to odontogenic signals from early tooth epithelium. Interestingly, the expression of markers that characterise tooth germ was seen in all three cell types after one or two days in culture. When such combinations of mouse tooth epithelium with, respectively, mouse embryonic stem cells, neural stem cells, and BMSC were transplanted into renal capsules for two weeks, only the latter combination showed the development of tooth. However it is not clear at present how efficient such a method is. In an attempt to purify cells with more morphogenetic potential, Yu *et al.* selected only the STRO-1-positive cells from both dental pulps and bone marrows of young rats[76] After recombining these two types of cells with epithelial apical bud cells, using absorbable gelatine sponge as a carrier, the pellets were transplanted into renal capsules of rats. Dentin-enamel structures of correct shape were found to be formed only with dental pulp-derived cells. Bone marrow cells in association with tooth epithelium do not produce ameloblast differentiation, although they did achieve atypical dentinogenesis. This result suggests that selecting the STRO-1-positive population of BMSC may not be advisable. Presumably this population lacks signals that can drive epithelial differentiation, although they can respond to epithelial signals themselves.

So far no epithelial counterpart has been identified that can serve for bioartificial tooth production, other than tooth epithelium. Trans-differentiation of epithelia has been reported for other applications. For example, oral mucosa epithelium has been employed with success to generate cornea.[84] Hu *et al.* made combinations of dissociated mouse cap-stage tooth epithelium mixed with BMSC from EGFP-transgenic mice with intact tooth mesenchyme.[85] After 20 days in organ culture tooth crowns were seen in two-thirds of experiments with BMSCs participating in the ameloblast layer. Double immunolocalisation revealed that among all BMSC present in the preparation, the c-kit-positive cells preferentially participated in tooth morphogenesis. Interestingly, some of the c-kit-positive cells were found to differentiate into odontoblasts too. The analysis for Y-chromosome in inter-gender associations demonstrated that BMSC cytodifferentiation into ameloblasts was unlikely to be due to cell fusion with epithelial cells.

It will be crucial to establish a method for implanting bioartificial tooth germs into the mouth. To determine whether it would be possible to implant tooth germs successfully into oral mucosa in such a way that they continue to develop, mouse cap-stage molar tooth germs were dissected and surgically implanted into the soft tissue of the diastema region of the maxilla of adult mice. Twenty-four days later the implanted tooth germs were found to have developed into teeth of normal size connected to the bone by a differentiating connective tissue, roots were formed and teeth erupted.[82] Bioengineered tooth germs were also implanted into the diastema. Mouse incisor tooth germs were put in organ culture for two days and then either implanted in the diastema directly or were allowed to grow in the renal capsule for two weeks and then implanted into the diastema. In both cases teeth were seen after 14 days with enamel, dentin, developing roots periodontal tissue, dental pulp and blood vessels, but no eruption was documented so far.[83]

To date the adequate conditions for the maturation of the pre-formed tooth germs can only be achieved by transplanting them into suitable anatomical locations in donors or hosts. Thus, further advancement is urgently needed in methods of organotypic culturing of dental cells combinations.

4. Challenges

Of the many challenges that remain there are some that currently remain big obstacles:

— Can dental stem cells be purified and expanded *in vitro* for clinical uses?
— Identification of easily expandable sources of epithelial and mesenchymal cells for tooth tissue engineering.
— GMP growth of both cell populations for testing (trials) in patients.

5. Conclusions

Dental stem cells are an easily accessible source of mesenchymal stem cells that can have wide-ranging uses in clinical applications.

Tooth tissue engineering continues to show promise. The fact that teeth are clinically important but easily accessible, non-essential organs makes them ideal targets in translational medicine.

References

1. Jernvall, J. and Thesleff, I. Reiterative signaling and patterning during mammalian tooth morphogenesis. *Mech. Dev.* **92**, 19–29 (2000).
2. Tucker, A. and Sharpe, P. The cutting-edge of mammalian development; how the embryo makes teeth. *Nat. Rev.* **5**, 499–508 (2004).
3. Graham, A. and Lumsden, A. Patterning the cranial neural crest. *Biochem. Soc. Symp.* **62**, 77–83 (1996).
4. Lumsden, A.G. Spatial organization of the epithelium and the role of neural crest cells in the initiation of the mammalian tooth germ. *Development (Cambridge, England)* **103**(Suppl.), 155–169 (1988).
5. Zhang, Y.D., Chen, Z., Song, Y.Q., Liu, C. and Chen, Y.P. Making a tooth: growth factors, transcription factors, and stem cells. *Cell Res.* **15**, 301–316 (2005).
6. Mitsiadis, T.A. and Smith, M.M. How do genes make teeth to order through development? *J. Exp. Zool. Part B* **306**, 177–182 (2006).
7. Wang, X.P., Suomalainen, M., Felszeghy, S., Zelarayan, L.C., Alonso, M.T., Plikus, M.V., Maas, R.L., Chuong, C.M., Schimmang, T. and Thesleff, I.

An integrated gene regulatory network controls stem cell proliferation in teeth. *PLoS Biol.* **5**, e159 (2007).

8. Yoshiba, N., Yoshiba, K., Stoetzel, C., Perrin-Schmitt, F., Cam, Y., Ruch, J.V., Hosoya, A., Ozawa, H. and Lesot, H. Differential regulation of TIMP-1, -2, and -3 mRNA and protein expressions during mouse incisor development. *Cell Tissue Res.* **324**, 97–104 (2006).
9. Yoshiba, N., Yoshiba, K., Stoetzel, C., Perrin-Schmitt, F., Cam, Y., Ruch, J.V. and Lesot, H. Temporospatial gene expression and protein localization of matrix metalloproteinases and their inhibitors during mouse molar tooth development. *Dev. Dyn.* **228**, 105–112 (2003).
10. Ten Cate, A. *Oral Histology: Development, Structure and Function* (Elsevier, Amsterdam, 2003).
11. Luan, X., Ito, Y. and Diekwisch, T.G. Evolution and development of Hertwig's epithelial root sheath. *Dev. Dyn.* **235**, 1167–1180 (2006).
12. Steele-Perkins, G., Butz, K.G., Lyons, G.E., Zeichner-David, M., Kim, H.J., Cho, M.I. and Gronostajski, R.M. Essential role for NFI-C/CTF transcription-replication factor in tooth root development. *Mol. Cell. Biol.* **23**, 1075–1084 (2003).
13. Shimono, M., Ishikawa, T., Ishikawa, H., Matsuzaki, H., Hashimoto, S., Muramatsu, T., Shima, K., Matsuzaka, K. and Inoue, T. Regulatory mechanisms of periodontal regeneration. *Microsc. Res. Tech.* **60**, 491–502 (2003).
14. Lumsden, A. Multipotent cells in the avian neural crest. *Trends Neurosci.* **12**, 81–83 (1989).
15. Tziafas, D. Basic mechanisms of cytodifferentiation and dentinogenesis during dental pulp repair. *Int. J Dev. Biol.* **39**, 281–290 (1995).
16. Tecles, O., Laurent, P., Zygouritsas, S., Burger, A.S., Camps, J., Dejou, J. and About, I. Activation of human dental pulp progenitor/stem cells in response to odontoblast injury. *Arch. Oral Biol.* **50**, 103–108 (2005).
17. Yamamura, T. Differentiation of pulpal cells and inductive influences of various matrices with reference to pulpal wound healing. *J. Dent. Res.* **64**, 530–540 (1985).
18. About, I., Bottero, M.J., de Denato, P., Camps, J., Franquin, J.C. and Mitsiadis, T.A. Human dentin production *in vitro*. *Exp. Cell Res.* **258**, 33–41 (2000).
19. Batouli, S., Miura, M., Brahim, J., Tsutsui, T.W., Fisher, L.W., Gronthos, S., Robey, P.G. and Shi, S. Comparison of stem-cell-mediated osteogenesis and dentinogenesis. *J. Dent. Res.* **82**, 976–981 (2003).
20. Gronthos, S., Brahim, J., Li, W., Fisher, L.W., Cherman, N., Boyde, A., DenBesten, P., Robey, P.G. and Shi, S. Stem cell properties of human dental pulp stem cells. *J. Dent. Res.* **81**, 531–535 (2002).

21. Gronthos, S., Mankani, M., Brahim, J., Robey, P.G. and Shi, S. Postnatal human dental pulp stem cells (DPSCs) *in vitro* and *in vivo*. *Proc. Natl. Acad. Sci. USA* **97**, 13625–13630 (2000).
22. Miura, M., Gronthos, S., Zhao, M., Lu, B., Fisher, L.W., Robey, P.G. and Shi, S. SHED: stem cells from human exfoliated deciduous teeth. *Proc. Natl. Acad. Sci. USA* **100**, 5807–5812 (2003).
23. Yamazaki, H., Tsuneto, M., Yoshino, M., Yamamura, K. and Hayashi, S. Potential of dental mesenchymal cells in developing teeth. *Stem Cells* **25**, 78–87 (2007).
24. Balic, A. and Mina, M. Analysis of developmental potentials of dental pulp *in vitro* using GFP transgenes. *Orthodont. Craniofac. Res.* **8**, 252–258 (2005).
25. Braut, A., Kollar, E.J. and Mina, M. Analysis of the odontogenic and osteogenic potentials of dental pulp *in vivo* using a Col1a1-2.3-GFP transgene. *Int. J. Dev. Biol.* **47**, 281–292 (2003).
26. Mina, M. and Braut, A. New insight into progenitor/stem cells in dental pulp using Col1a1-GFP transgenes. *Cells Tissues Organs* **176**, 120–133 (2004).
27. Yu, J., Deng, Z., Shi, J., Zhai, H., Nie, X., Zhuang, H., Li, Y. and Jin, Y. Differentiation of dental pulp stem cells into regular-shaped dentin-pulp complex induced by tooth germ cell conditioned medium. *Tissue Eng.* **12**, 3097–3105 (2006).
28. Zussman, W.V. Osteogenic activity of odontoblasts in transplanted tooth pulps. *J. Dent. Res.* **45**, 144–151 (1966).
29. Doherty, M.J., Ashton, B.A., Walsh, S., Beresford, J.N., Grant, M.E. and Canfield, A.E. Vascular pericytes express osteogenic potential *in vitro* and *in vivo*. *J. Bone Miner. Res.* **13**, 828–838 (1998).
30. Nehls, V. and Drenckhahn, D. The versatility of microvascular pericytes: from mesenchyme to smooth muscle? *Histochemistry* **99**, 1–12 (1993).
31. Shi, S. and Gronthos, S. Perivascular niche of postnatal mesenchymal stem cells in human bone marrow and dental pulp. *J. Bone Miner. Res.* **18**, 696–704 (2003).
32. Laino, G., Graziano, A., d'Aquino, R., Pirozzi, G., Lanza, V., Valiante, S., De Rosa, A., Naro, F., Vivarelli, E. and Papaccio, G. An approachable human adult stem cell source for hard-tissue engineering. *J. Cell. Physiol.* **206**, 693–701 (2006).
33. Laino, G., d'Aquino, R., Graziano, A., Lanza, V., Carinci, F., Naro, F., Pirozzi, G. and Papaccio, G. A new population of human adult dental pulp stem cells: a useful source of living autologous fibrous bone tissue (LAB). *J. Bone Miner. Res.* **20**, 1394–1402 (2005).
34. d'Aquino, R., Graziano, A., Sampaolesi, M., Laino, G., Pirozzi, G., De Rosa, A. and Papaccio, G. Human postnatal dental pulp cells co-differentiate into

osteoblasts and endotheliocytes: a pivotal synergy leading to adult bone tissue formation. *Cell Death Differ.* **14**, 1162–1171 (2007).
35. Iohara, K., Zheng, L., Ito, M., Tomokiyo, A., Matsushita, K. and Nakashima, M. Side population cells isolated from porcine dental pulp tissue with self-renewal and multipotency for dentinogenesis, chondrogenesis, adipogenesis, and neurogenesis. *Stem Cells* **24**, 2493–2503 (2006).
36. Huang, G.T., Shagramanova, K. and Chan, S.W. Formation of odontoblast-like cells from cultured human dental pulp cells on dentin *in vitro*. *J. Endodont.* **32**, 1066–1073 (2006).
37. Shi, S., Robey, P.G. and Gronthos, S. Comparison of human dental pulp and bone marrow stromal stem cells by cDNA microarray analysis. *Bone* **29**, 532–539 (2001).
38. De Bari, C., Dell'Accio, F., Vanlauwe, J., Eyckmans, J., Khan, I.M., Archer, C.W., Jones, E.A., McGonagle, D., Mitsiadis, T.A., Pitzalis, C. and Luyten, F.P. Mesenchymal multipotency of adult human periosteal cells demonstrated by single-cell lineage analysis. *Arthritis Rheum.* **54**, 1209–1221 (2006).
39. Zhang, W., Walboomers, X.F., Shi, S., Fan, M. and Jansen, J.A. Multilineage differentiation potential of stem cells derived from human dental pulp after cryopreservation. *Tissue Eng.* **12**, 2813–2823 (2006).
40. Papaccio, G., Graziano, A., d'Aquino, R., Graziano, M.F., Pirozzi, G., Menditti, D., De Rosa, A., Carinci, F. and Laino, G. Long-term cryopreservation of dental pulp stem cells (SBP-DPSCs) and their differentiated osteoblasts: a cell source for tissue repair. *J. Cell. Physiol.* **208**, 319–325 (2006).
41. Pierdomenico, L., Bonsi, L., Calvitti, M., Rondelli, D., Arpinati, M., Chirumbolo, G., Becchetti, E., Marchionni, C., Alviano, F., Fossati, V., Staffolani, N., Franchina, M., Grossi, A. and Bagnara, G.P. Multipotent mesenchymal stem cells with immunosuppressive activity can be easily isolated from dental pulp. *Transplantation* **80**, 836–842 (2005).
42. Nosrat, I.V., Smith, C.A., Mullally, P., Olson, L. and Nosrat, C.A. Dental pulp cells provide neurotrophic support for dopaminergic neurons and differentiate into neurons *in vitro*; implications for tissue engineering and repair in the nervous system. *Eur. J. Neurosci.* **19**, 2388–2398 (2004).
43. Arany, S., Nakata, A., Kameda, T., Koyota, S., Ueno, Y. and Sugiyama, T. Phenotype properties of a novel spontaneously immortalized odontoblast-lineage cell line. *Biochem. Biophys. Res. Commun.* **342**, 718–724 (2006).
44. McCulloch, C.A. Progenitor cell populations in the periodontal ligament of mice. *Anat. Rec.* **211**, 258–262 (1985).

45. Handa, K., Saito, M., Tsunoda, A., Yamauchi, M., Hattori, S., Sato, S., Toyoda, M., Teranaka, T. and Narayanan, A.S. Progenitor cells from dental follicle are able to form cementum matrix *in vivo*. *Connect. Tissue Res.* **43**, 406–408 (2002).
46. Ivanovski, S., Haase, H.R. and Bartold, P.M. Isolation and characterization of fibroblasts derived from regenerating human periodontal defects. *Arch. Oral Biol.* **46**, 679–688 (2001).
47. Jo, Y.Y., Lee, H.J., Kook, S.Y., Choung, H.W., Park, J.Y., Chung, J.H., Choung, Y.H., Kim, E.S., Yang, H.C. and Choung, P.H. Isolation and characterization of postnatal stem cells from human dental tissues. *Tissue Eng.* **13**, 767–773 (2007).
48. Techawattanawisal, W., Nakahama, K., Komaki, M., Abe, M., Takagi, Y. and Morita, I. Isolation of multipotent stem cells from adult rat periodontal ligament by neurosphere-forming culture system. *Biochem. Biophys. Res. Commun.* **357**, 917–923 (2007).
49. Seo, B.M., Miura, M., Gronthos, S., Bartold, P.M., Batouli, S., Brahim, J., Young, M., Robey, P.G., Wang, C.Y. and Shi, S. Investigation of multipotent postnatal stem cells from human periodontal ligament. *Lancet* **364**, 149–155 (2004).
50. Hasegawa, M., Yamato, M., Kikuchi, A., Okano, T. and Ishikawa, I. Human periodontal ligament cell sheets can regenerate periodontal ligament tissue in an athymic rat model. *Tissue Eng.* **11**, 469–478 (2005).
51. Seo, B.M., Miura, M., Sonoyama, W., Coppe, C., Stanyon, R. and Shi, S. Recovery of stem cells from cryopreserved periodontal ligament. *J. Dent. Res.* **84**, 907–912 (2005).
52. Ivanovski, S., Lichanska, A.M., d'Aniello, E., Xiao, Y. and Waters, M.J. Gene expression profiling of cells involved in periodontal regeneration. *Tissue Eng.* **13**, 393–404 (2007).
53. Kawanabe, N., Murakami, K. and Takano-Yamamoto, T. The presence of ABCG2-dependent side population cells in human periodontal ligaments. *Biochem. Biophys. Res. Commun.* **344**, 1278–1283 (2006).
54. Handa, K., Saito, M., Yamauchi, M., Kiyono, T., Sato, S., Teranaka, T. and Sampath Narayanan, A. Cementum matrix formation *in vivo* by cultured dental follicle cells. *Bone* **31**, 606–611 (2002).
55. Luan, X., Ito, Y., Dangaria, S. and Diekwisch, T.G. Dental follicle progenitor cell heterogeneity in the developing mouse periodontium. *Stem Cells Dev.* **15**, 595–608 (2006).
56. Kemoun, P., Laurencin-Dalicieux, S., Rue, J., Farges, J.C., Gennero, I., Conte-Auriol, F., Briand-Mesange, F., Gadelorge, M., Arzate, H., Narayanan, A.S., Brunel, G. and Salles, J.P. Human dental follicle

cells acquire cementoblast features under stimulation by BMP-2/-7 and enamel matrix derivatives (EMD) *in vitro*. *Cell Tissue Res.* **329**, 283–294 (2007).

57. D'Errico, J.A., Berry, J.E., Ouyang, H., Strayhorn, C.L., Windle, J.J. and Somerman, M.J. Employing a transgenic animal model to obtain cementoblasts *in vitro*. *J. Periodontol.* **71**, 63–72 (2000).
58. D'Errico, J.A., Ouyang, H., Berry, J.E., MacNeil, R.L., Strayhorn, C., Imperiale, M.J., Harris, N.L., Goldberg, H. and Somerman, M.J. Immortalized cementoblasts and periodontal ligament cells in culture. *Bone* **25**, 39–47 (1999).
59. Saito, M., Handa, K., Kiyono, T., Hattori, S., Yokoi, T., Tsubakimoto, T., Harada, H., Noguchi, T., Toyoda, M., Sato, S. and Teranaka, T. Immortalization of cementoblast progenitor cells with Bmi-1 and TERT. *J. Bone Miner. Res.* **20**, 50–57 (2005).
60. Yokoi, T., Saito, M., Kiyono, T., Iseki, S., Kosaka, K., Nishida, E., Tsubakimoto, T., Harada, H., Eto, K., Noguchi, T. and Teranaka, T. Establishment of immortalized dental follicle cells for generating periodontal ligament *in vivo*. *Cell Tissue Res.* **327**, 301–311 (2007).
61. Fujii, S., Maeda, H., Wada, N., Kano, Y. and Akamine, A. Establishing and characterizing human periodontal ligament fibroblasts immortalized by SV40T-antigen and hTERT gene transfer. *Cell Tissue Res.* **324**, 117–125 (2006).
62. Ohshima, H., Nakasone, N., Hashimoto, E., Sakai, H., Nakakura-Ohshima, K. and Harada, H. The eternal tooth germ is formed at the apical end of continuously growing teeth. *Arch. Oral Biol.* **50**, 153–157 (2005).
63. Thesleff, I., Wang, X.P. and Suomalainen, M. Regulation of epithelial stem cells in tooth regeneration. *C. R. Biol.* **330**, 561–564 (2007).
64. Kawano, S., Saito, M., Handa, K., Morotomi, T., Toyono, T., Seta, Y., Nakamura, N., Uchida, T., Toyoshima, K., Ohishi, M. and Harada, H. Characterization of dental epithelial progenitor cells derived from cervical-loop epithelium in a rat lower incisor. *J. Dent. Res.* **83**, 129–133 (2004).
65. Yokohama-Tamaki, T., Ohshima, H., Fujiwara, N., Takada, Y., Ichimori, Y., Wakisaka, S., Ohuchi, H. and Harada, H. Cessation of Fgf10 signaling, resulting in a defective dental epithelial stem cell compartment, leads to the transition from crown to root formation. *Development (Cambridge, England)* **133**, 1359–1366 (2006).
66. Harada, H., Kettunen, P., Jung, H.S., Mustonen, T., Wang, Y.A. and Thesleff, I. Localization of putative stem cells in dental epithelium and their association with Notch and FGF signaling. *J. Cell Biol.* **147**, 105–120 (1999).

67. Morotomi, T., Kawano, S., Toyono, T., Kitamura, C., Terashita, M., Uchida, T., Toyoshima, K. and Harada, H. *In vitro* differentiation of dental epithelial progenitor cells through epithelial-mesenchymal interactions. *Arch. Oral Biol.* **50**, 695–705 (2005).
68. Nakata, A., Kameda, T., Nagai, H., Ikegami, K., Duan, Y., Terada, K. and Sugiyama, T. Establishment and characterization of a spontaneously immortalized mouse ameloblast-lineage cell line. *Biochem. Biophys. Res. Commun.* **308**, 834–839 (2003).
69. Takahashi, S., Kawashima, N., Sakamoto, K., Nakata, A., Kameda, T., Sugiyama, T., Katsube, K. and Suda, H. Differentiation of an ameloblast-lineage cell line (ALC) is induced by Sonic hedgehog signaling. *Biochem. Biophys. Res. Commun.* **353**, 405–411 (2007).
70. Yan, Q., Zhang, Y., Li, W. and DenBesten, P.K. Differentiation of human ameloblast-lineage cells *in vitro*. *Eur. J. Oral Sci.* **114**(Suppl. 1), 154–158 (2006).
71. Komine, A., Suenaga, M., Nakao, K., Tsuji, T. and Tomooka, Y. Tooth regeneration from newly established cell lines from a molar tooth germ epithelium. *Biochem. Biophys. Res. Commun.* **355**, 758–763 (2007).
72. Sonoyama, W., Liu, Y., Fang, D., Yamaza, T., Seo, B.M., Zhang, C., Liu, H., Gronthos, S., Wang, C.Y., Shi, S. and Wang, S. Mesenchymal stem cell-mediated functional tooth regeneration in swine. *PLoS ONE* **1**, e79 (2006).
73. Diekwisch, T.G. The developmental biology of cementum. *Int. J. Dev. Biol.* **45**, 695–706 (2001).
74. Sonoyama, W., Seo, B.M., Yamaza, T. and Shi, S. Human Hertwig's epithelial root sheath cells play crucial roles in cementum formation. *J. Dent. Res.* **86**, 594–599 (2007).
75. Honda, M.J., Tsuchiya, S., Sumita, Y., Sagara, H. and Ueda, M. The sequential seeding of epithelial and mesenchymal cells for tissue-engineered tooth regeneration. *Biomaterials* **28**, 680–689 (2007).
76. Yu, J., Wang, Y., Deng, Z., Tang, L., Li, Y., Shi, J. and Jin, Y. Odontogenic capability: bone marrow stromal stem cells versus dental pulp stem cells. *Biol. Cell* **99**, 465–474 (2007).
77. Duailibi, M.T., Duailibi, S.E., Young, C.S., Bartlett, J.D., Vacanti, J.P. and Yelick, P.C. Bioengineered teeth from cultured rat tooth bud cells. *J. Dent. Res.* **83**, 523–528 (2004).
78. Young, C.S., Abukawa, H., Asrican, R., Ravens, M., Troulis, M.J., Kaban, L.B., Vacanti, J.P. and Yelick, P.C. Tissue-engineered hybrid tooth and bone. *Tissue Eng.* **11**, 1599–1610 (2005).
79. Young, C.S., Terada, S., Vacanti, J.P., Honda, M., Bartlett, J.D. and Yelick, P.C. Tissue engineering of complex tooth structures on biodegradable polymer scaffolds. *J. Dent. Res.* **81**, 695–700 (2002).

80. Hu, B., Nadiri, A., Bopp-Kuchler, S., Perrin-Schmitt, F. and Lesot, H. Dental epithelial histomorphogenesis *in vitro*. *J. Dent. Res.* **84**, 521–525 (2005).
81. Hu, B., Nadiri, A., Kuchler-Bopp, S., Perrin-Schmitt, F., Peters, H. and Lesot, H. Tissue engineering of tooth crown, root, and periodontium. *Tissue Eng.* **12**, 2069–2075 (2006).
82. Ohazama, A., Modino, S.A., Miletich, I. and Sharpe, P.T. Stem cell-based tissue engineering of murine teeth. *J. Dent. Res.* **83**, 518–522 (2004).
83. Nakao, K., Morita, R., Saji, Y., Ishida, K., Tomita, Y., Ogawa, M., Saitoh, M., Tomooka, Y. and Tsuji, T. The development of a bioengineered organ germ method. *Nat. Meth.* **4**, 227–230 (2007).
84. Nishida, K., Yamato, M., Hayashida, Y., Watanabe, K., Yamamoto, K., Adachi, E., Nagai, S., Kikuchi, A., Maeda, N., Watanabe, H., Okano, T. and Tano, Y. Corneal reconstruction with tissue-engineered cell sheets composed of autologous oral mucosal epithelium. *New Engl. J. Med.* **351**, 1187–1196 (2004).
85. Hu, B., Unda, F., Bopp-Kuchler, S., Jimenez, L., Wang, X.J., Haikel, Y., Wang, S.L. and Lesot, H. Bone marrow cells can give rise to ameloblast-like cells. *J. Dent. Res.* **85**, 416–421 (2006).

Chapter 31

Urogenital Repair

Anthony Atala

Abstract

Congenital abnormalities, trauma, infection, and cancer can all necessitate reconstructive surgery in the genitourinary tract. Currently, such surgeries may be performed with native non-urologic tissues, homologous tissues from a donor, heterologous tissues or substances, or artificial materials. However, these materials often lead to complications after reconstruction, including rejection of the implanted tissue. The field of tissue engineering may soon allow replacement of lost or deficient urologic tissues with anatomically and functionally equivalent ones that are derived from a small sample of the patient's own tissue. This would improve the outcome of reconstructive surgery in the genitourinary system and lead to new methods of treating these disorders.

Keywords: Stem Cells; Tissue Engineering; Bladder Reconstruction; Renal Replacement; Cell-Based Therapy.

Outline

1. Introduction
2. Tissue Engineering Strategies for Urogenital Repair
   - 2.1. The use of cells in urogenital tissue engineering applications
     - 2.1.1. *Stem cells*
     - 2.1.2. *Therapeutic cloning*
   - 2.2. Biomaterials for genitourinary tissue construction
   - 2.3. Vascularization of engineered tissue
3. Tissue Engineering of Specific Urologic and Genital Structures
   - 3.1. Urethra
   - 3.2. Bladder
     - 3.2.1. *Tissue expansion for bladder augmentation*
     - 3.2.2. *Seromuscular grafts and de-epithelialized bowel segments*

3.2.3. *Matrices for bladder regeneration*
3.2.4. *Bladder replacement using tissue engineering*
3.3. Ureters
4. Kidney
4.1. *Ex vivo* functioning renal units
4.2. Creation of functional renal structures *in vivo*
5. Genital Tissues
5.1. Reconstruction of corporal smooth muscle
5.2. Engineered penile prostheses
5.3. Female genital tissues
6. Other Applications of Genitourinary Tissue Engineering
6.1. Fetal tissue engineering
6.2. Injectable therapies
7. Conclusion
References

1. Introduction

Congenital disorders, cancer, trauma, infection, inflammation, iatrogenic injuries, or other conditions of the genitourinary system can lead to organ damage or complete loss of function. Both situations usually necessitate eventual reconstruction. Currently, reconstruction may be performed with native non-urologic tissues (skin, gastrointestinal segments, or mucosa from multiple body sites), homologous tissues from a donor (cadaver fascia, cadaver or living donor kidney), heterologous tissues or substances (bovine collagen), or artificial materials (silicone, polyurethane, Teflon). However, these materials often lead to complications after reconstruction, either because the implanted tissue is rejected, or because inherently different functional parameters cause a mismatch in the system. For example, replacement of bladder tissue with gastrointestinal segments can be problematic due to the opposite ways in which these two tissues handle solutes — urologic tissue normally excretes material, and gastrointestinal tissue generally absorbs the same materials. This mismatched state can lead to metabolic complications as well as infection and other issues.[1] Therefore, the replacement of lost or deficient urologic tissues with functionally equivalent ones would certainly improve the outcome of reconstructive surgery in the genitourinary system. This goal may soon be attainable with the use of tissue engineering techniques.

2. Tissue Engineering Strategies for Urogenital Repair

Tissue engineering uses the principles of cell transplantation, materials science, and biomedical engineering to develop biological substitutes that can restore and maintain normal function of damaged or lost tissues and organs. Tissue engineering may involve injection of functional cells into a nonfunctional site to stimulate regeneration. It can also involve the use of natural or synthetic matrices, often termed scaffolds, which encourage the body's natural ability to repair itself and assist in determination of the orientation and direction of new tissue growth. Often, tissue engineering uses a combination of both of these techniques. For example, matrices seeded with cells can be implanted into the body to encourage the growth or regeneration of functional tissue.

2.1. The use of cells in urogenital tissue engineering applications

Often, when cells are used for tissue engineering, donor tissue is dissociated into individual cells, which are then implanted directly into the host or expanded in culture, attached to a support matrix, and re-implanted after expansion. The implanted tissue can be heterologous, allogeneic, or autologous. Ideally, this approach allows lost tissue function to be restored or replaced *in toto* and with limited complications.[2–7]

However, an area of concern has been the source of cells for regeneration. The concept of creating engineered tissue constructs involves obtaining cells from the organ to be replaced and expanding these cells *in vitro* so that sufficient quantities are available for the implantation technique chosen. However, a major concern has been that, in certain cases, there may not be enough normal cells present in the diseased organ to begin this expansion process. This may not be the case; for example, one study has shown that cultured neuropathic bladder smooth muscle cells possess and maintain different characteristics than normal smooth muscle cells *in vitro*, as demonstrated by growth assays, contractility and adherence tests *in vitro*.[8] Despite these differences, when neuropathic smooth muscle cells were cultured *in vitro*, and then seeded onto matrices and implanted *in vivo*, the tissue engineered constructs showed the same properties as the constructs engineered with normal cells.[9] It is known that genetically normal progenitor cells, which are the reservoirs for new cell formation and are present even in diseased tissue, are programmed to give rise to normal tissue, regardless of whether they reside in normal or diseased tissues. Therefore, the stem cell niche and its role in normal tissue regeneration remains a fertile area of ongoing investigation.

2.1.1. *Stem cells*

Most current strategies for tissue engineering depend upon a sample of autologous cells from the diseased organ of the host. However, for many patients with extensive end-stage organ failure, a tissue biopsy may not yield enough normal cells for expansion and transplantation. In other instances, primary autologous human cells cannot be expanded from a particular organ, such as the pancreas. In these situations, pluripotent human embryonic stem cells are envisioned as an ideal source of cells, as they can differentiate into nearly any replacement tissue in the body.

Embryonic stem cells exhibit two remarkable properties: the ability to proliferate in an undifferentiated, but still pluripotent state (self-renewal), and the ability to differentiate into a large number of specialized cell types.[10] They can be isolated from the inner cell mass of the embryo during the blastocyst stage, which occurs five days post-fertilization. These cells have been maintained in the undifferentiated state for at least 80 passages when grown using current published protocols.[11] In addition, many protocols for differentiation into specific cell types in culture have been published. However, many uses of these cells are currently banned in the United States due to the ethical dilemmas that are associated with the manipulation of embryos in culture.

Adult stem cells have the advantage of avoiding some of the ethical issues associated with embryonic cells, and, unlike embryonic cells, they do not transdifferentiate into a malignant phenotype, so there is a diminished risk of teratoma formation should the cells be implanted *in vivo*. However, adult stem cells are limited for clinical use because expansion to the large quantities needed for tissue engineering is difficult.

Fetal stem cells derived from amniotic fluid and placentas have recently been described and represent a novel source of stem cells.[12,13] The cells express markers consistent with human embryonic stem cells, such as OCT4 and SSEA-4, but they do not form teratomas. The cells are multipotent and are able to differentiate into cells from all three germ layers. In addition, the cells have a high replicative potential and could be stored for future self-use, without the risks of rejection, and without ethical concerns.

2.1.2. *Therapeutic cloning*

Nuclear cloning, which has also been called nuclear transplantation and nuclear transfer, involves the introduction of a nucleus from a donor cell into an enucleated oocyte to create an embryo with a genetic makeup identical to that of the donor. Two types of nuclear cloning, reproductive cloning and therapeutic cloning, have been described. Banned in most countries for human applications, reproductive cloning is used to generate an embryo that has the identical genetic material as its cell source in order to produce offspring that are genetically identical to the donor. On the other hand, therapeutic cloning is used to generate early stage embryos that are used *in vitro*. They are explanted in culture to produce embryonic stem cell lines that are genetically identical to the source. These autologous

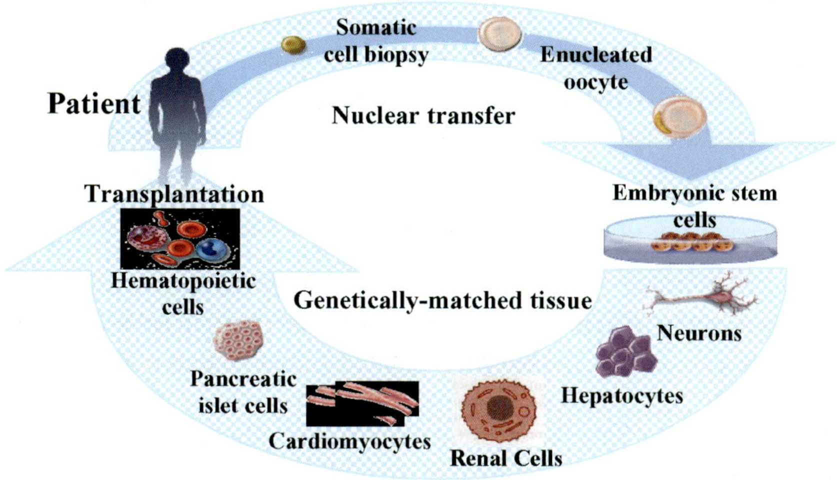

Fig. 1. Strategy for therapeutic cloning and tissue engineering.

stem cells have the potential to become almost any type of cell in the adult body, and thus would be useful in tissue and organ replacement applications (Fig. 1).[14] A particularly useful application for these cells would be in the treatment of end-stage kidney disease, for which there is limited availability of immunocompatible tissue transplants.

2.2. Biomaterials for genitourinary tissue construction

Biomaterials in genitourinary tissue engineering function as an artificial extracellular matrix (ECM) and are used to replace biologic and mechanical functions of native ECM found in tissues in the body. Biomaterials facilitate the localization and delivery of cells and/or bioactive factors (e.g. cell adhesion peptides, growth factors) to desired sites in the body as well as define a three-dimensional space for the formation of new tissues with appropriate structure. They also serve as a guide for the development of new tissues with appropriate function.[15,16] Direct injection of cell suspensions without such matrices has been used in some cases, but without this scaffold function, it is difficult to control the localization of transplanted cells.[17,18]

The ideal biomaterial should be biocompatible, promote cellular interaction and tissue development, and possess the proper mechanical and physical properties found in the tissue to be generated. Generally, three classes of biomaterials have been used for the engineering of genitourinary tissues: naturally derived materials, such as collagen and alginate, acellular tissue matrices, such as bladder submucosa and small-intestinal submucosa, and synthetic polymers, such as polyglycolic acid (PGA), polylactic acid (PLA), and poly(lactic-co-glycolic acid)

(PLGA). While synthetic polymers can be produced reproducibly on a large scale with controlled properties of strength, degradation rate, and microstructure, naturally derived materials and acellular tissue matrices have the potential advantage of biologic recognition, which can lessen host versus graft reactions.

2.3. Vascularization of engineered tissue

Limitations of nutrient and gas exchange currently restrict tissue-engineered implants to a volume of approximately 3 mm^3.[19] Therefore, to achieve the goals of engineering large complex tissues and organs, vascularization of the regenerating cells is essential.

Three approaches have been used to encourage the vascularization of bioengineered tissue. First, incorporation of angiogenic factors in the bioengineered tissue has been used in order to attract host capillaries and to enhance neovascularization of the implanted tissue. Second, some studies have investigated the effects of seeding EC with other cell types in the bioengineered tissue. Finally, prevascularization of the matrix prior to cell seeding has been attempted. There are many obstacles to overcome before large tissue-engineered solid organs are produced, but recent developments in angiogenesis research may provide important knowledge and essential materials to accomplish this goal.

3. Tissue Engineering of Specific Urologic and Genital Structures

3.1. Urethra

Various strategies have been proposed over the years for the regeneration of urethral tissue. Woven meshes of PGA (Dexon) have been used to reconstruct urethras in dogs.[20] Also, PGA has been used as a cell transplantation vehicle to engineer tubular urothelium *in vivo*.[2] Small-intestinal submucosa (SIS) without cells was used as an onlay patch graft for urethroplasty in rabbits.[21] Finally, a homologous graft of acellular urethral matrix was also used in a rabbit model.[22]

Bladder-derived acellular collagen matrix has proven to be a suitable graft for repair of urethral defects in rabbits. In the rabbit model, the neourethras created with these matrices demonstrated a normal urothelial luminal lining and organized muscle bundles shortly after repair.[21,22] These results were confirmed clinically in a series of patients with a history of failed hypospadias reconstruction, wherein the urethral defects were repaired with human bladder acellular collagen matrices (Fig. 2).[23,24] One the advantages of this material over nongenital tissue grafts currently used for urethroplasty (e.g. buccal mucosa) is that the material is "off the shelf." This eliminates the necessity of additional surgical procedures for graft

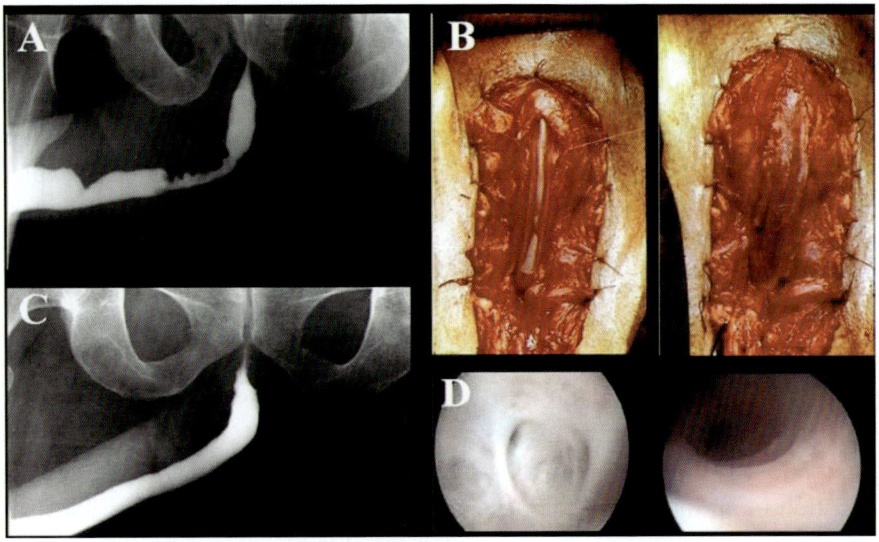

Fig. 2. Tissue engineering of the urethra using a collagen matrix. **(A)** Representative case of a patient with a bulbar stricture. **(B)** Urethral repair. Strictured tissue is excised, preserving the urethral plate on the left side, and matrix is anastomosed to the urethral plate in an onlay fashion on the right. **(C)** Urethrogram six months after repair. **(D)** Cystoscopic view of urethra before surgery on the left side and four months after repair on the right side.

harvesting, which may decrease operative time, as well as the potential morbidity due to the harvest procedure.

The above techniques, using non-seeded acellular matrices, were successfully applied experimentally and clinically for onlay urethral repairs. However, when tubularized urethral repairs were attempted experimentally, adequate urethral tissue regeneration was not achieved, and complications ensued, such as graft contracture and stricture formation.[25] Tubularized collagen matrices seeded with cells have performed better in animal studies. In a rabbit model, entire urethral segments were resected and urethroplasties were performed with tubularized collagen matrices either seeded with autologous cells or without cells. The tubularized collagen matrices seeded with autologous cells formed new tissue which was histologically similar to native urethral tissue.[26] The tubularized collagen matrices without cells lead to poor tissue development, fibrosis, and stricture formation.

3.2. Bladder

Currently, gastrointestinal segments are commonly used for bladder replacement or repair. However, gastrointestinal tissues are designed to absorb solutes that urinary tissue excretes, and due to this difference in function, multiple complications may ensue, such as infection, metabolic disturbances, urolithiasis, perforation, increased mucus production, and malignancy.[1,27,28] Because of the problems encountered with the use of gastrointestinal segments, numerous investigators have attempted alternative reconstructive procedures for bladder replacement or repair. The use of tissue expansion, seromuscular grafts, matrices for tissue regeneration, and tissue engineering with cell transplantation have been investigated.

3.2.1. *Tissue expansion for bladder augmentation*

A system of progressive dilation for ureters and bladders has been proposed as a method of bladder augmentation but has not yet been attempted clinically. Augmentation cystoplasty performed with dilated ureteral segments in animals has resulted in an increased bladder capacity ranging from 190% to 380%.[29,30] A system for the progressive expansion of native bladder tissue has also been used for augmenting bladder volumes in animals. Within 30 days after progressive dilation, the neoreservoir volume was expanded at least tenfold. Urodynamic studies showed normal compliance in all animals and microscopic examination of the expanded neoreservoir tissue showed a normal histology. A series of immunocytochemical studies demonstrated that the dilated bladder tissue maintained normal phenotypic characteristics.[30]

3.2.2. *Seromuscular grafts and de-epithelialized bowel segments*

Seromuscular grafts and de-epithelialized bowel segments, either alone or over a native urothelium, have also been attempted.[31–36] Keeping the urothelium intact avoids the complications associated with use of bowel in continuity with the urinary tract.[32,33] An example of this strategy is the combination of the techniques of autoaugmentation with those of enterocystoplasty. An autoaugmentation is performed and the diverticulum is covered with a demucosalized gastric or intestinal segment.[42]

3.2.3. *Matrices for bladder regeneration*

Non-seeded allogeneic acellular matrices have served as scaffolds for the ingrowth of host bladder wall components. The matrices are prepared by mechanically and chemically removing all cellular components from bladder tissue.[37–40] The matrices serve as vehicles for partial bladder regeneration, and relevant antigenicity is not evident. One example is SIS, a biodegradable, acellular, xenogeneic collagen-based tissue matrix. SIS was first used in the early 1980s as an acellular matrix for tissue replacement in the vascular field. It has been shown to promote regeneration of a variety of host tissues, including blood vessels and ligaments.[41] Animal studies have shown that the non-seeded SIS matrix used for bladder augmentation is able to regenerate *in vivo*.[42,43]

In multiple studies using various materials as non-seeded grafts for cystoplasty, the urothelial layer was able to regenerate normally, but the muscle layer, although present, was not fully developed.[37–39,43] Often the grafts contracted to 60%–70% of their original size with little increase in bladder capacity or compliance.[44,45] Studies involving acellular matrices that may provide the necessary environment to promote cell migration, growth, and differentiation are being conducted. Recently, bladder regeneration has been shown to be more reliable when the SIS was derived from the distal ileum.[42] With continued research in this area, these matrices may have a clinical role in bladder replacement in the future.

3.2.4. *Bladder replacement using tissue engineering*

Cell-seeded allogeneic acellular bladder matrices have been used for bladder augmentation in dogs. A group of experimental dogs underwent a trigone-sparing cystectomy and were randomly assigned to one of three groups. One group underwent closure of the trigone without a reconstructive procedure, another underwent reconstruction with a non-seeded bladder-shaped biodegradable scaffold, and the last underwent reconstruction using a bladder-shaped biodegradable scaffold that was seeded with autologous urothelial and smooth muscle cells.[46]

The cystectomy-only and non-seeded controls maintained average capacities of 22% and 46% of preoperative values, respectively. However, an average bladder capacity of 95% of the original precystectomy volume was achieved in the cell-seeded tissue engineered bladder replacements (Fig. 3). The subtotal cystectomy reservoirs that were not reconstructed and the polymer-only reconstructed bladders showed a marked decrease in bladder compliance (10% and 42% total compliance, respectively). The compliance of the cell-seeded tissue-engineered bladders was almost no different from preoperative values (106%). Histologically, the non-seeded scaffold bladders presented a pattern of normal urothelial cells with a thickened fibrotic submucosa and a thin layer of muscle fibers. The retrieved tissue-engineered bladders showed a normal cellular organization, consisting of a trilayer of urothelium, submucosa, and muscle.[46]

A clinical experience involving engineered bladder tissue for cystoplasty reconstruction was conducted starting in 1999. A small pilot study of seven patients was reported, using a collagen scaffold seeded with cells either with or without omentum coverage, or a combined PGA-collagen scaffold seeded with cells and omental coverage (see Fig. 4). The patients reconstructed with the engineered

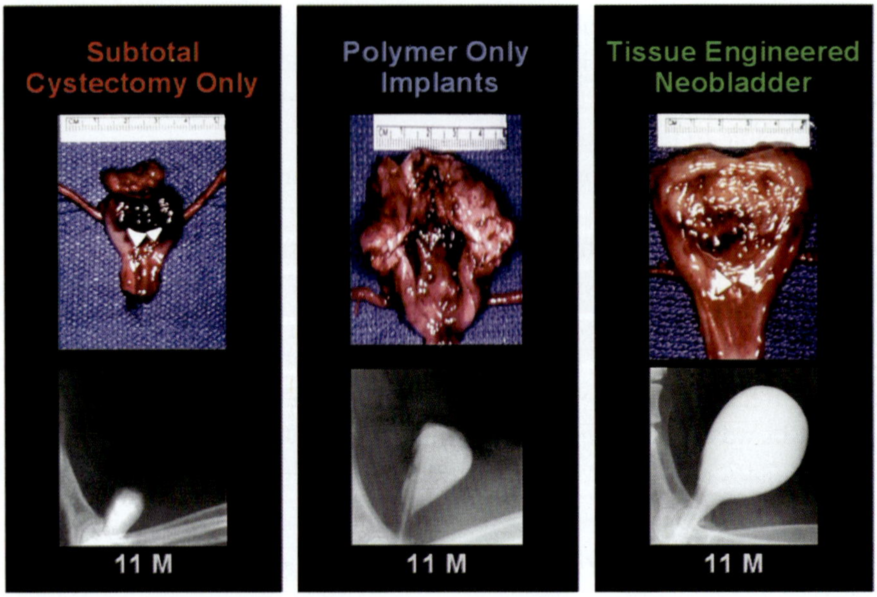

Fig. 3. Gross specimens and cystograms at 11 months of the cystectomy-only, non-seeded controls, and cell-seeded tissue engineered bladder replacements in dogs. The cystectomy-only bladder had a capacity of 22% of the preoperative value and a decrease in bladder compliance to 10% of the preoperative value. The non-seeded controls showed significant scarring with a capacity of 46% of the preoperative value and a decrease in bladder compliance to 42% of the preoperative value. An average bladder capacity of 95% of the original precystectomy volume was achieved in the cell-seeded tissue engineered bladder replacements and the compliance showed almost no difference from preoperative values that were measured when the native bladder was present (106%).

bladder tissue created with the PGA-collagen cell-seeded scaffolds showed increased compliance, decreased end-filling pressures, increased capacities and longer dry periods.[47] Although the experience is promising in terms of showing that engineered tissues can be implanted safely, it is just a start in terms of accomplishing the goal of engineering fully functional bladders. Further experimental and clinical work is being conducted.

3.3. Ureters

Ureteral non-seeded matrices have been used as a scaffold for the ingrowth of ureteral tissue in rats. These matrices promoted regeneration of the ureteral wall components.[48] Ureteral replacement with polytetrafluoroethylene (Teflon) grafts was also attempted in dogs, but with poor functional results.[49] In a more recent

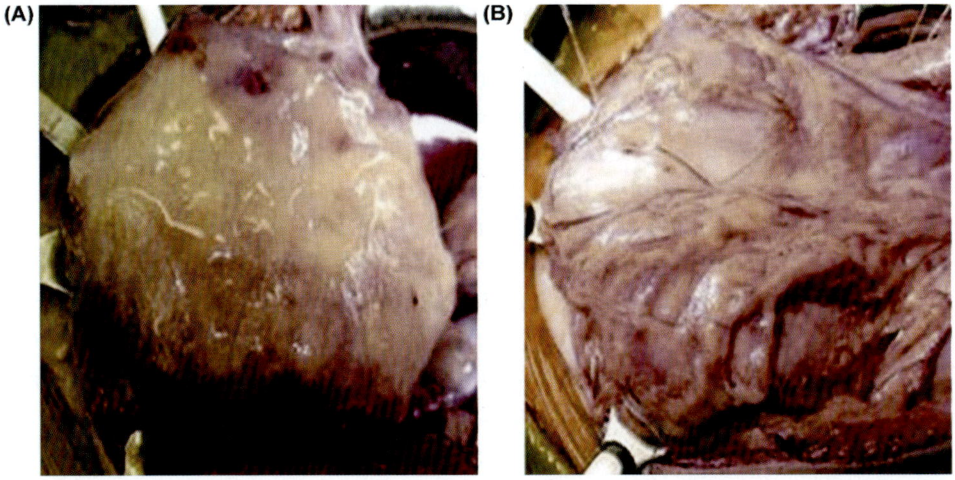

Fig. 4. Construction of an engineered human bladder. **(A)** The engineered bladder anastamosed to native bladder tissue with running 4-0 polyglycolic sutures. **(B)** Implanted bladder covered with fibrin glue and omentum.

study, non-seeded acellular collagen matrices were tubularized and used to replace 3 cm segments of canine ureters. The non-seeded acellular matrix tube was not able to successfully replace the segment of the ureter.[50]

Cell-seeded biodegradable polymer scaffolds have been used with more success to reconstruct ureteral tissues. In one study, urothelial and smooth muscle cells isolated from bladders and expanded *in vitro* were seeded onto tubular PGA scaffolds and implanted subcutaneously into athymic mice. After implantation, the urothelial cells proliferated to form a multilayered lining of the tubular structure, while the smooth muscle cells organized into multilayered structures surrounding the urothelial cells. Abundant angiogenesis was evident. The degradation of the polymer scaffolds resulted in the eventual formation of natural urothelial tissues. This approach was expanded to replacement of ureters in dogs.

4. Kidney

The kidney is the most challenging organ in the genitourinary system to reconstruct because of its extremely complex structure and function. Concepts for a bioartificial kidney are currently being explored. Some investigators are pursuing the replacement of isolated kidney function parameters using extracorporeal units, while others are aiming to replace total renal function with tissue-engineered bioartificial renal structures.

4.1. *Ex vivo* functioning renal units

Dialysis is currently the most common form of renal replacement therapy. However, the relatively high morbidity and mortality resulting from this process have spurred investigators to seek alternative solutions. In an attempt to assess the viability and physiologic functionality of a cell-seeded device to replace the filtration, transport, metabolic, and endocrinologic functions of the kidney, a synthetic hemofiltration device and a device which contained tissue-engineered porcine renal tubules were incorporated into an extracorporeal perfusion circuit, and this was introduced into acutely uremic dogs. Levels of potassium and blood urea nitrogen (BUN) were controlled during treatment with the device. The fractional reabsorption of sodium and water was possible. Active transport of potassium, bicarbonate, and glucose, and a gradual ability to excrete ammonia was observed. These results demonstrated the feasibility of an extracorporeal assist device that is reinforced by the use of proximal tubular cells.[51]

Using similar techniques, the development of a tissue-engineered bioartificial kidney consisting of a conventional hemofiltration cartridge in series with a renal tubule assist device containing human renal proximal tubule cells was used in patients with acute renal failure in the intensive care unit. The initial clinical experience with this bioartificial kidney suggests that renal tubule cell therapy may provide a dynamic and individualized treatment program as assessed by acute physiologic and biochemical indices.[52]

4.2. Creation of functional renal structures *in vivo*

Another approach to improve renal function involves the augmentation of renal tissue with kidney cells expanded *in vitro* and used for subsequent autologous transplantation. Most recently, an attempt was made to reconstitute renal epithelial cells for the generation of functional nephron units. Renal cells were harvested and expanded in culture. The cells were seeded onto a tubular device constructed from a polycarbonate membrane, which was connected at one end to a Silastic catheter that terminated in a reservoir. The device was implanted in athymic mice. Histologic examination of the implanted devices over time revealed extensive vascularization with formation of glomeruli and highly organized tubule-like structures. Immunocytochemical staining confirmed the renal phenotype. Additionally, yellow fluid was collected from inside the implant, and its creatinine and uric acid concentrations were consistent with the makeup of dilute urine. Further studies have shown the formation of renal structures in cows using nuclear transfer techniques (Fig. 5).[53] The expansion of this system to larger, three-dimensional structures is the next challenge awaiting researchers in the urogenital tissue engineering field.

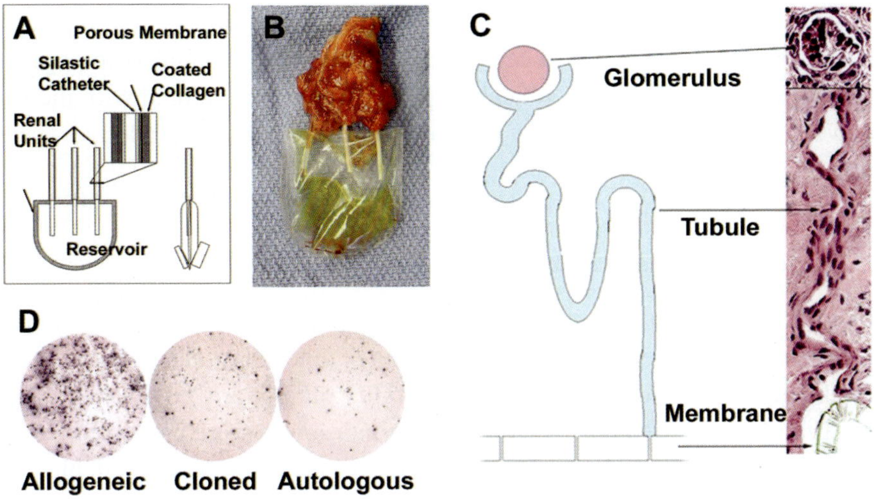

Fig. 5. Combining therapeutic cloning and tissue engineering to produce kidney tissue. **(A)** Illustration of the tissue-engineered renal unit. **(B)** Renal unit seeded with cloned cells, three months after implantation, showing the accumulation of urine-like fluid. **(C)** There was a clear unidirectional continuity between the mature glomeruli, their tubules, and the polycarbonate membrane. **(D)** Elispot analyses of the frequencies of T-cells that secrete IFN-gamma after primary and secondary stimulation with allogeneic renal cells, cloned renal cells, or nuclear donor fibroblasts.

5. Genital Tissues

5.1. Reconstruction of corporal smooth muscle

One of the major components of the phallus is corporal smooth muscle. The creation of autologous functional and structural corporal tissue *de novo* would be beneficial in cases of congenital abnormality of the genitals and in other situations where reconstruction is functionally and aesthetically necessary. In order to look at the functional parameters of engineered corpora, acellular corporal collagen matrices were obtained from donor rabbit penile tissue, and autologous corpus cavernosal smooth muscle and endothelial cells were harvested, expanded and seeded on the matrices. The entire rabbit corpora was removed and replaced with the engineered structures. The experimental corporal bodies demonstrated intact structural integrity by cavernosography and showed similar intracoporal pressures by cavernosometry when compared to the normal controls. Rabbits that received scaffolds without cells failed to achieve normal erectile function throughout the study period. However, mating activity in the animals with the cell-seeded corpora appeared normal by one month after implantation. The presence of sperm was confirmed during mating, and was present in all rabbits with the engineered corpora. The female rabbits that mated with the animals implanted with engineered corpora conceived and delivered healthy pups. Animals implanted with the matrix alone were unable to demonstrate normal mating activity and failed to ejaculate into the vagina.[54,55]

5.2. Engineered penile prostheses

Although silicone is an accepted biomaterial for penile prostheses, biocompatibility is a concern.[56,57] Use of a natural prosthesis composed of autologous cells may be advantageous. In a recent study, the feasibility of applying engineered cartilage rods *in situ* was investigated.[58] Autologous chondrocytes were harvested from rabbit ear and expanded in culture. The cells were seeded onto biodegradable poly-L-lactic acid-coated polyglycolic acid polymer rods and then implanted into the corporal spaces of rabbits. Examination at retrieval showed the presence of well-formed, milky-white cartilage structures within the corpora at one month and the polymer scaffolding had degraded by two months. There was no evidence of erosion or infection in any of the implantation sites. Subsequent studies were performed to assess the long-term functionality of the cartilage penile rods *in vivo*. To date, the animals have done well and can copulate and impregnate their female partners without problems.

5.3. Female genital tissues

Congenital malformations of the uterus may have profound implications clinically. Patients with cloacal exstrophy and intersex disorders may not have sufficient uterine tissue present for future reproduction. We investigated the possibility of engineering functional uterine tissue using autologous cells. Autologous rabbit uterine smooth muscle and epithelial cells were harvested and expanded in culture. These cells were seeded onto pre-configured uterine-shaped biodegradable polymer scaffolds, and these were used for subtotal uterine tissue replacement in the corresponding autologous animals. Upon retrieval six months after implantation, histological, immunocytochemical, and Western blot analyses confirmed the presence of normal uterine tissue components. Biomechanical analyses and organ bath studies showed that the functional characteristics of these tissues were similar to those of normal uterine tissue. Breeding studies using these engineered uteri are currently being performed.

Similarly, several pathologic conditions, including congenital malformations and malignancy, can adversely affect normal vaginal development or anatomy. To investigate tissue engineering methods of generating vaginal tissue for use in these situations, vaginal epithelial and smooth muscle cells of female rabbits were harvested, grown, and expanded in culture. These cells were seeded onto biodegradable polymer scaffolds, and the cell-seeded constructs were then implanted into nude mice for up to six weeks. Immunocytochemical, histological, and Western blot analyses confirmed the presence of vaginal tissue phenotypes. Electrical field stimulation studies in the tissue-engineered constructs showed similar functional properties to those of normal vaginal tissue. When these constructs were used for autologous total vaginal replacement, patent vaginal structures were noted in the tissue-engineered specimens, while the non-cell-seeded structures were noted to be stenotic.[59]

6. Other Applications of Genitourinary Tissue Engineering

6.1. Fetal tissue engineering

The prenatal diagnosis of fetal abnormalities is now more common and more accurate. Improvements in prenatal diagnosis have led to demand for novel interventions designed to reverse potentially life-threatening processes before birth. Having a ready supply of urologic-associated tissue for immediate surgical reconstruction of congenital malformations at birth may be advantageous. Theoretically, once the diagnosis of the pathologic condition is confirmed prenatally, a small tissue biopsy could then be obtained under ultrasound guidance. These biopsy materials could then be processed expanded *in vitro*. Using tissue engineering techniques, *in vitro*-reconstituted structures could then be readily available at the time of birth for reconstruction.

6.2. Injectable therapies

Both urinary incontinence and vesicoureteral reflux are common conditions affecting the genitourinary system. Currently, injectable bulking agents are one treatment used clinically for these conditions, but biocompatibility of current synthetic bulking agents is a concern. The ideal substance for endoscopic treatment of reflux and incontinence should be injectable, non-antigenic, non-migratory, volume stable, and safe for human use. Animal studies have shown that chondrocytes (cartilage cells) can be easily harvested and combined with alginate *in vitro* and the resulting suspension can be easily injected cystoscopically. The elastic cartilage tissue formed as a result of the injection is able to correct vesicoureteral reflux without any evidence of obstruction. This technology has been applied in humans for the correction of vesicoureteral reflux in children and for urinary incontinence in adults (Fig. 4).[60,61]

Using cell therapy techniques, the use of autologous smooth muscle cells has been explored for both urinary incontinence and vesicoureteral reflux applications. The potential use of injectable, cultured myoblasts for the treatment of stress urinary incontinence has also been investigated.[62,63] The use of injectable muscle precursor cells has also been investigated for use in the treatment of urinary incontinence due to irreversible urethral sphincter injury or maldevelopment.[64] A clinical trial involving the use of muscle-derived stem cells (MDSC) to treat stress urinary incontinence has also been performed with good results. Biopsies of skeletal muscle were obtained and autologous myoblasts and fibroblasts were cultured. Under ultrasound guidance, myoblasts were injected into the rhabdosphincter and fibroblasts mixed with collagen were injected into the submucosa. One year following injection, the thickness and function of the rhabdosphincter had significantly increased, and all patients were continent.[65] These are the first demonstrations of the replacement of both sphincter muscle tissue and its innervation by the injection of muscle precursor cells.

In addition, injectable muscle-based gene therapy and tissue engineering were combined to improve detrusor function in a bladder injury model, and may potentially be a novel treatment option for urinary incontinence.[66]

Patients with testicular dysfunction require androgen replacement for somatic development. Conventional treatment for testicular dysfunction consists of periodic intramuscular injections of chemically modified testosterone or application of a transdermal testosterone patch. However, long-term non-pulsatile testosterone therapy is not optimal and can cause multiple problems, including erythropoiesis and bone density changes.

A system was designed wherein Leydig cells were microencapsulated for controlled testosterone replacement. Purified Leydig cells were isolated and encapsulated in an alginate-poly-L-lysine solution. The encapsulated Leydig cells were injected into castrated animals, and serum testosterone was measured serially; the animals were able to maintain testosterone levels in the long term.[67] These studies suggest that microencapsulated Leydig cells may be able to replace or supplement testosterone in situations where anorchia or testicular failure is present.

7. Conclusion

Tissue engineering efforts are currently being undertaken for every type of tissue and organ within the urinary system. Most of the effort expended to engineer genitourinary tissues has occurred within the last decade. Tissue engineering techniques require a cell culture facility designed for human application. Personnel who have mastered the techniques of cell harvest, culture, and expansion as well as polymer design are essential for the successful application of this technology. Before these engineering techniques can be applied to humans, further studies need to be performed in many of the tissues described. Recent progress suggests that engineered urologic tissues and cell therapy may have clinical applicability.

References

1. McDougal, W.S. Metabolic complications of urinary intestinal diversion. *J. Urol.* **147**(5), 1199–1208 (1992).
2. Atala, A., Vacanti, J.P., Peters, C.A., Mandell, J., Retik, A.B. and Freeman, M.R. Formation of urothelial structures *in vivo* from dissociated cells attached to biodegradable polymer scaffolds *in vitro*. *J. Urol.* **148**(2 Pt. 2), 658–662 (1992).
3. Atala, A., Cima, L.G., Kim, W., Paige, K.T., Vacanti, J.P., Retik, A.B. *et al.* Injectable alginate seeded with chondrocytes as a potential treatment for vesicoureteral reflux. *J. Urol.* **150**(2 Pt. 2), 745–747 (1993).
4. Atala, A., Freeman, M.R., Vacanti, J.P., Shepard, J. and Retik, A.B. Implantation *in vivo* and retrieval of artificial structures consisting of rabbit and human urothelium and human bladder muscle. *J. Urol.* **150**(2 Pt. 2), 608–612 (1993).
5. Atala, A., Kim, W., Paige, K.T., Vacanti, C.A. and Retik, A.B. Endoscopic treatment of vesicoureteral reflux with a chondrocyte-alginate suspension. *J. Urol.* **152**(2 Pt. 2), 641–643 (1994); discussion 4.
6. Atala, A., Bauer, S.B., Soker, S., Yoo, J.J. and Retik, A.B. Tissue-engineered autologous bladders for patients needing cystoplasty [see comment]. *Lancet* **367**(9518), 1241–1246 (2006).
7. Atala, A. Tissue engineering, stem cells, and cloning for the regeneration of urologic organs. *Clin. Plast. Surg.* **30**(4), 649–667 (2003).
8. Lin, H.K., Cowan, R., Moore, P., Zhang, Y., Yang, Q., Peterson, J.A., Jr. *et al.* Characterization of neuropathic bladder smooth muscle cells in culture. *J. Urol.* **171**(3), 1348–1352 (2004).
9. Lai, J.Y., Yoon, C.Y., Yoo, J.J., Wulf, T. and Atala, A. Phenotypic and functional characterization of *in vivo* tissue engineered smooth muscle from normal and pathological bladders. *J. Urol.* **168**(4 Pt. 2), 1853–1857 (2002); discussion 8.
10. Brivanlou, A.H., Gage, F.H., Jaenisch, R., Jessell, T., Melton, D. and Rossant, J. Stem cells. Setting standards for human embryonic stem cells [see comment]. *Science* **300**(5621), 913–916 (2003).
11. Thomson, J.A., Itskovitz-Eldor, J., Shapiro, S.S., Waknitz, M.A., Swiergiel, J.J., Marshall, V.S. *et al.* Embryonic stem cell lines derived from human

blastocysts [see comment] [Erratum appears in *Science* **282**(5395), 1827 (1998)]. *Science* **282**(5391), 1145–1147 (1998).

12. De Coppi, P., Bartsch, G., Jr., Siddiqui, M.M., Xu, T., Santos, C.C., Perin, L. *et al.* Isolation of amniotic stem cell lines with potential for therapy [see comment]. *Nat. Biotechnol.* **25**(1), 100–106 (2007).
13. De Coppi, P., Callegari, A., Chiavegato, A., Gasparotto, L., Piccoli, M., Taiani, J. *et al.* Amniotic fluid and bone marrow derived mesenchymal stem cells can be converted to smooth muscle cells in the cryo-injured rat bladder and prevent compensatory hypertrophy of surviving smooth muscle cells. *J. Urol.* **177**(1), 369–376 (2007).
14. Hochedlinger, K., Rideout, W.M., Kyba, M., Daley, G.Q., Blelloch, R. and Jaenisch, R. Nuclear transplantation, embryonic stem cells and the potential for cell therapy. *Hematol. J.* **5**(Suppl. 3), S114–S117 (2004).
15. Kim, B.S., Baez, C.E. and Atala, A. Biomaterials for tissue engineering. *World J. Urol.* **18**(1), 2–9 (2000).
16. Kim, B.S. and Mooney, D.J. Development of biocompatible synthetic extracellular matrices for tissue engineering. *Trends Biotechnol.* **16**(5), 224–230 (1998).
17. Ponder, K.P., Gupta, S., Leland, F., Darlington, G., Finegold, M., DeMayo, J. *et al.* Mouse hepatocytes migrate to liver parenchyma and function indefinitely after intrasplenic transplantation. *Proc. Natl. Acad. Sci. USA* **88**(4), 1217–1221 (1991).
18. Brittberg, M., Lindahl, A., Nilsson, A., Ohlsson, C., Isaksson, O. and Peterson, L. Treatment of deep cartilage defects in the knee with autologous chondrocyte transplantation [see comment]. *N. Engl. J. Med.* **331**(14), 889–895 (1994).
19. Folkman, J. and Hochberg, M. Self-regulation of growth in three dimensions. *J. Exp. Med.* **138**(4), 745–753 (1973).
20. Olsen, L., Bowald, S., Busch, C., Carlsten, J. and Eriksson, I. Urethral reconstruction with a new synthetic absorbable device. An experimental study. *Scand. J. Urol. Nephrol.* **26**(4), 323–326 (1992).
21. Kropp, B.P., Ludlow, J.K., Spicer, D., Rippy, M.K., Badylak, S.F., Adams, M.C. *et al.* Rabbit urethral regeneration using small intestinal submucosa onlay grafts. *Urology* **52**(1), 138–142 (1998).
22. Sievert, K.D., Bakircioglu, M.E., Nunes, L., Tu, R., Dahiya, R. and Tanagho, E.A. Homologous acellular matrix graft for urethral reconstruction in the rabbit: histological and functional evaluation. *J. Urol.* **163**(6), 1958–1965 (2000).
23. Atala, A., Guzman, L. and Retik, A.B. A novel inert collagen matrix for hypospadias repair. *J. Urol.* **162**(3 Pt. 2), 1148–1151 (1999).

24. Chen, F., Yoo, J.J. and Atala, A. Acellular collagen matrix as a possible "off the shelf" biomaterial for urethral repair. *Urology* **54**(3), 407–410 (1999).
25. le Roux, P.J. Endoscopic urethroplasty with unseeded small intestinal submucosa collagen matrix grafts: a pilot study. *J. Urol.* **173**(1), 140–143 (2005).
26. Roger, E.D.F., James, J.Y. and Anthony, A. Urethral replacement using cell seeded tubularized collagen matrices. *J. Urol.* **168**(4), 1789–1793 (2002).
27. Kaefer, M., Hendren, W.H., Bauer, S.B., Goldenblatt, P., Peters, C.A., Atala, A. *et al.* Reservoir calculi: a comparison of reservoirs constructed from stomach and other enteric segments [see comment]. *J. Urol.* **160**(6 Pt. 1), 2187–2190 (1998).
28. Kaefer, M., Tobin, M.S., Hendren, W.H., Bauer, S.B., Peters, C.A., Atala, A. *et al.* Continent urinary diversion: the Children's Hospital experience. *J. Urol.* **157**(4), 1394–1399 (1997).
29. Lailas, N.G., Cilento, B. and Atala, A. Progressive ureteral dilation for subsequent ureterocystoplasty. *J. Urol.* **156**(3), 1151–1153 (1996).
30. Satar, N., Yoo, J.J. and Atala, A. Progressive dilation for bladder tissue expansion. *J. Urol.* **162**(3 Pt. 1), 829–831 (1999).
31. Blandy, J.P. Ileal pouch with transitional epithelium and anal sphincter as a continent urinary reservoir. *J. Urol.* **86,** 749–767 (1961).
32. Blandy, J.P. The feasibility of preparing an ideal substitute for the urinary bladder. *Ann. R. Coll. Surg. Engl.* **35**, 287–311 (1964).
33. Harada, N., Yano, H., Ohkawa, T., Misse, T., Kurita, T. and Nagahara, A. New surgical treatment of bladder tumours: mucosal denudation of the bladder. *Br. J. Urol.* **37**(5), 545–547 (1965).
34. Oesch, I. Neourothelium in bladder augmentation. An experimental study in rats. *Eur. Urol.* **14**(4), 328–329 (1988).
35. Salle, J.L., Fraga, J.C., Lucib, A., Lampertz, M., Jobim, G., Jobim, G. *et al.* Seromuscular enterocystoplasty in dogs. *J. Urol.* **144**(2 Pt. 2), 454–456 (1990); discussion 60.
36. Cheng, E., Rento, R., Grayhack, J.T., Oyasu, R. and McVary, K.T. Reversed seromuscular flaps in the urinary tract in dogs. *J. Urol.* **152**(6 Pt. 2), 2252–2257 (1994).
37. Yoo, J.J., Meng, J., Oberpenning, F. and Atala, A. Bladder augmentation using allogenic bladder submucosa seeded with cells. *Urology* **51**(2), 221–225 (1998).
38. Probst, M., Dahiya, R., Carrier, S. and Tanagho, E.A. Reproduction of functional smooth muscle tissue and partial bladder replacement. *Br. J. Urol.* **79**(4), 505–515 (1997).
39. Sutherland, R.S., Baskin, L.S., Hayward, S.W. and Cunha, G.R. Regeneration of bladder urothelium, smooth muscle, blood vessels and nerves into an acellular tissue matrix. *J. Urol.* **156**(2 Pt. 2), 571–577 (1996).

40. Piechota, H.J., Dahms, S.E., Nunes, L.S., Dahiya, R., Lue, T.F. and Tanagho, E.A. *In vitro* functional properties of the rat bladder regenerated by the bladder acellular matrix graft. *J. Urol.* **159**(5), 1717–1724 (1998).
41. Badylak, S.F., Lantz, G.C., Coffey, A. and Geddes, L.A. Small intestinal submucosa as a large diameter vascular graft in the dog. *J. Surg. Res.* **47**(1), 74–80 (1989).
42. Kropp, B.P., Cheng, E.Y., Lin, H.K. and Zhang, Y. Reliable and reproducible bladder regeneration using unseeded distal small intestinal submucosa. *J. Urol.* **172**(4 Pt. 2), 1710–1713 (2004).
43. Kropp, B.P., Rippy, M.K., Badylak, S.F., Adams, M.C., Keating, M.A., Rink, R.C. *et al.* Regenerative urinary bladder augmentation using small intestinal submucosa: urodynamic and histopathologic assessment in long-term canine bladder augmentations. *J. Urol.* **155**(6), 2098–2104 (1996).
44. Portis, A.J., Elbahnasy, A.M., Shalhav, A.L., Brewer, A., Humphrey, P., McDougall, E.M. *et al.* Laparoscopic augmentation cystoplasty with different biodegradable grafts in an animal model. *J. Urol.* **164**(4), 1405–1411 (2000).
45. Landman, J., Olweny, E., Sundaram, C.P., Andreoni, C., Collyer, W.C., Rehman, J. *et al.* Laparoscopic mid sagittal hemicystectomy and bladder reconstruction with small intestinal submucosa and reimplantation of ureter into small intestinal submucosa: one-year follow-up. *J. Urol.* **171**(6 Pt. 1), 2450–2455 (2004).
46. Oberpenning, F., Meng, J., Yoo, J.J. and Atala, A. *De novo* reconstitution of a functional mammalian urinary bladder by tissue engineering [see comment]. *Nature Biotechnol.* **17**(2), 149–155 (1999).
47. Atala, A., Bauer, S.B., Soker, S., Yoo, J.J. and Retik, A.B. Tissue-engineered autologous bladders for patients needing cystoplasty. *Lancet* **367**(9518), 1241–1246 (2006).
48. Dahms, S.E., Piechota, H.J., Nunes, L., Dahiya, R., Lue, T.F. and Tanagho, E.A. Free ureteral replacement in rats: regeneration of ureteral wall components in the acellular matrix graft. *Urology* **50**(5), 818–825 (1997).
49. Baltaci, S., Ozer, G., Ozer, E., Soygur, T., Bealti, O. and Anafarta, K. Failure of ureteral replacement with Gore-Tex tube grafts [see comment]. *Urology* **51**(3), 400–403 (1998).
50. Osman, Y., Shokeir, A., Gabr, M., El-Tabey, N., Mohsen, T. and El-Baz, M. Canine ureteral replacement with long acellular matrix tube: is it clinically applicable? *J. Urol.* **172**(3), 1151–1154 (2004).
51. Humes, H.D., Buffington, D.A., MacKay, S.M., Funke, A.J. and Weitzel, W.F. Replacement of renal function in uremic animals with a tissue-engineered kidney [see comment]. *Nat. Biotechnol.* **17**(5), 451–455 (1999).

52. Humes, H.D., Weitzel, W.F., Bartlett, R.H., Swaniker, F.C. and Paganini, E.P. Renal cell therapy is associated with dynamic and individualized responses in patients with acute renal failure. *Blood Purif.* **21**(1), 64–71 (2003).
53. Lanza, R.P., Chung, H.Y., Yoo, J.J., Wettstein, P.J., Blackwell, C., Borson, N. *et al.* Generation of histocompatible tissues using nuclear transplantation. [see comment]. *Nat. Biotechnol.* **20**(7), 689–696 (2002).
54. Yoo, J.J., Park, H.J. and Atala, A. Tissue-engineering applications for phallic reconstruction. *World J. Urol.* **18**(1), 62–66 (2000).
55. Yoo, J.J. and Atala, A. Tissue engineering of genitourinary organs. *Ernst Schering Res. Found. Workshop* **35**, 105–127 (2002).
56. Nukui, F., Okamoto, S., Nagata, M., Kurokawa, J. and Fukui, J. Complications and reimplantation of penile implants. *Int. J. Urol.* **4**(1), 52–54 (1997).
57. Thomalla, J.V., Thompson, S.T., Rowland, R.G. and Mulcahy, J.J. Infectious complications of penile prosthetic implants. *J. Urol.* **138**(1), 65–67 (1987).
58. Yoo, J.J., Park, H.J., Lee, I. and Atala, A. Autologous engineered cartilage rods for penile reconstruction. *J. Urol.* **162**(3 Pt. 2), 1119–1121 (1999).
59. De Filippo, R.E., Yoo, J.J. and Atala A. Engineering of vaginal tissue *in vivo*. *Tissue Eng.* **9**(2), 301–306 (2003).
60. Diamond, D.A. and Caldamone, A.A. Endoscopic correction of vesicoureteral reflux in children using autologous chondrocytes: preliminary results. *J. Urol.* **162**(3 Pt. 2), 1185–1188 (1999).
61. Bent, A.E., Tutrone, R.T., McLennan, M.T., Lloyd, L.K., Kennelly, M.J. and Badlani, G. Treatment of intrinsic sphincter deficiency using autologous ear chondrocytes as a bulking agent. *Neurourol. Urodyn.* **20**(2), 157–165 (2001).
62. Yokoyama, T., Huard, J. and Chancellor, M.B. Myoblast therapy for stress urinary incontinence and bladder dysfunction. *World J. Urol.* **18**(1), 56–61 (2000).
63. Chancellor, M.B., Yokoyama, T., Tirney, S., Mattes, C.E., Ozawa, H., Yoshimura, N. *et al.* Preliminary results of myoblast injection into the urethra and bladder wall: a possible method for the treatment of stress urinary incontinence and impaired detrusor contractility. *Neurourol. Urodyn.* **19**(3), 279–287 (2000).
64. Yiou, R., Yoo, J.J. and Atala, A. Restoration of functional motor units in a rat model of sphincter injury by muscle precursor cell autografts. *Transplantation* **76**(7), 1053–1060 (2003).
65. Strasser, H., Berjukow, S., Marksteiner, R., Margreiter, E., Hering, S., Bartsch, G. *et al.* Stem cell therapy for urinary stress incontinence. *Exp. Gerontol.* **39**(9), 1259–1265 (2004).

66. Huard, J., Yokoyama, T., Pruchnic, R., Qu, Z., Li, Y., Lee, J.Y. *et al.* Muscle-derived cell-mediated *ex vivo* gene therapy for urological dysfunction. *Gene Ther.* **9**(23), 1617–1626 (2002).
67. Machluf, M., Orsola, A., Boorjian, S., Kershen, R. and Atala, A. Microencapsulation of Leydig cells: a system for testosterone supplementation. *Endocrinology* **144**(11), 4975–4979 (2003).

PART VIII

CARDIAC REPAIR

Chapter 32

Basic Science

Sian E. Harding

Abstract

The lack of appropriate regenerative capacity of the heart has been the spur to early clinical trials using autologous stem cells (bone marrow-derived and skeletal myoblast) for cardiac repair. Associated *in vivo* and *in vitro* laboratory experiments have been essential in advancing understanding of mechanism of benefit or harm from these cells, and in suggesting other sources of stem cells for cardiac application. Skeletal myoblasts have dangers, in that lack of integration with myocardium produces an arrhythmic substrate. Bone marrow-derived stem cells, while safer, do not generate significant new cardiac muscle as part of their beneficial actions, suggesting an angiogenic mechanism or paracrine protection of existing cardiomyocytes. Other adult stem cells, primarily from heart but also from organs such as testis, are currently being characterised as source of new cardiac tissue. Embryonic stem cells reliably produce contracting cardiac muscle, and are a likely candidate for future repair when problems of teratoma formation and immune response have been solved. They are also more suited to tissue engineering application, because of their ready availability and potential for expansion. Study of the biology of stem cells is also generating new paradigms for understanding the intrinsic regenerative capacity of the heart.

Keywords: Heart; Stem Cells; Regeneration; Repair; Cardiomyocyte.

Outline

1. Introduction
    1.1. Adult stem cells — bone marrow-derived stem cells
    1.2. Adult stem cells — cardiac
    1.3. Adult stem cells — skeletal myoblasts
    1.4. Adult stem cells — others
    1.5. Embryonic stem cells
2. Conclusion
    References

1. Introduction

Heart disease has always been a prime driver in the search for regenerative strategies because of the clear perception that loss of myocardial mass is not reversed in any meaningful way. The deduction from this that the heart had no intrinsic regenerative capacity is now being questioned and, as we will see in this chapter, the paradigm may be reformulated to state that patients who present with heart disease have regenerative capacities which are not adequate to bring about sufficient cardiac repair for their immediate and future needs. There are a number of reasons why the investigation of cardiac regeneration is especially active at the moment, both social and scientific. First, the number of patients surviving with damaged hearts is increasing, due to successes in acutely reperfusing occluded blood vessels after myocardial infarction (MI). Second, the realisation that secondary damage occurs when power output is low due to inappropriately prolonged neurohormonal activation.[1] This implies that other cardiac disease states apart from direct ischaemic damage, for example genetic cardiomyopathies or valve disease, might also benefit from increased myocardial mass. It is interesting to note that blockade of these neurohormonal systems has been a mainstay in the prevention or retardation of heart failure development, but it is being appreciated that this strategy has reached a plateau of effectiveness. Adding new blockers to the beta-adrenoceptor antagonists, renin-angiotensin inhibitors, aldosterone antagonists and statins is unlikely to achieve a quantum leap in clinical effectiveness. However, on the positive side, the use of left ventricular assist devices has shown that temporary support of the failing heart can allow some recovery of intrinsic function even in end-stage heart failure patients[2] (see also Chapter 34 of this book). This suggests that even incomplete restoration of myocardial mass could provide sufficient power to allow partial recovery of the native tissue.

The third circumstance which promotes hope about cardiac repair at this time is of course the burgeoning science of stem cell biology. Clinical trials in MI patients using autologous stem cells, such as those derived from the bone marrow or skeletal muscle of the subject, were launched relatively soon after the demonstration of feasibility in animal models. This probably says more about the regulatory environment than the strength of the underlying scientific rationale, but important lessons have been learnt. These are described in more detail in a following chapter, but the situation at present is that the results have been sufficiently encouraging for trials with bone marrow-derived cells (BMCs) to continue[3] while those with skeletal myoblasts have lost impetus.[4] A new class of autologous cells, intrinsic cardiac progenitor cells, is being explored but the necessity for expansion in culture sets them apart from BMCs in ease of translation to the clinic. The same is true for other cells derived from adult tissue, fat or testes for example. Embryonic stem cell (ESC)-derived cardiomyocytes from human lines (hESC-CM) are another clear contender, and for these the number of hurdles to overcome is probably greatest. This chapter will discuss each of the candidates in turn. The basic science of stem cells has been addressed in the opening section of the book. Here the cells will be considered for a number of characteristics in relation to heart and cardiac repair: whether they differentiate into cardiomyocytes *in vitro*; how this is defined i.e. by cardiac markers or by contractile and electrical activity; whether they remain in or home to the cardiac niche *in vivo*, and under what circumstances; if there is evidence of functional benefit in animal models; additionally, whether there is evidence of potentially harmful consequences in animal models; if functional benefit is observed, is this associated with new muscle or must some other effect be postulated. Finally, in the context of the present volume, we will consider which cells have potential for tissue engineering solutions.

1.1. Adult stem cells — bone marrow-derived stem cells

Bone marrow is a complex mixture of stem cell types: clearly there are progenitors for haemopoietic stem (HSCs) but there also seem to be stem cell populations capable of contributing to regeneration of other organs. This has been shown in animal models, by transplantation of bone marrow from GFP or ROSA (β-galactosidase expressing) transgenic mice and subsequent examination of various tissues including the heart.[5] It has also been observed from sex-mismatched heart transplant patients, where the Y-chromosome has been detected by FISH staining.[6] In these, the number of Y-positive cardiomyocytes found in the implanted female heart in male recipients was low, but did increase with time[6] from approximately 0.1% to 0.5% over a year. It has been pointed out that a male pregnancy could also account for Y-positive cells in a female organ[7] but this increase with time would argue against that explanation. Taking the questions posed above in turn, is there evidence of transdifferentation of BMCs into cardiomyocytes *in vitro*? Mesenchymal stem cells (MSC) have been associated with *in vitro* transdifferentation, and it was reported that stimulation with 5-azacytidine could form beating cells with cardiac markers and action potential characteristics.[8] However, this has been difficult to reproduce.[9] *In vivo*, there is evidence for retention of injected BMCs in the heart, either after direct intramyocardial injection or perfusion, although the percentage retained is low.[10,11] Figure 1 illustrates the use of iron-GFP labelled bone marrow stromal cells to track retention after injection. There is also evidence that BMCs home particularly to the injured region of the heart and this is true both for native and transplanted BMCs.[12,13] (Interestingly, new results suggest that platelets

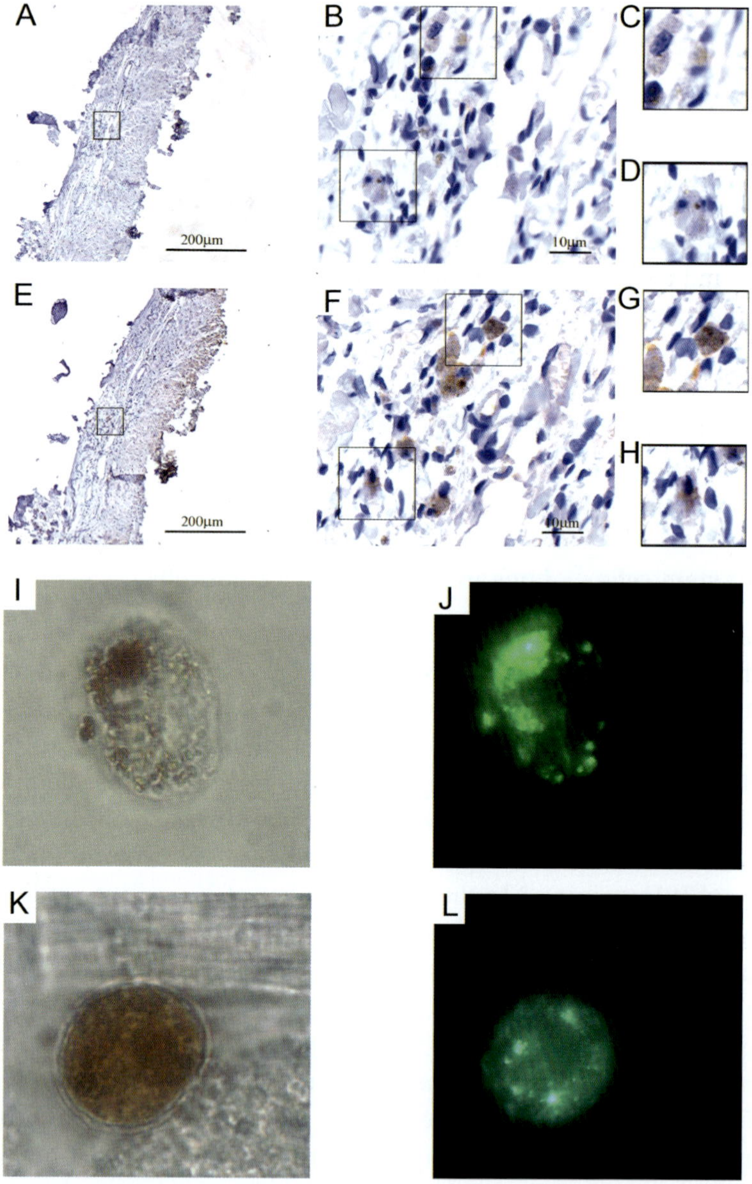

Fig. 1. Retention of injected bone marrow-derived stromal cells within rat heart. **(A–H)** Light microscopy of adjacent 5 μm sections taken from a heart that received Fe-BMSCs 16 weeks previously. **(A–D)** Sections stained with IgG isotype control and **(E–H)** sections stained for GFP. Higher magnification images **(C–D)**, **(G–H)** show that iron particles and GFP co-locate within cells. Light **(I** and **K)** and fluorescence **(J** and **L)** microscopy of iron particle and GFP double-labelled BMSCs pre-injection **(I** and **J)**, and after isolation **(K** and **L)** from hearts that had labelled BMSCs injected 20 weeks earlier.[10]

attracted to the site of injury may be a first step in recruiting the BMCs by secreted factors.[14]) From this resulted the idea that mobilisation and homing factors themselves might be sufficient to stimulate natural repair processes, or increase the success of cell implantation.[13] First clinical results with BMC mobilising factors such as G-CSF have not reliably replicated the effects of BMCs themselves, but more complex mixtures of factors or adjustments to timing may be required.[15] A large number of animal studies show greater or lesser functional improvements in contraction of injured heart (usually MI) after BMC treatment (e.g. Refs. 16–19). One caveat is that anecdotal evidence indicates that journals may have a (not unusual) bias for positive studies, and that negative or borderline series may not be attractive to journal editors. This may lead to overestimation of the reliability of the phenomenon. A source of great contention has been the mechanism of benefit. An initial study showed regeneration of tissue including myocytes within the infarct scar within nine days after BMC implantation.[18] It has since been difficult to show that BMCs transdifferentiate into cardiomyocytes, despite increasingly sophisticated experimental designs.[20–22] The consensus now is that this transdifferentation is a rare event, and is certainly not quantitatively accounting for the increase in contractility of the damaged ventricle. Even a cursory examination of the number of cells injected (usually around 10^8 for human[3]) shows that proliferation and highly efficient transdifferentation would be needed to replace the 10^9 cardiomyocytes lost in the infarct, given the loss after injection. This leaves the mechanism of benefit as yet undefined, with the most likely explanation being increased blood vessel formation rescuing viable but underperforming myocardium.[23] This would set an upper limit for benefit in man, since revascularisation is commonly performed anyway, but would indicate that patients in which revascularisation was not an option might be prime candidates for BMC therapy. Although clinical trials have shown short- and medium-term safety, it should be borne in mind that systemic pro-angiogenic strategies might support tumour growth. Another concern is the suggestion that cardiac disease in man may be associated with, or even be a consequence, of poor angiogenic potential of native cells. *In vitro* proliferative activity of endothelial progenitor cell populations, for example, was found to be impaired in cardiac patients,[24] and such impairment was a statistically independent risk factor for progression of disease.[25] Transplant from healthy volunteers is of course possible, but this obviates the advantage of the autologous source. Other explanations for BMC-related benefit, such as paracrine mechanisms to protect existing cardiomyocytes against apoptosis or rescue their contractile function are also being investigated.[26] Nevertheless, the lack of new muscle generation must remain a limitation to the use of BMCs.

1.2. Adult stem cells — cardiac

One of the most interesting results to have emerged in the cardiac field as a direct result of stem cell research is an acceptance of the view that there may be dividing cardiomyocytes in the adult heart. This had been suggested, but was strongly opposed since the conventional view was that no myogenesis occurred within the adult myocardium, with all growth and adaptation being ascribed to hypertrophy of existing myocytes.[27] In an attempt to provide an explanation for revival of myocardium by bone marrow cell transplantation in the absence of transdifferentiation into myocytes, other paradigms were considered. Paradoxically, not one but an apparent variety of cardiac stem cells have now been observed, with a range of different markers and characteristics. Sca-l-positive cells express CD31 but not c-kit, CD45 or CD34, and can express cardiac markers *in vitro* after treatment with 5-azacytidine.[28] They have been successfully used *in vivo* to regenerate mouse myocardium, and have some ability to home to injured tissue. Side population cells are similar in that they can extrude the Hoecsht 33342 dye, express Sca-1 and not c-kit, but differ in that they are not positive for CD31.[29] They produce the full cardiac phenotype when co-cultured with rat neonatal cardiomyocytes, and are effective after *in vivo* injection. C-kit-positive cells express cardiac transcription factors without 5-azacytidine treatment, but not later markers: they have also been observed to regenerate vascularised myocardium, and again can home to sites where repair is needed.[30] Importantly, similar cells have been found in human myocardium.[31] Islet-1 positive cells have no markers in common with the others, but can form cardiac myocytes when co-cultured with neonatal myocytes.[32] Possibly the most promising are the cardiospheres, since they have been obtained not only from mouse but from human subjects in the age and disease range of the patients for whom tissue repair is primarily intended. They are a mixed population having, at various times and on various phenotypes, many of the markers described for the other cells mentioned, including c-kit, CD31 and CD45.[33] They are capable of large-scale expansion *in vitro*, and demonstration of the cardiac phenotype including beating activity. *In vivo*, they have been shown to regenerate myocardium although the functional benefit was modest.[34] The autologous nature of these cells and their presence in the population of interest gives them a clear advantage for translation.

1.3. Adult stem cells — skeletal myoblasts

Stem cells from skeletal muscle were used early in cardiac clinical trials largely because of their autologous source, with biopsies from patients expanded in the laboratory before implantation.[35] The success of cardiomyoplasty, where skeletal muscle is wrapped around the heart and electrically conditioned, had also encouraged use of these cells.[36] Animal experiments had also suggested functional benefit was possible from this approach.[37] However, the initial trials had increased incidences of arrhythmia, which was controlled by implantable cardiodefibrillators (ICD).[38] It was found that the myoblasts have developed into myotubes with contractile activity within the heart, but were not coupling to cardiac muscle. A large scale trial (MAGIC) obviated the arrhythmic risk by using ICDs, but recruitment was slow and the trial was stopped early because of this.[4] It seems likely that the relative success and safety of the trials use bone marrow-derived stem cells has presently led to a loss of interest in the skeletal myoblast option.

1.4. Adult stem cells — others

There have been sporadic reports of a number of tissues, but one of the most advanced for cardiac applications are spermatogonial stem cells.[39] These have been isolated from mouse testis, and are the only self-renewing germ cells in that organ, retaining spermatogenesis throughout the animals' life span. These recapitulated many of the properties of ESCs, including embryoid body formation, differentiation into all three germ layers and the production of beating cardiomyocytes.

1.5. Embryonic stem cells

The embryonic stem cells (ESCs) represent the opposite end of the spectrum, in that their cardiomyogenic potential is assured but their translation into the clinic will be retarded by a number of real concerns. Although ESCs from mouse have been studied for many years, a population with similar characteristics was only identified in human embryos in 1998. Mouse and human embryonic stem cell (hESC) lines are similar in that they derive from the inner cell mass of the blastocyst, and have the properties of self-renewal in the undifferentiated state, coupled with subsequent ability to differentiate into all cell types of the body. They differ in the growth factors required to maintain pluripotency, and in the underlying signalling pathways that promote or retard differentiation: it has been suggested that this betokens a more fundamental difference in their nature.[40] Because of this, and because of the greater relevance for clinical use, this section will concentrate on hESC. Appearance of cardiac cells is rapid after induction of differentiation by withdrawal of basic fibroblast growth factor (bFGF) and fibroblast conditioned medium.[41–43] Early cardiac markers such as GATA4 and Nkx2.5 are identified within a few days of embryoid body formation, with cardiac myosin heavy chain (MHC) and cardiac actin observed soon after.[44] hESC-derived cardiac myocytes (hESC-CM) are readily identified by rhythmic beating activity, which occurs around day 9–15 after the beginning of differentiation.[44,45] hESC-CM are initially similar to immature cardiomyocytes, with a disorganised sarcomeric structure and spontaneous pacemaker activity. With time the sarcomeres become aligned as in adult myocytes (Fig. 2), and a variety of action potential configurations are observed, spanning the nodal, atrial and ventricular phenotypes.[46] Developments in both ion currents and the contribution of sarcoplasmic reticulum calcium stores bring hESC-CM close to the adult phenotype.[47,48] Experience with implantation of hESC-CM is not extensive, but it has been shown to be feasible in rat, guinea pig and pig heart.[49–51] For guinea pig and pig, the evidence for hESC-CM survival was the pacemaker activity, which was sufficient to trigger contraction when natural pacemakers were ablated (even with relatively small numbers of implanted embryoid bodies).[50,51] For rat, hESC-CM have been implanted into normal, briefly ischaemic or fully infarcted heart, and a number of points are becoming clear.[49,52] First, the ESC need to be differentiated before introduction: the cardiac niche itself did not drive hESC-CM differentiation and the risk for teratomas from undifferentiated cells was high.[53] Measures to increase the proportion of cardiomyocytes within the differentiated cells also improve the success of the graft.[52] Second, injury itself does not appear to promote homing or hESC-CM growth, but rather introduces damaging inflammatory mediators which decrease the size of the graft. A cocktail of protective agents such as IGF-1, pan-caspase inhibitors and NO blockers was injected with hESC-CM in one study: it is unclear how long the protective effect of these agents lasted over the four weeks before the heart was excised.[52] The additional pretreatment of the hESC-CM by heat shock had the possibility to confer a more lasting protection. Thirdly, the initial reports which suggested that implanted ESC have a certain immune privilege may be over-optimistic: implantation of hESC-CM in immunocompetent mice does not produce measurable grafting.[53] It is likely that the degree of differentiation required to reduce the risk of teratomas also increases the visibility of the hESC-CM to immune surveillance. Encouragingly, implantation of hESC-CM did preserve function of the infarcted heart, compared with deterioration over the same period when a non-cardiac hESC fraction was used.[52]

While hESC are the most efficient at producing cardiac muscle, they are also the most problematic to translate to the clinic. The strategy of reducing teratomas by using late stages of differentiation is promising, but it may be that genetic deselection of undifferentiated cells will be necessary to ensure complete safety in a clinical setting. Alterations in cell phenotype, such as appearance of immunogenic sialic acid residues,[54] due to animal components in the culture of ESC have largely been solved, with an array of human feeder lines and defined media available. Immune attack of the implanted hESC-CM is probably the most problematic area,

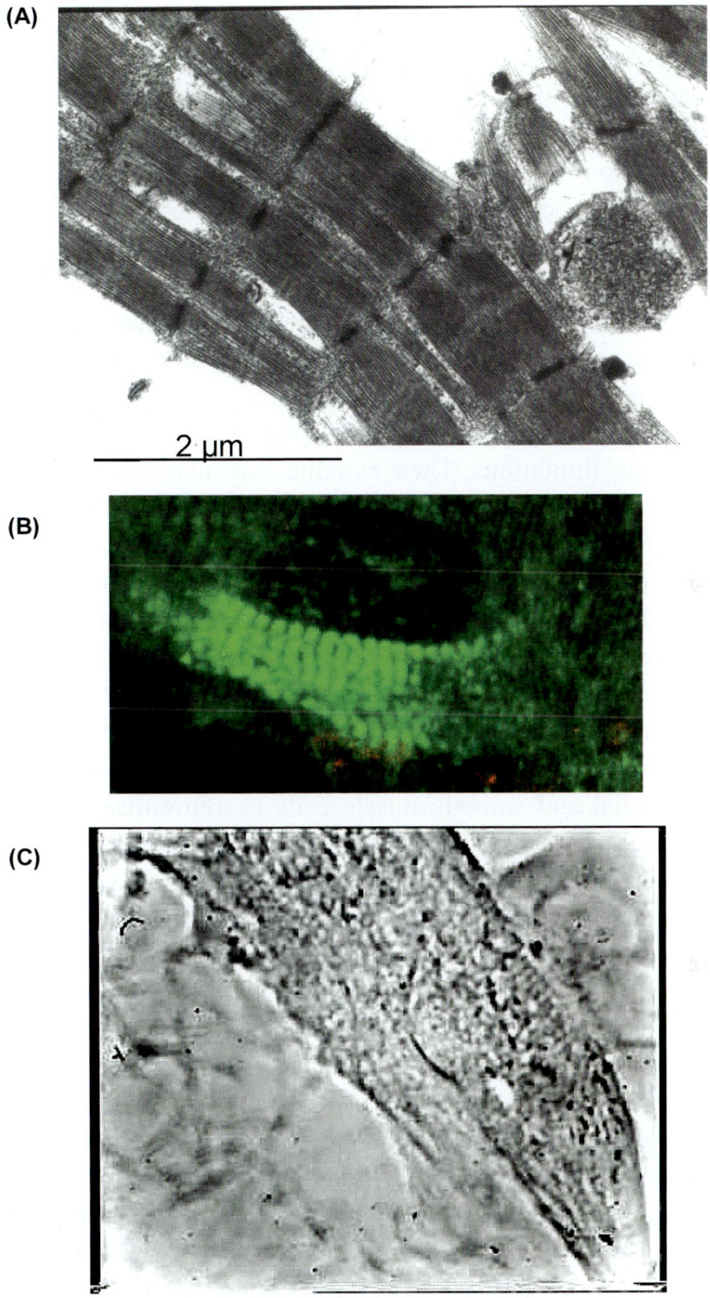

Fig. 2. **(A)** Transmission electron micrograph of human embryonic stem cell-derived cardiomyocyte, H7 line at day 76 of differentiation. **(B)** Areas of sarcomere alignment shown with MHC staining. **(C)** Areas of sarcomere alignment observed under light microscopy, on day 78 of differentiation.[45]

although it should be remembered that the degree of incompatibility will not be greater than for existing heart transplants. If, as suggested, approximately 150 ESC lines could match the major histocompatibility complex haplotypes in the UK, then matching might be better than for present donor-recipient pairs.[55] More sophisticated solutions such as producing the ESC lines by transfer of somatic nuclei into enucleated oocytes (therapeutic cloning) are technically challenging, and ethically problematic in some countries (see also Chapter 10 of this book). Ethical objections extend to the generation of ESC *per se* and it is not yet clear whether scientific solutions will be acceptable, for example, using single cells from a blastocyst (as in pre-implantation genetic diagnosis) to preserve the embryo while generating an ESC line.[56]

For tissue engineering, however, ESC have distinct advantages which may well outweigh their limitations. Their extreme expandability, especially coupled with advances in bioreactor design, means that they can be generated and prepared well in advance of need (unlike autologous cells). Similarly, experimental work is possible to optimise the combination of ESC with materials on the actual cell line that will finally be used in the product. They can be genetically manipulated to produce pure cardiomyocytes, and hESC-CM can potentially be driven into a more compatible mature phenotype before amalgamation with the chosen matrix. It is unclear, though, whether extreme genetic selection of the cardiomyocytes is necessary or even desirable. When this was attempted, it was found necessary to add back endothelial and smooth muscle cells to allow maximum vascularisation.[57] A strategy of partial purification may be more productive. Initial experiments to combine hESC with materials or scaffolds, or to create engineered tissue without added materials, are showing promise for creation of constructs.[58]

2. Conclusion

For cardiac repair in general, the immediate advantage of using autologous cells is undeniable. This chapter has outlined the scientific basis for use of various cell types, and has raised some concerns about the utility of straightforward autologous cell implantation into the heart. Additionally, this strategy is antithetical to commercial involvement in the development of widely applicable therapies. Lack of industry input could be argued to represent a beneficial shift towards academic-led medicine, but there is no doubt that the reduced investment levels will slow the introduction of tested therapies. This is particularly true for engineered tissue, since regulatory approval of a mixed material/cellular product is complex and industry is more experienced with the scale of pre-clinical testing needed. It should be remembered that it is not only the mechanism of benefit which needs to be considered carefully for a given cell type, but the route to clinical application.

One unexpected benefit from the mechanistic studies, though, has been the greater understanding of the nature of pluripotency, culminating in advances such as the development of ES-like cell lines from human fibroblasts by introduction of a limited number of "stemness" genes. This has the potential to bridge the gap between autologous and embryonic stem cells in designing patient-specific therapies.

References

1. Francis, G.S., McDonald, K.M. and Cohn, J.N. Neurohumoral activation in preclinical heart failure. Remodeling and the potential for intervention. *Circulation* **87**, IV90–IV96 (1993).
2. Birks, E.J., Tansley, P.D., Hardy, J., George, R.S., Bowles, C.T., Burke, M., Banner, N.R., Khaghani, A. and Yacoub, M.H. Left ventricular assist device and drug therapy for the reversal of heart failure. *N. Engl. J. Med.* **355**, 1873–1884 (2006).
3. Rosenzweig, A. Cardiac cell therapy — mixed results from mixed cells. *N. Engl. J. Med.* **355**, 1274–1277 (2006).
4. Cleland, J.G., Coletta, A.P., Abdellah, A.T., Nasir, M., Hobson, N., Freemantle, N. and Clark, A.L. Clinical trials update from the American Heart Association 2006: OAT, SALT 1 and 2, MAGIC, ABCD, PABA-CHF, IMPROVE-CHF, and percutaneous mitral annuloplasty. *Eur. J. Heart Fail.* **9**, 92–97 (2007).
5. Dedja, A., Zaglia, T., Dall'Olmo, L., Chioato, T., Thiene, G., Fabris, L., Ancona, E., Schiaffino, S., Ausoni, S. and Cozzi, E. Hybrid cardiomyocytes derived by cell fusion in heterotopic cardiac xenografts. *FASEB J.* **20**, 2534–2536 (2006).
6. Pfeiffer, P., Muller, P., Kazakov, A., Kindermann, I. and Bohm, M. Time-dependent cardiac chimerism in gender-mismatched heart transplantation patients. *J. Am. Coll. Cardiol.* **48**, 843–845 (2006).
7. Grounds, M.D., White, J.D., Rosenthal, N. and Bogoyevitch, M.A. The role of stem cells in skeletal and cardiac muscle repair. *J. Histochem. Cytochem.* **50**, 589–610 (2002).
8. Fukuda, K. Reprogramming of bone marrow mesenchymal stem cells into cardiomyocytes. *C. R. Biol.* **325**, 1027–1038 (2002).
9. Balana, B., Nicoletti, C., Zahanich, I., Graf, E.M., Christ, T., Boxberger, S. and Ravens, U. 5-Azacytidine induces changes in electrophysiological properties of human mesenchymal stem cells. *Cell Res.* **16**, 949–960 (2006).
10. Stuckey, D.J., Carr, C.A., Martin-Rendon, E., Tyler, D.J., Willmott, C., Cassidy, P.J., Hale, S.J., Schneider, J.E., Tatton, L., Harding, S.E., Radda, G.K., Watt, S. and Clarke, K. Iron particles for non-invasive monitoring of

bone marrow stromal cell engraftment into, and isolation of viable engrafted donor cells from, the heart. *Stem Cells* **24**(8), 1968–1975 (2006).

11. Hofmann, M., Wollert, K.C., Meyer, G.P., Menke, A., Arseniev, L., Hertenstein, B., Ganser, A., Knapp, W.H. and Drexler, H. Monitoring of bone marrow cell homing into the infarcted human myocardium. *Circulation* **111**, 2198–2202 (2005).
12. Jackson, K.A., Majka, S.M., Wang, H., Pocius, J., Hartley, C.J., Majesky, M.W., Entman, M.L., Michael, L.H., Hirschi, K.K. and Goodell, M.A. Regeneration of ischemic cardiac muscle and vascular endothelium by adult stem cells. *J. Clin. Invest.* **107**, 1395–1402 (2001).
13. Hisashi, Y., Tomita, S., Nakatani, T., Fukuhara, S., Yutani, C. and Kitamura, S. Granulocyte-colony stimulating factor enhanced the recruitment of bone marrow cells into the heart: time course evaluation of phenotypic differentiation in the doxorubicin-induced cardiomyopathic model. *Jap. J. Thorac. Cardiovasc. Surg.* **52**, 451–455 (2004).
14. Massberg, S., Konrad, I., Schurzinger, K., Lorenz, M., Schneider, S., Zohlnhoefer, D., Hoppe, K., Schiemann, M., Kennerknecht, E., Sauer, S., Schulz, C., Kerstan, S., Rudelius, M., Seidl, S., Sorge, F., Langer, H., Peluso, M., Goyal, P., Vestweber, D., Emambokus, N.R., Busch, D.H., Frampton, J. and Gawaz, M. Platelets secrete stromal cell-derived factor 1-alpha and recruit bone marrow-derived progenitor cells to arterial thrombi *in vivo*. *J. Exp. Med.* **203**, 1221–1233 (2006).
15. Ince, H. and Nienaber, C.A. Granulocyte-colony-stimulating factor in acute myocardial infarction: future perspectives after FIRSTLINE-AMI and REVIVAL-2. *Nat. Clin. Pract. Cardiovasc. Med.* **4**(Suppl. 1), S114–S118 (2007).
16. Yau, T.M., Tomita, S., Weisel, R.D., Jia, Z.Q., Tumiati, L.C., Mickle, D.A. and Li, R.K. Beneficial effect of autologous cell transplantation on infarcted heart function: comparison between bone marrow stromal cells and heart cells. *Ann. Thorac. Surg.* **75**, 169–176 (2003).
17. Saito, T., Kuang, J.Q., Lin, C.C. and Chiu, R.C. Transcoronary implantation of bone marrow stromal cells ameliorates cardiac function after myocardial infarction. *J. Thorac. Cardiovasc. Surg.* **126**, 114–123 (2003).
18. Orlic, D., Kajstura, J., Chimenti, S., Jakoniuk, I., Anderson, S.M., Li, B., Pickel, J., McKay, R., Nadal-Ginard, B., Bodine, D.M., Leri, A. and Anversa, P. Bone marrow cells regenerate infarcted myocardium. *Nature* **410**, 701–705 (2001).
19. Tomita, S., Li, R.K., Weisel, R.D., Mickle, D.A., Kim, E.J., Sakai, T. and Jia, Z.Q. Autologous transplantation of bone marrow cells improves damaged heart function. *Circulation* **100**, II247–II256 (1999).

20. Balsam, L.B., Wagers, A.J., Christensen, J.L., Kofidis, T., Weissman, I.L. and Robbins, R.C. Haematopoietic stem cells adopt mature haematopoietic fates in ischaemic myocardium. *Nature* **428**, 668–673 (2004).
21. Murry, C.E., Soonpaa, M.H., Reinecke, H., Nakajima, H., Nakajima, H.O., Rubart, M., Pasumarthi, K.B., Virag, J.I., Bartelmez, S.H., Poppa, V., Bradford, G., Dowell, J.D., Williams, D.A. and Field, L.J. Haematopoietic stem cells do not transdifferentiate into cardiac myocytes in myocardial infarcts. *Nature* **428**, 664–668 (2004).
22. Nygren, J.M., Jovinge, S., Breitbach, M., Sawen, P., Roll, W., Hescheler, J., Taneera, J., Fleischmann, B.K. and Jacobsen, S.E. Bone marrow-derived hematopoietic cells generate cardiomyocytes at a low frequency through cell fusion, but not transdifferentiation. *Nat. Med.* **10**, 494–501 (2004).
23. Tse, H.F., Kwong, Y.L., Chan, J.K., Lo, G., Ho, C.L. and Lau, C.P. Angiogenesis in ischaemic myocardium by intramyocardial autologous bone marrow mononuclear cell implantation. *Lancet* **361**, 47–49 (2003).
24. Vasa, M., Fichtlscherer, S., Aicher, A., Adler, K., Urbich, C., Martin, H., Zeiher, A.M. and Dimmeler, S. Number and migratory activity of circulating endothelial progenitor cells inversely correlate with risk factors for coronary artery disease. *Circ. Res.* **89**, E1–E7 (2001).
25. Schmidt-Lucke, C., Rossig, L., Fichtlscherer, S., Vasa, M., Britten, M., Kamper, U., Dimmeler, S. and Zeiher, A.M. Reduced number of circulating endothelial progenitor cells predicts future cardiovascular events: proof of concept for the clinical importance of endogenous vascular repair. *Circulation* **111**, 2981–2987 (2005).
26. Xu, M., Uemura, R., Dai, Y., Wang, Y., Pasha, Z. and Ashraf, M. *In vitro* and *in vivo* effects of bone marrow stem cells on cardiac structure and function. *J. Mol. Cell Cardiol.* **42**, 441–448 (2007).
27. Anversa, P. and Nadal-Ginard, B. Myocyte renewal and ventricular remodelling. *Nature* **415**, 240–243 (2002).
28. Oh, H., Chi, X., Bradfute, S.B., Mishina, Y., Pocius, J., Michael, L.H., Behringer, R.R., Schwartz, R.J., Entman, M.L. and Schneider, M.D. Cardiac muscle plasticity in adult and embryo by heart-derived progenitor cells. *Ann. N. Y. Acad. Sci.* **1015**, 182–189 (2004).
29. Mouquet, F., Pfister, O., Jain, M., Oikonomopoulos, A., Ngoy, S., Summer, R., Fine, A. and Liao, R. Restoration of cardiac progenitor cells after myocardial infarction by self-proliferation and selective homing of bone marrow-derived stem cells. *Circ. Res.* **97**, 1090–1092 (2005).
30. Beltrami, A.P., Barlucchi, L., Torella, D., Baker, M., Limana, F., Chimenti, S., Kasahara, H., Rota, M., Musso, E., Urbanek, K., Leri, A., Kajstura, J.,

Nadal-Ginard, B. and Anversa, P. Adult cardiac stem cells are multipotent and support myocardial regeneration. *Cell* **114**, 763–776 (2003).

31. Urbanek, K., Quaini, F., Tasca, G., Torella, D., Castaldo, C., Nadal-Ginard, B., Leri, A., Kajstura, J., Quaini, E. and Anversa, P. Intense myocyte formation from cardiac stem cells in human cardiac hypertrophy. *Proc. Natl. Acad. Sci. USA* **100**, 10440–10445 (2003).
32. Moretti, A., Lam, J., Evans, S.M. and Laugwitz, K.L. Biology of Isl1+ cardiac progenitor cells in development and disease. *Cell Mol. Life Sci.* **64**, 674–682 (2007).
33. Messina, E., De, A.L., Frati, G., Morrone, S., Chimenti, S., Fiordaliso, F., Salio, M., Battaglia, M., Latronico, M.V., Coletta, M., Vivarelli, E., Frati, L., Cossu, G. and Giacomello, A. Isolation and expansion of adult cardiac stem cells from human and murine heart. *Circ. Res.* **95**, 911–921 (2004).
34. Smith, R.R., Barile, L., Cho, H.C., Leppo, M.K., Hare, J.M., Messina, E., Giacomello, A., Abraham, M.R. and Marban, E. Regenerative potential of cardiosphere-derived cells expanded from percutaneous endomyocardial biopsy specimens. *Circulation* **115**, 896–908 (2007).
35. Pouzet, B., Vilquin, J.T., Hagege, A.A., Scorsin, M., Messas, E., Fiszman, M., Schwartz, K. and Menasche, P. Factors affecting functional outcome after autologous skeletal myoblast transplantation. *Ann. Thorac. Surg.* **71**, 844–850 (2001).
36. Jessup, M. Dynamic cardiomyoplasty: expectations and results. *J. Heart Lung Transplant.* **19**, S68–S72 (2000).
37. Scorsin, M., Hagege, A., Vilquin, J.T., Fiszman, M., Marotte, F., Samuel, J.L., Rappaport, L., Schwartz, K. and Menasche, P. Comparison of the effects of fetal cardiomyocyte and skeletal myoblast transplantation on postinfarction left ventricular function. *J. Thorac. Cardiovasc. Surg.* **119**, 1169–1175 (2000).
38. Menasche, P., Hagege, A.A., Vilquin, J.T., Desnos, M., Abergel, E., Pouzet, B., Bel, A., Sarateanu, S., Scorsin, M., Schwartz, K., Bruneval, P., Benbunan, M., Marolleau, J.P. and Duboc, D. Autologous skeletal myoblast transplantation for severe postinfarction left ventricular dysfunction. *J. Am. Coll. Cardiol.* **41**, 1078–1083 (2003).
39. Guan, K., Wagner, S., Unsold, B., Maier, L.S., Kaiser, D., Hemmerlein, B., Nayernia, K., Engel, W. and Hasenfuss, G. Generation of functional cardiomyocytes from adult mouse spermatogonial stem cells. *Circ. Res.* **100**, 1615–1625 (2007).
40. Vallier, L. and Pedersen, R.A. Human embryonic stem cells: an *in vitro* model to study mechanisms controlling pluripotency in early mammalian development. *Stem Cell Rev.* **1**, 119–130 (2005).

41. Mummery, C., Ward, D., van den Brink, C.E., Bird, S.D., Doevendans, P.A., Opthof, T., Brutel, dl.R., Tertoolen, L., van der, H.M. and Pera, M. Cardiomyocyte differentiation of mouse and human embryonic stem cells. *J. Anat.* **200**, 233–242 (2002).
42. Kehat, I., Kenyagin-Karsenti, D., Snir, M., Segev, H., Amit, M., Gepstein, A., Livne, E., Binah, O., Itskovitz-Eldor, J. and Gepstein, L. Human embryonic stem cells can differentiate into myocytes with structural and functional properties of cardiomyocytes. *J. Clin. Invest.* **108**, 407–414 (2001).
43. Reppel, M., Pillekamp, F., Brockmeier, K., Matzkies, M., Bekcioglu, A., Lipke, T., Nguemo, F., Bonnemeier, H. and Hescheler, J. The electrocardiogram of human embryonic stem cell-derived cardiomyocytes. *J. Electrocardiol.* **38**, 166–170 (2005).
44. Sachinidis, A., Fleischmann, B.K., Kolossov, E., Wartenberg, M., Sauer, H. and Hescheler, J. Cardiac specific differentiation of mouse embryonic stem cells. *Cardiovasc. Res.* **58**, 278–291 (2003).
45. Harding, S.E., Ali, N.N., Brito-Martins, M. and Gorelik, J. The human embryonic stem cell-derived cardiomyocyte as a pharmacological model. *Pharmacol. Ther.* **113**, 341–353 (2007).
46. Kehat, I., Gepstein, A., Spira, A., Itskovitz-Eldor, J. and Gepstein, L. High-resolution electrophysiological assessment of human embryonic stem cell-derived cardiomyocytes: a novel *in vitro* model for the study of conduction. *Circ. Res.* **91**, 659–661 (2002).
47. Liu, J., Fu, J.D., Siu, C.W. and Li, R.A. Functional sarcoplasmic reticulum for calcium-handling of human embryonic stem cell-derived cardiomyocytes: insights for driven maturation. *Stem Cells* **25**, 3038–3044 (2007).
48. Sartiani, L., Bettiol, E., Stillitano, F., Mugelli, A., Cerbai, E. and Jaconi, M.E. Developmental changes in cardiomyocytes differentiated from human embryonic stem cells: a molecular and electrophysiological approach. *Stem Cells* **25**, 1136–1144 (2007).
49. Laflamme, M.A., Gold, J., Xu, C., Hassanipour, M., Rosler, E., Police, S., Muskheli, V. and Murry, C.E. Formation of human myocardium in the rat heart from human embryonic stem cells. *Am. J. Pathol.* **167**, 663–671 (2005).
50. Xue, T., Cho, H.C., Akar, F.G., Tsang, S.Y., Jones, S.P., Marban, E., Tomaselli, G.F. and Li, R.A. Functional integration of electrically active cardiac derivatives from genetically engineered human embryonic stem cells with quiescent recipient ventricular cardiomyocytes: insights into the development of cell-based pacemakers. *Circulation* **111**, 11–20 (2005).
51. Kehat, I., Khimovich, L., Caspi, O., Gepstein, A., Shofti, R., Arbel, G., Huber, I., Satin, J., Itskovitz-Eldor, J. and Gepstein, L. Electromechanical

integration of cardiomyocytes derived from human embryonic stem cells. *Nat. Biotechnol.* **22**, 1282–1289 (2004).
52. Laflamme, M.A., Chen, K.Y., Naumova, A.V., Muskheli, V., Fugate, J.A., Dupras, S.K., Reinecke, H., Xu, C., Hassanipour, M., Police, S., O'Sullivan, C., Collins, L., Chen, Y., Minami, E., Gill, E.A., Ueno, S., Yuan, C., Gold, J. and Murry, C.E. Cardiomyocytes derived from human embryonic stem cells in pro-survival factors enhance function of infarcted rat hearts. *Nat. Biotechnol.* **25**, 1015–1024 (2007).
53. Nussbaum, J., Minami, E., Laflamme, M.A., Virag, J.A., Ware, C.B., Masino, A., Muskheli, V., Pabon, L., Reinecke, H. and Murry, C.E. Transplantation of undifferentiated murine embryonic stem cells in the heart: teratoma formation and immune response. *FASEB J.* **21**, 1345–1357 (2007).
54. Martin, M.J., Muotri, A., Gage, F. and Varki, A. Human embryonic stem cells express an immunogenic non-human sialic acid. *Nat. Med.* **11**, 228–232 (2005).
55. Taylor, C.J., Bolton, E.M., Pocock, S., Sharples, L.D., Pedersen, R.A. and Bradley, J.A. Banking on human embryonic stem cells: estimating the number of donor cell lines needed for HLA matching. *Lancet* **366**, 2019–2025 (2005).
56. Klimanskaya, I., Chung, Y., Becker, S., Lu, S.J. and Lanza, R. Human embryonic stem cell lines derived from single blastomeres. *Nature* **444**(7118), 481–485 (2006).
57. Caspi, O., Lesman, A., Basevitch, Y., Gepstein, A., Arbel, G., Habib, I.H., Gepstein, L. and Levenberg, S. Tissue engineering of vascularized cardiac muscle from human embryonic stem cells. *Circ. Res.* **100**, 263–272 (2007).
58. Jawad, H., Ali, N.N., Lyon, A.R., Chen, Q.Z., Harding, S.E. and Boccaccini, A.R. Myocardial tissue engineering: a review. *J. Tissue Eng. Regen. Med.* **1**, 327–342 (2007).

Chapter 33

Cardiac Repair Clinical Trials

Amanda Green and Eric Alton

Abstract

Cardiovascular disease is a major problem worldwide, and remains the leading cause of death within Europe. Clinical trials of stem cells have been conducted in the setting of acute myocardial infarction and chronic heart disease. At the time of writing there have been a number of different cell types, preparation and doses delivered by a number of different delivery routes to a variety of patients in small, mostly uncontrolled trials, generally with positive outcomes. Larger placebo-controlled trials are ongoing and we eagerly await their results. The challenges currently facing the field are to define the optimal target patient population, cell source, preparation, dose, delivery and retention of cells within the myocardium and the appropriate assays to detect meaningful clinical changes. Clearly not all of these questions can be answered in the clinical trial setting and it is imperative that basic science and translational research are conducted simultaneously to help guide clinical research. At present there appears to be no clear steer in answering these questions but we attempt to discuss the evidence to date and the future of stem cells in the management of cardiovascular disease.

Keywords: Clinical Trials; Stem Cells; Heart.

Outline

1. Introduction
2. A New Treatment Option
3. Acute Myocardial Infarction
    3.1. Skeletal myoblasts
    3.2. Autologous bone marrow cells
    3.3. Granulocyte cell stem cell factor (G-CSF)
4. Chronic Ischaemic Heart Disease
    4.1. Skeletal myoblasts
    4.2. Autologous bone marrow mononuclear cells
    4.3. G-CSF in chronic ischaemic heart disease

5. Which is the Optimal Method of Delivery?
6. The Future for Clinical Trials
References

1. Introduction

Cardiovascular disease (CVD) is a major problem worldwide, and remains the leading cause of death within Europe, nearly half resulting from coronary artery disease (CAD).[1] The World Health Organization (WHO) estimates this situation will worsen and predicts by 2020 over 25 million deaths from CVD worldwide.[2]

Irrespective of the aetiology, heart failure is characterised by cardiac hypertrophy, insufficient vascularisation and cardiomyocyte loss. The key events, leading to heart failure, are a loss of a critical quantity of functioning myocardial cells after prolonged stress or injury with CAD still remaining the main cause.[3] Although cumulative evidence has now demonstrated that cardiomyocytes are not terminally differentiated,[4] the natural regenerative process is inadequate to replace the massive number of up to one billion cardiomyocytes lost after an acute myocardial infarction (AMI),[5] thus resulting in chronic heart failure (CHF).

The heart is now viewed as a self-renewing organ in which myocyte regeneration occurs throughout life with cardiac stem cells replacing apoptotic cells. However, evidence from sex-mismatched heart transplant recipients has shown that small numbers of cells of recipient origin and therefore from a non-cardiac source, have become engrafted into the donor heart.[6] Animal studies have shown a naturally occurring reparative process that consists of up-regulation of progenitor cell release from bone marrow after myocardial infarction, homing of these cells to the injured tissue, and differentiation of these progenitor cells into endothelium and cardiac myocytes within the infarcted tissue. However in humans, it appears that this process does not naturally occur with high frequency, and thus the rate of repair is insufficient for clinical restoration of function. The goal of cell therapy is to increase the frequency of this process. Regeneration of functional myocardial tissue after an ischaemic insult could, therefore, be achieved either by implanting exogenous cells, which may transdifferentiate into the relevant tissues, or by secretion of cytokines which stimulate the endogenous mature cardiomyocytes and resident cardiac stem cells. The ultimate aims of treatment are to generate contractile cells that integrate functionally and structurally in the viable myocardium and to develop a network of capillaries and large-sized blood vessels for supply of nutrients and oxygen to the injured region.

2. A New Treatment Option

Whilst mortality associated with CAD may be decreasing, morbidity is increasing as more people survive an AMI; the major underlying aetiology of heart failure is now survival post-AMI rather than hypertension or valve disease.[7] There are two main patient populations who have been targeted with stem cell therapy. Firstly, those who have sustained an AMI, in an attempt to salvage ischaemic tissue and prevent the devastating LV-remodelling effects which are responsible for the development of heart failure. Secondly, many patients with a longstanding history of CAD, multiple coronary interventions and a number of prior coronary events and are no longer suitable for conventional revascularisation strategies such as percutaneous coronary interventions (PCI) or coronary artery bypass graphs (CABG) because of the advanced nature of their disease or significant co-morbidities. Estimates from the US classify over 100,000 of these so-called "no option" patients every year.[8] In these patients presenting with chronic heart failure (CHF), therapy is aimed at restoring hibernating myocardium into functional myocardial tissue. Although described as two distinct patient populations, in reality there is often significant overlap between these groups. Because of the success of pre-clinical studies, particularly the pivotal but controversial paper by Orlic *et al.*,[9] translation from "bench to bedside" has been rapid.

The first clinical trials were conducted with skeletal myoblasts (SM) but bone marrow still remains the most frequent source of cells used in clinical trials, which are either harvested directly from the bone marrow itself or are stimulated with drugs such as granulocyte-colony stimulating factor (G-CSF) and then harvested from peripheral blood. Bone marrow contains a complex assortment of progenitor cells, including haematopoietic stem cells (HSC), side population (SP) cells, mesenchymal stem cells (MSC) and stromal cells, a subset of which are multipotential adult progenitor cells (MAPC). The following paragraphs summarise the clinical trials to date, have been broken down into cell type and into acute and chronic patient populations, and are summarised in Table 1.

3. Acute Myocardial Infarction

Mortality after AMI improves with early reperfusion of the infarct-related artery. However, this is not always achieved and necrosis, scar formation and left ventricular remodelling often occur, which adversely affects the left ventricular function (LVF) and prognosis. The aim of cell therapy in this patient population is to reduce these detrimental effects and replace lost myocardium with functional tissues. A number of different cell types as well as delivery routes have been utilised in this setting.

3.1. Skeletal myoblasts

Menasché *et al.*[10] conducted the first Phase 1 trial in ten patients with an open-label, uncontrolled design, utilising autologous skeletal myoblasts (SM) delivered

Table 1. Cardiac stem cell clinical trials.

Study	N	Design and numbers	Time post-MI	Cell type	Cell preparation	Culture	Mean cell no.	Delivery	Safety	Efficacy
ACUTE MI										
Menasché *et al.*[10] (2003)	10	Open label	6.2 ± 1.9 years	SM	Muscle biopsy with complex processing	3 weeks	8.7×10^8 myoblasts	Epicardial with CABG	Arrhythmias Required ICD	11 months ↑ LVEF ↓ NYHA
Siminiak *et al.*[11] (2004)	10	Open label	Minimum of 90 days post-MI	SM	Muscle biopsy with complex processing	3 weeks	$4 \times 10^5 - 5 \times 10^7$ myoblasts	Epicardial with CABG	Arrhythmias	12 months ↑ LVEF ↓ NYHA
Siminiak *et al.*[12] (2005) POZNAN	10	Open label	Minimum of 90 days post-MI	SM	Muscle biopsy with complex processing	3 weeks	Up to 10^8 myoblasts	Transcoronary venous	Arrhythmias	6 months ↓ NYHA ↑ LVEF

(*Continued*)

Table 1. (*Continued*).

Study	N	Design and numbers	Time post-MI	Cell type	Cell preparation	Culture	Mean cell no.	Delivery	Safety	Efficacy
Strauer *et al.*[13] (2002)	10	Open label 10 BMC 10 control refusing cell therapy	7 ± 2 days post-AMI	BMC	Ficoll 40 ml	Overnight	2.8×10^7	Intracoronary	None	3 months ↑ Wall movement ↓ Infarct size ↑ SV ↓ LVESV ↑ Perfusion
Fernandez-Aviles *et al.*[14] (2004)	20	No	13.5 days	BMC	Ficoll 50 ml	Overnight	$5.0–1.25 \times 10^7$	Intracoronary	No adverse events	6 months ↑ LVEF ↓ LVESV ↑ Regional wall thickness
Schächinge *et al.*[15] TOPCARE–AMI (2004)	59	Randomised open label 30 CPC 29 BMC	4.9 days	CPC BMC	250 ml blood 50 ml Ficoll	3 days None	Not stated $5.5 \pm 3.9 \times 10^6$ CD34/CD45+	Intracoronary	No adverse events	1 year ↑ LVEF ↓ Infarct size
Wollert *et al.*[16] BOOST (2004)	60	Randomised controlled 30 medical management 30 BMC	4.8 days	MBC	130 ml gelatin-polysuccinate sedimentation	None	2.5×10^9 (SD 9.4×10^8)	Intracoronary	No adverse events	6 months ↑ LVEF 18 months follow-up Meyer *et al.*[54] no change in LVEF

(*Continued*)

Table 1. (*Continued*).

Study	N	Design and numbers	Time post-MI	Cell type	Cell preparation	Culture	Mean cell no.	Delivery	Safety	Efficacy
Janssens *et al.*[17] (2006)	67	RDBPCT 34 control 33 ABMC	24 h post-AMI and PCI + STENT	BMC	Ficoll	None	$3. \times 10^8$ (SD 1.3×10^8) nucleated cells 1.7×10^8 (SD 7.2×10^7) mononucleated cells	Intracoronary	1 death	4 months No change in LVEF
Schächinger *et al.*[18] REPAIR AMI (2006)	204	RDBPCT 103 placebo 101 BMC	3–7 days	BMC	Ficoll 50 ml	BMC	2.4×10^8	Intracoronary	None	1 year ABMC group ↓ Major adverse cardiovascular events after AMI (death, recurrence of MI, or re-hospitalisation
Lunde *et al.*[19] ASTAMI (2006)	100	Randomised controlled 50 medical management 50 ABMC	6 days post-AMI	BMC	Ficoll	None	6.8×10^7 median interquartile range 5.4×10^7 to 1.3×10^8	Intracoronary	None	6 months No change LVEF LVEDV Infarct size

(*Continued*)

Table 1. (*Continued*).

Study	*N*	Design and numbers	Time post-MI	Cell type	Cell preparation	Culture	Mean cell no.	Delivery	Safety	Efficacy
Ge *et al.*[20] TCT STAMI (2006)	20	Randomised controlled 10 BMC 10 bone marrow supernatant	24 h post-AMI and PCI + STENT	BMC	Ficoll 40 ml	None	4×10^7	Intracoronary	None	6 months ↑ LVEF ↑ Perfusion No change in LVED diameter
Meluzin *et al.*[21] (2006)	66	Randomised controlled 22 high dose (HD) 22 low dose (LD) 22 medical control	AMI post-PC1	BMMC	Buffy coated density gradient centrifugation	Overnight	High dose 1×10^8 Low dose 1×10^7	Intracoronary	1 Thrombus 1 Dissection 1 Infected bone marrow sample	3 months ↑ Regional wall motion in dose related fashion 12 months HD ↑ LVEF maintained No change in regional wall motion in LD or HD

(*Continued*)

Table 1. (*Continued*).

Study	*N*	Design and numbers	Time post-MI	Cell type	Cell preparation	Culture	Mean cell no.	Delivery	Safety	Efficacy
Ince *et al.*[22] (2005)	50	Open label randomised 25 control 25 GSCF	AMI post-PC1	GSCF	10 mg/kg GCSF 6 days	None	20 X ↑	Systemic	No adverse events	4 months ↑ LVEF ↑ Regional wall thickness and movement ↓ LVEDV ↑ Metabolic activity
Kuethe *et al.*[23] (2005)	23	Open label randomised 9 control refused GCSF 14 GSCF	48 hours AMI post-PC1	GSCF	10 mg/kg GCSF 7 days	None	Stopped when leucocytes above 5.0×10^{10}/L	Systemic	No adverse events	3 months ↑ Regional wall thickness and movement ↑ Perfusion
Zohlnhöfer *et al.*[24] REVIVAL-2 (2006)	114	RDBPCT	0.5 days	GCSF	10 mg/kg 5 days	None	5 × ↑ in WBC compared to placebo 14 × ↑ CD34+ cells	GCSF mobilised	Bone pain and muscle discomfort	4–6 months No change LVEF Restenosis Infarct size
Ripa *et al.*[25] STEMMI (2006)	78	RDBPCT	0.5 days	GCSF	10 mg/kg 6 days	None	5 × ↑ CD45–/CD34– 2 × ↑ IN CD45–/CD34–/CXCR4+ 10 × ↑ CD34–/CD34–VEGFR-2+	GCSF mobilised	None	No change LVEF Restenosis Infarct size

(*Continued*)

Table 1. (*Continued*).

Study	*N*	Design and numbers	Time post-MI	Cell type	Cell preparation	Culture	Mean cell no.	Delivery	Safety	Efficacy
Engelmann *et al.*[26] GCSF-STEMI (2006)	44	RDBPCT	0.25–7 days	GCSF	10 mg/kg 5 days	None	4 × ↑ in leucocytes compared to placebo. 23–29 fold ↑ in EPC's	GCSF mobilised	1 death 1 MI requires careful monitoring	3 months ↑ Perfusion no change in LVEF
CHRONIC										
Smit *et al.*[27] (2003)	5	Open label	Chronic	SM	Muscle biopsy and complex processing	17 days	$2 \pm 1.1\ 10^8$	NOGA endocardial	1/5 VT and ICD placement	3 months ↑ LVEF ↑ Wall thickness
Pagini *et al.*[28] (2003)	5	Open label	Chronic LVAD bridge to transplant and SM	SM	Muscle biopsy and complex processing	3 weeks	3.0×10^8	Epicardial + LVAD	1 death	3 transplant SM survival and differentiation into mature myofibres
Patel *et al.*[29] (2005)	20	Randomised 10 CABG 10 CABG + cells	Chronic	AC133+ BM	500–600 ml	None	2.2×10^7 CD34+	Epicardial with CABG	None	6 months ↑ LVEF
De la Fuente *et al.*[30] (2007)	10	Open label	Chronic	BMC	50 ml Ficoll	None	8.6×10^7	Endocardial helical needle	None	12 months ↑ LVEF

(*Continued*)

Table 1. (*Continued*).

Study	N	Design and numbers	Time post-MI	Cell type	Cell preparation	Culture	Mean cell no.	Delivery	Safety	Efficacy
Goussetis *et al.*[31] (2006)	8	Open label	Chronic	CD133+ CD133– CD34+ 300 ml BM	Labelled with 99m Tc-hexamethyl-propylenamine-oxime	None	$0.8 \pm 0.3 \times 10^7$ CD133+ $0.8 \pm 0.2 \times 10^7$ CD133– CD34+	Intracoronary	None	1 h distribution $n = 8$, 9.2% ± 3.6 24 h distribution $n = 4$, 6.8% ± 2.4
Losordo *et al.*[32] (2007)	24	RDBPCT dose escalation 3 active doses and control	Chronic	CD34+	5 mg/kg GCSF 5 days	Leuka-pheresis followed by overnight storage at 4° C	CD34+ cell/kg 5×10^4 1×10^5 5×10^5	NOGA and endocardial injections	13 pts ↑ Angina with GCSF	6 months Trends for ↓ Angina ↓ GTN ↓ CCC ↑ Exercise tolerance No dose response
Klein *et al.*[33] (2007)	10	Open label	Chronic	CD133+	Separated CliniMACS	None	$1.5–9.7 \times 10^6$	Epicardial with CABG	None	9 months ↓ NYHA ↓ CCC ↑ LVEF

(*Continued*)

Table 1. (*Continued*).

Study	*N*	Design and numbers	Time post-MI	Cell type	Cell preparation	Culture	Mean cell no.	Delivery	Safety	Efficacy
Hüttermann *et al.*[34] (2006)	12	Open label historical controls	Chronic ICM 9 DCM 7	CD34+	4×10 days dose sufficient to raise leukocyte count to 45,000–50,000/ml until day 10	None	Leukocyte count to 45,000–50,000/ml until day 10	Systemic	One death from ventricular fibrillation ↑ Angina ↑ Dyspnea	6 months ↓ NYHA ↑ 6 min walk test
Manginas *et al.*[35] (2007)	24	Open label retrospective medical management	Chronic	CD133+ CD133– CD34+	Ficoll	BMMNC incubated with monoclonal antibody against CD133 followed by CD34	CD133+ $1.7 \pm 0.5 \times 10^7$ CD133– CD43+ $8 \pm 4 \times 10^6$	Intracoronary	1/12 restenosis 1/12 increased atherosclerosis No increase compared to control	4–6 months ↑ LVEF ↓ LVEDV ↓ LVESV ↑ Perfusion
Stamm *et al.*[36] (2003)	6	Open label	Within 90 days	AC133+ from bone marrow aspirate	Ficoll Incubated ferrite conjugated monoclonal antibody against AC133 + for 30 min	Overnight at 4°C in 0.9% NaCl	AC133+ 1.5×10^6	Epicardial with CABG	Possible local inflammation	3–9 months 4/6 ↑ LVEF 5/6 ↑ Perfusion 6/6 ↑ NYHA

(*Continued*)

Table 1. (*Continued*).

Study	*N*	Design and numbers	Time post-MI	Cell type	Cell preparation	Culture	Mean cell no.	Delivery	Safety	Efficacy
Stamm *et al.*[37] (2007)	55	Open label 15 safety patients 20 CABG + cells 20 CABG control	MI at least 14 days previous	CD133+ 91–265 ml BMC	Ficoll Incubated Ferrite conjugated monoclonal antibody against CD133+	Overnight at 4°C in 0.9% NaCl	3×10^7 CD34+ 5.8×10^6AC133+	Epicardial with CABG	None	6 months ↑ LVEF ↑ Perfusion
Fuchs *et al.*[38] (2003)	10	Open label	Chronic	BMC	Not described	None	$3.3 \pm 2.8 \times 10^7$/ml 0.2 ml $\times 12$	NOGA and endocardial injection	2 re-admitted to hospital for recurrent chest pain	3 months ↓ CCC ↓ Stress induced ischaemia
Tse *et al.*[39] (2003)	8	Open label	Chronic	BMMC	Ficoll	None	$10^7 \times$ cells/ml	NOGA and endocardial injection	None	3 months ↑ Symptoms ↑ Perfusion ↑ LVEF
Perin *et al.*[40] (2004)	21	Open label 14 BMMC 7 medical control patients	Chronic	BMMC	Ficoll	None	3×10^7	NOGA and endocardial injection	1 sudden cardiac death	4 months ↑ LVEF ↓ ESLV 6–12 months ↑ SPECT ↑ MVO2

(*Continued*)

Table 1. (*Continued*).

Study	*N*	Design and numbers	Time post-MI	Cell type	Cell preparation	Culture	Mean cell no.	Delivery	Safety	Efficacy
Mocini *et al.*[41] (2005)	36	Open label 18 CABG+ BMC 18 CABG	Chronic MI < 6 months	BMC	400 ml	None	5.2×10^8	Epicardial with CABG	↑ Trop 1	3 months ↑ LVEF
Strauer *et al.*[42] (2005)	18	Open label 18 BMC historic controls	Chronic	80 ml BMC	Ficoll	Overnight	1.5 to 2.2×10^7	Intracoronary	None	3 months ↑ LVEF ↑ Wall movement ↓Infarct size ↑ VO_2 max ↑ Metabolic activity
Assmus *et al.*[43] TOPCARE-CHD (2006)	75	Randomised controlled crossover 23 no cell infusion 24 CPC 28 BMC	< 90 days	BMC	50 ml Ficoll	None	$2.1 \times 10^8 \pm 1.1 \times 10^8$ BMC	Intracoronary	3 of 135 infusion procedures local dissection of coronary artery	3 months BMC ↑ LVEF
				CPC	270 ml	3 days	$2.2 \times 10^7 \pm 1.1 \times 10^7$ CPC			No change with CPC

(*Continued*)

Table 1. (*Continued*).

Study	N	Design and numbers	Time post-MI	Cell type	Cell preparation	Culture	Mean cell no.	Delivery	Safety	Efficacy
Archundia *et al.*[44] (2005)	5	Open label compared to 10 matched controls receiving CABG	> 1 year	CD34+	300 mg G-CSF for 5 days	Apheresis selection of CD34+ cells frozen for 2 days	4×10^8 6×10^5 CD34+	Epicardial with CABG	None	6 months ↑ LVEF ↓ LVDV ↓ LVESV ↑ LV wall thickness 2/5 pts ↑ NYHA
Wang *et al.*[45] (2005)	29	Open label 13 GCSF 16 medical controls	Chronic	GCSF exercise	5 mg/kg GCSF 6 days and induced ischaemia	None	4 poor mobilisers CD34+ 5000 ± 700/ml blood 9 mobilisers 28,900 ± 5100 ml blood	GCSF mobilised	Possible adverse effect on LVEF	2 months GCSF group ↓ CCC ↓ Angina ↓ GTN No change Perfusion
Boyle *et al.*[46] (2005)	5	Open label	Chronic	CD34+	G-CSF	10 mg/kg 4 days and leuka-pheresis	CD43+ $6.7 \pm 1.8 \times 10^7$	Intracoronary	Bone pain 1/5 prolonged headache 1/5 ACS 2 months after 1/5 lentigo maliga in an area with previous solar keratosis	12 months ↓ CCC ↑ QOL ↑ Collateral blood flow

(*Continued*)

Table 1. (*Continued*).

Study	*N*	Design and numbers	Time post-MI	Cell type	Cell preparation	Culture	Mean cell no.	Delivery	Safety	Efficacy
Erbs *et al.*[47] (2005)	26	RDBPCT	Chronic occlusion post-PCI 10 days later cell/control	CPC 4 days	300 μg GCSF	4 days	$6.9 \pm 1.4 \times 10^7$	Intracoronary	None	3 months ↑ LVEF ↑ Flow reserve ↓ Hibernation ↓ Infarct size
Kang *et al.*[48] MAGIC Cell-3-DES (drug eluding stent) (2006)	96	Randomised controlled AMI PCI + DES 25 AMI cell infusion 25 AMI control 16 old MI cell infusion 16 old MI control	Old and new MI	PBSC	10 mg/kg GCSF 3 days	Apheresis	$1.4 \pm 0.5 \times 10^9$ mononuclear cells	Intracoronary	None	6 months AMI PBSCs ↑ LVEF ↓ LVEDV Old MI No change LVEF
Kang *et al.*[49] MAGIC 2007 Terminated early	30	Randomised controlled 10 control medical management 10 GCSF 10 GCSF and PBSC	Old and new MI	GCSF GCSF and PBSC	10 mg/kg GCSF 4 days	Apheresis for PBSC	PBSC $1 \times 10^9 \times 3$ patients 1.7×10^9 remaining patients	PBSCs intracoronary infusion	Terminated early GCSF alone ↑ restenosis	6 months PBSC ↑ LVEF Gone by 2 years GCSF alone ↑ restenosis

(*Continued*)

Table 1. (*Continued*).

Study	N	Design and numbers	Time post-MI	Cell type	Cell preparation	Culture	Mean cell no.	Delivery	Safety	Efficacy
Kang *et al.*[50] (2007)	50	Randomised controlled AMI sirolimus eluding stent (SES) or paclitaxel eluding stent (PES) 19 cells 17 medical management	Old and new MI	PBSCs	10 mg/kg GCSF 3 days	Apheresis	$1–2 \times 10^9$ mononuclear cells	Intracoronary	No ↑ Neointimal hyperplasia but may ↑ peri-stent tissue growth at stented segment especially with PES	6 months No aggravation in-stent neointimal hyperplasia or adverse effects or *de novo* lesions May↑ peri-stent tissue growth at stented segment especially with PES
Ripa *et al.*[51] (2006)	16	Historical control 32	Chronic	VEGF-A_{165} and GCSF	0.5 mg VEGF-A_{165} 10 mg/kg G-CSF 6 days	None	4 poor mobilisers CD34+ cells < 15.000/ml blood 60.2 ± 11.0/ml blood	NOGA and endocardial VEGF-A_{165} G-CSF	None	3 months No effect

Shaded rows represent randomised double-blind placebo-controlled trials (RDBPCT), skeletal myoblasts (SM), circulating progenitors cells (CPC), peripheral bone marrow stem cells (PBMC) autologous bone marrow cells (BMC, ABMC), autologous bone marrow mononuclear cells (ABMMC), granulocyte-colony stimulating factor (G-CSF), bone marrow (BM), vascular endothelial growth factor (VEGF), ↑ = increased ↓ = decreased, left ventricular end systolic volume (LVESV), left ventricular ejection fraction (LVEF), stroke volume (SV), left venticular end diastolic volume (LVEDV), acute myocardial infarction (AMI), myocardial infarction (MI), glycerine trinitrate (GTN), New York Heart Association class (NYHA), Canadian cardiovascular class (CCC), myocardial oxygen consumption (MVO_2), ischaemic cardiomypathy (ICM), dilated cardiomyopathy (DCM), drug eluding stent (DES), sirolimus eluding stent (SES), paclitaxel eluding stent (PES).

epicardially as an adjunct to coronary artery bypass grafts (CABG). All patients had a history of myocardial infarction (MI) with a left ventricular injection fraction (LVEF) < 35% and a residual akinetic scar. The age of infarct was very heterogeneous from one month to 19 years, and thus the regenerative capability and homing milieu were likely to be significantly different. Each patient received between 27 and 57 injections of cells, the dose varying from $5.0–11.5 \times 10^8$ cells. These were directed in and around the scar in an area distinct from grafts. Initial follow-up was 10.9 ± 4.5 months. LVEF increased significantly from 24 ± 1% preoperatively to 32 ± 1% postoperatively ($p < 0.02$). New York Heart Association (NYHA) class improved significantly from 2.7 ± 0.2 preoperatively to 1.6 ± 0.1 postoperatively ($p < 0.0001$). There was one death, which occurred early after the operation, and was not attributed to cell therapy. However, a significant adverse event was the development of multiple sustained ventricular arrhythmias on 24-hour Holter monitoring despite prophylactic amiodarone. Five of these patients went on to have internal cardiac defibrillators (ICD) implantation. Long-term follow-up of these patients (median 52 (range 18–58) months) after CABG and SM transplantation[52] demonstrated that the improvements in NYHA class and LVEF were maintained. It was also noted that there was a low incidence of hospitalisation for heart failure (0.13/patient/year) and the arrhythmic risk could be managed albeit by medical and or ICD implantation.

Siminiak *et al.*[11] also conducted an open-label, uncontrolled study of SM administered as an adjunct to CABG in ten patients with viable under-perfused myocardium. There was an increase in LVEF from 35.2% (25% to 40%) preoperatively to 42% (29% to 47%), four months postoperatively which was maintained at the 12-month follow-up. Although this group also noted ventricular arrhythmias these were managed medically without the need for ICD implantation. As for the Menasché study, it is unknown whether detected improvements in LVEF and NYHA class resulted from the application of SM cells or from the revascularisation resulting from surgery. However, in the subsequent POZNAN trial an additional ten patients received percutaneous transcoronary venous delivery of SM.[12] All patients were post-MI with a demonstrated area of non-viable myocardium, eight of which had previously undergone CABG. Again there was a large variability in the time from infarct and the dose of cells delivered. LVEF increased 3%–8% in six of the ten patients, with no change seen in the remaining subjects. In patients treated with amiodarone, only single ventricular extrasystoles were seen; however, in a patient not receiving amiodarone but with an ICD *in situ*, sustained episodes of VT occurred.

The clinical experience with SM cells remains limited; apart from the Myoblast Autologous Grafting in Ischaemic Cardiomyopathy (MAGIC) study all have been very small and uncontrolled. These studies have demonstrated the feasibility of SM cell delivery via a number of different routes but have also raised significant safety concerns regarding potential arrhythmogenic effects. In the majority of these studies SM have been administered in combination with either CABG and therefore any improvements in LV function are difficult to interpret and cannot be attributable to the haemodynamic effects of SM cells alone. There is no evidence of the formation of gap-junctions between the SM and resident cardiomyocytes; thus any reported improvements likely result from the transplanted cells strengthening the post-infarction scar with amelioration of LV remodelling rather than direct effects on contractility. The lack of electromechanical coupling may have also inadvertently lead to areas of electrical instability and the reported incidence of VT; thus support for the use of SM remains controversial.

3.2. Autologous bone marrow cells

Ge *et al.*[20] administered autologous bone marrow mononuclear cells (ABMMC) via the intracoronary route in ten patients within 24 hours of an AMI and successful PCI and stent implantation. Six months later, compared to the control group, who received intracoronary instillation of supernatant ($n = 10$), the ABMMC treated group had significant improvements in LVEF from a mean 53.8 (SD 9.2%) to 58.6 (9.9%) ($p < 0.05$) and unlike the control group they did not have an increase in left ventricular diameter (LVD) diameter. Myocardial perfusion defect scores improved from 21 ± 11 to 13 ± 10 ($p < 0.01$) in the ABMMC group but were unchanged in the control group 20 ± 14 to 17 ± 15. Schächinger *et al.*[15] in the Transplantation of Progenitor Cells and Regeneration Enhancement in Acute Myocardial Infarction (TOPCARE-AMI) Phase 1 trial studied 59 patients post-AMI and successful stenting of the culprit lesion. Patients were treated with either intracoronary delivery of bone-marrow derived progenitor cells consisting of haematopoietic progenitor, side populations and stromal cells or circulating blood-derived progenitor cells which were predominantly endothelial progenitor cells 4.9 ± 1.5 days post-AMI. At the four-month follow-up LVEF significantly increased (50 ± 10% to 58 ± 10%, $p < 0.001$), left ventricular end systolic volume (LVESV) decreased (54 ± 19 ml to 44 ± 20 ml, $p < 0.001$); at one year the improvement in LVEF was maintained.

The Bone Marrow Transfer to Enhance ST Elevation Infarct Regeneration (BOOST) Trial by Wollert *et al.*[16] was the first randomised controlled trial using autologous bone marrow. After sustaining an AMI and following acute reperfusion PCI, 60 patients were randomised either to receive optimal medical treatment or intracoronary unselected autologous bone marrow cells (ABMC) 4.8 days (SD 1.3) later into the culprit vessel. Six months later LVEF increased significantly in patients receiving ABMCs from 50 ± 10% to 56.7% versus 51.3 ± 9.3%

to 52 ± 12.4% in the control group ($p = 0.003$). There was no reported incidence of arrhythmias, or restenosis. These changes, however, were not sustained when patients were reviewed at 18 months, when the LVEF had increased by 0.7% in the control group and 3.1% in the group receiving ABMCs.[53] Two possible explanations for this are that the initial improvements in the ABMC group may have arisen as a result of a marked placebo effect, which has previously been noted in angiogenesis trials,[54,55] or that the long-term viability of transplanted cells is questionable and improvements are only temporary and do not translate into permanent myocardial repair.

Meluzin *et al.*[21] in the only dose-escalation study, randomised 66 patients into three groups post-AMI and PCI to either intracoronary delivery of high dose (10^8 BMMC) or low dose (10^7 BMMC) or control (no instillation). Three months later they found that both cell groups had improvements in regional wall motion compared to the control in a dose-related fashion. However, at 12 months this difference was no longer apparent. In the largest study to date Schächinger *et al.*[18] reported the results of the Reinfusion of Enriched Progenitor cells and Infarction Remodelling in Acute Myocardial Infarction (REPAIR AMI) trial. They randomised 204 patients with successful reperfused AMI to receive intracoronary infusion of ABMC, or placebo, three to seven days later. At 12 months in the ABMC group compared to the control there were significant improvements in rates of the pre-specified cumulative endpoint of death, recurrence of MI and rehospitalisation for heart failure ($p = 0.006$). At four months quantitative LV angiography was performed to assess LVEF; a significant if modest 2.5% improvement in LVEF ($p = 0.02$) was detected. An improvement of at least 5% has been suggested to be required to predict improvements in symptoms or mortality.[56] Subgroup analysis showed that those patients with lower baseline LVEF gained the most improvement. The 12-month LVEF data have not yet been published.

In contrast Lunde *et al.*[19] reported negative results from their Autologous Stem Cell Transplantation in Acute Myocardial Infarction (ASTAMI) randomised controlled trial. Forty-seven patients received intracoronary injections of ABMMC six days after an AMI and these were compared to a control group who received no infusion. Three non-invasive imaging modalities of single photon-emission computed tomography (SPECT), echocardiography (ECHO) and magnetic resonance imaging (MRI) were assessed. The study was powered to have an 80% chance of detecting a 5% increase in LVEF. At six months there was no significant difference between groups, and with arguably the most sensitive measure (MRI) the change in LVEF ($p = 0.05$) favoured the control group.

Janssens *et al.*[17] also did not detect an improvement in LVEF assessed by MRI at four-month follow-up of their double-blind placebo-controlled trial comparing intracoronary BMC ($n = 33$) to infusion of placebo ($n = 34$) in patients 24 hours post-AMI and successful PCI and stenting. However, in secondary outcome measures the BMC group, compared to controls, had a significant reduction of 28% in MI size ($p = 0.04$) and improvements in regional systolic function. There were no differences between the groups for perfusion and metabolism as assessed by PET. This group concluded that although cell therapy did not improve LV function after AMI, it could have had favourable effects on post-infarct remodelling. Interestingly this study administered cells 24 hours after optimal reperfusion. In the REPAIR AMI[18] study it was found that patients receiving cells five to seven days after MI had a greater improvement in LVEF than in those patients that received cells at an earlier time point. It has been suggested that the milieu immediately post-AMI may adversely affect cell homing and retention within in myocardium, potentially impacting on the efficacy in this trial. Clearly further studies are required to determine the optimal time, post-AMI, at which to deliver cells, and the most efficient delivery route, together with the longer-term sustainability of any noted improvements.

3.3. Granulocyte cell stem cell factor (G-CSF)

After AMI there is enhanced spontaneous mobilisation of bone marrow-derived stem cells.[57] The extent and duration of this mobilisation have also been shown to correlate with improvements in LVEF. Thus, Numaguchi *et al.*[58] assessed 51 patients who had had an AMI treated with primary PCI and successful stent implantation. They quantified the number of endothelial progenitor cells (EPC) identified as $CD45^{Low}$, $CD34^{+}$, $CD133^{+}$ and $VEGFR2^{+}$ by flow cytometry; the potential for differentiation was also shown by an up-regulation of CD31 and VEGF2. After seven days of culture they separated patients into two groups; those with >15 cells with $CD45^{Low}$, $CD34^{+}$ $CD133^{+}$ $VEGFR2^{+7}$ leukocytes (differentiated group) and those with <15 cells with $CD45^{Low}$, $CD34^{+}$ $CD133^{+}$ $VEGFR2^{+}/10^{7}$ leukocytes (non-differentiated group). Patients in the differentiated group, with higher numbers of circulating EPCs compared to the undifferentiated group, had a larger area of myocardial salvage ($p = 0.009$), a decrease in end-systolic volume ($p = 0.01$), and better recovery of LVEF ($p = 0.01$). Kang *et al.*[48] in the Myocardial Regeneration and Angiogenesis in Myocardial Infarction with G-CSF and Intra-Coronary Stem Cell (MAGIC CELL) study investigated the effects of G-CSF alone or G-CSF with intracoronary delivery of peripheral circulating progenitor cells (PCPCs) compared with a medical control. At six months the PCPC group had an increase in LVEF, although at two years this improvement was not maintained.[49] Of concern was the early termination of the study due to the increased rate of restenosis in the group receiving G-CSF alone. Thus, manipulating the number of circulating EPCs at the time of AMI may be a novel therapeutic target to salvage ischaemic damage. A number of small open-label uncontrolled studies have reported favourable effects on LVEF with the administration of G-CSF post-AMI.[22,23,59,60] However, some safety concerns were noted including an increase in restenosis rates, which may be explained by the rise in leukocytes numbers to leukaemia-like levels. Thus, it may be preferable to use strategies which do not induce massive inflammation, which could lead to plaque growth or destabilisation.

Three large randomised placebo-controlled studies have subsequently reported that although GCSF appeared safe, it did not improve LVEF or reduce infarct size[24–26] despite successfully increasing the number of circulating stem cells. These studies consistently showed that in the placebo group there was an improvement in LVEF from the time of primary PCI to follow-up three to six months later. This observed improvement ranged from 2% to 8%, probably due to recovery of stunned myocardium. Importantly, this highlights the critical need to include a true placebo group in all protocol designs where efficacy is reported as an outcome measure. So why, when it has been reported that an increased number of circulating stem cells have beneficial effects on LVEF, were these studies negative? It may be that the timing was wrong. Zohlnhöfer *et al.*[24] have suggested that the milieu five days post-AMI adversely affect the homing signals for cells. In the REPAIR-AMI trial the most beneficial effects were evident in those patients receiving stem cell transplantation more than five days post-AMI and there were no improvements in LVEF in those patients receiving stem cells within four days of their infarct.[18] Secondly, the wrong dose may have been used. Alternatively the type of cells mobilised may not be what is required. Thus, Ripa *et al.*[51] reported that patients randomised to G-CSF over seven days were exposed to 25×10^9 G-CSF mobilised CD34+ cells compared with 3×10^9 cells in the placebo group; in contrast these values were respectively 4.9×10^{11} and 2.0×10^{11} for mesenchymal stem cells. The fraction of CD34+ cells/leucocytes increased during G-CSF treatment compared to placebo, whereas the fraction of putative mesenchymal stem cells/leucocytes decreased because the total leucocyte number had increased more in the G-CSF group. An inverse association between the numbers of circulating mesenchymal stems cells and LVEF was found.

4. Chronic Ischaemic Heart Disease

These patients will generally have run through standard medical and surgical options and remain with a bleak outlook, both in terms of life expectancy and quality of life. The presenting symptom of some of these patients is recalcitrant angina, but in a large proportion the focus is heart failure of increasing severity.

4.1. Skeletal myoblasts

In a small pilot study of five patients with symptomatic heart failure post-MI, Smit *et al.*[27] delivered SM cells as sole therapy via NOGA electromechanical mapping and endocardial injections. Three months later one patient needed an ICD for non-sustained VT. Angiographic assessment of LVEF increased from 36 ± 11% to 41 ± 9% ($p = 0.009$); this improvement was not seen by nuclear or MRI assessments. At six months both angiographic and MRI assessments showed non-significant trends in improvement, but with such small patient numbers it is hard to draw conclusions. Pagani *et al.*[28] injected SM at the time of left ventricular assist device (LVAD) insertion as a bridge to transplantation. In the four patients transplanted, examination of the native heart revealed engraftment of the SM within the infracted myocardium. Both these were open-label, uncontrolled studies. The first randomised placebo-controlled study of the use of SM cells, the MAGIC study, was presented at the American Heart Association in November 2006.[61] This trial included 97 patients undergoing CABG after MI with moderate to severely reduced LVF. All patients had prophylactic ICDs implanted. The study was ended prematurely because the treatment was not superior to placebo for the primary outcome measure of LVEF; however, final results are still pending.

4.2. Autologous bone marrow mononuclear cells

A number of studies have been undertaken with unselected autologous bone marrow. Tse *et al.*[39] studied eight patients with stable angina, refractory to maximal medical therapy; 10^7 cells/ml were delivered percutaneously with endocardial injections using the Myostar catheter and NOGA mapping system. After three months patients reported improvements in symptoms, as well as more objective changes in myocardial perfusion. Left ventricular ejection fraction (LVEF) remained unchanged, but there was a 12% improvement in wall thickening, and a 6% improvement in target wall motion determined by magnetic resonance imaging (MRI).

Fuchs *et al.*[38] also administered unfractionated autologous bone marrow in ten patients not eligible for conventional revascularisation, but with evidence of reversible ischaemia. At three months there was a significant reduction in stress-induced ischaemia in the injected territories. In addition symptomatic angina was also significantly reduced. Perin *et al.*[40] reported on 21 patients with more severe chronic heart failure (mean LVEF 20%) who were enrolled in a non-randomised open label study (the first 14 patients received autologous bone marrow whilst the last seven patients had no intervention and were used as controls). At two months there was a significant reduction in total reversible defect (assessed by SPECT) and improvement in global left ventricular function within the treatment group and between groups, together with improvements in VO_{2max}. At four months there was an improvement in LVEF from 20% to 29%, a 31% relative increase. Encouragingly at 12 months SPECT had improved ($p = 0.01$) and the significant improvements in VO_2 max were maintained.

Strauer *et al.*[42] performed a cohort study of 18 consecutive patients, five months to 18.5 years following myocardial infarction, comparing them with a control group who fulfilled the entry criteria, but had refused cell therapy. Three months after coronary artery delivery of mononuclear cells, they reported a 30% reduction in infarct size, an improvement in LVEF of 15%, and an improvement of 57% in infarct wall movement velocity. They also noted an 11% improvement in $VO_{2max,}$ a 15% improvement in regional 18F-fluor-desoxy-glucose uptake in the infarct tissues, and concluded that the application of mononuclear cells had lead to functional and metabolic regeneration of infarcted and ischaemic tissue. The composition of the control group may have biased their results. In a larger randomised crossover trial (TOPCARE CHD), Assmus *et al.*[43] reported their results in 75 patients who had an MI at least three months previously. Patients were randomised to no cell infusion ($n = 23$), circulating progenitor cells (CPC, $n = 24$) or BMC ($n = 28$). After three months the patients who were initially randomised to no cell infusion were subsequently randomised to receive either CPC or BMC and the patients who initially received CPC or BMC crossed over to BMC and CPC respectively. The LVEF significantly increased (2.9%) in patients receiving BMC compared to those receiving CPC (–0.4%, $p = 0.003$) or no infusion (–1.2%, $p < 0.001$). There was also an increase in regional contractility in the area targeted by BMC. The crossover phase revealed that intracoronary infusion of BMC was associated with this improvement of LV function and contractility regardless of whether BMC were received as their primary or secondary therapy.

In addition to unfractionated BM a number of groups have also isolated specific cells. Stamm *et al.*[37] delivered epicardial CD133+ cells as an adjunct to CABG in 20 patients, compared to 20 patients in whom only CABG was performed. The CABG plus cell therapy group had significantly higher LVEF at six months compared to the CABG control (37.4% ± 8.4% preoperatively to 50.2% ± 8.5% at six months and 47.1% ± 6% at 18 months, compared to 37.9% ± 10.3% to 41.3% ± 9.1% at six months in the CABG only group). Between group comparisons showed an improvement in the cell group compared to the CABG group ($p = 0.03$). Perfusion improved more in the CABG plus cell therapy patients. Klein *et al.*[33] recently reported ten patients with ischaemic cardiomyopathy, where CD133+ cells were injected epicardially. All patients had improvement in cardiac function, both in New York Heart Association class, and LVEF. Patel *et al.*[29] randomised 20 patients into two groups; one received off-pump CABG (OPCAB), the other OPCAB and epicardial injection of CD34+ ABMMC. They used a large volume (30 ml in 1 ml aliquots) with a special needle with side holes to prevent leakage. Six months later, those patients receiving the addition of putative stem cells had improvements in LVEF from preoperative levels of 29% ± 0.6% to 46.1% ± 1.9% compared to OPCAB of 30.7% ± 2.5% to 37.2% ± 3.4%. The direct injections may have been a confounding variable, although in a preceding open-labelled pilot study this group had demonstrated a superior outcome with stem cells injections and OPCAB compared to serum injections and OPCAB. Although encouraging, cell transplant was performed at the time of CABG in these studies so it is difficult to interpret the effects of cell therapy alone.

Manginas *et al.*[35] reported on 24 patients with old myocardial infarction; 12 received intracoronary delivery of CD133+, CD133− CD34+ progenitor cells and 12 were followed up medically. In the treatment group LVED and LVESV were reduced, LVEF at rest improved from 27.2 ± 6.8% to 29.7 ± 7.3%, and perfusion was also improved. One patient developed restenosis and one progression of atherosclerosis at the site of cell delivery, but after 28 ± 8.7 months only one patient had deterioration in heart failure. In the control group there was deterioration in end diastolic and end systolic volumes, a non-significant reduction in LVEF and no change in perfusion.

4.3. G-CSF in chronic ischaemic heart disease

Wang *et al.*[45] found that treatment with G-CSF in patients with severe IHD improved symptoms but did not alter myocardial ischaemia. Indeed on MRI and SPECT, although not with Echo, there was a reduction in LVEF. This reduction may have been due to an adverse affect of G-CSF on the myocardium, or because there was a large variability of LVEF assessed by the three different assays utilised. Caution is required in future trials to assess whether there is a negative effect of G-CSF on LVEF in patients with severe IHD.

Losordo *et al.*[32] administered a five-day course of GCSF and then separated the CD34+ cells via leukoapheresis. No-option patients were randomised in a 3:1 ratio to receive CD34+ cells or placebo. Patients underwent NOGA electromechanical mapping followed by endocardial injections. All patients had an ICD or wore a life vest defibrillator. In the six-month follow-up there were no life-threatening ventricular arrhythmias. Angina and nitroglycerin use (GTN) and Canadian Cardiovascular Society (CCS) classification and exercise times showed trends for improvement in the CD34+ group compared to control; a larger Phase II trial is underway.

Huttmann *et al.*[34] administered G-CSF in two groups of heart failure patients, one with ischaemic cardiomyopathy (ICM) and another presenting with dilated cardiomyopathy (DCM). G-CSF was administered over a ten-day period interrupted by treatment-free intervals of equal length. Peak CD34+ levels remained constant throughout treatment periods. Nine of the 12 patients received the intended four full cycles and had statistically significant improved six-minute walk tests and an improvement of one in their NYHA class, compared to a historical control with ICM. There was one death, and occasional episodes of dyspnea and angina. Archundia *et al.*[44] mobilised peripheral mononuclear CD34+ rich cells and injected them epicardially in the peri-infarct zone at the time of CABG in patients who had an MI > 1 year previously. Compared to a matched group of patients just undergoing CABG there appeared to be beneficial effects of the addition of CD34+ enriched mononuclear cells. The cell group had improved LVEF, and decreased LVDV, improvements in NYHA class, and LV wall thickness was seen to increase significantly in 2/5 patients. Boyle *et al.*[46] administered a G-CSF mobilised CD34+ enriched mononuclear cells intracoronary in five patients with chronic ischaemic heart failure. Although they documented improved 12-month follow-up of collateral blood flow on angiography, there were two complications. One patient developed acute coronary syndrome and the other *lentigo maligna*. Ripa *et al.*[51] investigated three groups of 16 "no-option" patients. One group received intramyocardial injection of plasmid VEGF-A^{165} followed a week later by six days of G-CSF, one group received just intramyocardial injection of plasmid VEGF-A^{165} and the last group received injection of placebo. Although combined therapy resulted in a ten-fold increase in circulating CD34+ cells there was no improvements in perfusion, LVEF, exercise time, CCS classification, angina frequency or GTN use. Whether higher doses of VEGF or increasing the time of expression over one week would lead to benefit remain unanswered questions.

Although initial clinical results appear consistently encouraging, they must be interpreted with a certain degree of caution as to date the majority of clinical trials in the no-option patient population have been uncontrolled and involved small patient numbers. Importantly, however, no adverse events related to the administration of bone-marrow derived cells have been reported.

5. Which is the Optimal Method of Delivery?

No matter how good the cell, if it is not delivered to the target organ then therapy is unlikely to be successful. A number of different delivery routes have been used in the clinical trials published to date. Each method has advantages and disadvantages and each may lend itself more readily to a different patient population. For chronic patients with extensive ischaemia the routes used have been split between epicardial,[28,29,33,36,37,41,44] endocardial,[27,30,32,38–40,51] and intracoronary[31,35,42,43,46–50] injections and studies are currently underway with retrograde venous delivery at our institution.

Epicardial delivery has been used in patients with chronic disease and is most likely to guarantee delivery, as injections can be directly visualised and allows delivery of large cells such as SM, which cannot be delivered via the intracoronary route. In addition it may avoid the obstacle of microvascular obstruction which has been reported to have an incidence of up to 50% in patients post-AMI[17] and can be a significant obstacle to intracoronary delivery. Clearly, however, as sole therapy it precludes the inclusion of a control group because of ethical constraints. The alternative of delivering stem cells at the time of a planned CABG, allows controls to be included, but is bedeviled by the large proven confounding factor of the surgical revascularisation. The injections themselves may also be pro-angiogenic and, depending on volumes used, may also lead to local myocyte damage.

Intracoronary delivery has been used as the delivery route of choice in patients post-AMI. Although a number of the open-label studies reported positive results[13,15,16,20,21] larger placebo-controlled trials has been variable with both positive[18] and negative outcomes.[17] This follows a pattern seen in early Phase I angiogenesis gene therapy clinical trials, reporting positive results not borne out by larger placebo-controlled Phase II studies. One explanation for these negative results is that the rapid first passage through the coronary artery system provides a very short contact time between the product and the target cells and therefore reduces efficiency of delivery and retention. It is likely that to be effective cells must exit the vasculature into the myocardial interstitium. Goussetis *et al.*[31] performed an interesting pilot study where they labelled CD133+, CD133− and CD34+ cells with 99 m Tc-hexamethylpropylenamineoxime and infused these into the infarct-related artery of eight patients. Scintigraphic images at one and 24 hours post-delivery showed uptake of 9% and 7% of the delivered cells respectively. However other clinical data suggests that after intracoronary delivery only approximately 2% of infused BMC are retention in the heart.[62] The timing of delivery may also be critical after an AMI, the resultant secretion of growth factors such as VEGF and other cytokines causing vessels to become more leaky. This in turn might allow cells to cross the endothelial barrier more easily, and may explain homing of bone marrow cells to areas of infarction. However, the optimal time interval between an ischaemic event and cell delivery has not been clarified. As noted earlier, in the REPAIR AMI study it was found that patients receiving cells five to seven days after MI had a greater improvement in LVEF than in those patients that received cells at an earlier time point. It has been suggested that the milieu and inflammation immediately post-AMI may adversely affect cell homing, retention and survival within the myocardium. However in "no-option patients", two obvious differences exist in comparison to the acute setting. By definition these patients have inoperable coronary artery disease and it is, therefore, unlikely that their coronary arteries will be sufficiently patent to allow passage of an angioplasty catheter for delivery. Secondly, in this chronic setting the cytokine milieu is unlikely to be as active, and the vessels are consequently less likely to be permeable to cells. We, therefore, consider this an unlikely route for successful cell transfer in these patients.

Endocardial injections have also been utilised in association with electromagnetic mapping of the left ventricle. This technique has the theoretical advantages of targeting ischaemic, but viable myocardium, but delivery of cells is limited to a small radius around each needle track. The delivery technique is technically challenging, and unless the needle is exactly perpendicular to the endocardial surface there is likely to be a loss of cells both into the left ventricle and via backflow along the needle path. Whilst endocardial injections have been utilised in the clinical trial setting, their efficacy is far from assured.

Two methodologies have been described for delivery of products via the coronary venous system. Firstly low pressure delivery, which aims to increase contact time, with the vessels, without endothelial disruption of the vein.[63] Secondly high pressure delivery which causes disruption of the endothelial tight junctions so that cells are mechanically driven across from the venocapillary vasculature into the myocardial interstium resulting in a biological reservoir.[63] Both techniques involve catheterisation of the subclavian or femoral artery, cannulation of the coronary sinus, and then of a sub-selective coronary vein. Preclinical data suggest that the retrograde coronary venous approach is at least as efficient as endocardial injection[64,65] and covers a larger area of the myocardium, appealing in the patient population presenting with global ischaemia. Murad-Netto *et al.*[63] published the first case report of autologous bone marrow CD34+ cell infusion via the coronary veins in a patient 14 days after acute myocardial infarction. No complications were noted and the patient was discharged 48 hours after treatment. We are currently undertaking a randomised placebo-controlled clinical trial of autologous unselected bone marrow using this technique in end-stage ischaemic heart disease.

6. The Future for Clinical Trials

At the time of writing there have been a number of different cell types, preparation and doses delivered by a number of different routes to a variety of patients in small, mostly uncontrolled trials, generally with positive outcomes. However, many of the larger, controlled studies have been negative, a similar situation to that experienced by the angiogenesis gene therapy field a number of years ago. It is imperative that we learn lessons from the past, if the challenges for the future are to be overcome. Inadequately designed angiogenesis studies substantially set back progress in arguably what may have been, and may still be, a promising field.

Thus, an important factor that emerged from double-blind placebo-controlled angiogenesis gene therapy studies was the large placebo effect, the extent and prevalence of which surpassed expectations. Importantly, this effect was not limited to "soft" endpoints such as angina frequency, medication use or exercise capacity, but was also seen in objective measures of perfusion and function such as MRI and SPECT. The latter are likely related to the enhanced will to exercise created by the placebo effect. This strongly suggests that evaluation of efficacy is only possible in a double-blind trial format, and yet slightly disappointingly the majority of cell-based studies published to date remain open label whilst still reporting on efficacy.

Some may argue the move from bench to bedside has been premature, without adequate knowledge of the optimal cell type, method of handling and isolation and basic mechanisms of action. Further unanswered questions include the optimal patient population, timing related to the myocardial event, dose, optimal delivery route and appropriate assays and time points to detect clinically relevant advances. Although we are sympathetic to this view, we suggest that much will be learnt by clinical studies proceeding in parallel with the necessary preclinical experiments. Certain answers can only be gained in the clinic, including (i) optimal delivery route, (ii) assay optimisation and (iii) assessment of safety. An example of the latter is the electrical instability exhibited with SM in man[10] that was not detected in animal models.

There is no clear indication of the most suitable patient population, or timing of delivery. However, "no-option" patients with a poor prognosis have the best risk-benefit ratio and sub-group analysis, suggesting that the greatest benefit is seen in patients with the lowest baseline LVEF.[18] The effects of cell therapy seem to be restricted to the infarct-related territory, which may already be receiving adequate early revascularisation.[16] Thus, based on the evidence to date, patients with persistently lower LVEF several days after an AMI infarct may gain the most benefit, and delivery of cells may be optimally administered at slightly later time points than initially thought.[18]

Another key question is what is the optimal cell type? At present there appears to be no clear steer on this question. The balance lies between delivering a highly enriched population of a particular favoured cell type, with the risk this is not the one relevant to myocardial benefit, or a heterogeneous population for which the risk is that the key cell type is too diluted. A further likely relevant factor is that many in the field believe that even if a subpopulation can be defined, these cells may depend on a "soup" of growth factors secreted by other bone marrow cells. Further research is clearly needed to elucidate the mechanisms by which cells may impart their benefit, and in turn identifying which is the optimal cell type. Little data are available on how long the effect of stem cells needs to be maintained.

Thus, repeated delivery may offer additional benefits. Extended follow-up will be important in future trials, to assess whether benefits seen can be maintained; initial improvements in LVEF were often not seen at later follow-up.[27,49,54] Advances in cell labelling and tracking may also assist in optimising delivery and retention within the myocardium.

Clearly not all of these questions can be answered in the clinical trial setting and it is imperative that basic science and translational research are conducted simultaneously to ascertain the mechanisms of action, to elucidate further the optimal cell type, dose, delivery and timing of administration, and help guide clinical research. Whilst it is important that the field moves on enthusiastically, it is again worth harking back to the angiogenesis gene therapy studies. In general, these were poorly conducted, without knowledge of the most basic variables such as dose, delivery route and duration of action. The resulting, predictably unfavourable results, dampened enthusiasm for what may be just as, or a more useful approach, than stem cells. In order to avoid the same pitfall, there is the obvious need for clearly focused translational preclinical research running in parallel with, and well coordinated with, randomised, placebo-controlled trials. Finally, it is unlikely that this will be a "quick-fix"; the old adage about 20 years from bench to bedside will likely represent a minimum in this field, and patience will be a key virtue for both the community and the funders if this approach is to succeed.

References

1. Bleumink, G.S., Knetsch, A.M., Sturkenboom, M.C., Straus, S.M., Hofman, A., Deckers, J.W., Witteman, J.C. and Stricker, B.H. Quantifying the heart failure epidemic: prevalence, incidence rate, lifetime risk, prognosis of heart failure. The Rotterdam Study. *Eur. Heart J.* **25**(18), 1614–1619 (2004).
2. Mackay, J. and Mensah, G.A. *The Atlas of Heart Disease, Stroke*, 1st Ed. (World Health Organization, Geneva, 2004), pp. 1–112.
3. Gheorghiade, M. and Bonow, R.O. Chronic heart failure in the United States: a manifestation of coronary artery disease. *Circulation* **97**(3), 282–289 (1998).
4. Anversa, P. and Nadal-Ginard, B. Myocyte renewal and ventricular remodelling. *Nature* **415**(6868), 240–243 (2002).
5. Odorico, J., Zhang, S. and Pedersen, R. *Human Embryonic Stem Cells* (Bios Scientific Publishers, 2005).
6. Bertolini, F. and Pruneri, G. Chimerism of the transplanted heart. *N. Engl. J. Med.* **346**(18), 1410–1412 (2002).
7. Kocher, A.A., Schlechta, B., Gasparovicova, A., Wolner, E., Bonaros, N. and Laufer, G. Stem cells and cardiac regeneration. *Transplant. Int.* **20**(9), 731–746 (2007).

8. Lenzen, M.J., Boersma, E., Bertrand, M.E., Maier, W., Moris, C., Piscione, F., Sechtem, U., Stahle, E., Widimsky, P., de Jaegere, P., Scholte op Reimer, W.J., Mercado, N. and Wijns, W. European Society of Cardiology Management and outcome of patients with established coronary artery disease: the Euro Heart Survey on coronary revascularization. *Eur. Heart J.* **26**(12), 1169–1179 (2005).
9. Orlic, D., Kajstura, J., Chimenti, S., Jakoniuk, I., Anderson, S.M., Li, B., Pickel, J., McKay, R., Nadal-Ginard, B., Bodine, D.M., Leri, A. and Anversa, P. Bone marrow cells regenerate infarcted myocardium. *Nature* **410**(6829), 701–705 (2001).
10. Menasché, P., Hagège, A.A., Vilquin, J.T., Desnos, M., Abergel, E., Pouzet, B., Bel, A., Sarateanu, S., Scorsin, M., Schwartz, K., Bruneval, P., Benbunan, M., Marolleau, J.P. and Duboc, D. Autologous skeletal myoblast transplantation for severe postinfarction left ventricular dysfunction. *J. Am. Coll. Cardiol.* **41**(7), 1078–1083 (2003).
11. Siminiak, T., Kalawski, R., Fiszer, D., Jerzykowska, O., Rzeźniczak, J., Rozwadowska, N. and Kurpisz, M. Autologous skeletal myoblast transplantation for the treatment of postinfarction myocardial injury: phase I clinical study with 12 months of follow-up. *Am. Heart J.* **148**(3), 531–537 (2004).
12. Siminiak, T., Fiszer, D., Jerzykowska, O., Grygielska, B., Rozwadowska, N., Kałmucki, P. and Kurpisz, M. Percutaneous trans-coronary-venous transplantation of autologous skeletal myoblasts in the treatment of post-infarction myocardial contractility impairment: the POZNAN trial. *Eur. Heart J.* **26**(12), 1188–1195 (2005).
13. Strauer, B.E., Brehm, M., Zeus, T., Köstering, M., Hernandez, A., Sorg, R.V., Kögler, G. and Wernet, P. Repair of infarcted myocardium by autologous intracoronary mononuclear bone marrow cell transplantation in humans. *Circulation* **106**(15), 1913–1918 (2002).
14. Fernández-Avilés, F., San Román, J.A., García-Frade, J., Fernández, M.E., Peñarrubia, M.J., de la Fuente, L., Gómez-Bueno, M., Cantalapiedra, A., Fernández, J., Gutierrez, O., Sánchez, P.L., Hernández, C., Sanz, R., García-Sancho, J. and Sánchez, A. Experimental and clinical regenerative capability of human bone marrow cells after myocardial infarction. *Circ. Res.* **95**(7), 742–748 (2004).
15. Schächinger, V., Assmus, B., Britten, M.B., Honold, J., Lehmann, R., Teupe, C., Abolmaali, N.D., Vogl, T.J., Hofmann, W.K., Martin, H., Dimmeler, S. and Zeiher, A.M. Transplantation of progenitor cells and regeneration enhancement in acute myocardial infarction: final one-year results of the TOPCARE-AMI Trial. *J. Am. Coll. Cardiol.* **44**(8), 1690–1699 (2004).

16. Wollert, K.C., Meyer, G.P., Lotz, J., Ringes-Lichtenberg, S., Lippolt, P., Breidenbach, C., Fichtner, S., Korte, T., Hornig, B., Messinger, D., Arseniev, L., Hertenstein, B., Ganser, A. and Drexler, H. Intracoronary autologous bone-marrow cell transfer after myocardial infarction: the BOOST randomised controlled clinical trial. *Lancet* **364**(9429), 141–148 (2004).
17. Janssens, S., Dubois, C., Bogaert, J., Theunissen, K., Deroose, C., Desmet, W., Kalantzi, M., Herbots, L., Sinnaeve, P., Dens, J., Maertens, J., Rademakers, F., Dymarkowski, S., Gheysens, O., Van Cleemput, J., Bormans, G., Nuyts, J., Belmans, A., Mortelmans, L., Boogaerts, M. and Van de Werf, F. Autologous bone marrow-derived stem-cell transfer in patients with ST-segment elevation myocardial infarction: double-blind, randomised controlled trial. *Lancet* **367**(9505), 113–121 (2006).
18. Schächinger, V., Erbs, S., Elsässer, A., Haberbosch, W., Hambrecht, R., Hölschermann, H., Yu, J., Corti, R., Mathey, D.G., Hamm, C.W., Süselbeck, T., Werner, N., Haase, J., Neuzner, J., Germing, A., Mark, B., Assmus, B., Tonn, T., Dimmeler, S., Zeiher, A.M. and REPAIR-AMI Investigators. Improved clinical outcome after intracoronary administration of bone-marrow-derived progenitor cells in acute myocardial infarction: final 1-year results of the REPAIR-AMI trial. *Eur. Heart J.* **27**(23), 2775–2783 (2006).
19. Lunde, K., Solheim, S., Aakhus, S., Arnesen, H., Abdelnoor, M., Egeland, T., Endresen, K., Ilebekk, A., Mangschau, A., Fjeld, J.G., Smith, H.J., Taraldsrud, E., Grøgaard, H.K., Bjørnerheim, R., Brekke, M., Müller, C., Hopp, E., Ragnarsson, A., Brinchmann, J.E. and Forfang, K. Intracoronary injection of mononuclear bone marrow cells in acute myocardial infarction. *N. Engl. J. Med.* **355**(12), 1199–1209 (2006).
20. Ge, J., Li, Y., Qian, J., Shi, J., Wang, Q., Niu, Y., Fan, B., Liu, X., Zhang, S., Sun, A. and Zou, Y. Efficacy of emergent transcatheter transplantation of stem cells for treatment of acute myocardial infarction (TCT-STAMI). *Heart* **92**(12), 1764–1767 (2006).
21. Meluzín, J., Janoušek, S., Mayer, J., Groch, L., Horňáček, I., Hlinomaz, O., Kala, P., Panovský, R., Prášek, J., Kamínek, M., Staníček, J., Klabusay, M., Kořístek, Z., Navrátil, M., Dušek, L. and Vinklárková, J. Three-, 6-, and 12-month results of autologous transplantation of mononuclear bone marrow cells in patients with acute myocardial infarction. *Int. J. Cardiol.*, 29 August 2007 [Epub ahead of print].
22. Ince, H., Petzsch, M., Kleine, H.D., Schmidt, H., Rehders, T., Körber, T., Schümichen, C., Freund, M. and Nienaber, C.A. Preservation from left ventricular remodeling by front-integrated revascularization and stem cell liberation in evolving acute myocardial infarction by use of

granulocyte-colony-stimulating factor (FIRSTLINE-AMI). *Circulation* **112**(20), 3097–3106 (2005).

23. Kuethe, F., Figulla, H.R., Herzau, M., Voth, M., Fritzenwanger, M., Opfermann, T., Pachmann, K., Krack, A., Sayer, H.G., Gottschild, D. and Werner, G.S. Treatment with granulocyte colony-stimulating factor for mobilization of bone marrow cells in patients with acute myocardial infarction. *Am. Heart J.* **150**(1), 115 (2005).
24. Zohlnhöfer, D., Ott, I., Mehilli, J., Schömig, K., Michalk, F., Ibrahim, T., Meisetschläger, G., von Wedel, J., Bollwein, H., Seyfarth, M., Dirschinger, J., Schmitt, C., Schwaiger, M., Kastrati, A., Schömig, A. and REVIVAL-2 Investigators. Stem cell mobilization by granulocyte colony-stimulating factor in patients with acute myocardial infarction: a randomized controlled trial. *J. Am. Med. Assoc.* **295**(9), 1003–1010 (2006).
25. Ripa, R.S., Jørgensen, E., Wang, Y., Thune, J.J., Nilsson, J.C., Søndergaard, L., Johnsen, H.E., Køber, L., Grande, P. and Kastrup, J. Stem cell mobilization induced by subcutaneous granulocyte-colony stimulating factor to improve cardiac regeneration after acute ST-elevation myocardial infarction: result of the double-blind, randomized, placebo-controlled stem cells in myocardial infarction (STEMMI) trial. *Circulation* **113**(16), 1983–1992 (2006).
26. Engelmann, M.G., Theiss, H.D., Hennig-Theiss, C., Huber, A., Wintersperger, B.J., Werle-Ruedinger, A.E., Schoenberg, S.O., Steinbeck, G. and Franz, W.M. Autologous bone marrow stem cell mobilization induced by granulocyte colony-stimulating factor after subacute ST-segment elevation myocardial infarction undergoing late revascularization: final results from the G-CSF-STEMI (Granulocyte Colony-Stimulating Factor ST-Segment Elevation Myocardial Infarction) trial. *J. Am. Coll. Cardiol.* **48**(8), 1712–1721 (2006).
27. Smits, P.C., van Geuns, R.J., Poldermans, D., Bountioukos, M., Onderwater, E.E., Lee, C.H., Maat, A.P. and Serruys, P.W. Catheter-based intramyocardial injection of autologous skeletal myoblasts as a primary treatment of ischemic heart failure: clinical experience with six-month follow-up. *J. Am. Coll. Cardiol.* **42**(12), 2063–2069 (2003).
28. Pagani, F.D., DerSimonian, H., Zawadzka, A., Wetzel, K., Edge, A.S., Jacoby, D.B., Dinsmore, J.H., Wright, S., Aretz, T.H., Eisen, H.J. and Aaronson, K.D. Autologous skeletal myoblasts transplanted to ischemia-damaged myocardium in humans. Histological analysis of cell survival and differentiation. *J. Am. Coll. Cardiol.* **41**(5), 879–888 (2003).
29. Patel, A.N., Geffner, L., Vina, R.F., Saslavsky, J., Urschel, H.C. Jr., Kormos, R. and Benetti, F. Surgical treatment for congestive heart failure

with autologous adult stem cell transplantation: a prospective randomized study. *J. Thorac. Cardiovasc. Surg.* **130**(6), 1631–1638 (2005).

30. De la Fuente, L.M., Stertzer, S.H., Argentieri, J., Peñaloza, E., Miano, J., Koziner, B., Bilos, C. and Altman, P.A. Transendocardial autologous bone marrow in chronic myocardial infarction using a helical needle catheter: 1-year follow-up in an open-label, nonrandomized, single-center pilot study (the TABMMI study). *Am. Heart J.* **154**(1), 79.e1–7 (2007).
31. Goussetis, E., Manginas, A., Koutelou, M., Peristeri, I., Theodosaki, M., Kollaros, N., Leontiadis, E., Theodorakos, A., Paterakis, G., Karatasakis, G., Cokkinos, D.V. and Graphakos, S. Intracoronary infusion of CD133+ and CD133–CD34+ selected autologous bone marrow progenitor cells in patients with chronic ischemic cardiomyopathy: cell isolation, adherence to the infarcted area, and body distribution. *Stem Cells* **24**(10), 2279–2283 (2006).
32. Losordo, D.W., Schatz, R.A., White, C.J., Udelson, J.E., Veereshwarayya, V., Durgin, M., Poh, K.K., Weinstein, R., Kearney, M., Chaudhry, M., Burg, A., Eaton, L., Heyd, L., Thorne, T., Shturman, L., Hoffmeister, P., Story, K., Zak, V., Dowling, D., Traverse, J.H., Olson, R.E., Flanagan, J., Sodano, D., Murayama, T., Kawamoto, A. and Kusano, K.F. Intramyocardial transplantation of autologous CD34+ stem cells for intractable angina: a phase I/IIa double-blind, randomized controlled trial. *Circulation* **115**(25), 3165–3172 (2007).
33. Klein, H.M., Ghodsizad, A., Marktanner, R., Poll, L., Voelkel, T., Mohammad Hasani, M.R., Piechaczek, C., Feifel, N., Stockschlaeder, M., Burchardt, E.R., Kar, B.J., Gregoric, I. and Gams, E. Intramyocardial implantation of CD133+ stem cells improved cardiac function without bypass surgery. *Heart Surg. Forum* **10**(1), E66–9 (2007).
34. Hüttmann, A., Dührsen, U., Stypmann, J., Noppeney, R., Nückel, H., Neumann, T., Gutersohn, A., Nikol, S. and Erbel, R. Granulocyte colony-stimulating factor-induced blood stem cell mobilisation in patients with chronic heart failure — feasibility, safety and effects on exercise tolerance and cardiac function. *Basic Res. Cardiol.* **101**(1), 78–86 (2006).
35. Manginas, A., Goussetis, E., Koutelou, M., Karatasakis, G., Peristeri, I., Theodorakos, A., Leontiadis, E., Plessas, N., Theodosaki, M., Graphakos, S. and Cokkinos, D.V. Pilot study to evaluate the safety and feasibility of intracoronary CD133(+) and CD133(–) CD34(+) cell therapy in patients with nonviable anterior myocardial infarction. *Catheter Cardiovasc. Interv.* **69**(6), 773–781 (2007).
36. Stamm, C., Westphal, B., Kleine, H.D., Petzsch, M., Kittner, C., Klinge, H., Schümichen, C., Nienaber, C.A., Freund, M. and Steinhoff, G. Autologous

bone-marrow stem-cell transplantation for myocardial regeneration. *Lancet* **361**(9351), 45–46 (2003).

37. Stamm, C., Kleine, H.D., Choi, Y.H., Dunkelmann, S., Lauffs, J.A., Lorenzen, B., David, A., Liebold, A., Nienaber, C., Zurakowski, D., Freund, M. and Steinhoff, G. Intramyocardial delivery of CD133+ bone marrow cells and coronary artery bypass grafting for chronic ischemic heart disease: safety and efficacy studies. *J. Thorac. Cardiovasc Surg.* **133**(3), 717–725 (2007).
38. Fuchs, S., Satler, L.F., Kornowski, R., Okubagzi, P., Weisz, G., Baffour, R., Waksman, R., Weissman, N.J., Cerqueira, M., Leon, M.B. and Epstein, S.E. Catheter-based autologous bone marrow myocardial injection in no-option patients with advanced coronary artery disease: a feasibility study. *J. Am. Coll. Cardiol.* **41**(10), 1721–1724 (2003).
39. Tse, H.F., Kwong, Y.L., Chan, J.K., Lo, G., Ho, C.L. and Lau, C.P. Angiogenesis in ischaemic myocardium by intramyocardial autologous bone marrow mononuclear cell implantation. *Lancet* **361**(9351), 47–49 (2003).
40. Perin, E.C., Dohmann, H.F., Borojevic, R., Silva, S.A., Sousa, A.L., Mesquita, C.T., Rossi, M.I., Carvalho, A.C., Dutra, H.S., Dohmann, H.J., Silva, G.V., Belém, L., Vivacqua, R., Rangel, F.O., Esporcatte, R., Geng, Y.J., Vaughn, W.K., Assad, J.A., Mesquita, E.T. and Willerson, J.T. Transendocardial, autologous bone marrow cell transplantation for severe, chronic ischemic heart failure. *Circulation* **107**(18), 2294–2302 (2003).
41. Mocini, D., Staibano, M., Mele, L., Giannantoni, P., Menichella, G., Colivicchi, F., Sordini, P., Salera, P., Tubaro, M. and Santini, M. Autologous bone marrow mononuclear cell transplantation in patients undergoing coronary artery bypass grafting. *Am. Heart J.* **151**(1), 192–197 (2006).
42. Strauer, B.E., Brehm, M., Zeus, T., Bartsch, T., Schannwell, C., Antke, C., Sorg, R.V., Kögler, G., Wernet, P., Müller, H.W. andKöstering, M. Regeneration of human infarcted heart muscle by intracoronary autologous bone marrow cell transplantation in chronic coronary artery disease: the IACT Study. *J. Am. Coll. Cardiol.* **46**(9), 1651–1658 (2005).
43. Assmus, B., Honold, J., Schächinger, V., Britten, M.B., Fischer-Rasokat, U., Lehmann, R., Teupe, C., Pistorius, K., Martin, H., Abolmaali, N.D., Tonn, T., Dimmeler, S. and Zeiher, A.M. Transcoronary transplantation of progenitor cells after myocardial infarction. *N. Engl. J. Med.* **355**(12), 1222–1232 (2006).
44. Archundia, A., Aceves, J.L., López-Hernández, M., Alvarado, M., Rodriguez, E., Díaz Quiroz, G., Páez, A., Rojas, F.M., Montaño, L.F., Wollins, J., Welt, F., Shah, P., Soukas, P., Asahara, T. and Henry, T.D. Direct cardiac injection of G-CSF mobilized bone-marrow stem-cells improves

ventricular function in old myocardial infarction. *Life Sci.* **78**(3), 279–283 (2005).

45. Wang, Y., Tägil, K., Ripa, R.S., Nilsson, J.C., Carstensen, S., Jørgensen, E., Søndergaard, L., Hesse, B., Johnsen, H.E. and Kastrup, J. Effect of mobilization of bone marrow stem cells by granulocyte colony stimulating factor on clinical symptoms, left ventricular perfusion and function in patients with severe chronic ischemic heart disease. *Int. J. Cardiol.* **100**(3), 477–483 (2005).
46. Boyle, A.J., Whitbourn, R., Schlicht, S., Krum, H., Kocher, A., Nandurkar, H., Bergmann, S., Daniell, M., O'Day, J., Skerrett, D., Haylock, D., Gilbert, R.E. and Itescu, S. Intra-coronary high-dose CD34+ stem cells in patients with chronic ischemic heart disease: a 12-month follow-up. *Int. J. Cardiol.* **109**(1), 21–27 (2006).
47. Erbs, S., Linke, A., Adams, V., Lenk, K., Thiele, H., Diederich, K.W., Emmrich, F., Kluge, R., Kendziorra, K., Sabri, O., Schuler, G. and Hambrecht, R. Transplantation of blood-derived progenitor cells after recanalization of chronic coronary artery occlusion: first randomized and placebo-controlled study. *Circ. Res.* **97**(8), 756–762 (2005).
48. Kang, H.J., Lee, H.Y., Na, S.H., Chang, S.A., Park, K.W., Kim, H.K., Kim, S.Y., Chang, H.J., Lee, W., Kang, W.J., Koo, B.K., Kim, Y.J., Lee, D.S., Sohn, D.W., Han, K.S., Oh, B.H., Park, Y.B. and Kim, H.S. Differential effect of intracoronary infusion of mobilized peripheral blood stem cells by granulocyte colony-stimulating factor on left ventricular function and remodeling in patients with acute myocardial infarction versus old myocardial infarction: the MAGIC Cell-3-DES randomized, controlled trial. *Circulation* **114**(1 Suppl.), I145–51 (2006).
49. Kang, H.J., Kim, H.S., Koo, B.K., Kim, Y.J., Lee, D., Sohn, D.W., Oh, B.H. and Park, Y.B. Intracoronary infusion of the mobilized peripheral blood stem cell by G-CSF is better than mobilization alone by G-CSF for improvement of cardiac function and remodeling: 2-year follow-up results of the Myocardial Regeneration and Angiogenesis in Myocardial Infarction with G-CSF and Intra-Coronary Stem Cell Infusion (MAGIC Cell) 1 trial. *Am. Heart J.* **153**(2), 237.e1–8 (2007).
50. Kang, H.J., Kim, Y.S., Koo, B.K., Park, K.W., Lee, H.Y., Sohn, D.W., Oh, B.H., Park, Y.B. and Kim, H.S. Effects of stem cell therapy with G-CSF on coronary artery after drug-eluting stent implantation in patients with acute myocardial infarction. *Heart*, 10 December 2007 [Epub ahead of print].
51. Ripa, R.S., Wang, Y., Jørgensen, E., Johnsen, H.E., Hesse, B. and Kastrup, J. Intramyocardial injection of vascular endothelial growth factor-A165 plasmid followed by granulocyte-colony stimulating factor to induce angiogenesis

in patients with severe chronic ischaemic heart disease. *Eur. Heart J.* **27**(15), 1785–1792 (2006).
52. Hagège, A.A., Marolleau, J.P., Vilquin, J.T., Alhéritière, A., Peyrard, S., Duboc, D., Abergel, E., Messas, E., Mousseaux, E., Schwartz, K., Desnos, M. and Menasché, P. Skeletal myoblast transplantation in ischemic heart failure: long-term follow-up of the first phase I cohort of patients. *Circulation* **114** (1 Suppl.), I108–13 (2006).
53. Meyer, G.P., Wollert, K.C., Lotz, J., Steffens, J., Lippolt, P., Fichtner, S., Hecker, H., Schaefer, A., Arseniev, L., Hertenstein, B., Ganser, A. and Drexler, H. Intracoronary bone marrow cell transfer after myocardial infarction: eighteen months' follow-up data from the randomized, controlled BOOST (BOne marrOw transfer to enhance ST-elevation infarct regeneration) trial. *Circulation* **113**(10), 1287–1294 (2006).
54. Henry, T.D., Annex, B.H., McKendall, G.R., Azrin, M.A., Lopez, J.J., Giordano, F.J., Shah, P.K., Willerson, J.T., Benza, R.L., Berman, D.S., Gibson, C.M., Bajamonde, A., Rundle, A.C., Fine, J., McCluskey, E.R. and VIVA Investigators. The VIVA trial: vascular endothelial growth factor in ischemia for vascular angiogenesis. *Circulation* **107**(10), 1359–1365 (2003).
55. Simons, M., Annex, B.H., Laham, R.J., Kleiman, N., Henry, T., Dauerman, H., Udelson, J.E., Gervino, E.V., Pike, M., Whitehouse, M.J., Moon, T. and Chronos, N.A. Pharmacological treatment of coronary artery disease with recombinant fibroblast growth factor-2: double-blind, randomized, controlled clinical trial. *Circulation* **105**(7), 788–793 (2002).
56. Cintron, G. *et al.* Prognostic significance of serial changes in left ventricular ejection fraction in patients with congestive heart failure. The V-HeFT VA Cooperative Studies Group. *Circulation* **87**(Suppl. 6), VI17–VI23.
57. Leone, A.M., Rutella, S., Bonanno, G., Contemi, A.M., de Ritis, D.G., Giannico, M.B., Rebuzzi, A.G., Leone, G. and Crea, F. Endogenous G-CSF and CD34+ cell mobilization after acute myocardial infarction. *Int. J. Cardiol.* **111**(2), 202–208 (2006).
58. Numaguchi, Y., Sone, T., Okumura, K., Ishii, M., Morita, Y., Kubota, R., Yokouchi, K., Imai, H., Harada, M., Osanai, H., Kondo, T. and Murohara, T. The impact of the capability of circulating progenitor cell to differentiate on myocardial salvage in patients with primary acute myocardial infarction. *Circulation* **114**(1 Suppl.), I114–9 (2006).
59. Kang, H.J., Kim, H.S., Zhang, S.Y., Park, K.W., Cho, H.J., Koo, B.K., Kim, Y.J., Soo Lee, D., Sohn, D.W., Han, K.S., Oh, B.H., Lee, M.M. and Park, Y.B. Effects of intracoronary infusion of peripheral blood stem-cells mobilised with granulocyte-colony stimulating factor on left ventricular systolic function and restenosis after coronary stenting in myocardial

infarction: the MAGIC cell randomised clinical trial. *Lancet* **363**(9411), 751–756 (2004).
60. Valgimigli, M., Rigolin, G.M., Cittanti, C., Malagutti, P., Curello, S., Percoco, G., Bugli, A.M., Della Porta, M., Bragotti, L.Z., Ansani, L., Mauro, E., Lanfranchi, A., Giganti, M., Feggi, L., Castoldi, G. and Ferrari, R. Use of granulocyte-colony stimulating factor during acute myocardial infarction to enhance bone marrow stem cell mobilization in humans: clinical and angiographic safety profile. *Eur. Heart J.* **26**(18), 1838–1845 (2005).
61. Cleland, J.G., Coletta, A.P., Abdellah, A.T., Nasir, M., Hobson, N., Freemantle, N. and Clark, A.L. Clinical trials update from the American Heart Association 2006: OAT, SALT 1 and 2, MAGIC, ABCD, PABA-CHF, IMPROVE-CHF, and percutaneous mitral annuloplasty. *Eur. J. Heart Fail.* **9**(1), 92–97 (2007).
62. Hofmann, M., Wollert, K.C., Meyer, G.P., Menke, A., Arseniev, L., Hertenstein, B., Ganser, A., Knapp, W.H. and Drexler, H. Monitoring of bone marrow cell homing into the infracted human myocardium. *Circulation* **111**(17), 2198–2202 (2005).
63 Murad-Netto, S., Moura, R., Romeo, L.J., Manoel Neto, A., Duarte, N., Barreto, F., Jensen, A., Viña, R.F., Vraslovik, F., Oberdan, A., Benetti, F., Saslavsky, J., Viña, M.F. and Amino, J.G. Stem cell therapy with retrograde coronary perfusion in acute myocardial infarction. A new technique. *Arq. Bras. Cardiol.* **83**(4), 352–354, 349–351 (2004).
64. Hou, D., Maclaughlin, F., Thiesse, M., Panchal, V.R., Bekkers, B.C., Wilson, E.A., Rogers, P.I., Coleman, M.C. and March, K.L. Widespread regional myocardial transfection by plasmid encoding Del-1 following retrograde coronary venous delivery. *Catheter Cardiovasc. Interv.* **58**(2), 207–211 (2003).
65. Fearon, W.F., Ikeno, F., Bailey, L.R., Hiatt, B.L., Herity, N.A., Carter, A.J., Fitzgerald, P.J., Rezaee, M., Yeung, A.C. and Yock, P.G. Evaluation of high-pressure retrograde coronary venous delivery of FGF-2 protein. *Catheter Cardiovasc. Interv.* **61**(3), 422–428 (2004).

Chapter 34

Myocardial Recovery Following LVAD Support

Robert S. George and Emma J. Birks

Abstract

The use of left ventricular assist devices (LVADs) has been increasing in patients with end-stage heart failure in order to bridge them to transplantation and these devices are now also being used as "destination therapy". Some of these patients have demonstrated signs of recovery of the native heart and this recovery has been sufficient in some cases to allow explantation of the device and hence a new concept of bridge-to-recovery has been described. The rate of successful recovery leading to device explantation, however, has been only 5%–24% until recently when we showed it to be possible to increase the frequency and durability of myocardial recovery using drug therapy combined with LVAD unloading. This chapter summarises different types of LVADs and describes the clinical implications of myocardial recovery following device implantation, monitoring of myocardial recovery and gives the up-to-date findings of the histological and molecular changes in recovery.

Keywords: Left Ventricular Assist Device; Myocardial Recovery; Remodelling; Reverse Remodelling; Quality of Life; Bridge to Recovery; Bridge to Transplantation; Extracellular Matrix; Harefield Bridge-to-Recovery Protocol; Metalloproteinases; Excitation-Contraction Coupling; SERCA2a; Na^+/Ca^{2+} Exchanger.

Outline

1. Introduction
2. Overview of LVADs
3. LVAD and Myocardial Recovery — Clinical Implications
   3.1. Monitoring recovery
   3.2. Improvements in exercise capacity
   3.3. Haemodynamic improvements

3.4. Electrocardiographic improvements
3.5. Quality of life
3.6. Device explantation for recovery
4. Remodelling versus Reverse Remodelling
4.1. The effects of unloading on myocyte size and structural proteins
4.2. Reverse remodelling and calcium regulation
5. Conclusion
References

1. Introduction

Heart failure is a major problem associated with high morbidity and mortality. Medical therapy with angiotensin converting enzymes (ACE) inhibitors, β-blockers, angiotensin-2 inhibitors and aldosterone antagonists together with resynchronisation therapy has improved the survival of many with heart failure, but there remains a large group of patients who, despite optimal medical therapy, are in New York Heart Association (NYHA) Class III/IV heart failure with a very poor prognosis.

Unfortunately the numbers of useable donor hearts available to perform heart transplantation for these patients in advanced heart failure has significantly decreased over recent years and is totally inadequate for the population who require heart transplantation (Fig. 1).[1] The decrease in donors means that an

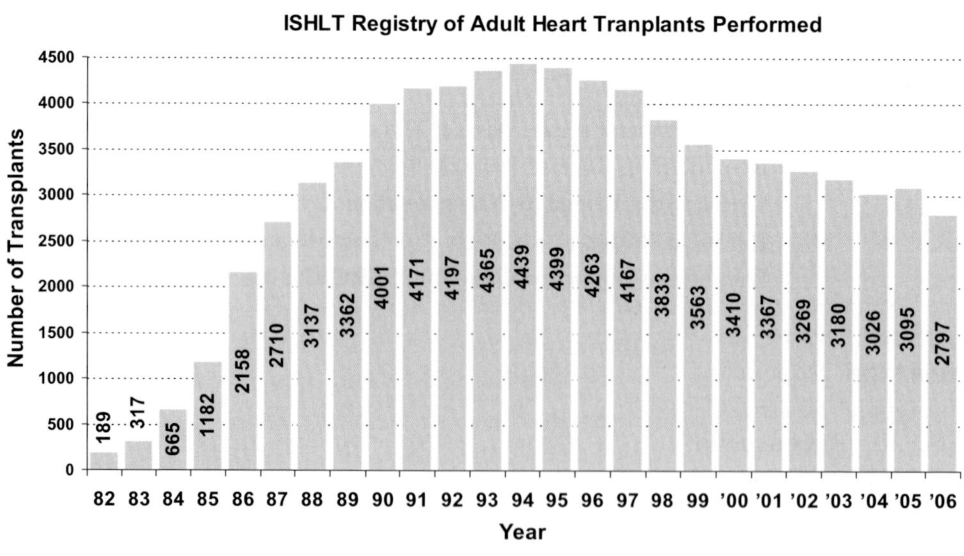

Fig. 1. The ISHLT registry of all adult heart transplantation since 1982 until 31 December 2006.[1]

increasing number of patients are requiring support with a left ventricular assist device (LVAD) for survival when their clinical status deteriorates to bridge them to cardiac transplantation.[2–6]

LVADs are artificial hearts that assist the circulation; they are rapidly evolving and are being inserted into an increasing number of patients with advanced heart failure. LVADs are not only life saving in these deteriorating patients with advanced heart failure that would otherwise die before a donor heart became available, but also improve secondary organ function for transplantation, reduce pulmonary hypertension and allow for improvement of nutritional status.

The future use of these devices, particularly as survival continues to increase, is likely to extend to their wider use as destination therapy, i.e. when the device is inserted lifelong as an alternative to transplantation. The Randomised Evaluation of Mechanical Assistance for the Treatment of Congestive Heart Failure (REMATCH) trial, which randomised patients with advanced NYHA Class IV heart failure to a pulsatile first generation pump or optimal medical therapy, was an influential study that initially demonstrated the efficacy of permanent device therapy.[6,7]

There is also now compelling evidence that with LVAD unloading recovery of the patient's myocardial function can also occur, allowing device removal and avoiding the need for transplantation, immunosuppression and its associated complications and leaving the patient with an excellent quality of life. This also means that the precious resource of a donor organ can be used for another needy individual. This indication, known as "bridge to recovery" (BTR) is a newer and expanding indication and will be the focus of this chapter. Unloading of the left ventricle (LV) using an LVAD for prolonged periods of time can result in regression of dilatation and hypertrophy associated with a series of changes in myocardial gene expression involving genes encoding contractile and structural proteins as well as calcium handling proteins, ion channels and extracellular matrix (ECM) components.

2. Overview of LVADs

Key features of long-term mechanical circulatory support are reliability, durability, biocompatibility, functionality and affordability. Blood-biomaterial interaction, the coagulation system and the immune response are all challenging biological barriers to the development of long term mechanical circulatory support.[8–14] These have been partially defeated by developing and using specific materials and specific blood-contacting corrosion resistance surfaces which have certain structural strength and almost no toxicity levels.[15]

Blood pressure and flow can be generated by positive displacement (pulsatile), rotary (non-pulsatile), or magnetohydrodynamic pumps. The last two types have been developed on the background of large sizes of the pulsatile devices and

Table 1. Descriptions of some of the pulsatile and non-pulsatile devices with their clinical use and their anti-coagulation requirement.

Pump type	Device description	Clinical experience	Reference(s)
Positive Displacement (Pulsatile)			
Thoratec Paracorporeal (PVAD)	Has a 65 ml stroke volume (SV) blood pumping chamber with two mechanical valves. Can produce a beat rate of 40–110 with a flow rate of 1.3–7.2 l/min using alternating positive and negative air pressures.	Approved in USA, Europe, Japan, Canada and Australia for bridge-to-transplant (BTT) and post-cardiotomy recovery. Can be used on either left (LVAD), right (RVAD), or both ventricles as a biventricular assist device (BiVAD). INR to be maintained between 2.5 and 3.5. Until March 2005, 2900 patients were implanted with PVAD. Worldwide, survival from implant to transplantation or recovery was 64.8% with LVAD, 56.6% with BiVAD, and 31.2% with RVAD. Also been used as BTR in 3.8% with LVAD, 5.6% with BiVAD, and 8.3% with RVAD. The five-year transplant-free survival was 77% for this subgroup.	

(*Continued*)

Table 1. *(Continued)*

Pump type	Device description	Clinical experience	Reference(s)
Thoratec Intracorporeal (IVAD)	Implantable version of PVAD, hence: — a smooth polished titanium housing for implantability, — a reduced weight (399 vs. 417 g), — a reduced volume (252 vs. 318 ml), — a narrower percutaneous lead (9 vs. 20 mm). In addition, it has an optical sensor that determines whether the pump is full or empty.	Approved by the FDA in August 2004. INR to be maintained between 2.5 and 3.5. Until March 2005, 86 patients had IVAD of which 63 were for BTT.	
HeartMate XVE — two models: First generation — Implantable Pneumatic (IP) Current — Vented Electric (VE) which is modified to XVE	Made of titanium with polyurethane diaphragm and a pusher-plate actuator. May be powered pneumatically and electrically and operates on two modes: auto and fixed. A porcine valve (25 mm) is suspended in the inflow cannula. Outflow graft is made of 20 mm Dacron. Pumps between 4 and 10 l/min with a maximum SV of 83 ml.	Since its first application in 1986, 5000 have been implanted where 65% experienced a successful outcome as BTT. Anticoagulation is limited to aspirin. REMATCH trial started in 1998 as a multicentre study and compared the outcome of using HeartMate LVAD with optimal medical management. Showed improved one-year survival (52% vs. 25%, $p = 0.002$) and an improved quality of life (QOL).	6, 7, 13, 16–20

(Continued)

Table 1. *(Continued)*

Pump type	Device description	Clinical experience	Reference(s)
	It has a unique blood pumping surface which consists of titanium microspheres and a fibrillar textured surface that promotes pseudointima and resists thrombogenesis, although it has also been shown to be immunologically active.	Also been used as a bridge-to-recovery (BTR) (Harefield BTR Protocol) in association with pharmacological therapy. Explantation rate was 75% and five-year survival exceeded 90%. Since FDA approval in 2002 for destination therapy (DT), 246 patients were implanted of which 67% were discharged home with longest duration of support of 86 days to date.	
Novacor LVAS	Uses an implanted pump drive unit (PDU) with the following characteristics: — symmetrical dual pusher plate, — 70 ml SV, — sac-type blood pump with smooth blood-containing surface coupled to a pulsed-solenoid energy converter drive that drives the blood pump to match the physiological requirements, — two porcine valves, each in the outflow/inflow cannulae that control unidirectional flow of blood.	INR to be maintained between 2.0 and 3.0. Since 1984 to date, more than 1700 have been implanted with the majority for BTT. BTR and DT represent a much smaller experience.	21

(Continued)

Table 1. (*Continued*)

Pump type	Device description	Clinical experience	Reference(s)
Rotary Devices (Continuous Flow Pump)			
Jarvik 2000	This device is placed in the LV with no inlet cannula. No valves, internal compliance chamber, or an externalised vent. Uses a brushless direct-current motor which is contained within the titanium housing and creates electromagnetic force necessary to rotate the impeller. Operates in a range of 8000–12000 rotations per minute (rpm) generating flow of up to 8 l/min. Power consumption is only 3–7 watts and supplied by a 12-volt battery. Power cable can come out abdominally or post-auricular through a skull-mounted pedestal.	INR to be kept 2.5–3.5. Has been used for BTT and destination therapy. Harefield has also used it for BTR purposes (unpublished data).	

(*Continued*)

Table 1. *(Continued)*

Pump type	Device description	Clinical experience	Reference(s)
HeartMate II	Small, lightweight device is designed to accommodate a broad patient population — body surface area (BSA) as low as 1.2 m^2. Flexible inflow and outflow conduits accommodate anatomical changes. Consists of an internal blood pump (a 12 mm diameter straight tube made of titanium alloy) that can produce up to 10 l/min of blood. The tube incorporates the hydraulic components of the pump that include the inlet stator, a pump rotor (incorporates a pump magnet), and an outlet stator. Shares the same hardware platform and system components as the HeartMate XVE LVAD.	INR kept between 2.0 and 3.0 Has been used for BTT and BTR purposes Following the successful completion of its Phase I Pilot Study in 2004, Thoratec was granted approval to begin a Phase II Pivotal Trial in February 2005. The HeartMate II Pivotal Trial is a multi-centre evaluation of the HeartMate II LVAD for advanced-stage heart failure patients. To date, more than 1000 patients have been implanted with the HeartMate II.	20, 22–24
Micromed-Debakey	Weighs 93 g. Can provide up to 10 l/min of blood.	INR to be maintained between 2.5 and 3.5; 326 implanted worldwide for BTT.	

the relatively high incidence of device-related infection and malfunction of the pulsatile pumps. Table 1 lists some of the devices with their brief description and the associated clinical experience.

3. LVAD and Myocardial Recovery — Clinical Implications

In addition to beneficial effects at the cellular and molecular level, unloading can result in significant improvements in haemodynamic, echocardiographic and clinical parameters in heart failure which may be sufficient to allow device explantation. However the success rate of recovery leading to device explanation has been noted in only 5%–24% of patients[25–31] with a number of patients developing failure again or dying within the first two years.[29,31] To improve the incidence of recovery and rate of device explantation, we evolved and used a form of combination therapy Harefield Bridge-to-Recovery Protocol using LVADs combined with drugs known to enhance reverse remodelling followed by the use of the β_2-adrenoceptor agonist, clenbuterol, which has been shown to induce physiological hypertrophy and improve ventricular function.[20,32–36] Using this therapy patients moved from advanced NYHA Class IV heart failure to Class I and reversal of end-stage heart failure secondary to idiopathic dilated cardiomyopathy was achieved in more than 75% of patients with a five-year survival exceeding 90%.[20]

3.1. Monitoring recovery

Recovery critically depends on a safe, accurate and reproducible method of monitoring myocardial recovery during the period of LVAD support. Previously described weaning protocols have not assessed the true myocardial response to device cessation and have based their explantation criteria on measurements taken whilst the device was switched on.[25,37,38] Those groups have generally relied on assessing patients whilst the device is on, or after only reducing the speed of the device and hence reducing its power, or after only momentarily discontinuing the device.[30,37,38] This has only provided a rough estimate of the underlying myocardial function. Full device cessation on the other hand allows clinicians to assess the true capacity of the unaided myocardium to support the circulation and importantly to respond to increased demand through loading and exercise. In the Harefield Bridge-to-Recovery Protocol, switching off the device completely for 15 minutes or more followed by exercise with continuous monitoring of several clinical, haemodynamic and echocardiographic parameters was an important part and provided a more robust method to assess the true underlying function and inotropic reserve.[39] This prospective study of the efficacy of device discontinuation showed that there was an immediate drop after device discontinuation in mean arterial blood pressure (MAP), a rise in heart rate (HR), a reduction in ejection fraction (EF) and increases in ventricular dimensions. These changes tended to be more marked in patients that did not recover. After a six-minute walk, those who recovered had a significant rise in HR and EF and a non-significant increase in MAP, whereas in those that did not recover there was a significant drop in MAP compensated by a rise in HR. These acute haemodynamic changes reflect the heart's response to acute loading and were more marked in the bridge-to-transplant (BTT) group.

Complete unprotected device discontinuation could increase the risk of within-pump thrombosis. To prevent this potentially serious complication, full heparinisation (10,000 units of unfractionated heparin) prior to turning the pump off, coupled with pneumatic hand pumping in devices such as the HeartMate I XVE was used. In non-pulsatile pumps where the flow is continuous, reducing the pump speed to a rate where there is no forward or back flow, i.e. an "off-pump" equivalent study is an alternative in assessing recovery as performed by us with axial flow pumps and as suggested by Myers *et al.*[40] Alternatively, Ferrari *et al.* have used a minimally invasive method to determine the predictive power of *E*max, the slope of the LV end-systolic pressure–volume relationship and an afterload and preload-independent index of LV contractility, for bridging assist device recipients to myocardial recovery.[41]

3.2. Improvements in exercise capacity

The above recovery protocol also included testing of exercise capacity by performing cardiopulmonary exercise tests both on the pump and with the pump turned off after heparinisation.[20] Following explantation, exercise capacity further improved as reflected by an increase in oxygen consumption to a mean of 26.6 ml/kg/min at two years. This was better than that observed after cardiac transplantation[42] and could be due to the fact that the recovery patients retained innervation or possibly be due to the effect of temporary treatment with clenbuterol. Furthermore, following device explantation most of the patients showed sustained improvements in their echocardiographic parameters and maximal oxygen consumption (Fig. 2).[20]

3.3. Haemodynamic improvements

Following LVAD implantation a significant reduction in pulmonary artery systolic pressure generally occurs.[43–45] Haemodynamic studies performed at the time of device explantation (measured with the pump off for 15 minutes) revealed an improved pulmonary capillary wedge pressure (PCWP) (9.0 ± 4.1 mmHg vs. 24.9 ± 7 mmHg on inotropic therapy before implantation, $p = 0.004$) and cardiac

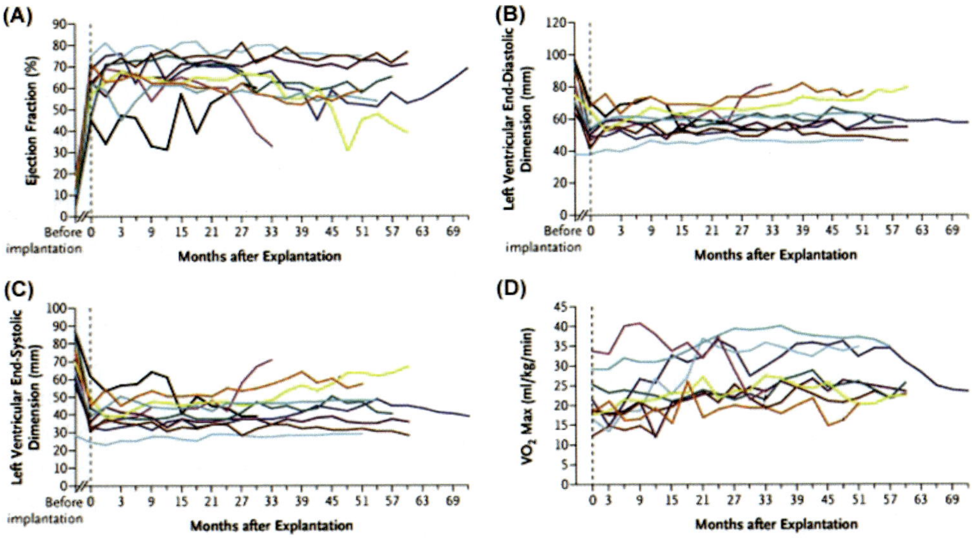

Fig. 2. Ejection fraction (**A**), left ventricular end-diastolic diameter (**B**), and left ventricular end-systolic diameter (**C**) before implantation and after explantation, and maximal oxygen consumption (VO_2 Max) (**D**) before and after explantation.

index (CI) (2.8 ± 0.7 l/min/m^2 vs. 1.8 ± 0.7 l/min/ m^2 pre-device implantation).[20] In addition the pulmonary-artery oxygen saturation improved to 66.9 ± 4.8%. At three months after device explantation the right atrial pressure (RAP) was 6.2 ± 2.1 mmHg, PCWP was 12.8 ± 6.9 mmHg, left ventricular end-diastolic pressure was 12.9 ± 5.9 mmHg, cardiac output (CO) was 4.9 ± 2.1 l/min, CI of 2.4 ± 1.0 l/min/m^2, and pulmonary-artery oxygen saturation was 69.8 ± 29.9%. Repeat catheterisation one year after explantation, revealed an RAP of 5.1 ± 3.3 mmHg, PCWP of 9.5 ± 6.2 mmHg, left ventricular end-diastolic pressure of 9.3 ± 5.5 mmHg, CO of 4.9 ± 2.1 l/min, CI of 2.4 ± 1.2 l/min/m^2, and pulmonary-artery oxygen saturation of 73.5 ± 32%.

3.4. Electrocardiographic improvements

Electrophysiological disturbances are important manifestations in heart failure. ECG analysis one week post-LVAD implantation in patients being bridged to transplantation showed significant reduction in the heart rate and a decrease in QT_C intervals compared to pre-implant ECGs; 91 vs. 107 bpm and 445 ± 9 vs. 479 ± 10 ms, respectively.[46] In addition, the magnitude of myocyte contraction was significantly higher in the LVAD-supported hearts, while the time to maximal contraction and the action potential duration at 50% repolarisation (ADP50) was decreased (529 ± 154 vs. 863 ± 37 ms).[46]

3.5. Quality of life

QOL of patients receiving LVAD has been extensively studied especially in those who were bridged to transplantation. Longer term quality of life (QOL) outcome in patients who have had an LVAD explanted due to myocardial recovery was recently investigated.[47] Using SF36 questionnaires the QOL of transplanted patients (Tx) was compared to those bridged to transplantation following transplantation (BTT) and those bridged to recovery following explantation (BTR). After an average of 3.8 years since transplantation both physical health dimensions (PHD) and mental health dimensions (MHD) were better in the BTT and BTR groups compared to the Tx group (PHD score of 64.5 ± 23.2 and 71.9 ± 21 vs. 41.4 ± 48 respectively, and MHD of 71.4 ± 21.1 and 78 ± 16.08 vs. 39.4 ± 44, respectively) with the mental QOL reaching statistical significance. Similarly Dew *et al.* showed that patients are most likely to have compliance and psychiatric problems early post-transplant.[48] These mental QOL changes could be attributed to the use of immunosuppressants, in particular cyclosporine which is known to induce neurotoxicity and hence impaired brain cognitive function.[49,50] These findings raise an important question of the significance of LVAD in affecting patients QOL outcome post-transplantation.

Similarly QOL following recovery was assessed using the Minnesota Living with Heart Failure Questionnaire.[20] The mean score (scores can range from 0 to 105, with higher scores indicating a worse quality of life) was 12.1 ± 11.7 three years following explantation.

3.6. Device explantation for recovery

After recovery, explantation of LVADs has traditionally comprised median sternotomy and laparotomy[51,52] with the associated extensive dissection of adhesions that can increase morbidity and mortality. To avoid these complications, Professor Yacoub developed and evaluated a minimally invasive approach.[53] In this approach the apex of the heart is exposed through an anterolateral mini thoracotomy and two other separate small incisions are made: an epigastric incision to expose the device and a limited anterior thoracotomy through the second intercostal space to expose the aortic anastomosis of the outflow cannula. After establishing femoro-femoral cardiopulmonary bypass the heart is fibrillated to allow explantation of the device inflow cannula. Following deairing the apex is repaired by direct suture. The heart is then defibrillated and cardiopulmonary bypass is discontinued. The outflow graft is divided as close as possible to the aortic anastomosis and the residual graft stump, which usually measures 2 to 3 mm in length, is oversewn followed by device explantation, along with the intrathoracic part of the Dacron polyester fabric graft. In our early series of device explantation for myocardial recovery, 11 out of 12 patients remained alive and their post-device removal intensive care unit stay was 6.2 ± 3.3 days.[53]

4. Remodelling versus Reverse Remodelling

LV remodelling describes the dynamic process of progressive LV dilatation. Remodelling could be triggered by mechanical stretch secondary to either volume overload or pressure overload, ischaemia and infarction, cardiotoxic agents, and predetermined genetic phenotype abnormalities, however many of the triggers are poorly understood. This structural remodelling of the cardiac interstitium represents a major determinant of pathological hypertrophy as it accounts for abnormal myocardial stiffness and impaired coronary reserve, thereby leading to ventricular diastolic and systolic dysfunction and ultimately the appearance of symptomatic heart failure. The molecular, cellular, biochemical and structural changes occurring in remodelling have been extensively studied.[54–59] However, the exact mechanisms involved remain poorly defined. At the structural level there is an increase in both LV mass and volume and alteration in the 3D structure of the LV.[57,60–62] The pathogenesis of heart failure will determine whether these structural alterations are global and symmetrical or asymmetrical as in the case of LV dilatation secondary to myocardial infarction.[63]

One intriguing feature of remodelling is that it can occasionally reverse itself, i.e. *"reverse remodelling"*, in response to specific therapeutic interventions.[56,59,64–68] This phenomenon has attracted the attention of scientists and clinicians who have used multiple strategies to try to reverse the deleterious damage that occurs in heart failure.[69] Recently there has been increasing evidence that prolonged near complete unloading of the LV using LVADs can produce certain degrees of reverse remodelling.[20,25,26,70] Whether reverse remodelling is due to a "U-turn" of the pathological mechanisms that occur in remodelling or the generation of new pathways remains unclear.

4.1. The effects of unloading on myocyte size and structural proteins

There is sufficient evidence that cardiomyocytes change their phenotype in heart failure with marked enlargement in their size.[59,71–74] This represents an important hallmark in the remodelling process. Hematoxylin and eosion (H&E) staining of longitudinal and cross-sectional myocardial samples taken from patients with end-stage heart failure demonstrated an increase in the size of cardiomyocytes with loading followed by reduction in myocyte size after long-term LVAD support

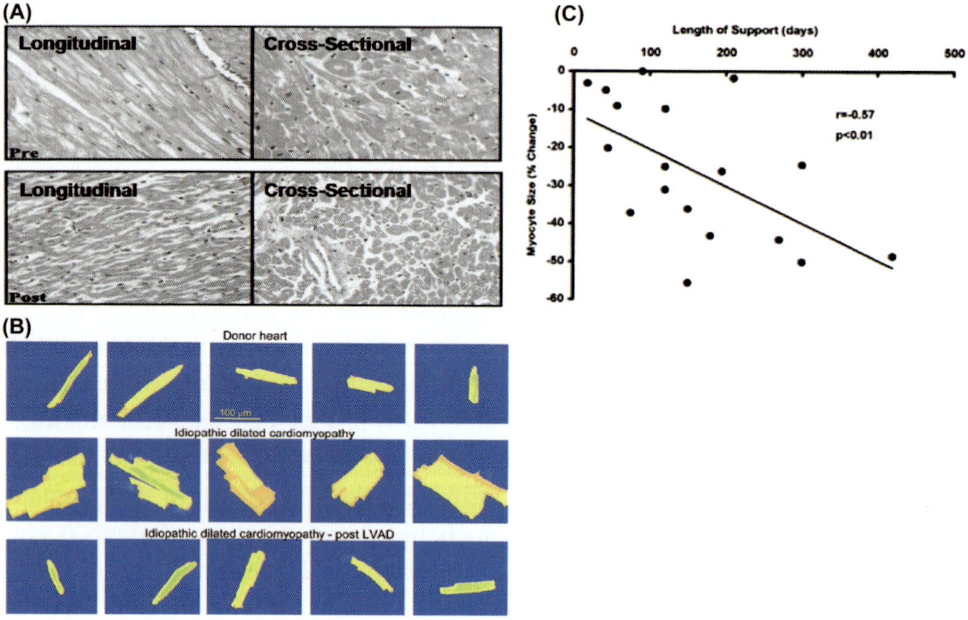

Fig. 3. (**A**) Myocardial longitudinal and cross-sectional samples taken from a patient pre- and post-LVAD support. The paired samples demonstrate reduction in myocyte size after long-term LVAD support.[75] (**B**) *Isolated* ventricular myocytes from patients at the time of insertion of a LVAD (middle horizontal panel) showing severe hypertrophy as compared to normal controls (top horizontal panel). Bottom panel shows "normalization" of the size of the myocytes post-LVAD.[59] (**C**) The significant correlation between duration of LVAD support and the percentage reduction in myocyte size ($r = -0.57$, $p < 0.01$).[74]

(Fig. 3).[75–77] It has been suggested that enlargement of isolated myocytes might be more pronounced in the long axis dimension compared to the width or depth.[59] Interestingly, Bruckner *et al.* showed a significant correlation between the length of LVAD support and the percentage change in myocyte size (Fig. 3C).[74]

Alterations in the size of the cardiomyocytes are associated with a deterioration in contractile function and cardiac performance[78–82] and induction of specific gene programmes involving several groups of gene.[62,83–87] In remodelling there is an enhancement in the secretory activity of the fibroblasts which leads to both reparative and reactive fibrosis which is related to changes in total collagen content, an increase in collagen I and III concentration and a change in collagen cross-linking in the cardiac ECM. All these changes have been shown to be important features of remodelling.[57,59,88] However, the responses of collagen concentration and cross-linking to LVAD support are controversial. Klotz *et al.* demonstrated an increase in collagen cross-linking and the ratio of collagen types I to III after 145 ± 33 days of LVAD support and hence an increase in myocardial stiffness.[89] Xydas *et al.*, on the other hand, showed a decrease in collagen deposition from time of LVAD implantation to time of transplantation in 17 patients who were bridged to transplantation.[68] Similarly, Bruckner *et al.* demonstrated that long-term mechanical circulatory support significantly reduced collagen content and fibrosis (Fig. 4).[74] Bruggink *et al.* suggested a biphasic pattern for the collagen turnover and the volume of ECM.[90] They showed that the initial response of the cardiac ECM response to LVAD (within 200 days of support) was an increase in types I and III collagen turnover and after 400 days of support there was a reduction in both the collagen and the volume of the ECM. Therefore, it could be assumed that the pattern of ECM and collagen response to LVAD is correlated to the length of support.

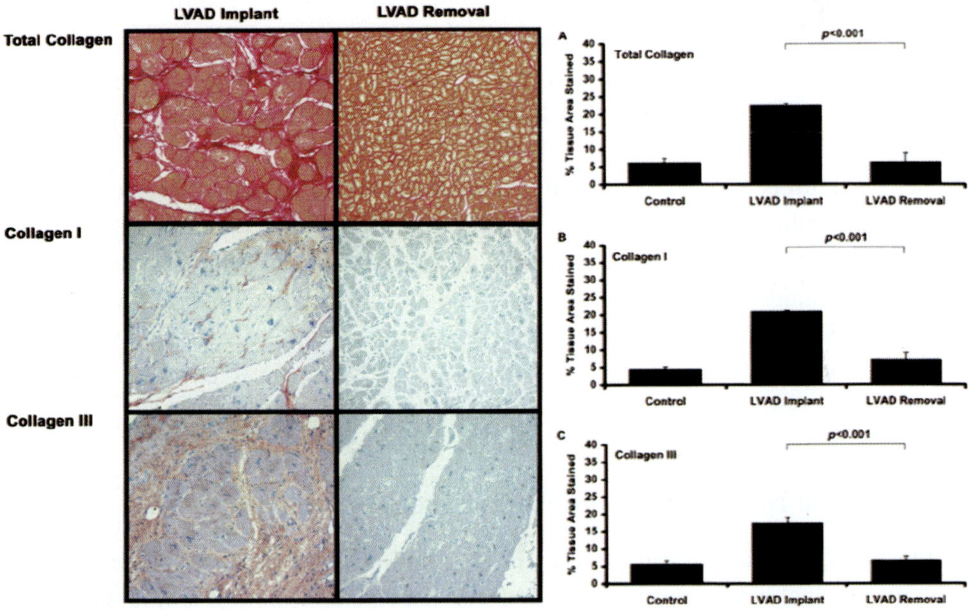

Fig. 4. Effect of LVAD support on collagen content. Total collagen, collagen I, and collagen III immunostaining of failing myocardium at the time of LVAD implant and end of support. Representative myocardial sections of failing myocardium from an individual patient at the time of LVAD implant (left panel) and removal (right panel) were stained for total collagen **(A)**, collagen I **(B)**, and collagen III **(C)**. Semi-quantitative immunohistochemistry revealed a significant reduction in total collagen content, collagen I, and collagen III contents following LVAD support. There was no difference in the total collagen, collagen I and III contents in the post-explant group compared to the controls.[74]

Using real-time polymerase chain reaction (PCR), Felkin *et al.* demonstrated that patients in deteriorating heart failure requiring LVAD support have elevated matrix metalloproteinases-1 and -8 (MMP-1 and -8) mRNA levels compared to patients with stable heart failure undergoing elective heart transplantation.[91] In addition, tissue inhibitor of metalloproteinases (TIMP-1 to -4) mRNA levels were also higher in the deteriorating patients requiring LVAD support with only TIMP-4 reaching statistical significance.[91] Li *et al.* looked at the protein expression of MMPs and TIMPs after LVAD support and showed that MMP-1 and -9 were decreased, TIMP-1 and -3 were increased, and there was no change in MMP-2 and -3 and TIMP-2 and -4 after LVAD support.[92]

Myocardial cytoskeletal proteins are necessary to maintain the structural and functional integrity of the myofibrillar apparatus as well as being essential cellular organelles. Changes in cytoskeletal elements have been observed in patients with heart failure.[93–95] Analysing paired myocardial samples collected at implantation and explantation using real-time PCR, we described that myocardial recovery is associated with a specific pattern of changes in the mRNA content of some sarcomeric, non-sarcomeric, and membrane-associated genes.[96] This was also confirmed at the protein level.[97] These changes paralleled improvements in haemodynamic function in LVAD patients showing clinical myocardial recovery.

Dystrophin is another structural protein that provides support for the myocyte and the cardiomyocyte membrane by linking N-terminus with the dystrophin-associated protein complex and sarcolemmal at the C-terminus. Studies have shown that mutations in the N-terminus of dystrophin are responsible for X-linked dilated cardiomyopathy.[98,99] Vatta *et al.* demonstrated that with prolonged unloading with an LVAD most patients with reduced N-terminus expression before implantation exhibited an increase in expression.[95]

4.2. Reverse remodelling and calcium regulation

In the heart, excitation-contraction (EC) coupling is the central mechanism by which electrical activation is translated into cardiac contraction and its initiation in ventricular myocytes is highly localised to the T-tubule network, a complex system of interconnected membrane structures continuous with the extracellular space and is integrally involved in multiple cellular processes including membrane transport, excitability, and cellular signalling. T-tubules are also the site of most junctional complexes formed between sarcolemma containing L-type calcium (Ca^{2+}) channels and junctional sarcoplasmic reticulum with Ca^{2+} release channels or ryanodine receptors (RyR). In heart failure, several proteins involved in this finely concerted regulation are changed with respect to their expression, phosphorylation status, and function leading to remodelling of EC coupling.

In the failing ventricular myocytes T-tubules are depleted and dilated and associated with changes in the density of a variety of proteins in both surface and T-tubular sarcolemma but with preservation of the protein composition of junctional complexes.[100,101] This subcellular remodelling contributes to abnormal EC coupling in heart failure. In addition, the action potential is initially prolonged in heart failure mainly due to calcium flux (which acts as a compensatory mechanism) but later becomes maladaptive due to decreased calcium response.[102] Terracciano *et al.*, studied the effects of LVAD and myocardial recovery on sarcoplasmic reticulum (SR) Ca^{2+} content and EC coupling by studying isolated cardiomyocytes from the LVAD core at the time of LVAD implantation and from myocardial samples at the time of explantation in patients who recovered and at the time of transplantation in those who did not recover and were transplanted.[103,104] They showed that recovered patients had an increase in their SR Ca^{2+} content and an increased gain in EC coupling resulting in larger Ca^{2+} transient and stronger cell contraction (Fig. 5). They speculated that the increased levels of SR Ca^{2+} could be due to one of the following: (1) increased SR Ca^{2+} uptake, (2) reduced SR Ca^{2+} leak as a consequence of a more stable ryanodine receptor complex, or possibly, (3) altered function of the sodium-calcium (Na^+/Ca^{2+}) exchanger.

Interestingly this study showed that reduction in cell capacitance and cell size were not necessarily associated with clinical recovery.[103] Further studies of the effects of LVAD on the Na^+/Ca^{2+} exchanger suggest that unloading increases the function of the exchanger,[105] an important player which is over expressed in heart failure.[106] Furthermore, an animal model suggested the enhancement in Ca^{2+} handling, SR Ca^{2+} content which resulted in increased amplitude of the Ca^{2+} transients, improved EC coupling, and the improvement of myocardial metabolism might be related to clenbuterol.[107]

It is important to note that Ca^{2+} regulation in the SR is regulated via sarcoplasmic reticulum calcium ATPase (SERCA2a), an energy requiring pump. In normal hearts this pump maintains the store for contraction as well as allowing the heart to relax rapidly in order to have sufficient time to fill before the next beat. In heart failure, contraction is poor and relaxation is slow and incomplete. This is due to the decrease in SERCA2a levels.[108] Therefore it could be speculated that increasing the levels of SERCA2a could increase the probability of reverse remodelling and improve calcium regulation. Currently studies are being conducted where AAV6 adenovirus will be used as vector to deliver the SERCA2a gene in patients who have an LVAD implanted for end-stage heart failure.

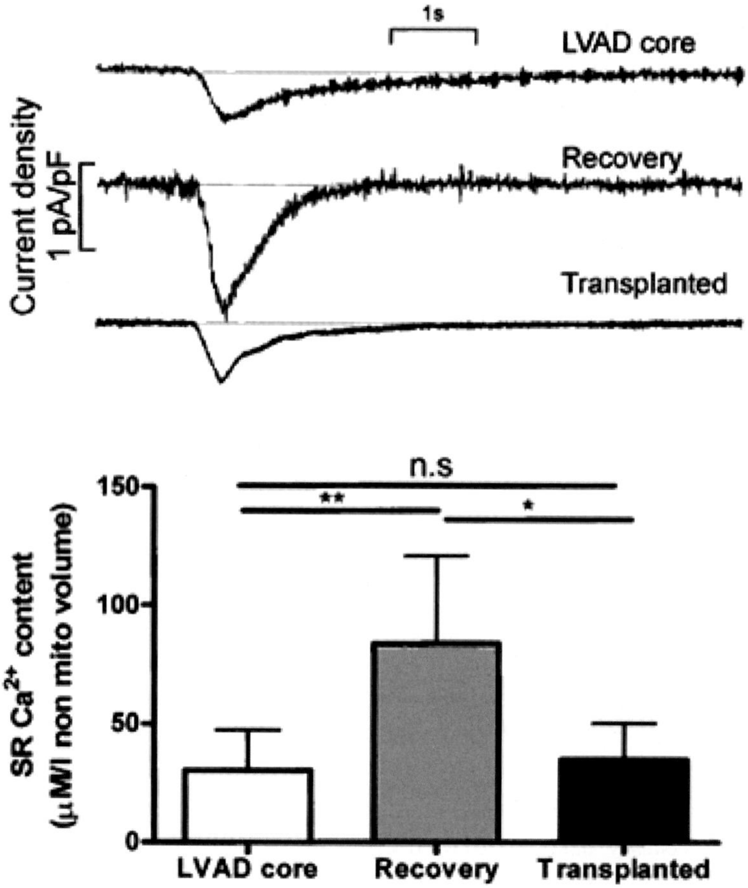

Fig. 5. SR Ca^{2+} content from myocytes isolated from LVAD cores and tissue from explanted (recovery) and transplanted hearts (without recovery). Cells were voltage-clamped at their resting membrane potentials. The SR Ca^{2+} content was significantly higher in the recovered patients and was associated with an improved current density.

5. Conclusion

LVADs are increasingly being implanted to treat deteriorating patients with advanced heart failure. Although they have mainly been implanted as a bridge to transplantation, and in the future are likely to be increasingly implanted as chronic therapy, there is now compelling evidence that reversal of advanced heart failure resulting in myocardial recovery sufficient to allow explantation of the device can occur. Adjunctive therapy such as a combination of pharmacologic therapy and LVAD unloading can increase the incidence of recovery. Studying changes in the myocardium and serum at the time of LVAD implantation and removal following myocardial recovery provides a unique opportunity to determine the mechanisms involved in the reversal of human heart failure.

References

1. Taylor, D.O., Edwards, L.B., Boucek, M.M., Trulock, E.P., Aurora, P., Christie, J., Dobbels, F., Rahmel, A.O., Keck, B.M. and Hertz, M.I. Registry of the International Society for Heart and Lung Transplantation: twenty-fourth official adult heart transplant report — 2007. *J. Heart Lung Transplant.* **26**(8), 769–781 (2007).
2. Frazier, O.H., Macris, M.P., Myers, T.J., Duncan, J.M., Radovancevic, B., Parnis, S.M. and Cooley, D.A. Improved survival after extended bridge to cardiac transplantation. *Ann. Thorac. Surg.* **57**(6), 1416–1422 (1994).
3. Oz, M.C. *et al.* Bridge experience with long-term implantable left ventricular assist devices. Are they an alternative to transplantation? *Circulation* **95**(7), 1844–1852 (1997).
4. DeRose, J.J., Jr. *et al.* Implantable left ventricular assist devices provide an excellent outpatient bridge to transplantation and recovery. *J. Am. Coll. Cardiol.* **30**(7), 1773–1777 (1997).
5. Piccione, W. Bridge to transplant with HeartMate device. *J. Card. Surg.* **16**, 272–279 (2001).
6. Rose, E.A., Gelijns, A.C., Moskowitz, A.J., Heitjan, D.F., Stevenson, L.W., Dembitsky, W., Long, J.W., Ascheim, D.D., Tierney, A.R., Levitan, R.G., Watson, J.T., Meier, P., Ronan, N.S., Shapiro, P.A., Lazar, R.M., Miller, L.W., Gupta, L., Frazier, O.H., Desvigne-Nickens, P., Oz, M.C. and Poirier, V.L. Long-term mechanical left ventricular assistance for end-stage heart failure. *N. Engl. J. Med.* **345**(20), 1435–1443 (2001).
7. Rose, E.A., Moskowitz, A.J., Packer, M., Sollano, J.A., Williams, D.L., Tierney, A.R., Heitjan, D.F., Meier, P., Ascheim, D.D., Levitan, R.G., Weinberg, A.D., Stevenson, L.W., Shapiro, P.A., Lazar, R.M., Watson, J.T., Goldstein, D.J., Gelijns, A.C. and Mar for the REMATCH investigators. The REMATCH trial: rationale, design, and end points. Randomized evaluation of mechanical assistance for the treatment of congestive heart failure. *Ann. Thorac. Surg.* **67**(3), 723–730 (1999).
8. Dee, K.C., Puleo, D.A. and Bizios, R. Biocompatibility. In: *An Introduction to Tissue-Biomaterial Interactions* (Wiley, Hoboken, 2002).
9. Didisheim, P. Current concepts of thrombosis and infection in artificial organs. *ASAIO J.* **40**(2), 230–237 (1994).
10. Koster, A., Sanger, S., Hansen, R., Sodian, R., Mertzlufft, F., Harke, C., Kuppe, H., Hetzer, R. and Loebe, M. Prevalence and persistence of

heparin/platelet factor 4 antibodies in patients with heparin coated and non-coated ventricular assist devices. *ASAIO J.* **46**(3), 319–322 (2000).

11. Loebe, M., Koster, A., Sanger, S., Potapov, E.V., Kuppe, H., Noon, G.P. and Hetzer, R. Inflammatory response after implantation of a left ventricular assist device: comparison between the axial flow MicroMed DeBakey VAD and the pulsatile Novacor device. *ASAIO J.* **47**(3), 272–274 (2001).
12. Menconi, M.J., Pockwinse, S., Owen, T.A., Dasse, K.A., Stein, G.S. and Lian, J.B. Properties of blood-contacting surfaces of clinically implanted cardiac assist devices: gene expression, matrix composition, and ultrastructural characterization of cellular linings. *J. Cell. Biochem.* **57**(3), 557–573 (1995).
13. Spanier, T., Oz, M., Levin, H., Weinberg, A., Stamatis, K., Stern, D., Rose, E. and Schmidt, A.M. Activation of coagulation and fibrinolytic pathways in patients with left ventricular assist devices. *J. Thorac. Cardiovasc. Surg.* **112**(4), 1090–1097 (1996).
14. Schuster, M., Kocher, A., Lietz, K., Ankersmit, J., John, R., Edwards, N., Oz, M. and Itescu, S. Induction of CD40 ligand expression in human T cells by biomaterials derived from left ventricular assist device surface. *Transplant. Proc.* **33**(1–2), 1960–1961 (2001).
15. Burgreen, G.W., Loree, H.M. II, Bourque, K., Dague, C., Poirier, V.L., Farrar, D., Hampton, E., Wu, Z.J., Gempp, T.M. and Schöb, R. Computational fluid dynamics analysis of a maglev centrifugal left ventricular assist device. *Artif. Organs* **28**(10), 874–880 (2004).
16. Long, J.W. Advanced mechanical circulatory support with the HeartMate left ventricular assist device in the year 2000. *Ann. Thorac. Surg.* **71**(3 Suppl.), S176–S182 (2001).
17. Spanier, T.B., Chen, J.M., Oz, M.C., Stern, D.M., Rose, E.A. and Schmidt, A.M. Time-dependent cellular population of textured-surface left ventricular assist devices contributes to the development of a biphasic systemic procoagulant response. *J. Thorac. Cardiovasc. Surg.* **118**(3), 404–413 (1999).
18. Ankersmit, H.J., Edwards, N.M., Schuster, M., John, R., Kocher, A., Rose, E.A., Oz, M., Itescu, S. Quantitative changes in T-cell populations after left ventricular assist device implantation: relationship to T-cell apoptosis and soluble CD95. *Circulation* **100**(19 Suppl.), II211–II215 (1999).
19. Ankersmit, H.J., Tugulea, S., Spanier, T., Weinberg, A.D., Artrip, J.H., Burke, E.M., Flannery, M., Mancini, D., Rose, E.A., Edwards, N.M., Oz, M.C. and Itescu, S. Activation-induced T-cell death and immune dysfunction after implantation of left-ventricular assist device. *Lancet* **354**(9178), 550–555 (1999).

20. Birks, E.J., Tansley, P.D., Hardy, J.P., George, R.S., Bowles, C.T., Burke, M., Banner, N.R., Khaghani, A. and Yacoub, M.H. Left ventricular assist device and drug therapy for the reversal of heart failure. *N. Engl. J. Med.* **355**, 1873–1884 (2006).
21. Wheeldon, D.R., LaForge, D.H., Lee, J., Jansen, P.G., Jassawalla, J.S. and Portner, P.M. Novacor left ventricular assist system long-term performance: comparison of clinical experience with demonstrated *in vitro* reliability. *ASAIO J.* **48**(5), 546–551 (2002).
22. Butler, K.C. and Farrar, D.J. Long-term reliability testing of the HeartMate II LVAD. *ASAIO J.* **51**(2), 32A (2005).
23. Frazier, O.H., Delgado, R.M., III, Kar, B., Patel, V., Gregoric, I.D. and Myers, T.J. First clinical use of the redesigned HeartMate® II left ventricular assist system in the United States: a case report. *Tex. Heart Inst. J.* **31**(2), 157–159 (2004).
24. Pagani, F.D., Strueber, M., Naka, Y., Kormos, R.L., Long, J.W., Dowling, R.D., Massey, H.T., Farrar, D.J. and Frazier, O.H. Initial multicenter clinical results with the HeartMate® II axial flow left ventricular assist device. *J. Heart Lung Transplant.* **24**(2 Suppl. 1), S74 (2005).
25. Dandel, M., Weng, Y., Siniawski, H., Potapov, E., Lehmkuhl, H., Hetzer, R., Muller, J.H., Weng, Y., Meyer, R. and Dandel, M. Long-term results in patients with idiopathic dilated cardiomyopathy after weaning from left ventricular assist devices. *Circulation* **112**, I-37–I-45 (2005).
26. Frazier, O.H. and Myers, T.J. Left ventricular assist system as a bridge to myocardial recovery. *Ann. Thorac. Surg.* **68**(2), 734–741 (1999).
27. Mancini, D.M., Beniaminovitz, A., Levin, H., Catanese, K., Flannery, M., DiTullio, M., Savin, S., Cordisco, M.E., Rose, R. and Oz, M. Low incidence of myocardial recovery after left ventricular assist device implantation in patients with chronic heart failure. *Circulation* **98**(22), 2383–2389 (1998).
28. Frazier, O.H., Delgado, R.M., III, Scroggins, N., Odegaard, P. and Kar, B. Mechanical bridging to improvement in severe acute "nonischemic, nonmyocarditis" heart failure. *Congest. Heart Fail.* **10**(2), 109–113 (2004).
29. Hetzer, R., Muller, J.H., Weng, Y.G., Loebe, M. and Wallukat, G. Midterm follow-up of patients who underwent removal of a left ventricular assist device after cardiac recovery from end-stage dilated cardiomyopathy. *J. Thorac. Cardiovasc. Surg.* **120**(5), 843–853 (2000).
30. Hetzer, R., Muller, J.H., Weng, Y., Meyer, R. and Dandel, M. Bridging-to-recovery. *Ann. Thorac. Surg.* **71**(Suppl. 3), S109–S113 (2001).
31. Mann, D.L. and Willerson, J.T. Left ventricular assist devices and the failing heart: a bridge to recovery, a permanent assist device, or a bridge too far? *Circulation* **98**, 2367–2369 (1998).

32. Petrou, M., Clarke, S., Morrison, K., Bowles, C., Dunn, M. and Yacoub, M. Clenbuterol increases stroke power and contractile speed of skeletal muscle for cardiac assist. *Circulation* **99**(5), 713–720 (1999).
33. Petrou, M., Wynne, D.G., Boheler, K.R. and Yacoub, M.H. Clenbuterol induces hypertrophy of the latimus dorsi muscle and heart in the rat with molecular and phenotypic changes. *Circulation* **92**(Suppl. II), II-483–II-489 (1995).
34. Hon, J.K., Steendijk, P., Perou, M., Wong, K. and Yacoub, M.H. Influence of clenbuterol treatment during six weeks of chronic right ventricular pressure overload as studied with pressure–volume analysis. *J. Thorac. Cardiovasc. Surg.* **122**(4), 767–774 (2001).
35. Wong, K., Boheler, K.R., Petrou, M. and Yacoub, M.H. Pharmacological modulation pressure overload cardiac hypertrophy changes in ventricular function, extracellular matrix, and gene expression. *Circulation* **96**, 2239–2246 (1997).
36. Wong, K., Boheler, K.R., Bishop, J., Petrou, M. and Yacoub, M.H. Clenbuterol induces cardiac hypertrophy with normal functional, morphological and molecular features. *Cardiovasc. Res.* **37**, 115–122 (1998).
37. Maybaum, S., Stockwell, P., Naka, Y., Catanese, K., Flannery, M., Fisher, P., Oz, M. and Mancini, D. Assessment of myocardial recovery in a patient with acute myocarditis supported with a left ventricular assist device: a case report. *J. Heart Lung Transplant.* **22**(2), 202–209 (2003).
38. Slaughter, M., Silver, M., Farrar, D., Tatooles, A. and Pappas, P. A new method of monitoring recovery and weaning the Thoratec LVAD. *Ann. Thorac. Surg.* **71**, 215–218 (2001).
39. George, R.S., Yacoub, M.H., Tasca, G., Webb, C., Bowles, C.T., Tansley, P., Hardy, J.H., Dreyfus, G., Khaghani, A. and Birks, E.J. Hemodynamic and echocardiographic responses to acute interruption of left ventricular assist device support: relevance to assessment of myocardial recovery. *J. Heart Lung Transplant.* **26**(10), 967–973 (2007).
40. Myers, T., Frazier, O., Mesina, H., Radovancevic, B. and Gregoric, I. Hemodynamics and patient safety during pump-off studies of an axial-flow left ventricular assist device. *J. Heart Lung Transplant.* **25**(4), 379–383 (2006).
41. Ferrari, M., Kadipasaoglu, K., Croitoru, M., Conger, J., Myers, T., Gregoric, I., Radovancevic, B., Letsou, G. and Frazier, O.J. Evaluation of myocardial function in patients with end-stage heart failure during support with the Jarvik 2000 left ventricular assist device. *J. Heart Lung Transplant.* **24**(2), 226–228 (2005).

42. Banner, N.R., Patel, N., Cox, A.P., Patton, H.E., Lachno, D.R. and Yacoub, M.H. Altered sympathoadrenal response to dynamic exercise in cardiac transplant recipients. *Cardiovasc. Res.* **23**(11), 965–972 (1989).
43. Etz, C.D., Welp, H.A., Tjan, T.D.T., Hoffmeier, A., Weigang, E., Scheld, H.H. and Schmid, C. Medically refractory pulmonary hypertension: treatment with nonpulsatile left ventricular assist devices. *Ann. Thorac. Surg.* **83**(5), 1697–1705 (2007).
44. George, R.S., Birks, E.J., Radley-Smith, R.C., Khaghani, A. and Yacoub, M. Bridge to transplantation with a left ventricular assist device for systemic ventricular failure after mustard procedure. *Ann. Thorac. Surg.* **83**, 308–310 (2007).
45. Salzberg, S.P., Lachat, M.L., von Harbou, K., Zünd, G. and Turina, M.I. Normalization of high pulmonary vascular resistance with LVAD support in heart transplantation candidates. *Eur. J. Cardiothorac. Surg.* **27**(2), 222–225 (2005).
46. Harding, J.D., Piacentino, V., III, Gaughan, J.P., Houser, S.R. and Margulies, K.B. Electrophysiological alterations after mechanical circulatory support in patients with advanced cardiac failure. *Circulation* **104**(11), 1241–1247 (2001).
47. George, R.S., Yacoub, M.H., Bowles, C.T., Hipkin, M., Rogers, P., Hallas, C., Banner, N.R., Dreyfus, G., Khaghani, A. and Birks, E.J. Quality of life following LVAD removal for myocardial recovery. *J. Heart Lung Transplant.* **27**(2), 165–172 (2008).
48. Dew, M., Kormos, R., Roth, L., Murali, S., DiMartini, A. and Griffith, B. Early post-transplant medical compliance and mental health predict physical morbidity and mortality one to three years after heart transplantation. *J. Heart Lung Transplant.* **18**, 549–562 (1999).
49. Grimm, M., Yeganehfar, W., Laufer, G., Madl, C., Kramer, L., Eisenhuber, E., Simon, P., Kupilik, N., Schreiner, W., Pacher, R., Bunzel, B., Wolner, E. and Grimm, G. Cyclosporine may affect improvement of cognitive brain function after successful cardiac transplantation. *Circulation* **94**, 1339–1345 (1996).
50. Steinbüchel, N., Limm, H., Leopold, C. and Carr, D. Assessment of health-related quality-of-life in patients after heart transplantation under therapy with tacrolimus or cyclosporine. *Transpl. Int.* **13**(Suppl. 1), S609–S614 (2000).
51. Borovetz, H.S., Ramasamy, N., Zerbe, T.R. and Portner, P.M. Evaluation of an implantable ventricular assist system for humans with chronic refractory heart failure: device explant protocol. *ASAIO J.* **41**, 42–48 (1995).

52. Oz, M.C., Levin, H.R. and Rose, E.A. Technique for removal of left ventricular assist devices. *Ann. Thorac. Surg.* **58**, 257–258 (1994).
53. Tansley, P. and Yacoub, M. Minimally invasive explantation of implantable left ventricular assist devices. *J. Thorac. Cardiovasc. Surg.* **124**(1), 189–191 (2002).
54. Dimmeler, S., Zeiher, A.M. and Schneider, M.D. Unchain my heart: the scientific foundations of cardiac repair. *J. Clin. Invest.* **115**(3), 572–583 (2005).
55. Iwanaga, Y., Hoshijima, M., Gu, Y., Iwatate, M., Dieterle, T., Ikeda, Y., Date, M., Chrast, J., Matsuzaki, M., Peterson, K.L., Chien, K.R. and Ross, J. Jr. Chronic phospholamban inhibition prevents progressive cardiac dysfunction and pathological remodeling after infarction in rats. *J. Clin. Invest.* **113**(5), 727–736 (2004).
56. Lowes, B.D., Gilbert, E.M., Abraham, W.T., Minobe, W.A., Larrabee, P., Ferguson, D., Wolfel, E.E., Lindenfeld, J., Tsvetkova, T., Robertson, A.D., Quaife, R.A. and Bristow, M.R. Myocardial gene expression in dilated cardiomyopathy treated with beta-blocking agents. *N. Engl. J. Med.* **346**(18), 1357–1365 (2002).
57. Swynghedauw, B. Molecular mechanisms of myocardial remodeling. *Phys. Rev.* **79**, 215–262 (1999).
58. Starling, R. Inducible nitric oxide synthase in severe human heart failure: impact of mechanical unloading. *J. Am. Coll. Cardiol.* **34**(9), 1425–1427 (2005).
59. Yacoub, M. A novel strategy to maximize the efficacy of left ventricular assist devices as a bridge to recovery. *Eur. Heart J.* **22**(7), 534–540 (2001).
60. Bozkurt, B. Medical and surgical therapy for cardiac remodeling. *Curr. Opin. Cardiol.* **14**, 196–205 (1999).
61. Cohen, J. Structural basis for heart failure. Ventricular remodelling and its pharmacological inhibition. *Circulation* **91**, 2504–2507 (1995).
62. Gaballa, M.A. and Goldman, S. Ventricular remodelling in heart failure. *J. Card. Fail.* **8**(Suppl. 6), S476–S485 (2002).
63. Opie, L.A., Commerford, P.J., Gersh, B.J. and Pfefer, M.A. Controversies in ventricular remodelling. *Lancet* **367**(9507), 356–367 (2006).
64. Margulies, K. Reversal mechanisms of left ventricular remodeling: lessons from left ventricular assist device experiments. *J. Card. Fail.* **8**(Suppl. 6), S500–S505 (2002).
65. Margulies, K.B., Matiwala, S., Cornejo, C., Olsen, H., Craven, W.A. and Bednarik, D. Mixed messages: transcription patterns in failing and recovering human myocardium. *Circ. Res.* **96**(5), 592–599 (2005).
66. Pieske, B. Reverse remodelling in heart failure — fact or fiction? *Eur. Heart J. Suppl.* **6**(Suppl. D), D66–D78 (2004).

67. Sutton, M.J. and Keane, M.G. Reverse remodelling in heart failure with cardiac resynchronisation therapy. *Heart* **93**, 167–171 (2007).
68. Xydas, S., Rosen, R., Ng, C., Mercando, M., Cohen, J., DiTullio, M., Magnano, A., Marboe, C.C., Mancini, D.M., Naka, Y., Oz, M.C. and Maybaum, S. Mechanical unloading leads to echocardiographic, electrocardiographic, neurohormonal, and histologic recovery. *Circulation* **25**(1), 7–15 (2006).
69. Poole-Wilson, P.A. and Katz, A.M. Future strategies of reverse remodelling: summation. *J. Card. Fail.* **8**(6 Suppl.), S556–S561 (2002).
70. Levin, H.R., Oz, M.C., Chen, J.M., Packer, M., Rose, E.A. and Burkhoff, D. Reversal of chronic ventricular dilation in patients with end-stage cardiomyopathy by prolonged mechanical unloading. *Circulation* **91**(11), 2717–2720 (1995).
71. Onodera, T., Tamura, T., Said, S., McCune, S.A. and Gerdes, A.M. Maladaptive remodelling of cardiac myocyte shape begins long before failure in hypertension. *Hypertension* **32**, 753–757 (1998).
72. Gerdes, A.M. and Cappaso, J.M. Structural remodelling and mechanical dysfunction of cardiac myocytes in heart failure. *J. Mol. Cell. Biol.* **27**, 849–856 (1995).
73. Barbone, A., Holmes, J.W., Heerdt, P.M., The, A.H.S., Naka, Y., Joshi, N., Daines, M., Marks, A.R., Oz, M.C. and Burkhoff, D. Comparison of right and left ventricular responses to left ventricular assist device support in patients with severe heart failure. *Circulation* **104**, 670–675 (2001).
74. Bruckner, B.A., Stetson, S.J., Perez-Verdia, A., Youker, K.A., Radovancevic, B., Connelly, J.H., Koerner, M.M., Entman, M.E., Frazier, O.H., Noon, G.P. and Torre-Amione, G. Regression of fibrosis and hypertrophy in failing myocardium following mechanical circulatory support. *J. Heart Lung Transplant.* **20**(4), 457–464 (2001).
75. Khan, N., Delgado, R.M., Radovancevic, B., Torre-Amione, G., Abrams, J., Miller, K., Myers, T., Okerberg, K., Stetson, S.J., Gregoric, I., Hernandez, A. and Frazier, O.H. Dobutamine stress echocardiography predicts myocardial improvement in patients supported by left ventricular assist devices (LVADs): hemodynamic and histologic evidence of improvement before LVAD explantation. *J. Heart Lung Transplant.* **22**(2), 137–146 (2003).
76. Dipla, K., Mattiello, J.A., Jeevanandam, V., Houser, S.R. and Margulies, K.B. Myocyte recovery after mechanical circulatory support in humans with end-stage heart failure. *Circulation* **97**, 2316–2322 (1998).
77. Zafeiridis, A., Jeevanandam, V., Houser, S.R. and Margulies, K.B. Regression of cellular hypertrophy after left ventricular assist device support. *Circulation* **98**, 656–662 (1998).

78. Davies, C.H., Davia, K., Bennett, J.G., Pepper, J.R., Poole-Wilson, P.A. and Harding, S.E. Reduced contraction and altered frequency response of isolated ventricular myocytes from patients with heart failure. *Circulation* **92**, 2540–2549 (1995).
79. Houser, S.R. and Lakatta, E.G. Function of the cardiac myocyte in the conundrum of end-stage, dilated human heart failure. *Circulation* **341**, 1276–1283 (1999).
80. Ginsburg, R., Bristow, M.R., Billingham, M.E., Stinson, E.B., Schroeder, J.S. and Harrison, D.C. Study of the normal and failing isolated human heart: decreased response of failing heart to isoproterenol. *Am. Heart J.* **106**(3), 535–540 (1983).
81. Marian, A. Pathogenesis of diverse clinical and pathological phenotypes in hypertrophic cardiomyopathy. *Lancet* **335**(9197), 58–60 (2000).
82. Zile, M.R., Koide, M., Sato, H., Ishiguro, Y., Conrad, C.H., Buckley, J.M., Morgan, J.P. and Cooper, G., IV. Role of microtubules in the contractile dysfunction of hypertrophied myocardium. *J. Am. Coll. Cardiol.* **33**(1), 250–260 (1999).
83. Chien, K. Stress pathways and heart failure. *Cell* **98**, 555–558 (1999).
84. Corbett, J.M., Why, H.J., Wheeler, C.H., Richardson, P.J., Archard, L.C., Yacoub, M.H. and Dunn, M.J. Cardiac protein abnormalities in cardiomyopathy detected by two dimensional polyacrylamide gel electrophoresis. *Electrophoresis* **19**, 2031–2042 (1998).
85. Francis, G. Cellular and subcellular basis for remodelling: summation. *J. Card. Fail.* **8**(Suppl. 6), S450–S451 (2002).
86. Latif, N. *et al.* Upregulation of the Bcl-2 family of proteins in end-stage heart failure. *J. Am. Coll. Cardiol.* **35**, 1769–1777 (2000).
87. Schneider, M.D. and Schwartz, R.J. Chips ahoy: gene expression in hearts surveyed by high-density microarrays. *Circulation* **102**, 3020–3027 (2000).
88. Weber, K. Targeting pathological remodelling: concepts of cardioprotection and reparation. *Circulation* **102**, 1342–1345 (2000).
89. Klotz, S., Foronjy, R.F., Dickstein, M.L., Gu, A., Garrelds, I.M., Jan Danser, A.H., Oz, M.C., D'Armiento, J. and Burkhoff, D. Mechanical unloading during left ventricular assist device support increases left ventricular collagen cross-linking and myocardial stiffness. *Circulation* **112**, 364–374 (2005).
90. Bruggink, A.H., van Oosterhout, M.F., de Jonge, N., Ivangh, B., van Kuik, J., Voorbij, R.H., Cleutjens, J.P., Gmelig-Meyling, F.H. and de Weger, R.A. Reverse remodeling of the myocardial extracellular matrix after prolonged left ventricular assist device support follows a biphasic pattern. *J. Heart Lung Transplant.* **25**(9), 1091–1098 (2006).

91. Felkin, L.E., Birks, E.J., George, R., Wong, S., Khaghani, A., Yacoub, M.H. and Barton, P.J.R. A quantitative gene expression profile of matrix metalloproteinases (MMPS) and their inhibitors (TIMPS) in the myocardium of patients with deteriorating heart failure requiring left ventricular assist device support. *J. Heart Lung Transplant.* **25**(12), 1413–1419 (2006).
92. Li, Y.Y., Feng, Y., McTiernan, C.F., Pei, W., Moravec, C.S., Wang, P., Rosenblum, W., Kormos, R.L. and Feldman, A.M. Downregulation of matrix metalloproteinases and reduction in collagen damage in the failing human heart after support with left ventricular assist devices. *Circulation* **104**, 1147–1152 (2001).
93. Hein, S., Scholz, D., Fujitani, N., Rennollet, H., Brand, T., Friedl, A. and Schaper, J. Altered expression of titin and contractile proteins in failing human myocardium. *J. Mol. Cell. Card.* **26**(10), 1291–1306 (1994).
94. Di Somma, S., Marotta, M., Salvatore, G., Cudemo, G., Cuda, G., De Vivo, F., Di Benedetto, M.P., Ciaramella, F., Caputo, G. and de Divitiis, O. Changes in myocardial cytoskeletal intermediate filaments and myocyte contractile dysfunction in dilated cardiomyopathy: an *in vivo* study in humans. *Heart* **84**(6), 659–667 (2000).
95. Vatta, M., Stetson, S., Perez-Verdia, A., Entman, M.L., Noon, G.P., Torre-Amione, G., Bowles, N.E. and Towbin, J.A. Molecular remodelling of dystrophin in patients with end-stage cardiomyopathies and reversal in patients on assistance-device therapy. *Lancet* **359**(9310), 936–941 (2002).
96. Birks, E.J., Hall, J.L., Barton, P.J., Grindle, S., Latif, N., Hardy, J.P., Rider, J.E., Banner, N.R., Khaghani, A., Miller, L.W. and Yacoub, M.H. Gene profiling changes in cytoskeletal proteins during clinical recovery after left ventricular-assist device support. *Circulation* **112**(9 Suppl.), I57–I64 (2005).
97. Latif, N., Yacoub, M.H., George, R., Barton, P.J.R. and Birks, E.J. Changes in sarcomeric and non-sarcomeric cytoskeletal proteins and focal adhesion molecules during myocardial recovery after left ventricular assist device support. *J. Heart Lung Transplant.* **26**(3), 230–235 (2007).
98. Towbin, J.A., Hejtmancik, J.F., Brink, P., Gelb, B., Zhu, X.M., Chamberlain, J.S., McCabe, E.R. and Swift, M. X-linked dilated cardiomyopathy. Molecular genetic evidence of linkage to the Duchenne muscular dystrophy (dystrophin) gene at the Xp21 locus. *Circulation* **87**(6), 1854–1865 (1993).
99. Ortiz-Lopez, R., Li, H., Su, J., Goytia, V. and Towbin, J.A. Evidence for a dystrophin missense mutation as a cause of X-linked dilated cardiomyopathy. *Circulation* **95**(10), 2434–2440 (1997).
100. Kostin, S., Scolz, D., Shimada, T., Maeno, Y., Mollnau, H., Hein, S. and Schaper, J. The internal and external protein scaffold of the T-tubular system in cardiomyocytes. *Cell Tissue Res.* **294**(3), 449–460 (1998).

101. Balijepallia, R.C., Lokuta, A.J., Maertzb, N.A., Bucka, J.M., Haworthc, R.A., Valdiviab, H.H. and Kamp, T.J. Depletion of T-tubules and specific subcellular changes in sarcolemmal proteins in tachycardia-induced heart failure. *Cardiovasc. Res.* **59**(1), 67–77 (2003).
102. Wickenden, A.D., Kaprielian, R., Kassiri, Z., Tsoporis, J.N., Tsushima, R., Fishman, G.I. and Backx, P.H. The role of action potential prolongation and altered intracellular calcium handling in the pathogenesis of heart failure. *Cardiovasc. Res.* **37**(2), 312–323 (1998).
103. Terracciano, C.M.N., Hardy, J., Birks, E.J., Khaghani, A., Banner, N.R. and Yacoub, M.H. Clinical recovery from end-stage heart failure using left ventricular assist device and pharmacologic therapy correlates with increased sarcoplasmic reticulum calcium content, but not with regression of cellular hypertrophy. *Circulation* **109**, 2263–2265 (2004).
104. Terracciano, C.M.N., Harding, S.E., Tansley, P.T., Birks, E.J., Barton, P.J.R. and Yacoub, M.H. Changes in sarcolemmal Ca entry and sarcoplasmic reticulum Ca content in ventricular myocytes from patients with end-stage heart failure following myocardial recovery after combined pharmacological and ventricular assist device therapy. *Eur. Heart J.* **24**, 1329–1339 (2003).
105. Terracciano, C.M., Koban, M.U., Soppa, G.K., Siedlecka, U., Lee, J., Stagg, M.A. and Yacoub, M.H. The role of the cardiac Na^+/Ca^{2+} exchanger in reverse remodeling: relevance for LVAD-recovery. *Ann. N. Y. Acad. Sci.* **1099**, 349–360 (2007).
106. Stagg, M.A., Malik, A.H., MacLead, K.T. and Terracciano, C.M. The effects of overexpression of the Na^+/Ca^{2+} exchanger on calcium regulation in hypertrophied mouse cardiac myocytes. *Cell Calcium* **36**(2), 111–118 (2004).
107. Soppa, G.K.R., Smolenski, R.T., Latif, N., Yuen, A.H.Y., Malik, A., Karbowska, J., Kochan, Z., Terracciano, C.M.N. and Yacoub, M.H. Effects of chronic administration of clenbuterol on function and metabolism of adult rat cardiac muscle. *Am. J. Physiol. Heart Circ. Physiol.* **288**, H1468–H1476 (2005).
108. Scmidt, U., Hajjar, R.J., Helm, P.A., Kim, C.S., Doye, A.A. and Gwathmey, J.K. Contribution of abnormal sarcoplasmic reticulum ATPase activity to systolic and diastolic dysfunction in human heart failure. *J. Mol. Cell. Cardiol.* **30**, 1929–1937 (1998).

PART IX

OSTEOARTICULAR REPAIR

Chapter 35

Animal Models

Elizabeth A. Horner, Jennifer Kirkham and Xuebin B. Yang

Abstract

In vivo animal models are currently the gold standard for testing the capacity of stem/progenitor cells, smart biomaterials and novel growth factors for successful tissue engineering. *In vitro* models ultimately fail to provide the appropriate physiologically relevant microenvironment and hence animal models are an essential pre-requisite in the translation of any new therapy to the clinic. The aim of this chapter is to consider the available animal models commonly in use for tissue engineering, with a particular focus upon bone and cartilage research. Factors driving the choice of a given animal model are reviewed, according to the requirements of experimental design, hypothesis and the specific parameters to be tested. A number of animal models, together with their respective advantages and limitations are described, ranging from relatively simple experimental designs such as the subcutaneous implant and muscle pouch models, through to the diffusion chamber model and chorioallantoic membrane assay, to the more complex *in vivo* bioreactors and (arguably the most clinically relevant) bone and cartilage defect models. The need to consider the ethical issues of using animal models and the principles of reduction, replacement and refinement are emphasised in selecting the final experimental model of choice.

Keywords: Animal Model; *In Vivo*; Bone; Cartilage; Bioreactor; Tissue Engineering; Subcutaneous Implant; Intramuscular Implant; Diffusion Chamber; CAM

Outline

1. Introduction
2. The Choice of Animal
    2.1. Species
    2.2. Size
    2.3. Age

3. *In Vivo* Models for Bone and Cartilage Tissue Engineering
   3.1. The ectopic subcutaneous implant model
   3.2. The intramuscular implant model
   3.3. The diffusion chamber model
   3.4. The chorioallantoic membrane assay
   3.5. *In vivo* bioreactors
   3.6. *In vivo* defect models
      3.6.1. *Bone defect models*
      3.6.2. *The cartilage defect model*
4. Conclusion
   References

1. Introduction

In developing a clinical therapy from an original hypothesis through to stage 3 clinical trials, there is a need for the use of animal models to test whether what has been achieved *in vitro* can also be recapitulated *in vivo*. Animal models for tissue engineering vary from simple subcutaneous implants to models for testing complex tissue regeneration *in situ*. Living tissues require a balanced and multifactorial portfolio of cells and biomolecules acting in concert to maintain homeostasis and affect biological repair. For example, the cascade of growth factors that occurs following a bone fracture in order to promote healing is complex and subject to strict tempero-spatial control.[1] The ideal *in vivo* model allows the researcher to mimic appropriate clinical conditions (both physiological and pathological) and to create a permissive microenvironment that provides physiologically relevant nutrients, relative humidity, gaseous concentrations and growth factors. *In vivo* models can also potentially allow the testing of angiogenesis and biomechanical loading which cannot be easily replicated using *in vitro* techniques.

In the simplest animal models for tissue engineering, biomaterial scaffolds can be implanted subcutaneously for biocompatibility testing (immune and inflammatory responses, cytotoxicity), investigations of the cell-material interface and determination of material degradation rate. Furthermore, the capacity for repairing damaged tissues or regeneration of a tissue/organ (e.g. bone, cartilage) using cells (e.g. stem cells) in combination with/without growth factors and/or scaffolds can be investigated. However, it is crucial to choose the most suitable animal species and to carefully design the model in order to optimally test the desired parameters according to the specific tissue to be regenerated. In addition, the investigator must always attempt to minimise the suffering of the animal and consider the replacement, reduction and refinement (3Rs) of using experimental animals in research. In some countries, the investigator is required to attend special training and obtain an appropriate license before carrying out *in vivo* work.

2. The Choice of Animal

Reasons for selecting one animal over another for tissue engineering research are related to ethical issues, size of animal, physiology/anatomy and similarity to humans and cost implications.

2.1. Species

To date, almost all species of common laboratory mammals and some birds have been used to study tissue engineering of the skeleton.[2] Each species offers advantages and disadvantages and thus must be carefully considered prior to any study. The species that is most similar to the human, in terms of bone biology, is of course the primate. However, the use of primates, which are not readily available in some countries, has major ethical implications due to their elevated position within the phylogenetic table. Therefore, other mammals (e.g. pig, sheep, goat, rabbit, rat, mouse, etc.) are more commonly used.

One major disadvantage in the use of animal models for the study of human tissue regeneration is the variation in physiology between species. Animal models do, however, provide an indicative insight into procedures and techniques that may or may not work in the clinic. The use of mice and rats has a clear advantage because of the in-depth available knowledge of their skeletal biology (including bone loading). On the other hand, rodents do not have the same skeletal physiology as humans; rat growth plates often do not close (and so keep growing) until very late in life[3] and lack Haversian systems and remodelling.[2,4,5] However, when testing the capacity of human cells for tissue engineering *in vivo*, there is no alternative other than using rodents as such studies can only be carried out using genetically immuno-compromised animals which are currently restricted to mouse or rat.

2.2. Size

The advantage of using a large animal (e.g. pig, sheep, goat) as an experimental model is that a larger volume of tissue can be engineered. In addition, these larger species often exert similar forces to those of humans on any implant. Thus, large animals are often used for testing the biomechanical strength of any new material. However, large animal studies can be expensive, have a more restrictive sample size, a longer study period and are frequently of greater ethical concern.

Whilst the use of a smaller animal, such as mouse, rat and rabbit, negates these problems to some extent, the size of the bioengineered tissue and biomechanical stimuli are often correspondingly small. This places constraints when attempting to extrapolate and to investigate specific parameters such as nutrient delivery into the centre of the scaffold and the mechanical strength of the construct, which can be of importance especially in a critical bone defect. It must also be borne in mind that other, post-surgical, factors may also differ between species. For example, there is interspecies variation in the effects of simple anti-inflammatories and antibiotics.[2] However, van der Donk *et al.*[6] reported that bone in-growth into chambers placed on the tibias of rats and goats were similar if the defect is corrected to take into account body size, even though the general bone physiology and timing of repair in animals may be different. Finally, the use of small animal models in tissue engineering has been necessarily limited due to the requirement for special micro-surgical skills and the limited size of regenerated tissue for post-surgical analysis.

2.3. Age

The age of any experimental animal must also be considered when selecting an appropriate model. For example, a study investigating the age differences of bone marrow haematopoietic stem cells showed that a mouse of one year correlated to a human of 50 years.[7] A young immature skeleton has a far greater natural reparative capability than older bone or cartilage.[2,8–10] Therefore, the value of carrying out any tissue engineering investigations using very young animals is probably limited as this would rarely reflect the situation encountered in the clinic. Surprisingly, Gan[11] reported that there were apparently no age-related changes in fracture repair in nude mice. Further work is needed to understand the full significance of this and in extrapolating this to the human situation.

3. *In Vivo* Models for Bone and Cartilage Tissue Engineering

There are several animal models that have been used for studies of skeletal regeneration/repair with varying degrees of success. These include those that involve the production of bone or cartilage either subcutaneously or within muscle pouches of the animal and the most complex, but arguably the most clinically significant, critical defect models. Many researchers have utilised subcutaneous implant models or muscle pouch models in preliminary investigations to screen new scaffolds, cells and growth factors, before progressing to the defect model with their best candidates.[12–14]

3.1. The ectopic subcutaneous implant model

The ectopic subcutaneous implant model is the least invasive animal model used in tissue engineering research. Briefly, a small implant of scaffold with or without cells and/or growth factors can be implanted directly under the animal's skin, usually around the dorsal area of the shoulders or haunches (Fig. 1). This technique provides a permissive environment that can supply the nutrients, growth factors, gases/waste exchange and angiogenesis required for repair. The model has been employed in a number of different species, including the mouse,[15,16] rat[17] and rabbit[18] and has shown very promising results for the bioengineering of bone[19,20] and elastic cartilage[21] especially as different morphologies of engineered tissue can be produced.

Using this technique, new bone has been generated with dimensions of up to approximately 21 mm^3.[16,18] Isogai *et al.* went further, producing different types of

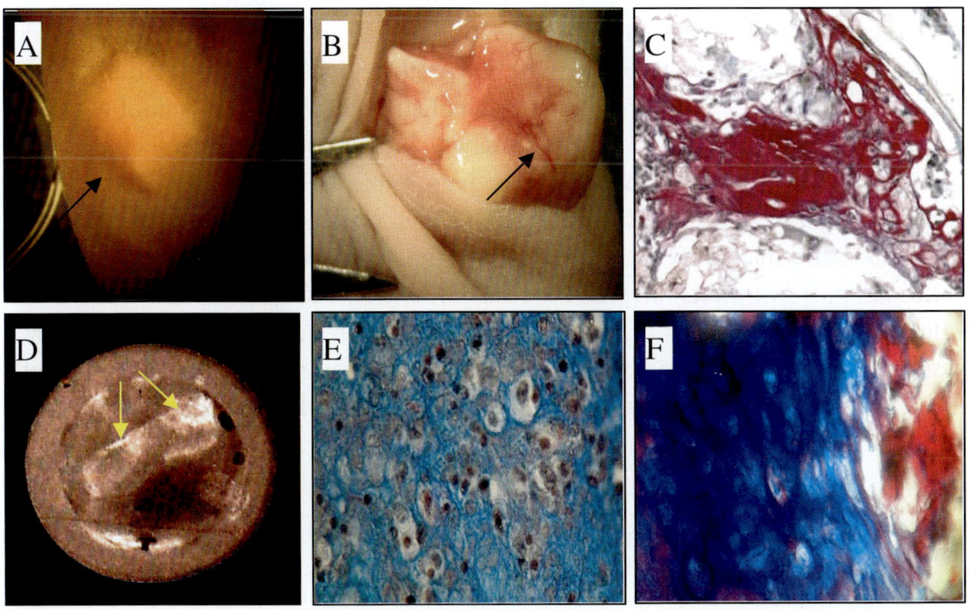

Fig. 1. Subcutaneously implant model **(A–C)** and diffusion chamber model **(D–F)** *in vivo*. **(A)** Human bone marrow stromal cells and biomaterial construct implanted subcutaneously in nude mice; **(B)** new tissue formation and vascular supply to the implants (arrow); **(C)** Sirius red staining showed new bone formation within the pleiotrophin absorbed PLGA porous scaffold (reproduced from Ref. 32 with permission from the American Society for Bone and Mineral Research); **(D)** X-ray images showed high density bone nodule formation; **(E)** Alcian blue staining showed cartilage matrix formation within the diffusion chamber;[27] and **(F)** Alcian blue/Sirius red staining showed bone and cartilage matrix formation within the diffusion chamber.[20]

tissue (bone, cartilage, tendon) in 24 weeks using scaffolds implanted with the relevant cell type in the dorsal subcutaneous space of the athymic mouse.[16] However, the ectopic subcutaneous implant model is also subject to limitations and constraints (especially for investigations of bone regeneration) as newly-formed tissue may be re-absorbed with time due to lack of appropriate daily mechanical stimulation.

3.2. The intramuscular implant model

An intramuscular implant model can provide a suitable physiological and mechanical environment for implanted cells to proliferate, differentiate and form new tissues. In brief, scaffolds with or without cells and/or growth factors can be implanted directly within the selected muscle (often the calf muscle). The intramuscular implant model not only supplies implanted cells with a suitable environment for appropriate access to nutrients, growth factors, gaseous/waste exchange and angiogenesis for tissue engineering applications, but also provides stimuli from biomechanical loading to both cells and scaffold during daily movement. Muscle contraction also promotes and facilitates the movement of nutrients, gases and waste products through the newly forming tissue. This technique has been used in rodents,[22,23] rabbits[24] and even man.[25] The intramuscular implant model is slightly more invasive than the subcutaneous implant model, but given the availability of a good blood supply and the additional advantage of mechanical loading, it is a better model for research aimed towards tissue regeneration. There is however, one note of caution when interpreting results obtained when stem cell constructs are used in this model. Muscle is known to contain stem cells itself,[26] which may exaggerate the efficacy of any observed regeneration and could confound the final results.

3.3. The diffusion chamber model

A diffusion chamber (Fig. 1), made of a microporous membrane, is a simple model that allows an enclosed, permissive environment to be generated within a host animal while preventing any host tissue participation that might otherwise confuse the findings.[27] This model therefore restricts investigation of test parameters to the implanted cells only. The test scaffolds are sealed in the diffusion chamber with or without growth factors. The experimental cells are then injected into the diffusion chamber via an injection port and the device implanted in the intraperitoneal cavity of the animal. The filters on the diffusion chamber allow free exchange of nutrients and waste but effectively isolate the experimental cells from the host tissues. This ensures that any tissue formed in the diffusion chamber is derived exclusively from the test — not the host — cells. The diffusion chamber model has been successfully used to test the regenerative capacity of both bone[27–31] and cartilage.[30,32,33] Figure 1 shows histology of successful bone and cartilage regeneration using this model. Because of its design, this model does not have any vascular/angiogenic involvement, which tends to mitigate strongly against endochondral ossification during bone formation. On the other hand, this very same characteristic offers great advantages for tissue engineering of cartilage, where a diffusion chamber creates a more hypoxic environment reliant upon diffusion and no angiogenesis.

A modification of this technique has been used by Miller and co-workers, who have used a chamber that is open-ended on one side but is otherwise impervious. The chamber, containing inserted scaffolds or morselised bone, was laid against the side of an ovine rib with the periosteum dissected, exposing the cell-rich cambium layer. This technique allowed migration of cells from the cambium layer into the chamber, leading to osteogenesis.[34,35]

3.4. The chorioallantoic membrane assay

The chorioallantoic membrane (CAM) of growing chick embryos (Fig. 2) has been widely used as an assay system for angiogenesis and as an explant tissue culture system.[36] It allows the culture of tissues and scaffolds on the chorioallantoic membrane of growing chick embryos.[20,37] The blood vessels of the CAM surround and nourish the scaffolds and since there is no immunologic or inflammatory response, human stromal/stem cells seeded onto the scaffold can be tested. This

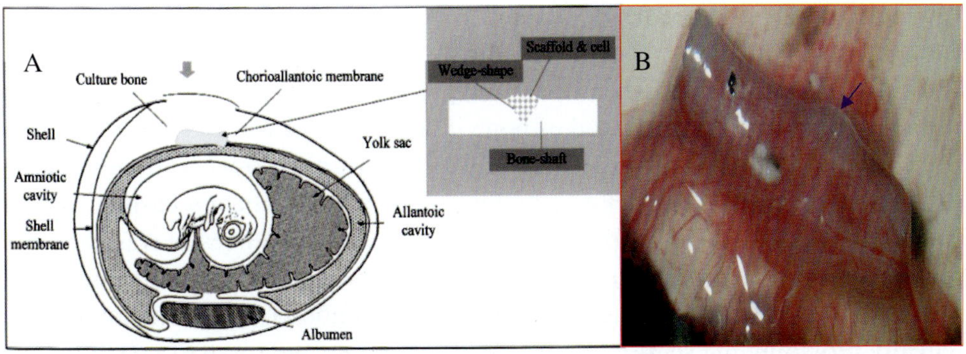

Fig. 2. Chorioallantoic membrane (CAM) assay. **(A)** CAM model showing how a chicken femur bone defect filled with cells-scaffolds construct is cultured on the chorioallantoic membrane. **(B)** Retrieved construct showing angiogenesis and vascular supply to the construct after seven days.[20]

ex vivo system is the least invasive and most cost-effective means of providing a suitable physiological microenvironment for investigation of basic scientific questions (e.g. angiogenesis, biocompatibility, cytotoxicity of biomaterial and fast screening of the capacity of tissue regeneration using human cells, novel growth factors and biomaterials).

Chick femurs are excised from day 18 embryos and a wedge-shaped segmental defect is created in the middle of the femur, into which the experimental scaffold construct can be used to fill the defect site. The chick bone and scaffold/cells/growth factor constructs are then placed directly onto the chorioallantoic membrane of ten-day-old eggs (through a 1 cm^2 square section cut into the shell) and incubated at 37°C in an automatic incubator. Yang *et al.* have used human bone marrow stromal cell/BMP-2 encapsulated 3D porous scaffold constructs on day 10 chick chorioallantoic membrane, which showed vascular invasion and bone/cartilage matrix formation in a seven-day period.[20]

3.5. *In vivo* bioreactors

Recently, two animal models have been developed to meet the demands for skeletal tissue engineering that are distinct from all of those described above. These models rely on the host animal's own reparative capacities, utilising either the environment of the periosteum[38] or circulating stem cells.[39,40]

Stevens *et al.* reported a novel *in vivo* bioreactor using the regenerative capacity of the periosteum. The periosteum of a rabbit tibia was revealed and a hypodermic needle carefully inserted beneath so that it was directly adjacent to the bone. A quick-gelating agent was then introduced to elevate the periosteum and create an oval-shaped pocket between the periosteum and bone as a test site for new bone formation. The newly formed bone was well organised with a similar biomechanical strength to that of the parent bone, but was still a separate entity, allowing for easy removal when desired (Fig. 3). The study also demonstrated that cartilage could be formed instead of bone if angiogenesis was inhibited.[38]

Using a very different approach, Holt, Halpern and co-workers designed a self-contained bioreactor enclosed within a non-porous silicone casing with or without a vascular supply. In this model, the epigastric blood vessel was carefully revealed and cut to produce a ligated vascular pedicle. The vascular pedicle was then carefully inserted into a scaffold encased by a non-porous membrane, theoretically providing the scaffold with a vascular supply, permissive growth factors (BMP-2) and circulating stem cells for tissue engineering. Stem cells were reported to have migrated into the bioreactor and initiated bone formation.[40] This model has also been used to investigate the interactions of ectopic bone formation and mammary tumour cells.[39]

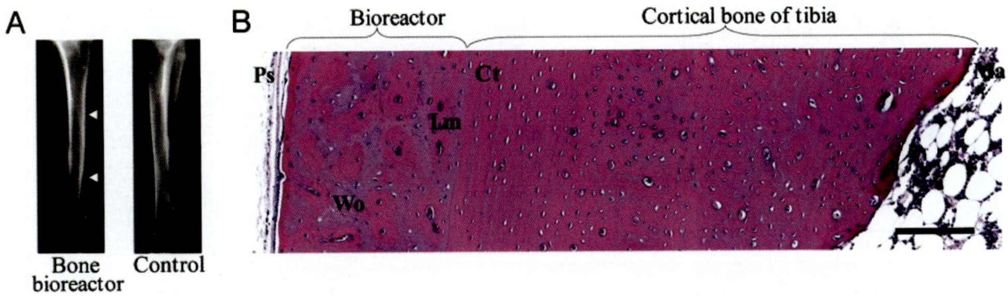

Fig. 3. *In vivo* bone bioreactor. **(A)** Radiograph of tibia with the bioreactor and contralateral limb after six weeks. Arrowheads indicate top and bottom of bioreactor. **(B)** H&E-stained cross-section of the bone bioreactor, adjacent cortical bone, and marrow cavity after six weeks. Ps, periosteum; bone, Wo, woven; Lm, lamellar; Ct, cortical; Ma, marrow (bar, 300 μm).[38] With permission from the *Proc. Natl. Acad. Sci. USA*. Copyright (2005) National Academy of Sciences, USA.

3.6. *In vivo* defect models

The production of a site-specific defect followed by tissue regeneration offers distinct advantages over other animal models for tissue engineering. Firstly, the appropriate clinically relevant environmental factors (nutrients, growth factors, biomechanical loading, etc.) are present within the defect site. Secondly the experimental scaffold, cells and growth factors implanted can be tested under conditions that more faithfully mimic those required clinically. However, the main disadvantage of defect models (either for cartilage or bone) is the lack of a chronic diseased state.

3.6.1. *Bone defect models*

There are several *in vivo* bone defect models that are currently available. The model selected for any given experimental strategy will depend on the end goal of the research and the test parameters required, such as whether a weight bearing or non-weight bearing bone is the clinical target.

For non-weight bearing testing, the bone defect can be made within the calvaria, rib or mandible, which have relatively low mechanical forces.[41] A full-thickness calvarial bone defect can be created using a drilling burr or trephine (with a choice of different diameters according to the required defect size which in turn relates to the animal species used) in a slow-speed dental hand piece supplemented with 0.9% sterile saline irrigation taking care not to damage the dura or superior sagittal sinus. The bone defect is then reconstructed (Fig. 4) using the test material/scaffold in combination with/without cells and/or growth factors.[42,43]

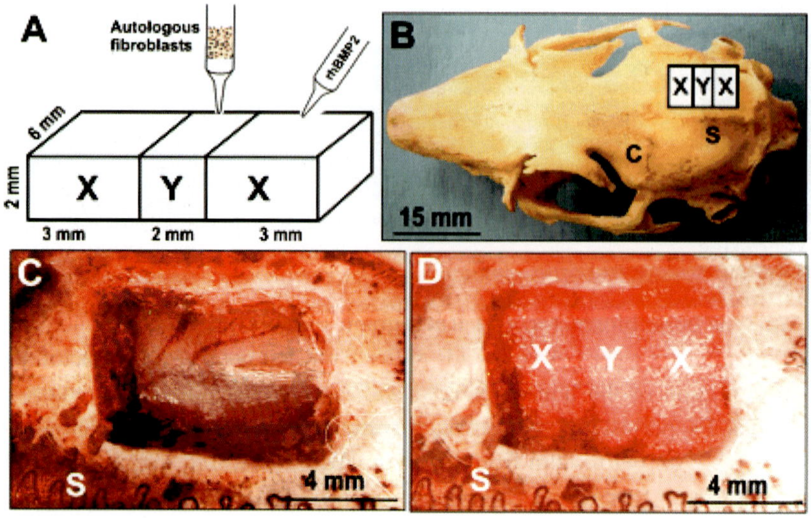

Fig. 4. Calvarial bone defect model. **(A)** Dermal fibroblast-seeded gelatin scaffold (Y) is sandwiched between two microporous collagen sponges loaded with rhBMP2 (X). **(B)** Location of surgically created calvarial defect. **(C)** Surgical creation of full-thickness calvarial defect ($6 \times 2 \times 8$ mm^3). **(D)** The surgically created calvarial defect was filled with a composite tissue construct (XYX).[42] With permission from the *J. Dent. Res.* Copyright (2004) International and American Associations for Dental Research, USA.

Most scaffold materials in such use can be moulded to fit the bone defect and do not require further fixation. The periosteum can then be closed to cover the defect.[44]

Where weight bearing is an important criterion during testing, the defect can be made in the femur or tibia. These models are appropriate for testing almost all parameters that are relevant within the clinic (e.g. fixation, mechanical loading). The use of non-healing segmental defect models (including fracture/osteotomy models) has long been established for testing tissue ultimately destined for repair of the long bones.[45,46] The size of defect can vary from a partial thickness defect[47,48] to a full diameter defect.[49–51] It is also necessary to take into account whether the desired outcome is regeneration/formation/repair of cortical and/or cancellous bone.[47] A femur segmental bone defect can be made using a saw supplemented with 0.9% sterile saline irrigation. The final size of the defect will differ according to the animal species chosen. The bone defect is then reconstructed with the test material (Fig. 5) in combination with/without cells and/or growth factors and fixed using external[46,52] or internal fixation.[49,50]

The biomechanical properties of the selected scaffold are crucial for optimum bone regeneration in a weight bearing bone defect. Scaffolds need to be

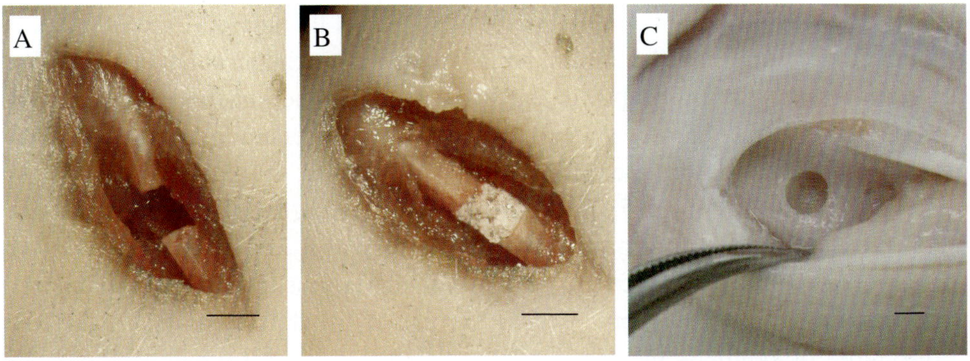

Fig. 5. Segmental bone defect model (**A** and **B**) and articular cartilage defect model (**C**). (**A**) A 2 mm bone defect created in a mouse femur. (**B**) The mouse femur defect was repaired by a porous poly(-lactic acid) scaffold and intramedullar pin fixation. (**C**) A 3 mm articular cartilage defect has been created on the knee joint of a rabbit (scale bars: 2 mm).

initially strong enough to allow weight-bearing during the first stage of the regenerative process. However, if the scaffold is too rigid and so fails to transmit load appropriately, it can slow the regenerative process and bone remodelling as bone requires daily loading for optimal homeostasis. A bone defect model may allow testing of the scaffold alongside fixation of the bone. If the combination of scaffold and fixator is too rigid, this can lead to load shielding from the healthy bone causing bone fragility in the surrounding area.[53] There has been very little research to date on bone regeneration and remodelling following removal of any fixator though Zhu *et al.*[50] showed that fixator removal did increase organised bone remodelling. These factors must be taken into account when evaluating results and extrapolating forward to the clinical situation.

3.6.2. *The cartilage defect model*

An *in vivo* cartilage defect model needs to mimic the correct clinical conditions and provide a microenvironment (no angiogenesis, appropriate mechanical stimuli and nutrient/gaseous/waste based upon diffusion exchange) optimised for chondrogenesis and regeneration of articular cartilage. Cartilage defects can be created on articular cartilage surfaces using either a scalpel blade/cork borer (chondral defect) or bone trephine (osteochondral defect). The anatomical location of any defect must be carefully considered in respect of whether the site is to be weight bearing or non-weight bearing, as well as the size and type of the defect according to the experimental design and the animal species being used. Cartilage defects are usually one of three main types: (1) partial thickness, (2) full thickness without cutting into the subchondral bone, and (3) osteochondral defects, with the defect traversing the full depth of cartilage and into the subchondral bone. An osteochondral defect allows more efficient tissue regeneration due to the in-growth of stem cells from the bone marrow.[54] However, the partial or full thickness cartilage defect (Fig. 5) is often used for testing implanted cells and/or materials with or without growth factors. The cartilage defect can be filled with stem/stromal cells, chondrocytes and/or biomaterials scaffolds in combination with or without growth factors.[55–57] The surface of the defect can be covered by periosteum[58] or synthetic membranes.[59]

The use of animal models for testing the efficacy of cartilage repair is limited due to the anatomical differences of human and animal cartilage. Hunziker,[60] pointed out that cartilage thickness and overall cell volume densities vary greatly across the different species. Furthermore, the need for repair of a joint is usually preceded by chronic degenerative disease (e.g. osteoarthritis, osteochondrosis). The presence of chronic degeneration, with multiple underlying causes is neglected in most animal models, and yet is the largest challenge to any reparative attempt.

4. Conclusion

To date, a number of different animal models have been used in tissue engineering strategies in attempts to mimic clinically relevant conditions and provide an appropriate permissive microenvironment (e.g. angiogenesis, mechanical stimuli and physiological/pathological conditions) for tissue regeneration. However, each model has its own advantages and limitations. It is concluded that when considering the use of an *in vivo* animal model for research in skeletal tissue regeneration, species, size and age of the animal must all be taken into account, in addition to its suitability in respect of the specific hypothesis being tested. Finally, the need to consider the ethical issues and work to the best practice of replacement, reduction and refinement while maintaining true statistical viability for any given study is essential in the selection of a suitable animal model.

References

1. Lieberman, J.R., Daluiski, A. and Einhorn, T.A. The role of growth factors in the repair of bone. Biology and clinical applications. *J. Bone Joint Surg. Am.* **84-A**, 1032–1044 (2002).
2. Nunamaker, D.M. Experimental models of fracture repair. *Clin. Orthop. Relat. Res.* **355**(Suppl.), S56–S65 (1998).

3. Kilborn, S.H., Trudel, G. and Uhthoff, H. Review of growth plate closure compared with age at sexual maturity and lifespan in laboratory animals. *Contemp. Top. Lab. Anim. Sci.* **41**, 21–26 (2002).
4. Jee, W.S. and Yao, W. Overview: animal models of osteopenia and osteoporosis. *J. Musculoskelet. Neuronal. Interact.* **1**, 193–207 (2001).
5. Devine, M.J., Mierisch, C.M., Jang, E., Anderson, P.C. and Balian, G. Transplanted bone marrow cells localize to fracture callus in a mouse model. *J. Orthop. Res.* **20**, 1232–1239 (2002).
6. van der Donk, S., Buma, P., Aspenberg, P. and Schreurs, B.W., Similarity of bone ingrowth in rats and goats: a bone chamber study. *Comp. Med.* **51**, 336–340 (2001).
7. Kucia, M., Dawn, B., Hunt, G., Guo, Y., Wysoczynski, M., Majka, M., Ratajczak, J., Rezzoug, F., Ildstad, S.T., Bolli, R. and Ratajczak, M.Z. Cells expressing early cardiac markers reside in the bone marrow and are mobilized into the peripheral blood after myocardial infarction. *Circ. Res.* **95**, 1191–1199 (2004).
8. Key, J.A. The effect of a local calcium depot on osteogensis and healing of fractures. *J. Bone Joint Surg. Am.* **16**, 176–184 (1934).
9. Hak, D.J., Makino, T., Niikura, T., Hazelwood, S.J., Curtiss, S. and Reddi, A.H. Recombinant human BMP-7 effectively prevents non-union in both young and old rats. *J. Orthop. Res.* **24**, 11–20 (2006).
10. Huibregtse, B.A., Johnstone, B., Goldberg, V.M. and Caplan, A.I. Effect of age and sampling site on the chondro-osteogenic potential of rabbit marrow-derived mesenchymal progenitor cells. *J. Orthop. Res.* **18**, 18–24 (2000).
11. Gan, J.H.Y. *Characterisation of Bone Defect Models in Immunodeficient Animals* (The University of New South Wales, 2005).
12. Huang, Q., Goh, J.C., Hutmacher, D.W. and Lee, E.H. *In vivo* mesenchymal cell recruitment by a scaffold loaded with transforming growth factor beta1 and the potential for *in situ* chondrogenesis. *Tissue Eng.* **8**, 469–482 (2002).
13. Cui, Q., Xiao, Z., Li, X., Saleh, K.J. and Balian, G. Use of genetically engineered bone-marrow stem cells to treat femoral defects: an experimental study. *J. Bone Joint Surg. Am.* **88**, 167–172 (2006).
14. Ueda, H., Nakamura, T., Yamamoto, M., Nagata, N., Fukuda, S., Tabata, Y. and Shimizu, Y. Repairing of rabbit skull defect by dehydrothermally crosslinked collagen sponges incorporating transforming growth factor [beta]1. *J. Control Release* **88**, 55–64 (2003).
15. Zhu, S.J., Choi, B.H., Huh, J.Y., Jung, J.H., Kim, B.Y. and Lee, S.H. A comparative qualitative histological analysis of tissue-engineered bone using bone marrow mesenchymal stem cells, alveolar bone cells, and periosteal

cells. *Oral Surg. Oral Med. Oral Pathol. Oral Radiol. Endod.* **101**, 164–169 (2006).

16. Isogai, N., Landis, W., Kim, T.H., Gerstenfeld, L.C., Upton, J. and Vacanti, J.P. Formation of phalanges and small joints by tissue-engineering. *J. Bone Joint Surg. Am.* **81**, 306–316 (1999).
17. Dennis, J.E., Haynesworth, S.E., Young, R.G. and Caplan, A.I. Osteogenesis in marrow-derived mesenchymal cell porous ceramic composites transplanted subcutaneously: effect of fibronectin and laminin on cell retention and rate of osteogenic expression. *Cell Transplant.* **1**, 23–32 (1992).
18. Eyckmans, J. and Luyten, F.P. Species specificity of ectopic bone formation using periosteum-derived mesenchymal progenitor cells. *Tissue Eng.* **12**, 2203–2214 (2006).
19. Yang, X.B., Webb, D., Blaker, J., Boccaccini, A.R., Maquet, V., Cooper, C. and Oreffo, R.O.C. Evaluation of human bone marrow stromal cell growth on biodegradable polymer/bioglass(R) composites. *Biochem. Biophys. Res. Commun.* **342**, 1098–1107 (2006).
20. Yang, X.B., Whitaker, M.J., Sebald, W., Clarke, N., Howdle, S.M., Shakesheff, K.M. and Oreffo, R.O.C. Human osteoprogenitor bone formation using encapsulated bone morphogenetic protein 2 in porous polymer scaffolds. *Tissue Eng.* **10**, 1037–1045 (2004).
21. Duda, G.N., Haisch, A., Endres, M., Gebert, C., Schroeder, D., Hoffmann, J.E. and Sittinger, M. Mechanical quality of tissue engineered cartilage: results after 6 and 12 weeks *in vivo*. *J. Biomed. Mater. Res.* **53**, 673–677 (2000).
22. Feeley, B.T., Conduah, A.H., Sugiyama, O., Krenek, L., Chen, I.S. and Lieberman, J.R. *In vivo* molecular imaging of adenoviral versus lentiviral gene therapy in two bone formation models. *J. Orthop. Res.* **24**, 1709–1721 (2006).
23. Fujii, T., Ueno, T., Kagawa, T., Sakata, Y. and Sugahara, T. Comparison of bone formation ingrafted periosteum harvested from tibia and calvaria. *Microsc. Res. Tech.* **69**, 580–584 (2006).
24. Saito, M., Shimizu, H., Beppu, M. and Takagi, M. The role of beta-tricalcium phosphate in vascularized periosteum. *J. Orthop. Sci.* **5**, 275–282 (2000).
25. Warnke, P.H., Wiltfang, J., Springer, I., Acil, Y., Bolte, H., Kosmahl, M., Russo, P.A., Sherry, E., Lutzen, U., Wolfart, S. and Terheyden, H. Man as living bioreactor: fate of an exogenously prepared customized tissue-engineered mandible. *Biomaterials* **27**, 3163–3167 (2006).
26. Bosch, P., Musgrave, D.S., Lee, J.Y., Cummins, J., Shuler, F., Ghivizzani, S.C., Evans, C., Robbins, P.D. and Huard, J. Osteoprogenitor cells within skeletal muscle. *J. Orthop. Res.* **18**, 933–944 (2000).

27. Partridge, K., Yang, X., Clarke, N.M.P., Okubo, Y., Bessho, K., Sebald, W., Howdle, S.M., Shakesheff, K.M. and Oreffo, R.O.C. Adenoviral BMP-2 gene transfer in mesenchymal stem cells: *in vitro* and *in vivo* bone formation on biodegradable polymer scaffolds. *Biochem. Biophys. Res. Commun.* **292**, 144–152 (2002).
28. Howard, D., Partridge, K., Yang, X., Clarke, N.M.P., Okubo, Y., Bessho, K., Howdle, S.M., Shakesheff, K.M. and Oreffo, R.O.C. Immunoselection and adenoviral genetic modulation of human osteoprogenitors: *in vivo* bone formation on PLA scaffold. *Biochem. Biophys. Res. Commun.* **299**, 208–215 (2002).
29. Gundle, R., Joyner, C.J. and Triffitt, J.T. Human bone tissue formation in diffusion chamber culture *in vivo* by bone-derived cells and marrow stromal fibroblastic cells. *Bone* **16**, 597–601 (1995).
30. Ashton, B.A., Allen, T.D., Howlett, C.R., Eaglesom, C.C., Hattori, A. and Owen, M. Formation of bone and cartilage by marrow stromal cells in diffusion chambers *in vivo*. *Clin. Orthop. Relat. Res.* **151**, 294–307 (1980).
31. Yang, X.B., Bhatnagar, R.S., Li, S. and Oreffo, R.O. Biomimetic collagen scaffolds for human bone cell growth and differentiation. *Tissue Eng.* **10**, 1148–1159 (2004).
32. Yang, X., Tare, R.S., Partridge, K.A., Roach, H.I., Clarke, N.M., Howdle, S.M., Shakesheff, K.M. and Oreffo, R.O. Induction of human osteoprogenitor chemotaxis, proliferation, differentiation, and bone formation by osteoblast stimulating factor-1/pleiotrophin: osteoconductive biomimetic scaffolds for tissue engineering. *J. Bone Miner. Res.* **18**, 47–57 (2003).
33. Nawata, M., Wakitani, S., Nakaya, H., Tanigami, A., Seki, T., Nakamura, Y., Saito, N., Sano, K., Hidaka, E. and Takaoka, K. Use of bone morphogenetic protein 2 and diffusion chambers to engineer cartilage tissue for the repair of defects in articular cartilage. *Arthritis Rheum.* **52**, 155–163 (2005).
34. Cheng, M.H., Brey, E.M., Allori, A., Satterfield, W.C., Chang, D.W., Patrick, C.W., Jr., and Miller, M.J. Ovine model for engineering bone segments. *Tissue Eng.* **11**, 214–225 (2005).
35. Thomson, R.C., Mikos, A.G., Beahm, E., Lemon, J.C., Satterfield, W.C., Aufdemorte, T.B. and Miller, M.J. Guided tissue fabrication from periosteum using preformed biodegradable polymer scaffolds. *Biomaterials* **20**, 2007–2018 (1999).
36. Roach, H.I., Baker, J.E. and Clarke, N.M. Initiation of the bony epiphysis in long bones: chronology of interactions between the vascular system and the chondrocytes. *J. Bone Miner. Res.* **13**, 950–961 (1998).
37. Borges, J., Tegtmeier, F.T., Padron, N.T., Mueller, M.C., Lang, E.M. and Stark, G.B. Chorioallantoic membrane angiogenesis model for tissue engineering: a new twist on a classic model. *Tissue Eng.* **9**, 441–450 (2003).

38. Stevens, M.M., Marini, R.P., Schaefer, D., Aronson, J., Langer, R. and Shastri, V.P. *In vivo* engineering of organs: the bone bioreactor. *Proc. Natl. Acad. Sci. USA* **102**, 11450–11455 (2005).
39. Halpern, J., Lynch, C., Fleming, J., Hamming, D., Martin, M., Schwartz, H., Matrisian, L. and Holt, G. The application of a murine bone bioreactor as a model of tumor: bone interaction. *Clin. Exp. Metastasis* **23**, 345–356 (2006).
40. Holt, G.E., Halpern, K.J.L., Dovan, T.T., Hamming, D. and Schwartz, H.S. Evolution of an *in vivo* bioreactor. *J. Orthop. Res.* **23**, 916–923 (2005).
41. Hillam, R., Jackson, M., Goodship, A. and Skerry, T. Comparison of physiological strains in the human skull and tibia. *Bone* **19**, 686–686 (1996).
42. Hong, L. and Mao, J.J. Tissue-engineered rabbit cranial suture from autologous fibroblasts and BMP2. *J. Dent. Res.* **83**, 751–756 (2004).
43. Peltola, M.J., Aitasalo, K.M., Suonpaa, J.T., Yli-Urpo, A. and Laippala, P.J. *In vivo* model for frontal sinus and calvarial bone defect obliteration with bioactive glass S53P4 and hydroxyapatite. *J. Biomed. Mater. Res.* **58**, 261–269 (2001).
44. Chen, T.M., Shih, C., Lin, T.F. and Lin, F.H. Reconstruction of calvarial bone defects using an osteoconductive material and post-implantation hyperbaric oxygen treatment. *Mater. Sci. Eng. C* **24**, 855–860 (2004).
45. Bonnarens, F. and Einhorn, T.A. Production of a standard closed fracture in laboratory animal bone. *J. Orthop. Res.* **2**, 97–101 (1984).
46. Einhorn, T.A., Lane, J.M., Burstein, A.H., Kopman, C.R. and Vigorita, V.J. The healing of segmental bone defects induced by demineralized bone matrix. A radiographic and biomechanical study. *J. Bone Joint Surg. Am.* **66**, 274–279 (1984).
47. Martin, R.B., Chapman, M.W., Sharkey, N.A., Zissimos, S.L., Bay, B. and Shors, E.C. Bone ingrowth and mechanical properties of coralline hydroxyapatite one year after implantation. *Biomaterials* **14**, 341–348 (1993).
48. Nuss, K.M., Auer, J.A., Boos, A. and von Rechenberg, B. An animal model in sheep for biocompatibility testing of biomaterials in cancellous bones. *BMC Musculoskelet. Disord.* **7**, 67 (2006).
49. Cook, S.D., Wolfe, M.W., Salkeld, S.L. and Rueger, D.C. Effect of recombinant human osteogenic protein-1 on healing of segmental defects in non-human primates. *J. Bone Joint Surg. Am.* **77**, 734–750 (1995).
50. Zhu, L., Liu, W., Cui, L. and Cao, Y. Tissue-engineered bone repair of goat-femur defects with osteogenically induced bone marrow stromal cells. *Tissue Eng.* **12**, 423–433 (2006).
51. Kokubu, T., Hak, D.J., Hazelwood, S.J. and Reddi, A.H. Development of an atrophic nonunion model and comparison to a closed healing fracture in rat femur. *J. Orthop. Res.* **21**, 503–510 (2003).

52. Sciadini, M.F. and Johnson, K.D. Evaluation of recombinant human bone morphogenetic protein-2 as a bone-graft substitute in a canine segmental defect model. *J. Orthop. Res.* **18**, 289–302 (2000).
53. Perren, S.M. Evolution of the internal fixation of long bone fractures. The scientific basis of biological internal fixation: choosing a new balance between stability and biology. *J. Bone Joint Surg. Br.* **84-B**, 1093–1110 (2002).
54. Shapiro, F., Koide, S. and Glimcher, M.J. Cell origin and differentiation in the repair of full-thickness defects of articular cartilage. *J. Bone Joint Surg. Am.* **75**, 532–553 (1993).
55. Huang, X., Yang, D., Yan, W., Shi, Z., Feng, J., Gao, Y., Weng, W. and Yan, S. Osteochondral repair using the combination of fibroblast growth factor and amorphous calcium phosphate/poly(L-lactic acid) hybrid materials. *Biomaterials* **28**, 3091–3100 (2007).
56. Holland, T.A., Bodde, E.W.H., Cuijpers, V.M.J.I., Baggett, L.S., Tabata, Y., Mikos, A.G. and Jansen, J.A. Degradable hydrogel scaffolds for *in vivo* delivery of single and dual growth factors in cartilage repair. *Osteoarthr. Cartil.* **15**, 187–197 (2007).
57. Hunziker, E.B. and Rosenberg, L.C. Repair of partial-thickness defects in articular cartilage: cell recruitment from the synovial membrane. *J. Bone Joint Surg. Am.* **78**, 721–733 (1996).
58. Pulliainen, O., Vasara, A.I., Hyttinen, M.M., Tiitu, V., Valonen, P., Kellomaki, M., Jurvelin, J.S., Peterson, L., Lindahl, A., Kiviranta, I. and Lammi, M.J. Poly-L-D-lactic acid scaffold in the repair of porcine knee cartilage lesions. *Tissue Eng.* **13**, 1347–1355 (2007).
59. Bartlett, W., Skinner, J.A., Gooding, C.R., Carrington, R.W.J., Flanagan, A.M., Briggs, T.W.R. and Bentley, G. Autologous chondrocyte implantation versus matrix-induced autologous chondrocyte implantation for osteochondral defects of the knee: a prospective, randomised study. *J. Bone Joint Surg. Br.* **87-B**, 640–645 (2005).
60. Hunziker, E.B., Biologic repair of articular cartilage: defect models in experimental animals and matrix requirements. *Clin. Orthop. Relat. Res.* **367**(Suppl.), S135–S146 (1999).

Chapter 36

In Vitro 3D Human Tissue Models for Osteochondral Diseases

Sourabh Ghosh and David L. Kaplan

Abstract

The use of animal models is essential in bridging *in vitro* studies with clinical outcomes for tissue repair and regeneration. However, *in vitro* cell culture systems and animal studies have significant limitations that impede translation of the research findings to successful clinical outcomes. This limitation also applies to the study and treatment of human osteochondral tissues and associated diseases. An alternative approach, based on a combination of tissue engineering and knowledge of developmental biology, can be employed to establish relevant *in vitro* 3D human disease model systems for osteochondral needs. To address this goal, tissue-specific microenvironments to simulate hydrodynamic fluid flow, biomechanical functions and cell biology have been established *in vitro*. As a result, more physiologically relevant human tissue engineered disease model systems can be established to provide better input into a variety of osteochondral diseases, as well as to prescreen treatment options.

Keywords: Tissue Engineering; Disease Models; Osteochondral; Osteoporosis; Osteoarthritis.

Outline

1. Introduction
2. Factors Governing the Simulation of Tissue Microenvironments
   2.1. Role of scaffold
   2.2 Role of bioreactor
   2.3. Cells
   2.4. Soluble factors
3. Osteochondral Tissue
   3.1. State of the art: 3D osteochondral tissue systems

3.2. Model of osteoarthritis
3.3. Gravitation models for bone loss
3.4. Osteoporosis model
4. Conclusions
References

1. Introduction

Tissue engineering research has made significant progress toward tissue constructs for the repair or replacement of lost functions in diseased or damaged organs. One important area of current research is the combination of tissue engineering principles and knowledge from developmental biology to establish relatively simple *in vitro* disease model systems, in order to gain insight into mechanisms of disease origins, pathological conditions and to screen treatment options. The successful development of engineered human tissue models, which mimic their *in vivo* counterparts, offers the potential to advance our understanding of human disease processes. This insight includes understanding intra- and intercellular signaling pathways in order to develop novel treatment modalities. There remains a need to develop *in vitro* engineered tissue systems which can simulate the three-dimensional (3D) morphology and microenvironment of the target tissue, including both normal and pathological states. Aside from appropriate cells and scaffolding, the environmental inputs to these systems include the appropriate biomechanical forces, hydrodynamic fluid transport of nutrients and metabolic waste products and growth factors.

Conventional two-dimensional (2D) *in vitro* cell culture systems provide a convenient, relatively inexpensive and high throughput platform for biochemical analysis. Accumulating evidence suggests that three-dimensional (3D) cell culture models are fundamentally superior to 2D systems with respect to mimicking physiologically relevant conditions.[1,2] Cells respond to 3D architectures in ways that can not be replicated in standard 2D cell culture formats, especially with respect to complex cues related to homotypic or heterotypic cell adhesion, mechanical forces and biochemical signals. For example, the production of milk protein by mammary epithelial cells needed complex cell-cell, cell-ECM organization, and rat mammary epithelial cells failed to produce milk protein, whey acidic protein, when cultured in monolayers.[3] Even when scattered single cells were cultured on reconstituted basement membrane (matrigel) the milk protein was not generated. However, when mammary epithelial cells formed into multicellular architectures with hollow lumen (acini), milk protein was produced in response of lactogenic stimulation.[4,5] Basal epidermal cells do not express involucrin, a marker of late-stage differentiation, when they are cultured in monolayers or on basement membrane, but in non-adherent culture they express involucrin, the major precursor of the cornified envelop involved in epidermal wound healing.[6]

3D tumor systems engineered *in vitro* showed that the phenotype and functional characteristics of tumor cells is significantly changed. Melanoma antigen-specific cytotoxic T-lymphocytes recognized and successfully killed cancer cells, as anticipated, when they were cultured as monolayers. In contrast, antigen recognition and functional activities of T-lymphocytes were significantly inhibited when tumor cells were grown in multicellular 3D architectures, a potential escape mechanism of cancer cells from the immune recognition process.[7,8] 3D aggregates of melanoma cells modulated gene expression patterns compared to conventional monolayer culture.[9] Interestingly, many of the genes that are upregulated in 3D tumor tissue models are also upregulated in melanoma tumors *in vivo*; these genes are known to play an important role in melanoma progression. These studies suggest that commonly used monolayer cell culture systems may not provide the best representation of cellular microenvironments present in human tissues, and thus their utility toward the study of human diseases is reduced.

Current clinical research depends on the use of animal models to understand the mechanism of human diseases at cellular and molecular levels.[10,11] Transgenic or genetically engineered animal models of diseases play an instrumental role in the design of therapeutic options for human diseases and for the development of clinical protocols, particularly when human experimentation is not feasible. In the post-genomic era, the contribution of animal models to human clinical research requires some re-evaluation,[12] due to the differences in genomics[13] as well as the costs associated with the use of animals. Fundamental difference between mouse and human tissue systems with respect to disease thresholds, telomere regulation mechanisms,[14] and other issues question the relevance of such animal models. Unfortunately many new investigational drug candidates fail in Phase I and II clinical trials because commonly used animal models are insufficient to represent the human tissue systems. This high failure rate necessitates the search for better drug screening assays and simple human cell-based disease models to better predict tolerability and efficacy. A relatively simple *in vitro* tissue engineered human disease model can be used to generate novel early-stage diagnostic tools, facilitating inexpensive and rapid testing of new therapeutic approaches *in vitro*. The treatment of chronic-debilitating diseases such as cancer, arthritis and osteoporosis are potentially addressable with such an approach. These diseases present significant social and financial burdens to society and more effective predictive treatment tools are urgently needed.

To address this need, new models for the *in vitro* study of diseases through the use of human tissue engineered model systems are emerging.[8,15,16] The potential impact using this approach is immense. Based on knowledge of the human genome, understanding of stem cells and cell biology, and insight into engineering aspects required to maintain 3D tissue systems, these *in vitro* systems are contributing to this progress. Related research arenas are also expanding rapidly; for example, novel 3D biomaterial scaffolds as extracellular matrix analogs and a comprehensive resource of well-characterized affinity reagents for analysis of the human proteome.[17] The establishment of human cell-based *in vitro* engineered disease model systems could represent a paradigm shift from inadequate conventional monolayer cell cultures, or animal models, towards more physiologically tissue-relevant, patient-specific human tissue approaches.

In this chapter, we focus on how tissue engineering principles can be used to establish relatively simple, yet relevant, *in vitro* disease model systems for the understanding and study of human osteochondral pathologies (Table 1).

2. Factors Governing the Simulation of Tissue Microenvironments

In tissue engineering, a scaffold is usually biodegradable to offer full restoration of native tissue structure and function. The scaffold also provides the architecture for extracellular matrix deposition. Cells can be collected from a patient by biopsy and then cultured in monolayers to increase numbers prior to seeding on the scaffold. Specialized "bioreactors" are usually used to provide nutrients, oxygen and physiologically relevant environments. When the cells multiply to fill the scaffold and deposit new extracellular matrix to resemble native tissue, the tissue construct can be removed from the bioreactor and implanted into the body for restoration of native tissue function. Hence, the proper combination of: (a) selected population of undifferentiated cells or specialized cells, (b) a biomaterial having appropriate chemistry, structure and morphology, and (c) a bioreactor to deliver appropriate physiological cues are required to give rise to *in vitro* engineered tissues (discussed in detail in Chapters 24 and 25).

2.1. Role of scaffold

For simulating a tissue microenvironment, fabrication of suitable scaffolds is critical in order to control cell adhesion, cell proliferation towards tissue-specific outcomes, to regulate expression of a specific phenotype and to encourage extracellular matrix deposition in a predictable and controlled fashion. The physical architecture of a scaffold can control cell functions by regulating diffusion of nutrients, metabolic waste products and influencing cell-cell interactions.

Table 1. Selected *in vitro* model systems for the study of osteochondral disease pathologies.

Disease	Cells or tissues	Culture conditions (scaffolds/bioreactor)	Soluble factors	Outcomes	Ref.
Osteoarthritis	Healthy non-inflamed porcine synovial tissue co-cultured with OA cartilage fragments.	Static culture.		Progressive breakdown of cartilage matrix by the resident chondrocytes.	129
	Cartilage explants co-cultured with synovial membrane-derived cells expressing IGF-1 and IL-1Ra.	Monolayer, static culture.	IGF-1 and IL-1Ra	Anabolic action of IGF-1 and the catabolic blocking of IL-1Ra partially restored cartilage matrix.	125
	Articular cartilage and synovial explants from canine cadaveric tissues or tissues excised during surgery.	Presence or absence of synovium; presence or absence of IL-1β; static culture.	IL-1β	GAG and collagen expression were compared in normal and OA cartilage co-culture model.	121
	Bovine articular cartilage explants.	Cyclic loading (continuous, uniaxial, unconfined compression applied, 1 Hz, peak stress 0.1–0.5 MPa (6% strain) for 1–16 hrs).		Mechanical loading increased expression and activation of MMPs-2, -9, expression of tissue inhibitors of MMPs was unaffected.	133

(*Continued*)

Table 1. *(Continued)*

Disease	Cells or tissues	Culture conditions (scaffolds/bioreactor)	Soluble factors	Outcomes	Ref.
Bone loss during spaceflight	Human bone marrow-derived mesenchymal stem cells seeded on polystyrene microcarrier beads.	Rotary Cell Culture System, Synthecon, 7-day culture.		Reduced expression of collagen-I, focal adhesion kinase, MAP kinase, reduced osteoblastic differentiation.	157
	Mouse osteoblast cells.	Rotary Cell Culture System, Synthecon.	Osteoblasts conditioned media.	Osteoblasts conditioned media subjected to microgravity increased osteoclastogenesis, bone resorption in mouse bone marrow cultures. OPG-treatment inhibited osteoclastogenesis.	163
Osteoporosis	Human osteoblasts.		Low molecular weight heparin.	Inhibition of osteoblast growth.	169

The surface chemistry of scaffold can affect cell adhesion, cell morphology and cellular activity by regulating adsorption of proteins and influencing cellular differentiation. Traditional approaches to fabricate 3D tissues have focused on modulating the non-adhesive properties of the culture substrate[18–21] or on the use of traditional scaffold fabrication techniques, such as solvent casting, temperature-induced phase separation, non-woven fiber meshes,[22] particle-leaching techniques[23] and gas foaming.[24] These latter approaches can be used to form porous interconnected polymer architectures, hence process driven strategies. Thus by using these methods complex architectures with tunable micro- and macro-scale features can be difficult to achieve. Recently, significant expansion in the design of biomaterials has been attained using rapid prototyping microfabrication manufacturing techniques, such as stereolithography, selective laser sintering,[25,26] depositing molten polymers[27] or by stacking large numbers of polymer films.[28] Such micro-patterned scaffolds with highly interconnected pores may be useful for an improved understanding of role of architectural complexity in tissue-specific differentiation. These approaches enable the fabrication of controlled multicellular aggregate structures required for the formation of complex microenvironments.[29]

Although many biomaterials have been explored for engineering osteochondral constructs, such as collagen and polyglycolic/polylactic acids, none of these satisfy all of the requirements for a functional system. In our recent studies we have demonstrated that silk fibroin-based protein biomaterial scaffolds offer such attributes. Protein-based polymers, and in particular silks, can address these needs due to the following combination of properties.

(i) Biocompatibility: Silk fibroin is biocompatible and has been used as surgical sutures for decades as an FDA approved biomaterial. Comprehensive studies, both *in vitro* and *in vivo*, demonstrate that silk fibroin is less immunogenic and inflammatory than other commonly used polymeric degradable biomaterials.[30–32] Further, unlike bovine-derived collagens, silks have no known bioburdens.

(ii) Mechanical Properties and Stability: Silk fibroin exhibits remarkable mechanical strength and toughness, as well as compressive strength, which exceeds other commonly used degradable polymeric biomaterials.[33] In addition, thermal stability is a hallmark of silk fibroin materials; they can be autoclaved without loss of mechanical integrity.[30] Silks achieve these outcomes without chemical or photo-initiated crosslinking, unlike other polymeric biomaterials such as collagen. These properties are derived from the β-sheet secondary structures, which are stable physical crosslinks formed via hydrogen bonding and hydrophobic interactions via inter- and intra-chain interactions. Silk biomaterials are digestable by proteases.

(iii) Ease of Modification: Silk fibroin is easily modified with selective cell binding or cell activation features. The use of unmodified silk fibroin from the silkworm, *Bombyx mori*, which does not contain any specific cell-binding domains and is predominately hydrophobic, suffices for most cell-based studies. However, chemical decoration with arginine-glycine-aspartate peptides (RGD), bone morphogenic protein (BMP-2) and other cell modulating factors has been demonstrated using facile carbodiimide chemical coupling.[34,35] This approach provides options to functionally modify the protein for selective needs. Silk fibroin mats decorated with hydroxyapaptite and BMP-2 supported the differentiation of human bone marrow-derived mesenchymal stem cells (hMSCs) to osteogenic cells, with enhanced calcium deposition and upregulation of osteogenic markers. Inducement of bone markers from hMSCs was also found when BMP-2 was covalently immobilized versus adsorbed on silk fibroin films, based on enhanced alkaline phosphatase, calcium deposition, expression of collagen type I, bone sialoprotein, osteopontin and BMP-2.[36,37]

(iv) Slow Degradability: Silk can be designed to degrade relatively fast (weeks) to very slow (years) depending on the mode of processing,[27] so that degradation rate can be controlled to match tissue growth.[33,38,39] Silk films and hydrogel systems offer versatile platforms for controlled release of soluble factors[40–43] for tissue engineering.

(v) Aqueous Processing: The formation of silk-based biomaterials (sponges, films, gels) can be accomplished in all water environments, thus allowing the incorporation of bioactive compounds, such as cell signaling factors, without loss of function. For example, electrospun matrices with embedded BMP-2 have been generated and used to improve bone formation as described earlier.[37]

2.2. Role of Bioreactor

Bioreactor systems provide controlled physicochemical culture conditions necessary to support engineered tissues (see also other chapters in this book). Uniform cell seeding is important and the most commonly used methods are based on pipetting a dense cell suspension onto porous scaffolds. This approach is not highly controlled. Stirred-flask bioreactors can enhance control of seeding if the scaffold is relatively thin and porous; however, often higher cell densities are found on the surfaces ("donut effect"), due to insufficient penetration or survival

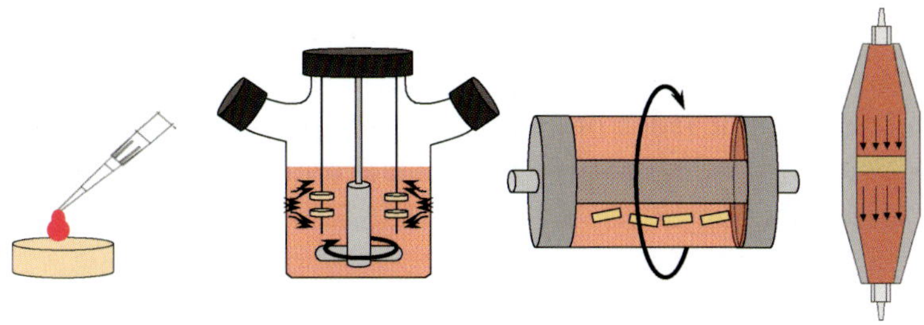

	Static cell seeding method	Stirrer Flask Bioreactor	Rotating wall Vessel Bioreactor	Perfusion Bioreactor
Cell seeding uniformity	User dependent High variablility	Improved uniformity than static	Improved uniformity than stirred flask	Improved uniformity than other systems
Fluid flow pattern	None-Gravitational settling	Turbulent flow-around the scaffold	Laminar flow-around the scaffold	Laminar flow-through the scaffold
Mass transfer	Molecular diffusion	Turbulent convection	Convection	Convection due to continuous recirculation
Medium exchange	Batch wise	Batch wise	Batch wise	Continuous recirculation

Fig. 1. Advantages and disadvantages of static and dynamic seeding: dynamic seeding can generate uniform tissue microenvironments.

of cells in interior regions. Dynamic laminar flow in rotating wall vessels is associated with reduced shear and enhanced uniformity of cell seeding and mass transport (Fig. 1). Perfusion bioreactors are effective and reproducible for uniform seeding and in generating well-defined physicochemical culture environments, important for the formation of tissue engineered disease models with clinically relevant sizes.

Bioreactors that apply dynamic mechanical loading to simulate physiological loading stresses of a load-bearing joint can be used to establish *in vitro* engineered tissue models based on human cartilage or bone. Any osteochondral joint *in vivo* is continuously subjected to intermittent cyclic stress. These stresses can be classified into two categories: compressive stress or shear stress. In native tissue, compressive stress is supported by an incompressible fluid component, whereas shear stress is supported by the matrix architecture. Dramatic effects on cartilage[44–47] and bone[48,49] tissue development have been demonstrated in response of dynamic stress, but these responses depend upon specific magnitude and/or frequency applied, as well as the stage of tissue development. Dynamic compression can enhance cartilage ECM production, whereas fluid flow-induced shear has been also shown to promote osteopontin and osteocalcin gene expression in MSCs.[48,49] Human chondrocyte-based *in vitro* models were used to show that dynamic compression can reverse interleukin-1β induced nitric oxide and prostaglandin (PGE2) release.[50] Hence, application of mechanical loading can increase the relevance of *in vitro* models for the joint environment and the proper selection of loading regime to simulate a diseased joint can provide insight into disease pathology.[51]

2.3. Cells

Selection, characterization and handing of cells are crucial to cell vitality and cell-cell and cell-ECM interactions. Depending on the expression of cell surface adhesion receptors cells respond selectively to the scaffold surface chemistry to transduce these surface signals to intracellular transcriptional processes. These processes need to be clearly understood for simulating targeted 3D tissue microenvironments.

When primary cells are isolated from a diseased organ and used to develop an *in vitro* tissue model there are risks of losing typical physiology, senescence related events, and de- or trans-differentiation.[52] Primary cells can be cultured only up to a certain number of passages ("Hayflick limit"), after which they stop dividing and enter a "senescent" state.[53] Similar to other somatic cells, adult human mesenchymal progenitor cells also undergo telomere shortening with each cell division, despite the ability of telomerase to repair telomeres after each cell division, which eventually causes a cessation in cell proliferation. Although MSCs are present in the human body throughout life, the total number reduces with age, and often also depends upon the site of extraction and the systemic disease pathology.[54] Incorporation of basic fibroblast growth factor (bFGF) in the culture medium can delay, but does not eliminate, cellular senescence. Long-term growth kinetics and osteogenic differentiation potential of hMSCs have been studied and it was found that the cells averaged 38 ± 4 population doublings following consecutive subculture and cryopreservation before reaching senescence.[55] Retroviral transduction of the telomerase gene into adult hMSCs extends the proliferative ability maintaining the undifferentiated, multi-lineage differentiation potential.[56] Embryonic stem cells, derived from the inner cell mass of the embryonic blastocyst, can be cultured indefinitely *in vitro* without loss of differentiation potential. Maintaining their undifferentiated phenotype or inducing a lineage-specific differentiation path can be challenging.

Gene transfer technology can generate genetically modified cells to simulate disease tissue models *in vitro*. A patient's cells can be harvested, expanded, and a specific cell population can be transfected prior to development of *in vitro* engineered tissue to screen functionality before re-implantation. Compared to primary cells, if "immortalized" cell lines are used to prepare tissue *in vitro*, these cells can grow continuously. Conventional methods for gene delivery include both *in vitro* and *in vivo* gene transfer based on viral vectors or DNA-polymer matrix complexes. During *in vivo* gene transfer, a vector carrying the therapeutic DNA is directly implanted or injected into the patient. However, it is difficult to target gene delivery to a specific cell population and cells of surrounding tissues are often transfected. This approach is also dependent on a substantial number of host cells being available for transfection but it is often limited by inefficient transient delivery and poor specificity.[57,58] Immortalized cell lines generated by infecting primary hMSCs with human papilloma virus Type16 with E6/E7 genes within a viral vector, maintained their multi-lineage differentiation potential toward osteogenic, chondrogenic, and adipogenic lineages, while exceeding the lifespan of normal adult hMSCs for up to a year in culture.[59]

Cell lines are useful for developing *in vitro* diseased tissue models but their application for clinical translation must be carefully monitored due to the genetic modifications which can give rise to questions of relevancy, and malignancy. However, immortalized cells are not always tumorigenic cells.[60] Such systems provide insight about osteochondral disease; for example, human chondrocytes are modified by transferring genes encoding the interleukin-1 receptor antagonist (IL-1RA), which blocks IL-1 induced degeneration during arthritis.[61] When these transformed chondrocytes were implanted onto the surface of osteoarthritic cartilage in organ culture, they were able to protect cartilage matrix from IL-1 induced degradation. RNA interference strategies (reviewed in Ref. 62), such as selective knockdown of cathepsin B in chondrocytes *in vitro* resulted in a 70% decrease in expression of cathepsin B, and led to reduced cartilage destruction. This result indicated that cathepsin B could be useful as a potential target in osteoarthritis treatment.[63] There remain critical gaps in our understanding of disease mechanisms and gene transfer technologies,[64] and tissue engineered *in vitro* disease models can help to fill this crucial gap.

Exciting recent finding suggests that a higher proportion of the mammalian genome is transcribed into RNA than previously thought. More than one third of mammalian coding transcripts are silenced by microRNAs. MicroRNAs (miRNAs) constitute a class of small non-coding RNAs that appear to have a central role in controlling gene expression. Tissue engineering based *in vitro* initiatives to explore such mechanisms have potential to expand fundamental understanding about contributions of these mechanisms in cell biology for many pathological conditions.[65]

2.4. Soluble factors

A clear understanding of the action of growth factors, cytokines, chemokines, protease inhibitors, and kinases can shed light on the signaling cues in tissue regeneration or degeneration processes. These soluble factors can either be exogenously added, immobilized within permeable polymeric structures, chemically bonded with a polymer scaffold, or delivered via genetically modified cells (as described earlier). The incorporation of regulatory molecules can play a role in tissue engineering by providing appropriate stimulation to the cells in the proper developmental sequence, with appropriate dose and duration. While this remains a challenging goal due to the lack of insight into the native processes involved, *in vitro* tissue engineered systems offer a direct route to understand some of these complexities due to the controlled nature of these systems and the ability to more systemically explore variants both in terms of concentration, time of delivery and combinations of factors.[66]

Transforming growth factor-β (TGF-β) has a growth-stimulatory role on mesenchymal cells and a growth-inhibitory role on epithelial cells.[67] Epidermal growth factor (EGF) and other members of the EGF family are chemotactic and mitogenic for epithelial cells. bFGF can regulate the synthesis and deposition of various ECM components. Basic FGF is also a potent mitogen for vascular endothelial cells. VEGF induces the formation and maintenance of blood vessels, endothelial cell growth, cell migration, and inhibition of apoptosis. VEGF-A, -B, and -E play important roles in adult vascular angiogenesis, while VEGF-C and –D regulate lymphatic vessel growth. Proper concentration and sequential delivery of growth factors will be critical for simulating a tissue-specific microenvironment. For example, VEGF-related angiogenic effects depended on the microenvironmental concentration of VEGF, but not on total exogenously added VEGF.[68]

Silk fibroin scaffolds can be fabricated with a wide range of degradation rates, by varying β-sheet (crystalline) content, for controlled delivery of growth factors.[69] We have encapsulated gradients of different chondrogenic or osteogenic inducers in these protein scaffolds to regulate hMSCs and to mimic their tissue-specific environment. Controlled sequential delivery of growth factors to hMSC culture can induce optimal matrix development as seen by enhanced matrix ingrowth and collagen type I production.[70]

3. Osteochondral Tissue

The main function of articular cartilage is to provide low friction, efficient load-bearing and load distribution between bones. Cartilage is composed of one cell type — chondrocytes, embedded within a highly specialized extracellular matrix of collagens (type-II, IX, XI), proteoglycans, noncollageneous proteins and 70% water. The permanently charged macromolecules, proteoglycans and glycosaminoglycans, provide hydration and electrostatic load, whereas the collagen network provides tension; this special tissue composition provides a homeostatic balance between physicochemical properties of tissue loading and the biosynthetic activity of chondrocytes. Despite the relatively simple avascular structure of cartilage, the inability of articular cartilage to repair itself when this tissue is damaged remains a clinical and scientific challenge. Orthopaedic clinicians have placed a variety of "model tissues" into or between damaged articular surfaces (e.g. placement of fascia, bladder, periosteum, perichondrium) and have used a broad spectrum of operations (drilling, abrasion, resection, or microfracture of ulcerated regions of articular cartilage, and subchondral bone) with the intention of improving the repair response. All of these strategies have been for the most part unsuccessful. For cartilage tissue engineering, chondrocytes must be expanded *in vitro* to provide sufficient numbers of cells, however, human chondrocytes have a limited capacity to proliferate, which is dependent on the age of the donor.[71] Tissue engineering of cartilage-like tissues have been extensively reported,[27,72–76] and offers the potential to restore cartilage defects and to serve as a starting point for cartilage-related *in vitro* disease models.

Bone has an excellent capacity to self-repair or regenerate after trauma. The regeneration process is impaired when there is movement in the area of the defect which disrupts the interfacial region of the repair site, or in case of several pathologic events ranging from osteoporosis, periodontal disease, congenital inherited disease (e.g. *Osteogenesis imperfecta*), rheumatoid arthritis and others. Bone tissue is composed of cellular components and extracellular matrix (organic and calcified/mineralized matrix). The cellular components of bone include osteoblasts, osteoclasts, osteocytes, stem cells (both hematopoietic and mesenchymal), bone marrow stroma (composed of various cell types such as adipocytes, fibroblasts), nerve, and endothelial cells. Osteoblasts are terminally differentiated cells responsible for the production of inorganic bone matrix and regulation of mineralization. Osteoclasts are large multinucleated cells that resorb bone. Bone formation by osteoblasts and resorption by osteoclasts result in continuous bone remodeling. Over the last decade, engineered bone-like tissues have been formed from biocompatible materials and osteogenic cells.[23,32,34,36,37,48] These systems can be used as *in vitro* models to study bone-related diseases.

We have investigated the effect of scaffold microarchitecture with hMSCs for osteogenic tissue formation. Upon exposure to osteogenic factors, the hMSCs produced mineralized nodules in porous silk scaffolds; however, the size and distribution of these nodules was influenced by the initial pore structure of the scaffold. Specifically, in scaffolds with pore sizes of 100 to 200 μm the bone nodules were small and highly connected, resembling cortical bone. When pore sizes of 400 to 500 μm were used the sizes of the nodules increased, and they exhibited an open structure that more closely resembled that of native trabecular bone.[23] When non-mineralized scaffolds were used, hMSC-seeded collagen sponges exhibited considerably less mineralized deposits than silk scaffolds because they degraded too quickly and were unable to provide the proper architecture for tissue development due to loss of transport.[32]

3.1. State of the art: 3D osteochondral tissue systems

Custom-designed tissue engineered osteochondral grafts of specific sizes and shapes is an option for the repair of defects where both cartilage and subchondral trabecular bone are damaged. These osteochondral grafts, coupled with biochemical cues to integrate with native tissue, can be implanted in the area of the defect in load bearing cartilage and underlying bone. These osteochondral grafts could also serve as controlled *in vitro* models, simulating microenvironments of an osteoarthritic joint, in order to study pathophysiological interactions between physical forces and soluble factors on the engineered cartilage. Such *in vitro* models could be used as 3D human tissue model systems for drug screening to facilitate the identification and validation of effective drugs for osteoarthritis.

Attempts have been made to fabricate such osteochondral grafts by using different cell populations such as primary bovine chondrocytes and bovine periosteal cells,[77] human fetal chondrocytes and human fetal osteoblasts,[78] and adult human articular chondrocytes and hMSCs.[79] hMSCs are particularly advantageous for the formation of *in vitro* tissue engineered osteochondral constructs due to their ability to differentiate into both chondrogenic and osteogenic lineages. Undifferentiated hMSCs can be separately cultured for inducing differentiation to bone or cartilage, and then the two constructs can be sutured or glued together and cultivated in culture medium containing supplements essential to the development of both lineages.[80–82] An osteochondral construct using one scaffold of poly-lactic acid has been reported.[83] First separately cultured hMSCs in a medium supplemented with TGF-β1 was used to support chondrogenesis. In parallel hMSCs were cultured in a monolayer with factors inducing osteogenesis (β-glycerophosphate and dexamethasone). These two systems were combined, to form a bilayer construct and were cultured in a medium containing both chondrogenic and osteogenic factors, resulting in composites consisting of both a hyaline cartilage-like layer and a dense bone-like component, with a well-developed transition zone between the two components.

The generation of large-sized (8 mm diameter × 4 mm length) tissue engineered osteochondral grafts has also been reported,[84,85] using silk-based porous scaffolds and dynamic culture conditions (Fig. 2). Enhanced mass transfer in rotating bioreactors compared to static culture environments, and osteogenic effects of supplementation with rhBMP-2 resulted in improved mineralization and abundant osteogenic matrix formation. Cartilage-bone integration was most enhanced in

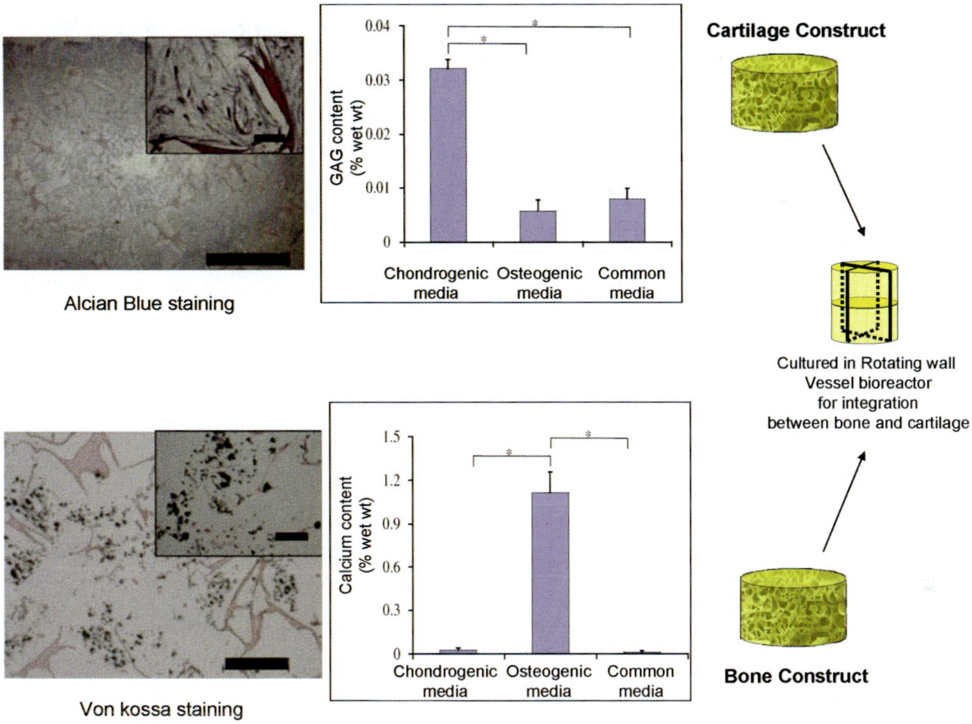

Fig. 2. Strategy to generate oseochondral tissue constructs. Cartilagenous constructs were characterized by Alcian Blue staining for glucosaminoglycan deposition. Bone was characterized by Von Kossa staining. Culturing in RWV bioreactors resulted in enhanced integration between bone and cartilage.

composites cultured in chondrogenic medium, where the interfacial region of composites showed a network of collagen spanning the cartilage and bone regions.

3.2. Model of osteoarthritis

Osteoarthritis (OA) is not a life threatening disease, but onset of arthritis leads to excruciating joint pain, swelling, and often causes significant reduction in the quality of life. The etiology of osteoarthritis is poorly understood but appears to result from the imbalance between anabolic (synthesis) and catabolic (resorptive) activities of chondrocytes. Under normal physiological conditions a fine balance is maintained to support the structural and functional integrity of cartilage. When this balance is lost, the loss of ECM components, such as proteoglycans and the disruption of collagenous fibrillar networks, initiate the tissue degeneration process leading to chondrocyte apoptosis and deterioration of cartilage morphology.[86]

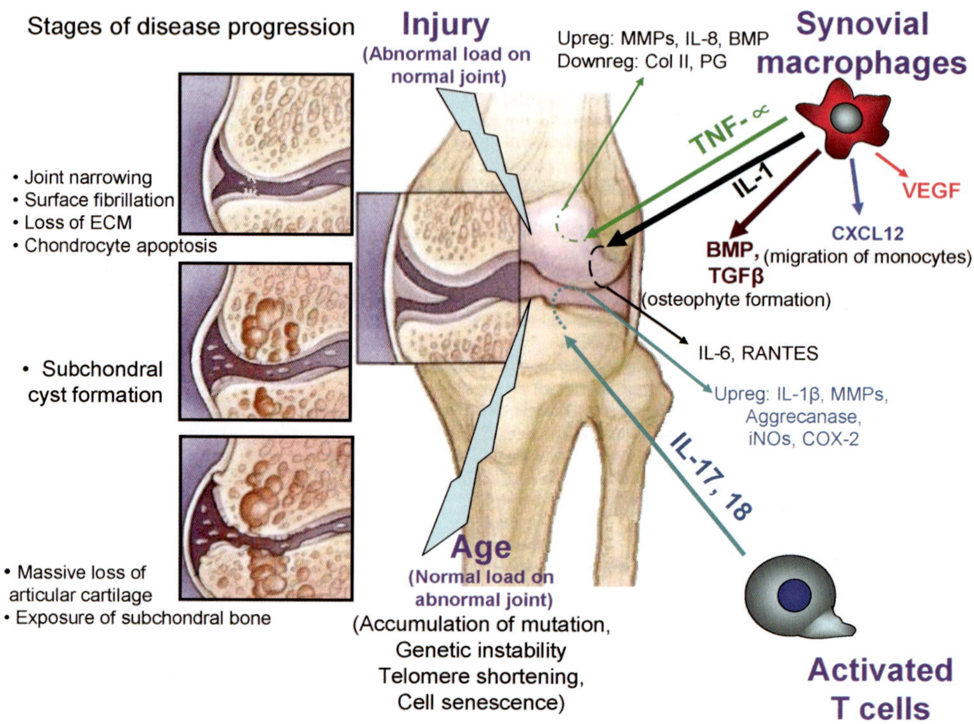

Fig. 3. Pathophysiology of osteoarthritis disease progression: an engineered *in vitro* OA disease model should address the involvement of articular cartilage, subchondral bone, ligaments and synovial membrane (adapted from Ref. 172).

OA probably comprises a combination of several interconnected factors: age, mechanical events (micro-injury, mechanical stress, obesity), programmed cell death (apoptosis), or genetic predisposition (mutation of matrix molecules matrix degradation, local inflammatory processes and increased cytokine activity) (Fig. 3). Primary OA is predominantly a degenerative disease of articular cartilage, which is caused by either normal forces applied on abnormal cartilage, or abnormal forces acting on normal cartilage. If the patient has a prior history of trauma that was left untreated, it is classified as secondary OA. Traumatic arthritis can occur at any age, if the defect is left untreated, but the prevalence of primary (idiopathic) OA gradually increases with age. With age, chondrocytes lose their ability to maintain metabolic homeostasis, mostly due to cell senescence, cumulative oxidative damage and genetic mutations, telomere shortening, and increased expression of senescence-associated β-galactosidase.[87,88] Injury caused by excessive or normal mechanical impact or stress could be a potential initiator of OA, which can induce cartilage catabolism and degradation of the matrix,[89] which in turn generates fragments of proteoglycans, collagens and fibronectin that can stimulate production of matrix-degrading proteinases.[90,91]

Inflammation within cartilage or in surrounding synovial tissue can cause severe changes in the expression of proinflammatory soluble factors, which can potentially initiate OA-related degradation. IL-1 has been defined as "catabolin," as IL-1 had been found to degrade the cartilage matrix by stimulating chondrocytes to produce proteinase, nitric oxide (NO), PGE2, metalloproteinases (MMPs) and to inhibit the synthesis of proteoglycans and type II collagen. TNF-α and IL-1 can synergistically induce cartilage damage, the inhibition of proteoglycan synthesis,[92] and the increased production of nitric oxide synthase, leading to an increase of NO radicals which can down regulate matrix synthesis and upregulate matrix degradation via the activation of MMPs. Initially it was thought that MMPs solely degraded collagen (MMP-13 cleaves collagen type II), elastins and other ECM components. Aggrecans can be cleaved by MMPs (aggrecanase), but it is currently thought that the enzymes ADAMTS-4 and ADAMTS-5 (a disintegrin and metalloproteinase with thrombospondin motifs) are mostly responsible for the initial cleavage of aggrecan.[93–95] Many attempts have been made to identify causative genetic factors associated in development of OA.[96,97]

Significant efforts have been made in the last 30 years to develop disease models for understanding the pathogenesis of OA, either by surgical intervention or by chemical induction. However, these models only partially simulate the complexity of human OA pathology. Small animal models (e.g. mouse, rabbit) to large models (e.g. horse) have been used to form a partial-thickness cartilage defect or osteochondral model defect. The defect size and reproducibility can be challenging as articular cartilage thickness is different between species, and no animal model is directly comparable to the human.[98] The selection of an appropriate animal age is also debatable, as there might be differences in structure and function of cartilage from juvenile and senescent animal models.[99] In general, commonly used OA animal models (reviewed in Refs. 100 and 101) can be categorized as follows:

(i) Spontaneous degeneration of unknown, but genetic origin — Normal aging animals[102] or genetically modified animals, which have no history of injury have been used as spontaneous OA models. Some genetically modified mouse strains (e.g. STR/ort or C57BL mice)[103] show spontaneous OA development at an early age.
(ii) Degeneration due to a mutation in a cartilage extracellular matrix gene — Point mutations in ECM molecules of articular cartilage have been reported.[104] The structure of ECM molecules can be altered by mutations in enzymes that cause post-translational modifications of collagen and side chains of proteoglycans. Defects in collagen type-II

(Col2A1 with large internal deletions),[105] overexpression of the Col2A1 gene,[106] and complete knockout of Col2A1[107] in experimental animal models demonstrated some characteristics of OA. Animal models with mutations in other ECM components (e.g. proteoglycans)[108] or β_1 integrins[109] are also available which show modulation in cartilage metabolism.

(iii) Surgically induced instability, which leads to degeneration — Injury, obesity or genetic predisposition can cause muscle weakness, joint instability and misalignment, which can promote OA. Surgically induced OA disease models include various combinations of menisectomy, collateral and/or anterior cruciate ligaments or myectomy.[110–112] Anterior cruciate ligament transection can simulate characteristics of injury-induced arthritis, osteoblasts proliferate along the joint margin to form osteophytes and synovial cells proliferate causing synovial cell hyperplasia. However, this model does not simulate primary OA.[113–115]

(iv) Chemically induced OA — Intra-articular injection of iodoacetate or papain[116] can cause a constellation of changes in load-bearing joint physiology, such as inhibition of chondrocyte metabolism, where injection of immunotoxins[117] causes selective joint denervation. Degenerative joint pathology can also be induced by intra-articular injection of collagenase, which causes instability of ligaments and tendons, and subsequently leads to OA pathology. However no direct damage to cartilage is observed in this model. Attempts have been made to improve this model by incorporating other cell populations, such as macrophages.[118]

These animal models provide significant understanding about degenerative joint diseases and have been used to demonstrate high levels of collagen synthesis associated with OA.[119] Animal models intra-articular injection of IL-1 resulted in proteoglycan loss while the inhibition of IL-1 via IL-1 receptor antagonist (IL-1RA) slowed progression of cartilage loss.[120] IL-1RA is a secreted protein that binds to interleukin-1 receptor (IL-1R), thereby blocking IL-1R downstream signaling, and inhibiting the pro-inflammatory properties of IL-1a and IL-1b. Although these animal models of OA have characteristics somewhat similar to the human pathology, none has proven to be a complete model of OA.

In light of the above challenges, the design of a simple *in vitro* disease model of OA may require the involvement of complex cross-talk among articular cartilage, subchondral bone, ligaments and synovial membrane. A hallmark of OA includes fibrosis in the synovium, changes in composition of synovial fluid, changes at the subchondral bone (early changes like increased trabecular bone and stiffness, late-stage changes like bone outgrowth-cyst and osteophytes formation).[121]

Although OA is separate from rheumatoid arthritis (RA) in terms of inflammation and immune activation, more than 50% of OA patients have inflammation in their synovial membrane, which underlines the importance of incorporating T-cells and/or macrophages in the *in vitro* disease model.[122,123] These cells may be unrelated to initiation of OA, but probably contribute to disease progression. To establish human cell based *in vitro* OA disease models, one choice would be human cartilage biopsies collected during partial/total knee joint replacement. These tissues would only represent the late-stage disease. In order to study human cartilage at earlier stages of OA, human cartilage can be obtained from tissue banks or at the time of autopsy of a donor who had not been diagnosed with OA (clear absence of osteophytes and signs of joint space narrowing). Controlled degeneration can be induced on this asymptomatic "normal" cartilage tissue. However, such studies can be confounding, as found with cartilage samples collected from a tissue bank where different stages of degeneration taken from the same joint was used to represent "early stage" (i.e. intact cartilage) and "late stage" (i.e. fibrillated cartilage) OA.[124] Two markers of anabolism were selected (the C-propeptide of collagen and epitope 846 of aggrecan) as well as markers of degradation. At the same early stage of the disease, ankle OA cartilage was more reactive than knee OA cartilage, indicating that there may be major differences between cartilage sources.

Dual expression of IL-1RA and insulin-like growth factor (IGF-I) were studied in a horse OA model.[125] Cartilage explants were exposed to the environment of synovial membrane-derived cells cultured in monolayers and expressing IGF-1 and IL-1RA. The data confirmed that combining the anabolic action of IGF-1 and the catabolic blocking of IL-1RA protected and partially restored the cartilage matrix. Genetic transfer of TIMP-1 into bovine chondrocytes resulted in resistance to the catabolic effects of IL-1, including reduced MMP activity and a decreased loss of collagen type II.[126] Lorenzo *et al.*[127] compared the rate of biosynthesis of extracellular noncollagenous cartilage matrix macromolecules in "normal or early stag" (cartilage obtained from surgery for sarcomas) and "late stages" of OA (cartilage from total knee replacement patient). The early-stage OA samples were further subdivided into those with some fibrillation or those showing no fibrillation, by visual assessment. Non-fibrillated samples showed a lower level of matrix production than fibrillated samples. Both early- and late-stage OA cartilage samples showed increased synthesis of cartilage oligomeric protein, fibronectin, and cartilage intermediate-layer protein, compared to normal cartilage tissue. Collagen synthesis was significantly higher in late-stage OA. Altered composition and pattern of matrix synthesis may indicate that the joint actually attempts to undergo metabolic alterations in early stages of the disease before there is nominal fibrillation of the tissue. Further confusing the situation, chondrocytes from normal but aged patients secrete protein patterns resembling that of chondrocytes from OA patients but not young individuals.[128]

Attempts to establish *in vitro* models of OA started as early as the 1970s. Cocultures of healthy non-inflamed porcine synovial tissue with cartilage fragments were studied and there was a progressive breakdown of cartilage matrix by the resident chondrocytes.[129] It was speculated that the synovial tissue produced a soluble factor which stimulated chondrocytes to break down the surrounding cartilage matrix. Later, tissue engineering principles using fibrin matrices were used to establish *in vitro* models of RA.[130] Culturing chondrocytes on polymeric scaffolds delivering proteolytic enzymes, such as matrix metalloproteinases (MMPs) and aggrecanase can be used to establish *in vitro* OA tissue models. The occurrence of point mutations related to extracellular matrix macromolecules of arthritic articular cartilage are documented.[131] Therefore, engineered tissues composed of cells with these mutations, or having mutations in enzymes that can cause post-translational modifications of collagen, might be useful to establish *in vitro* OA models.

Computational fluid dynamic modeling has been used to assess the fluid-induced shear stresses and mass transport within the porous architecture of the 3D scaffold.[132] Static compression decreased the synthesis of cartilage matrix proteins, while dynamic compression may increase matrix synthetic activity.[133] A bioreactor system applying controlled regimes of repetitive loading and specific combinations of inflammatory cytokines, such as, TNF-α, IL-1β or other OA-related pro-inflammatory cytokines, can be used to investigate the response of engineered tissue to an environment simulating an OA joint.

There is a lack of studies regarding the role of OA cartilage causing inflammation. Cartilage fragments are frequently found in inflamed synovial membranes of OA patients. Normally HLA-DR-negative (a member of the major histocompatibility complex, class-II) chondrocytes become positive in OA, which suggests that they may act as antigen-presenting cells.[134,135] Compared with RA, OA patients show a strong response of peripheral blood and synovial fluid T-cells to chondrocytes and to fibroblast-seeded membranes, but not to epithelial cell seeded membranes. Interestingly, T-cells from normal donors do not show significant responses to any membrane preparations.[136] Future investigations should be focused on developing a clear understanding of the cross-talk between OA diseased tissue and immune cells.

3.3. Gravitational models for bone loss

During spaceflight astronauts experience multiple physiological changes, including musculoskeletal tissue loss, cardiovascular and neuro-vestibular malfunction, and suppression of immune responses (reviewed by Refs. 137–140). Around the

1960s, bioanalysis studies on astronauts of Vostok 2 and 3 space flights indicated increased excretion of urinary calcium.[141,142] Post-mortem reports of three Russian cosmonauts who died during their return to Earth in 1971 after a 24-day mission indicated greater packing of the bone crystal lattice, enhanced mineralization of bone tissues, and increased micro-hardness of bone, indicative of loss of calcium. Later space research demonstrated that excretion of calcium through the urine steadily increased during exposure to microgravity, but was gradually recovered after return to earth.

Due to the limited opportunities for in-flight experimentation, a commonly used animal model was the "hind-limb suspension model", where young or mature rats are suspended by their tails, with the hind-limb suspended for a longer duration of time. This model has been used to demonstrate reduced bone mineral density in mechanically unloaded hind-limbs, with reduced osteocalcin, osteoblast surface area, hypercalcemia, and reduced levels of the active metabolite of vitamin D. These results contrast with the maintenance of bone density at control levels in the mechanically loaded forelimbs. However, this model has limited representation for microgravity.[143–145]

NASA has developed several types of rotating wall vessel (RWV) bioreactors to simulate microgravity and to study changes in osteochondral tissues during space flight due to bone loss and skeletal degeneration. RWV bioreactors can be classified as: (a) high aspect ratio vessels (HARV), (b) slow-turning lateral vessel (STLV), and (c) rotating wall perfusion vessel (RWPV) (Fig. 4). RWV bioreactors can generate human cell-based tissue models to help design improved strategies to prevent degeneration of bone and muscle.[146] In RWV bioreactors, scaffold-cell constructs are in a perpetual floating condition, in a low-shear freefall environment, which is maintained by balancing gravitational settling with centrifugation caused by the vessel rotation and fluid drag. However, there is a lack of agreement regarding the terminology used for describing this "zero-gravity situation," or "simulated microgravity," "modeled microgravity," "microgravity analog," or most recently "vector-less gravity." It is probably impossible to fully recapitulate microgravity on earth using ground-based RWV bioreactors, however, these attempts to simulate such conditions offer an important step forward. Generally in rotating bioreactors, eight to 12 tissue constructs are cultured per vessel, in a rotational flow field. The rotational speed of the vessel is gradually adjusted to balance gravity, buoyancy and drag forces in such manner that each construct remains freely suspended. For example, to culture chondrocyte-seeded constructs for six weeks in STLV, the speed can be increased at regular intervals from 10 to 45 rpms to compensate for the gradual increase in construct mass due to ECM formation over time.

Bovine articular chondrocytes were seeded on polyglycolic acid scaffolds in spinner flasks and cultured for three months, and then cultured in RWPV bioreactors

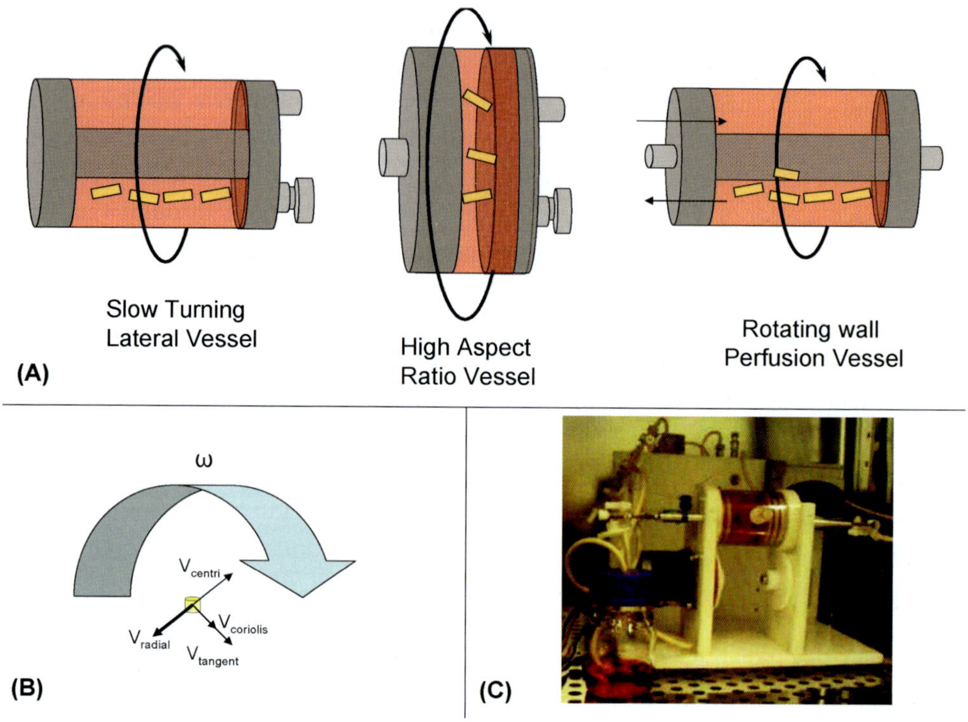

Fig. 4. **(A)** Three different types of rotating wall vessel bioreactors, **(B)** working principle of RWV bioreactor. A tissue construct floating in a fluid medium will travel in a small circular path at the same angular rate (ω) as the vessel rotation due to balancing of centrifugal (*V*centri), coriolis (*V*coriolis) and gravitational forces. The gravity vector has been resolved into its radial (*V*radial) and tangential (*V*tangent) vectors. **(C)** A model of a perfusion vessel bioreactor.

for four months, either on the *Mir* space station at 10^{-4}–10^{-6} g, or on Earth at 1 g.[147] The engineered cartilage tissue consisted of viable, metabolically active cells and was structurally and functionally cartilaginous, but constructs grown on *Mir* were smaller, contained less glucosaminoglycan and were mechanically inferior to those grown on earth. Microgravity-based studies have provided some insight into mechanisms of bone loss in space, such as due to a decrease in bone formation by osteoblasts, whose differentiation and function are negatively affected in zero-gravity. Culturing osteoblasts in rotating wall vessel bioreactors resulted in a reduction in osteoblast-specific markers, such as alkaline phosphatase, osteocalcin, and Runt-related transcription factor 2 (Runx2).[148–152]

Cytoskeletal disruption has been correlated with inhibition of osteoblastic gene expression. RhoA GTPase has a role in stress fiber formation in fibroblastic cells. RhoA mediates stress fiber formation through phosphorylation and inactivation of actin-disrupting protein cofilin and myosin phosphatase. Significant alterations in cellular morphology, such as disruption of the actin cytoskeleton and reduced focal adhesions, have been reported in microgravity situations with many different cell types, such as osteoblastic MC3T3-E1 cells.[153] It would be interesting to investigate whether gene transfer-mediated overexpression of constitutively active RhoA can lead to recovery of stress fibers, and induce osteogenic differentiation in engineered *in vitro* tissue models in microgravity situations. RhoA also plays a crucial role in lineage specific differentiation of hMSCs. Inactivation of RhoA leads to adipocytic differentiation, whereas constitutively active RhoA induces osteoblastic differentiation. Activation of the p38 MAPK and peroxisome proliferators activated receptor g2 (PPARg2), accompanied by suppression of Runx2 expression, were shown to induce differentiation of hMSCs to adipocytes.[154] Hence, reduced activation of RhoA during microgravity and spaceflight likely contributes to altered hMSCs differentiation.

Collagen-I, the major ECM component of bone, is secreted by osteoblasts during differentiation. Alterations in collagen-I expression are associated with post-menopausal osteoporosis, as well as disruption of collagen fibrils in the bone matrix following ovariectomy in mice.[155] The C-terminal peptide of type I collagen was secreted from astronauts duration a 180-day space flight.[156] Collagen-I biosynthesis is down regulated when hMSCs were cultured in simulated microgravity conditions for seven days.[157] Collagen/integrin signaling is linked to bone morphogenic protein signaling for osteoinduction or osteogenic differentiation.[158] Thus blocking collagen fibril formation, or inhibiting interaction of collagen-I with its integrin receptor, $\alpha 2\beta 1$, can inhibit complete osteoblastic differentiation.[159]

Microgravity can also indirectly stimulate osteoclast formation. Animal studies have resulted in confusing results. Osteoclast number and activity remained unchanged in initial studies using bones from animals sent on space flight,[160] whereas studies of less than one-week duration showed an increase in osteoclastic populations using different bone sites in pregnant rats.[161,162] Although osteoclast-induced bone resorption activity could be transiently elevated between the first and second week of microgravity exposure, during longer periods of weightlessness bone turnover decreases. Microgravity can indirectly stimulate osteoclast formation and activity by regulating secretion of key regulatory factors such as receptor activator for nuclear factor κ B ligand (RANKL) and osteoprotegerin (OPG). Conditioned media from osteoblasts subjected to microgravity increased osteoclastogenesis and bone resorption in mouse bone marrow cultures. In these osteoblasts, the RANKL/OPG ratio was higher than that observed in gravity-based cultures.[162] Treatment with high concentrations of OPG inhibited osteoclastogenesis and bone resorption in the presence of conditioned media from osteoblasts cultured under microgravity. Microgravity induced a time-dependent increase in extracellular signal-regulated kinase (ERK-1/2) phosphorylation. Apoptosis was similar among the various experimental groups, while it increased under microgravity after treatment with a MEK-1/2 inhibitor (MAPK/ERK kinase), suggesting a protective role by ERK-1/2.[163]

Ground-based studies using RWV bioreactors simulating microgravity have demonstrated that in space bone loss could be at least partially due to a decrease in bone formation by osteoblasts, whose differentiation and function are negatively affected, under these conditions. The bone loss may also be related to the activation of osteoclasts. Further studies are required to gather insight into the observed bone remodeling process and 3D tissue models offer an important option toward this goal.

3.4. Osteoporosis model

Osteoporosis is a metabolic or aging-related disease in which low bone mineral density causes bone fragility. This occurs as a result of aging (often occurring in post-menopausal women), disuse of skeletal system, or the chronic use of glucocorticoids and calcineurin inhibitors. The pathophysiology that underlies reduced bone formation during normal aging processes has remained unclear due to lack of good animal models.

Investigations using *in vivo* osteoporosis models fail to cover all causative aspects of the disease, probably due to the complexity of different factors involved. Studies using the transgenic SAMP6 mouse, with accelerated cell senescence intended to simulate aging process, indicate that reduced osteoblastogenesis partly addresses mechanisms of osteoporosis.[164] One of the main mechanisms by which estrogen deficiency causes bone loss is by stimulating osteoclast formation.[165] Osteoclast precursor cells can be activated by macrophage colony stimulating factor and a tumor necrosis factor-related factor known as receptor activator of NF-KB ligand (RANKL). TNF-α-transgenic mice or thyrotropin receptor-knockout mice are a high-turnover osteoporosis model, but do not address low-turnover osteoporosis. Analysis of a murine genetic model of osteoporosis[166] demonstrated that genetic variation within the 15-lipoxygenase (*Alox15*) gene affected peak bone mass, thus inhibiting this enzyme could improve bone mass.

Heparin-induced osteoporosis is a well-known complication that follows high dose heparin therapy during the prenatal period.[167,168] An *in vitro* osteoporosis model addressed this issue by incubating osteoblasts in the presence of four different low-molecular-weight heparins (nadroparin, enoxaparin, dalteparin, certoparin), showing significant inhibition of osteoblast growth.[169]

Compared with the pathological role of T-cells in RA and OA, the role of T-cells in non-inflammatory metabolic bone diseases such as post-menopausal osteoporosis has recently been proposed.[170,171] However, the mechanisms are poorly understood due to lack of efficient *in vitro* and *in vivo* models. A simple *in vitro* engineered tissue model may provide a platform with which to address how oestrogen can regulate T-cell immune responses in post-menopausal osteoporosis.

4. Conclusions

In the last decade, tissue engineering, which originated from the clinical need of organ transplants and repairs, has morphed into new venues of research. One critical goal is to have reliable 3D human tissue models to study diseases such as those described herein. The field is aimed at developing *in vitro* tissue equivalents by culturing primary, progenitor or genetically modified cells on 3D degradable polymeric matrices. If these *in vitro* engineered tissues can be used as reproducible and viable "human cell-based disease model systems," this may provide an important step forward for the translation of basic science to clinically meaningful therapies. It is tempting to speculate that in the near future these engineered human tissue models may provide a new way to study disease *in vitro*, and help in designing tailor-made patient-specific treatments under highly controlled conditions. This outcome would fill a critical gap that presently exists between cell-based assays and human clinical trials.

References

1. Birgersdotter, A., Sandberg, R. and Ernberg, I. Gene expression perturbation *in vitro* — a growing case for three-dimensional (3D) culture systems. *Semin. Cancer Biol.* **15**, 405–412 (2005).
2. Pampaloni, F., Reynaud, E.G. and Stelzer, E. The third dimension bridges the gap between cell culture and live tissue. *Nat. Rev. Mol. Cell. Biol.* **8**(10), 839–845 (2007).
3. Lin, C.Q. and Bissell, M.J. Mulifaceted regulation of cell differentiation by extracellular matrix. *FASEB J.* **7**, 737–743 (1993).
4. Petersen, O.W., Ronnov-Jessen, L., Howlett, A.R. and Bissell, M.J. Interaction with basement membrane serves to rapidly distinguish growth and differentiation pattern of normal and malignant human breast epithelial cells. *Proc. Natl. Acad. Sci. USA* **89**, 9064–9068 (1992).
5. Shaw, K.R.M., Wrobel, C.R. and Brugge, J.S. Use of three-dimensional basement membrane cultures to model oncogene-induced changes in

mammary epithelial morphogenesis. *J. Mammary Gland Biol. Neoplasia* **9**(4), 297–310 (2004).
6. Adams, J.C. and Watt, F.M. Fibronectin inhibits the terminal differentiation of human keratinocytes. *Nature* **340**, 307–309 (1989).
7. Ghosh, S., Rosenthal, R., Zajac, P., Weber, W.P., Oertli, D., Heberer, M., Martin, I., Spagnoli, G.C. and Reschner, A. Culture of melanoma cells in 3-dimensional architectures results in impaired immunorecognition by cytotoxic T lymphocytes specific for Melan-A/MART-1 tumor-associated antigen. *Ann. Surg.* **242**(6), 851–858 (2005).
8. Feder-Mengus, C., Ghosh, S., Weber, W.P., Wyler, S., Zajac, P., Terracciano, L., Oertli, D., Heberer, M., Martin, I., Spagnoli, G.C. and Reschner, A. Multiple mechanisms underlie defective recognition of melanoma cells cultured in three-dimensional architectures by antigen-specific cytotoxic T lymphocytes. *Br. J. Cancer* **96**(7), 1072–1082 (2007).
9. Ghosh, S., Spagnoli, G.C., Martin, I., Ploegert, S., Demougin, P., Heberer, M. and Reschner, A. Three-dimensional culture of melanoma cells profoundly affects gene expression profile: a high density oligonucleotide array study. *J. Cell Physiol.* **204**, 522–531 (2005).
10. Lieschke, G.J. and Currie, P.D. Animal models of human disease: zebrafish swim into view. *Nat. Rev. Genet.* **8**, 353–367 (2007).
11. Rosenthal, N. and Brown, S. The mouse ascending: perspectives for human-disease Models. *Nat. Cell Biol.* **9**(9), 993–999 (2007).
12. Pound, P., Ebrahim, S., Sandercock, P., Bracken, M.B. and Roberts, I. Where is the evidence that animal research benefits humans? *Br. Med. J.* **328**, 514–517 (2004).
13. Guigo, R., Dermitzakis, E.T., Agarwal, P., Ponting, C.P., Parra, G., Reymond, A., Abril, J.F., Keibler, E., Lyle, R., Ucla, C., Antonarakis, S.E. and Brent, M.R. Comparison of mouse and human genomes followed by experimental verification yields an estimated 1019 additional genes. *Proc. Natl. Acad. Sci. USA* **100**(3), 1140–1145 (2003).
14. Wright, W.E. and Shay, J.W. Telomere dynamics in cancer progression and prevention: fundamental differences in human and mouse telomere biology. *Nat. Med.* **6**(8), 849–851 (2000).
15. Garlick, J.A. Engineering skin to study human disease — tissue models for cancer biology and wound repair. *Adv. Biochem. Eng. Biotechnol.* **103**, 207–239 (2006).
16. Fischbach, C., Chen, R., Matsumoto, T., Schmelzle, T., Brugge, J.S., Polverini, P.J. and Mooney, D.J. Engineering tumors with 3D scaffolds. *Nat. Methods* **4**, 855–860 (2007).

17. Taussig, M.J., Stoevesandt, O., Borrebaeck, C.A.K., Bradbury, A.R., Cahill, D., Cambillau, C., Daruvar, A., Dübel, S., Eichler, J., Frank, R., Gibson, T.J., Gloriam, D., Gold, L., Herberg, F.W., Hermjakob, H., Hoheisel, J.D., Joos, T.O., Kallioniemi, O., Koegl, M., Konthur, Z., Korn, B., Kremmer, B., Krobitsch, S., Landegren, U., van der Maarel, S., McCafferty, J., Muyldermans, S., Nygren, S., Palcy, S., Plückthun, A., Polic, B., Przybylski, M., Saviranta, P., Sawyer, A., Sherman, D.J., Skerra, A., Templin, M., Ueffing, M. and Uhlén, M. ProteomeBinders: planning a European resource of affinity reagents for analysis of the human proteome. *Nat. Meth.* **4,** 13–17 (2007).
18. Carlsson, J. and Yuhas, J.M. Liquid-overlay culture of cellular spheroids. *Recent Results Cancer Res.* **95**,1–23 (1984).
19. Mueller-Klieser, W. Multicellular spheroids: a review on cellular aggregates in cancer research. *J. Cancer Res. Clin. Oncol.* **113**, 101–122 (1987).
20. Sutherland, R.M. Cell and environment interactions in tumor microregions: the multicell spheroid model. *Science* **240**, 177–184 (1988).
21. Albrecht, D.R., Underhill, G.H., Wassermann, T.B., Sah, R.L. and Bhatia, S.N. Probing the role of multicellular organization in three-dimensional microenvironments. *Nat. Meth.* **3**(5), 369–375 (2006).
22. Freed, L.E., *Biodegradable Polymer Scaffolds for Tissue Engineering*, Vol. 12 (Biotechnology, New York, 1994), pp. 689–693.
23. Hofmann, S., Hagenmuller, H., Koch, A.M., Muller, R., Vunjak-Novakovic, G., Kaplan, D.L., Merkle, H.P. and Meinel, L. Control of *in vitro* tissue-engineered bone-like structures using human mesenchymal stem cells and porous silk scaffolds. *Biomaterials* **28**, 1152–1162 (2007).
24. Nazarov, R., Jin, H.J. and Kaplan, D.L. Porous 3D scaffolds from regenerated silk fibroin. *Biomacromolecules* **5**, 718–726 (2004).
25. Hutmacher, D.W. and Cool, S. Concepts of scaffold-based tissue engineering — the rationale to use solid free-form fabrication techniques. *J. Cell. Mol. Med.* **11**(4), 654–669 (2007).
26. Yeong, W., Chua, C., Leong, K. and Chandrasekaran, M. Rapid prototyping in tissue engineering: challenges and potential. *Trend Biotechnol.* **22**(12)**,** 643–652 (2004).
27. Miot, S., Woodfield, T., Daniels, A.U., Suetterlin, R., Peterschmitt, I., Heberer, M., van Blitterswijk, C.A., Riesle, J. and Martin, I. Effects of scaffold composition and architecture on human nasal chondrocyte redifferentiation and cartilaginous matrix deposition. *Biomaterials* **26**, 2479–2489 (2005).
28. Bettinger, C.J., Weinberg, E.J., Wang, Y., Borenstein J.T. and Langer, R. Three-dimensional microfluidic tissue engineering scaffolds using a flexible biodegradable polymer. *Adv. Mater.* **18**, 165–169 (2006).

29. Lutolf, M.P. and Hubbell, J.A. Synthetic biomaterials as instructive extracellular microenvironments for morphogenesis in tissue engineering. *Nat. Biotechnol.* **23**(1), 47–55 (2005).
30. Altman, G.H., Diaz, F., Jakuba, C., Calabro, T., Horan, R.L., Chen, J., Lu, H., Richmond, J. and Kaplan, D.L. Silk-based biomaterials. *Biomaterials* **24**, 401–416 (2003).
31. Panilaitis, B., Altman, G.H., Chen, J., Jin, H.J., Karageorgiou, V. and Kaplan, D.L. Macrophage responses to silk. *Biomaterials* **24**(18), 3079–3085 (2003).
32. Meinel, L., Karageorgiou, V., Hofmann, S., Fajardo, R., Snyder, B., Li, C., Zichner, L., Langer, R., Vunjak-Novakovic, G. and Kaplan, D.L. Engineering bone-like tissue *in vitro* using human bone marrow stem cells and silk scaffolds. *J. Biomed. Mater. Res. A.* **71**(1), 25–34 (2004).
33. Kim, U.J., Park, J., Joo, H., Wada, M. and Kaplan, D.L. Three-dimensional aqueous-derived biomaterial scaffolds from silk fibroin. *Biomaterials* **26**, 2775–2785 (2005).
34. Sofia, S., McCarthy, M.B., Gronowicz, G. and Kaplan, D.L. Functionalized silk-based biomaterials for bone formation. *J. Biomed. Mater. Res.* **54**(1), 139–148 (2001).
35. Bini, E., Foo, C., Huang, J., Karageorgiou, V., Kitchel, B. and Kaplan, D.L. RGD-functionalized bioengineered spider dragline silk biomaterial. *Biomacromolecules* **7**(11), 3139–3145 (2006).
36. Karageorgiou, V., Meinel, L., Hofmann, S., Malhotra, A., Volloch, V. and Kaplan, D.L. Bone morphogenetic protein-2 decorated silk fibroin films induce osteogenic differentiation of human bone marrow stromal cells. *J. Biomed. Mater. Res. A* **71**(3), 528–537 (2004).
37. Li, C., Vepari, C., Jin, H.J., Kim, H.J. and Kaplan, D.L. Electrospun silk-BMP-2 scaffolds for bone tissue engineering. *Biomaterials* **27**(16), 3115–3124 (2006).
38. Jin, H.J., Park, J., Karageorgiou, V., Kim, U.J., Valluzzi, R., Cebe, P. and Kaplan, D.L. Water-stable silk films with reduced β-sheet content. *Adv. Funct. Mater.* **15**(8), 1241–1247 (2005).
39. Horan, R.L., Antle, K., Collette, A.L., Wang, Y., Huang, J., Moreau, J.E., Volloch, V., Kaplan, D.L. and Altman, G.H. *In vitro* degradation of silk fibroin. *Biomaterials* **26**, 3385–3393 (2005).
40. Wang, X., Hu, X., Daley, A., Rabotyagova, O., Cebe, P. and Kaplan, D.L. Nanolayer biomaterial coatings of silk fibroin for controlled release. *J. Control. Release* **121**, 190–199 (2007).
41. Hofmann, S., Foo, C.W., Rossetti, C., Textor, M., Merkle, H.P., Kaplan, D.L. and Meinel, L. Silk fibroin as an organic polymer for controlled drug delivery. *J. Control. Release* **111**, 219–227 (2006).

42. Bayraktar, O., Malay, O., Ozgarip, Y. and Batgun, A. Silk fibroin as a novel coating material for controlled release of theophylline. *Eur. J. Pharm. Biopharm.* **60**, 373–381 (2005).
43. Rujiravanit, R., Kruaykitanon, S., Jamieson, A.M. and Tokura, S. Preparation of cross-linked chitosan/silk fibroin blend films for drug delivery system. *Macromol. Biosci.* **3**(10), 604–611 (2003).
44. Buschmann, M.D., Gluzband, Y.A., Grodzinsky A.J. and Hunziker, E.B. Mechanical compression modulates matrix biosynthesis in chondrocyte/agarose culture. *J. Cell Sci.* **108**(4), 1497–1508 (1995).
45. Demarteau, O., Wendt, D., Braccini, A., Jakob, M., Schafer, D., Heberer M. and Martin, I. Dynamic compression of cartilage constructs engineered from expanded human articular chondrocytes. *Biochem. Biophys. Res. Commun.* **310**, 580–588 (2003).
46. Grad, S., Lee, C.R., Gorna, K., Gogolewski, S., Wimmer, M.A. and Alini, M. Surface motion upregulates superficial zone protein and hyaluronan production in chondrocyte-seeded three-dimensional scaffolds. *Tissue Eng.* **11,** 249–256 (2005).
47. Seidel, J.O., Pei, M., Gray, M.L., Langer, R., Freed, L.E. and Vunjak-Novakovic, G., Long-term culture of tissue engineered cartilage in a perfused chamber with mechanical stimulation. *Biorheology* **41**(3–4), 445–458 (2004).
48. Mauney, J.R., Sjostorm, S., Blumberg, J., Horan, R., O'Leary, J.P., Vunjak-Novakovic, G., Volloch, V. and Kaplan, D.L. Mechanical stimulation promotes osteogenic differentiation of human bone marrow stromal cells on 3-D partially demineralized bone scaffolds *in vitro*. *Calcif. Tissue Int.* **74**(5), 458–468 (2004).
49. Li, Y.J., Batra, N.N., You, L., Meier, S.C., Coe, I.A., Yellowley, C.E. *et al.* Oscillatory fluid flow affects human marrow stromal cell proliferation and differentiation. *J. Orthop. Res.* **22**, 1283–1289 (2004).
50. Chowdhury, T.T., Bader, D.L. and Lee, D.A. Dynamic compression counteracts IL-1β induced iNOS and COX-2 activity by human chondrocytes cultured in agarose constructs. *Biorheology* **43**(3–4), 413–429 (2006).
51. Piscoya, J., Fermor, B., Kraus, V., Stabler, T. and Guilak, F. The influence of mechanical compression on the induction of osteoarthritis-related biomarkers in articular cartilage explants. *Osteoarthr. Cartil.* **13**(12), 1092–1099 (2005).
52. Benya, P.D. and Shaffer, J.D. Dedifferentiated chondrocytes re-express the differentiated collagen phenotype when cultured in agarose gels. *Cell* **30**, 215–224 (1982).
53. Swim, H.E. and Parker, R.F., Culture characteristics of human fibroblasts propagated serially. *Am. J. Hygiene* **66**, 235–243 (1957).

54. Majors, A.K., Boehm, C.A., Nitto, H., Midura, R.J. and Muschler, G.F., Characterization of human bone marrow stromal cells with respect to osteoblastic differentiation. *J. Orthop. Res.* **15**, 546–557 (1997).
55. Bruder, S.P., Jaiswal, N. and Haynesworth, S.E. Growth kinetics, self-renewal, and the osteogenic potential of purified human mesenchymal stem cells during extensive subcultivation and following cryopreservation. *J. Cell Biochem.* **64**, 278–294 (1997).
56. Simonsen, J.L., Rosada, C., Serakinci, N., Justesen, J., Stenderup, K., Rattan, S.I., Jensen, T.G. and Kassem, M. Telomerase expression extends the proliferative life-span and maintains the osteogenic potential of human bone marrow stromal cells. *Nat. Biotechnol.* **20**, 592–596 (2002).
57. Gersbach, C.A., Phillips, J.E. and Garcia, A.J. Genetic engineering for skeletal regenerative medicine. *Ann. Rev. Biomed. Eng.* **9**, 87–119 (2007).
58. Fang, J., Zhu, Y.Y., Smiley, E., Bonadio, J., Rouleau, J.P. and Goldstein, S.A. Stimulation of new bone formation by direct transfer of osteogenic plasmid genes. *Proc. Natl. Acad. Sci. USA* **93**(12), 5753–5758 (1996).
59. Osyczka, A.M., Noth, U., O'Connor, J., Caterson, E.J., Yoon, K., Danielson, K.G. and Tuan, R.S. Multilineage differentiation of adult human bone marrow progenitor cells transduced with human papilloma virus type 16 E6/E7 genes. *Calcif. Tissue Int.* **71**, 447–458 (2002).
60. Morales, C.P., Holt, S.E., Ouelette, M., Kaur, K.J., Yan, Y., Wilson, K.S., White, M.A., Wright, W.E. and Shay, J.W. Absence of cancer-associated changes in human fibroblasts immortalized with telomerase. *Nat. Genet.* **21**, 115–118 (1999).
61. Baragi, V.M., Renkievicz, R.R., Jordan, H., Bonadio, J., Hartman, J.W. and Roessler, B.J. Transplantation of transduced chondrocytes protects articular cartilage from interleukin-1 induced extracellular matrix degradation. *J. Clin. Invest.* **96**, 2454–2460 (1995).
62. Li, Y., Quarles, L.D., Zhou, H. and Xiao, Z.S. RNA interference and its application in bone-related diseases. *Biochem. Biophys. Res. Commun.* **361**, 817–821 (2007).
63. Zwicky, R., Muntener, K., Goldring, M.B. and Baici, A. Cathepsin B expression and down-regulation by gene silencing and antisense DNA in human chondrocytes. *Biochem. J.* **367**, 209–217 (2002).
64. Friedmann, T. A new serious adverse event in a gene therapy study. *Mol. Ther.* **15**(11), 1899–1900 (2007).
65. Georges, M., Coppieters, W. and Charlier, C. Polymorphic miRNA-mediated gene regulation: contribution to phenotypic variation and disease. *Curr. Opin. Genet. Dev.* **17**, 166–176 (2007).

66. Simmons, C., Alsberg, E., Hsiong, S., Kim, W. and Mooney, D. Dual growth factor delivery and controlled scaffold degradation enhance *in vivo* bone formation by transplanted bone marrow stromal cells. *Bone* **35**(2), 562–569 (2004).
67. Rod, A., Rahimi, E. and Leof, B. TGF-β signaling: a tale of two responses. *J. Cell. Biochem.* **102**(3), 593–608 (2007).
68. Ozawa, C.R., Banfi, A., Glazer, N.L., Thurston, G., Springer, M.L., Kraft, P.E., McDonald, D. and Blau, H.M. Microenvironmental VEGF concentration, not total dose, determines a threshold between normal and aberrant angiogenesis. *J. Clin. Invest.* **113**, 516–527 (2004).
69. Uebersax, L., Mattotti, M., Papaloïzos, M., Merkle, H.P., Gander, B. and Meinel, L. Silk fibroin matrices for the controlled release of nerve growth factor (NGF). *Biomaterials* **28**(30), 4449–4460 (2007).
70. Moreau, J., Chen, J., Kaplan, D. and Altman, G. Sequential growth factor stimulation of bone marrow stromal cells in extended culture. *Tissue Eng.* **12**(10), 2905–2912 (2006).
71. Verbruggen, G., Cornelissen, M., Almqvist, K.F., Wang, L., Elewaut, D., Broddelez, C., de Ridder, L. and Veys, E.M. Influence of aging on the synthesis and morphology of the aggrecans synthesized by differentiated human articular chondrocytes. *Osteoarthr. Cartil.* **8**(3), 170–179 (2000).
72. Hofmann, S., Knecht, S., Langer, R., Kaplan, D.L., Vunjak-Novakovic, G., Merkle, H.P. and Meinel, L. Cartilage-like tissue engineering using silk scaffolds and mesenchymal stem cells. *Tissue Eng.* **12**(10), 2729–2738 (2006).
73. Wang, Y., Blasioli, D.J., Kim, H.J., Kim, H.S. and Kaplan, D.L. Cartilage tissue engineering with silk scaffolds and human articular chondrocytes. *Biomaterials* **27**(25), 4434–4442 (2006).
74. Kim, T.K., Sharma, B., Williams, C.G., Ruffner, M.A., Malik, A., McFarland, E.G. and Elisseeff, J.H. Experimental model for cartilage tissue engineering to regenerate the zonal organization of articular cartilage. *Osteoarthr. Cartil.* **11**(9), 653–664 (2003).
75. Hwang, N.S., Kim, M.S., Sampattavanich, S., Baek, J.H., Zhang, Z. and Elisseeff, J. Effects of three-dimensional culture and growth factors on the chondrogenic differentiation of murine embryonic stem cells. *Stem Cells* **24**(2), 284–291 (2006).
76. Li, W., Tuli, R., Okafor, C., Derfoul, A., Danielson, K.G., Hall, D.J. and Tuan, R.S. A three-dimensional nanofibrous scaffold for cartilage tissue engineering using human mesenchymal stem cells. *Biomaterials* **26**, 599–609 (2005).
77. Schaefer, D., Martin, I., Shastri, P., Padera, R.F., Langer, R., Freed, L.E. and Vunjak-Novakovic, G. *In vitro* generation of osteochondral composites. *Biomaterials* **21**, 2599–3003 (2000).

78. Mahmoudifar, N. and Doran, P.M. Tissue engineering of human cartilage and osteochondral composites using recirculation bioreactors. *Biomaterials* **26**(34), 7012–7024 (2005).
79. Pound, J.C., Green, D.W., Chaudhuri, J.B., Mann, S., Roach, H.I., Richard O. and Oreffo, C. Strategies to promote chondrogenesis and osteogenesis from human bone marrow cells and articular chondrocytes encapsulated in polysaccharide templates. *Tissue Eng.* **12**(10), 2789–2799 (2006).
80. Wakitani, S., Goto, T., Pineda, S.J., Young, R.G., Mansour, J.M., Caplan, A.I. and Goldberg, V.M. Mesenchymal cell-based repair of large, full-thickness defects of articular cartilage. *J. Bone Joint Surg. Am.* **76**, 579–583 (1994).
81. Schaefer, D., Martin, I., Jundt, G., Seidel, J., Heberer, M., Grodzinsky, A., Bergin, I., Vunjak-Novakovic, G. and Freed, L.E. Tissue-engineered composites for the repair of large osteochondral defects. *Arthritis Rheum.* **46**, 2524–2529 (2002).
82. Angele, P., Kujat, R., Nerlich, M., Yoo, J., Goldberg, V. and Johnstone, B. Engineering of osteochondral tissue with bone marrow mesenchymal progenitor cells in a derivatized hyaluronan-gelatin composite sponge. *Tissue Eng.* **5**, 545–550 (1999).
83. Tuli, R., Nandi, S., Li, W.J., Tuli, S., Huang, X., Manner, P.A. and Tuan R. Human mesenchymal progenitor cell-based tissue engineering of a single-unit osteochondral construct. *Tissue Eng.* **10**, 1169–1179 (2004).
84. Augst, A., Marolt, D., Freed, L.E., Vepari, C., Meinel, L., Farley, M., Fajardo, R., Patel, N., Gray, M., Kaplan, D.L. and Vunjak-Novakovic, D. Effects of chondrogenic and osteogenic regulatory factors on composite constructs grown bone-cartilage using human mesenchymal stem cells, silk scaffolds and bioreactors. *J. R. Soc. Interface*, in press (2008).
85. Marolt, D., Augst, A., Freed, L.E., Vepari, C., Fajardo, R., Patel, N., Gray, M., Farley, M., Kaplan, D.L. and Vunjak-Novakovic, G. Bone and cartilage tissue constructs grown using human bone marrow stromal cells, silk scaffolds and rotating bioreactors. *Biomaterials* **27**, 6138–6149 (2006).
86. Nesic, D., Whiteside, R., Brittberg, M., Wendt, D., Martin I. and Mainil-Varlet, P. Cartilage tissue engineering for degenerative joint disease. *Adv. Drug Deliv. Rev.* **58**(2), 300–322 (2006).
87. Busuttil, R.A., Dolle, M., Campisi, J. and Vijga, J. Genomic instability, aging, and cellular senescence. *Ann. N. Y. Acad. Sci.* **1019**, 245–255 (2004).
88. Price, J.S., Waters, J.G., Darrah, C., Pennington, C., Edwards, D.R., Donell, S.T. and Clark, I.M. The role of chondrocyte senescence in osteoarthritis. *Aging Cell* **1**, 57–65 (2002).
89. Patwari, P., Cook, M.N., DiMicco, M.A., Blake, S.M., James, I.E., Kumar, S., Cole, A.A., Lark, M.W. and Grodzinsky, A.J. Proteoglycan degradation

after injurious compression of bovine and human articular cartilage *in vitro*: interaction with exogenous cytokines. *Arthritis Rheum.* **48**, 1292–1301 (2003).

90. Forsyth, C.B., Pulai, J. and Loeser, R.F. Fibronectin fragments and blocking antibodies to α2β1 and α5β1 integrins stimulates mitogen-activated protein kinase signaling and increase colagenase-3 production in human articular condrocytes. *Arthritis Rheum.* **46**, 2368–2376 (2002).
91. Stanton, H., Ung, L. and Fosang, A.J. The 45 kDa collagen-binding fragments of fibronectin induces matric metalloproteinase-13 synthesis by chondrocytes and aggrecan degradation by aggrecanases. *Biol. Chem. J.* **364**, 181–190 (2002).
92. Henderson, B. and Pettipher, E.R. Arthritogenic actions of recombinant IL-1 and tumour necrosis factor alpha in the rabbit: evidence for synergistic interactions between cytokines *in vivo*. *Clin. Exp. Immunol.* **75**, 306–310 (1989).
93. Arner, E.C., Pratta, M.A., Trzaskos, J.M., Decicco, C.P. and Tortorella, M.D. Generation and characterization of aggrecanase: a soluble, cartilage- derived aggrecan-degrading activity. *J. Biol. Chem.* **274**, 6594–6601 (1999).
94. Abbaszade, I., Liu, R.Q., Yang, F., Rosenfeld, S.A., Ross, O.H., Link, J.R. *et al.* Cloning and characterization of ADAMTS11, an aggrecanase from the ADAMTS family. *J. Biol. Chem.* **274**, 23443–23450 (1999).
95. Glasson, S.S., Askew, R., Sheppard, B., Carito, B., Blanchet, T. and Ma, H.L. Deletion of active ADAMTS5 prevents cartilage degradation in a murine model of osteoarthritis. *Nature* **434**, 644–648 (2005).
96. Holderbaum, D., Haggi, T.M. and Moskowitz, R.W. Genetics and osteoarthritis: exposing the iceberg. *Arthritis Rheum.* **42**, 397–405 (1999).
97. Aigner, T., Fundel, K., Saas, J., Gebhard, P.M., Haag, J., Weiss, J., Zien, A., Obermayr, F., Zimmer, R. and Bartnik, E. Large-scale gene expression profiling reveals major pathogenetic pathways of cartilage degeneration in osteoarthritis. *Arthritis Rheum.* **54**(11), 3533–3544 (2006).
98. An, Y.H. and Friedman, R.J. Animal models of articular cartilage defect. In: *Animal Models in Orthopaedic Research* (CRC Press LLC, Boca Raton FL, 1999).
99. Aigner, T., Rose, J., Martin, J. and Buckwalter, J. Aging theories of primary osteoarthritis: from epidemiology to molecular biology. *Rejuvenation Res.* **7**(2), 134–145 (2004).
100. Van der Berg, W.B., Lessons from animal models of osteoarthritis. *Curr. Opin. Rheumatol.* **101**, 452–456 (2001).
101. Li, Y. and Olsen, B.R. Murine models of human genetic skeletal disorders. *Matrix Biol.* **16**, 49–52 (1997).

102. Anderson-MacKenzie, J.M., Billingham, M.E and Bailey, A.J. Collagen remodeling in the anterior cruciate ligament associated with developing spontaneous murine osteoarthritis. *Biochem. Biophys. Res. Commun.* **258,** 763–767 (1999).
103. Mason, R.M., Chambers, M.G., Flannely, J., Gaffen, J.D., Dudhia, J. and Bayliss, M.T. The STR/ort mouse and its use as a model of osteoarthritis. *Osteoarthr. Cartil.* **9**, 85–91 (2001).
104. So, C.L., Kaluarachchi, K., Tam, P.P. and Cheah, K.S. Impact of mutations of cartilage matrix genes on matrix structure, gene activity and chondrogenesis. *Osteoarthr. Cartil.* **9**(Suppl. A), S160–S173 (2001).
105. Vandenberg, P., Khillan, J.S., Prockop, D.J. *et al.* Expression of a partially deleted gene of human type II procollagen (COL2A1) in transgenic mice produces a chondrodysplasia. *Proc. Natl. Acad. Sci. USA* **88**, 7640–7644 (1991).
106. Garofalo, S., Metsaranta, M., Ellard J. *et al.* Assembly of cartilage collagen fibrils is disrupted by overexpression of normal type II collagen in transgenic mice. *Proc. Natl. Acad. Sci. USA* **90**, 3825–3829 (1993).
107. Hyttinen, M.M., Toyras, J., Lapvetelainen, T. *et al.* Inactivation of one allele of the type II collagen gene alters the collagen network in murine articular cartilage and make cartilage softer. *Ann. Rheum. Dis.* **60**, 262–268 (2001).
108. Wadhwa, S., Embree, M.C., Kilts, T. *et al.* Accelerated osteoarthritis in the temporomandibular joint of biglycan/fibromodulin double-deficient mice. *Osteoarthr. Cartil.* **13**, 817–827 (2005).
109. Zenmyo, M., Meharra, J., Kuhn, K. and Lotz, M. The alpha-1 integrin deficient mouse as a model of accelerated aging dependent OA development. *Osteoarthr. Cartil.* **9**(Suppl. B), s57 (2001).
110. Quasnichka, H.L., Anderson-MacKenzie, J.M., Tarlton, J.F. *et al.* Cruciate ligament laxity and femoral intercondylar notch narrowing in early-stage knee osteoarthritis. *Arthritis Rheum.* **52**, 3100–3109 (2005).
111. Kamekura, S., Hoshi, K., Shimoaka, T. *et al.* Osteoarthritis development in novel experimental mouse models induced by knee joint instability. *Osteoarthr. Cartil.* **13**, 632–641 (2005).
112. Sandy, J.D., Adams, M.E., Billingham, M.E.J., Plaas, A. and Muir, H. *In vivo* and *in vitro* stimulation of chondrocyte biosynthetic activity in early experimental osteoarthritis. *Arthritis Rheum.* **27**, 388–397 (1984).
113. Telhag, H. and Lindberg, L.A. A method for inducing osteoarthritic changes in rabbits' knees. *Clin. Orthop. Relat. Res.* **86**, 214–223 (1972).
114. Stoop, R., Buma, P., van der Kraan, P.M., Hollander, A.P., Billinghust, R.C. and Meijers, T.H. Type II collagen degradation in articular cartilage fibrillation

after anterior cruciate ligament transaction in rats. *Osteoarthr. Cartil.* **9**, 308–315 (2001).

115. Appleton, C.T.G., Pitelka, V., Henry, J. and Beier, F. Global analyses of gene expression in early experimental osteoarthritis. *Arthritis Rheum.* **56**, 1854–1868 (2007).
116. van der Kraan, P.M., Vitters, E.L., van de Putte, L.B.A. and van der Berg, W. Development of osteoarthritic lesions in mice by "metabolic" and "mechanical" alterations in the knee joints. *Am. J. Pathol.* **135**(6), 1001–1014 (1989).
117. Salo, P.T., Theriault, E. and Wiley, R.G. Selective ablation of rat knee joint innervation with injected immunotoxin: a potential new model for the study of neurpathic arthritis. *J. Orthop. Res.* **15**, 622–628 (1997).
118. Blom, A.B., Lent, P.L., Libregts, S., Holthuysen, A.E., van der Kraan, P.M., Rooijen, N. and van der Berg, W.B. Crucial role of macrophages in matrix metalloproteinase-mediated cartilage destruction during experimental osteoarthritis. *Arthritis Rheum.* **56**, 147–157 (2007).
119. Floman, Y., Eyre, D.R. and Glimcher, M.J. Induction of osteoarthrosis in the rabbit knee joint: biochemical studies on the articular cartilage. *Clin. Orthop. Relat. Res.* **147**, 278–286 (1980).
120. Goldring, S.R. and Goldring, M.B. The role of cytokines in cartilage matrix degeneration in osteoarthritis. *Clin. Orthop. Relat. Res.* **427**(Suppl.), S27–S36 (2004).
121. Cook, J.L., Kuroki, K., Stoker, A., Streppa, H. and Fox, D.B. Review of *in vitro* models and development and initial validation of a novel co-culture model for the study of osteoarthritis. *Curr. Rheumatol. Rev.* **3**(3), 172–182 (2007).
122. Sakkas, L.I. and Platsoucas, C.D. The role of T cells in the pathogenesis of osteoarthritis. *Arthritis Rheum.* **56**(2), 409–424 (2007).
123. Haywood, L., McWilliums, D.F., Pearson, C.I., Gill, S.E., Ganesan, A. and Wilson, D. Inflammation and angiogenesis in osteoarthritis. *Arthritis Rheum.* **48**, 2173–2177 (2003).
124. Aurich, M., Squires, G.R., Reiner, A., Mollenhauer, J.A., Kuettner, K.E. and Poole, A.R. Differential matrix degradation and turnover in early cartilage lesions of human knee and ankle joints. *Arthritis Rheum.* **52**, 112–119 (2005).
125. Haupt, J.L., Frisbie, D.D., McIlwraith, C.W., Robbins, P.D., Ghivizzani, S., Evans, C.H. and Nixon, A.J. Dual transduction of insulin-like growth factor-I and interleukin-1 receptor antagonist protein controls cartilage degradation in an osteoarthritic culture model. *J. Orthop. Res.* **23**, 118–126 (2005).

126. Kafienah, W., Al-Fayez, F., Hollander, A.P. and Barker, M.D. Inhibition of cartilage degradation: a combined tissue engineering and gene therapy approach. *Arthritis Rheum.* **48**, 709–718 (2003).
127. Lorenzo, P., Bayliss, M. and Heinegard, D. Altered patterns and synthesis of extracellular matrix macromolecules in early osteoarthritis. *Matrix Biol.* **23**, 381–391 (2004).
128. Dozin, B., Malpeli, M., Camardella, L., Cancedda, R. and Pietrangelo, A. Response of young, aged and osteoarthritic human articular chondrocytes to inflammatory cytokines: molecular and cellular aspects. *Matrix Biol.* **21**, 449–459 (2002).
129. Fell, H.B. and Jubb, R.W. The effect of synovial tissue on the breakdown of articular cartilage in organ culture. *Arthritis Rheum.* **20**, 1359–1371 (1977).
130. Schultz, O., Keyszer, G., Zacher, J., Sittinger, M. and Burmester, G.R. Development of *in vitro* model systems for destructive joint diseases: novel strategies for establishing inflammatory pannus. *Arthritis Rheum.* **40**(8), 1420–1428 (1997).
131. Ala-Kokko, L., Baldwin, C.T., Moskowitz, R.W. and Prockop, D.J. Single base mutation in the type II procollagen gene (COL2A1) as a cause of primary osteoarthritis associated with a mild chondrodysplasia. *Proc. Natl. Acad. Sci. USA* **87**, 6565–6568 (1990).
132. Raimondi, M.T., Boschetti, F., Falcone, L., Fiore, G.B., Remuzzi, A., Marinoni, E., Marazzi, M. and Pietrabissa, R. Mechanobiology of engineered cartilage cultured under a quantified fluid-dynamic environment. *Biomech. Model Mechanobiol.* **1**, 69–82 (2002).
133. Blain, E.J., Gilbert, S.J., Wardale, R.J., Capper, S.J., Mason, D.J. and Duance, V.C. Up-regulation of matrix metalloproteinase expression and activation following cyclical compressive loading of articular cartilage *in vitro*. *Arch. Biochem. Biophys.* **396**, 49–55 (2001).
134. Lance, E.M., Kimura, L.H. and Manibog, C.N. The expression of major histocompatibility antigens on human artiular chondrocytes. *Clin. Orthop. Relat. Res.* **291**, 266–282 (1993).
135. Sakata, M., Masuko-Hongo, K., Nakamura, H., Onuma, H., Tsuruha, J.I. and Aoki, H. Osteoarthritic articular chondrocytes simulate autologous T cell responses *in vitro*. *Clin. Exp. Rheumatol.* **21**, 704–710 (2003).
136. Alsalameh, S., Mollenhauer, J., Hain, N., Stock, K.P., Kalden, J.R. and Burmster, G.R. Cellular immune response toward human articular chondrocytes: T cell reactivities against chondrocytes and fibroblast membranes in destructive joint diseases. *Arthritis Rheum.* **33**, 1477–1486 (1990).
137. Zayzafoon, M., Meyers, V.E. and McDonald, J.M. Microgravity: the immune response and bone. *Immunol. Rev.* **208**, 267–280 (2005).

138. Whedon, G.D. and Rambaut, P.C. Effects of long-duration space flight on calcium metabolism: review of human studies from Skylab to the present. *Acta Astronaut.* **58**, 59–81 (2006).
139. Ziambaras, K., Civitelli, R. and Papavasiliou, S.S. Weightlessness and skeleton homeostasis. *Hormones* **4**(1), 18–27 (2005).
140. Cogoli, A. Gravitational physiology of human immune cells: a review of *in vivo, ex vivo* and *in vitro* studies. *J. Gravit. Physiol.* **3**, 1–9 (1996).
141. David, H. Russians discuss space radiation in conference. *Missiles Rockets* **21**, 34–38 (1963).
142. Mack, P.B., LaChance, P.A., Vose, V.P. and Vogt, F.B. Bone demineralization of foot and hand on Gemini-Titan IV, V, and VII astronauts during orbital flight. *Am. J. Roentgenol.* **100**, 503–511 (1967).
143. Van Loon, J.J., Bervoets, D.J., Burger, E. and Veldhuijzen, J.P. Decreased mineralization and increased calcium release in isolated fetal mouse long bones under near weightlessness. *J. Bone Miner. Res.* **10**, 550–557 (1995).
144. Matsumoto, D., Nakayama, K., Kodama, Y., Fuse, H., Nakamura, T. and Fukumoto, S. Effect of mechanical unloading and reloading on periosteal bone formation and gene expression in tail-suspended rapidly growing rats. *Bone* **22**, 89S–93S (1998).
145. Grano, M., Mori, G., Minielli, G., Barou, O., Collucci, S., Gianelli, G. and Vico, L. Rat hindlimb unloading by tail suspension reduces osteoblast differentiation, induces IL-6 secretion, and increases bone resorption in *ex vivo* cultures. *Calcif. Tissue Int.* **70**, 176–185 (2002).
146. Martin, Y. and Vermette, P. Bioreactors for tissue mass culture: design, characterization, and recent advances. *Biomaterials* **26**, 7481–7503 (2005).
147. Freed, L.E., Langer, R., Martin, I., Pellis, N.R. and Vunjak-Novakovic, G. Tissue engineering of cartilage in space. *Proc. Natl. Acad. Sci. USA* **94**, 13885–13890 (1997).
148. Nakamura, H., Kumei, Y., Morita, S., Shimokawa, H., Ohya, K. and Shinomya, K. Suppression of osteoblastic phenotypes and modulation of pro- and anti-apoptotic features in normal human osteoblastic cells under a vector averaged gravity condition. *J. Med. Dent. Sci.* **50**, 167–176 (2003).
149. Ontiveros, C. and McCabe, L.R. Simulated microgravity suppresses osteoblast phenotype, Runx2 levels and AP-1 transactivation. *J. Cell. Biochem.* **88**, 427–437 (2003).
150. Zayzafoon, M., Gathings, W.E. and McDonald, J.M. Modeled microgravity inhibits osteogenic differentiation of human mesechymal stem cells and increase adipogenesis. *Endocrinology* **145**, 2421–2432 (2004).
151. Caillot-Augusseau, A., Vico, L., Heer, M., Voroviev, D., Souberbille, J.C., Zitterman, A., Alexandre, C. and Lafage-Proust, M.H. Space flight is associated

with rapid decreases of undercarboxylated osteocalcin and increases of markers of bone resorption without changes in their circadian variations: observations in two cosmonauts. *Clin. Chem.* **46**, 1136–1143 (2000).

152. Pardo, S.J., Patel, M.J., Sykes, M.C., Platt, M.O., Boyd, N.L., Sorescu, G.P., Xu, M., van Loon, J.J., Wang, M.D. and Jo, H. Simulated microgravity using the random positioning machine inhibits differentiation and alters gene expression profiles of 2T3 preosteoblasts. *Am. J. Physiol. Cell. Physiol.* **288**, C1211–C1221 (2005).
153. Hughes-Fulford, M. and Lewis, M.L. Effects of microgravity on osteoblast growth activation. *Exp. Cell. Res.* **224**, 103–109 (1996).
154. Lecka-Czernik, B. Inhibition of Osf2/Cbfa1 expression and terminal osteoblast differentiation by PPARgamma2. *J. Cell. Biochem.* **74**, 357–371 (1999).
155. Kafantari, H., Kounadi, E., Fatouros, M., Milonakis, M. and Tzaphlidou, M. Structural alterations in rat skin and bone collagen fibrils induced by ovariectomy. *Bone* **26**, 349–353 (2000).
156. Caillot-Augusseau, A., Lafage-Proust, M.H., Soler, C., Pernod, J., Dubois, F. and Alexandre, C. Bone formation and resorption biological markers in cosmonauts during and after a 180-day space flight (Euromir 95). *Clin. Chem.* **44**, 578–585 (1998).
157. Meyers, V.E., Zayzafoon, S., Gonda, S.R., Gathering, W.E. and McDonald, J.M. Modeled microgravity disrupts collagen-I/integrin signaling during osteoblasic differentiation of human mesenchymal stem cells. *J. Cell. Biochem.* **93**, 697–707 (2004).
158. Kundu, A.K. and Putnam, A.J. Vitronectin and collagen I differentially regulate osteogenesis in mesenchymal stem cells. *Biochem. Biophys. Res. Commun.* **347**(1), 347–357 (2006).
159. Takeuchi, Y., Suzawa, M., Kikuchi, T., Nishida, E., Fujita, T. and Matsumoto, T. Differentiation and transformating growth factor-beta receptor downregulation by collagen-alpha2beta1 integrin interaction is mediated by focal adhesion kinase and its downstream signals in murine osteoblastic cells. *J. Biol. Chem.* **272**, 29309–29310 (1997).
160. Jee, W.S.S., Wronski, T.J., Morey, E.R. and Kimmel, D.B. Effects of spaceflight on trabecular bone in rats. *Am. J. Physiol.* **244**, 310–314 (1983).
161. Vico, L., Chappard, D. and Alexandre, C. Effect of weightlessness on bone mass and osteoclast number in pregnant rats after a five-day spaceflight (Cosmos 1514). *Bone* **8**, 95–103 (1987).
162. Vico, L., Chappard, D., Palle, S., Bakulin, A.V., Novikov, V.E. and Alexandre, C. Trabecular bone remodeling after seven days of weightlessness exposure (BIOCOSMOS 1667). *Am. J. Physiol.* **255**, R243–R247 (1988).

163. Rucci, N., Rufo, A., Alamanou, M. and Teti, A. Modeled microgravity stimulates osteoclastogenesis and bone resorption by increasing osteoblast RANKL/OPG ratio. *J. Cell. Biochem.* **100**, 464–473 (2007).
164. Jilka, R.L., Weinstein, R., Takahashi, K., Parfitt, A.M. and Manolagas, S.C. Linkage of decreased bone mass with impaired osteoblastogenesis in a murine model of accelerated senescence. *J. Clin. Invest.* **97**, 1732–1740 (1996).
165. Roggia, C., Gao, Y. and Cenci, S. Up-regulation of TNF-producing T cells in the bone marrow: a key mechanism by which estrogen deficiency induces bone loss *in vivo*. *Proc. Natl. Acad. Sci. USA* **98**, 13960–13965 (2001).
166. Klein, R.F., Allard, J., Avnur, Z., Nikolcheva, T., Rotstein, D., Carlos, A.S., Shea, M., Waters, R.V., Belknap, J.K. and Peltz, G. Regulation of bone mass in mice by the lipoxygenase gene Alox15. *Science* **303**, 229–232 (2004).
167. Dahlman, T.C. Osteoporotic fractures and the recurrence of thromboembolism during pregnancy and the puerperium in 184 women undergoing thromboprophylaxis with heparin. *Am. J. Obstet. Gynaecol.* **168**, 1265–1269 (1993).
168. Backos, M., Rai, R., Thomas, E., Murphy, M., Dore, C. and Regan, L. Bone density changes in pregnant women treated with heparin: a prospective, longitudinal study. *Hum. Reprod.* **14**(11), 2876–2880 (1999).
169. Kock, H.J. and Handschin, A.E. Osteoblast growth inhibition by unfractionated heparin and by low molecular weight heparins: An *in-vitro* investigation. *Clin. Appl. Thromb. Hemost.* **8**, 251–255 (2002).
170. Cenci, S., Toraldo, G., Weitzmann, M.N., Roggia, C., Gao, Y., Qian, W.P., Sierra, O. and Pacifici, R. Estrogen deficiency induces bone loss by increasing T cell proliferation and lifespan through IFN-γ-induced class II transactivator. *Proc. Natl. Acad. Sci. USA* **100**, 10405–10410 (2003).
171. Teitelbaum, S.L. Postmenopausal osteoporosis, T cells, and immune dysfunction. *Proc. Natl. Acad. Sci. USA* **101**, 16711–16712 (2004).
172. Rosneck, J., Higuera, C., Tadross, N., Krebs, V. and Barsoum, W.K. Managing knee osteoarthritis before and after arthroplasty. *Cleveland Clin. J. Med.* **74**(9), 663–671 (2007).

Chapter 37

Application of Tissue Engineering for Craniofacial Reconstruction

Deepak M. Gupta, Matthew D. Kwan, Bethany J. Slater and Michael T. Longaker

Abstract

The potential applications of tissue engineering for craniofacial tissue reconstruction are numerous. However, much remains to be accomplished if translation of research findings is to change the way craniofacial surgeons manage tissue defects presented in the clinical realm. Currently, research in craniofacial tissue engineering aims to address concerns stemming from several shortcomings of strategies that intend to replace or repair tissue defects rather than regenerate them. The majority of this research is conducted in three areas employing various animal models. First, an increased understanding of the mechanisms underlying craniosynostosis/cranial suture biology (i.e. suture fusion or patency) and tissue interactions that may exist between dura mater and the overlying suture complex may impart insight into the molecular mechanisms underlying bone formation. Signaling cascades that have received significant attention in this area include BMPs, TGF-β and FGFs. Second, distraction osteogenesis, a form of endogenous bone tissue engineering, has served to further highlight not only the cellular and molecular components needed for successful bone formation, but also the interaction of mechanical forces with the microenvironment of osseous regeneration. Finally, cell-based therapies have received widespread attention, as multipotent mesenchymal cells have been shown to differentiate towards a multitude of lineages and cell types in a number of experimental designs and applications both *in vitro* and *in vivo*. As such, they may be suitable for application in craniofacial tissue engineering and reconstruction. These three models of tissue induction and *de novo* tissue formation form the basis of regenerative medicine for craniofacial reconstruction, and promise to reveal factors in recapitulating nature, which may be used in applications of tissue engineering to craniofacial reconstruction.

Keywords: Tissue Engineering; Regenerative Medicine; Craniosynostosis; Distraction Osteogenesis; Craniofacial.

Outline

1. Introduction
2. Craniosynostosis: A Case for Mechanisms Underlying Bone Formation
3. Distraction Osteogenesis: Endogenous Tissue Engineering
4. Cellular-Based Tissue Engineering: Regenerative Medicine
5. Conclusion

References

1. Introduction

Craniofacial reconstruction has represented a major challenge for surgeons since before plastic and reconstructive surgery even became a distinct specialty. Paul Tessier is commonly regarded as the "father" of the field of craniofacial surgery. In 1967, he presented his surgical approaches to correct a number of "deformities" involving the orbits at a conference in Rome. His presentations described a series of osteotomies on the face.[1,2] Other major facial deformities were characterized by LeFort (early 20th century) and Gillies (World War II). Far removed from these events, it was only less than a decade ago, in 1997, that craniofacial surgery was officially recognized by the Accreditation Committee for Graduate Medical Education (ACGME) as a subspecialty of plastic and reconstructive surgery. How the field of craniofacial surgery evolved during the intervening eras was remarkably complicated, but one thing is certain: the evolution is by no measure over. On the contrary, it is occurring at an increasing rate.

From trauma and tumor resection to degenerative and congenital disease, common clinical scenarios for surgeons on the front lines of craniofacial practice have also evolved to include professions not limited to audiology, dental medicine, genetics, neurosurgery, nursing, nutrition, occupational and physical therapy, ophthalmology, oral surgery, otolaryngology, pediatrics, psychiatry, respiratory care, social work, and speech pathology. The diversity of backgrounds from which craniofacial care is administered reflects both the complexity of patient care, as well as the depth of burden. Further, data from the United States Healthcare Cost and Utilization Project indicates the number of craniotomies/craniectomies has been rising since 2001.[3] Added to these procedures, post-traumatic facial reconstructions and other skeletal-related procedures account for over one billion dollars annually in direct medical costs in the United States alone. Currently, the aforementioned procedures employ the means of autologous, allogeneic and prosthetic materials to achieve osseous reconstruction and replacement. A number of inorganic/alloplastic materials including metal alloys, glass, plaster of Paris, polymethylmethacrylate have also been employed but often fail to achieve complete restoration of form and function. Each method has inherent limitations, such as donor site morbidity, graft resorption, contour irregularities, structural failure, infection and others. The number of therapeutic techniques reflects the inadequacy of each, as well as the significant need to develop improved reconstructive strategies for the skeleton of the craniofacium. As a result of this socioeconomic and biomedical burden, clinical practice is increasingly demanding an alternative to tissue replacement.

This chapter will focus on three main areas of research in the field of craniofacial surgery that have promised to yield novel treatment strategies by way of tissue engineering which may ultimately translate into improved clinical outcomes: craniosynostosis and its underlying developmental and molecular mechanisms, distraction osteogenesis (DO) and its underlying mechanical and molecular mechanisms, and cell-based therapies for bone tissue engineering. Seminal findings and recent advances which will help to navigate future investigations are presented here.

2. Craniosynostosis: A Case for Mechanisms Underlying Bone Formation

Craniosynostosis, or the premature fusion of one or more cranial sutures, results in cranial deformity, restriction of the growing brain, and increased intracranial pressure. Research in this area has focused on molecular mechanisms controlling whether cranial sutures remain patent or proceed to fuse. Although craniosynostosis can be viewed as a state of pathological bone formation, an understanding of the biology underlying cranial suture biology has provided further insight into potential regulators of bone formation. These results further serve to propel the development of molecular-based treatments for craniosynostosis by providing clues to investigators in disparate research fields about underlying mechanisms of osteogenesis. Although craniosynostosis is usually an isolated condition without known molecular etiologies, many syndromic forms with specific genetic mutations have been described, including Crouzon syndrome, Apert syndrome, Pfeiffer syndrome, and Saethre-Chotzen syndrome. Over the last two decades, a number of mutations have been identified that are responsible for the major craniosynostosis syndromes. Currently, craniosynostosis is treated with complex surgical procedures employing suturectomies and cranial vault remodeling techniques. For the young patient, these are often physiologically taxing operations. Ultimately, the goal of tissue engineering in this field is to develop molecular-based, minimally invasive approaches, to serve as an adjunct or alternative to current strategies for the treatment of premature cranial suture fusion.

With increased understanding of both normal cranial suture biology as well as the pathological processes resulting in craniosynostosis, several signaling factors and growth factors have been implicated in cranial suture development, including bone morphogenetic proteins (BMPs), transforming growth factor-β (TGF-β), and fibroblast growth factors (FGFs). These molecules help to regulate many different events that are essential for bone formation and accordingly have a potential role in tissue engineering.

The TGF-β superfamily is comprised of potent cytokines implicated in a wide range of cellular processes.[4] Many investigators have demonstrated the role of TGF-β isoforms in cranial suture biology and their potential involvement in the pathogenesis of craniosynostosis. Recently, mutations of *Tgf-β receptor-1* (*R1*) and *Tgf-βR2* were found to be associated with human syndromes of craniosynostosis.[4] Roth *et al.* investigated the spatial and temporal expression of TGF-β isoforms during normal cranial suture development and fusion in the murine model.[5] They found intense immunoreactivity for the isoforms (namely β2 isoform) during posterior frontal (PF) suture fusion with no change in immunoreactivity in the patent sagittal suture. In addition, the authors injected recombinant TGF-β2 locally into the sutures and found initiation of premature fusion of the PF suture as well as ectopic bone formation in the sagittal suture, thereby implicating it in a regulatory role in calvarial osteogenesis. Similarly, Opperman and colleagues demonstrated that the reactivity for TGF-β3 declined, while that of TGF-β1 and -β2 maintained continued expression during the process of PF suture fusion.[6] Given the role of TGF-β in cranial suture fusion, several investigators used techniques to abrogate expression in order to prevent fusion. Using a dominant negative *Tgf-βR2* adenovirus in an *in vitro* mouse organ culture system, Mehrara *et al.* showed nearly complete patency of the PF suture compared to fusion of the control sutures and sham infected sutures.[7] Mooney and colleagues also performed an *in vivo* study with anti-TGF-β2 therapy in familial craniosynostotic rabbits to successfully prevent refusion and encourage normal calvarial vault growth after suturectomy.[8]

Investigations of the role of BMP signaling in cranial suture biology have underscored the balance between BMP agonism and antagonism in dictating bone formation that may apply to a variety of craniofacial tissue engineering settings. Urist first discovered in 1965 that demineralized bone matrix could induce ectopic bone formation when placed subcutaneously in rats.[9] As an extension of this study, BMPs were "discovered" to underlie his findings and have formed the foundation for many studies of its role in osteogenesis. Today, BMPs are thought to recruit mesenchymal stem cells and induce them to undergo osteogenic differentiation.[10] A number of BMPs have been used in many animal and clinical trials for various tissue engineering purposes, particularly BMP-2.[11] More recently, BMP antagonists have also been investigated as possible targets in regenerative medicine, particularly in craniosynostosis and tissue engineering. Warren *et al.* have shown that the BMP antagonist, Noggin, was expressed in suture mesenchyme of patent, but not fusing, sutures.[12] To further explore this observation, the authors injected Noggin-encoded adenovirus into PF sutures of three-day-old mice and subsequently noted pathologic patency and exuberant growth at these sutures on histological analysis. In addition, Cooper *et al.* demonstrated decreased incidence of postoperative resynostosis when craniosynostotic rabbits were treated with local application of Noggin-loaded collagen gels after suturectomy.[13] These findings from cranial suture models have further been extended to the tissue engineering setting. In a recent report, Wan *et al.* studied the role of Noggin during osteogenesis mediated by calvarial osteoblasts when implanted into critical-sized calvarial defects. In this study, the investigators primarily demonstrated that abrogation of Noggin's action by RNA interference resulted in significant upregulation of cell-mediated osteogenesis within the healing defect. In theory, when Noggin's role as a "brake" of the osteogenic "accelerators" was removed, BMP signaling was left relatively unopposed, and accelerated bone healing was noted in this animal model.[14] These findings have potential implications for clinical craniofacial tissue engineering beyond the addition of exogenous BMP.

The significance of FGF signaling in suture development has also previously been established, given that mutations in their receptors have long been established in human craniosynostosis syndromes.[15–20] Of the four known FGF receptors, mutations in three have been linked to premature pathologic fusion. Given these findings, significant work has been done to elucidate the function of the receptors and ligands in this family that may lead to bone formation in craniosynostoses. For example, Moore and colleagues have previously demonstrated that FGF signaling was necessary for normal calvarial and suture development. In their study, they demonstrated that reduction of fibroblast growth factor-2 (FGF-2) signaling by introduction of beads coated with a neutralizing antibody decreased differentiation of chick cranial mesenchyme.[21] In contrast, when recombinant human fibroblast growth factor-2 (rhFGF-2) was added to cranial mesenchyme of embryonic mice, these cells proliferated less and showed upregulation of osteopontin and osteonectin, markers of osteoblast differentiation.[22,23] Greenwald *et al.* have also investigated the role of FGF signaling in cranial suture biology by performing targeted *in utero* injections of a dominant-negative fibroblast growth factor receptor-1 (FGFR-1) construct in rats.[24] Briefly, when this construct was injected into dural tissue underlying the normally fusing posterior frontal cranial suture *in vivo*, bone formation in this area and fusion were inhibited. Further study of FGF signaling in cranial suture biology has been conducted by Zhou and colleagues.

In their study, a transgenic mouse carrying the FGFR-1 Pro252Arg mutation underlying Pfeiffer syndrome exhibited pathologic sagittal and coronal synostoses, facial asymmetry and midface hypoplasia.[25] When these findings were studied in a tissue engineering setting, the results were consistent; when osteoblasts with this mutation were examined *in vitro*, they exhibited an accelerated capacity to deposit mineralized extracellular matrix and elevated markers of terminal osteogenic differentiation such as osteopontin, osteocalcin and bone sialoprotein.[25] FGF-2 signaling in calvarial osteogenesis has also been studied by our own group. Recently, we have shown that FGF-2 is also likely to be one molecular player underlying the observed regenerative capacity engendered by immature dura mater.[26] In our study, dura mater cells and calvarial osteoblasts from adult and/or juvenile animals were placed in co-culture. When juvenile dura mater cells were co-cultured with osteoblasts, the latter proliferated faster than when cultured alone or with dura mater cells derived from adult animals. When conditioned media from juvenile dura mater cells were placed on osteoblasts cultured alone, the same effect was observed, but not with conditioned media collected from cultures of dura mater cells originating from adult animals. Upon probing, it was found that media from both co-culture and conditioning with juvenile dura mater cells contained elevated levels of FGF-2. When a neutralizing antibody to FGF-2 was added, the observed effects on proliferation were blunted; the juvenile dura mater cells and their conditioned media now performed similarly to adult dura mater cells and their conditioned media. Conversely, when osteoblasts were cultured alone but supplemented with exogenous FGF-2, they again exhibited accelerated proliferation similar to culture with immature dura mater cells or their conditioned media. Finally, our laboratory has documented a link between FGF and BMP signaling in the setting of normal cranial suture biology. Noggin was found to be tightly coordinated with FGF-2 activity, as inducing high levels of FGF-2 *in vitro* and *in vivo* in the PF suture resulted in downregulation of Noggin.[12] These data underscore an important signaling pathway that may underlie the regenerative properties of immature dura mater and also have implications for craniofacial tissue engineering. It still remains to be determined how FGF-2 may be regulated and exactly how it may be used to develop novel therapeutic strategies for tissue engineering in the clinical setting.

Thus, advances in understanding cranial suture biology and craniosynostosis have led to increased knowledge of the molecular underpinnings of bone formation in these cases. These studies have elucidated several important morphogens and signaling pathways that may be applied to craniofacial tissue engineering applications. More specifically, abrogation or increased expression of these factors may be useful in tissue engineering modalities and consequently facilitate novel, less invasive, and more effective treatment options for both children with craniosynostoses and other patients needing reconstruction of the craniofacial skeleton.

3. Distraction Osteogenesis: Endogenous Tissue Engineering

An overview of the fundamentals of craniofacial tissue engineering would be incomplete without addressing the ideal milieu for successful regeneration of bone. DO has served to further highlight not only the cellular and molecular components needed for successful bone formation, but the interaction of mechanical forces with the microenvironment of osseous regeneration. DO, first described by Alexander Codivilla in 1905 for limb lengthening and later codified by Gavril Ilizarov in the 1950s, represents an endogenous form of bone tissue engineering.[27–29] It involves the application of constant but gradual mechanical vectors to separate two osteogenic fronts, resulting in the formation of new intervening bone. Briefly, from animal studies and clinical experience, investigators have previously documented that the osteotomy site is ideally widened at a rate of 1–2 mm per day ("adaptive"). More rapid rates result in fibrous non-union, while slower rates can cause premature bony union ("maladaptive"). Usually, distraction is performed once or twice per day. Following craniofacial distraction, the resulting callus is allowed to mature over a period of multiple weeks to months, during the so-called consolidation phase. It is thought that the mechanical forces applied by distraction are translated into signals that guide mesenchymal cells in the bony gap to differentiate along an osteogenic lineage.

Since McCarthy's description of DO to lengthen mandibles, this treatment modality has expanded to commonly treat an array of skeletal deficiencies or hypoplasias in the craniofacial region.[30] Current craniofacial applications of DO include the treatment of mandibular deficits resulting from traumatic mechanisms or extirpation of malignancies. The host of craniofacial anomalies resulting in micrognathia-glossoptosis, including Pierre-Robin syndrome, Nager syndrome, and mandibulofacial dystosis, have also been shown to benefit from mandibular DO, especially in correcting the airway obstruction inherent to these syndromes.[31] DO has also been expanded as an alternative to frontofacial monobloc advancements in the treatment of frontal and facial retrusion. Proponents of DO cite a lower incidence of frontal bone infection and decreased mortality rates, while achieving increased cranial vault volume, correction of exorbitism, and alleviation of upper airway obstruction.[32] Undoubtedly, the development of distractor devices which allow for application of multidirectional vectors has contributed to the ability of surgeons to regenerate the complex shaped bones of the craniofacial region.

Given the significant osseous regeneration that results from DO, efforts have been directed at characterizing the association between stress and strain patterns with bone formation. Correlating tensile force measurements with histology, Loboa *et al.* found that the maximal bone formation was observed during active distraction, the period of greatest strain.[33] Loboa went on to describe the forces of distraction, using finite element analysis models created from three-dimensional computed tomography image data of rat mandibles at different phases of DO.[34] This model described patterns of moderate hydrostatic stress within the gap, predictive of intramembranous ossification, and patterns of mild compressive stress in the periphery, consistent with endochondral ossification. These data derived from finite element analysis were consistent with previous histological findings.

The question remains, however, how the application of forces is translated into cellular signals that promote new bone formation. Work by Tong *et al.* points to the role of focal adhesion kinase (FAK), a regulator of the integrin-mediated signal transduction cascade, in DO.[35] In a rat model of mandibular DO, Tong and colleagues demonstrated immunolocalization of FAK in regions of new bone formation secondary to distraction, which was absent in the control groups where new bone formation occurred without distraction. Similarly, recent work has also co-localized c-SRC, a kinase involved with activation of the mechanical transduction complex (p130), in regions of bone regeneration secondary to DO.[36] To further elucidate the effect of mechanical forces on cell biology, Bradley and colleagues utilized an *in vitro* microdistractor device.[37] MC3T3 preosteoblasts were seeded on three-dimensional, collagen type I gels and subjected either to linear distraction or cycles of distraction and compression. Interestingly, Bradley and colleagues found that distraction forces alone increased cellular proliferation, while cycling of distraction and compression forces increased alkaline phosphatase and osteocalcin expression, markers of osteogenic differentiation.

Given that DO remains a relatively lengthy process most directly attributed to the many weeks required for consolidation of the newly formed callus, substantial efforts have been directed towards accelerating this process. One strategy has been to identify potential growth factors involved in bone regeneration in this setting. A multitude of growth factors, including TGF-β1, insulin-like growth factor-1 (IGF-1), FGFs, and BMPs have all been shown to be endogenously expressed during distraction and to accelerate bone formation when exogenously applied to distracted regions in various animal models. An increased understanding of the roles of these factors could be applied towards acceleration of bone formation during DO.

Because of the crucial role that angiogenesis plays during bone development and postnatal fracture repair, our laboratory has examined the role of blood vessel formation in achieving successful bone formation during distraction. Fang *et al.* utilized TNP-470, a synthetic fumagillin analogue that is known to inhibit capillary formation *in vivo* and prevent endothelial cell proliferation *in vitro*.[38–40] Fang and colleagues demonstrated *in vitro* that exposure to TNP-470 did not affect osteoblast proliferation or differentiation. However, TNP-470 did abrogate endothelial cell proliferation *in vitro*. When TNP-470 was applied *in vivo* in a rat model of mandibular distraction, fibrous non-union resulted with a distinct lack of angiogenesis demonstrated upon histological analysis. These findings of failed bone formation and the lack of angiogenesis paralleled that for animals treated with acute distraction. These data together suggest an interplay between application of controlled mechanical forces, angiogenesis, and robust bone formation.

While adequate vascular perfusion is needed for successful wound healing, it is known that ischemia itself is a potent biological stimulus.[41] Using a rat model, Gurtner and colleagues demonstrated the role of ischemia during the process of mandibular DO.[42] Doppler flow analysis revealed relative ischemia in the distracted region during the period of midactivation, as compared to control mandibles which only underwent osteotomy. Gurtner and colleagues then systemically administered endothelial progenitor cells and demonstrated a significantly greater number of these progenitor cells homing to distracted mandibles than the control mandibles at time points during activation and consolidation. These data serve to underscore the role that ischemic signals in the distracted region may serve in regulating trafficking of progenitor cells to that region.

Investigators have also taken great interest in the potential role of mesenchymal cells in enhancing the bone formation during DO. In animal studies where bone marrow mesenchymal cells were injected into regions of mandibular distraction, increased bone formation was observed relative to sites treated with saline injection.[43,44] These findings suggest that the microenvironment of distraction sites acts to stimulate osteogenic differentiation of progenitor cells. However, further work is needed to determine whether the addition of progenitor cells, whether locally or systemically, could lead to accelerated rates of distraction in the clinical realm.

DO has thus proved to be a powerful modality for reconstruction of skeletal deficits in the craniofacial region. More importantly, DO has served to emphasize the link between mechanical force and skeletal regeneration. While progress is being made, further work is needed to elucidate the mechanisms translating discrete force vectors into molecular signals that stimulate competent cells to deposit bone matrix. From clinical experience and laboratory investigations, DO has served to emphasize the critical factors involved with robust bone regeneration, namely a competent cell population, pro-osteogenic molecular signals, angiogenesis, and appropriate mechanical forces.

4. Cellular-Based Tissue Engineering: Regenerative Medicine

While DO is certainly a powerful technique used to engineer new bone *in vivo* in some clinical settings, its applications remain limited in certain clinical scenarios. In some clinical cases of craniosynostosis, craniofacial hypoplasia and injuries secondary to facial trauma, DO may lead to suboptimal results as these cases often represent irregular defects and/or loss of tissue with which to conduct *de novo* endogenous tissue engineering *in situ*. Furthermore, DO is not without its own morbidities, such as soft-tissue infection, osteomyelitis, pintract infection or loosening, and patient discomfort.[45] In these cases, alternatives to repair or regeneration that replace tissue often fall short of the desired clinical outcome.

As briefly mentioned in the introduction, autogenous, allogeneic and prosthetic materials have been the most frequently employed strategies to reconstruct the craniofacial skeleton.[46–54] Amongst many craniofacial surgeons, autogenous bone grafting is regarded as the clinical gold standard. The quality of bone in these situations is ideal and concerns regarding immunologic rejection, structural failure and infection are minimized. However, autogenous bone grafting is limited by the amount of bone available at the donor site and the morbidity imparted to this site during its harvest, which can often be significant.[55] Allogeneic techniques often employ the use of demineralized bone matrix which may evade concerns of insufficient amount of autogenous bone. However, this technique is not without its disadvantages; use of demineralized bone matrix is associated with potential concerns of infection, immunologic rejection, resorption and graft-versus-host disease.[53,56] Prosthetic materials that have assisted in craniofacial reconstruction include metal alloys, glass, Plaster of Paris, polymethylmethacrylate and others, but these are also accompanied by their own disadvantages including risk of infection, contour irregularity and foreign body non-integration. Despite the development of these techniques and materials to address bony defects, there is an unabated demand for improved techniques to reconstruct the craniofacial skeleton. The number of approaches available underscores the lack of an ideal solution and significant burden posed by craniofacial defects.

Nonetheless, DO has become the treatment modality of choice in several clinical scenarios. Before the early 1990s, this technique was not commonplace in the craniofacial realm, but a number of research studies and practical clinical reports have garnered support for this extremely effective mode of endogenous tissue engineering and reconstruction. In the same vein, research on cell-based therapies has received widespread attention largely over the last decade and perhaps holds the same potential to transform clinical practice in the future. This approach holds implications to regenerate tissue in a novel way where previous approaches have failed or resulted in suboptimal tissue replacement.

The term, "cell-based therapy" has represented many areas of research within regenerative medicine. The breadth of this term is likely due to large-scale efforts to define the ideal cellular building block to be used in regenerative strategies. For example, numerous research reports have touted embryonic stem cells (ESCs) as the ideal cellular source for tissue engineering applications due to their vast biologic potential.[57] ESCs represent a cellular building block that is pluripotent and potentially immune-compatible, perhaps two of the most highly desired traits of the ideal cellular source. ESCs are not without their disadvantages though; while their biology may potentially be sound for tissue engineering applications, political and ethical debate surrounds their research and use and makes their clinical involvement in the near future unrealistic.[58–60] Somatic cell nuclear transfer (SCNT), or "therapeutic cloning", also results in the generation of pluripotent, potentially immune-matched ESCs. This technology, however, is different from reproductive cloning in that this cell mass is not implanted in a uterus to generate an embryo; instead, genetic material from the donor is cultured in an oocyte not past the 100-cell stage to generate stem cells with genetically identical information. These "autologous stem cells" would then have the potential to become a multitude of cell types in the adult body, and thus would be useful in tissue engineering and organ replacement applications theoretically without the disadvantages associated with traditional transplantation such as immunologic rejection, the need for immunosuppression, or graft-versus-host disease. SCNT research, however, has not evaded similar ethical debate and is unlikely to be clinically available in the foreseeable future.[61] Finally, gene therapy and genetically modified adult cells also continue to engender significant debate given adverse outcomes in clinical trials and calls for a moratorium, making their use in therapeutic modalities in the clinical setting unlikely in the near term.[62–65]

As an alternative to these highly debated sources of cells for tissue engineering applications, postnatal progenitor cells have received widespread attention recently as an attractive candidate for use in cell-based strategies. These multipotent mesenchymal cells have been shown to differentiate towards a multitude of lineages and cell types in a number of experimental designs and applications both *in vitro* and *in vivo*. Furthermore, they are relatively abundant and readily available. They also avoid concern for immunologic rejection when used autologously. Thus, they may be suitable for application in craniofacial tissue engineering and reconstruction.

The majority of early work with postnatal progenitors focused on mesenchymal cells derived from bone marrow aspirates. As mentioned above, these cells have also been shown to differentiate towards numerous lineages including bone, cartilage, muscle, fat, ligament, tendon and stroma to regenerate tissues throughout the body.[66–68] One seminal and early work that illustrated the *in vitro* capacity of these cells to differentiate to several lineages was that of Pittenger and colleagues. In this study, they demonstrated the ability of bone marrow-derived mesenchymal cells to undergo osteogenic, chondrogenic and adipogenic differentiation under appropriate culture conditions.[66,69] Furthermore, these findings have subsequently been validated *in vivo* in a number of animal models. In one report, these cells were placed in a fibrin glue scaffold and delivered to non-healing rabbit calvarial defects. In this study, the bone marrow-derived mesenchymal cells led to osseous healing of the parietal bone defects. The cells demonstrated similar integration into surrounding corticocancellous bone when compared with implanted osteoblasts.[70,71] The newly regenerated bone also demonstrated comparable stiffness and strength in both bone marrow-derived mesenchymal and osteoblast groups. Finally, both groups performed better than defects that were not filled with a cellular-scaffold construct. In a similar study, bone marrow-derived mesenchymal cells have been used to reconstruct non-healing porcine orbital defects.[72] As these findings provide ongoing enthusiasm for their potential in craniofacial tissue engineering applications, it is important to also recognize that there are several shortcomings associated with the use of bone marrow-derived mesenchymal cells that may preclude them from becoming the ideal cellular building block. Bone marrow-derived mesenchymal cells have been reported by some investigators to require selective sera and growth factor supplementation. Furthermore, investigators have also argued that multipotent mesenchymal cells are a small minority within the nucleated cell fraction derived from bone marrow aspirates; in some reports, this frequency is as low as one in 27,000 cells.[73] Harvest of bone marrow aspirate is also necessarily accompanied by donor site morbidity and potential anesthesia-related complications, which can be significant. Finally, age-related changes in the cellular biology of these bone marrow-derived progenitors may play a significant role and have not been completely elucidated in the literature.[68,73–75] As such, a search for an alternate source of postnatal progenitor cells has been extended beyond the bone marrow in recent years.

As mentioned above, the marrow compartment of long bones undergoes age-related changes with adipose tissue that replaces stroma, and perhaps reflective of this reciprocal relationship between fat and bone, increasing attention has been paid to adult adipose tissue as a source for multipotent mesenchymal cells for bone tissue engineering. Adipose-derived stromal cells (ASCs) have been isolated from adult adipose tissue compartments and have been shown to differentiate towards multiple lineages that may apply to craniofacial reconstruction.

Initially demonstrated by the work of Zuk and colleagues, ASCs have been shown to have the potential to differentiate towards osteogenic, chondrogenic, adipogenic, and myogenic lineages under the appropriate culture conditions *in vitro*.[76–78] Furthermore, ASCs have demonstrated the potential to differentiate to a variety of lineages *in vivo* in a number of animal models including murine, porcine and lagomorph. ASCs also have inherent advantages that may make them attractive; ASCs are more readily accessible and relatively more abundant than their bone marrow-derived counterparts.[73] ASCs are also relatively easy to expand in culture and reports of special serum requirements or growth factor supplementation do not exist.

Bone tissue engineering is perhaps the most widely investigated application of ASCs. In addition to *in vitro* osteogenic differentiation of multipotent ASCs discussed above, these cells have been shown to form bone in craniofacial tissue engineering applications in several animal models. Briefly, in the murine model, ASCs have been shown to form bone when implanted on a polyglycolic acid scaffold subcutaneously.[79] These findings have been extended to settings with human cells implanted in immunodeficient mice when loaded on hydroxyapatite/tricalcium phosphate.[80] In the craniofacial skeleton, murine ASCs have demonstrated the ability to heal critical-sized calvarial defects when implanted on apatite coated poly (lactic-co-glycolic) acid scaffolds.[81] These undifferentiated cells performed comparably to calvarial osteoblasts and bone marrow-derived stromal cells without genetic manipulation or exogenous growth factors. Furthermore, these cells contributed to over 92% of the cellular composition by chromosomal detection to the osseous, histologically mature regenerate.[81] As illustrated, ASCs hold significant potential for their use in the clinical management of craniofacial bone reconstruction and regeneration. This is further corroborated by a case report from Germany where a seven-year-old child underwent calvarial reconstruction and regeneration with autologous adipose tissue-derived stromal cells and bone chips harvested from the iliac crest. In this case, surgeons demonstrated near complete continuity in the region of the calvarial defect only three months postoperatively.[82] This seminal finding supports the proof of principle proposed by documented *in vitro* experiments and *in vivo* animal models.

To summarize, significant progress has already been made with cellular building blocks for craniofacial bone tissue engineering applications. While this biotechnology has largely not entered the clinical human setting, numerous successful cases of bone regeneration mediated by ASCs have been documented in the field of veterinary care namely in the equine and canine models. The majority of these cases has focused on the appendicular skeleton, but still provides an important proof of principle that ASCs may be utilized for craniofacial bone tissue engineering in the human setting. Similar to the past decade, the next ten years promise advancements in the field of cell-mediated tissue engineering; for example, several reports have alluded to elucidation of cellular subsets within the heterogeneous mix of adipose-derived stroma and this exciting investigation holds significant potential for the clinical management of craniofacial repair. Craniofacial tissue engineering by stem cells, however, will also be dependent on parallel advances in other fields such as materials science and bioengineering. The successful delivery of progenitor cells to tissue engineer new bone *in vivo* will depend on properties engendered by specific scaffolds. For example, in our own studies, we advocate the use of scaffolds which are osteoinductive, biocompatible and biodegradable in a controlled manner while maintaining structural integrity. Other investigations propose scaffolds on which cells are highly proliferative or specified to other lineages by means of cytokine signaling. Scaffolds present many options and each should be considered with respect to their advantages and disadvantages.[53,54,83–87]

5. Conclusion

To supplement healing and perhaps develop new tissue replacements, researchers have studied endogenous models of tissue induction and *de novo* tissue formation. Regenerative medicine promises the potential of engineering tissue with multipotent building blocks and appropriate molecular and environmental cues without the disadvantages of the more traditional approaches. More recently added to these components, consideration of polymer scaffolds and mechanical variables has led to work by material scientists and bioengineers to add to the body of literature produced by cell, developmental, and molecular biologists. Already, this interdisciplinary approach has revealed some of the factors in recapitulating nature, which may be used in applications of tissue engineering to craniofacial reconstruction.

References

1. Tessier, P. Total facial osteotomy. Crouzon's syndrome, Apert's syndrome: Oxycephaly, scaphocephaly, turricephaly. *Ann. Chir. Plast.* **12**, 273–286 (1967).
2. Tessier, P., Guiot, G., Rougerie, J. *et al.* Cranio-naso-orbito-facial osteotomies. Hypertelorism. *Ann. Chir. Plast.* **12**, 103–118 (1967).
3. HCUP. Healthcare Cost and Utilization Project (Agency for Healthcare Research and Quality, Rockville, MD, 2005).

4. Loeys, B.L., Chen, J., Neptune, E.R. *et al.* A syndrome of altered cardiovascular, craniofacial, neurocognitive and skeletal development caused by mutations in TGFBR1 or TGFBR2. *Nat. Genet.* **37**, 275–281 (2005).
5. Roth, D.A., Longaker, M.T., McCarthy, J.G. *et al.* Studies in cranial suture biology: part I. Increased immunoreactivity for TGF-beta isoforms (β1, β2, and β3) during rat cranial suture fusion. *J. Bone Miner. Res.* **12**, 311–321 (1997).
6. Opperman, L.A., Nolen, A.A. and Ogle, R.C. TGF-beta 1, TGF-beta 2, and TGF-beta 3 exhibit distinct patterns of expression during cranial suture formation and obliteration *in vivo* and *in vitro*. *J. Bone Miner. Res.* **12**, 301–310 (1997).
7. Mehrara, B.J., Spector, J.A., Greenwald, J.A. *et al.* Adenovirus-mediated transmission of a dominant negative transforming growth factor-beta receptor inhibits *in vitro* mouse cranial suture fusion. *Plast. Reconstruct. Surg.* **110**, 506–514 (2002).
8. Mooney, M.P., Losken, H.W., Moursi, A.M. *et al.* Postoperative anti-Tgf-beta2 antibody therapy improves intracranial volume and craniofacial growth in craniosynostotic rabbits. *J. Craniofac. Surg.* **18**, 336–346; discussion 347 (2007).
9. Urist, M.R. Bone: formation by autoinduction. *Science* **150**, 893–899 (1965).
10. Yoon, S.T. and Boden, S.D. Osteoinductive molecules in orthopaedics: basic science and preclinical studies. *Clin. Orthop. Relat. Res.* **395**, 33–43 (2002).
11. Salgado, A.J., Coutinho, O.P. and Reis, R.L. Bone tissue engineering: State of the art and future trends. *Macromol. Biosci.* **4**, 743–765 (2004).
12. Warren, S.M., Brunet, L.J., Harland, R.M. *et al.* The BMP antagonist noggin regulates cranial suture fusion. *Nature* **422**, 625–629 (2003).
13. Cooper, G.M., Curry, C., Barbano, T.E. *et al.* Noggin inhibits postoperative resynostosis in craniosynostotic rabbits. *J. Bone Miner. Res.* **22**, 1046–1054 (2007).
14. Wan, D.C., Pomerantz, J.H., Brunet, L.J. *et al.* Noggin suppression enhances *in vitro* osteogenesis and accelerates *in vivo* bone formation. *J. Biol. Chem.* **282**, 26450–26459 (2007).
15. Wilkie, A.O.M., Slaney, S.F., Oldridge, M. *et al.* Apert syndrome results from localized mutations of FGFR2 and is allelic with Crouzon syndrome. *Nat. Genet.* **9**, 165–171 (1995).
16. Robin, N.H., Feldman, G.J., Mitchell, H.F. *et al.* Linkage of Pfeiffer syndrome to chromosome 8 centromere and evidence for genetic heterogeneity. *Hum. Mol. Genet.* **3**, 2153–2158 (1994).

17. Muenke, M., Schell, U., Hehr, A. *et al.* A common mutation in the fibroblast growth factor receptor 1 gene in Pfeiffer syndrome. *Nat. Genet.* **8**, 269–274 (1994).
18. Bellus, G.A., Gaudenz, K., Zackai, E.H. *et al.* Identical mutations in three different fibroblast growth factor receptor genes in autosomal dominant craniosynostosis syndromes. *Nat. Genet.* **14**, 174–176 (1996).
19. Muenke, M., Gripp, K.W., McDonald-McGinn, D.M. *et al.* A unique point mutation in the fibroblast growth factor receptor 3-gene (FGFR3) defines a new craniosynostosis syndrome. *Am. J. Hum. Genet.* **60**, 555–564 (1997).
20. Tavormina, P.L., Shiang, R., Thompson, L.M. *et al.* Thanatophoric dysplasia (types I and II) caused by distinct mutations in fibroblast growth factor receptor 3. *Nat. Genet.* **9**, 321–328 (1995).
21. Moore, R., Ferretti, P., Copp, A. *et al.* Blocking endogenous FGF-2 activity prevents cranial osteogenesis. *Dev. Biol.* **243**, 99–114 (2002).
22. Iseki, S., Wilkie, A.O., Heath, J.K. *et al.* Fgfr2 and osteopontin domains in the developing skull vault are mutually exclusive and can be altered by locally applied FGF2. *Development* **124**, 3375–3384 (1997).
23. Iseki, S., Wilkie, A.O. and Morriss-Kay, G.M. Fgfr1 and Fgfr2 have distinct differentiation- and proliferation-related roles in the developing mouse skull vault. *Development* **126**, 5611–5620 (1999).
24. Greenwald, J.A., Mehrara, B.J., Spector, J.A. *et al. In vivo* modulation of FGF biological activity alters cranial suture fate. *Am. J. Pathol.* **158**, 441–452 (2001).
25. Zhou, Y.X., Xu, X., Chen, L. *et al.* A Pro250Arg substitution in mouse Fgfr1 causes increased expression of Cbfa1 and premature fusion of calvarial sutures. *Hum. Mol. Genet.* **9**, 2001–2008 (2000).
26. Li, S., Quarto, N. and Longaker, M.T. Dura mater-derived FGF-2 mediates mitogenic signaling in calvarial osteoblasts. *Am. J. Physiol. Cell Physiol.* **293**, C1834–1842 (2007).
27. Codivilla, A. On the means of lengthening in the lower limbs, the muscles and tissues which are shortened through deformity. *Am. J. Orthop. Surg.* **2**, 353–369 (1905).
28. Ilizarov, G.A. The tension-stress effect on the genesis and growth of tissues: part I. The influence of stability of fixation and soft-tissue preservation. *Clin. Orthop.* **238**, 249–281 (1989).
29. Ilizarov, G.A. The tension-stress effect on the genesis and growth of tissues: part II. The influence of the rate and frequency of distraction. *Clin. Orthop.* **239**, 263–285 (1989).

30. McCarthy, J.G., Schreiber, J., Karp, N. *et al.* Lengthening the human mandible by gradual distraction. *Plast. Reconstr. Surg.* **89**, 1–8 (1992).
31. Imola, M.J., Hamlar, D.D., Thatcher, G. *et al.* The versatility of distraction osteogenesis in craniofacial surgery. *Arch. Facial Plast. Surg.* **4**, 8–19 (2002).
32. Arnaud, E., Marchac, D. and Renier, D. Reduction of morbidity of the frontofacial monobloc advancement in children by the use of internal distraction. *Plast. Reconstr. Surg.* **120**, 1009–1026 (2007).
33. Loboa, E.G., Fang, T.D., Warren, S.M. *et al.* Mechanobiology of mandibular distraction osteogenesis: experimental analyses with a rat model. *Bone* **34**, 336–343 (2004).
34. Loboa, E.G., Fang, T.D., Parker, D.W. *et al.* Mechanobiology of mandibular distraction osteogenesis: finite element analyses with a rat model. *J. Orthop. Res.* **23**, 663–670 (2005).
35. Tong, L., Buchman, S.R., Ignelzi Jr., M.A. *et al.* Focal adhesion kinase expression during mandibular distraction osteogenesis: evidence for mechanotransduction. *Plast. Reconstr. Surg.* **111**, 211–222 (2003).
36. Rhee, S.T. and Buchman, S.R. Colocalization of c-Src (pp60src) and bone morphogenetic protein 2/4 expression during mandibular distraction osteogenesis: *in vivo* evidence of their role within an integrin-mediated mechanotransduction pathway. *Ann. Plast. Surg.* **55**, 207–215 (2005).
37. Gabbay, J.S., Zuk, P.A., Tahernia, A. *et al.* *In vitro* microdistraction of preosteoblasts: distraction promotes proliferation and oscillation promotes differentiation. *Tissue Eng.* **12**, 3055–3065 (2006).
38. Ingber, D., Fujita, T., Kishimoto, S. *et al.* Synthetic analogues of fumagillin that inhibit angiogenesis and suppress tumour growth. *Nature* **348**, 555–557 (1990).
39. Kusaka, M., Sudo, K., Matsutani, E. *et al.* Cytostatic inhibition of endothelial cell growth by the angiogenesis inhibitor TNP-470 (AGM-1470). *Br. J. Cancer* **69**, 212–216 (1994).
40. Fang, T.D., Salim, A., Xia, W. *et al.* Angiogenesis is required for successful bone induction during distraction osteogenesis. *J. Bone Miner. Res.* **20**, 1114–1124 (2005).
41. Ceradini, D.J., Kulkarni, A.R., Callaghan, M.J. *et al.* Progenitor cell trafficking is regulated by hypoxic gradients through HIF-1 induction of SDF-1. *Nat. Med.* **10**, 858–864 (2004).
42. Cetrulo Jr., C.L., Knox, K.R., Brown, D.J. *et al.* Stem cells and distraction osteogenesis: endothelial progenitor cells home to the ischemic generate in

activation and consolidation. *Plast. Reconstr. Surg.* **116**, 1053–1064 (2005); discussion 1065–1067.

43. Qi, M., Hu, J., Zou, S. *et al.* Mandibular distraction osteogenesis enhanced by bone marrow mesenchymal stem cells in rats. *J. Craniomaxillofac. Surg.* **34**, 283–289 (2006).
44. Shao, Z., Liu, B., Peng, Q. *et al.* Transplantation of osteoblast-like cells to the distracted callus in the rabbit mandible. *Plast. Reconstr. Surg.* **119**, 500–507 (2007).
45. Mofid, M.M., Inoue, N., Atabey, A. *et al.* Callus stimulation in distraction osteogenesis. *Plast. Reconstr. Surg.* **109**, 1621–1629 (2002).
46. Shenaq, S.M. Reconstruction of complex cranial and craniofacial defects utilizing iliac crest-internal oblique microsurgical free flap. *Microsurgery* **9**, 154–158 (1988).
47. Bruens, M.L., Pieterman, H., de Wijn, J.R. *et al.* Porous polymethylmethacrylate as bone substitute in the craniofacial area. *J. Craniofac. Surg.* **14**, 63–68 (2003).
48. Nicholson, J.W. Glass-ionomers in medicine and dentistry. *Proc. Inst. Mech. Eng. [H]* **212**, 121–126 (1998).
49. Moss, S.D., Joganic, E., Manwaring, K.H. *et al.* Transplanted demineralized bone graft in cranial reconstructive surgery. *Pediatr. Neurosurg.* **23**, 199–204 (1995); discussion 204–205.
50. Goodrich, J.T., Argamaso, R. and Hall, C.D. Split-thickness bone grafts in complex craniofacial reconstructions. *Pediatr. Neurosurg.* **18**, 195–201 (1992).
51. Marchac, D. Split-rib grafts in craniofacial surgery. *Plast. Reconstr. Surg.* **69**, 566–567 (1982).
52. Dean, D., Topham, N.S., Rimnac, C. *et al.* Osseointegration of preformed polymethylmethacrylate craniofacial prostheses coated with bone marrow-impregnated poly (DL-lactic-co-glycolic acid) foam. *Plast. Reconstr. Surg.* **104**, 705–712 (1999).
53. Bostrom, R. and Mikos, A. (eds.). *Tissue Engineering of Bone* (Birkhauser, Boston, 1997), pp. 215–234.
54. Cho, B.C., Kim, J.Y., Lee, J.H. *et al.* The bone regenerative effect of chitosan microsphere-encapsulated growth hormone on bony consolidation in mandibular distraction osteogenesis in a dog model. *J. Craniofac. Surg.* **15**, 299–311 (2004); discussion 312–313.
55. Silber, J.S., Anderson, D.G., Daffner, S.D. *et al.* Donor site morbidity after anterior iliac crest bone harvest for single-level anterior cervical discectomy and fusion. *Spine* **28**, 134–139 (2003).

56. Mulliken, J.B. and Glowacki, J. Induced osteogenesis for repair and construction in the craniofacial region. *Plast. Reconstr. Surg.* **65**, 553–560 (1980).
57. Thomson, J.A., Itskovitz-Eldor, J., Shapiro, S.S. *et al.* Embryonic stem cell lines derived from human blastocysts. *Science* **282**, 1145–1147 (1998).
58. Weissman, I.L. Stem cells — scientific, medical, and political issues. *N. Engl. J. Med.* **346**, 1576–1579 (2002).
59. Bahadur, G. The moral status of the embryo: the human embryo in the UK Human Fertilisation and Embryology (Research Purposes) Regulation 2001 debate. *Reprod. Biomed. Online* **7**, 12–16 (2003).
60. Chin, J.J. Ethical issues in stem cell research. *Med. J. Malaysia* **58**(Suppl. A), 111–118 (2003).
61. Hwang, W.S., Roh, S.I., Lee, B.C. *et al.* Patient-specific embryonic stem cells derived from human SCNT blastocysts. *Science* **308**, 1777–1783 (2005).
62. Dixon, N. Cancer scare hits gene cures: a second major setback for medicine's most pioneering treatment has split the scientific community. Could a moratorium do more harm than good? *New Sci.* **176**, 4–5 (2002).
63. Grilley, B.J. and Gee, A.P. Gene transfer: regulatory issues and their impact on the clinical investigator and the good manufacturing production facility. *Cytotherapy* **5**, 197–207 (2003).
64. Smith, L. and Byers, J.F. Gene therapy in the post-Gelsinger era. *JONAS Health Law Ethics Regul.* **4**, 104–110 (2002).
65. Verma, I.M. A voluntary moratorium? *Mol. Ther.* **7**, 141 (2003).
66. Pittenger, M.F., Mackay, A.M., Beck, S.C. *et al.* Multilineage potential of adult human mesenchymal stem cells. *Science* **284**, 143–147 (1999).
67. Prockop, D.J. Marrow stromal cells as stem cells for non-hematopoietic tissues. *Science* **276**, 71–74 (1997).
68. Haynesworth, S.E., Baber, M.A. and Caplan, A.I. Cell surface antigens on human marrow-derived mesenchymal cells are detected by monoclonal antibodies. *Bone* **13**, 69–80 (1992).
69. Pittenger, M.F., Mosca, J.D. and McIntosh, K.R. Human mesenchymal stem cells: progenitor cells for cartilage, bone, fat and stroma. *Curr. Top. Microbiol. Immunol.* **251**, 3–11 (2000).
70. Schantz, J.T., Teoh, S.H., Lim, T.C. *et al.* Repair of calvarial defects with customized tissue-engineered bone grafts I. Evaluation of osteogenesis in a three-dimensional culture system. *Tissue Eng.* **9**(Suppl. 1), S113–S126 (2003).

71. Schantz, J.T., Hutmacher, D.W., Lam, C.X. *et al.* Repair of calvarial defects with customised tissue-engineered bone grafts II. Evaluation of cellular efficiency and efficacy *in vivo*. *Tissue Eng.* **9**(Suppl. 1), S127–S139 (2003).
72. Rohner, D., Hutmacher, D.W., Cheng, T.K. *et al. In vivo* efficacy of bone-marrow-coated polycaprolactone scaffolds for the reconstruction of orbital defects in the pig. *J. Biomed. Mater. Res. B Appl. Biomater.* **66**, 574–580 (2003).
73. De Ugarte, D.A., Morizono, K., Elbarbary, A. *et al.* Comparison of multi-lineage cells from human adipose tissue and bone marrow. *Cells Tissues Organs* **174**, 101–109 (2003).
74. Mendes, S.C., Tibbe, J.M., Veenhof, M. *et al.* Bone tissue-engineered implants using human bone marrow stromal cells: effect of culture conditions and donor age. *Tissue Eng.* **8**, 911–920 (2002).
75. Banfi, A., Muraglia, A., Dozin, B. *et al.* Proliferation kinetics and differentiation potential of *ex vivo* expanded human bone marrow stromal cells: implications for their use in cell therapy. *Exp. Hematol.* **28**, 707–715 (2000).
76. Zuk, P.A., Zhu, M., Ashjian, P. *et al.* Human adipose tissue is a source of multipotent stem cells. *Mol. Biol. Cell* **13**, 4279–4295 (2002).
77. Zuk, P.A., Zhu, M., Mizuno, H. *et al.* Multilineage cells from human adipose tissue: implications for cell-based therapies. *Tissue Eng.* **7**, 211–228 (2001).
78. Shi, Y., Nacamuli, R.P., Salim, A. *et al.* The osteogenic potential of adipose-derived mesenchymal cells is maintained with aging. *Plast. Reconstr. Surg.* **116**(6), 1686–1696 (2005).
79. Lee, J.A., Parrett, B.M., Conejero, J.A. *et al.* Biological alchemy: engineering bone and fat from fat-derived stem cells. *Ann. Plast. Surg.* **50**, 610–617 (2003).
80. Hicok, K.C., Du Laney, T.V., Zhou, Y.S. *et al.* Human adipose-derived adult stem cells produce osteoid *in vivo*. *Tissue Eng.* **10**, 371–380 (2004).
81. Cowan, C.M., Shi, Y.Y., Aalami, O.O. *et al.* Adipose-derived adult stromal cells heal critical-size mouse calvarial defects. *Nat. Biotechnol.* **22**, 560–567 (2004).
82. Lendeckel, S., Jodicke, A., Christophis, P. *et al.* Autologous stem cells (adipose) and fibrin glue used to treat widespread traumatic calvarial defects: case report. *J. Craniomaxillofac. Surg.* **32**, 370–373 (2004).
83. Saadeh, P.B., Khosla, R.K., Mehrara, B.J. *et al.* Repair of a critical size defect in the rat mandible using allogenic type I collagen. *J. Craniofac. Surg.* **12**, 573–579 (2001).

84. Seol, Y.J., Lee, J.Y., Park, Y.J. *et al.* Chitosan sponges as tissue engineering scaffolds for bone formation. *Biotechnol. Lett.* **26**, 1037–1041 (2004).
85. Bumgardner, J.D., Wiser, R., Gerard, P.D. *et al.* Chitosan: potential use as a bioactive coating for orthopaedic and craniofacial/dental implants. *J. Biomater. Sci. Polym. Ed.* **14**, 423–438 (2003).
86. Solchaga, L.A., Gao, J., Dennis, J.E. *et al.* Treatment of osteochondral defects with autologous bone marrow in a hyaluronan-based delivery vehicle. *Tissue Eng.* **8**, 333–347 (2002).
87. Lanza, R.P., Langer, R.S. and Vacanti, J. *Principles of Tissue Engineering*, 2nd Ed., Vol. xli (Academic Press, San Diego, CA, 2000), 995 pp.

Chapter 38

Clinical Trials

Anne K. Haudenschild and Marc H. Hedrick

Abstract

Osteoarticular repair involves replacing damaged cartilage and subchondral bone in an articulating, load bearing joint. The human and financial impact of osteoarticular defects is immense. In current clinical practice the most common non-surgical methods for treatment of joint degradation are symptomatic only. For patients who fail conservative treatment, surgical intervention is required. Standard surgical interventions addressing both bone and cartilage damage include osteotomies, mosaicplasty and joint replacements. Cell based treatments of cartilage defects have emerged as promising strategies, but their FDA approved use is limited. Clinical trials that employ tissue-engineering methods have so far focused predominantly on the regeneration of cartilage in damaged knees. Emerging strategies use autologous cells in combination with bioresorbable delivery scaffolds to provide initial mechanical support, homogenous three-dimensional cell distribution, improved tissue differentiation, and suitable handling properties for delivery into patients. The invasive surgical nature of osteochondral repair, the challenges of "blinding" in scaffold implantation procedures, and ethical considerations presents considerable difficulties in implementing large prospective clinical studies to evaluate tissue engineered constructs. Despite these obstacles, the convergence of technological advances in the fields of cell culture, biomaterials, biologics, and surgical techniques makes this an exciting and highly competitive field.

Keywords: Osteoarticular Repair; Cell Therapy; Osteoarthritis; Clinical Trial; Stem Cell.

Outline

1. Introduction to Osteoarticular Repair
2. Impact

3. Current Medical Treatment of OA
  3.1. Pharmacological intervention
  3.2. Surgical intervention
  3.3. Cell therapy
  3.4. Shortcomings of current treatments
4. Current Trials
  4.1. ChondroCelect® [TiGenix]
  4.2. Chondrogen™ [Osiris]
  4.3. BST-CarGel® [BioSyntech Canada Inc]
5. Promising Future Technologies
6. Discussion
  References

1. Introduction to Osteoarticular Repair

Osteoarticular repair involves replacing damaged cartilage and subchondral bone in an articulating, load bearing joint. The primary cause of osteoarticular defects is from osteoarthritis (OA), a chronic degenerative disorder and one of the oldest and most common forms of arthritis.[1,2] Commonly referred to as the "wear-and-tear" type of arthritis, OA is characterized by fibrillation of the cartilage surface accompanied by a net reduction in proteoglycan content of the cartilage. Because the negative charge of proteoglycan chains provides the osmotic pressure necessary to hydrate the tissue, there is a net loss of tissue water content and of the accompanying pressure necessary to withstand physiological loads. A loss of lubrication in the articulating surfaces causes fibrillation of the cartilage and thus exacerbates the degeneration. Inflammation of the surrounding joint capsule can occur as breakdown products from the cartilage are released into the synovial space, and the cells lining the joint attempt to remove them. New bone outgrowths, called osteophytes, form on the margins of the joints in an attempt to compensate for the mechanically compromised articular cartilage surfaces. These defects in both the cartilage and underlying bony surfaces of the joint, together with the inflammation, are both painful and debilitating. OA commonly affects the hands, feet, spine, and the large weight-bearing joints, such as the hips and knees (see also Chapter 13 of this book).

A secondary cause of osteoarticular defects is mechanical injury caused by impact loading from sports related injuries or accidental falls. The increased loading of the joint surfaces of these events result in focal defects that can adopt the same pathology as OA over time due to frictional wear from decreased lubrication of the joint, and high shear forces.[3,4]

2. Impact

Both the human and financial impact of osteoarticular defects is immense. Regardless of the original cause, OA directly impacts the quality of life of 10% of the population in the United States over the age of 60 totaling nearly 21 million people and is second only to cardiovascular disease in producing chronic disability.[5] In the UK, the total costs topped four billion pounds and comprise 1.1% of the GNP[6] and in France, the direct costs of osteoarthritis exceeded 1.6 billion Euros in 2002, and contributed to approximately 1.7% of expenses of the French health insurance system.[7]

The World Health Organization estimates that OA accounts for 25% of visits to primary care physicians, and half of all NSAID (Non-Steroidal Anti-Inflammatory Drugs) prescriptions. The total cost of arthritis and other rheumatic conditions in 2003 were US$128 billion — equivalent to 1.2% of the US gross domestic product.[8,9] The increase in the prevalence of symptomatic OA with obesity[10] and age pushes the estimated number of OA affected adults to 59 million by 2020.[11]

3. Current Medical Treatment of OA

Currently, therapeutic options are limited to conservative management, and invasive surgical procedures. All management options are designed to address the pain symptoms associated with OA, and no treatment option offers treatment for the pathology or improvements in the function of the deteriorating cartilage and bone.[12,13] This gap in therapeutic options results in a large unmet medical need for more efficacious, less invasive treatment options.

3.1. Pharmacological intervention

Disease modifying pharmacological drugs are currently in the development phase and hold tremendous promise.[14] However, in today's clinical practice the most common non-surgical method for treatment of joint degradation is symptomatic only. NSAIDs are the most commonly prescribed medications for the treatment of OA, there is little evidence that NSAIDs actually inhibit inflammation or delay the progression of OA. The general consensus is that NSAIDs provide temporary relief of pain caused by OA and that other pain relievers, like acetaminophen, have fewer or no side effects relative to NSAIDs when used appropriately.

A second line of treatment to reduce OA symptoms are intra-articular injections of either corticosteroids or hyaluronic acid.[15] While corticosteroid injections provide pain relief for weeks to months, no evidence has shown that they successfully modify the progression of the articular cartilage degeneration caused by OA. Similarly, hyaluronic acid injections have been shown to provide short-term relief of pain but require multiple injections and patients should not expect pain relief for several months, if at all.[16,17]

3.2. Surgical intervention

For patients who fail conservative treatment, surgical intervention is required. While surgical interventions are more aggressive in nature, they typically result in little to no pain in treated joints for decades after surgery.[18] The standard surgical interventions addressing both the bone and cartilage damage in an osteoarticular defect include osteotomies, mosaicplasty and joint replacements. There is also a range of repair strategies available to surgeons focusing on regenerating only the cartilage portion of an osteoarticular defect. These techniques include marrow stimulating techniques like microfracture, and cell therapy strategies including autologous chondrocyte implantation.

Microfracture is one of the most minimally-invasive surgical techniques to restore cartilage, and is usually done arthroscopically. It relies on small fractures made through the damaged cartilage and into the adjacent bone.[19,20] This causes bleeding from the bone into the joint, and the resulting influx of cells and growth factors heal the fractured areas and also contribute to the healing of the cartilage surface. This technique has the advantage of being fast and relatively inexpensive, and has met with reasonable clinical success.[21] However, the resulting cartilage is often fibrillar and may not integrate well with the surrounding healthy cartilage.

Osteotomies are more invasive surgical procedures in which a wedge of bone is removed from the joint to shift the body weight off the portion of the joint structure with arthritic damage to the other side of the knee, where the cartilage is still healthy. Osteotomies are helpful because they do not generally require much hardware, making them a more economical surgical option, and they often allow the involved joint to function normally for many years.

Mosaicplasties are more invasive surgical procedures in which healthy cartilage and underlying bone are transferred from a normal area of the joint to a damaged area.[22–24] Osteochondral autograft transfer system (OATS) is a similar technique. The repair tissue has a mosaic tile type appearance. There are the disadvantages of harvesting donor tissue from a healthy region of the joint which can cause additional morbidity.[25,26] In addition, the process of tissue harvest causes cell death at the perimeter of the osteochondral tissue graft, and this can affect both the stability of the graft and its ability to integrate at the new site.[27]

Joint replacement is one of the most common and successful operations in modern orthopedic surgery resulting in pain relief in 95% of its patients at 15 years.[28,29] Good to excellent results are typically achieved in more than 92% of patients treated with femoral condylar implantations.[30] On the down side, joint replacement is major surgery and comes with a major price tag. The joint must be exposed and dislocated, the cartilaginous joint surface and some bone tissue is then removed from the bone ends. The prosthetic components are implanted using an interference fit with the expectation of bone ingrowths, or using bone cement as a grout to hold the metal components into the bone.

3.3. Cell therapy

Cell based treatments of cartilage defects have emerged as promising strategies, but their FDA approved use is limited to focal chondral defects. Focal defects are distinct from osteoarthritis, but when untreated they can contribute to progressive cartilage deterioration through alteration in joint geometry and consequent changes in local forces. There are currently no FDA approved cell therapies to treat osteoarthritis. The most successful cell based therapy for focal defects is Genzyme's Carticel® procedure. In Carticel®, a small piece of healthy cartilage is harvested from a non-essential donor site, and cells are expanded *in vivo* to increase their numbers. The cartilage defect is then covered with a piece of periosteal tissue in a second surgical procedure, and a liquid suspension of the expanded chondrocytes is injected into the defect under the periosteal flap.

3.4. Shortcomings of current treatments

Approved pharmacological treatments of osteoarthritis primarily address symptoms, but they do not delay progression of the underlying disease, nor do they restore damaged cartilage or underlying bone. In addition, there are undesirable side effects. All NSAIDs act by inhibiting the formation of prostaglandins, which play a central role in inflammation and pain; however, these drugs have adverse effects on the gastrointestinal tract. Certain types of NSAIDs (for example COX-2 selective inhibitors) carry such an elevated risk for cardiovascular disease, resulting in their withdrawal from the market despite their success in reducing joint pain.

Surgical interventions are also limited by current technology. Both osteotomies and mosaicplasties create new areas of damage through load transfers or harvesting of healthy tissue, respectfully. Total joint replacements often limit the full range of motion and only last an average of 15 to 20 years before they wear out, loosen or break. With more people living longer and younger patients receiving surgery, multiple surgeries are required for increasing numbers of patients.

Cell based therapies are similar in concept to tissue-engineering, but they also have inherent limitations. Genzyme pioneered the commercialization of cell-based cartilage repair with their Carticel® product more than a decade ago, but this technology leaves room for improvement. The shortcomings of autologous chondrocyte therapy are due to the limited usefulness for chronic indications such as osteoarthritis. In addition, the *in vitro* expansion and dedifferentiation of the articular cartilage-derived cells is a major limitation resulting in multiple surgeries, long lead times for cell expansion and inconsistent cellular product.[31] Perhaps because of these limitations, full recovery of joint function and complete relief of pain are not always attained, and some patients do not respond to this approach.[32] The long term benefits to patients and to halting the progression of joint degradation still need to be determined.[33,34] Failure may be a result of inferior repair tissue that consists of a mixture of fibrocartilage and hyaline cartilage or a lack of new tissue integration with the existing cartilage within the joint space. Despite the successful long-term tissue repair obtained using autologous cell based approaches in the joint environment,[35–37] the potential role and success of autologous cell implantation in an OA joint is unclear, as is the impact of the influence of this procedure on the development and disease progression.

4. Current Trials

Of the nearly 200 ongoing clinical trials for treatment of osteochondral defects[38] the overwhelming majority are pharmacological in nature. The clinical trials that employ tissue-engineering methods have so far focused predominantly on the regeneration of cartilage in damaged knees. These approaches have had some success, and are summarized below.

4.1. ChondroCelect® [TiGenix]

One such trial involves the use of cell characterization in an attempt to improve upon Genzyme's current protocol for autologous chondrocyte repair. TiGenix has recently completed a phase III randomized controlled trial based on the use of proprietary genetic marker technology to improve the selection, characterization, and expansion of cartilage forming cell populations for autologous chondrocyte transplantation procedures. Researchers examined the genetic profiles of individual cartilage cells and found a specific gene expression profile for stabile hyaline cartilage formation.[39] It was demonstrated that 150 (positive) markers were linked to building stable hyaline cartilage and 60 (negative) markers were linked to fibrous tissue formation, which is an undesirable outcome in cartilage repair. Using a subset of these markers, cells are scored and selected for implantation in a process is called Characterized Chondrocyte Implantation (trademarked ChondroCelect™).[40]

After earlier success in animal models,[41] a multicenter, open-label randomized controlled trial was initiated comparing repair of symptomatic cartilage lesions of the knee with ChondroCelect™ to microfracture. ChondroCelect™ formed a regenerated tissue superior to the repair tissue formed following microfracture, as assessed by histomorphometry ($p = 0.003$) and overall histology ($p = 0.012$) of biopsies taken 12 months after treatment. The repair tissue in patients treated with ChondroCelect™ was found to be less fibrous and to display features indicative of more durable hyaline-like cartilage. To date, TiGenix is the only company to have successfully completed a GCP-controlled, prospective, randomized multicentre clinical trial for a cell-based therapy product intended for cartilage repair.

4.2. Chondrogen™ [Osiris]

One of the largest challenges faced in tissue engineering is that of cell sourcing. An adequate number of cells capable of forming the desired tissue type is required but difficult to obtain from mature tissues due to problems associated with donor tissue disease, limited harvesting sites, donor site morbidity, and the limited expansion capacity of fully differentiated adult cells.[42]

Human adult stem cells have two characteristics that have gained a great deal of interest from the tissue engineering community: their pluripotency-meaning that they can form a variety of desired cell phenotypes and their ability to self-renew.[43] These properties make stem cells a potential replacement for expanded chondrocytes in an autologous "cartilage" repair procedure. "Adult" stem cells can be isolated from multiple sources, including bone marrow, fat, or synovium.[44,45]

Osiris initiated a recent phase I/II, double-blind study evaluating the safety and exploratory effectiveness of a preparation of adult stem cells (trade name Chondrogen™) formulated for direct injection into the knee for the regeneration of meniscus. At the six-month time point, Chondrogen™ met its primary end point, demonstrating product safety but did not demonstrate a statistically significant increase in the volume of meniscus as compared to placebo. However, an improvement in baseline cartilage and joint condition was noted in about 30% of patients treated with Chondrogen that was not seen in patients that received placebo. Patients will be followed for safety and additional preliminary efficacy, such as pain, cartilage damage, and changes in the meniscus for two years under the current study protocol.

4.3. BST-CarGel® [BioSyntech Canada Inc]

In an attempt to circumvent the cell sourcing issue all together and produce an off-the-shelf solution, scaffold only approaches to tissue engineering problems have also come into the clinical realm.

One such attempt to improve cartilage function using a scaffold only approach is currently being studied using a thrombogenic and adhesive polymer chitosan (trademark BST-CarGel®), which is being used to stabilize a blood clot and provide a scaffold for cartilage repair.[46] This approach has previously showed some success in improving the quality of microfracture defect repair using a sheep model at six months postoperatively, and in rabbit trochlear defects.[47–49] A randomized, comparative multicenter clinical trial was initiated to investigate whether the treatment of damaged cartilage in the knee with BST-CarGel® will increase the amount and quality of cartilage repair tissue when compared with microfracture alone. The effect of decreasing cartilage related pain and improving cartilage related function in the knee will be assessed by the degree of lesion filling by repair tissue at 12 months.

5. Promising Future Technologies

The current situation is that individual surgeons in the operating room are continually improving their approach to healing osteoarticular defects. However, to bring an idea to clinical trials requires the backing of large corporate sponsors and hundreds of millions of dollars — clearly beyond the reach of individual surgeons or even hospitals. When discussing future promising technologies and products in current clinical trials it becomes necessary to list these companies and their products.

This being said, there are a number of industry competitors with products currently in the marketplace and under development for the treatment of osteoarticular defects. Many of these are geared towards addressing the current treatment shortcomings, and the distinguishing factors for market success will include ease of use, cost and time for production, and the long term efficacy of treatment.

A number of strategies under development use autologous cells in combination with bioresorbable delivery scaffolds. The major functions of a delivery scaffold are to provide initial mechanical support, homogenous three-dimensional cell distribution, improved tissue differentiation, and suitable handling properties for delivery into patients.[50]

Genzyme Tissue Repair's Carticel® represents the market leader in autologous cell based therapy at present with over 10,000 patients treated since it gained FDA approval in 1995. Since then, a less invasive second generation

Matrix-assisted Autologous Chondrocyte Implantation (MACI) product was developed based on a collagen bilayer seeded with autologous chondrocytes. MACI outperformed autologous chondrocyte implantation with a type I/type III collagen derived cover in a recent prospective, randomized study[51] and MACI is currently being manufactured and sold in Europe and Australia with plans to enter the US market.

Another advocate of collagen based scaffolds is ARS ARTHRO AG® who offers the CaReS®-Cartilage Repair System based on patient specific autologous cartilage cells embedded in a 3D mechanically stable collagen type I hydrogel derived from rat tail.

BioTissue Technologies has developed the Autologous Chondrocyte Graft BioSeed®-C for the treatment of defective joint cartilage based on autologous chondrocytes embedded in a three-dimensional bioresorbable two-component gel-polymer scaffold comprised of fibrin and polyglycolic/polylactic acid (polyglactin, vicryl).[52] Clinical outcome was assessed in 40 patients with a two-year clinical follow-up after implantation and showed significant improvement after implantation of BioSeed-C in focal osteoarthritic defects. The results suggest that implanting BioSeed-C is an effective treatment option for the regeneration of posttraumatic and/or osteoarthritic defects of the knee.[52]

Johnson & Johnson's orthopedic division Depuy has developed a treatment that can be delivered without *ex vivo* cell expansion and relies on mechanical fragmentation of cartilage tissue sufficient to mobilize embedded chondrocytes via increased tissue surface area. Direct treatment of full-thickness chondral defects in goats using cartilage fragments on a resorbable scaffold produced hyaline-like repair tissue at six months.[53] Thus, delivery of chondrocytes in the form of cartilage tissue fragments in conjunction with appropriate polymeric scaffolds provides a novel and promising intraoperative approach for cell based cartilage repair.

Smith & Nephew is also in pursuit of an expansion free system and has recently acquired OsteoBiologics, Inc. in an attempt to expand its thriving implant and endoscopic divisions to include the research, development and manufacturing of bioabsorbable tissue-engineered scaffolds and instrumentation to identify and repair osteoarticular defects. OsteoBiologics's, TruFit™ CB is an innovative bioabsorbable bone graft substitutes (BGS) and one-step arthroscopic procedure for repairing bone and articular cartilage defects marketed in Europe.

6. Discussion

Much of the progress that has been made in the field of orthopedics has been by individual surgeons trying their best with different ideas. Brittberg's initial work on autologous cell implantation led to products like Carticel® and pioneered the field of commercialized cell therapy.[35] As more complex treatments are developed to treat more challenging orthopedic problems, research needs to switch from individual surgeons proving concepts to centrally organized and initiated double-blind clinical trials. While this sounds logical, the invasive surgical nature of osteochondral repair presents considerable difficulties in designing and implementing such studies.

The imminent advent of disease modifying treatments for osteoarthritis demands that we take this approach. Orthopaedic trials may be subject to an interim analysis initially looking at the clinical end points perhaps at one and two years and then reporting differences in radiological progression at four and five years. The challenge will be to identify early markers that predict a good outcome, and the initial studies need to be large to do this.[54] Orthopedists interested in early treatment of osteoarthritis need to look at the way cardiologists have developed their large intervention studies in coronary artery disease. High cholesterol levels has been identified as indicative of cardiovascular disease long before clinical manifestation of the disease.[54] Studies on the lipid lowering agents or treatment for thrombolysis have hundreds of patients followed up for long periods of time.[55,56] Similarly, in the field of cancer biology there are early detection markers of disease. We need to ask ourselves why these large controlled studies are not done in orthopaedics and why there are no equivalent indicators of cartilage degradation. Without these types of diagnostic tools, orthopedic surgeons are faced treating more difficult end-stage disease pathologies. While these large studies sound ideal, the invasive surgical nature of osteochondral repair as well as the challenges of "blinding" in device or scaffold implantation procedures presents considerable difficulties in designing and implementing such studies. Additionally, ethical considerations may not allow the use of placebo procedures due to the invasive nature and substantial risk to the patient.

Despite these obstacles, significant improvement in the field is evident. Current osteoarticular repair strategies rely on surgical interventions to stimulate the body's inherent ability for repair. The new and emerging technologies attempt to augment this repair while minimizing the side effects. The ideal tissue-engineered product for osteoarticular repair will be a multi-layered construct with a base layer similar to subchondral bone, and a surface layer with properties similar to cartilage. It would allow immediate fixation into the osteochondral defect, tight integration into both the bone and cartilage surrounding healthy tissue, appropriate mechanical properties to withstand physiological loads, minimal rehabilitation time for the patient, and a long time to failure.

The ideal tissue-engineered osteochondral construct does not exist today, but technological advances in the fields of cell culture, biomaterials, and biologics are converging to hopefully make it a possibility in the near future.

References

1. Felson, D.T. *et al.* Osteoarthritis: new insights. Part 1: the disease and its risk factors. *Ann. Int. Med.* **133**(8), 635–646 (2000).
2. Zhang, W. *et al.* OARSI recommendations for the management of hip and knee osteoarthritis, part I: critical appraisal of existing treatment guidelines and systematic review of current research evidence. *Osteoarthr. Cartil.* **15**(9), 981–1000 (2007).
3. Carter, D.R. *et al.* Influences of mechanical stress on prenatal and postnatal skeletal development. *Clin. Orthop. Relat. Res.* **219**, 237–250 (1987).
4. Griffin, T.M. and Guilak, F. The role of mechanical loading in the onset and progression of osteoarthritis. *Exerc. Sport Sci. Rev.* **33**(4), 195–200 (2005).
5. Altman, R.D. and Moskowitz, R. Intraarticular sodium hyaluronate (Hyalgan) in the treatment of patients with osteoarthritis of the knee: a randomized clinical trial. Hyalgan Study Group. *J. Rheumatol.* **25**(11), 2203–2212 (1998).
6. March, L.M. and Bachmeier, C.J. Economics of osteoarthritis: a global perspective. *Baillieres Clin. Rheumatol.* **11**(4), 817–834 (1997).
7. Le Pen, C., Reygrobellet, C. and Gerentes, I. Financial cost of osteoarthritis in France. The "COART" France study. *Joint Bone Spine* **72**(6), 567–570 (2005).
8. *Prevalence of Doctor-Diagnosed Arthritis and Arthritis-Attributable Activity Limitation — United States, 2003–2005.* CDC and Publications, http://www.cdc.gov/mmwr/preview/mmwrhtml/mm5601a2.htm?s_cid=mm5601a2_e (2006).
9. *National and State Medical Expenditures and Lost Earnings Attributable to Arthtiris and Other Rheumatic Conditions — United States 2003.* CDC and Publication, http://www.cdc.gov/mmwr/preview/mmwrhtml/mm5540a2.htm?s_cid=mm5540a2_e (2006).
10. Christensen, R. *et al.* Effect of weight reduction in obese patients diagnosed with knee osteoarthritis: a systematic review and meta-analysis. *Ann. Rheum. Dis.* **66**(4), 433–439 (2007).
11. Lawrence, R.C. *et al.* Estimates of the prevalence of arthritis and selected musculoskeletal disorders in the United States. *Arthritis Rheum.* **41**(5), 778–799 (1998).

12. Goldberg, S.H., Von Feldt, J.M. and Lonner, J.H. Pharmacologic therapy for osteoarthritis. *Am. J. Orthop.* **31**(12), 673–680 (2002).
13. Singh, G. Treatment options for osteoarthritis. *Surg. Technol. Int.* **11**, 287–292 (2003).
14. Moskowitz, R.W. and Hooper, M. State-of-the-art disease-modifying osteoarthritis drugs. *Curr. Rheumatol. Rep.* **7**(1), 15–21 (2005).
15. Leopold, S.S. *et al.* Corticosteroid compared with hyaluronic acid injections for the treatment of osteoarthritis of the knee. A prospective, randomized trial. *J. Bone Joint Surg. Am.* **85-A**(7), 1197–1203 (2003).
16. Jubb, R.W. *et al.* A one-year, randomised, placebo (saline) controlled clinical trial of 500–730 kDa sodium hyaluronate (Hyalgan) on the radiological change in osteoarthritis of the knee. *Int. J. Clin. Pract.* **57**(6), 467–474 (2003).
17. Karlsson, J., Sjogren, L.S. and Lohmander, L.S. Comparison of two hyaluronan drugs and placebo in patients with knee osteoarthritis. A controlled, randomized, double-blind, parallel-design multicentre study. *Rheumatology (Oxford)* **41**(11), 1240–1248 (2002).
18. Dennis, D.A., Trends in total knee arthroplasty. *Orthopedics* **29**(9 Suppl.), S13–S16 (2006).
19. Frisbie, D.D. *et al.* Arthroscopic subchondral bone plate microfracture technique augments healing of large chondral defects in the radial carpal bone and medial femoral condyle of horses. *Vet. Surg.* **28**(4), 242–255 (1999).
20. Steadman, J.R. *et al.* The microfracture technic in the management of complete cartilage defects in the knee joint. *Orthopade* **28**(1), 26–32 (1999).
21. Steadman, J.R. *et al.* Outcomes of microfracture for traumatic chondral defects of the knee: average 11-year follow-up. *Arthroscopy* **19**(5), 477–484 (2003).
22. Kock, N. *et al.* Histological evaluation of a mosaicplasty of the femoral condyle-retrieval specimens obtained after total knee arthroplasty — a case report. *Acta Orthop. Scand.* **75**(4), 505–508 (2004).
23. Hangody, L. *et al.* Mosaicplasty for the treatment of articular defects of the knee and ankle. *Clin. Orthop. Relat. Res.* **391** (Suppl.), S328–S336 (2001).
24. Hangody, L. *et al.* Arthroscopic autogenous osteochondral mosaicplasty for the treatment of femoral condylar articular defects. A preliminary report. *Knee Surg. Sports Traumatol. Arthrosc.* **5**(4), 262–267 (1997).
25. Feczko, P. *et al.* Experimental results of donor site filling for autologous osteochondral mosaicplasty. *Arthroscopy* **19**(7), 755–761 (2003).
26. Iwasaki, N. *et al.* Donor site evaluation after autologous osteochondral mosaicplasty for cartilaginous lesions of the elbow joint. *Am. J. Sports Med.* **35**, 2096–2100 (2007).

27. Huntley, J.S. *et al.* Chondrocyte death associated with human femoral osteochondral harvest as performed for mosaicplasty. *J. Bone Joint Surg. Am.* **87**(2), 351–360 (2005).
28. Murray, D.W., Goodfellow, J.W. and O'Connor, J.J. The Oxford medial unicompartmental arthroplasty: a ten-year survival study. *J. Bone Joint Surg. Br.* **80**(6), 983–989 (1998).
29. Price, A.J. *et al.* A history of Oxford unicompartmental knee arthroplasty. *Orthopedics* **30**(5 Suppl.), 7–10 (2007).
30. Bartha, L. *et al.* Autologous osteochondral mosaicplasty grafting. *J. Orthop. Sports Phys. Ther.* **36**(10), 739–750 (2006).
31. De Bari, C., Dell'Accio, F. and Luyten, F.P. Failure of *in vitro*-differentiated mesenchymal stem cells from the synovial membrane to form ectopic stable cartilage *in vivo. Arthritis Rheum.* **50**(1), 142–150 (2004).
32. Nehrer, S., Spector, M. and Minas, T. Histologic analysis of tissue after failed cartilage repair procedures. *Clin. Orthop. Relat. Res.* **365**, 149–162 (1999).
33. Knutsen, G. *et al.* A randomized trial comparing autologous chondrocyte implantation with microfracture. Findings at five years. *J. Bone Joint Surg. Am.* **89**(10), 2105–2112 (2007).
34. Knutsen, G. *et al.* Autologous chondrocyte implantation compared with microfracture in the knee. A randomized trial. *J. Bone Joint Surg. Am.* **86-A**(3), 455–464 (2004).
35. Brittberg, M. *et al.* Treatment of deep cartilage defects in the knee with autologous chondrocyte transplantation. *N. Engl. J. Med.* **331**(14), 889–895 (1994).
36. Brittberg, M. *et al.* Autologous chondrocytes used for articular cartilage repair: an update. *Clin. Orthop. Relat. Res.* **391**(Suppl.), S337–S348 (2001).
37. Peterson, L. *et al.* Two- to nine-year outcome after autologous chondrocyte transplantation of the knee. *Clin. Orthop. Relat. Res.* **374**, 212–234 (2000).
38. NIH. *Clinical Trials Website*, http://clinicaltrials.gov/ (2007).
39. Dell'Accio, F., De Bari, C. and Luyten, F.P. Molecular markers predictive of the capacity of expanded human articular chondrocytes to form stable cartilage *in vivo. Arthritis Rheum.* **44**(7), 1608–1619 **(2001)**.
40. Dell'Accio, F., De Bari, C. and Luyten, F.P. Microenvironment and phenotypic stability specify tissue formation by human articular cartilage-derived cells *in vivo. Exp. Cell Res.* **287**(1), 16–27 (2003).
41. Dell'Accio, F. *et al.* Expanded phenotypically stable chondrocytes persist in the repair tissue and contribute to cartilage matrix formation and structural integration in a goat model of autologous chondrocyte implantation. *J. Orthop. Res.* **21**(1), 123–131 (2003).

42. Sharma, B. and Elisseeff, J.H. Engineering structurally organized cartilage and bone tissues. *Ann. Biomed. Eng.* **32**(1), 148–159 (2004).
43. Bianco, P. and Robey, P.G. Stem cells in tissue engineering. *Nature* **414**(6859), 118–121 (2001).
44. Moseley, T.A., Zhu, M. and Hedrick, M.H. Adipose-derived stem and progenitor cells as fillers in plastic and reconstructive surgery. *Plast. Reconstr. Surg.* **118**(3 Suppl.), 121S–128S (2006).
45. Zuk, P.A. *et al.* Human adipose tissue is a source of multipotent stem cells. *Mol. Biol. Cell.* **13**(12), 4279–4295 (2002).
46. Hoemann, C.D. *et al. Neocartilage Formation In Vitro and In Vivo Using an Injectable Chitosan-Based Cell-Delivery Vehicle that Gels and Adheres In Situ* (Trans Federation European Connective Tissue Societies, Patras, Greece, 2000).
47. Chevrier, A. *et al.* Chitosan-glycerol phosphate/blood implants increase cell recruitment, transient vascularization and subchondral bone remodeling in drilled cartilage defects. *Osteoarthr. Cartil.* **15**(3), 316–327 (2007).
48. Hoemann, C.D. *et al.* Chitosan-glycerol phosphate/blood implants improve hyaline cartilage repair in ovine microfracture defects. *J. Bone Joint Surg. Am.* **87**(12), 2671–2686 (2005).
49. Hoemann, C.D. *et al.* Chitosan-glycerol phosphate/blood implants elicit hyaline cartilage repair integrated with porous subchondral bone in microdrilled rabbit defects. *Osteoarthr. Cartil.* **15**(1), 78–89 (2007).
50. Sittinger, M., Hutmacher, D.W. and Risbud, M.V. Current strategies for cell delivery in cartilage and bone regeneration. *Curr. Opin. Biotechnol.* **15**(5), 411–418 (2004).
51. Bartlett, W. *et al.* Autologous chondrocyte implantation versus matrix-induced autologous chondrocyte implantation for osteochondral defects of the knee: a prospective, randomised study. *J. Bone Joint Surg. Br.* **87**(5), 640–645 (2005).
52. Ossendorf, C. *et al.* Treatment of posttraumatic and focal osteoarthritic cartilage defects of the knee with autologous polymer-based three-dimensional chondrocyte grafts: two-year clinical results. *Arthritis Res. Ther.* **9**(2), R41 (2007).
53. Lu, Y. *et al.* Minced cartilage without cell culture serves as an effective intraoperative cell source for cartilage repair. *J. Orthop. Res.* **24**(6), 1261–1270 (2006).
54. March, L.M. and Brooks, P.M. Clinical trials in osteoarthritis. *Ann. Rheum. Dis.* **55**(8), 491–493 (1996).

55. Shepherd, J. *et al.* Prevention of coronary heart disease with pravastatin in men with hypercholesterolemia. West of Scotland Coronary Prevention Study Group. *N. Engl. J. Med.* **333**(20), 1301–1307 (1995).
56. An international randomized trial comparing four thrombolytic strategies for acute myocardial infarction. The GUSTO investigators. *N. Engl. J. Med.* **329**(10), 673–682 (1993).

Chapter 39

Tissue Engineering for the Respiratory Epithelium: Is There a Future for Stem Cell Therapy in the Lung?

Valérie Besnard and Jeffrey A. Whitsett

Abstract

Recent advances in stem cell biology provide the conceptual framework for the development of cell-based therapies for life-threatening diseases affecting many organs, including the lung. Because of its complexity and structure, cell-based therapy for the lung faces significant technical challenges. Therapeutic goals span a spectra of expectations that might include: (1) regeneration of functional lung tissue, (2) replacement of specific cells affected by inherited or acquired diseases with genetically altered progenitor cells, (3) provision of cells capable of enhancing repair or influencing oncogenesis directly or indirectly, and (4) introduction of cells capable of expressing therapeutic molecules for local or systemic delivery. The technical hurdles required for accomplishing each of these goals are distinct and of various heights. None are trivial. Knowledge of the cellular and molecular basis for specification and differentiation of stem/progenitor cells will be required for the successful application of cell-based therapies for the lung. This chapter reviews concepts derived from study of lung morphogenesis and repair as well as stem cell biology that will be relevant to the development of novel therapies for pulmonary diseases in the future.

Keywords: Pulmonary; Tissue Engineering; Morphogenesis; Stem Cells; Cell-Based Therapy.

Outline

1. Introduction: The Challenges Facing Cell-Based Therapy for Treatment of Lung Disease

2. Lung Morphogenesis
  2.1. Integration of signaling and transcriptional pathways during lung formation
  2.2. Multiple cell types in the lung
  2.3. Concept of unique cellular niches within the lung
  2.4. Epithelial cell plasticity
  2.5. Cell proliferation
3. Sources of Stem Cells
  3.1. Definition of stem cells
  3.2. Programming of embryonic stem cells
  3.3. Potential utility of Mesenchymal Stem Cells (MSCs)
4. Do Bone Marrow-Derived Cells Contribute to Repair?
  4.1. Tumorigenic and mutational potentials of stem cells
  4.2. Delivery of stem cells into the lung
    4.2.1. *Vascular delivery of stem cells*
    4.2.2. *Intratracheal delivery of stem cells*
    4.2.3. *Recruitment of MSCs to tumors*
5. Use of Intrinsic Pulmonary Progenitor Cells
6. The Hope of Stem Cell Therapy for Treatment of Lung Disease
7. Conclusion
References

1. Introduction: The Challenges Facing Cell-Based Therapy for Treatment of Lung Disease

The lung is a remarkably complex organ that evolved for the adaptation of vertebrates to terrestrial life. While embryonic development proceeds normally in the absence of lung tissue, life after birth is entirely dependent upon respiration, a process that, in turn, is dependent upon the structure and function of the lung. The respiratory tract consists of distinct anatomic regions from the nasal passages, pharynx, larynx, trachea, bronchi, lobar bronchi, bronchioles and peripheral airways that directly inhaled gases to the alveoli. Gas exchange occurs across alveolar epithelial and capillary endothelial cells. Ventilation is driven by mechanical forces dependent upon neuromuscular activity that is precisely controlled by neurosensory inputs to maintain normal pCO_2, pO_2, and pH. There is little evidence that the lung serves roles other than gas exchange. Lung function is entirely dependent on its remarkable structure that exchanges millions of liters of environmental gases throughout our lifetime. Unlike other organs, for example endocrine organs that synthesize and secrete hormones critical for growth and metabolism of many target

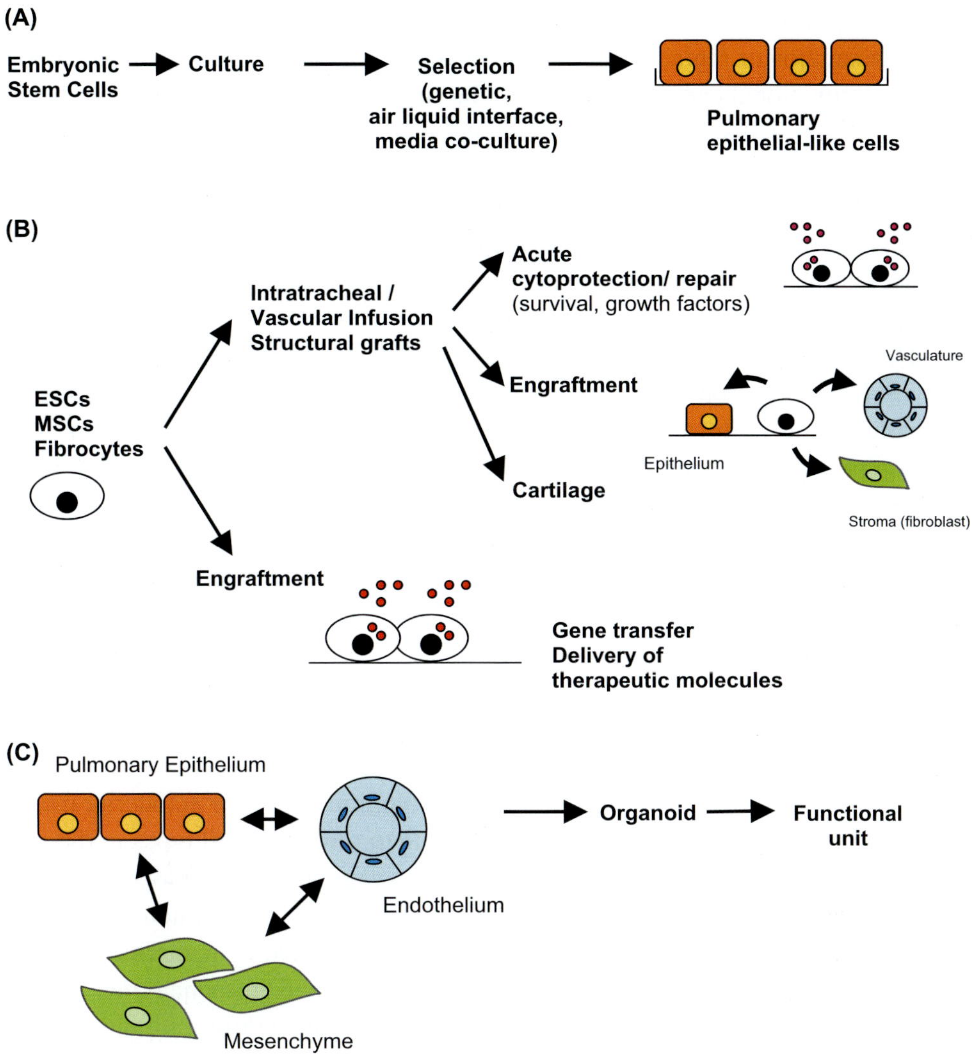

Fig. 1. Strategies for cell-based therapies for the lung. Cell-based therapies for treatment of lung disease are being actively studied using distinct strategies and all are in experimental stages. **(A)** Embryonic stem cells culture can be induced in response to selective culture conditions to generate differentiated respiratory epithelial cells. **(B)** Stem cells may be delivered to the lung by systemic or intratracheal injection to (1) participate in cytoprotection and repair after acute injury, (2) replace cells affected by genetic or acquired diseases with genetically altered progenitor cells, and (3) introduce cells capable of expressing therapeutic molecules for local or systemic delivery or provision of tissue (cartilage) for reconstruction of airway structures. **(C)** Stem cells capable of differentiation into epithelial, endothelial and mesenchymal cells may used to generate pulmonary organoids that may be used for engraftment into the lung.

organs, respiration is an intrinsic property defined by lung structure and mechanics. Regeneration or replacement of functional respiratory tissues will depend upon creating a functional organ that enhances the surface area across which oxygen and carbon dioxide are exchanged. Achieving these goals presents formidable technical challenges for regenerative medicine. Repair/regeneration of tissues in conducting airways (for example, the repair or replacement of the larynx, trachea or bronchial cartilage), or the engraftment of therapeutic cells to the pulmonary epithelium or endothelium may represent a more achievable target for the application of regenerative medicine for the respiratory tract (Fig. 1).

This chapter will consider issues regarding the application of cell-based therapy for prevention and/or treatment of lung diseases, with focus to some of the cellular and molecular processes mediating lung formation and repair. The development of new therapeutic strategies using tissue engineering will depend upon this knowledge of pulmonary morphogenesis and repair. Understanding these processes will be useful in guiding studies in which the principles of cell biology can be applied to enhance the treatment of life-threatening pulmonary disorders.

2. Lung Morphogenesis

The lung is a highly complex organ that contains multiple cell types that are derived from the three layers of the early embryo, including the ectoderm, mesoderm, and endoderm. All germ layers contribute to renew all cell types in the respiratory tract. Lung morphogenesis begins as evagination of endodermal cells along the anterior foregut (Fig. 2). Epithelial cells of the lung primordia proliferate and migrate into the splanchnic mesenchyme influenced by paracrine and perhaps cell-cell signaling between the epithelium and mesenchyme. These interactions are regulated by numerous pathways modulated by retinoids, SHH, FGFs, Wnts, BMPs and TGF-β family members, that instruct cell proliferation, migration, and differentiation.[1]

Understanding the early events directing the differentiation of embryonic cells to pulmonary lineages will be useful in the programming of embryonic or other stem cells to become cells with pulmonary characteristics. Embryonic cells must first be programmed to endoderm and then respiratory epithelial cells. There is increasing evidence that FGF signaling from the cardiac mesoderm occurs as pulmonary epithelial cells are distinguished from other organs along the foregut endoderm. Receipt of high levels of FGF signaling may serve to distinguish lung cells from others destined to form the gastrointestinal tract, liver and pancreas.[2,3] The commitment, restriction and differentiation of endodermal cells to form pulmonary epithelial cell lineages, are first marked by the expression of the transcription factor

NKX2.1 (previously termed thyroid transcription factor-1 or TTF-1).[4] While formation of the trachea and main bronchi occurs in *Nkx2.1*-deleted mice, branching morphogenesis and differentiation of various pulmonary epithelial cell types fails to occur in these mice.[5] *Nkx2.1*$^{-/-}$ mice have a tracheal-esophageal fistula, with failure of separation of the trachea and esophagus, and nearly complete lack of peripheral lung morphogenesis. Lung tubules as well as the stromal and vascular components typical of the normal lung are present in *Nkx2.1*$^{-/-}$ mice. Thereafter, the activity of *Nkx2.1* is influenced by signaling molecules, receptors, and its partnerships with other transcription factors and co-factors that regulate gene expression and differentiation in the various cell types that form the respiratory epithelium.[6] NKX2.1 interacts directly or indirectly with a number of transcription factors, including FOXA2, FOXA1, GATA-6, NFATc3, NF-1, SOX and ETS family members (to name a few) to regulate gene expression and cell differentiation in the respiratory epithelium.[6]

2.1. Integration of signaling and transcriptional pathways during lung formation

A number of signaling molecules mediate the autocrine and paracrine signals that are precisely regulated during lung formation. Wnt-β-catenin,[7,8] FGF,[9–11] SHH,[12,13] BMP4,[14] Notch[15] and retinoic acid[16] pathways play critical roles during formation of the lung. The sites and functions of ligands and receptors that determine the multiple signaling centers required for precise temporal-spatial control of cellular behaviors during lung morphogenesis are highly complex and relatively poorly understood at present. Elucidation of these pathways will be relevant to understanding the control of proliferation and differentiation of embryonic stem cells or other "stem" or "progenitor" cells that might be reprogrammed to generate the pulmonary progenitor cells that may be used for cell-based therapies for repair or regeneration of the damaged lung.

2.2. Multiple cell types in the lung

Lung morphogenesis is precisely orchestrated by inductive and instructive interactions among multiple cells. The respiratory epithelium itself consists of many distinct cell types that are characteristic of the proximal (conducting or cartilaginous airways), as compared to peripheral airways and alveolar (gas exchange) regions of the lung (Fig. 2). The extent of the respiratory tract covered by distinct cell types varies greatly among species and during development. Conducting airways are lined by a pseudostratified epithelium consisting primarily of ciliated, basal, goblet and non-ciliated cells (Clara cells). Neuroendocrine cells are relatively a rare cell type

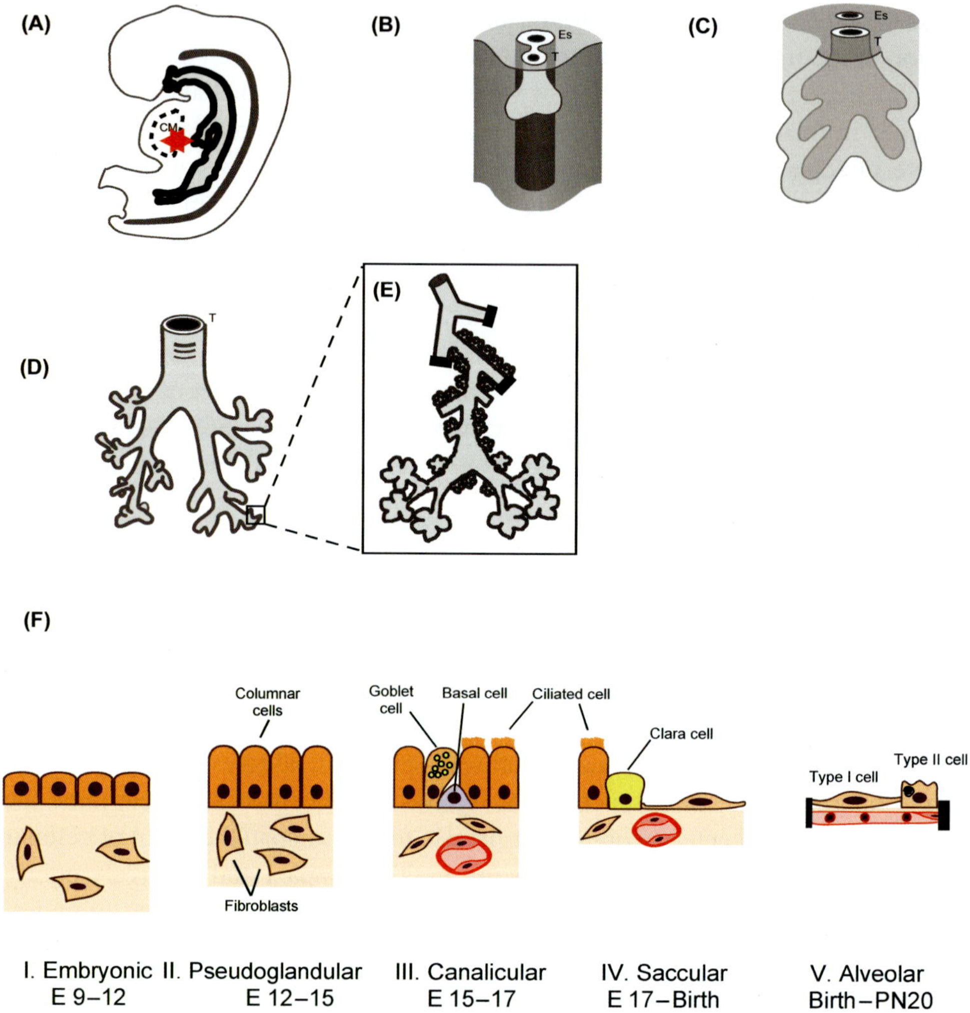

Fig. 2. Complexity of lung structures during development. **(A–E)** Schematic representation of developmental stages of lung development in the mouse. **(A)** FGF signaling from the cardiac mesoderm (CM) participates in the induction of the lung bud formation. **(B)** Lung buds form along the anterior region of the foregut endoderm and migrates into the splanchnic mesenchyme as the primordial trachea and two main stem bronchi are formed. The tracheal and bronchial stalks extend, the latter forming the main bronchi, a process completed by approximately E10.5 in the mouse. **(C)** Stereotypic branching and budding occur as the bronchial and bronchiolar tubules form between approximately E11 to 16 in the mouse and are accompanied by the development of an extensive blood supply from both bronchial and pulmonary circulations. **(D)** Cell proliferation and branching morphogenesis take place until E16.5 when proximal-distal patterning and cell differentiation begins to establish the various epithelial cell lineages lining the airways. **(E)** Alveolarization is associated with rapid cellular proliferation and the successive subdivisions of the pulmonary

present either as isolated cells or as organized clusters termed neuroepithelial bodies (NEB) in conducting airways. Smaller non-cartilaginous airways are lined primarily by ciliated and Clara cells. Squamous type I and cuboidal type II cells comprise the alveolar surfaces of the peripheral lung. After birth, the diversity of cells lining the respiratory tract is strongly influenced by acute and chronic injury. The pattern of stromal tissue also varies greatly along the proximal-distal axis of the lung. Proximal regions of the lung are supported by cartilage, nerves, larger pulmonary arteries, veins and lymphatics as well as tracheal-bronchial glands. Pulmonary vessels (arteries, veins, capillaries and lymphatics) are precisely aligned with the lung tubules and are supported by smooth muscle and other stromal cells. In alveolar regions, mesenchymal cell populations are minimized and endothelial cells in the microvascular network come in close contact with the epithelial cells supported by a network of elastin to produce the gas exchange area in the alveoli. Autocrine-paracrine and direct cell-cell interactions among the various cell types control proliferation and differentiation required to precisely determine the location of specific cells during lung formation. During embryonic and early postnatal development, mitotic activity is generally very high, and proliferating cells differentiate into multiple cell types. In contrast, proliferative rates in the mature lung are remarkably low, cell turnover occurring over many months in the absence of lung injury.

2.3. Concept of unique cellular niches within the lung

Epithelial cell types are distinguished primarily by anatomic, morphologic, biochemical, and genetic features. Antibodies and RNA probes have been developed that are useful for immunohistochemistry, cell sorting and *in situ* hybridization for the identification of specific cell types. Cell survival and proliferative capacity differ amongst various regions within the lung. In conducting airways, the epithelium

---

saccules. **(F)** Multiple cell types are distinguished along the proximal/distal axis during lung development. I. Undifferentiated cuboidal cells line the lung tubules in the embryonic period (E 9–12). II. For the pseudoglandular stage, the respiratory epithelium becomes cuboidal-columnar. III. A pseudo-stratified columnar epithelium consisting of basal, ciliated, goblet, Clara (non-ciliated secretory cells), intermediate cells, and others forms during the canalicular stage. IV. At the saccular stage, the pseudostratified epithelium of the proximal airways contains mostly ciliated cells, Clara-like secretory cells, basal and goblet cells. Bronchioles open into alveoli through the bronchioalveolar ducts. Alveoli are lined by type II epithelial cell and squamous type I cells the latter being closely apposed to alveolar capillaries. V. During the alveolar period, the type II cells proliferate and differentiate into type I cells, increasing the surface area available for gas exchange. Thus the lung is lined by a remarkable diversity of cells that are spatially and developmentally controlled.

is repaired primarily by proliferation of basal cells and non-ciliated epithelial cells. After severe lung injury, cells protected within the tracheal-bronchial gland ducts can migrate onto the airway surfaces, proliferate and redifferentiate to repair the respiratory epithelium.[17] So-called "label retaining cells" or LRCs are present in distinct anatomic microenvironments where they are relatively protected, allowing them to provide progenitor cells for repair of the respiratory epithelium. Basal cells, cells within tracheal-bronchial glands, and subsets of non-ciliated epithelial cells are capable of rapid migration, proliferation and differentiation during repair of the conducting airways. While ciliated cells become squamous and migrate rapidly to repair the injured airway, they appear to be terminally differentiated and do not contribute to cell proliferation following epithelial cell injury.[18] In conducting airways, differences in proliferation and progenitor cell behavior occur in the regions surrounding neuroepithelial bodies near the bifurcations of airways. Mitotic activity may be particularly important in the bronchoalveolar duct junctions (BADJ), a region proposed to harbor subsets of bronchoalveolar stem cells or BASC cells that may have unique properties during lung repair or oncogenesis.[19,20] In the peripheral lung, type II epithelial cells proliferate following injury and differentiate into type I epithelial cells. Selected regions of cytoprotection and repair are observed along blood vessels stroma and pleural surfaces, sites that may serve in regeneration of tissue following acute and chronic lung injury. There is evidence that many type II cells can be recruited into the cell cycle following injury.[21,22] Whether distinct activities of subsets of type II epithelial cells contribute uniquely to various repair processes or to oncogenesis remains to be clarified. While epithelial-mesenchymal transitions are well established during formation of the kidney[23] and other organs, there is no evidence that this occurs during normal lung morphogenesis on the basis of lineage tracing experiments.[24,25] Recent studies provide evidence that during severe injury, the epithelial-mesenchymal transition may be involved in formation of fibrotic lesions, suggesting that the precise cellular and extracellular environments in which epithelial cells are found may influence their behavior and differentiated phenotypes.[26]

2.4. Epithelial cell plasticity

Early in foregut endoderm differentiation, there is considerable cell plasticity among foregut epithelial cells. Expression of β-catenin-LEF caused differentiation of embryonic pulmonary epithelial cells into cells with gastrointestinal tract selective markers.[27] Later in lung development, mesenchyme obtained from the peripheral lung can reprogram epithelial cells from the trachea to differentiate with peripheral alveolar cell types.[28] While these processes occur during defined developmental stage, foregut endodermal cells display remarkable plasticity during the embryonic period. Recent findings, that embryonic stem cells (ESCs) can be grown under conditions that induce respiratory epithelial cell-like properties suggest that embryonic stem cells are capable of being engineered into cells with lung-like phenotypes, presumably recapitulating regulatory processes that direct normal differentiation of respiratory epithelial cells during its formation. Culture conditions including nutrients, hormones, matrices and co-cultures with various pulmonary cells may provide some of the processes that occur during normal lung development. Understanding these developmental programs may be useful in developing strategies for engineering stem cells capable of contributing to lung repair.

2.5. Cell proliferation

In general, cell proliferation in the mature lung is remarkably quiescent. Proliferative rates of all types of pulmonary cells are remarkably slow in stark contrast to the rapid and ongoing turnover of cells lining the gastrointestinal tract or skin, wherein stem cells proliferate, differentiate, migrate and senesce within days. Rates of cell proliferation in the lung are low unless subjected to injury, chronic disease or resection, after which the lung is capable of remarkable proliferative responses. Perhaps the most dramatic of these responses is the response to ipsilateral lung resection that is followed by rapid proliferation of many cell types in the remaining lung tissue that serves to regenerate the lung volumes within several weeks after resection.[29] Since the respiratory tract is continuously exposed to pathogens, including viruses, bacteria, fungus, toxic particles, and toxicants, a remarkable system of innate defense has evolved to maintain pulmonary homeostasis throughout life. This is accomplished by the intrinsic integrity of the epithelial lining, mucocillary clearance, production of innate host defense molecules, and the instruction of acquired immune defenses that serve to maintain lung sterility after birth. Since many viruses and pathogens have developed strategies for targeting specific cells, vertebrates have also developed strategies to maintain cells capable of regeneration. Variation of surface molecules and distinct capacities for proliferation and differentiation of cells in protected anatomic regions provides multiple progenitor cells capable of repairing the lung. Under most circumstances, lung injury is followed by rapid repair with maintenance and restoration of normal lung structure and function. However, following catastrophic injury and/or chronic injury or exposure to toxicants (for example, smoking), metaplasia, dysplasia, and goblet cell hyperplasia are associated with fibrosis and infiltration that underlie the pathogenesis of severe chronic lung diseases including chronic obstructive lung disease, asthma, emphysema, and idiopathic pulmonary fibrosis.

3. Sources of Stem Cells

3.1. Definition of stem cells

"Stem cells" is a term used to define cells that are capable of both self-renewal and production of daughter cells with capacity to differentiate and contribute to the formation or repair of a tissue. Stem cells able to provide cells characteristic of all three germ layers, ectoderm, mesoderm and endoderm are generally termed "pluripotent" stem cells, exemplified by embryonic stem cells (ESCs) that derived from the inner cell mass of early embryos. These cells are capable of proliferation as well as contributing to the formation of multiple organs or the entire embryo. The term "stem cell" is used with varying precision amongst fields of study, for example, being more precisely defined for embryonic or marrow stem cells but perhaps more loosely in organs in which precise cell lineage relationships are not known. Hematopoietic stem cells (HSCs) in the marrow are considered as such when a single stem cell can regenerate the entire hematopoietic system after marrow ablation and transplantation. Bone marrow-derived mesenchymal stem cells are considered "multipotent" when exhibiting capacity for differentiation into two of the germ layers. Distinct stem cell properties have been described for hematopoietic stem cells (HSCs), mesenchymal stem cells (MSCs), circulating endothelial progenitor cells (EPCs) that are capable of differentiating into endothelial-like cells *in vitro,*[30] side population (SP) cells, and multipotent adult progenitor cells (MAPCs) (for review[31] and various contributors in this book). Capacities of various stem/progenitor cells to contribute to the diverse cell types present in the lung are not well defined at present.

3.2. Programming of embryonic stem cells

Embryonic stem cells may provide a source of cells that when subjected to various cell culture conditions, can differentiate towards respiratory epithelial cell-like phenotypes. Differentiation of ES cells into respiratory epithelial cells requires that cells acquire endodermal and pulmonary characteristics. Surfactant protein C (pro-SPC) is a useful cell marker for alveolar type II cells. Both Foxj1 and β-tubulin have been useful in identifying ciliated cells. Non-ciliated respiratory epithelial cells express Clara cell protein (CCSP). These markers have been used to demonstrate that ES cells have been differentiated into Clara-like cells[32] or alveolar type II cells[33,34] during culture. It is unclear, however, whether other differentiated functions have been recapitulated in these preparations. Nevertheless, there is increasing evidence that ES cells can be cultured under conditions to enhance expression of pulmonary epithelial characteristics. Culture of ES cells at an air/liquid interface produced a complex, highly differentiated epithelium with properties of conducting airways, including ciliated, basal, and secretory-like cells.[32] Similarly, co-culture of ES cells with embryonic lung mesenchyme enhanced the differentiation into cells with respiratory epithelial cell characteristics.[33] Nuclear extracts obtained from mouse lung epithelial (MLE)-12 cells (an SV40 large T-antigen immortalized cell line) increased pulmonary cell differentiation of embryonic stem cells *in vitro*.[34] Differentiated alveolar type II-like cells were produced after genetically modified ES cells were grown under conditions that selected for the survival of respiratory epithelial cell phenotype.[35] Differentiation of ES cells into respiratory epithelial-like cells has been associated with the expression of pro-SP-C, CFTR, α1AT, cytokeratin, ZO-1 and others.[35–37] It remains unknown whether these "differentiated" cells can be reintroduced into the lung to contribute to normal lung function or repair. Notably, immortalized or transformed cell lines have been isolated and cultured under conditions in which they express markers specific for respiratory cell type, but these cell lines do not recapitulate gene expression or other biological functions comparable to endogenous lung epithelial cells. Engineering ES cells capable of engrafting, proliferating, differentiating and repopulating the respiratory epithelium represents a considerable technical challenge.

The use of ES cells is the subject of important social and political controversies related to the destruction of embryos required to produce pluripotent ES cells. Recently, reprogramming of somatic cells into pluripotent ES cells was achieved in the mouse[38,39] and in Rhesus macaque.[40] Skin fibroblast nuclei were introduced into mature oocytes by somatic cell nuclear transfer and were able to differentiate into cell types representative of all three germ layers.[40–43] These recent advances provide a new strategy for the production of ES cells for cell-based therapies. The success of exogenous stem cell-based therapies for the lung and other organs will depend on the ability to isolate, culture, genetically modify cells capable of engrafting, proliferating and differentiating into cells that can contribute to organ function. Because of their capacity for self-renewal and pluripotency, it may be possible to generate large numbers of ES cells *in vitro* for cell-based therapy. Reprogramming skin cells into pluripotent ES-like cells may overcome social and religious barriers to the use of ES cells, however, the abilities of these reprogrammed cells to safely contribute to organ function *in vivo* remains unclear. The ability to produce ES cells directly from the patient and to genetically modify the cells for therapy may bypass immunologic barriers to cell-based therapies. Specific lung cell promoters, for example SFTPC or SFTPB, may be useful tools for introducing genes for treatment of genetic disease affecting the respiratory epithelium.[35,44] Mesenchymal stem cells (MSCs), umbilical cord blood cells (UCBs) and embryonic stem cells (ESCs) represent alternative sources of potential stem cells. At present, however, adult bone marrow-derived stem cells are present in few numbers, have a limited proliferation and differentiation capacities and have not been effectively engrafted into the lung.

3.3. Potential utility of mesenchymal stem cells

MSCs have been derived from bone marrow stroma, umbilical cord blood (UCB),[45] whole blood and other organs (see also other chapters in this book). Adult bone marrow-derived MSCs have limited capacity to proliferate *in vitro* but are able to grow and differentiate into bone, cartilage, fibrocytes and other cell types.[46] However, MSCs and hematopoietic stem cells (HSCs) isolated from umbilical cord blood (UCB) contain pluripotent hematopoietic stem cells that are readily isolated at birth. UCB stem cells have clinical and logistical advantages for cell-based therapy and are routinely used for bone marrow transplantation. UCB cells are collected without risk to the donor and can be readily stored. UCB cells are amenable to HLA matching for allogenic transplantation. MSCs are also present in UCB and are relatively abundant (0.1%)[47] compared to adult bone marrow, wherein the frequency of the MSCs ranges from 0.01% to 0.001% of total nucleated cells.[31] UCB-derived MSCs produce multipotent colony, and have longer telomeres than adult BM-derived MSCs perhaps providing an advantage for longevity as progenitor cells.[48,49] MSC derived from UCB can be expanded *in vitro* and can differentiate into multiple cell types (chondrocyte, osteocyte, myocyte and adipocyte lineages).[46,50,51] "Fibrocytes" are bone marrow-derived mesenchymal progenitors that express cell surface markers indicating hematopoetic origin but express collagen and other matrix proteins characteristics of fibroblasts.[52,53] Circulating fibrocytes migrate to wounds where they contribute to populations of fibroblasts, lipocytes and myofibroblasts.[54] Circulating epithelial progenitor cells (CEPCs) have been identified in bone marrow and blood by expression of CD45, cytokeratin-5 and CXCR4.[55] CEPCs are recruited to the airways in response to a gradient of CXCL12 following lung injury. FGF-7 enhanced the migration of CEPCs to proximal airways following injury.[56] CEPCs enhanced re-epithelialization of tracheal transplants *in vitro*, inhibiting squamous metaplasia resulting from the excessive proliferation of the resident epithelial progenitor cells during repair of the airway epithelium.[55]

4. Do Bone Marrow-Derived Cells Contribute to Repair?

Stem cell-based therapies may be designed to acutely enhance repair by endogenous cells or alternatively to provide progenitor cells (or genetically altered progenitor cells) capable of engraftment and permanently contributing to the lung.

Long-term engraftment and proliferation of stem cells for repair following injury has not been accomplished in the laboratory and faces many technical hurdles. Evaluation for Y-chromosome-containing type II pneumocytes using FISH technology indicated that repopulation by bone marrow-derived stem cells or their progeny may occur at a low frequency in the lungs.[57,58] Early studies of potential bone marrow contribution to lung repair reported that bone marrow-derived cells from a male donor engrafted and accounted for 20% of the alveolar epithelial cells into a female lung.[59] Similarly, in a model of bleomycin-induced lung injury, MSCs were proposed to differentiate into alveolar type I cells.[60] However, engrafted MSCs did not proliferate or differentiate into pulmonary epithelial cells. Their presence in the lung was considered a result of cell fusion between the MSCs and endogenous epithelial cells or false colocalization of cell markers due to insufficient technical resolution.[61,62] While it is well known that hematopoietic cells migrate to the lung and repopulate alveolar macrophages following bone marrow transplantation, engraftment of bone marrow-derived cells capable of proliferation and differentiation in the lung has not been accomplished to date.

Introduction of MSCs into the lung or other organs has been proposed as a therapy that does not depend on long term engraftment. For example, in the kidney, in a rat model of ischemia/reperfusion, MSCs infusion enhances the restoration of renal function.[63] Recovery after MSCs infusion was mediated via paracrine actions with decreased expression of proinflammatory cytokines (IL-1β, TNF-α, IFN-γ), and increased anti-inflammatory cytokines (IL-10 and bFGF, TGF-α) and Bcl-2.[64] Pulmonary hypertension and endothelial function improved after engraftment of MSCs into the lung in a process likely related to paracrine effects rather than replacement of endothelial cells.[65,66]

4.1. Tumorigenic and mutational potentials of stem cells

Stem cell-based therapies present significant safety concerns including immune rejection and tumorigenesis. Stem cells may acquire chromosomal abnormalities during prolonged culture.[67] Mutations can enhance proliferation, cloning efficiency and inhibit apoptosis. While these traits are useful characteristics in culture, these abnormalities may influence cell differentiation and increase the potential for malignancy. Because stem cells have characteristics in common with cancer cells, including capacity of self-renewal and production of amplifying cells, control of stem cells proliferation may be required at sites of engraftment. Recent studies demonstrated the risk of using bone marrow cells or mesenchymal cells and their contribution to tumorigenesis.[68–71] Likewise, proliferation of fibrocytes raises concerns regarding their potential contribution to lung pathology and pulmonary fibrosis. The ability of fibrocytes to differentiate into mature mesenchymal cells *in vivo* may enhance rather than ameliorate fibrogenesis and vascular remodeling.[72–74] Further characterization of the fibrocyte-derived stromal cells will be necessary to determine whether they will participate therapeutically or pathologically. Genetic modifications of stem cells will require the use of integrating viruses or DNA raising concerns regarding the potential for insertional mutagenesis. The use of retroviruses to treat severe combined immunodeficiency (SCID-X1) was associated with lymphoma related to the insertion of the vector near an oncogenic locus.[75] Murine retroviruses insert preferentially into promoters of active genes.[76] Thus, it will be important to evaluate the risks of insertional mutagenesis in the genetic engineering of stem cells used in treatment of lung disease.

4.2. Delivery of stem cells into the lung

4.2.1. *Vascular delivery of stem cells*

The efficient recruitment or delivery of stem cells into the lung represents another challenge for the application of stem cell therapy to the lung. There is at present little evidence that exogenous cells readily engraft and permanently contribute to normal lung function. While normal and tumor cells readily enter the pulmonary vasculature and can embolize or metastasize to the lung, engraftment appears to occur with very low frequencies. Previous studies with MSCs or marrow-derived cells demonstrated extremely low efficiency of integration into the normal lung epithelium or vasculature after intravenous injection. In most studies, engraftment of bone marrow and mesenchymal stem cells into the lung does not occur unless the lung was severely injured. The mechanisms underlying this observation are unclear. Secretion of growth factors and cytokines may enhance recruitment of MSCs to sites of injury. *In vitro* experiments demonstrated that cell suspensions from the injured lung contain chemoattractants and growth-stimulating factors including hyaluronan (CD44), osteopontin and SDF-1α.[77] Repair of the lung after bleomycin injury was enhanced after intravenous administration of bone marrow derived cells and was associated with an increased expression of anti-inflammatory cytokines (IL-10, IL-1ra, IL-13), G-CSF and GM-CSF.[78]

4.2.2. *Intratracheal delivery of stem cells*

Since the surface of the lung is directly accessible via the trachea, stem cells may be directly injected intratracheally. Intratracheal injection of bone marrow-derived mesenchymal cells decreased bacterial and bleomycin-induced lung injury.[79,80] The beneficial effects of MSC were independent of the ability of the cells to engraft in the lung and were proposed to be mediated by paracrine effects modulating inflammation. Intratracheal injection of alveolar type II cells decreased fibrosis after induction by bleomycin.[81] Engrafted alveolar type II cells did not contribute to alveolar epithelium repopulation but reduced collagen deposition and fibrosis development. The remarkable structural integrity of the respiratory epithelium is likely to limit the ability to introduce exogenous cells and may require depletion of endogenous cells to provide "space" for engraftment. The success of bone marrow transplant depends on marrow ablation creating space for engraftment. Inhibition of c-kit, the receptor for the stem cell factor, by ACK2 antibody treatment depleted 99% of the host HSCs enhancing the engraftment of donor HSCs in the host stem cell niches.[82] Development of safe methodologies for providing the cellular space required for the engraftment of stem cells into the lung will likely be required for success of cell-based therapies dependent upon engraftment of cells into the respiratory epithelium.

4.2.3. *Recruitment of MSCs to tumors*

The use of genetically modified mesenchymal cells has been proposed for treatment of pulmonary tumors.[83–85] "Umbilical cord matrix stem cells" (UCMS) transplanted into SCID mice bearing lung tumors, migrated to pulmonary carcinomas,[86] indicating that mesenchymal cells may represent a vehicle for delivery of therapeutic molecules to the lung. Mesenchymal cells expressing therapeutic proteins (IFN-γ, NK4, CX3CL1) decreased the growth of pulmonary tumors.[86–90] The enhancement of recruitment of MSCs to pulmonary tumors represents a promising application of cell-based therapies for the lung.

5. Use of Intrinsic Pulmonary Progenitor Cells

Repair of the respiratory epithelium occurs rapidly to maintain alveolar-capillary permeability and pulmonary homeostasis. Failure to repair is associated with loss of alveolar capillary integrity that can cause respiratory failure. Repair of the respiratory epithelium primarily is dependent upon endogenous cells that line distinct regions of conducting and peripheral airways. Cells that survive injury rapidly spread and migrate to cover the epithelium surfaces. While the mature respiratory epithelium does not undergo rapid turnover,[91] the respiratory epithelium responds rapidly to acute injury and many cell types are capable of re-entering the cell cycle during the repair process.

The identity and properties of endogenous stem/progenitor cells in the lung are of considerable interest and are under active investigation. At present, few markers or characteristics are useful in precisely identifying pulmonary progenitor cells. In many tissues, "endogenous stem cells" are generally considered to represent a rare subset of non-proliferating cells that are normally maintained in undifferentiated non-proliferative state. However, findings in some organs (e.g. repair of pancreatic beta cells populations by differentiation of ductal or intra-islet pancreatic precursor cells) indicate that differentiated endogenous cells replicate and serve as pluripotent progenitor cells. Numerous cell types, mostly basal, non-ciliated epithelial and alveolar type II cells, have proliferative capacities and are able to differentiate into distinct cell types during repair. Subsets of "stem cells" have been described in the lung including side population cells, label retaining cells and BASCs. "Side population" (SP) cells are defined by their ability to actively efflux DNA-binding dye (Hoechst 33342). SP cells were identified in the embryonic and adult lung in distinct anatomic regions including the trachea and the alveolar epithelium.[92–95] SP cells are CD45 negative (non-hematopoietic), have Hoechst efflux capacity, verapamil-sensitivity, low auto-fluorescence and are enriched in vimentin mRNA.[96] Analysis of the stem cell-like subpopulations demonstrated that SP cells present in the tracheal compartment represent a heterogenous cell population that varies in clonogenicity, whereas alveolar SP cells are more homogenous. The alveolar SP and type II cells are equally susceptible to bleomycin and do not appear to have unique capacity to repopulate the alveolar epithelium.[96] Some subsets of "stem cells" in various organs have been defined by their low rate of proliferation and termed label retaining cells (LRCs) because of their prolonged retention of BrdU or ^{3}H thymidine after labeling. LRCs turn over slowly, are capable of self-renewal and may be an important source of progenitor cells that can further proliferate and differentiate during repair of the lung. LRCs populations are enriched in the necks of submucosal glands and in tracheal/bronchial folds in non-cartilaginous regions of the conducting airways.[17,97–99] Other potential populations of stem cells with label retaining capacity are located near neuroepithelial bodies at branch points of conducting airways.[100] Stem cells within terminal bronchioles adjacent to the bronchoalveolar duct junction, termed bronchoalveolar stem cells (BASCs) may represent another subset of cells with unique proliferative and differentiation capacities.[19,20] BASCs were defined as CCSP- and pro-SPC-co-expressing cells that resist naphthalene injury and repopulate terminal bronchioles. These cells are not abundant and it is not clear at present whether the properties of self-renewal and production of daughter cells of BASCs are distinct from more abundant cells (e.g. Clara cells, alveolar type II cells) that are known to contribute substantially to repair of terminal airways and the alveoli. Lineage tracing experiments will be useful in defining the ability of these "stem cells" to renew and differentiate into various cell populations.

While rare and aberrant subsets of epithelial cells have been identified in the lung, repair of the respiratory epithelium that occurs following the rapid infection or other injury, is not dependent upon rare subsets of cells like BASCs or LRCs

PART X

LUNG REPAIR

but mediated by the proliferation and re-differentiation following injury. In conducting airways, *in vitro* and *in vivo* experiments have demonstrated the ability of basal cells, Clara cells or other secretory cells to proliferate and differentiate into other respiratory epithelial cell types.[19,98–105] Ciliated cells contribute to repair by undergoing squamous metaplasia and re-differentiating to Clara cells following injury. Lineage tracing experiment indicate that ciliated cells are not proliferative and do not serve a role as progenitor cells after injury to airway epithelium.[18,106] Evidence indicating that many endogenous cells play a role in lung repair is derived from experiments in animals models exposed to toxicants (naphthalene, SO_2). Purified basal cells and/or Clara cells re-establish a complex respiratory epithelium consisting of numerous cell types *in vitro*[107,108] including ciliated, non-ciliated epithelial cells, and goblet cells. In the peripheral lung, type I cells are sensitive to injury and do not proliferate. Repair of the peripheral lung depends upon alveolar epithelial type II cells. Proliferation of alveolar type II epithelial cells occurs two to three days following the O_2 injury. Subsets of these cells differentiate into type I cells to complete repair.[21,22] Thus, under physiologic conditions, repair is accomplished primarily by the proliferation of endogenous progenitor cells present in the various compartment of the respiratory epithelium. Identification of endogenous lung "stem" or "progenitor" cells, characterization of their proliferation and differentiation potentials and the development of processes to genetically repair progenitor cells capable of permanent engraftment and proliferation represent important scientific goals that will enable stem cell therapies for pulmonary diseases.

6. The Hope of Stem Cell Therapy for Treatment of Lung Disease

Successful cell-based therapy for many genetic diseases affecting epithelial cell function (e.g. cystic fibrosis, genetic defects in genes controlling surfactant homeostasis and others) may require the permanent introduction of cells capable of maintaining stem cell activity as well as the ability to contribute to highly differentiated cell functions. Differentiated respiratory epithelial cells have been produced on various biomatrices, tracheal grafts, air-liquid interface cultures and after transplantation under the renal or testicular capsules of SCID mice[32,109–111] supporting the feasibility of direct cell differentiation useful for correction of pulmonary disease. Selectable markers may be used to enhance epithelial differentiation during preparation of stem cells. For example, ES cells transfected with a *SFTPC* promoter/NEO^r fusion gene were selected with gencyclovir, enhancing the numbers of cells with respiratory epithelial cell characteristics.[35] Use of selection strategies and the expression of regulatory genes known to play an important role in lung epithelial specification (for review, see Ref. 6) may be useful in differentiating stem/progenitor cells into further specific cell types that will be useful for pulmonary disorders. Likewise, elucidation of the signaling and transcriptional networks controlling lung epithelial cell differentiation will be useful in engineering stem cells for therapy of lung diseases in the future.

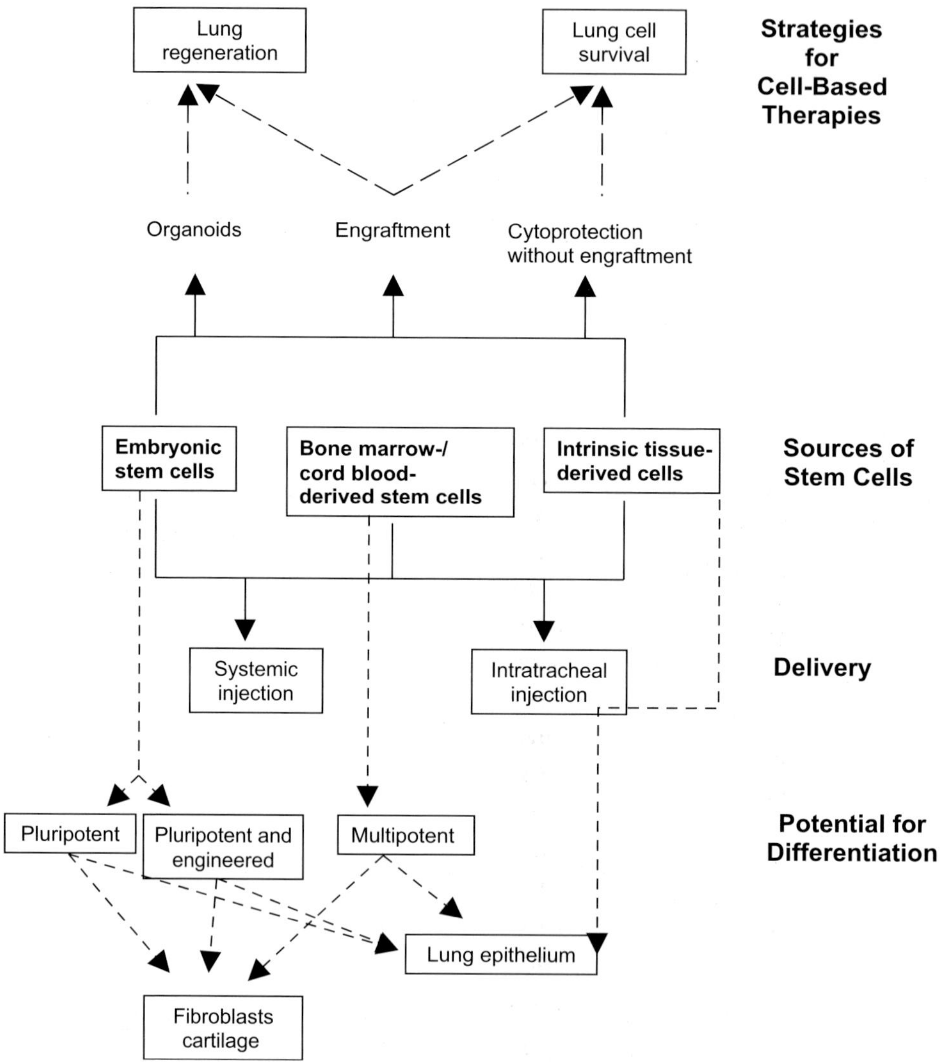

Fig. 3. Concepts regarding the use of stem cells for treatment of lung diseases. Stem cell-based therapies seek to enhance lung function via systemic or intratracheal delivery of exogenous stem/progenitor cells or the provision of progenitor cells (or genetically altered progenitor cells) to influence disease processes or contribute to the repair of lung tissues. Participation of various stem/progenitor cells in repair may occur through engraftment or release of therapeutic factors that protect the lung or enhance repair.

7. Conclusion

Knowledge regarding the pathophysiology, genetics and cell biology of life threatening diseases of the lung is expanding rapidly. Likewise, our understanding of development, lung morphogenesis and stem cell biology is accelerating at a rapid pace. Together, these technical and scientific advances raise the hopes that novel therapies may be developed for life-threatening disorders in many organs including the lung. Major barriers to successful application of cell-based therapies for the lung remain, however. The ability to engineer lung tissue capable of enhancing respiratory function remains conceptual rather than actual (Fig. 3). Nevertheless, isolation and culture of cells able to differentiate into pulmonary cell types has been achieved *in vitro*. Various stem and progenitor cells have been identified and can be introduced into the various tissue compartments of the lung providing the scientific basis for the development of cell-based therapies for pulmonary diseases in the future.

References

1. Cardoso, W.V. and Lu, J. Regulation of early lung morphogenesis: questions, facts and controversies. *Development* **133**(9), 1611–1624 (2006).
2. Bottcher, R.T. and Niehrs, C. Fibroblast growth factor signaling during early vertebrate development. *Endocr. Rev.* **26**(1), 63–77 (2005).
3. Serls, A.E. *et al.* Different thresholds of fibroblast growth factors pattern the ventral foregut into liver and lung. *Development* **132**(1), 35–47 (2005).
4. Lazzaro, D. *et al.* The transcription factor TTF-1 is expressed at the onset of thyroid and lung morphogenesis and in restricted regions of the foetal brain. *Development* **113**(4), 1093–1104 (1991).
5. Kimura, S. *et al.* The T/ebp null mouse: thyroid-specific enhancer-binding protein is essential for the organogenesis of the thyroid, lung, ventral forebrain, and pituitary. *Genes Dev.* **10**(1), 60–69 (1996).
6. Maeda, Y., Dave, V. and Whitsett, J.A. Transcriptional control of lung morphogenesis. *Physiol. Rev.* **87**(1), 219–244 (2007).
7. Mucenski, M.L. *et al.* Beta-catenin regulates differentiation of respiratory epithelial cells *in vivo*. *Am. J. Physiol. Lung Cell. Mol. Physiol.* **289**(6), L971–L979 (2005).
8. Mucenski, M.L. *et al.* Beta-catenin is required for specification of proximal/distal cell fate during lung morphogenesis. *J. Biol. Chem.* **278**(41), 40231–40238 (2003).
9. Cardoso, W.V. *et al.* FGF-1 and FGF-7 induce distinct patterns of growth and differentiation in embryonic lung epithelium. *Dev. Dyn.* **208**(3), 398–405 (1997).

10. Shannon, J.M. and Hyatt, B.A. Epithelial-mesenchymal interactions in the developing lung. *Annu. Rev. Physiol.* **66**, 625–645 (2004).
11. Warburton, D. *et al.* The molecular basis of lung morphogenesis. *Mech. Dev.* **92**(1), 55–81 (2000).
12. Litingtung, Y. *et al.* Sonic hedgehog is essential to foregut development. *Nat. Genet.* **20**(1), 58–61 (1998).
13. Pepicelli, C.V., Lewis, P.M. and McMahon, A.P. Sonic hedgehog regulates branching morphogenesis in the mammalian lung. *Curr. Biol.* **8**(19), 1083–1086 (1998).
14. Bellusci, S. *et al.* Evidence from normal expression and targeted misexpression that bone morphogenetic protein (Bmp-4) plays a role in mouse embryonic lung morphogenesis. *Development* **122**(6), 1693–1702 (1996).
15. Taichman, D.B. *et al.* Notch1 and Jagged1 expression by the developing pulmonary vasculature. *Dev. Dyn.* **225**(2), 166–175 (2002).
16. Mendelsohn, C. *et al.* Function of the retinoic acid receptors (RARs) during development (II). Multiple abnormalities at various stages of organogenesis in RAR double mutants. *Development* **120**(10), 2749–2771 (1994).
17. Engelhardt, J.F., Stem cell niches in the mouse airway. *Am. J. Respir. Cell. Mol. Biol.* **24**(6), 649–652 (2001).
18. Evans, M.J. *et al.* Renewal of the terminal bronchiolar epithelium in the rat following exposure to NO_2 or O_3. *Lab. Invest.* **35**(3), 246–257 (1976).
19. Giangreco, A., Reynolds, S.D. and Stripp, B.R. Terminal bronchioles harbor a unique airway stem cell population that localizes to the bronchoalveolar duct junction. *Am. J. Pathol.* **161**(1), 173–182 (2002).
20. Kim, C.F. *et al.* Identification of bronchioalveolar stem cells in normal lung and lung cancer. *Cell* **121**(6), 823–835 (2005).
21. Evans, M.J. *et al.* Acute kinetic response and renewal of the alveolar epithelium following injury by nitrogen dioxide. *Chest* **65**(Suppl.), 62S–65S (1974).
22. Evans, M.J. *et al.* Transformation of alveolar type 2 cells to type 1 cells following exposure to NO_2. *Exp. Mol. Pathol.* **22**(1), 142–150 (1975).
23. Davies, J.A. Mesenchyme to epithelium transition during development of the mammalian kidney tubule. *Acta Anat. (Basel)* **156**(3), 187–201 (1996).
24. Perl, A.K. *et al.* Conditional recombination reveals distinct subsets of epithelial cells in trachea, bronchi, and alveoli. *Am. J. Respir. Cell. Mol. Biol.* **33**(5), 455–462 (2005).
25. Perl, A.K. *et al.* Early restriction of peripheral and proximal cell lineages during formation of the lung. *Proc. Natl. Acad. Sci. USA* **99**(16), 10482–10487 (2002).

26. Willis, B.C. and Borok, Z. TGF-beta-induced EMT: mechanisms and implications for fibrotic lung disease. *Am. J. Physiol. Lung Cell. Mol. Physiol.* **293**(3), L525–L534 (2007).
27. Okubo, T. and Hogan, B.L. Hyperactive Wnt signaling changes the developmental potential of embryonic lung endoderm. *J. Biol.* **3**(3), 11 (2004).
28. Shannon, J.M. Induction of alveolar type II cell differentiation in fetal tracheal epithelium by grafted distal lung mesenchyme. *Dev. Biol.* **166**(2), 600–614 (1994).
29. Hsia, C.C. Signals and mechanisms of compensatory lung growth. *J. Appl. Physiol.* **97**(5), 1992–1998 (2004).
30. Doyle, B., Metharom, P. and Caplice, N.M. Endothelial progenitor cells. *Endothelium* **13**(6), 403–410 (2006).
31. Pittenger, M.F. *et al.* Multilineage potential of adult human mesenchymal stem cells. *Science* **284**(5411), 143–147 (1999).
32. Coraux, C. *et al.* Embryonic stem cells generate airway epithelial tissue. *Am. J. Respir. Cell. Mol. Biol.* **32**(2), 87–92 (2005).
33. Van Vranken, B.E. *et al.* Coculture of embryonic stem cells with pulmonary mesenchyme: a microenvironment that promotes differentiation of pulmonary epithelium. *Tissue Eng.* **11**(7–8), 1177–1187 (2005).
34. Qin, M. *et al.* Cell extract-derived differentiation of embryonic stem cells. *Stem Cells* **23**(6), 712–718 (2005).
35. Wang, D. *et al.* A pure population of lung alveolar epithelial type II cells derived from human embryonic stem cells. *Proc. Natl. Acad. Sci. USA* **104**(11), 4449–4454 (2007).
36. Popov, B.V. *et al.* Lung epithelial cells induce endodermal differentiation in mouse mesenchymal bone marrow stem cells by paracrine mechanism. *Tissue Eng.* **13**(10), 2441–2450 (2007).
37. Samadikuchaksaraei, A. *et al.* Derivation of distal airway epithelium from human embryonic stem cells. *Tissue Eng.* **12**(4), 867–875 (2006).
38. Brambrink, T. *et al.* ES cells derived from cloned and fertilized blastocysts are transcriptionally and functionally indistinguishable. *Proc. Natl. Acad. Sci. USA* **103**(4), 933–938 (2006).
39. Wakayama, S. *et al.* Efficient establishment of mouse embryonic stem cell lines from single blastomeres and polar bodies. *Stem Cells* **25**(4), 986–993 (2007).
40. Byrne, J.A. *et al.* Producing primate embryonic stem cells by somatic cell nuclear transfer. *Nature* **450**(7169), 497–502 (2007).
41. Takahashi, K. *et al.* Induction of pluripotent stem cells from adult human fibroblasts by defined factors. *Cell* **131**(5), 861–872 (2007).

42. Takahashi, K. and Yamanaka, S. Induction of pluripotent stem cells from mouse embryonic and adult fibroblast cultures by defined factors. *Cell* **126**(4), 663–676 (2006).
43. Yu, J. *et al.* Induced pluripotent stem cell lines derived from human somatic cells. *Science* **318**(5858), 1917–1920 (2007).
44. Yan, C. *et al.* Gene delivery by the hSP-B promoter to lung alveolar type II epithelial cells in LAL-knockout mice through bone marrow mesenchymal stem cells. *Gene Ther.* **14**(20), 1461–1470 (2007).
45. Romanov, Y.A., Svintsitskaya, V.A. and Smirnov, V.N. Searching for alternative sources of postnatal human mesenchymal stem cells: candidate MSC-like cells from umbilical cord. *Stem Cells* **21**(1), 105–110 (2003).
46. Fuchs, J.R. *et al.* Cartilage engineering from ovine umbilical cord blood mesenchymal progenitor cells. *Stem Cells* **23**(7), 958–964 (2005).
47. Campagnoli, C. *et al.* Identification of mesenchymal stem/progenitor cells in human first-trimester fetal blood, liver, and bone marrow. *Blood* **98**(8), 2396–2402 (2001).
48. Broxmeyer, H.E. *et al.* Human umbilical cord blood as a potential source of transplantable hematopoietic stem/progenitor cells. *Proc. Natl. Acad. Sci. USA* **86**(10), 3828–3832 (1989).
49. Nakahata, T. and Ogawa, M. Hemopoietic colony-forming cells in umbilical cord blood with extensive capability to generate mono- and multi-potential hemopoietic progenitors. *J. Clin. Invest.* **70**(6), 1324–1328 (1982).
50. Hutson, E.L., Boyer, S. and Genever, P.G. Rapid isolation, expansion, and differentiation of osteoprogenitors from full-term umbilical cord blood. *Tissue Eng.* **11**(9–10), 1407–1420 (2005).
51. Kim, J.W. *et al.* Mesenchymal progenitor cells in the human umbilical cord. *Ann. Hematol.* **83**(12), 733–738 (2004).
52. Bucala, R. *et al.* Circulating fibrocytes define a new leukocyte subpopulation that mediates tissue repair. *Mol. Med.* **1**(1), 71–81 (1994).
53. Lama, V.N. and Phan, S.H. The extrapulmonary origin of fibroblasts: stem/progenitor cells and beyond. *Proc. Am. Thorac. Soc.* **3**(4), 373–376 (2006).
54. Bellini, A. and Mattoli, S. The role of the fibrocyte, a bone marrow-derived mesenchymal progenitor, in reactive and reparative fibroses. *Lab. Invest.* **87**(9), 858–870 (2007).
55. Gomperts, B.N. *et al.* Circulating progenitor epithelial cells traffic via CXCR4/CXCL12 in response to airway injury. *J. Immunol.* **176**(3), 1916–1927 (2006).
56. Gomperts, B.N. *et al.* Keratinocyte growth factor improves repair in the injured tracheal epithelium. *Am. J. Respir. Cell. Mol. Biol.* **37**(1), 48–56 (2007).

57. Suratt, B.T. *et al.* Human pulmonary chimerism after hematopoietic stem cell transplantation. *Am. J. Respir. Crit. Care Med.* **168**(3), 318–322 (2003).
58. Zander, D.S. *et al.* Bone marrow-derived stem-cell repopulation contributes minimally to the Type II pneumocyte pool in transplanted human lungs. *Transplantation* **80**(2), 206–212 (2005).
59. Krause, D.S. *et al.* Multi-organ, multi-lineage engraftment by a single bone marrow-derived stem cell. *Cell* **105**(3), 369–377 (2001).
60. Kotton, D.N. *et al.* Bone marrow-derived cells as progenitors of lung alveolar epithelium. *Development* **128**(24), 5181–5188 (2001).
61. Kotton, D.N., Fabian, A.J. and Mulligan, R.C. Failure of bone marrow to reconstitute lung epithelium. *Am. J. Respir. Cell. Mol. Biol.* **33**(4), 328–334 (2005).
62. Chang, J.C. *et al.* Evidence that bone marrow cells do not contribute to the alveolar epithelium. *Am. J. Respir. Cell. Mol. Biol.* **33**(4), 335–342 (2005).
63. Lange, C. *et al.* Administered mesenchymal stem cells enhance recovery from ischemia/reperfusion-induced acute renal failure in rats. *Kidney Int.* **68**(4), 1613–1617 (2005).
64. Togel, F. *et al.* Administered mesenchymal stem cells protect against ischemic acute renal failure through differentiation-independent mechanisms. *Am. J. Physiol. Renal Physiol.* **289**(1), F31–F42 (2005).
65. Baber, S.R. *et al.* Intratracheal mesenchymal stem cell administration attenuates monocrotaline-induced pulmonary hypertension and endothelial dysfunction. *Am. J. Physiol. Heart Circ. Physiol.* **292**(2), H1120– H1128 (2007).
66. Zhao, Y.D. *et al.* Rescue of monocrotaline-induced pulmonary arterial hypertension using bone marrow-derived endothelial-like progenitor cells: efficacy of combined cell and eNOS gene therapy in established disease. *Circ. Res.* **96**(4), 442–450 (2005).
67. Allegrucci, C. *et al.* Restriction landmark genome scanning identifies culture-induced DNA methylation instability in the human embryonic stem cell epigenome. *Hum. Mol. Genet.* **16**(10), 1253–1268 (2007).
68. Aguilar, S. *et al.* Murine but not human mesenchymal stem cells generate osteosarcoma-like lesions in the lung. *Stem Cells* **25**(6), 1586–1594 (2007).
69. Cogle, C.R. *et al.* Bone marrow contributes to epithelial cancers in mice and humans as developmental mimicry. *Stem Cells* **25**(8), 1881–1887 (2007).
70. Direkze, N.C. *et al.* Bone marrow contribution to tumor-associated myofibroblasts and fibroblasts. *Cancer Res.* **64**(23), 8492–8495 (2004).
71. Houghton, J. *et al.* Gastric cancer originating from bone marrow-derived cells. *Science* **306**(5701), 1568–1571 (2004).
72. Ishida, Y. *et al.* Essential roles of the CC chemokine ligand 3-CC chemokine receptor 5 axis in bleomycin-induced pulmonary fibrosis through regulation

of macrophage and fibrocyte infiltration. *Am. J. Pathol.* **170**(3), 843–854 (2007).

73. Phillips, R.J. *et al.* Circulating fibrocytes traffic to the lungs in response to CXCL12 and mediate fibrosis. *J. Clin. Invest.* **114**(3), 438–446 (2004).
74. Frid, M.G. *et al.* Hypoxia-induced pulmonary vascular remodeling requires recruitment of circulating mesenchymal precursors of a monocyte/macrophage lineage. *Am. J. Pathol.* **168**(2), 659–669 (2006).
75. Hacein-Bey-Abina, S. *et al.* LMO2-associated clonal T cell proliferation in two patients after gene therapy for SCID-X1. *Science* **302**(5644), 415–419 (2003).
76. Baum, C. *et al.* Mutagenesis and oncogenesis by chromosomal insertion of gene transfer vectors. *Hum. Gene Ther.* **17**(3), 253–263 (2006).
77. Imai, K. *et al.* Selective transendothelial migration of hematopoietic progenitor cells: a role in homing of progenitor cells. *Blood* **93**(1), 149–156 (1999).
78. Cashen, A.F., Lazarus, H.M. and Devine, S.M. Mobilizing stem cells from normal donors: is it possible to improve upon G-CSF? *Bone Marrow Transplant.* **39**(10), 577–588 (2007).
79. Gupta, N. *et al.* Intrapulmonary delivery of bone marrow-derived mesenchymal stem cells improves survival and attenuates endotoxin-induced acute lung injury in mice. *J. Immunol.* **179**(3), 1855–1863 (2007).
80. Rojas, M. *et al.* Bone marrow-derived mesenchymal stem cells in repair of the injured lung. *Am. J. Respir. Cell. Mol. Biol.* **33**(2), 145–152 (2005).
81. Serrano-Mollar, A. *et al.* Intratracheal transplantation of alveolar type II cells reverse bleomycin-induced lung fibrosis. *Am. J. Respir. Crit. Care Med.* **176**(12), 1261–1268 (2007).
82. Czechowicz, A. *et al.* Efficient transplantation via antibody-based clearance of hematopoietic stem cell niches. *Science* **318**(5854), 1296–1299 (2007).
83. Pereboeva, L. *et al.* Approaches to utilize mesenchymal progenitor cells as cellular vehicles. *Stem Cells* **21**(4), 389–404 (2003).
84. Studeny, M. *et al.* Bone marrow-derived mesenchymal stem cells as vehicles for interferon-beta delivery into tumors. *Cancer Res.* **62**(13), 3603–3608 (2002).
85. Studeny, M. *et al.* Mesenchymal stem cells: potential precursors for tumor stroma and targeted-delivery vehicles for anticancer agents. *J. Natl. Cancer Inst.* **96**(21), 1593–1603 (2004).
86. Rachakatla, R.S. *et al.* Development of human umbilical cord matrix stem cell-based gene therapy for experimental lung tumors. *Cancer Gene Ther.* **14**(10), 828–835 (2007).
87. Hakkarainen, T. *et al.* Human mesenchymal stem cells lack tumor tropism but enhance the antitumor activity of oncolytic adenoviruses in orthotopic lung and breast tumors. *Hum. Gene Ther.* **18**(7), 627–641 (2007).

88. Kanehira, M. *et al.* Targeted delivery of NK4 to multiple lung tumors by bone marrow-derived mesenchymal stem cells. *Cancer Gene Ther.* **14**(11), 894–903 (2007).
89. Stoff-Khalili, M.A. *et al.* Mesenchymal stem cells as a vehicle for targeted delivery of CRAds to lung metastases of breast carcinoma. *Breast Cancer Res. Treat.* **105**(2), 157–167 (2007).
90. Xin, H. *et al.* Targeted delivery of CX3CL1 to multiple lung tumors by mesenchymal stem cells. *Stem Cells* **25**(7), 1618–1626 (2007).
91. Breuer, R. *et al.* Cell kinetics of normal adult hamster bronchial epithelium in the steady state. *Am. J. Respir. Cell. Mol. Biol.* **2**(1), 51–58 (1990).
92. Giangreco, A. *et al.* Molecular phenotype of airway side population cells. *Am. J. Physiol. Lung Cell. Mol. Physiol.* **286**(4), L624– L630 (2004).
93. Liang, S.X. *et al.* Gene expression profiling and localization of Hoechst-effluxing $CD45^-$ and $CD45^+$ cells in the embryonic mouse lung. *Physiol. Genomics* **23**(2), 172–181 (2005).
94. Summer, R. *et al.* Embryonic lung side population cells are hematopoietic and vascular precursors. *Am. J. Respir. Cell. Mol. Biol.* **33**(1), 32–40 (2005).
95. Summer, R. *et al.* Side population cells and Bcrp1 expression in lung. *Am. J. Physiol. Lung Cell. Mol. Physiol.* **285**(1), L97–L104 (2003).
96. Reynolds, S.D. *et al.* Molecular and functional properties of lung SP cells. *Am. J. Physiol. Lung Cell. Mol. Physiol.* **292**(4), L972–L983 (2007).
97. Borthwick, D.W. *et al.* Evidence for stem-cell niches in the tracheal epithelium. *Am. J. Respir. Cell. Mol. Biol.* **24**(6), 662–670 (2001).
98. Reynolds, S.D. *et al.* Neuroepithelial bodies of pulmonary airways serve as a reservoir of progenitor cells capable of epithelial regeneration. *Am. J. Pathol.* **156**(1), 269–278 (2000).
99. Reynolds, S.D. *et al.* Conditional clara cell ablation reveals a self-renewing progenitor function of pulmonary neuroendocrine cells. *Am. J. Physiol. Lung Cell. Mol. Physiol.* **278**(6), L1256– L1263 (2000).
100. Hong, K.U. *et al.* Clara cell secretory protein-expressing cells of the airway neuroepithelial body microenvironment include a label-retaining subset and are critical for epithelial renewal after progenitor cell depletion. *Am. J. Respir. Cell. Mol. Biol.* **24**(6), 671–681 (2001).
101. Hong, K.U. *et al.* Basal cells are a multipotent progenitor capable of renewing the bronchial epithelium. *Am. J. Pathol.* **164**(2), 577–588 (2004).
102. Lawson, G.W. *et al.* Mouse strain modulates the role of the ciliated cell in acute tracheobronchial airway injury-distal airways. *Am. J. Pathol.* **160**(1), 315–327 (2002).
103. Plopper, C.G. *et al.* Relationship of cytochrome P-450 activity to Clara cell cytotoxicity. I. Histopathologic comparison of the respiratory tract of mice,

rats and hamsters after parenteral administration of naphthalene. *J. Pharmacol. Exp. Ther.* **261**(1), 353–363 (1992).

104. Van Winkle, L.S. *et al.* Cellular response in naphthalene-induced Clara cell injury and bronchiolar epithelial repair in mice. *Am. J. Physiol.* **269**(6 Pt. 1), L800–L818 (1995).
105. Van Winkle, L.S. *et al.* Early events in naphthalene-induced acute Clara cell toxicity: comparison of membrane permeability and ultrastructure. *Am. J. Respir. Cell. Mol. Biol.* **21**(1), 44–53 (1999).
106. Rawlins, E.L. *et al.* Lung development and repair: contribution of the ciliated lineage. *Proc. Natl. Acad. Sci. USA* **104**(2), 410–417 (2007).
107. Bongso, A. *et al.* Isolation and culture of inner cell mass cells from human blastocysts. *Hum. Reprod.* **9**(11), 2110–2117 (1994).
108. Schoch, K.G. *et al.* A subset of mouse tracheal epithelial basal cells generates large colonies *in vitro*. *Am. J. Physiol. Lung Cell. Mol. Physiol.* **286**(4), L631–L642 (2004).
109. Schwarz, M.A. *et al.* Angiogenesis and morphogenesis of murine fetal distal lung in an allograft model. *Am. J. Physiol. Lung Cell. Mol. Physiol.* **278**(5), L1000– L1007 (2000).
110. Vaughan, M.B. *et al.* A three-dimensional model of differentiation of immortalized human bronchial epithelial cells. *Differentiation* **74**(4), 141–148 (2006).
111. Vu, T.H., Alemayehu, Y. and Werb, Z. New insights into saccular development and vascular formation in lung allografts under the renal capsule. *Mech. Dev.* **120**(3), 305–313 (2003).

Chapter 40

The Artificial Lung

Andreas Nikolas Maurer and Georg Matheis

Abstract

Membrane ventilators are medical devices that provide extrapulmonary gas exchange, thus enabling novel treatment protocols for lung failure. These protocols prioritize lung protection over pulmonary gas exchange. In the intensive care setting, these devices are applied from several days up to a month to bridge patients to lung healing in acute respiratory failure, exacerbated COPD, and weaning failure. In these clinical settings the durability of current membrane ventilators is sufficient. However, in long-term applications periodical exchange of the membrane ventilator is required due to biofilm formation and clotting. This failure mode can possibly be avoided by development of long term artificial lungs. In the near future, vascular access, safety and control features for membrane ventilators will be a priority. In the mid and long term artificial lungs will combine artificial and biological components, to mimic biological surfaces, thus improving long-term function. In the far future, organoid structures, either as capillary microfluidic systems or three-dimensional structures, generated from stem cells, could serve as bioartificial lungs.

Keywords: Extrapulmonary Gas Exchange; Membrane Ventilator; Bioartificial Lung; Cell Seeding; Tissue Engineering.

Outline

1. Introduction
2. Mechanical Ventilation
3. Development of Membrane Ventilators (Artificial Lungs, Medical Devices)
4. Near Future Tasks to Enhance Membrane Ventilators
5. From Medical Devices to Biohybrid Lungs
6. Organoid Structures
7. Conclusion

References

1. Introduction

The human lung serves multiple functions ranging from gas exchange and metabolic functions, to a breathing pump together with the corresponding muscles. The gas exchange surface is lined with a layer of endothelial cells on the blood contacting side and a layer of pneumocytes on the air contacting side. Gas exchange takes place in minute dimensions of alveoli, where pulmonary capillaries with a diameter as small as 4 μm[1] form a vascular mesh over the lung's alveoli. While accomplishing its tasks perfectly in healthy conditions, the lung's performance is susceptible to acute or chronic disease.

Acute lung failure is due to trauma, inflammation and many other conditions. It is unpredictable, has no targeted treatment and a high mortality. Chronic lung disease including chronic obstructive pulmonary disease (COPD), cystic fibrosis and numerous other disease entities comprise the third largest mortality rate after cardiovascular diseases and cancer, with nine million deaths worldwide (WHO). COPD is a preventable and treatable disease with some significant extrapulmonary effects, such as right heart failure, that contribute to the severity in individual patients. Its pulmonary component is characterized by airflow limitation that is usually progressive, and is associated with an abnormal inflammatory response of the lung to noxious particles or gases.[2] Due to poor lung function, patients suffer from the need for substantially increased work of breathing and pathologic blood gas levels, leading to reduced general status and severely impaired quality of life. In many cases chronically elevated blood carbon dioxide levels cause constant breathing reflexes which lead to high energy consumption just for maintaining the vitally important gas exchange, and finally lead to exhaustion of the respiratory muscles.

Despite temporal symptomatic relief with mechanical ventilation or O_2 insufflation, there is no destination therapy available to maintain acceptable quality of life (QOL) in patients with advanced disease progression. This is due to the lack of efficient drugs to stop COPD progression. New treatment options are necessary, especially for late-stage patients where quality of life is impaired more severely than in end-stage heart failure patients. Future support solutions such as artificial lungs (i.e. medical devices), drug-device combinations or, in the long run, regenerative medicine solutions including tissue engineering, may create options in the treatment of late-stage chronic lung disease. These options will include support of the patient's gas exchange and breathing pump functions by means of an extrapulmonary gas exchanger, or in the future the replacement or regeneration of damaged lung tissue by means of tissue engineering and regenerative medicine.

2. Mechanical Ventilation

Mechanical ventilation remains the signature tool of critical care, and has greatly contributed to the tremendous progress in the treatment of critically ill patients. In almost all cases current mechanical ventilation protocols provide sufficient gas exchange to keep patients alive.

However, a growing body of evidence has recently evolved on positive pressure ventilation as a double edged sword. Ventilator associated lung injury (VALI) summarizes the barotrauma, volutrauma and biotrauma induced by positive pressure ventilation. Remote organ injury as a consequence of ventilator-induced lung injury has also been shown.[3,4] A vicious circle is initiated when the failing lung is forced to perform with unphysiological positive pressure instead of being allowed to partially rest and heal. It is mandatory to create conditions that prioritize lung healing instead of gas exchange performance. A basic treatment principle in medicine needs to be introduced in the treatment of acute lung failure. Mechanical ventilation in the treatment of acute lung failure can cause further lung injury, aggravate existent lung damage and, furthermore, cause multi-organ failure. Mechanical ventilation in acute lung failure today is confronted with a conflict between the need for gas exchange and lung protection. The limits of conventional mechanical ventilation create complications such as hypercapnia and respiratory acidosis. Non-compliance with lung protective protocols[5] leads to progressive lung damage by high respiratory pressure and volume. This calls for a treatment mode that reduces the pulmonary and extra-pulmonary pathogenicity of mechanical ventilation.

3. Development of Membrane Ventilators (Artificial Lungs, Medical Devices)

Membrane ventilators focus on partial gas exchange and resulting breathing pump support[6–9] whereas oxygenators serve to provide full gas exchange replacement during cardiac surgery[10] or for extended full support in extracorporeal membrane oxygenation (ECMO). Essentially we need solutions to provide adequate gas exchange and acid base status while maximizing pulmonary and remote organ protection. Extrapulmonary support of gas exchange by using membrane ventilators is a possible alternative. This allows gas exchange independent of impaired respiratory function. For the past 50 years numerous physicians and scientists have proposed concepts that enable patients to breathe outside the natural lung.[11,12] Most of these concepts have failed because of a lack of technologies. They were complex and invasive, labor-intensive and costly, and ultimately resulted in numerous complications.

Only in recent years technologies have evolved that allow the provision of focused respiratory assist with durable, single use devices. Membrane ventilators are lung assist devices that enable physicians to provide the desired level of ventilation without creating ventilator-associated lung injury (VALI). The work of breathing is supported by extrapulmonary CO_2 elimination. This enables novel protective ventilation strategies, and allows the separation of ventilation (extrapulmonary CO_2 elimination) and oxygenation.

A paradigm shift from prioritizing pulmonary gas exchange towards lung protection is enabled by extrapulmonary ventilation.[8,13–16] Extrapulmonary ventilation will allow the development of advanced modes of protective ventilation with or without the use of mechanical ventilation. This scenario is suitable for acute lung failure in the critical care unit. However, the risk-benefit ratio of extrapulmonary ventilation needs to be better than that of conventional ventilation to serve as a real alternative. This will require least invasive small veno-venous cannula and low flow blood pumps without hemolysis or polymorphonuclear neutrophil (PMN) activating shear stress.

The need for visionary permanent lung assist devices for homecare and destination therapy goes far beyond the specifications of devices for the acute critical care patient population. First clinical experience in bridge to lung transplant applications has been generated in intensive care environments.[17,18] Membrane ventilators are connected via arterio-venous femoral cannulation without a blood pump[17] or in veno-venous attachment mode with a centrifugal blood pump.[19] Pumped veno-venous systems will be more neutral for the circulatory system[20–22] compared with arterio-venous access.

4. Near Future Tasks to Enhance Membrane Ventilators

No chronically implantable, diffusion membrane-based device is available which remains functional (does mass transport across the membrane) over a prolonged time with appropriate low dose anticoagulant or antiplatelet drugs such as aspirin. Biofouling, which is due to protein and/or cell adsorption on the foreign surface as well as thrombus formation and filtration effects on the membrane mesh, accounts for a limited life span of such devices. The gas permeability of the hollow fiber membranes is reduced over time by deposition of blood constituents. A stable long-term blood contact surface would be a prerequisite for the development of an implantable artificial lung. Work on a paracorporeal solution is in progress (TAL, thoracic artificial lung),[23] but implantable devices will remain a vision for many years; surgical implantation is complex and invasive, and there is no possibility to do maintenance on the implanted devices. One option to permanently support the patient's breathing pump/gas exchange is to regularly exchange the membrane ventilator[17,18] which has already been demonstrated in bridge to lung transplant cases, that were supported by extracorporal CO_2 removal for more than 143 days without sedation and mechanical ventilation. Extracorporeal or paracorporeal devices remain the only option for the majority of artificial lung applications for the foreseeable future. Cannulas can be in the patient for even longer, but infections pose a risk.

The next step in the development of new artificial lungs will take place in the field of further enhancement of medical device technology. The position of artificial devices is likely to be paracorporeal and ambulatory.[24] Coating with hydrophilic substances or heparinization of blood contacting surfaces is a way to reduce protein adsorption and thrombosis,[25–27] but long-term application (up to 30 days) shows a decreasing performance due to biofilm in most of the patients. This performance reduction results in diminished gas transfer capabilities and elevated trans-module pressure. One of the near-future options is to enhance the hemocompatibility of artificial lungs by means of anticoagulation strategies. The gas exchange surface can be redesigned with newly available antithrombogenic capabilities.[28] New coating strategies have been developed and findings from other applications may be transferred to the field of artificial lungs.[29] Science lacks good models for coagulation and biofouling, since the mechanisms underlying the body's reaction to the foreign blood contacting surfaces is very complex. Whereas adhesion of single proteins can be monitored in laboratory experiments, and *ex vivo* experiments can be performed in short-time assays, there is no means to test hemocompatible surface coating in long-term experiments other than large animal models. More efficient and safer anticoagulation strategies are being evaluated in clinical trials and can be transferred to the field of artificial lungs.[30,31] For chronic lung assist devices, dimensions have to be reduced, and permanent access to the vascular system has to be established to enable patients' mobility. Instead of aterio-venous cannulation of femoral blood vessels, this can be achieved by application of vascular grafts with reusable connectors that are attached close to the lung/heart system. Alternatively, inflow attachment to the proximal pulmonary artery and outflow attachment to distal pulmonary artery[6] can be performed.

Sweep gas from ambient air can be used to avoid the need for oxygen cylinders and compressors. New safety systems are necessary that monitor the performance of the lung and can interrupt blood flow in case of emboli or blood leak. Safety features like online monitoring of blood gases and blood flow through the device have been developed.

5. From Medical Devices to Biohybrid Lungs

Biohybrid lungs may become a first stepping stone towards introducing regenerative medicine techniques in the treatment of chronic lung disease. Imitating the delicate structures of the human lung will not be possible in the foreseeable future, but the cells that are necessary to create a single cell layer are on hand. Endothelialization is a critical component of any extracorporeal or intracorporeal organ assist technology with biohybrid components. To generate a biohybrid organ for extracorporeal lung support, endothelialization of blood contacting surfaces would be a key milestone. In a second iteration, a layer of endothelium on the blood side and a layer of type II pneumocytes[32–34] that eventually differentiate in type I pneumocytes on the epithelial side would be necessary to create a non-compliant biohybrid lung. The generation of a sufficient amount of cell numbers of those two types of cells as well as rigid structures for cellular seeding is the prerequisite to establish biohybrid lungs. The supporting structure could be hollow fiber membranes that have sufficient media exchange capacity to support the cells on the epithelial side of the membrane. The conventional hollow fiber structures, which are applied in membrane ventilators, can be used as scaffolds to generate biohybrid constructs by means of cellular seeding. The rigid structure of the hollow fibers accounts for the security of the devices, but material properties, especially at the cell-material interface, must be engineered to keep cells in their phenotypic state and prevent the cells from releasing inflammatory mediators into the blood stream.

Endothelialization of the blood contacting surfaces of the membrane ventilator is a first step to replace avital coatings with a layer of cells that prevent immune reactions and avoid biofilm formation. This will prevent deterioration of gas exchange function of the artificial lung over extended periods. Coating blood contacting surfaces with a layer of autologous or allogenous endothelial cells has been a goal for more than 30 years,[35] and has been applied to several medical devices. Clinical data indicate initial positive effects of endothelialization especially in small-vessel prostheses[36] and heart valves.[37] Culturing systems have to be established, that allow the generation of high numbers of cells. Technologies that facilitate the propagation of cells are in the process of evaluation.[38,39] In addition to traditional cell culture techniques in flasks, three-dimensional cell culture on microcarriers, alginate embedding and revolving disc bioreactors are promising options to increase cell culture efficiency and the number of cells that can be generated. Access to different sources of stem or progenitor cells and cell culture methods is constantly growing, but it remains a challenge to cellularize devices. Several hundreds of millions of cells are needed to coat the artificial lung surfaces of about one meter squared. For endothelialization, cells isolated from biopsies of superficial veins are cultured for several passages until the needed amount of cells is generated. Subsequently cells are seeded on a scaffold that might be precoated with proteins or growth factors. Cells are then maintained *in vitro* on the implant for a designated time in which the cells can adhere to the graft. Blood flow can be simulated *in vitro* using pulsatile pumps where medium is pumped over the blood contacting surfaces. In this phase, the cells adapt to the pulsatile flow conditions that they will encounter in the patient's circulatory system.[40] Several approaches were developed for clinical application. But the number of patients treated with endothelial cell seeded implants is still limited. Broader application is still limited due to concern over the interpretation of the method's benefit, high unit prices caused by complex logistics, labor-intensive workflows and high regulatory demands on documentation and quality control. These issues might be obviated by new developments in allogene usage of cellular material and automatization of laboratory procedures and installation of on-site facilities for cell propagation and cell seeding in hospital areas.

Several approaches arose during the last few years that were intended to create surfaces which facilitate the self-endothelialization of blood contacting surfaces of medical devices. The aim is the creation of functional surfaces that move products towards cellularization and still are designed as "off the shelf products". These approaches should overcome obstacles such as donor-site morbidity, complicated logistics of biopsies and cells as well as labor-intensive personalized cell culture steps. Those approaches aim to capture molecules that are specific for binding endothelial progenitor cells (EPCs) to the blood contacting surface. EPCs circulate in the blood stream[41] and are a source of progenitor cells that are constantly produced in the bone marrow and secreted into the blood stream. EPCs are involved in neovascularization and vascular repair. If a surface is coated with capture molecules, EPCs can adhere predominantly on the surface and differentiation to endothelial cells may enable *in situ* cellularization of the surface. If this can be shown, reproduced, and scaled, the drawbacks of complex *in vitro* tissue engineering can be avoided. First approaches in other fields of intravascular therapies including peripheral vascular stenting and vascular prostheses[36,42,43] are close to clinical evaluation. These approaches utilize capture molecules like antibodies CD 34,[42,44,45] aptamers[46,47] or cell adhesion peptides.[48] It is possible to increase the number of EPCs[49] but it remains open how number and quality of cells correlate with the success of *in situ* endothelialization.

It is uncertain how long it will take to seed large surfaces *in situ*, and it is not known if unwanted conditions occur while surfaces are not yet fully covered with an endothelial layer. Furthermore, the cells will have to be differentiated towards the endothelial phenotype in order to create the desired cellular layer.

For the cellularization of the epithelial layer, cells can be generated from umbilical cord blood cells and introduced into the hollow fibers via the sweep gas tubing. The supply of nutrients to these cells through the wall of the hollow fiber membrane will be a challenge.

6. Organoid Structures

The vertebrate lung has unique mechanical properties, since nowhere else are mechanical durability and layers thin enough for efficient gas transfer such contradictory prerequisites. Therefore near-future developments may consider or even copy the evolutionary superior construction plan of the avian lung, which consists of more rigid structures. The avian parabronchial lung differs morphologically from the mammalian bronchealveolar lung in certain key aspects.[50,51] While the arrangement of the airway system of the mammalian lung displays iterating, commonly dichotomous bifurcation which terminate in blind-ended air conduits, in the avian lung an anastomotic system exists.[52] In complete contrast to the tidally ventilated mammalian respiratory system, where inhaled air is mixed with residual air in the respiratory airways, the avian lung is a flow-through system.[53] Through the system of air conduits and bronchial system, the avian-specific air sacs ventilate the lung continuously and unidirectionally like a pair of bellows.[53–55] Gas exchange takes place in the parabronchial tissue mantle, which is the fundamental functional part of the avian lung.[56,57]

As a blueprint for medtech or biohybrid devices the avian flow-through lungs appear to be the superior configuration over dead-end ventilation, not only in efficiency but also in size and ease of construction.[58–63]

There are contradictory prerequisites for stability and gas exchange performance. While allowing a maximum of specific gas exchange capability, the membrane thickness would have to be decreased to a nearly infinitely thin molecular mesh. This mesh would be susceptible to physical forces caused by blood pressure or movement of the artificial organ resulting in life threatening emboli. Second, the lung is a highly organised, fine structured organ with capillaries that transport blood on the one side and tiny structures that transport gas on the other side. Those two fluidic systems occur in a close and intimate manner. The alveolar structure brings the two cell types "endothelium" and "epithelium" very close to each other. This takes place in an extremely organized manner that determines the function of the lung as a gas exchange organ. The lung also functions as a metabolic organ, and cells from the immune system are capable of moving in and out of the lung.

Current biomaterials are not capable of dividing two cavities with stable barriers a few microns in thickness while providing continuous mass exchange (gas/fluid).[64] All reported approaches aimed to imitate the alveolar structures by sponge-like biomaterials seeded with cells, typically of embryonic origin.[65–67] Histologic examination revealed not only the development of alveolar structures but also an increasing vascularization over time. It needs to be shown that these organoid structures form two fluidic systems in a scaffold and that directed branching occurs. Genetic control of branching morphogenesis is complex[68,69] and to imitate these developmental processes *in vitro* is a prerequisite for the generation of tissue engineered parts of the lung.[67] The construction of such delicate structures might be dependent on sufficiently small structuring processes. Solutions will come from the field of rapid prototyping (when resolution of that technique will improve) or from microoptics and microproduction technologies. This shows that while access to cellular sources is progressing, structure engineering is needed to direct cells in the desired shape. Approaches with intermediate steps that use accelularized lungs or other highly vascularized tissues as seeding scaffolds[70] fill the scaffold gap. Subsequently such constructs will have to be introduced into the human lung to replace damaged areas. Therefore the newly developed structure must be connected to both blood vessels and the airway system of the lung flap.

Microstructured devices that mimic the scale and function of the alveolar-capillary structures of the lung could be an intermediate step for the development of organoid lung replacement. First described in the 1970s,[71] new technologies are capable of realizing microfluidic systems with technologies of microsystem engineering. Several academic groups are working on the design of microchannel assemblies as artificial blood-capillary beds for artificial organs.[72,73] The coagulation processes and the size of white blood cells within microchannels is a technical challenge, but experience from the application in hollow fiber modules[26] can be transferred to the new field. Evaluation of coating or cell seeding could be carried out.[74] Modeling and mass transfer through the microchannel material[75] can be investigated.

7. Conclusion

Further development in the field of artificial lungs will be dominated by progress in material science in the coming years. In the long run, biotechnology and stem cell technology as advanced methods for the generation of medical devices with biological activity will be applied to enhance product capabilities.

The current situation and the immense knowledge gained in academic research open the door for the application of such techniques to improve product quality and biocompatibility. Risks include regulatory hurdles and methods that are not clinically practical or too complex. Furthermore, potentially lower gas transfer capability or cellular instability may perpetuate the superiority of the current bioinactive medical devices.

The medical need in acute and even more so in chronic lung failure is a strong driver for further research on drug development, medical devices and regenerative medicine. A blend of medtech and biotech in a development process that comprises many steps may be a pathway with great potential, and a stepping stone towards true regenerative solutions.

References

1. Milnor, W.R. *Cardiovascular Physiology* (Oxford University Press, New York, 1990).
2. Rabe, K.F. *et al.* Global strategy for the diagnosis, management, and prevention of chronic obstructive pulmonary disease: GOLD executive summary. *Am. J. Respir. Crit. Care Med.* **176**(6), 532–555 (2007).
3. Imai, Y. and Slutsky, A.S. High-frequency oscillatory ventilation and ventilator-induced lung injury. *Crit. Care Med.* **33**(3 Suppl.), S129–S134 (2005).
4. Slutsky, A.S. and Imai, Y. Ventilator-induced lung injury, cytokines, PEEP, and mortality: implications for practice and for clinical trials. *Intensive Care Med.* **29**(8), 1218–1221 (2003).
5. Ventilation with lower tidal volumes as compared with traditional tidal volumes for acute lung injury and the acute respiratory distress syndrome. The Acute Respiratory Distress Syndrome Network. *N. Engl. J. Med.* **342**(18), 1301–1308 (2000).
6. Zwischenberger, J.B. and Alpard, S.K. Artificial lungs: a new inspiration. *Perfusion* **17**(4), 253–268 (2002).
7. Matheis, G. New technologies for respiratory assist. *Perfusion* **18**(4), 245–251 (2003).
8. Kopp, R., Dembinski, R. and Kuhlen, R. Role of extracorporeal lung assist in the treatment of acute respiratory failure. *Minerva Anestesiol.* **72**(6), 587–595 (2006).
9. Tatsumi, E. Artificial lungs: current state and trends of clinical use and research and development. *J. Artif. Organs* **10**(1), 1–5 (2007).
10. Lim, M.W. The history of extracorporeal oxygenators. *Anaesthesia* **61**(10), 984–995 (2006).

11. Kolobow, T. *et al.*, An alternative to breathing. *J. Thorac. Cardiovasc. Surg.* **75**(2), 261–266 (1978).
12. Gattinoni, L. *et al.* Control of intermittent positive pressure breathing (IPPB) by extracorporeal removal of carbon dioxide. *Br. J. Anaesth.* **50**(8), 753–758 (1978).
13. Walles, T. Clinical experience with the iLA Membrane Ventilator pumpless extracorporeal lung-assist device. *Expert Rev. Med. Devices* **4**(3), 297–305 (2007).
14. Elliot, S.C. *et al.* Pumpless extracorporeal carbon dioxide removal for life-threatening asthma. *Crit. Care Med.* **35**(3), 945–948 (2007).
15. Mallick, A. *et al.* Extracorporeal carbon dioxide removal using the Novalung in a patient with intracranial bleeding. *Anaesthesia* **62**(1), 72–74 (2007).
16. von Mach, M.A. *et al.* An update on interventional lung assist devices and their role in acute respiratory distress syndrome. *Lung* **184**(3), 169–175 (2006).
17. Fischer, S. *et al.* Bridge to lung transplantation with the novel pumpless interventional lung assist device NovaLung. *J. Thorac. Cardiovasc. Surg.* **131**(3), 719–723 (2006).
18. Dschietzig, T., Laule, M. and Baumann, G. Long-term treatment of severe respiratory failure with an extracorporeal lung assist — a case report. *Intensivmed. Notfallmed.* **42**(4), 365–370 (2005).
19. Fischer, S. *et al.* Bridge to lung transplantation with the extracorporeal membrane ventilator Novalung in the veno-venous mode: the initial Hannover experience. *ASAIO J.* **53**(2), 168–170 (2007).
20. Wang, D. *et al.* An ambulatory pulmonary and right heart assist device (OxyRVAD) in an ovine survival model. *J. Heart Lung Transplant.* **26**(10), 974–979 (2007).
21. Wang, D. *et al.* OxyRVAD for total right heart and respiratory support. *Conf. Proc. IEEE Eng. Med. Biol. Soc.* **1**, 5390–5391 (2006).
22. Asakawa, Y. *et al.* Development of an implantable oxygenator with cross-flow pump. *ASAIO J.* **52**(3), 291–295 (2006).
23. Perlman, C.E. and Mockros, L.F. Hemodynamic consequences of thoracic artificial lung attachment configuration: a computational model. *ASAIO J.* **53**(1), 50–64 (2007).
24. Lick, S.D. and Zwischenberger, J.B. Artificial lung: bench toward bedside. *ASAIO J.* **50**(1), 2–5 (2004).
25. Wendel, H.P. and Ziemer, G. Coating-techniques to improve the hemocompatibility of artificial devices used for extracorporeal circulation. *Eur. J. Cardiothorac. Surg.* **16**(3), 342–350 (1999).

26. Thorslund, S. *et al.* Bioactive heparin immobilized onto microfluidic channels in poly(dimethylsiloxane) results in hydrophilic surface properties. *Coll. Surf. B Biointerf.* **46**(4), 240–247 (2005).
27. Sandhu, S. and Luthra, A. New biointeracting materials. *Med. Dev. Technol.* **13**(8), 10–14, 16 (2002).
28. Statz, A.R. *et al.* New peptidomimetic polymers for antifouling surfaces. *J. Am. Chem. Soc.* **127**(22), 7972–7973 (2005).
29. Jordan, S.W. and Chaikof, E.L. Novel thromboresistant materials. *J. Vasc. Surg.* **45**(Suppl. A), A104–115 (2007).
30. Bates, S.M. and Weitz, J.I. New anticoagulants: beyond heparin, low-molecular-weight heparin and warfarin. *Br. J. Pharmacol.* **144**(8), 1017–1028 (2005).
31. Hirsh, J., O'Donnell, M. and Eikelboom, J.W. Beyond unfractionated heparin and warfarin: current and future advances. *Circulation* **116**(5), 552–560 (2007).
32. Lane, S., Rippon, H.J. and Bishop, A.E. Stem cells in lung repair and regeneration. *Regen. Med.* **2**(4), 407–415 (2007).
33. Bishop, A.E. and Rippon, H.J. Stem cells — potential for repairing damaged lungs and growing human lungs for transplant. *Expert. Opin. Biol. Ther.* **6**(8), 751–758 (2006).
34. Samadikuchaksaraei, A. *et al.* Derivation of distal airway epithelium from human embryonic stem cells. *Tissue Eng.* **12**(4), 867–875 (2006).
35. Herring, M.B. *et al.* Seeding arterial prostheses with vascular endothelium. The nature of the lining. *Ann. Surg.* **190**(1), 84–90 (1979).
36. Zilla, P., Deutsch, M. and Meinhart, J. Endothelial cell transplantation. *Semin. Vasc. Surg.* **12**(1), 52–63 (1999).
37. Lichtenberg, A. *et al.* Preclinical testing of tissue-engineered heart valves re-endothelialized under simulated physiological conditions. *Circulation* **114**(1 Suppl.), I559–1565 (2006).
38. Chen, H.C. and Hu, Y.C. Bioreactors for tissue engineering. *Biotechnol. Lett.* **28**(18), 1415–1423 (2006).
39. King, J.A. and Miller, W.M. Bioreactor development for stem cell expansion and controlled differentiation. *Curr. Opin. Chem. Biol.* **11**(4), 394–398 (2007).
40. Papadaki, M. and Eskin, S.G. Effects of fluid shear stress on gene regulation of vascular cells. *Biotechnol. Prog.* **13**(3), 209–221 (1997).
41. Urbich, C. and Dimmeler, S. Endothelial progenitor cells: characterization and role in vascular biology. *Circ. Res.* **95**(4), 343–353 (2004).
42. Ong, A.T. *et al.* How to accelerate the endothelialization of stents. *Arch. Mal. Coeur. Vaiss.* **98**(2), 123–126 (2005).

43. Bordenave, L. *et al. In vitro* endothelialized ePTFE prostheses: clinical update 20 years after the first realization. *Clin. Hemorheol. Microcirc.* **33**(3), 227–234 (2005).
44. Rotmans, J.I. *et al. In vivo* cell seeding with anti-CD34 antibodies successfully accelerates endothelialization but stimulates intimal hyperplasia in porcine arteriovenous expanded polytetrafluoroethylene grafts. *Circulation* **112**(1), 12–28 (2005).
45. Aoki, J. *et al.* Endothelial progenitor cell capture by stents coated with antibody against CD34: The HEALING-FIM (Healthy Endothelial Accelerated Lining Inhibits Neointimal Growth-First In Man) registry. *J. Am. Coll. Cardiol.* **45**(10), 1574–1579 (2005).
46. Guo, K. *et al.* Aptamer-based capture molecules as a novel coating strategy to promote cell adhesion. *J. Cell. Mol. Med.* **9**(3), 731–736 (2005).
47. Hoffmann, J. *et al.* Immobilized DNA aptamers used as potent attractors for porcine endothelial precursor cells. *J. Biomed. Mater. Res. A* **84A**(3), 614–621 (2007).
48. Blindt, R. *et al.* A novel drug-eluting stent coated with an integrin-binding cyclic Arg-Gly-Asp peptide inhibits neointimal hyperplasia by recruiting endothelial progenitor cells. *J. Am. Coll. Cardiol.* **47**(9), 1786–1795 (2006).
49. Aicher, A., Zeiher, A.M. and Dimmeler, S. Mobilizing endothelial progenitor cells. *Hypertension* **45**(3), 321–325 (2005).
50. Duncker, H.R. Structure of the avian respiratory tract. *Respir. Physiol.* **22**(1–2), 1–19 (1974).
51. Maina, J.N. Comparative respiratory morphology: themes and principles in the design and construction of the gas exchangers. *Anat. Rec.* **261**(1), 25–44 (2000).
52. Woodward, J.D. and Maina, J.N. A 3D digital reconstruction of the components of the gas exchange tissue of the lung of the muscovy duck, *Cairina moschata*. *J. Anat.* **206**(5), 477–492 (2005).
53. Maina, J.N. Some recent advances on the study and understanding of the functional design of the avian lung: morphological and morphometric perspectives. *Biol. Rev. Camb. Philos. Soc.* **77**(1), 97–152 (2002).
54. Fedde, M.R. Relationship of structure and function of the avian respiratory system to disease susceptibility. *Poult. Sci.* **77**(8), 1130–1138 (1998).
55. Scheid, P., Slama, H. and Piiper, J. Mechanisms of unidirectional flow in parabronchi of avian lungs: measurements in duck lung preparations. *Respir. Physiol.* **14**(1), 83–95 (1972).

56. Maina, J.N. Scanning electron microscope study of the spatial organization of the air and blood conducting components of the avian lung (*Gallus gallus* variant *domesticus*). *Anat. Rec.* **222**(2), 145–153 (1988).
57. Scheid, P. Mechanisms of gas exchange in bird lungs. *Rev. Physiol. Biochem. Pharmacol.* **86**, 137–186 (1979).
58. West, J.B., Watson, R.R. and Fu, Z. The human lung: did evolution get it wrong? *Eur. Respir. J.* **29**(1), 11–17 (2007).
59. Maina, J.N. Structure, function and evolution of the gas exchangers: comparative perspectives. *J. Anat.* **201**(4), 281–304 (2002).
60. Maina, J.N. and West, J.B. Thin and strong! The bioengineering dilemma in the structural and functional design of the blood-gas barrier. *Physiol. Rev.* **85**(3), 811–844 (2005).
61. Powell, F.L. and Hopkins, S.R. Comparative physiology of lung complexity: implications for gas exchange. *News Physiol. Sci.* **19**, 55–60 (2004).
62. Reese, S., Dalamani, G. and Kaspers, B. The avian lung-associated immune system: a review. *Vet. Res.* **37**(3), 311–324 (2006).
63. West, J.B. Thoughts on the pulmonary blood-gas barrier. *Am. J. Physiol. Lung Cell Mol. Physiol.* **285**(3), L501–L513 (2003).
64. Sachlos, E. and Czernuszka, J.T. Making tissue engineering scaffolds work. Review: the application of solid freeform fabrication technology to the production of tissue engineering scaffolds. *Eur. Cell Mater.* **5**, 29–39 (2003); discussion 39–40.
65. Chen, P. *et al.* Formation of lung alveolar-like structures in collagen-glycosaminoglycan scaffolds *in vitro*. *Tissue Eng.* **11**(9–10), 1436–1448 (2005).
66. Andrade, C.F. *et al.* Cell-based tissue engineering for lung regeneration. *Am. J. Physiol. Lung Cell Mol. Physiol.* **292**(2), L510–L518 (2007).
67. Mondrinos, M.J. *et al.* Engineering three-dimensional pulmonary tissue constructs. *Tissue Eng.* **12**(4), 717–728 (2006).
68. Cardoso, W.V. and Lu, J. Regulation of early lung morphogenesis: questions, facts and controversies. *Development* **133**(9), 1611–1624 (2006).
69. Metzger, R.J. and Krasnow, M.A. Genetic control of branching morphogenesis. *Science* **284**(5420), 1635–1639 (1999).
70. Lwebuga-Mukasa, J.S., Ingbar, D.H. and Madri, J.A. Repopulation of a human alveolar matrix by adult rat type II pneumocytes *in vitro*. A novel system for type II pneumocyte culture. *Exp. Cell Res.* **162**(2), 423–435 (1986).
71. Hung, T.K., Borovetz, H.S. and Weissman, M.H. Transport and flow phenomena in a microchannel membrane oxygenator. *Ann. Biomed. Eng.* **5**(4), 343–361 (1977).

72. Lee, J. *et al.* Microchannel technologies for artificial lungs. *ASAIO J.* **52**(2), 66A (2006).
73. Gilbert, R.J. *et al.* Computational and functional evaluation of a microfluidic blood flow device. *ASAIO J.* **53**(4), 447–455 (2007).
74. Kurotobi, K. *et al.* Short term evaluation of material blood compatibility using a microchannel array. *J. Mater. Sci. Mater. Med.* **18**(6), 1175–1184 (2007).
75. Wickramasinghe, S.R. *et al.* Designing blood oxygenators. *Ann. N. Y. Acad. Sci.* **984**, 502–514 (2003).

Index

acquired immunity 201
adaptive immune system 202
adipocyte 143, 146, 149, 150, 152, 155, 156
adipose tissue 143–145, 151, 153, 154, 156, 158
  brown 144, 145, 151
  stem cell 143–158
  white 144, 145, 151
adrenaline 382
adult 505, 507, 508
  mesenchymal stem cell 86
adult stem cell 17, 18, 20–22, 25, 29, 31, 32, 619, 620, 622
  fate 185, 186
affinity sensor 376
aggregation 313–327
alginate 410, 413, 414
  encapsulation 508
allogeneic 561, 563–572, 574, 575
amniotic fluid 658
angiogenesis 679, 683
animal model 763–767, 770, 771, 774
antigen-presenting cell (APC) 204, 205, 207
application 469, 470, 472, 477, 480
articular cartilage 475, 477, 478
artificial lung 887–889, 891, 895
autologous 679, 680, 683, 684, 688, 689
autologous chrondrocyte implantation (ACI) 475, 476

B-cell 202–205, 207
bioactive glass 279–281, 283, 284, 286, 291, 294, 298–301
bioactive material 279, 280
biocompatibility 373, 393
biodegradable polymer 279–285, 298, 299, 302, 303
bioengineered niche 175
biomaterial 313, 314, 318, 325, 659, 668
biomimetic 336
bioprocess 215–217, 219, 220, 223, 226
bioprocessing 130–132, 137
bioreactor 508, 509, 511
bladder replacement 661–664
blastocyst 64, 65, 68–70, 74–76, 685, 688
blastomere 65, 70
B-lymphocyte 202, 203
bone defect model 771–773
bone marrow 505, 507, 561–566, 568–571, 573, 574
  mesenchymal stem cell 143, 148
  stem cell 505, 679–683, 685
  stromal stem cell 22–26
bone regeneration
bone tissue engineering 511
bovine serum 234
bridge to recovery 733, 735, 736, 738–741, 744
bridge to transplantation 736–740, 742, 744, 750
Burns 593–597, 599–601, 603, 608
business model 495, 499

cantilever 391, 392
cardiac 677, 679–681, 683–686, 688
  cell 507
  myocyte 151
  repair 677, 679–681, 688
  stem cell 684
  tissue engineering
cardiomyocyte 679, 681, 683–688
cartilage defect model 763, 773
cell 15–33, 313–327
cell-based therapeutic 537–540, 542, 546, 548, 550–552
cell based therapy 861, 862, 864, 871, 872, 877
cell culture reagent 236, 237
cell expansion methodology 507
cell-instructive 255, 259, 260

Cell Line Master File (CLMF) 243–245
cell-responsive 255, 259–263, 271, 272
cell source 505
cell therapy 200, 619, 620, 622–625, 846, 847, 852
cell tracking 445–449, 451, 456, 457
cellular implant 471
chimera 50, 52, 54
chitinase-3-like protein-1 413, 416
chondrocyte 143, 151, 152, 401, 410, 412–414, 416
chorioallantoic membrane (CAM) assay 769
chronic wound 593, 595, 603, 605
clinical
  application 505, 512
  trial 695, 697, 698, 719–721, 723, 843, 848–850, 852
  utility 480
clinically relevant 474, 480
  number 505, 509
collagenase 143, 145–147
colony forming unit
  endothelial cell 104
  fibroblast 101
commercial 469, 470, 472, 476, 477, 481–483
  feasibility 469, 483
commercialization 482, 483
comparative genome hybridisation (CGH) 234
composite material 279–281, 283, 284, 288, 296–301
conditioned culture media 506
contamination 230, 233, 236–239, 242–244
controlled differentiation 505
cord blood 569, 570
  banking 127, 129, 132, 133
  stem cell 123, 124, 127, 134–137
craniofacial 821–827, 829–834
craniosynostosis 821, 823–826, 830
Creutzfeldt-Jakob disease (CJD) 237
crosscontamination 233
cryopreservation 235, 241
cryoprotectant 242
cytokine 394
  profile 146, 150
cytoplasmic hybrid 47, 55, 56

decellularized tissue 472
dendrimer 333, 335, 340
dendritic cell (DC) 200, 204, 205, 207, 209
dental stem cell 646
derivation 63, 65, 68–71, 73, 75–77
design of experiment (DoE) 216, 218–221, 223, 224, 226, 392
differentiated cell 505, 512
differentiation 15, 16, 19, 21–25, 28, 30, 31, 83–89, 146–157
diffusion chamber 763, 767–769
disease 781–786, 789–791, 793, 795–800, 804, 805
distraction osteogenesis 821, 823, 827
dog 563–565, 571–573
donor 242, 246
  selection 242
dopamine 382, 383
Duchenne muscular dystrophy 571–573
dynamic cultivation system 509
dynamic range 376, 379

embryoid body (EB) 207
  formation 505, 507
embryonic 503, 505
embryonic stem cell (ESC) 15, 17, 18, 20, 22, 23, 25–32, 83–85, 88, 199, 200, 205–210, 232, 238, 503, 508, 509, 679, 681, 685–689
  cardiomyocyte 681, 686–688
encapsulated cell 508
encapsulation 503, 507, 508
endodermal 152
endothelial cell 143, 151, 153
endothelial colony-forming cell 104, 105
endothelial progenitor cell (EPC) 95–97, 103–106, 108, 109, 683, 893
endothelialization 892, 893
enhance mesoderm formation 506
EnzFET 380
epithelial 150, 153

ethics committee 50, 54, 56
excitation-contraction coupling 748
extracellular matrix (ECM) 735, 746, 747
artificial 255, 259–264, 266, 267, 269–272
extracorporeal device 477

Fas ligand (FasL) 204, 206
fast phase liquid chromatography 410, 413
feeder cell 72–74, 234, 235
fluorescence resonance energy transfer (FRET) 387, 388, 390, 393
foetal 505, 507
mesenchymal stem cell 83, 87
fullerene 335, 338
functioning renal unit 666

gene therapy
glucose 375, 376, 380, 382, 383, 386, 392, 394
glutaminase 380
glutamine 375, 376, 380
Good Cell Culture Practice (GCCP) 234, 235
Good Manufacturing Practice (GMP) 243, 245
graft-versus-host disease (GVHD) 203, 207, 210, 563, 565–571, 574–576
graft-versus-tumor effect 561, 568–571
green fluorescent protein (GFP) 386, 393
growth factor 234, 238
guidance 539, 542, 543, 554, 555

Harefield Bridge-to-Recovery Protocol 741
hazard 234, 243
heart 695–697, 710, 711, 713, 715–721
hemangioblast 103, 105–109
hematopoietic cell transplantation 561, 562, 564–576
hematopoietic stem cell 17, 20, 22–26, 95–101, 103, 105, 106, 108, 109, 176, 183, 561–564, 566, 569, 571–573, 576
hepatocyte 148, 150, 152
hollow fiber membrane 890, 892
homogeneous population 505
human cell line 232, 233
human embryonic stem cell (hES) 63–77, 230, 232, 235, 237, 241, 243, 246, 247
hESCs (H1) 508
human leukocyte antigen (HLA) 203, 204, 206, 210
hybrid 260, 263–269
hydrogel 186–189, 255, 259–261, 263, 266, 268, 270, 272
hydrogen peroxide 380, 382

IgA 202, 203
IgE 202
IgG 202
IgM 202
immune privileged 201, 204, 206, 208, 209
immune rejection 199, 200, 207–209
immunogenesis 679, 686
immunogenicity 147, 148
immunophenotype 143, 146, 148, 153
immunosuppression 200, 201, 206, 208–210
indolamine-2,3-dioxygenase (IDO) 204, 206, 209
industrial perspective 469
industry 3, 4, 6–8, 469, 477, 483
inflammation 482
inflammatory cytokine 471
informed consent 48, 50, 52, 54–56
innate immune response 201
innate immune system 202
inner cell mass 63–65, 68, 69, 72, 74–76
interferon-γ (IFNγ) 202, 206, 207, 209, 210
international collaboration 43, 51, 53, 54, 56, 58
International Conference on Harmonisation (ICH) 246
International Stem Cell Forum 231, 246, 247
International Stem Cell Initiative 246, 247
interstitial fluid flow 528, 529
intervertebral disc 401, 402, 410, 412, 413, 415, 416
intramuscular implant 768
investment model 495
in vitro fertilisation 63–67, 72–74
in vivo bioreactor 763, 770
ion sensitive field effect transistor (ISFET) 379–381

iridium oxide 335, 341, 379
iron oxide 335, 341

karyology 234

lactate 376, 383, 394
left ventricular assist device 733, 735–737, 739–750
limit of detection (LoD) 376
liposome 335
liver
  assist device 469, 477–480
  disease 619–624
  repair 619

MAGIC 685
magnetic resonance imaging (MRI) 443, 445, 448–450, 452–457, 508
  contrast agent 443, 449, 450, 452
major challenges 512
major histocompatibility complex (MHC) 203–210
manufactured epidermis 482
marker 24–27, 30, 31
mass transport 517, 518, 532
master cell bank 235, 236
materno-foetal tolerance 204
Matrigel™ 235
matrix 657, 659–665, 667
mechanical stress
mechanical ventilation 888–891
media 65, 68, 69, 72–74, 234–236, 238, 239, 244, 246, 247
membrane fouling 408, 410, 411, 413
mesenchymal stem cell (MSC) 83, 86–88, 95–97, 100–102, 105–109, 143, 145, 146, 148, 152–156, 158, 205, 209, 210, 505, 508, 561, 563, 574–576, 681
meso-angioblast 106, 108
metabolic activity 401, 410, 412, 415
metalloproteinase 748
microarray 234
microdialysis 401, 403–416
microelectrode 382
microencapsulation technology 508
microfabrication 354, 359
microfluidics 361–363
microwell array 187
mixed lymphocyte reaction (MLR) 143, 147, 209
molecular beacon 387, 393
monitoring 401–403, 407–410, 412, 413, 415
morphogenesis 861, 864–866, 868, 879
multidisciplinary 470, 471, 483
multipotent marrow stromal cell 95–97, 100–102, 105–109
myocardial infarction 680, 683
myocardial recovery 733, 741, 742, 744, 745, 748–751
myocyte 156

Na^+/Ca^{2+} exchanger 749
nanocomposite 333, 335–338
nanofiber 336, 337, 339
nanofibrillar 269, 270
nanosensor 375, 392
nanotechnology 333, 334, 336, 343
nanotube 333, 335, 338
natural killer (NK) cell 203, 204, 206, 207
neonatal 505
neuronal 143, 154
niche 175–189
nitric oxide 381, 394
non-invasive 421, 424, 438
nuclear imaging 453

optical 445, 447, 448, 456
  coherence tomography 340
  fibre 387, 392
optimal bioreactor 509
osmotic pressure 407, 409
osteoarthritis 785, 791, 794–796, 844, 845, 847, 848, 852
osteoarticular repair 843, 844, 852
osteoblast 143, 149, 155, 157, 282, 288, 291, 294–297, 299, 301, 303
osteochondral 781, 784, 785, 787, 789, 791, 792, 794, 797, 801
osteogenic lineage differentiation 506
osteoporosis 783, 786, 793, 803–805
oxygen 74–76, 376, 378, 381, 383, 386, 392, 394

paracrine 679, 683
patent landscape 491, 492, 498, 499
PEBBLES 392
penile prosthesis 668
peptide 336, 337, 342
peptide-polymer 266–268
perfusion bioreactor 517, 522, 524–528
peripheral blood mononuclear cell 569, 571
placenta 658
planar waveguide 387
plasticity 570, 571, 576
pluripotency 685, 689, 888, 892
poly(ethyleneglycol) 393
poly(propylene fumarate) 338
porcine hepatocyte 480
positron emission tomography (PET) 446, 447, 449, 453, 454, 456
preimplantation genetic diagnosis 70, 71
process map 243, 537, 539, 553, 554
product life-cycle 538–541, 547, 555
protein
- adsorption 381, 393
- marker 411
- polymer 263–265, 269

public private partnership 492, 495
pulmonary 861, 863–871, 873–875, 877–879
pumping method 408, 409, 415

quality of life 735, 737, 744
quantum dot 333–335, 340, 342, 385, 386
quartz crystal microbalance (QCM) 391

Raman spectroscopy 376, 385, 390, 421, 423, 424, 426–428, 432–434, 438
Randles cell 384
ratiometric 385
reconstructive surgery 593, 595, 596, 603, 605, 608
regeneration 3–5, 7, 8, 10
regenerative medicine 123, 135, 137, 821, 825, 830, 831, 834
registry 246
regulation 537–539, 541, 542, 544, 546, 547, 550, 551–555
regulatory framework 492, 497
relative recovery 405, 408, 410, 411, 413, 415
remodelling 741, 745, 746, 748, 749
- reverse 741, 745, 748, 749

reproductive cloning 658, 659, 667
review 363
risk assessment 243

Sauerbrey equation 391
scaffold 279–288, 295–304, 421–426, 430, 434, 435, 437, 438, 470, 472, 477, 478, 657, 659, 662–665, 667–669
scaleable process 505
scale-up principle 509
self-assembly 265, 269, 271, 336, 337, 339
SERCA2a 749
serotonin 382
serum 234, 235, 237, 238
single nucleotide polymorphism (SNP) 381
skeletal muscle 145
skeletal myoblast 679, 680, 684, 685
skin 469, 481–483, 593–597, 599–613
- construct 469, 482

smooth muscle 148, 153
soft lithography 350, 351, 359, 361
somatic cell nuclear transfer 47
somatic stem cell 83, 84, 86
stability 230, 233, 234, 236
standardisation 229, 230, 241
standards 77
static culture 509
stem cell 15–29, 31–33, 175–183, 185–189, 349, 351, 352, 353, 355–357, 363, 470–472, 474–476, 480, 503, 505–507, 509, 512, 657–659, 670, 695–698, 710, 713–715, 718, 720, 722, 723, 849, 861, 863–865, 868–879
- bank 229, 232, 236, 241, 243, 248
- bioprocessing 503, 509, 512
- cryopreservation 123, 129, 131
- expansion
- line 229, 232–235, 237, 238, 242, 243, 246, 247
- technology 6, 10

storage 230, 239, 241, 242, 245
subcutaneous implant 763, 764, 766–768
surface modification 321

surface plasmon resonance (SPR) 376, 389, 390

T-cell 202–207, 209, 210
teratoma 679, 686
testes 679, 681, 685
therapeutic
  cloning 658, 659, 667, 688
  nuclear transfer 200
therapy 234, 236, 238, 242, 243, 246–248
three-dimensional microenvironment 530
three-dimensional setting 507
three-dimensional structure 507
tissue 232, 233, 242–244
  formation 401–403, 413–416
  monitoring 423
  replacement and repair 4, 7, 8, 10
tissue engineered blood vessel (TEBV) 474, 475
tissue engineering 255, 256, 259, 260, 263, 265, 266, 268, 269, 271, 272, 655–663, 666–670, 763–771, 774, 781, 782, 784, 788, 791–793, 800, 805, 821, 823–827, 830–834, 861, 864
T-lymphocyte 202
tooth
  regeneration 633
  repair 639
traceability 232, 235–237, 244, 245
translational
  activity 504
  research 504
transplantation 200, 203, 205, 208, 209, 210, 561–570, 572–575
tropism 243
two-dimensional culture 507

UK Stem Cell Bank 231–233, 235, 236, 242, 244, 247
umbilical cord cell 505
urethral repair 661
urinary incontinence 669, 670

vascular 469, 472–475
venture capital 491–495, 499
vesicoureteral reflux 669, 670
viral testing 237, 243
virus 236, 237, 242, 243
vitrification 241, 242

World Health Organization (WHO) 231, 246

xenogeneic cell therapy 480
X-ray micro-computed tomography 421, 423, 425, 429